Learning and Experiencing Cryptography with CrypTool and SageMath

For a listing of recent titles in the *Artech Computer Security Library*,
turn to the back of this book.

Learning and Experiencing Cryptography with CrypTool and SageMath

Bernhard Esslinger

ARTECH
HOUSE

BOSTON | LONDON
artechhouse.com

Library of Congress Cataloging-in-Publication Data
A catalog record of this book is available from the U.S. Library of Congress.

British Library Cataloguing in Publication Data
A catalog record for this book is available from the British British Library.

ISBN 978-1-68569-017-5

Cover design by Joi Garron

Accompanying software for this book can be found at:
https://www.cryptool.org/en/documentation/ctbook.

685 Canton Street
Norwood, MA 02062

10 9 8 7 6 5 4 3 2 1

Contents

Preface

The rapid spread of the internet has led to intensified research in the technologies involved, especially within the area of cryptography where a good deal of new knowledge has arisen.

This book provides a thorough overview of classical and modern cryptography. In particular, it also guides you very specifically to try it out. The CrypTool (CT) programs are used for this purpose, as well as sample code written for the computer-algebra system SageMath. Both CrypTool and SageMath are open-source and free.

Another special feature is the selection of topics and the clear statements for users. On the one hand, the theory is presented, but it also emphasizes which procedures are really reliable and which official recommendations exist where.

To our knowledge, this book contains the first concrete summary in book form of what concrete tasks the science of historical cryptology consists of.

This book is written for both laymen and beginners, as well as for students and practitioners who would like to delve deeper into this field. Anyone who enjoys prime numbers or wants to know what modern lattice methods are will find very understandable information here. For a large number of ciphers, you can find in tabular form what are currently the best attacks on them.

The first chapter of this book explains the principles of symmetric and asymmetric *encryption* and discusses definitions for their resistibility.

Because of didactic reasons the second chapter gives an exhaustive overview of *paper-and-pencil encryption methods* and explains a typical example of a precomputer machine cipher that later became embroiled in scandal.

Chapter 3 gives a comprehensive overview about *historical cryptology*, a new research area, which deals with the practical problems of cryptanalyzing and contextualizing encrypted historical documents.

A major part of this book is dedicated to the fascinating topic of *prime numbers* (Chapter 4).

Then, Chapter 5 introduces *modular arithmetic* and *elementary number theory* using numerous examples. Here, the features of the *RSA procedure* are a key aspect.

Chapter 6 provides insights into the mathematical ideas and concepts behind *modern asymmetric cryptography* including a new geometric illustration of the processes involved in RSA encryption.

Chapter 7 gives a very brief overview about the status of attacks against modern *hash algorithms* and is then briefly devoted to *digital signatures and public-key infrastructures*, which are an essential component of e-business applications.

Chapter 8 describes *elliptic curves*, which are an alternative cryptosystem to RSA and are particularly well suited for use on smart cards.

Chapter 9 introduces *modern symmetric cryptography*. Boolean algebra is the foundation for most modern, symmetric encryption algorithms as these algorithms operate on bit streams and bit groups. Principal construction methods are described and implemented in SageMath. Compared to the other chapters in this book, this is the most mathematical one.

Chapter 10 introduces *homomorphic crypto functions*: Homomorphic encryption allows one to run calculations on encrypted data. This is a modern research topic that gets special attention in the course of cloud computing.

Chapter 11 gives a very easy *lightweight introduction to lattices*, an area that enables quantum-computer-resistant methods.

Chapter 12 describes *results for solving discrete logarithms and factoring*. This chapter provides a broad picture of the current best algorithms for (a) computing discrete logarithms in various groups, for (b) the status of the factorization problem, and for (c) elliptic curves. This survey was put together as a reaction to a provocative talk at the Black Hat conference 2013, which caused some uncertainty by incorrectly extrapolating progress at finite fields of small characteristics to the fields used in the real world.

Chapter 13 about the *future of cryptography* discusses threats for currently used cryptographic methods and introduces alternative research approaches (post-quantum cryptography) to achieve long-term security of cryptographic schemes.

The individual main chapters have been written by various authors and are mostly self-contained. The main author contributed to all chapters and is responsible for any mistakes left. The contents covered are accompanied by numerous practical examples and SageMath code.

At the end of each chapter you will find references. The sections have been enriched with many footnotes. Within the footnotes you can see where the described functions can be called and tried within the different CrypTool versions, within SageMath, or within OpenSSL.

Whereas the CrypTool e-learning programs motivate and teach you how to use cryptography in practice, the book also provides a deeper understanding of the mathematical algorithms used, trying to do it in an instructive way.

The best overview of all the functions available in CrypTool programs can be found at the website https://www.cryptool.org/en/documentation/functionvolume.

Within the appendix at the end of this book, you can gain an overview about the four different CrypTool variants via:

- The functions from the CrypTool 1 menus (CT1);
- The functions within the CrypTool 2 templates (CT2);
- The JCrypTool functions (JCT);
- The CrypTool-Online applications (CTO).

The programs of this book and some specific additions can be found at the CrypTool website: https://www.cryptool.org/en/documentation/ctbook/. There are detailed introductions to SageMath and OpenSSL with many examples. SageMath is placed there in a broader context (LaTeX, Python, Jupyter). Another addition is the 90-page document *CUDA Tutorial – Cryptanalysis of Classical Ciphers Using*

Modern GPUs and CUDA, which contains a practical introduction to writing CUDA programs on Linux and Windows.

As with the e-learning program CrypTool, the quality of the book grows with the suggestions and proposals for improvement from you, the reader. We look gratefully forward to your feedback.

Acknowledgments

There are many people involved in the creation and publication of a comprehensive book about cryptography.

At this point I'd like to thank explicitly the following people who in particular contributed to the CrypTool project. They applied their very special talents and showed really great engagement:

- Mr. Henrik Koy
- Mr. Jörg-Cornelius Schneider
- Mr. Florian Marchal
- Dr. Peer Wichmann
- Mr. Dominik Schadow
- Mr. Simon Leischnig
- Dr. Nils Kopal
- Staff of Prof. Johannes Buchmann, Prof. Claudia Eckert, Prof. Alexander May, Prof. Torben Weis, and especially Prof. Arno Wacker

The students must be mentioned who contributed through their far over 100 bachelor's and master's theses. Also, I want to thank the many people not mentioned here for their focused work (mostly carried out in their spare time).

Thanks to Bart Wessel and George Lasry for information about the differences between the C-52/CX-52 models. Thanks to Georg Illies for pointing me to Pari-GP. Thanks to Lars Fischer for his help with fast Pari-GP code for primitive roots. Thanks to Volker Simon for writing the SageMath Example 5.36. Thanks to Minh Van Nguyen from Australia for his always quick, professional, and exhaustive help with the first SageMath code samples. It's a pity that he is no longer reachable ... Many thanks to Klaus Pommerening, who handed over the script of his lecture about symmetric cryptography to the CrypTool project and who shared our love for SageMath. We then first extended his script together. Subsequently, the editor took over Chapter 9 on his own. R.I.P. Prof. Pommerening—we lost an admirable person.

The contributors to this book would like to take this opportunity to thank their colleagues in the particular companies and at the universities of Bochum, Darmstadt, Frankfurt, Gießen, Karlsruhe, Lausanne, Munich, Paris, and Siegen.

A special thank you to Dr. Doris Behrendt, who took over the laborious task to bring two books of 500+ pages to KOMA-Script, to clean up and modernize the TeX sources written by different authors over years, and in addition critically read the content.

Thanks also to the readers who sent us feedback, and especially to Olaf Ostwald, Helmut Witten, and Prof. Ralph-Hardo Schulz for constructively proofreading many parts of this book. And to Herbert Voss, who helped us when things got difficult in LaTeX. And finally, many thanks to the publisher's staff and their external reviewer, all of whom have been very helpful in keeping everything focused.

I hope that many readers have fun with this book and that they get out of it more interest and a greater understanding of this modern but also very ancient topic.

Introduction

This section shows how the book and the programs work together.

The CrypTool Book

The chapters of this book are largely self-contained and can be read and understood independently of the CrypTool programs. In the following, we often abbreviate "CrypTool" by CT.

Chapters 6 ("Modern Asymmetric Cryptography"), 8 ("Elliptic Curves"), 9 ("Modern Asymmetric Cryptography"), 10 ("Homomorphic Ciphers"), and 12 ("Results for Solving Discrete Logarithms and for Factoring") require a deeper knowledge of mathematics, while the other chapters should be understandable with a high school diploma.

The authors have tried to describe cryptography for a broad audience—without being mathematically incorrect, but with various links to get practical experience. We believe that this didactic approach is the best way to promote awareness of IT security and the willingness to use standardized modern cryptography.

This book provides a thorough overview of classical and modern cryptography and also guides you to try it out using the following free programs.

The Programs CrypTool 1, CrypTool 2, and JCrypTool

CrypTool 1 (CT1) and its successor versions CrypTool 2 (CT2) and JCrypTool (JCT) are used worldwide for training in companies and teaching in schools and universities. CrypTool 1 is an educational program for Windows that allows you to use and analyze cryptographic procedures within a unified graphical user interface. The comprehensive online help in CrypTool 1 contains both instructions on how to use the program and explanations of the methods themselves (both not as detailed and in a different structure as in this book).

CT2 also runs on Windows and now has a significantly larger range of cryptanalytic functions than CT1. JCT runs on Windows, Linux, and macOS, and now includes many things not included in CT1.

The setups of these standalone desktop programs are downloaded more than 10,000 times a month.

The Programs on CrypTool-Online (CTO)

The CrypTool-Online website (http://www.cryptool-online.org or https://www.cryptool.org/en/cto/), where you can try out and use cryptographic methods in a browser on a PC, tablet, or smartphone, is another part of the CT project.

The scope of CTO is not yet as broad as that of the standalone CT1, CT2, and JCT programs. However, as CTO is what people are using more and more as a first contact, a lot of effort is going into the CTO development. So we redesigned the backbone and frontend system using modern web technology to provide a fast, consistent, and responsive look and feel. CTO also includes plugins using WebAssembly (wasm) such as a Python IDE, Msieve, or OpenSSL. Using WebAssembly makes this functionality run in a browser almost as fast as native applications. Another modern technology offered is models for cryptanalysis trained with machine learning algorithms (deep learning, neural networks). See Section A.4.

Besides the classic ciphers, the most popular plugins in CTO are "RSA step-by-step," "RSA visual and more," "AES step-by-step," "AES animation," and the "Password meter."

MysteryTwister

MTW is the abbreviation for MysteryTwister (https://www.mysterytwister.org), an international cryptography contest ("cipher contest by cryptool"), which is also based on the CT project. Here you can find cryptographic puzzles in four categories, a high-score list, and a moderated forum.

As of July 2023 more than 10,000 users are participating, and more than 360 challenges are offered (301 of them are solved by at least one participant).

The SageMath Computer-Algebra System

SageMath is a comprehensive open-source CAS package that can be used to easily program the mathematical methods explained in this book. A special feature of this CAS is that the scripting language is Python (version 3 since SageMath 9). Thus, in a Sage script, you have at your disposal not only the mathematical commands of SageMath, but also all the functions of the Python language. SageMath is increasingly becoming the standard CAS system at universities. Since SageMath 8, there is also a version for Windows that runs in a Bash shell.

CHAPTER 1

Ciphers and Attacks Against Them

For centuries, plaintext messages were encrypted by the military, by diplomats, and by alchemists, and much less frequently by businesses and the general population. The goal of cryptography was to protect the privacy between sender and receiver. Since the 1970s, further goals have been added to achieve integrity, authenticity, and non-repudiation, and also to compute on encrypted data in the cloud or to achieve quantum-computer resistance.

The science that deals with encryption is called *cryptology*—divided into the branches of cryptography (designing secure encryption procedures) and cryptanalysis (breaking encryption procedures). In reality, however, these branches are closely interrelated and the terms cryptography and cryptology are often used interchangeably. Therefore, cryptology is currently subdivided into fields like symmetric cryptography, public-key cryptography, hardware and embedded systems cryptography, theoretical cryptology, and real-world crypto [1].

The importance of cryptology continues to grow as our society becomes more and more dependent on information technology. Although cryptology and information security are interdisciplinary fields of research, mathematics now plays the largest role in cryptology. Finally, learning about cryptology can also be fun and entertaining.

The special thing about this book is that you can always try out the procedures right away—by using the links (in the footnotes) to the programs from the CrypTool project, from OpenSSL, or from SageMath. All these programs are open-source.

In this book, the basics are covered in great detail, then from the very extensive field of cryptology certain (current) topics are selected (like RSA, ECC, or lattices). This makes this book accessible to a wide audience, not just only for those interested in the natural sciences.

This chapter introduces the topic in a more descriptive way without using mathematics. To do so, it uses modern methods (RSA, AES) as examples. Then we dive deepen, for example, the property, how many possible keys (key space) different methods have (Section 1.6) and what are the best attacks against known methods (Section 1.7). Recommended books are presented in Section 1.10. In Section 1.11 you will find screenshots of how to use AES in various programs. Classic methods are presented in Chapters 2 and 3.

The purpose of encryption is to change data (plaintext messages) in such a way that only an authorized recipient is able to reconstruct the plaintext. This allows us to transmit encrypted data without worrying about it getting into unauthorized hands. Authorized recipients possess a secret information—called the key—which allows them to decrypt the data while it remains hidden from everyone else. An attacker cannot only try to break a cipher: She still can disturb the connection (e.g., denial-of-service attack) or tap metadata (who is communicating when with whom).

Plaintext is the data processed as input by the encryption method. This data can be text, but also binary data such as an image or an executable file. The encryption method is called a cipher. The output is called ciphertext. With modern ciphers the output is always binary data. Figure 1.1 shows this notation graphically.

1.1 Importance of Cryptology

With the use of the internet and wireless communication, encryption technologies are used (mostly transparently) by everyone. Cryptographic algorithms secure ATMs and the privacy of messengers, allow anonymity for voters, but also help criminals. Cryptography is dual-use, as are many human innovations.

However, cryptography is not only used today, but has been for centuries by governments, the military, and diplomats. The side with a better command of these technologies could exert more influence on politics and war with the help of secret services. This book touches on history only twice: when introducing the earlier cipher methods for didactical reasons in Chapter 2, and in Chapter 3 when explaining the real application of earlier methods. You can gain an understanding of how important cryptology was and still is by considering the following two examples: the BBC documentary film *War of the Letters* [2] and the debates around the so-called crypto wars.

The next two sections discuss the differences between symmetric (see Section 1.2) and asymmetric (see Section 1.3) methods for encryption.

1.2 Symmetric Encryption

For *symmetric* encryption, both the sender and recipient must be in possession of a common (secret) key that they have exchanged before actually starting to communicate (over another channel, out of the band). The sender uses this key

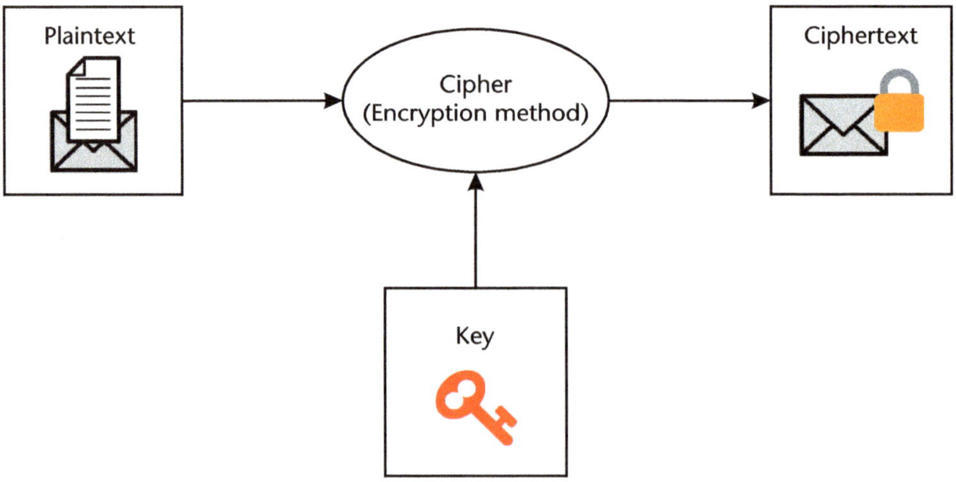

Figure 1.1 Common notations when using ciphers.

to encrypt the message and the recipient uses it to decrypt it. This is shown in Figure 1.2.

All classical ciphers are of the symmetric type. Examples can be found within the CT programs, in Chapter 2 of this book, or in [3]. In this section, however, we want to consider only modern symmetric mechanisms.

The main advantage of symmetric algorithms is the high speed with which data can be encrypted and decrypted. The main disadvantage is the high effort needed for key distribution. In order to communicate with one another confidentially, the sender and recipient must have exchanged a key using a secure channel before actually starting to communicate. Spontaneous communication between individuals who have never met therefore seems virtually impossible. If everyone wants to communicate with everyone else spontaneously at any time in a network of n subscribers, each subscriber must have previously exchanged a key with each of the other $n - 1$ subscribers. A total of $n(n - 1)/2$ keys must therefore be exchanged.

The current standard for modern symmetric ciphers is the Advanced Encryption Standard (AES).

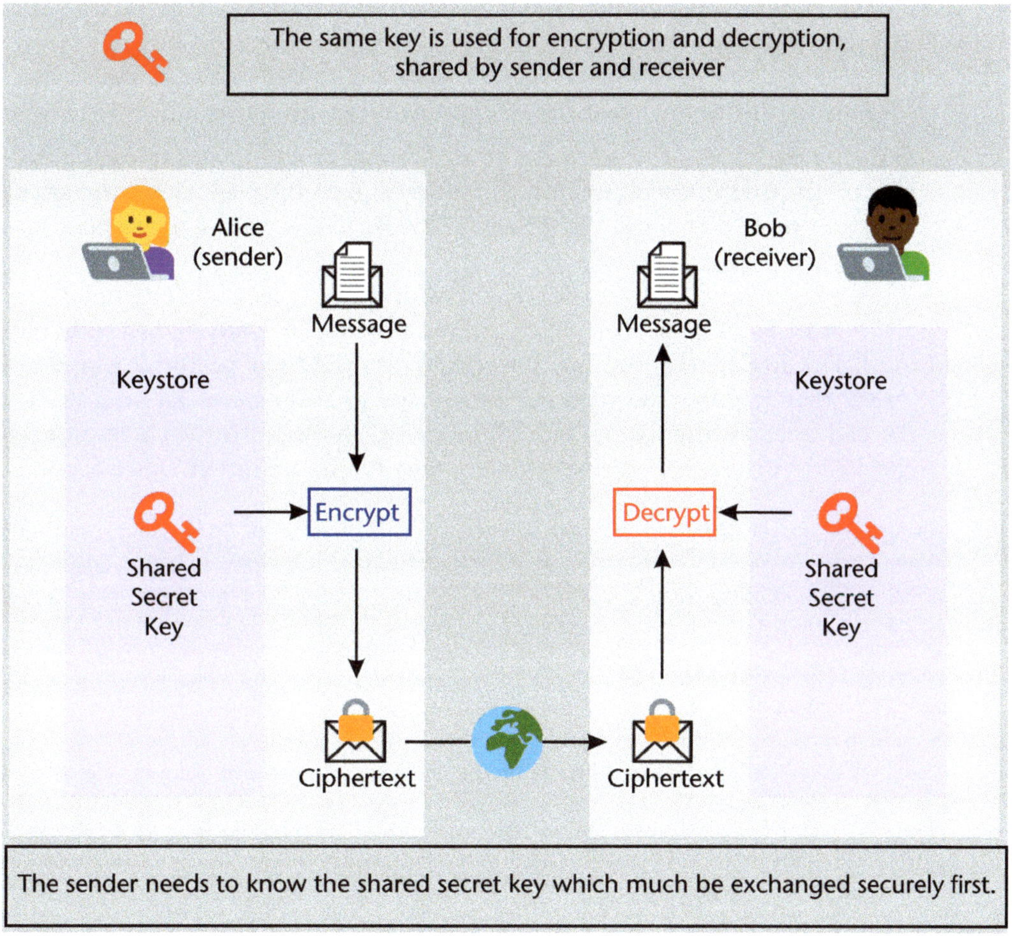

Figure 1.2 Symmetric or secret-key encryption.

1.2.1 AES[1]

Before AES, the most well-known modern symmetric encryption procedure was the Data Encryption Standard (DES). The DES algorithm was developed by IBM in collaboration with the National Security Agency (NSA), and was published as a standard in 1975. Despite the fact that the procedure is relatively old, no effective attack on it has yet been detected (what "effective" exactly means depends on the security definition—see Section 1.8). The most effective way of attacking DES consists of testing (almost) all possible keys until the right one is found (*brute-force attack*). Due to the relatively short key length of effectively 56 bit (64 bits, which however include 8 parity bits),[2] numerous messages encrypted using DES have in the past been broken. Therefore, the procedure cannot be considered secure any longer. Alternatives to the DES procedure include Triple-DES (TDES, 3DES) and especially AES.

The standard among symmetric methods today is AES. The associated Rijndael algorithm was declared the winner of the AES competition on October 2nd, 2000, and thus succeeds the DES procedure. Since then, the AES has been subjected to extensive research and has so far resisted all practical attempts at attack.

Further information about AES can be found in Section 9.2.7. Section 1.11 presents how the AES is animated in CTO, and how the AES is executed in CT2 and with OpenSSL.

1.2.2 Current Status of Brute-Force Attacks on Symmetric Algorithms

The current status of brute-force attacks on symmetric encryption algorithms can be explained with the attack on the block cipher RC5-64. A key length of 64 bit means at most $2^{64} = 18{,}446{,}744{,}073{,}709{,}551{,}616$ or about 18 quintillion (U.S.) $(= 18 \cdot 10^{18})$ keys to check.

Brute-force (exhaustive search, trial-and-error) means to completely examine all keys of the key space, which means no special analysis methods have to be used. The attacker knows only the ciphertext, and so he performs a ciphertext-only attack that requires the weakest knowledge prerequiste of all attacks. Therefore, the ciphertext is decrypted with all possible keys[3] and for each resulting text it is checked to determine whether this is a meaningful plaintext.[4] (See Section 1.6.)

1. - Using CTO in the browser, AES can be seen in two plugins: as "AES Animation" https://www.cryptool.org/en/cto/aes-animation and via "AES (step-by-step)" https://www.cryptool.org/en/cto/aes-step-by-step.
   - Using CT1 Indiv. Procedures ▷ Visualization of Algorithms ▷ AES you can find three visualizations for this cipher.
   - Using the search string `AES` in CT2 Startcenter ▷ Templates you can find a plugin performing AES step by step.
2. As a unit in formulas, we write "bit" in lower case and without the plural "s." See Section B.2.
3. - Using CT1 Analysis ▷ Symmetric Encryption (modern) you can perform brute-force attacks of modern symmetric algorithms.
   - Using CT2 Templates ▷ Cryptanalysis ▷ Modern you also can perform brute-force attacks. The KeySearcher is a highly powerful component used within these templates, which can distribute the calculations to many different computers.
4. If the plaintext is written in a natural language and at least 100 bytes long, this check also can be performed automatically. To achieve a result in an appropriate time with a single PC you should mark only at bits of the key as unknown. On a current PC in 2022, CT1 tries for AES 24 bit in about 20 seconds, but with 32 bit it takes 1:45 h. Compare screenshots in Section 1.6.

Companies like RSA Security provided so-called cipher challenges in order to quantify the security offered by well-known symmetric ciphers such as DES, 3DES, or RC5 [4, 5]. They offered prizes for those who managed to decipher ciphertexts, encrypted with different algorithms and different key lengths, and to unveil the symmetric key (under controlled conditions).[5]

It is well-known that the old standard algorithm DES with a fixed key length of 56 bit is no longer secure: This was already demonstrated in January 1999 by the Electronic Frontier Foundation (EFF). With their specialized computer Deep Crack they cracked a DES-encrypted message within less than a day.

The currently known record for strong symmetric algorithms unveiled a key that was 64-bit long. The algorithm used was RC5, a block cipher with variable key size.

The RC5-64 challenge was solved in July 2002 by the distributed.net team after 5 years [6]. In total 331,252 individuals cooperated over the internet to find the key. More than 15 quintillion ($= 15 \cdot 10^{18}$) keys were checked until the right key was found. This was about 85% of the whole search space.

Therefore, symmetric algorithms using keys of size 64 bit are (even if they have no cryptographic weakness) no longer appropriate to keep sensitive data private.

The BSI requires a security level of 120 bits for modern symmetric ciphers that will be used after 2022 (see [7], page 17f). Not only is AES-128 recommended, but details like suitable block modes and padding methods are also specified.

1.3 Asymmetric Encryption

In the case of *asymmetric* encryption (also called public-key encryption), each participant has their own pair of keys consisting of a *secret* key (called *private* key) and a *public* key. The public key, as its name implies, is made public—for example, within a certificate (see Section 7.5.2) or in a key directory on the internet (this type of billboard is also called a directory or sometimes public-key ring).

Figure 1.3 shows the process of asymmetric encryption and decryption.

If Alice[6] wants to communicate with Bob, she looks for Bob's public key and uses it to encrypt her message (plaintext) for him. She then sends this ciphertext to Bob, who is able to decrypt it again using his private key. As only Bob knows his private key, only he can decrypt messages addressed to him. Even Alice who sends the message cannot restore the plaintext from the (encrypted) message she has sent. In reality, asymmetric methods are not used to encrypt the whole message but only a session key (see Section 1.4). Asymmetric ciphers are designed in a way that the public key cannot be used to derive the private key from it.

Such a procedure can be demonstrated using a series of thief-proof letter boxes. If I have composed a message, I then look for the letter box of the recipient and post

5. Unfortunately, in May 2007 RSA Inc. announced that they will not confirm the correctness of the not-yet-solved RC5-72 challenge. Alternatively, a wide spectrum of both simple and complex, and both symmetric and asymmetric crypto riddles are included in the international cipher contest *MysteryTwister*: https://www.mysterytwister.org.

6. In order to describe cryptographic protocols, participants are often named Alice, Bob, ... (see [8, p. 23]). Alice and Bob perform all 2-person-protocols where Alice will initiate the protocol and Bob answers. The attackers are named Eve (eavesdropper) and Mallory (malicious active attacker).

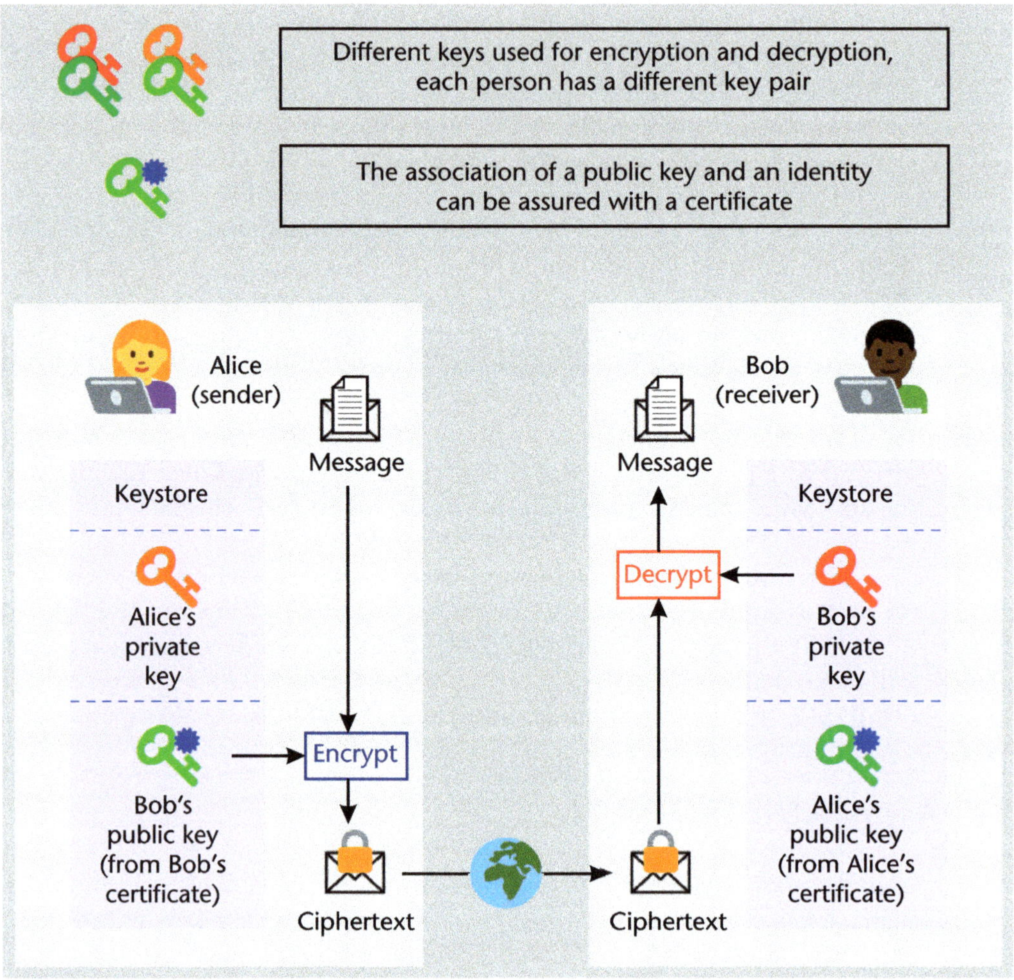

Figure 1.3 Asymmetric or public-key encryption.

the letter through it. After that, I can no longer read or change the message myself, because only the legitimate recipient has the key for the letter box.

The advantage of asymmetric procedures is the easier key management. Let's look again at a network with n subscribers. In order to ensure that each participant can establish an encrypted connection to each participant, each participant must possess a pair of keys. We therefore need $2n$ keys or n pairs of keys. Furthermore, no secure channel is needed before messages are transmitted, because all the information required in order to communicate confidentially can be sent openly. In this case, you simply have to pay attention to the accuracy (integrity and authenticity) of the public key. Nevertheless, the requirements for the key generation are not trivial. What could go wrong is explained, for example, in Section 5.12.5.4. Besides that, nowadays also (public-key) infrastructures themselves are targets of cyberattacks. A disadvantage of pure asymmetric procedures is that they take a lot longer to perform than symmetric ones (see Section 1.4).

The most well-known asymmetric procedure is the RSA algorithm,[7] named after its developers Ronald Rivest, Adi Shamir, and Leonard Adleman. The RSA algorithm was published in 1978. The concept of asymmetric encryption was first introduced by Whitfield Diffie and Martin Hellman in 1976. It is worth noting that the concept was known at the secret services Government Communications Headquarters (GCHQ) and National Security Agency (NSA) several years prior to its independent rediscovery by Diffie and Hellman. Today, the ElGamal procedures also play a decisive role, particularly the Schnorr variant in the Digital Signature Algorithm.

The German Federal Office for Information Security (BSI) requires a security level of 120 bit for processes used beyond 2022. Applied to RSA, the corresponding technical guideline recommends a key length of 3,000 bit (see [7], page 18, comment on Table 1.2).

1.4 Hybrid Procedures[8]

In order to benefit from the advantages of symmetric and asymmetric techniques together, hybrid procedures are usually used (for encryption) in practice.

In this case the bulk data is encrypted using symmetric procedures. The key used for this is a secret session key generated by the sender randomly that is only used for this message. This session key is then encrypted using the asymmetric procedure and transmitted to the recipient together with the message. Recipients can determine the session key using their private keys and then use the session key to decrypt the message.

In this way, we can benefit from the feasible key management of asymmetric procedures (using public/private keys) and we benefit from the efficiency of symmetric procedures to encrypt large quantities of data (using secret keys).

1.5 Kerckhoffs' Principle

In 1883, the Dutch cryptographer Auguste Kerckhoffs formulated six principles for the construction of secure military encryption procedures. The second one, Kerckhoffs' principle or Kerckhoffs' maxim, is now regarded as the principle of modern cryptography. It states that an encryption scheme should be secure even if everything about the scheme is known except the key used. Kerckhoffs' principle is often contrasted with "security through obscurity," in which the encryption algorithm must also be kept secret.

7. The RSA algorithm is extensively described within this book in Section 5.10. The topical research results concerning RSA are described in Section 5.12. In Section 6.5 the RSA algorithm is more deeply reasoned from number theory: The RSA plane is a model to illustrate the processes in this algorithm using pictures of rectangles.

8. - Using CT1 Encrypt/Decrypt ▷ Hybrid you can follow the single steps and its dependencies with concrete numbers. The variant with RSA as the asymmetric algorithm is graphically visualized; the variant with ECC uses the standard dialogs. In both hybrid cases AES is used as the symmetric algorithm.
- Using JCT Algorithm Perspective ▷ Hybrid Ciphers also offers hybrid methods like ECIES.

Kerckhoffs' principle was reinterpreted several times. For example, Claude Shannon formulated that one should design encryption systems under the assumption that an enemy knows the system exactly from the very beginning (Shannon's maxim).

1.6 Key Spaces: A Theoretical and Practical View

For good encryption procedures used today, the time needed to break an encryption is so long that it is almost impossible to do so. Such procedures are considered (practically) secure—from an algorithm's point of view. After the knowledge gathered by Edward Snowden, there were many discussions debating whether encryption is secure. In [9] is the result of an evaluation, which cryptographic algorithms can be relied on—but only according to current knowledge. The article investigates: Which cryptosystems can—despite the reveal of the NSA/GCHQ attacks—still be considered as secure? Where have systems been intentionally weakened? How can we create a secure cryptographic future? What is the difference between math and implementation?

The key space of a cipher is an important indicator for the security of a cipher. In a monoalphabetic substitution (MASC; also called simple substitution) for instance, using an alphabet of length of k, the key space is $k!$. For AES-128 it is 2^{128}.

A (sufficiently) large key space (approx. 2^{100}) is a necessary prerequisite for a secure cipher, but not a sufficient condition: The MASC has a large key space (with an alphabet of 26 characters approx. $2^{88.4}$ that corresponds to the number of possible ciphertext alphabets), but it has been cracked with frequency analysis for centuries.

The key space is used to calculate the effort required for a brute-force (BF) attack (i.e., for the systematic testing of all possible keys). If the key space is so small that an attacker can carry out a complete BF attack, the procedure is broken—not only theoretically but also practically.

In the case of a BF attack, the attacker decrypts the ciphertext (or parts of it) with every possible key (see Section 1.2.2). Then the found plaintext is evaluated. How surprisingly well fitness algorithms can recognize correct natural texts can be seen in Figures 1.4[9] and 1.5.[10] CT1 uses similar fitness functions as the solvers and analyzers in CT2.

Whether an attacker really has to try the maximal, theoretical key space is questionable, at least with the older ciphers. For this reason, the *practical key space* introduced by Ralph Simpson for historic cipher devices and the *work factor*, which is also known as *attack time*, are considered.

1.6.1 Key Spaces of Historic Cipher Devices

Key spaces of historic cipher devices are often reported in the popular press as a gargantuan number designed to impress the reader about the incredible strength of the encryption. This is often a lead-in to the story of the amazing ingenuity of

9. CT1 Analysis ▷ Symmetric Encryption (modern) ▷ AES (CBC).
10. CT2 Templates ▷ Cryptanalysis ▷ Modern ▷ AES Known-Plaintext Analysis (2).

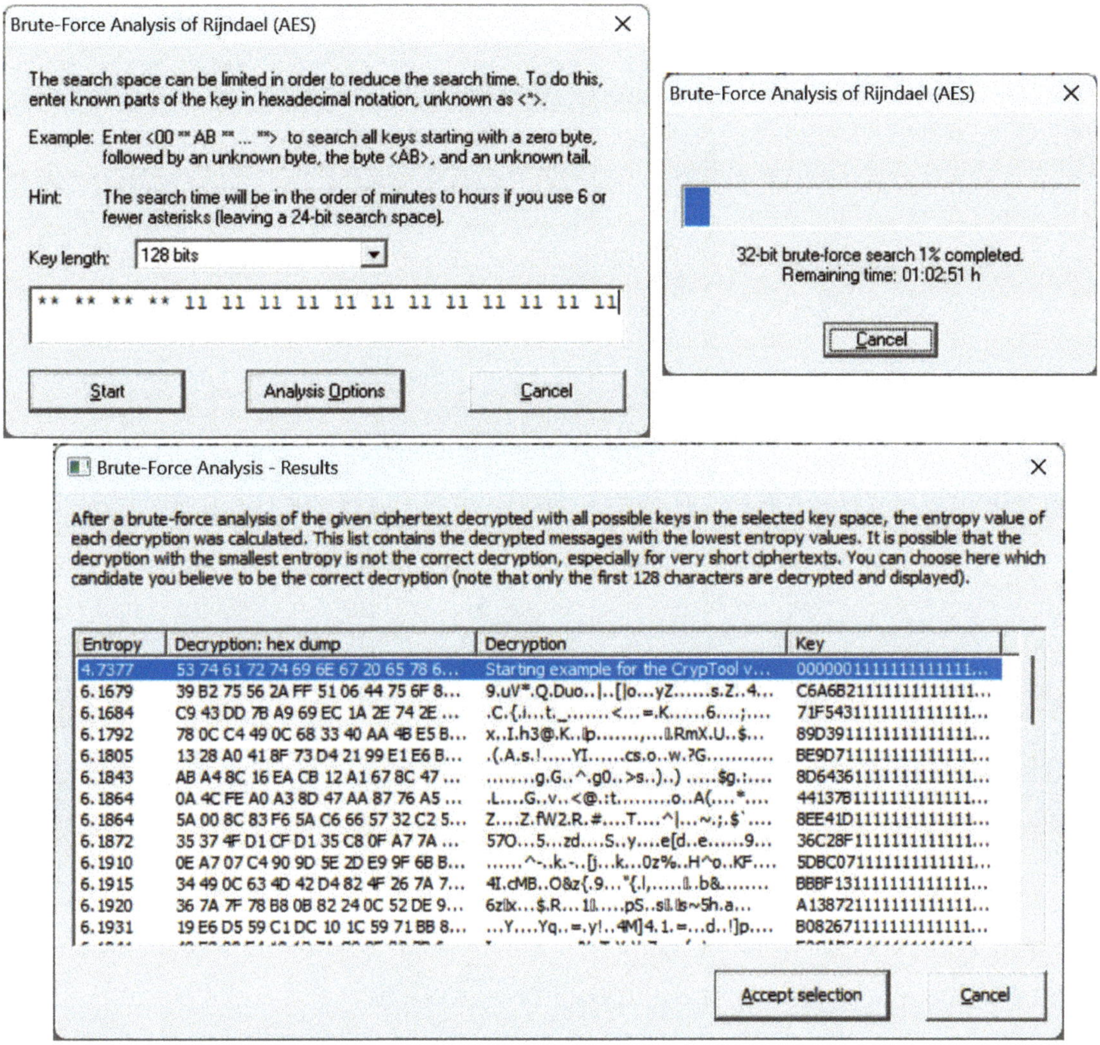

Figure 1.4 Brute-force analysis of AES in CT1 with partly known key.

the codebreakers who broke that encryption. Of course, they were all eventually broken.

For instance, the key space for the infamous Enigma I machine is larger than the number of atoms in the universe. According to Table 1.1, the theoretical key space of the Enigma is around $3 \cdot 10^{114}$, while the number of atoms in the universe is around 10^{77} (according to Table 4.13).

There are two main problems with key spaces of historic cipher devices. The first problem is that key space can be a misleading measure for the strength of the encryption. The reason for the confusion on this point arises because the key space of a modern symmetric cipher system, in contrast, usually provides a good measure for the strength of the encryption. But historic devices are mechanical or electromechanical, which results in limitations on the randomness of the encryptions. This means that methods can be developed to break that encryption without the need for brute force. Remember, key space is only a measure of the brute force

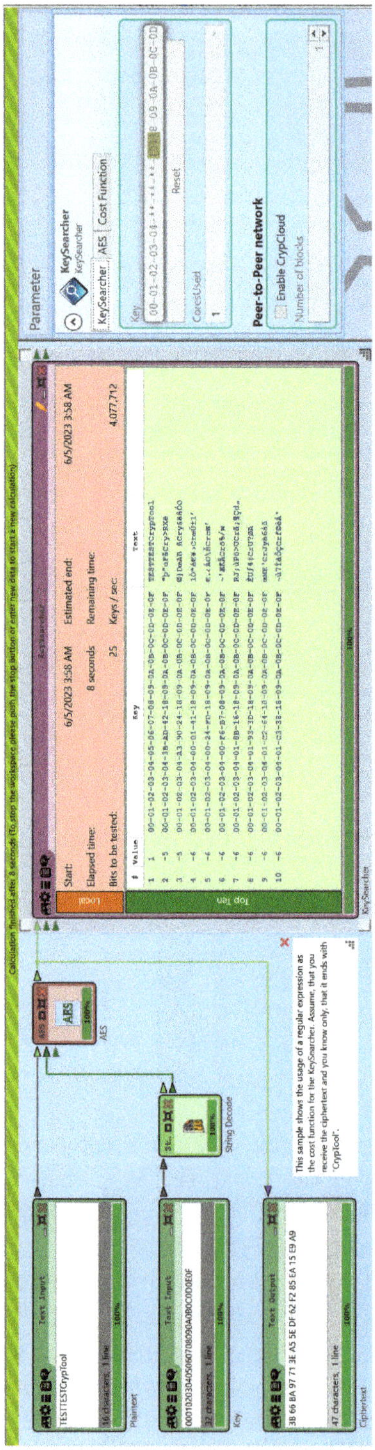

Figure 1.5 Brute-force analysis of AES in CT2 with partly-known plaintext.

required to break an encryption, without taking into account any methods used by cryptanalysts to shortcut (many) parts of that key space.

The second problem with key spaces of historic devices is due to the wild variations often reported for the very same device. This variation is usually due to differences in base assumptions, but those assumptions are not always stated.

Another thing to consider about key spaces is that cryptanalysis methods for some historic devices were not developed for many decades or even centuries after their invention. As with all things crypto-related, cryptanalysis methods are not necessarily made public. As an example, the Vigenère disk, which was invented in 1466, was reported by *Scientific American* magazine to be unbreakable in 1917. This article was published the same year that Joseph Mauborgne, U.S. Army Chief Signal Officer, boasted that his cryptographers could decrypt the Vigenère disk faster than the enemy could decrypt their own messages.

Despite the problems highlighted, a study of the key spaces of historic cipher devices is a useful tool to better understand the mind of the cipher inventor, user, and codebreaker. So with modern methods, we can discount and malign the value of key spaces of historic devices, but that alone would miss the point of understanding why historical decisions were made based on the strength of the encryption implied by these large key spaces.

1.6.2 Which Key Space Assumptions Should Be Used

After selecting a common set of assumptions, the key spaces of historic devices need to be calculated so they can be compared. Since the key space quoted most often originated from the NSA document [10] about the Enigma, that set of assumptions was used to develop the chart of historic key spaces (Table 1.1). The NSA document was written by Ray Miller and first published in 1995. In this document, Miller describes a maximum and a practical key space, but unfortunately he did not explicitly define the used assumptions.

1.6.2.1 Maximum Key Space vs Practical Key Space vs Work Factor

Miller used the term *maximum key space* for the theoretical maximum number of settings that would need to be tested for a brute-force attack. He assumed that the enemy captured the device, as per Kerckhoffs' principle, but any field-replaceable parts are unknown or could be changed, such as the rotors and reflectors. So all possible wirings of rotors and reflectors would have to be cryptanalyzed and any number of possible plugboard cables could be used.

The *practical key space* is also a theoretical number of settings but assumes that the captured machine and all field-replaceable parts are known and being used. This means that the wiring of the rotors and reflector are known but the rotors selected to be inserted into the machine and the order of those rotors are not known. This also means the reflector adds no cryptographic strength at all, since its wiring is known. Also, any user-imposed limitations are known and exploited, such as the Germans in WW2 mostly used 10 plugboard cables. These factors all help to reduce the practical key space compared to the maximum key space.

Another term (not used by Miller, but closely related to the key space) is *work factor*. This is the amount of work effort really required to break an encryption.

This number is usually smaller than the practical key space because any known cryptanalysis techniques are used as shortcuts. For the Enigma, this means that Rejewski's method of separating the cryptanalysis of the plugboard from the rotors and reflector greatly reduced the total number of settings that needed to be tested. Some of these cryptanalysis techniques were not known at the time of use or were not known by the users of these cipher devices.

Work factor is a concept more commonly used for the modern cipher systems. For the historical devices, there is very little available on work factors. It depends on the size of the message or number of messages captured. And it depends on the state of the cryptanalytic techniques that could be applied. For example: Although the Enigma machine has a huge theoretical key space, the Turing-Welchman Bombe only had to check about 422,000 settings in order to break the Enigma.[11] This work factor is what is called "attack time" when comparing the best attacks against modern ciphers in Table 1.3. For DES the work factor is drastically smaller (2^{43}) than the practical key space, and for AES it is around 2 bits smaller ($2^{254.4}$).

1.6.2.2 Key Space Assumptions Defined

The objective is to have one common set of assumptions to compare all the historic cipher devices and to use the assumptions that seem to have the most popular acceptance. Since Miller did not explicitely state his assumptions, they had to be reverse-engineered. A careful reading of the NSA document yields the following assumptions.

The maximum key space, as calculated by Miller, has three assumptions:

1. The base machine is captured and known to the enemy (per Kerckhoffs' principle);
2. Field-replaceable parts can be changed, so are not known (e.g., rotor and reflector wiring);
3. A "message setting" will be sent with each message, separate from the fixed machine setting.

The practical key space, as calculated by Miller, has four assumptions:

1. The base machine is captured and known to the enemy (per Kerckhoffs' principle);
2. Field-replaceable parts are also captured and known;
3. User-imposed limitations are known (e.g., always using 10 plugboard cables);

11. Why only 422,000? The British Bombe only tested for rotor order and rotor settings; ring settings and plugboard settings were then manually determined. With three rotors chosen from five, there are $5 \cdot 4 \cdot 3 = 60$ possible rotor orders. German procedures, however, did not allow any three rotor order to be repeated in the same month, which reduced the 60 possible orders at the beginning of the month to 30 by the end of the month. In addition, the Germans did not permit any individual rotor to be in the same position on the following day, reducing the 60 possible rotor orders to 32. Combined, these two rules reduced the possible orders to 32 at the beginning of the month, declining to 16 at the end of the month, or on average 24 rotor orders. This average rotor order multiplied with the 26^3 rotor settings yielded to $24 \cdot 17{,}576 = 421{,}824$ settings tested by the Bombe for a full run.

4. A "message setting" will be sent with each message, separate from the fixed machine setting.

1.6.2.3 Explanation of the NSA Key Space Assumptions

These assumptions detailed above seem reasonable and straightforward, except for possibly the last assumption of both the maximum and practical key spaces: A "message setting" will be sent with the message, separate from the fixed machine setting. The meaning and effect of this assumption requires further explanation.

For the Enigma, all possible wirings of the rotors are included in the maximum key space. Also, Miller includes the rotational starting positions of the rotors. Including the rotor starting position in the key space—besides it is already accounted for in all possible wirings—can be considered as redundant.

For instance, if a rotor is in position "A" and a particular wiring scheme is determined to be correct, that same wiring scheme could be advanced one position and now this new wiring scheme works when the rotor is moved to position "B." So all wiring schemes should yield 26 correct solutions as you rotate the rotor through the 26 positions. It seems you should just ignore the rotor starting position for the three rotors, which accounts for a contribution to the key space of 26^3. For this reason, many others have reported the Enigma key space without this factor.

We don't go deeper here into Enigma. There are many books and articles about this rotor machine and its history. A good summary of its design (flaws) and another approach calculating its relevant key space can be found in [11].

By including the rotor setting in the key space, Miller was allowing for a slightly larger key space that would break all daily messages after cryptanalysis of the first message. All subsequent messages using the same machine setting could then be decrypted in real time, just as the enemy would decrypt their own message.

Miller's rationale of the rotor position applies to all the rotor-based historic cipher devices, including the mechanical devices, like the Hagelin M-209. For this machine, all possible pin settings on each rotor are analyzed and included in the key space. So knowledge of the rotor rotational position is not necessary to break a message. The pin settings are part of the machine setting and fixed for the day, and the rotor setting is part of the message setting, which changes with every message. Again, just like in the case of the Enigma, the rotor positions must be known to break all daily messages in real time.

1.6.3 Conclusion of Key Spaces of Historic Cipher Devices

Having a clearly defined set of assumptions for key spaces, the key spaces could be calculated accordingly.

Table 1.1 lists 34 historic and 4 modern cipher systems, showing the maximum and practical key spaces for each one, using that same set of assumptions. This table was first presented to the International Conference on Cryptographic History (ICCH) group [12] by Ralph Simpson in Decmber 2022. The key spaces for some of these devices have not been previously reported, such as the Hebern, Japanese Purple machine, NEMA, KL-7, Transvertex HC-9, Russian VIC, and Hagelin CD-57. Most of the other historic cipher devices required new calculations to match the maximum and practical assumptions listed above.

Table 1.1 Key Space Sizes for 34 Historic and 4 Modern Cipher Systems

Year	*Cipher*	*Maximum Key Space*		*Practical Key Space*	
600 BCE	Monoalphabetic substitution	$4.03 \cdot 10^{26}$	2^{88}	$4.03 \cdot 10^{26}$	2^{88}
50 BCE	Caesar	$2.50 \cdot 10^{1}$	2^{5}	$2.50 \cdot 10^{1}$	2^{5}
1466	Vigenère (repeating keyword – 15 char.)	$1.68 \cdot 10^{21}$	2^{71}	$1.68 \cdot 10^{21}$	2^{71}
1586	Vigenère (autokey – 314 char. message)	$2.00 \cdot 10^{444}$	2^{1476}	$2.00 \cdot 10^{444}$	2^{1476}
1854	Playfair	$6.20 \cdot 10^{23}$	2^{79}	$6.20 \cdot 10^{23}$	2^{79}
1860s	Wheatstone Cryptograph	$4.03 \cdot 10^{26}$	2^{88}	$4.03 \cdot 10^{26}$	2^{88}
1912	Lugagne Transpositeur	$1.30 \cdot 10^{532}$	2^{1768}	$1.32 \cdot 10^{13}$	2^{44}
1912	M-94 cylinder cipher	$3.45 \cdot 10^{666}$	2^{2214}	$3.88 \cdot 10^{26}$	2^{88}
1916	M-138A strip cipher	$3.69 \cdot 10^{799}$	2^{2656}	$1.95 \cdot 10^{59}$	2^{197}
1918	ADFGX	$4.19 \cdot 10^{47}$	2^{158}	$4.19 \cdot 10^{47}$	2^{158}
1918	ADFGVX	$1.01 \cdot 10^{64}$	2^{213}	$1.01 \cdot 10^{64}$	2^{213}
1922	Hebern 5-rotor	$1.27 \cdot 10^{140}$	2^{466}	$4.56 \cdot 10^{10}$	2^{35}
1924	Kryha	$2.02 \cdot 10^{53}$	2^{177}	$1.78 \cdot 10^{29}$	2^{97}
1926	Enigma Swiss K	$1.60 \cdot 10^{101}$	2^{336}	$1.85 \cdot 10^{9}$	2^{31}
1930	Lugagne Le Sphinx	$1.30 \cdot 10^{532}$	2^{1768}	$2.43 \cdot 10^{24}$	2^{81}
1931	Abwehr Enigma G	$7.17 \cdot 10^{121}$	2^{405}	$4.82 \cdot 10^{10}$	2^{35}
1932	Enigma I	$3.28 \cdot 10^{114}$	2^{380}	$4.31 \cdot 10^{22}$	2^{75}
1937	SIGABA	$1.82 \cdot 10^{285}$	2^{941}	$5.95 \cdot 10^{28}$	2^{96}
1939	Japanese Purple	$3.81 \cdot 10^{59}$	2^{198}	$1.45 \cdot 10^{31}$	2^{104}
1939	Japanese JN-25 codebook (100 words)	$1.00 \cdot 10^{12}$	2^{40}	$8.25 \cdot 10^{10}$	2^{36}
1941	Lorenz SZ40/SZ42	$1.05 \cdot 10^{170}$	2^{565}	$1.05 \cdot 10^{170}$	2^{565}
1941	SG-41 "Hitler Mill"	$4.24 \cdot 10^{51}$	2^{171}	$4.24 \cdot 10^{51}$	2^{171}
1942	M-209 pin & lug	$6.16 \cdot 10^{60}$	2^{202}	$6.02 \cdot 10^{58}$	2^{195}
1942	Enigma M4	$2.33 \cdot 10^{145}$	2^{483}	$3.13 \cdot 10^{25}$	2^{85}
1942	T-52d Geheimschreiber	$7.23 \cdot 10^{213}$	2^{710}	$8.11 \cdot 10^{23}$	2^{79}
1943	Typex Mark 22	$1.82 \cdot 10^{195}$	2^{649}	$5.51 \cdot 10^{54}$	2^{182}
1947	NEMA	$5.99 \cdot 10^{164}$	2^{551}	$1.83 \cdot 10^{19}$	2^{64}
1952	Hagelin C-52	$1.68 \cdot 10^{117}$	2^{389}	$7.17 \cdot 10^{57}$	2^{192}
1952	Hagelin CX-52	$1.17 \cdot 10^{123}$	2^{409}	$1.10 \cdot 10^{104}$	2^{346}
1952	KL-7	$5.87 \cdot 10^{431}$	2^{1434}	$1.70 \cdot 10^{34}$	2^{114}
1950s	Transvertex HC-9	$2.96 \cdot 10^{71}$	2^{237}	$4.39 \cdot 10^{69}$	2^{231}
1953	VIC paper & pencil	$9.09 \cdot 10^{40}$	2^{136}	$1.00 \cdot 10^{27}$	2^{90}
1956	Fialka	$2.82 \cdot 10^{458}$	2^{1523}	$6.24 \cdot 10^{77}$	2^{258}
1957	Hagelin CD-57	$1.52 \cdot 10^{103}$	2^{343}	$1.49 \cdot 10^{60}$	2^{200}
1976	DES (56 bit)	$7.21 \cdot 10^{16}$	2^{56}	$7.21 \cdot 10^{16}$	2^{56}
1977	RSA-4096	$2.22 \cdot 10^{1225}$	2^{4071}	$2.22 \cdot 10^{1225}$	2^{4071}
1992	AT&T TSD 3600-E Clipper chip	$1.21 \cdot 10^{24}$	2^{80}	$1.21 \cdot 10^{24}$	2^{80}
2001	AES-256	$1.16 \cdot 10^{77}$	2^{256}	$1.16 \cdot 10^{77}$	2^{256}

Courtesy of Ralph Simpson.

It is important to remember that these key spaces are still not a good sole indicator of the cryptographic strength of the encryption method—examples for these criticisms are monoalphabetic substitution (2^{88}), Enigma I (2^{75}), and Playfair (2^{79}). But using a common set of assumptions will at least add a level of consistency among all these disparate devices.

1.7 Best Known Attacks on Given Ciphers

Tables 1.2 and 1.3 contain the best attacks known today for well-known classical and modern ciphers. For modern procedures, the effort (number of steps or attack time) is also given in Table 1.3. To our knowledge, this is the first time such a complete table is created.

For symmetric ciphers, the key space derived from the key length is an important indicator (see Section 1.6). It is used to calculate the effort required for a BF attack, the maximum effort that an attacker can have.

The following applies to AES-128 (see Table 1.3): The key length is 128 bits. The key space is 2^{128} and so is the theoretical attack time. The best known attack (biclique attack) reduces this maximum effort to $2^{126.1}$ steps. This difference of around 2 in the exponent means that the attack is about 4 times faster than a BF attack on average. This shows that AES is vulnerable in principle, but this attack is not at all relevant to practical security.

1.7.1 Best Known Attacks Against Classical Ciphers

The historical ciphers shown in Table 1.2 represent different periods in the history of cryptography, ranging from simple Caesar ciphers to more complex machine-assisted systems like Enigma. These selections are based on their historical significance. The attack types and methods shown in the table are the currently best known computerized methods for attacking these ciphers. All of the hand ciphers are vulnerable to simulated annealing and hill climbing. Composed ciphers, in our example here ADFGVX, need more sophisticated methods. With ADFGVX, a divide-and-conquer attack can be used to break substitution and transposition independently. Also noteworthy is SIGABA, since it can be attacked with a meet-in-the-middle attack. Additionally, all shown hand ciphers (substitution, transposition, and composed ciphers) can today be attacked in a pure ciphertext-only scenario. An exception are nomenclature ciphers, since the nomenclature elements (code words) can often only be decrypted when having either the original key or enough context to deduce them. Also, the chances of successfully attacking cipher machines, such as the Enigma and Typex, are enhanced when a crib (a partially known plaintext) is available. Only attacks on SIGABA still require the complete plaintext to be successful.

1.7.2 Best Known Attacks Against Modern Ciphers

Table 1.3 presents a selection of modern ciphers and the best attacks against them. The table includes historically significant ciphers such as DES and FEAL, ISO standards like AES, Camellia, and SNOW 2, national standards like GOST and SM4, and ciphers that were actively used in industrial solutions such as KeeLoq and A5.1. Cipher names typically encompass a family of encryption methods rather than referring to a single algorithm. These algorithms usually differ in the size of the key used and, in the case of block ciphers, the size of the data block. It is important to note that the best attacks against various versions of a cipher may differ. For the sake of brevity, we provide a single example from each cipher family and present the most successful attack against it.

In the right-most column of Table 1.3, the term "attack time" is used. "Time" is an established term used in modern cryptography. In order to understand what the attack time—as a measure for the resistability of a cipher—means, see Section 1.8 which introduces attack costs and different attack types.

Table 1.2 Best Known Attacks Against 17 Historical Ciphers

Cipher	*Attack Requirements*	*(Best) Cryptanalysis Methods*	*References*
Substitution ciphers			
Caesar	PCO	Brute force, frequency analysis	[13]
Monoalphabetic substitution	PCO	Hill climbing, frequency analysis	[13]
Homophonic substitution	PCO	Hill climbing / simulated annealing	[14]
Nomenclatures	PCO	Manual (deduced by context; or nomenclature available)	[15, 16]
Polyalphabetic substitution	PCO	Hill climbing / simulated annealing / (Friedman + Kasiski)	[13]
Playfair	PCO; crib	Simulated annealing	[17, 18]
Code books	PCO; crib/KP	Manual (deduced by context; availability of similar code book)	[19]
Chaocipher	PCO	Hill climbing / simulated annealing	[20]
Transposition ciphers			
Scytale	PCO	Brute force	[13]
Columnar transposition	PCO	Brute force (short keys) / hill climbing / simulated annealing	[21]
Double columnar transposition	PCO	Hill climbing / simulated annealing; IDP attack	[22]
Composed			
ADFGVX	PCO	DAC + hill climbing / simulated annealing	[23]
Machines			
Enigma	PCO, crib	DAC; hill climbing / simulated annealing; Turing Bombe	[24, 25, 26]
Typex	PCO, crib	DAC; hill climbing / simulated annealing; Turing Bombe	[24, 25, 26]
SZ42	PCO, crib	Testery methods and hill climbing	[27]
M209	PCO, crib	Simulated annealing / hill climbing	[28, 29]
SIGABA	KP	Meet in the middle; hill climbing / simulated annealing	[30, 31]

PCO = pure ciphertext-only, KP = known-plaintext, DAC = divide and conquer.

1.8 Attack Types and Security Definitions

If you are interested in the definitions used in modern cryptography, this section explains them with the fewest amount of mathematics as possible. Also, the relationship between the various definitions is declared—something which often falls short in courses. We believe that only understanding the differences between the various concepts enables learners to grasp the idea and apply it correctly later.

1.8.1 Attack Parameters

In cryptography, a security parameter is a way of measuring of how hard it is for an adversary to break a cryptographic scheme. Attack parameters describe the conditions available for the attacker.

Table 1.3 Best Known Attacks Against 36 Modern Ciphers

Cipher	*Attack Types*	*(Best) Cryptanalysis Methods*	*Attack Time*
Block ciphers			
DES	Single key. KPA. Full	Linear [32, 33]	2^{43}
3DES (TDEA). 3-key version [34]	Single key. KPA. Full	Meet-in-the-middle [34]	2^{112}
AES-128 (Rijndael) [35]	Single key. CCA. Full	Biclique [36]	$2^{126.1}$
Camellia-128 [37]	Single key. CPA. 11/18 rounds	Truncated differential [38]	$2^{121.3}$
MISTY1 [39]	Single key. CPA. Full	Integral [40, 41]	$2^{107.9}$
KASUMI [42]	Related-key. CCA. Full	Boomerang [43]	2^{32}
HIGHT [44]	Single key. CCA. Full	Biclique [45]	$2^{126.4}$
CAST-128 [46]	Single key. CPA. 9/16 rounds	Differential [47]	2^{73}
SEED-128 [48]	Single key. CPA. 8/16 rounds	Differential [49]	2^{122}
PRESENT [50]	Single key. CPA. 26/31 rounds	Truncated differential [51]	2^{70}
CLEFIA-128 [52]	Single key. CPA. 14/18 rounds	Truncated differential [38]	2^{108}
LEA-128 [53]	Single key. CPA. 13/24 rounds	Differential [54]	2^{127}
SM4 [55]	Single key. KPA. 24/32 rounds	Linear [56]	$2^{126.6}$
GOST 28147-89 [57] (Magma)	Single key. CPA. Full	Guess then truncated differential [58]	2^{179}
GOST R 34.12-2015 (Kuznechik) [59]	Single key. CCA. 5/10 rounds	Meet-in-the-middle [60]	2^{140}
KeeLoq [61]	Single key. KPA. Full	Slide and meet-in-the-middle [62]	$2^{44.5}$
Simon64/128 [63]	Single key. KPA. 31/44 rounds	Multidimensional linear [64]	2^{120}
Speck64/128 [63]	Single key. CPA. 20/27 rounds	Differential [65]	$2^{93.56}$
FEAL-32 [66]	Single key. CPA. 31/32 rounds	Differential [67]	2^{63}
Twofish-128 [68]	Single key. CPA. 7/16 rounds	Saturation [69]	2^{126}
Stream ciphers			
RC4	Variable-key. Plaintext recovery. COA	Statistical [70]	2^{31}
A5/1 [71]	Single key. KPA. Full	Time-memory-data trade-off [72]	2^{24}
A5/2 [73]	Single key. KPA. Full	Time-memory-data trade-off [72]	2^{16}
Chacha [74]	Single key. KPA. Chosen IV. 7/20 rounds	Differential [75]	2^{255}
Salsa20 [76]	Single key. KPA. Chosen IV. 8/20 rounds	Differential [75]	2^{255}
Crypto-1 [77]	Single key. KPA. Full	Algebraic [78]	2^{32}
Grain-128 [79]	Single key. KPA. Chosen IV. Full	Dynamic cube attack [80]	2^{74}
Trivium [81]	Single key. KPA. Chosen IV. 799/1152 rounds	Dynamic cube attack [82], see also footnote 1	2^{62}
Rabbit [83]	Not known	See also footnote 2	
Enocoro 128v2 [84]	Distinguishing. KPA. Chosen IV. 22/96 rounds	Higher order differential [85]	2^{16}
SNOW 2-128 [86]	Single key. KPA. Chosen IV. 14/32 rounds	Cube [87]	$2^{162.86}$
MUGI [88]	Distinguishing. KPA. Chosen IV. 21/32 rounds	Differential [89]	$2^{61.59}$
ZUC 1.6 [90]	Not known	See also footnote 3	

Table 1.3 *Continued*

Cipher	*Attack Types*	*(Best) Cryptanalysis Methods*	*Attack Time*
Public-key encryption			
RSA [91]	Single key. COA. For RSA-250 (829-bit number)	Number field sieve [92, 93], see also footnote 4	$2^{68.5}$
ElGamal [94]	Single key. CCA	Trivial algebraic	Instant
NTRUEncrypt [95]	Single key. COA	Hybrid [96] (Lattice reduction and combinatorial search)	PB, see also footnote 5

PB = parameter-based.

1. Another attack claiming to break 855 rounds [97] of Trivium has been questioned in [98].
2. We are not aware of any attacks faster than brute force. Rabbit has four initialization rounds. The values within the cipher become balanced after two rounds [83], hence there is a trivial distinguishing attack against at least one round of the cipher.
3. There exist attacks against earlier versions of the cipher. The cryptanalysis of the final version made by the designers is secret to the best of our knowledge.
4. Our upper-bound estimation: In [93], the attack time is given as 2, 700 core years of computations using Intel Xeon Gold 6130 CPU (each 2.1 GHz). To convert this attack time to the RSA-250 encryptions, we would need to know how much time is required on average to apply one encryption on the mentioned processor. For a rough estimate, we assume that one encryption requires less time than one integer operation as tested in [99].
5. The actual attack time depends on the specific parameter choices. See [100] for more details.

Attack definition. Before proceeding to the discussion about various attack types (see Section 1.8.1), it's essential to clarify the concept of an attack against a modern cipher. We start this explanation with Kerckhoffs's principle (see Section 1.5). This principle emphasizes that a cryptosystem should be secure even if all the system details, excluding the secret key, are known to the attacker.

However, the principle brings up the term "secure." To formulate the definition of security, we use ideas about the *infeasibility of distinguishing*—see Sections 1.8.2 and 1.8.3. In a nutshell, a cryptographic attack is an algorithm that aims to demonstrate the lack of security in a given cryptosystem.

Attack costs. When analyzing how difficult it is to apply a cryptographic attack, the *computational complexity* of the corresponding algorithm is evaluated. The computational complexity is the amount of resources needed to run the algorithm. There are typically three main resources considered: time, memory, and data.

- Time complexity of the attack, or just attack time, is an estimated upper limit of the number of operations required to successfully break a cipher. Time is the primary resource taken into account. If "computational complexity" is mentioned without further specification, it typically refers to "time complexity."
- Memory complexity is the storage space needed to execute the attack.
- The data complexity refers to the amount of data (plaintext, ciphertext, or both) that the attacker needs access to in order to carry out the attack.

Attack time. The *attack time* is generally expressed in the number of a particular cipher's encryptions. This is done in order to demonstrate by which factor the corresponding attack is faster than the brute-force attack. As discussed in Section 1.2.2, the key-space size has a direct relation to the attack time of the brute force. Testing each of the keys requires the corresponding encryption algorithm to run once completely. So if the key (in binary representation) length is L, and all possible variants of the key lead to different ciphertexts, then the key space size is 2^L. It means that in order to certainly break a cipher, 2^L encryptions are always enough. This determines the attack time of the exhaustive search.

Different attacks may not require running the encryption algorithm itself, but to perform other computational operations. In this case, an estimation is done on how many of such operations require the time equivalent to the time of one encryption. Then the whole number of operations needed to apply the attack is divided over the number of operations equating to a single encryption. This results in the time complexity for the current attack measured in encryptions.

Security parameter. A cryptographic attack is considered to be successful if it requires less costs than defined by the security parameter set by the designers of a cryptosystem. A security parameter measures the level of difficulty for an adversary to break a cryptographic scheme. It is often expressed in bits. For example, one can say that a certain scheme offers κ-bit security if the attack time is of $O(2^\kappa)$ encryptions. The $O()$ notation (also called big O notation or Bachmann–Landau notation or asymptotic notation) describes an upper bound on the time complexity of an algorithm. Essentially, it gives the worst-case scenario for how the run time grows as the input size increases. Here we don't need the big O notation, which is used to describe the limiting behavior of a function when the argument tends towards a particular value. But here in the table, we use the concrete versions of the ciphers and provide the complexities with a constant argument.

In the context of symmetric encryption schemes, the security parameter is typically equal to the key size. This is because the brute-force attack sets the minimum limit for the security parameter. However, the security parameter can be lower than the key size if an attack faster than the brute force is known at the stage of the design of a cipher. This is a common situation for public-key encryption schemes.

Goal. In modern cryptology, different classifications of cryptanalytical attacks exist. By the goal of the attacker we differentiate between *key-recovery attacks* and *distinguishing attacks*. The key-recovery attacks aim to obtain the actual encryption or decryption key, compromising the security of the cryptographic system completely. On the other hand, distinguishing attacks focus on the ability to differentiate encrypted data from truly random data, indicating deviations or weaknesses in the cipher that may lead to key-recovery attacks.

Single/multiple keys. Cryptanalytic attacks also vary based on the attacker's ability to observe different numbers of encryption instances related to distinct keys. *Single-key attacks* assume access to the ciphertexts encrypted under the same key. *Variable-key attacks* assume access to ciphertexts encrypted under multiple unknown keys. This often mirrors real-world situations where a cipher's user must change the key after a certain number of encryptions. If an attacker gains access to several corresponding ciphertexts, he can use this information as an advantage in attempting to break any of the corresponding encryptions. *Related-key attacks* assume that an attacker has knowledge of a certain mathematical relationship that exists between different secret keys and that she can observe the corresponding ciphertexts. Although at first glance, such a scenario can be seen as too unrealistic, several cryptosystems were broken using related-key attacks in the real world (e.g., [43]).

Access to data (ciphertext-plaintext pairs). The cryptographic attacks can be divided into the following four main categories based on the type of access to the ciphertext and plaintext (assuming the key is always unknown):

- Ciphertext-only attacks (COA) assume access only to ciphertexts without knowledge of corresponding plaintexts;
- Known-plaintext attacks (KPA) involve pairs of known plaintext and their corresponding ciphertext, aiming to recover the secret key;
- Chosen-plaintext attacks (CPA) allow the attacker to choose arbitrary plaintexts and obtain their ciphertexts, providing flexibility in analyzing the encryption algorithm;
- Chosen-ciphertext attacks (CCA) enable the attacker to choose arbitrary ciphertexts and obtain their plaintexts, possessing the power to manipulate ciphertexts during decryption.

Additionally, attacks differ based on specific mathematical methods, such as differential cryptanalysis (analyzing how differences between inputs of the ciphers affect resultant differences between outputs), linear cryptanalysis (exploiting linear relationships in the encryption process), meet-in-the-middle, biclique, integral, boomerang, cube, and other attacks. All these methods are unique, so we refer to the provided references for a comprehensive explanation.

1.8.2 Indistinguishability Security Definitions

The attack types CPA and CCA have a direct relationship with the cryptographic security definitions IND-CPA, IND-CCA1, and IND-CCA2. These definitions play a crucial role in the provable security branch of cryptography. This field focuses on proving mathematically the security of the cryptographic schemes. This is achieved by demonstrating that breaking a certain scheme would require solving a problem that is widely known to be difficult, such as factoring large numbers or computing discrete logarithms.

Indistinguishability under chosen-plaintext attack (IND-CPA). In this model, an attacker is allowed to choose arbitrary plaintexts and obtain the corresponding ciphertexts from the encryption oracle as many times as he needs. Then the adversary chooses two distinct challenge messages and sends them to the encryption oracle, which returns a ciphertext of just one of them called challenge ciphertext. After that, the attacker is allowed to perform any number of additional computations and encryptions. An encryption scheme is considered secure if the attacker can't guess to which plaintext the challenge refers to with the probability higher than $|1/2 + \eta|$ where η is negligible. Clearly, the attacker cannot choose the same messages for the challenge for which he gets the ciphertexts from the oracle. This security definition can be applied to both symmetric and asymmetric encryption schemes, although formally they are described differently [101]. However, in case of deterministic asymmetric encryption schemes, an attacker has access to the public key, which means that he can easily distinguish which ciphertext was produced by which message by encrypting the messages by himself. Therefore, the definition is only applied to probabilistic public-key encryption schemes where randomness is used in the encryption process. This implies that the same message encrypted several times under the probabilistic encryption scheme results in different ciphertexts.

Indistinguishability under chosen-ciphertext attack, also known as nonadaptive or lunchtime attack (IND-CCA1). This security definition imposes a higher level of security than IND-CPA. In this model, an attacker can choose both the plaintexts and obtain their corresponding ciphertexts from the oracle, and also decrypt arbitrary ciphertexts and get the corresponding plaintexts. The further procedure is similar to the IND-CPA case. However, in the case of IND-CCA1 after the adversary gets the challenge the decryption oracle becomes unavailable.

Indistinguishability under adaptive chosen-ciphertext attack (IND-CCA2). This is the strongest definition providing the highest level of security. It allows the attacker to continue to interact with the decryption oracle even after the challenge ciphertext is received.

When considering modern cryptographic encryption primitives, selecting the best attack is not a straightforward task. In Table 1.3, we have kept the information concise and prioritized key-recovery attacks requiring minimal computation and being faster than brute-force, which is a universal attack method against any encryption algorithm. By this prioritizing, we have left out other complexities such as data and memory costs (e.g., number of required plaintext-ciphertext pairs).

Single-key scenarios are typically targeted, except for two exceptions in our table: the related-key attack against Kasumi cipher and the variable-key attack against RC4. If the full cipher is not compromised, we aim to select attacks that break as many rounds as possible. We only refer to distinguishing attacks against MUGI and Enocoro as we are not aware of any published key-recovery attacks.

1.8.3 Security Definitions

Modern cryptography is heavily based on mathematical theory and computer science practice. Cryptographic algorithms are designed around computational hardness assumptions, making such algorithms hard to break in practice by any adversary.

There are different approaches (categories) to define the security of cryptosystems.

Most commonly, two fundamental approaches are used for formally defining the security of an encryption scheme [102]:

- The first one is semantic security, which implies that it is infeasible for an attacker to learn any information about the plaintext from the ciphertext;

- The second definition determines security as the infeasibility of distinguishing between encryptions of two given messages.

In both definitions of security, the term "infeasible" rather than "impossible" is used. This is because generic attacks exist against almost every known encryption scheme (with the exception of the one-time-pad). One such universal attack, namely a brute force, was discussed in Section 1.2.2. Brute-force attacks can be extended to time-memory trade-off (TMTO) attacks, a broader class of attacks, which in certain cases allow to reduce the key-recovery time by increasing the memory cost. See Table 1.3 for an in-depth discussion of different attack types.

Another main category in literature defines security depending on the adversary's capabilities (e.g., *Cryptography 101* [103, Chap. 1.2.2]):

Computational, conditional, or practical security. A cipher is *computationally* secure if it is theoretically possible to break such a system, but it is infeasible to do so by any known practical means. Theoretical advances (e.g., improvements in integer factorization algorithms) and faster computing technology require these solutions to be continually adapted.

Even using the best known algorithm for breaking it will require so many resources (e.g., 1,000,000 years) that essentially the cryptosystem is secure.

So this concept is based on assumptions of the adversary's limited computing power and the current state of science.

A typical example of a pragmatically secure procedure is AES: No practicable attack is known on it. Even so, AES is theoretically broken, which just means it can be broken with less effort than a brute-force attack. This effort is still unrealistically high. See Section 1.7.

Information-theoretical or unconditional security. A cipher is considered *unconditionally* secure if its security is guaranteed no matter how many resources (time, space) the attacker has. Even if the adversary has unlimited resources he is unable to gain any meaningful data from a ciphertext.

The only information-theoretically secure schemes that provably cannot be broken even with unlimited computing power are the *one-time pad* (OTP) or variants of it.

Figure 1.6 shows that it may be impossible to determine the correct plaintext from a OTP (if the OTP method has been applied correctly and if all keys have the same likelihood). The example in this figure uses an 8-character long given ciphertext: `11 1B 1E 18 00 04 0A 15`. The hex values correspond to the ASCII values of the letters: For example, the letter `C` has the numerical value `67` (decimal), which is `43` in hex representation.

There are many meaningful words with eight letters and for each there is a correct key. So an attacker cannot determine alone from the ciphertext which is the correct key and which is the correct plaintext word. In other words, with different keys the same ciphertext can lead to different meaningful plaintexts and so, in this case, it cannot be distinguished which plaintext is the correct one.[12]

12. The OTP procedure is discussed in more detail in Section 2.2.4 in item "One-time pad."
Also see Figure 9.12, where a corresponding example with text strings is built with SageMath, and the XOR method is explained.

Key (hex)	Revealed plaintext (hex)	Revealed plaintext (txt)
52 49 47 48 54 4B 45 59	43 52 59 50 54 4F 4F 4C	CRYPTOOL
41 5A 50 57 52 45 47 54	50 41 4E 4F 52 41 4D 41	PANORAMA
54 77 7B 68 68 65 64 61	45 4C 45 50 48 41 4E 54	ELEPHANT
50 55 5B 55 4F 4A 4F 46	41 4E 45 4D 4F 4E 45 53	ANEMONES
52 53 5F 55 50 4D 45 5B	43 48 41 4D 50 49 4F 4E	CHAMPION
42 58 5B 56 41 56 43 5A	53 43 45 4E 41 52 49 4F	SCENARIO

Figure 1.6 Illustration of the information-theoretically secure OTP scheme.[13]

As the OTP is information-theoretically secure it derives its security solely from information theory and is secure even with unlimited computing power at the adversary's disposal. However, OTP has several practical disadvantages (the key must be used only once, must be randomly selected, and must be at least as long as the message being protected), which means that it is hardly used except in closed environments such as for the hot wire between Moscow and Washington.

Two more security concepts are sometimes used:

- *Provable security.* This means that breaking such a cryptographic system is as difficult as solving some supposedly difficult problem, such as discrete logarithm computation, discrete square root computation, or very large integer factorization.

  Example: Currently we know that RSA is at most as difficult as factorization, but we cannot prove that it's exactly as difficult as factorization. So RSA has no proven minimum security. Or in other words, we cannot prove that if RSA (the cryptosystem) is broken, then factorization (the hard mathematical problem) can be solved.

  The Rabin cryptosystem was the first cryptosystem that could be proven to be computationally equivalent to a hard problem (integer factorization).
- *Ad-hoc security.* A cryptographic system has this security feature if it is not worth trying to break the system because the effort to do so is more expensive than the value of the data that would be obtained by doing so. Or an attack can't be done in sufficiently short time (see [104]).

  Example: This may apply if a message relevant to the stock market will be published tomorrow, and you would need a year to break it.

13. Source of the four photos: https://pixabay.com/.

1.9 Algorithm Types and Self-Made Ciphers

Here, two aspects of crypto procedures are mentioned briefly, which are often not discussed early enough: types of algorithms and the thinking up of new algorithms.

1.9.1 Types of Algorithms

Algorithms can be categorized as follows:

- *Random-based.* Algorithms can be divided up into *deterministic and heuristic* methods. Often students only become aware of deterministic methods, where the output is uniquely determined by the input. On the other hand, heuristic methods make decisions using random values and the results are only correct with a certain probability. One can differentiate even more precisely between randomized algorithms, and probabilistic and heuristic methods, but these subtleties are not important for understanding the contrast to deterministic methods.

  Random looms large in cryptographic methods. Keys have to be selected randomly, which means that at least for the key generation "random" is necessary. In addition, some methods, especially from cryptanalysis, are heuristic.
- *Constant-based.* Many modern methods (especially hash methods and symmetric encryption) use numeric constants. Their values should be plausible, and they shouldn't contain back doors. Numbers fulfilling this requirement are called nothing-up-my-sleeve numbers.

1.9.2 New Algorithms

It happens again and again that someone without deeper knowledge of adequate design concepts comes up with a "new" encryption procedure. However, reality shows that this is not a good idea. That's why people usually learn early not to design their own cryptosystem if they hope that the fact that it is not known will protect them. There are many reasons for this, including that it only takes one disgruntled employee or any other malicious actor to reveal the secrets that make the scheme secure. Designing secure cryptographic schemes is extremely difficult. It is incredibly easy to create something that looks secure, but actually leaks information.

Offering prize money and just single ciphertexts is unprofessional—serious researchers have little time and will not spend any effort on it (perhaps they give it to students as an exercise for didactic reasons). Modern best practice is that if you want to create a new encryption scheme, first publish it with a detailed explanation of how it works, its advantages, and any evidence of its security. Then you can see if anyone can find any weaknesses. This is not a quick process—you should expect it to take years.

1.10 Further References and Recommended Resources

Here are some good cryptography books that can serve as useful background on various topics in order from beginners (history) to intermediate (applied) to advanced (theory-focused):

- David Kahn: *The Codebreakers*, 1995.
- Elonka Dunin and Klaus Schmeh: *Codebreaking: A Practical Guide*, 2nd ed, 2023.
- Simon Singh: *The Code Book*, 2000 [105].
- Bruce Schneier, *Applied Cryptography, Protocols, Algorithms, and Source Code in C*, 2nd ed, 1996 [8].
- Christof Paar and Jan Pelzl: *Understanding Cryptography*, 2009 [106].
- David Wong: *Real-World Cryptography*, 2020 [107] (our favorite).
- Jean-Philippe Aumasson: *Serious Cryptography*, 2017 [108].
- Mike Rosulek: *The Joy of Cryptography*, 2021.
- Niels Ferguson, Bruce Schneier, and Tadayoshi Kohno: *Cryptography Engineering*, 2010.
- Dan Boneh and Victor Shoup: *A Graduate Course in Applied Cryptography*, v0.6, 2023.
- Mark Stamp and Richard M. Low: *Applied Cryptanalysis: Breaking Ciphers in the Real World*, 2007 [109].
- Rolf Oppliger, *Cryptography 101*, 2021 [103].
- Jonathan Katz and Yehuda Lindell: *Introduction to Modern Cryptography*, 3rd ed, 2020.
- Douglas R. Stinson: *Cryptography – Theory and Practice*, 3rd ed, 2006 [110].

Besides the information in these books and in the following chapters, there is also a good number of websites and the online help of all CrypTool variants that contain many details about encryption methods.

The book by Bruce Schneier [8] offers an easy overview of the different encryption algorithms. For a more in-depth introduction, in addition to the book by Rolf Oppliger [103], we also recommend the books by David Wong [107], Jean-Philippe Aumasson [108], and Douglas R. Stinson [110].

1.11 AES Visualizations/Implementations

AES is now probably the most widely used modern encryption algorithm worldwide. AES is a secure, standardized, symmetrical process that encrypts data, for example, in Wi-Fi and browser connections. The AES-192 and AES-256 variants are approved for top-class government documents in the United States.

In the following sections, first an AES animation is presented in CTO; and then AES is executed directly—once in CT2 and twice with OpenSSL (once on the command line of the operating system and once in the OpenSSL WebAssembly plugin in CTO).

1.11.1 AES Animation in CTO[14]

Figure 1.7 shows that the modern encryption algorithm receives both inputs (the key and the plaintext) in binary form and creates the output in binary form. Like most modern (block) ciphers, the algorithm contains a key scheduling part where from the given key (also called session key, master key, or cipher key) the round keys are generated, and another part where then the actual encryption is carried out using the generated round keys.

Figures 1.7 to 1.8 are taken from the AES animation in CrypTool-Online (CTO). Figure 1.9 is from CT1, but the image is also part of the CTO animation.

1.11.2 AES in CT2

After these visualizations, we want—in a concrete example—to encrypt a plaintext of length 128 bits (one block) with a 128-bit key with AES in CBC mode. From the

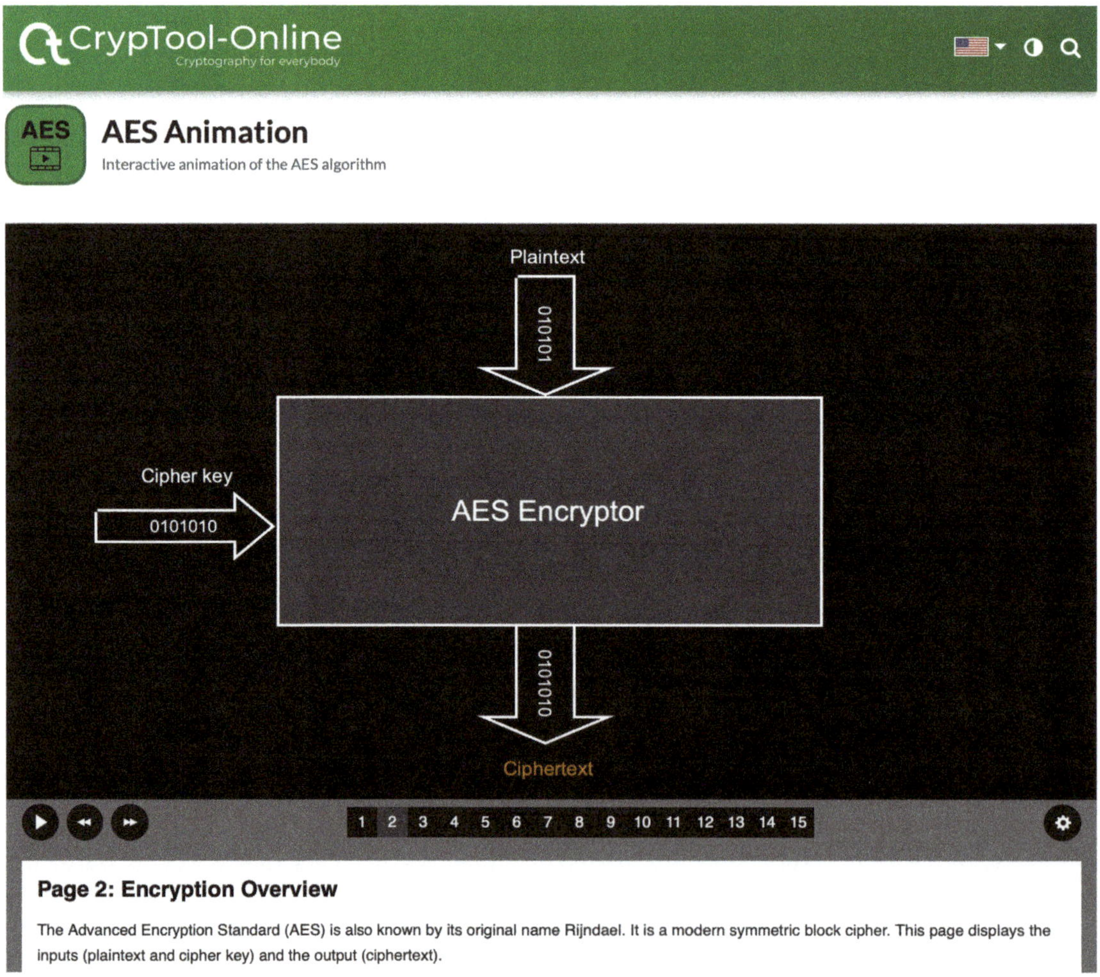

Figure 1.7 AES visualization from CTO (part 1).

14. https://www.cryptool.org/en/cto/aes-animation.

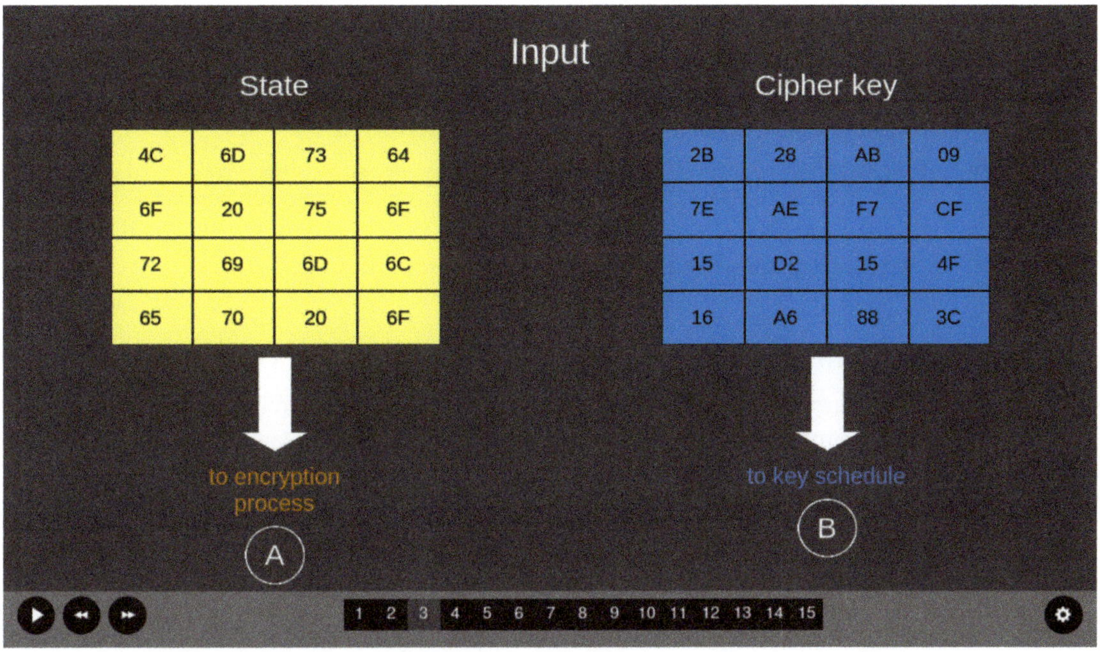

Figure 1.8 AES visualization from CTO (part 2).

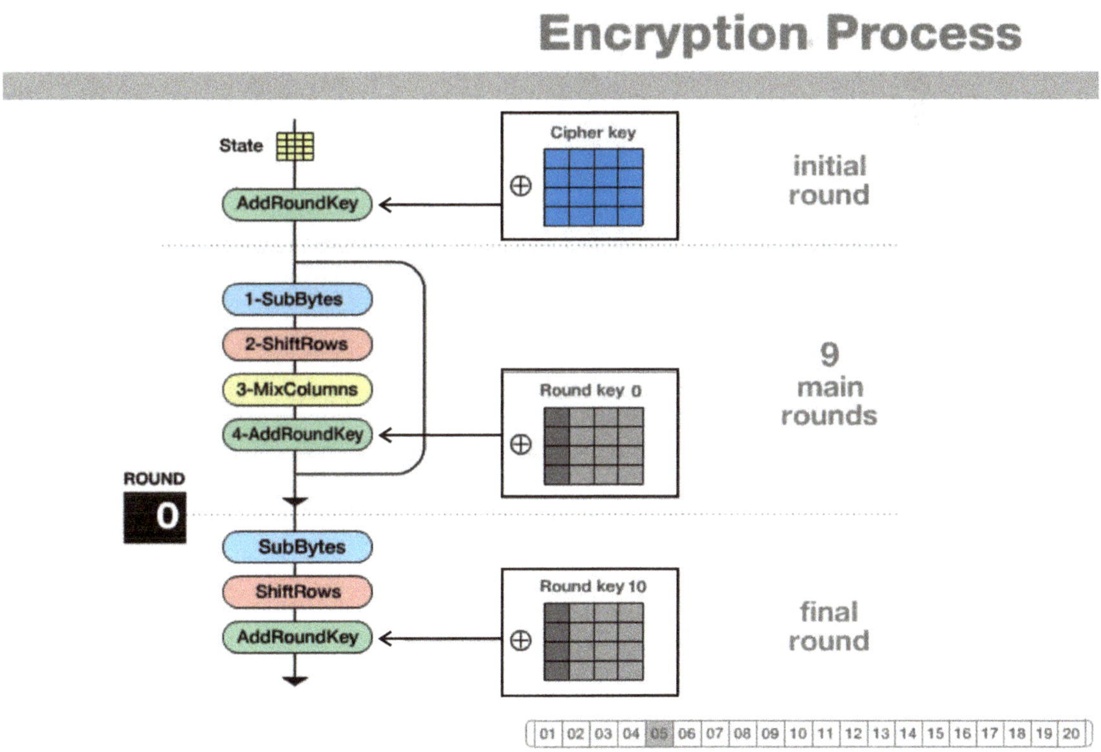

Figure 1.9 AES visualization by Enrique Zabala from CT1.

received ciphertext we are only interested in the first block (if the plaintext doesn't fill up a complete block, for the sake of simplicity, here we use zero padding).

For demonstration, we do it once with CT2 and twice with OpenSSL.[15]

The plaintext `AESTEST1USINGCT2` is converted to hex (41 45 53 54 45 53 54 31 55 53 49 4E 47 43 54 32). Using this and the key 3243F6A8885A308D313198A2E0370734 the AES component creates the ciphertext, which is in hex: B1 13 D6 47 DB 75 C6 D8 47 FD 8B 92 9A 29 DE 08.

Figure 1.10 shows the encryption of one block in CT2.[16]

1.11.3 AES with OpenSSL at the Command Line of the Operating System

OpenSSL Example 1.1 achieves the same result as CT2 with `OpenSSL` from the (Windows) command line.

OpenSSL Example 1.1: AES Encryption (Of Exactly One Block and Without Padding)

```
>openssl enc -e -aes-128-cbc -K 3243F6A8885A308D313198A2E0370734 -iv 00▸
  ▸00000000000000000000000000000 -in klartext-1.hex  -out klartext-1.▸
  ▸hex.enc
>dir
06.07.2016  12:43                  16 key.hex
20.07.2016  20:19                  16 klartext-1.hex
20.07.2016  20:37                  32 klartext-1.hex.enc
```

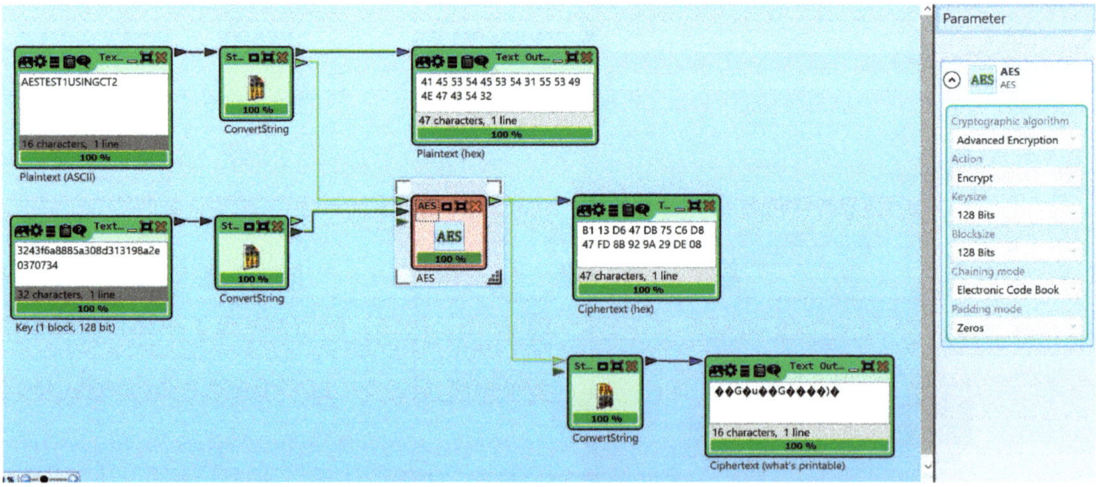

Figure 1.10 AES encryption (here exactly 1 block and without padding) in CT2.

15. OpenSSL is a widespread free open-source crypto library that contains the command line tool `openssl`.
    Using OpenSSL you can try out the functionality on many operating systems.
    You can find an introduction into the CLI `openssl` (e.g. at https://www.cryptool.org/en/documentation/ctbook/).
16. This is similar to the following template: CT2 Templates ▹ Cryptography ▹ Modern ▹ Symmetric ▹ AES Cipher (Text Input).

Note: As OpenSSL Example 1.2 shows, with a little effort, pipes, and the tool xxd, this can be achieved also in a Bash shell and without using temporary files:[17]

OpenSSL Example 1.2: AES Encryption (Without Temporary Files) With Bash

```
$ echo 0: 41 45 53 54 45 53 54 31 55 53 49 4E 47 43 54 32 | xxd -r | ▸
  ▸openssl enc -e -aes-128-cbc -nopad -K 3243F6A8885A308D313198A2E03707▸
  ▸34 -iv 00000000000000000000000000000000  | xxd -p
b113d647db75c6d847fd8b929a29de08

$ echo -n AESTEST1USINGCT2 | openssl enc -e -aes-128-cbc -nopad -K 3243▸
  ▸F6A8885A308D313198A2E0370734 -iv 00000000000000000000000000000000  |▸
  ▸ xxd -p
b113d647db75c6d847fd8b929a29de08
```

1.11.4 AES with OpenSSL within CTO[18]

As CTO has integrated a WebAssembly-based version of OpenSSL, this also can be done locally in your browser without the need to install OpenSSL. While Linux systems mostly have OpenSSL on board, Windows systems or smart phones don't. For such systems this plugin is helpful.

For the example in Figure 1.11 we store the message AESTEST1USINGCT2 in a file called "klartext-1.hex." Then we upload this file from the file system of the operating system into a virtual file system in the browser: This upload is done in the tab "Files" of the OpenSSL plugin. Then in the OpenSSL plugin the same `openssl` command is executed as before in the terminal (see Section 1.11.3). And if you download the resulting file klartext-1.hex.enc and compare it with the result from the terminal, you see both are identical.

1.12 Educational Examples for Symmetric Ciphers Using SageMath

Section 1.12.1 shows the SageMath implementation of a cipher (called Mini-AES) stripped for didactic purposes. Further publications with ciphers reduced for didactic reasons are listed in Section 1.12.2.

1.12.1 Mini-AES

The SageMath module `crypto/block_cipher/miniaes.py` supports Mini-AES to allow students to explore the inner working of a modern block cipher.

Mini-AES, originally described in [111], is a simplified variant of AES to be used for cryptography education.

Here is a short list about how Mini-AES was simplified compared to AES:

17. xxd creates a hex dump of a given file or of standard input. With the option "-r" it converts hex dump back to its original binary form.
18. `https://www.cryptool.org/en/cto/openssl`.

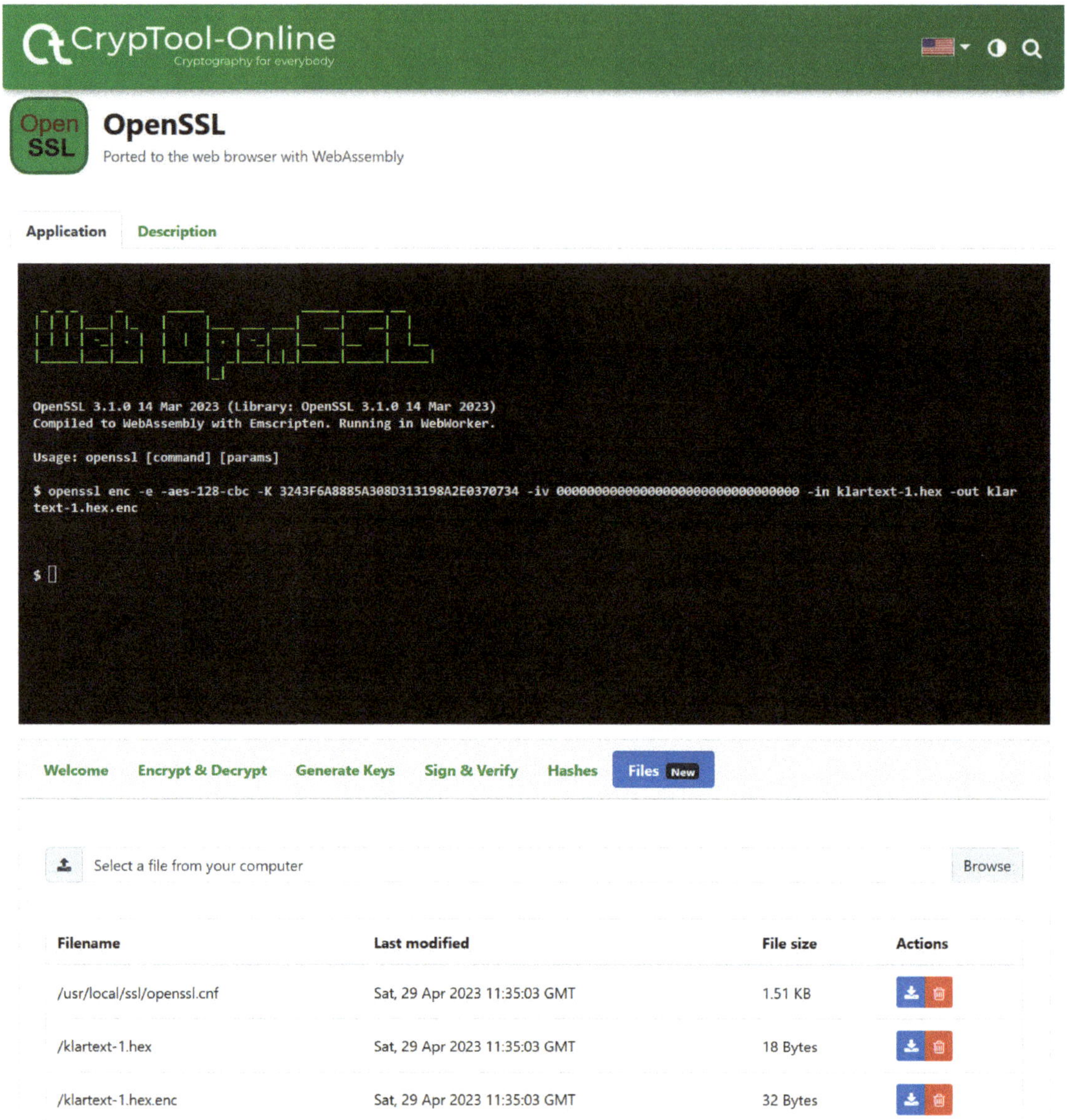

Figure 1.11 AES encryption using OpenSSL in the browser.

- The AES has a block size of 128 bits, and supports key sizes of 128, 192, and 256 bits. The number of rounds is 10, 12, or 14 for the three different key sizes, respectively.

  Mini-AES has a 16-bit block size, a 16-bit key size, and 2 rounds.

- The 128-bit block of the AES is expressed as a matrix of 4×4 bytes, in contrast to Mini-AES expressing its 16-bit block as a matrix of 2×2 nibbles (half-bytes).

- The AES key schedule takes the 128-bit secret key and expresses it as a group of four 32-bit words.

> The Mini-AES key schedule takes the 16-bit secret key and expresses it as a group of four nibbles (4-bit words).

How to use Mini-AES is exhaustively described at this SageMath reference page: https://doc.sagemath.org/html/en/reference/cryptography/sage/crypto/block_cipher/miniaes.html.

SageMath Example 1.1 was originally taken from the release tour of SageMath 4.1[19] and calls the implementation of the Mini-AES.

SageMath Example 1.1: Encryption and Decryption with Mini-AES

```
print("\n# CHAP01 -- Sage-Script-SAMPLE 010: =========")

# (1) Encrypting a plaintext using Mini-AES
from sage.crypto.block_cipher.miniaes import MiniAES
maes = MiniAES()
K = FiniteField(16, "x")
MS = MatrixSpace(K, 2, 2)

P = MS([K("x^3 + x"), K("x^2 + 1"), K("x^2 + x"), K("x^3 + x^2")]); ▸
  ▸print("(1) P:\n",P, sep="")

key = MS([K("x^3 + x^2"), K("x^3 + x"), K("x^3 + x^2 + x"), K("x^2 + x ▸
  ▸+ 1")]); print("key:\n",key, sep="")

C = maes.encrypt(P, key); print("C:\n",C, sep="")

# decryption process
plaintxt = maes.decrypt(C, key)
print(plaintxt == P)

# (2) Working directly with binary strings
maes = MiniAES()
bin = BinaryStrings()
key = bin.encoding("KE"); print("\n(2) key:\n",key, sep="")

P = bin.encoding("Encrypt this secret message!"); print("P:\n",P,sep▸
  ▸="")
C = maes(P, key, algorithm="encrypt"); print("C:\n",C,sep="")
plaintxt = maes(C, key, algorithm="decrypt")
print(plaintxt == P)

# 3) Or working with integers n such that 0 <= n <= 15:
maes = MiniAES()
P = [n for n in range(16)]; print("\n(3) P:\n",P, sep="")
key = [2, 3, 11, 0]; print("key:\n",key, sep="")

P = maes.integer_to_binary(P)
key = maes.integer_to_binary(key)
C = maes(P, key, algorithm="encrypt"); print("C:\n",C, sep="")
plaintxt = maes(C, key, algorithm="decrypt")
print(plaintxt == P)
```

19. See `https://mvngu.wordpress.com/2009/07/12/sage-4-1-released/`.
Further example code for Mini-AES can be found in [112].

Further details concerning cryptosystems within SageMath (e.g., about the Simplified Data Encryption Standard (SDES)) can be found in the thesis of Minh Van Nguyen [113].

1.12.2 Symmetric Ciphers for Educational Purposes

Compared to public-key ciphers based on mathematics, the structure of AES and most other modern symmetric ciphers (like DES, IDEA, or Present), is very complex and cannot be explained as easily as RSA.

So, simplified variants of modern symmetric ciphers were developed for educational purposes in order to allow beginners to perform encryption and decryption by hand and gain a better understanding of how the algorithms work in detail. These simplified variants also help to understand and apply the corresponding cryptanalysis methods.[20]

The most well-known of these variants are SDES[21] and Simplified-AES (S-AES)[22] by Ed Schaefer and his students [115], and Mini-AES (see Section 1.12.1):

- Edward F. Schaefer: *A Simplified Data Encryption Standard Algorithm* [116].
- Raphael Chung-Wei Phan: *Mini Advanced Encryption Standard (Mini-AES): A Testbed for Cryptanalysis Students* [111].
- Raphael Chung-Wei Phan: *Impossible Differential Cryptanalysis of Mini-AES* [117].
- Mohammad A. Musa, Edward F. Schaefer, Stephen Wedig: *A Simplified AES Algorithm and Its Linear and Differential Cryptanalyses* [118].
- Nick Hoffman: *A Simplified Idea Algorithm* [119].
- S. Davod. Mansoori, H. Khaleghei Bizaki: *On the Vulnerability of Simplified AES Algorithm Against Linear Cryptanalysis* [120].

References

[1] International Association for Cryptologic Research (IACR), *IACR*, https://www.iacr.org/.

[2] *War of the Letters*, BBC documentary film (in German, *Krieg der Buchstaben*), 1994, https://www.youtube.com/watch?v=yfw1JLI8aWk.

[3] Nichols, R. K., *Classical Cryptography Course*, Volumes 1 and 2, Tech. Rep., 12 lessons, Aegean ParkPress 1996, https://www.cryptogram.org/resource-area/crypto-lessons-tutorials-lanaki/.

20. A very good starting point to learn cryptanalysis is the book from Mark Stamp [109]. Also good, but very high-level and concentrating on analyzing symmetric block ciphers only, is the article from Bruce Schneier [114].

    Several of the cipher challenges at MysteryTwister (`https://www.mysterytwister.org`) are also well suited for educational purposes.

21. If you double-click on the title of the icon of the SDES component in CT2 you can see a visualization of the SDES algorithm, showing how the bits of the given data flow through the whole algorithm. A corresponding screenshot: https://www.facebook.com/CrypTool2/photos/a.505204806238612.1073741827.243959195696509/597354423690316.

22. See the template: `CT2 Templates` ▷ `Cryptography` ▷ `Modern` ▷ `Symmetric` ▷ `S-AES`.

[4] *The RSA Secret Key Challenge*, RSA Labs (formerly RSA Security), https://web.archive.org/web/20170417095446/http://www.emc.com/emc-plus/rsa-labs/historical/the-rsa-laboratories-secret-key-challenge.htm.

[5] *DES Challenge*, RSA Labs (former RSA Security), https://web.archive.org/web/20061210141223/http://www.rsasecurity.com/rsalabs/node.asp?id=2108.

[6] *Press and Articles Related to Project RC5-64*, https://www.distributed.net/Press-room_press-rc5-64.

[7] BSI, *Technical Guideline TR-02102-1, Cryptographic Mechanisms: Recommendations and Key Lengths (Version 2022-01)*, Tech. Rep. 2022, https://www.bsi.bund.de/SharedDocs/Downloads/EN/BSI/Publications/TechGuidelines/TG02102/BSI-TR-02102-1.pdf.

[8] Schneier, B., *Applied Cryptography, Protocols, Algorithms, and Source Code in C*, Second Edition, Wiley, 1996.

[9] Esslinger, B., J. Schneider, and V. Simon. "Krypto + NSA = ? – Kryptografische Folgerungen aus der NSA-Affäre," in *KES Zeitschrift für Informationssicherheit*, March 2014, pp. 70–77, https://www.cryptool.org/assets/ctp/documents/krypto_nsa.pdf.

[10] Miller, A. R., *The Cryptographic Mathematics of Enigma*, Revised Edition, 2019, Center for Cryptologic History, National Security Agency, 1995, https://www.nsa.gov/portals/75/documents/about/cryptologic-heritage/historical-figures-publications/publications/wwii/CryptoMathEnigma_Miller.pdf.

[11] Ostwald, O., *Cryptographic Design Flaws of Early Enigma*, 2023, https://cryptocellar.org/enigma/files/enigma-design-flaws.pdf.

[12] International Conference on Cryptologic History (ICCH), https://www.cryptologichistory.org/.

[13] Kopal, N, "Solving Classical Ciphers with CrypTool 2," in *Proceedings of the 1st International Conference on Historical Cryptology*, 2018, pp. 29–38.

[14] Kopal, N, "Cryptanalysis of Homophonic Substitution Ciphers Using Simulated Annealing with Fixed Temperature," in *Proceedings of the 2nd International Conference on Historical Cryptology*, 2019, pp. 107–116.

[15] Lasry, G., N. Biermann, and S. Tomokiyo, "Deciphering Mary Stuart's Lost Letters from 1578–1584," *Cryptologia*, Vol. 47, No. 2, 2023, pp. 101–202.

[16] Lasry, G., B. Megyesi, and N. Kopal, "Deciphering Papal Ciphers from the 16th to the 18th Century," *Cryptologia*, Vol. 45, No. 6, 2021, pp. 479–540, https://www.tandfonline.com/doi/full/10.1080/01611194.2020.1755915.

[17] Dunin, E., et al., "How We Set New World Records in Breaking Playfair Ciphertexts," *Cryptologia*, Vol. 46, No. 4, 2022, pp. 302–322.

[18] Lasry, G., "Solving a 40-Letter Playfair Challenge with CrypTool 2," in *Proceedings of the 2nd International Conference on Historical Cryptology*, 2019, pp. 23–26.

[19] Lasry, G., "Deciphering German Diplomatic and Naval Attaché Messages from 1914–1915," in *Proceedings of the 1st International Conference on Historical Cryptology*, 2018, pp. 55–64.

[20] Lasry, G., et al., "Cryptanalysis of Chaocipher and Solution of Exhibit 6," *Cryptologia*, Vol. 40, No. 6, 2016, pp. 487–514.

[21] Lasry, G., N. Kopal, and A. Wacker, "Cryptanalysis of Columnar Transposition Cipher with Long Keys," *Cryptologia*, Vol. 40, No. 4, 2016, pp. 374–398.

[22] Lasry, G., N. Kopal, and A. Wacker, "Solving the Double Transposition Challenge with a Divide-and-Conquer Approach," *Cryptologia*, Vol. 38, No. 3, 2014, pp. 197–214.

[23] Lasry, G., et al., "Deciphering ADFGVX Messages from the Eastern Front of World War I," *Cryptologia*, Vol. 41, No. 2, 2017, pp. 101–136.

[24] Gillogly, J. J., "Ciphertext-Only Cryptanalysis of Enigma," *Cryptologia*, Vol. 19, No. 4, 1995, pp. 405–413.

[25] Lasry, G., N. Kopal, and A. Wacker, "Cryptanalysis of Enigma Double Indicators with Hill Climbing," *Cryptologia*, Vol. 43, No. 4, 2019, pp. 267–292.

[26] Ostwald, O., and F. Weierud, "Modern Breaking of Enigma Ciphertexts," *Cryptologia*, Vol. 41, No. 5, 2017, pp. 395–421.

[27] Lasry, G., "Solving a Tunny Challenge with Computerized 'Testery' Methods," in *Proceedings of the 3rd International Conference on Historical Cryptology*, 2020.

[28] Lasry, G., N. Kopal, and A. Wacker, "Automated Known-Plaintext Cryptanalysis of Short Hagelin M-209 Messages," *Cryptologia*, Vol. 40, No. 1, 2016, pp. 49–69.

[29] Lasry, G., N. Kopal, and A. Wacker, "Ciphertext-Only Cryptanalysis of Short Hagelin M-209 Ciphertexts," *Cryptologia*, Vol. 42, No. 6, 2018, pp. 485–513.

[30] Lasry, G., "A Practical Meet-in-the-Middle Attack on SIGABA," in *Proceedings of the 2nd International Conference on Historical Cryptology*, 2019, pp. 23–26.

[31] Lasry, G., "Cracking SIGABA in Less than 24 Hours on a Consumer PC," *Cryptologia*, Vol. 47, No. 1, 2023, pp. 1–37.

[32] Matsui, M., "Linear Cryptanalysis Method for DES Cipher," in *Advances in Cryptology—EUROCRYPT'93: Workshop on the Theory and Application of Cryptographic Techniques*, Lofthus, Norway, May 23–27, 1993, Springer, pp. 386–397.

[33] Junod, P., "On the Complexity of Matsui's Attack," in *Selected Areas in Cryptography: 8th Annual International Workshop, SAC 2001*, Toronto, Ontario, Canada, August 16–17, 2001, Springer, pp. 199–211.

[34] Merkle, R. C., and M. E. Hellman, "On the Security of Multiple Encryption," *Communications of the ACM*, Vol. 24, No. 7, 1981, pp. 465–467.

[35] Daemen, J., and V. Rijmen, *The Design of Rijndael*, Volume 2, Springer, 2002.

[36] Bogdanov, A., D. Khovratovich, and C. Rechberger, "Biclique Cryptanalysis of the Full AES," in *Advances in Cryptology–ASIACRYPT 2011: 17th International Conference on the Theory andApplication of Cryptology and Information Security*, Seoul, South Korea, December 4–8, 2011, Springer, pp. 344–371.

[37] Aoki, K., et al., "Camellia: A 128-Bit Block Cipher Suitable for Multiple Platforms—Design and Analysis," in *Selected Areas in Cryptography: 7th Annual International Workshop, SAC 2000*, Waterloo, Ontario, Canada, August 14–15, 2000, Springer, pp. 39–56.

[38] Li, L., et al., "Meet-in-the-Middle Technique for Truncated Differential and Its Applications to CLEFIA and Camellia," in *Fast Software Encryption: 22nd International Workshop, FSE 2015*, Istanbul, Turkey, March 8–11, 2015, Revised Selected Papers, Springer, pp. 48–70.

[39] Matsui, M., "New Block Encryption Algorithm MISTY," in *Fast Software Encryption: 4th International Workshop, FSE'97*, Haifa, Israel, Springer, pp. 54–68.

[40] Bar-On, A., and N. Keller, "A 2^{70}Attack on the Full MISTY1," in *Advances in Cryptology–CRYPTO 2016: 36th Annual International Cryptology Conference*, Springer: Santa Barbara, California, August 14–18, 2016, pp. 435–456.

[41] Todo, Y., "Integral Cryptanalysis on Full MISTY1," *Journal of Cryptology*, Vol. 30, No. 3, 2017, pp. 920–959.

[42] ETSI (2014-10), *Universal Mobile Telecommunications System (UMTS); LTE; 3G Security; Specification of the 3GPP Confidentiality and Integrity Algorithms; Document 2: Kasumi Specification* (3GPP TS 35.202 version 12.0.0 Release 12), https://www.etsi.org/deliver/etsi_ts/135200_135299/135202/07.00.00_60/ts_135202v070000p.pdf.

[43] Dunkelman, O., N. Keller, and A. Shamir, "A Practical-Time Related-Key Attack on the KASUMI Cryptosystem Used in GSM and 3G Telephony," in *Advances in Cryptology–CRYPTO 2010: 30th Annual Cryptology Conference*, Santa Barbara, California, August 15–19, 2010, Springer, pp. 393–410.

[44] Hong, D., et al., "HIGHT: A New Block Cipher Suitable for Low-Resource Device," in *Cryptographic Hardware and Embedded Systems—CHES 2006: 8th International Workshop*, Yokohama, Japan, 2006, Springer, pp. 46–59.

[45] Hong, D., B. Koo, and D. Kwon, "Biclique Attack on the Full HIGHT," in *Information Security and Cryptology-ICISC 2011: 14th International Conference*, Seoul, Korea, November 30–December 2, 2011, Revised Selected Papers 14, Springer, pp. 365–374.

[46] Adams, C., *RFC2144: The CAST-128 Encryption Algorithm*, 1997, https://www.rfc-editor.org/rfc/r fc2144.

[47] Wang, S., T. Cui, and M. Wang, "Improved Differential Cryptanalysis of CAST-128 and CAST-256," in *Information Security and Cryptology: 12th International Conference, Inscrypt 2016*, Beijing, China, November 4–6, 2016, Springer, pp. 18–32.

[48] Lee, H. J., et al., *RFC4009: The SEED Encryption Algorithm*, 2005, https://www.rfc-editor.org/rfc/rfc4269.

[49] Sung, J., "Differential Cryptanalysis of Eight-Round SEED," *Information Processing Letters*, Vol. 111, No. 10, 2011, pp. 474–478.

[50] Bogdanov, A., et al., "PRESENT: An Ultra-Lightweight Block Cipher," in *Cryptographic Hardware and Embedded Systems-CHES 2007: 9th International Workshop*, Vienna, Austria, September 10–13, 2007, Springer, pp. 450–466.

[51] Blondeau, C., and K. Nyberg. "Links between Truncated Differential and Multidimensional Linear Properties of Block Ciphers and Underlying Attack Complexities," *Eurocrypt*, Vol. 14, 2014, pp. 165–182.

[52] Shirai, T., et al.,"The 128-Bit Blockcipher CLEFIA,"in *Fast Software Encryption, 14th International Workshop, FSE 2007, LNCS 4593*, 2007, pp. 181–195. 2007, http://citeseerx.ist.psu.edu/viewdoc/download?doi=10.1.1.111.9703&rep=rep1&type=pdf#page=191.

[53] Hong, D., et al. "LEA: A 128-Bit Block Cipher for Fast Encryption on Common Processors," in *Information Security Applications: 14th International Workshop, WISA 2013*, Jeju Island, Korea, August 19–21, 2013, pp. 3–27.

[54] Dwivedi, A. D., and G. Srivastava, "Differential Cryptanalysis of Round-Reduced LEA," *IEEE Access*, Vol. 6, 2018, pp. 79105–79113.

[55] Diffie, W., and G. Ledin, "SMS4 Encryption Algorithm for Wireless Networks," *Cryptology ePrint Archive*, 2008.

[56] Liu, Y., et al., "New Linear Cryptanalysis of Chinese Commercial Block Cipher Standard SM4," *Security and Communication Networks*, 2017.

[57] Zabotin, I. A., G. P. Glazkov, and V. B. Isaeva, "Cryptographic Protection for Information Processing Systems, Government Standard of the USSR, GOST 28147-89," *Government Committee of the USSR for Standards*, 1989.

[58] Courtois, N. T., "An Improved Differential Attack on Full GOST," in *The New Codebreakers: Essays Dedicated to David Kahn on the Occasion of His 85th Birthday* (P. Y. A. Ryan, D. Naccache, and J.-J. Quisquater, eds.), Berlin: Springer, 2016, pp. 282–303.

[59] Federal Agency on Technical Regulation and Metrology (GOST), *GOST R 34.12-2015: Block Cipher "Kuznyechik,"* https://www.rfc-editor.org/rfc/rfc7801.

[60] AlTawy, R., and A. M Youssef, "A Meet in the Middle Attack on Reduced Round Kuznyechik," *IEICE Transactions on Fundamentals of Electronics, Communications and Computer Sciences*, Vol. 98, No. 10, 2015, pp. 2194–2198.

[61] Bruwer, F. J., W. Smit, and G. J. Kuhn, *Microchips and Remote Control Devices Comprising Same*, U.S. Patent 5,517,187, May 1996.

[62] Indesteege, S., et al., "A Practical Attack on KeeLoq," in *Advances in Cryptology–EUROCRYPT 2008: 27th Annual International Conference on the Theory and Applications of Cryptographic Techniques*, Istanbul, Turkey, April 13–17, 2008, Springer, 2008, pp. 1–18.

[63] Beaulieu, R., et al., "The SIMON and SPECK Lightweight Block Ciphers," in *Proceedings of the 52nd Annual Design Automation Conference*, 2015, pp. 1–6.

[64] Chen, H., and X. Wang, "Improved Linear Hull Attack on Round-Reduced Simon with Dynamic Key-Guessing Techniques," in *Fast Software Encryption: 23rd International Conference, FSE 2016*, Bochum, Germany, March 20–23, 2016, Springer, pp. 428–449.

[65] Song, L., Z. Huang, and Q. Yang, "Automatic Differential Analysis of ARX Block Ciphers with Application to SPECK and LEA," in *Information Security and Privacy: 21st Australasian Conference, ACISP 2016*, Melbourne, Victoria, Australia, July 4–6, 2016, Springer, pp. 379–394.

[66] Miyaguchi, S., "The FEAL Cipher Family," in *Advances in Cryptology-CRYPTO'90: Proceedings 10*, Springer. 1991, pp. 628–638.

[67] Biham, E., and A. Shamir, "Differential Cryptanalysis of Feal and N-Hash," in *Advances in Cryptology—EUROCRYPT'91: Workshop on the Theory and Application of Cryptographic Techniques, Proceedings 10*, Brighton, UK, April 8–11, 1991, Springer, pp. 1–16.

[68] Schneier, B., et al., "Twofish: A 128-Bit Block Cipher," *NIST AES Proposal*, Vol. 15, No. 1, 1998, pp. 23–91.

[69] Lucks, S., "The Saturation Attack—A Bait for Twofish," in *Fast Software Encryption: 8th International Workshop, FSE 2001*, Yokohama, Japan, Springer, pp. 1–15.

[70] Vanhoef, M., and F. Piessens, "All Your Biases Belong to Us: Breaking RC4 in WPA-TKIP and TLS" in *24th USENIXSecurity Symposium (USENIXSecurity 15)*, 2015, pp. 97–112.

[71] Briceno, M., "A Pedagogical Implementation of A5/1," 1995, http://www. scard.org.

[72] Barkan, E., E. Biham, and N. Keller, "Instant Ciphertext-Only Cryptanalysis of GSM Encrypted Communication," in *Advances in Cryptology-CRYPTO 2003: 23rd Annual International Cryptology Conference, Proceedings 23*, Santa Barbara, California, August 17–21, 2003, Springer, pp. 600–616.

[73] Briceno, M., I. Goldberg, and D. Wagner, "A Pedagogical Implementation of the GSM A5/1 and A5/2 'Voice Privacy' Encryption Algorithms," 1999, *http://www.scard.org, mirror at http://cryptome.org/gsm-a512.htm* 26.

[74] Bernstein, D. J., et al., "ChaCha, a Variant of Salsa20," *Workshop Record of SASC*, Vol. 8, Citeseer, 2008, pp. 3–5.

[75] Aumasson, J.-P., et al., "New Features of Latin Dances: Analysis of Salsa, ChaCha, and Rumba," in *Fast Software Encryption: 15th International Workshop, FSE 2008, Revised Selected Papers 15*, Lausanne, Switzerland, February 10–13, 2008, Springer, pp. 470–488.

[76] Bernstein, D. J., "The Salsa20 Family of Stream Ciphers," *New Stream Cipher Designs: the eSTREAM Finalists*, 2008, pp. 84–97.

[77] Nohl, K., "Mifare, Little Security, Despite Obscurity," in *The 24th Congress of the Chaos Computer Club in Berlin*, December 2007.

[78] Courtois, N. T., K. Nohl, and S. O'Neil, "Algebraic Attacks on the Crypto-1 Stream Cipher in MiFare Classic and Oyster Cards," *Cryptology ePrint Archive*, 2008, https://eprint.iacr.org/2008/166.pdf.

[79] Hell, M., et al., "A Stream Cipher Proposal: Grain-128," in *2006 IEEE International Symposium on Information Theory*, IEEE, 2006, pp. 1614–1618.

[80] Fu, X., et al., "Determining the Nonexistent Terms of Non-linear Multivariate Polynomials: How to Break Grain-128 More Efficiently," *IACR Cryptol ePrint Archive*, 2017, p. 412.

[81] De Canniere, C., and B. Preneel, "Trivium," *New Stream Cipher Designs: The eSTREAM Finalists*, 2008, pp. 244–266.

[82] Fouque, P.-A., and T. Vannet, "Improving Key Recovery to 784 and 799 Rounds of Trivium Using Optimized Cube Attacks," in *Fast Software Encryption: 20th International Workshop, FSE 2013*, Singapore, March 11–13, 2013, Springer, pp. 502–517.

[83] Boesgaard, M., et al. "Rabbit: A New High-Performance Stream Cipher," in *Fast Software Encryption: 10th International Workshop, FSE 2003, Revised Papers 10*, Lund, Sweden, February 24–26, 2003, Springer, pp. 307–329.

[84] Watanabe, D., et al. "Update On Enocoro Stream Cipher," in *2010 International Symposium On Information Theory & Its Applications*, IEEE, 2010, pp. 778–783.

[85] Shibayama, N., and Y. Igarashi, "A New Higher Order Differential of Enocoro-128v2," in *2021 Ninth International Symposium on Computing and Networking Workshops (CANDARW)*, IEEE, 2021, pp. 379–384.

[86] Ekdahl, P., and T. Johansson, "A New Version of the Stream Cipher SNOW," in *Selected Areas in Cryptography: 9th Annual International Workshop, SAC 2002*, St. John's, Newfoundland, Canada, August 15–16, 2002, Springer, pp. 47–61.

[87] Funabiki, Y., et al., "Several MILP-Aided Attacks Against SNOW 2.0," in *Cryptology and Network Security: 17th International Conference, CANS 2018, Proceedings*, Naples, Italy, September 30–October 3, 2018, Springer, pp. 394–413.

[88] Watanabe, D., etal., "A New Key Stream Generator MUGI," in *Fast Software Encryption: 9th International Workshop, FSE 2002*, Leuven, Belgium, February 4–6, 2002, Springer, pp. 179–194.

[89] Watanabe, D., et al., *MUGI Psuedorandom Number Generator, Self Evaluation*, Tech. Rep., Hitachi Ltd., 2001, http://www.sdl.hitachi.co.jp/crypto/mugi/index-e.html.

[90] ETSI/SAGESpecification, *Specification of the 3GPP Confidentiality and Integrity Algorithms 128-EEA3 & 128-EIA3, Document 2: ZUC Specification, Version: 1.6*, 2011.

[91] Rivest, R. L., A. Shamir, and L. Adleman,"A Method for Obtaining Digital Signatures and Public-Key Cryptosystems," *Communications of the ACM*, Vol. 21, No. 2, 1978, pp. 120–126.

[92] Boudot, F., et al., "Comparing the Difficulty of Factorization and Discrete Logarithm: A 240-Digit Experiment," in *Advances in Cryptology–CRYPTO 2020: 40th Annual International Cryptology Conference, CRYPTO 2020, Proceedings, Part II 40*, Santa Barbara, CA, August 17–21, 2020, Springer, pp. 62–91.

[93] Boudot, F., et al., "The State of the Art in Integer Factoring and Breaking Public-Key Cryptography," *IEEE Security & Privacy*, Vol. 20, No. 2, 2022, pp. 80–86.

[94] ElGamal, T., "A Public Key Cryptosystem and a Signature Scheme Based on Discrete Logarithms," *IEEE Transactions on Information Theory*, Vol. 31, No. 4, 1985, pp. 469–472.

[95] Hoffstein, J., et al., "Practical Lattice-Based Cryptography: NTRUEncrypt and NTRUSign," in *The LLL Algorithm: Survey and Applications* (P. O. Nguyen, and V. Vallee, eds.), Berlin: Springer-Verlag, 2009, pp. 349–390.

[96] Howgrave-Graham, N., "A Hybrid Lattice-Reduction and Meet-in-the-Middle Attack against NTRU," in *Advances in Cryptology-CRYPTO 2007: 27th Annual International Cryptology Conference, Proceedings 27*, Santa Barbara, CA, August 19–23, 2007, Springer, pp. 150–169.

[97] Fu, X., et al., "A key-Recovery Attack on 855-Round Trivium," in *Advances in Cryptology–CRYPTO 2018: 38th Annual International Cryptology Conference, Proceedings, Part II 38*, Santa Barbara, CA, August 19–23, 2018, Springer, pp. 160–184.

[98] Hao, Y., et al., "Observations on the Dynamic Cube Attack of 855-Round TRIVIUM from Crypto'18," *Cryptology ePrint Archive*, 2018.

[99] PassMark Software, *CPU Benchmarks. Intel Xeon Gold 6130, 2.10GHz*, https://www.cpubenchmark.net/cpu.php?cpu=Intel+Xeon+Gold+6130+%40+2.10GHz&id=3126.

[100] Hoffstein, J., et al., "Choosing Parameters for NTRUEncrypt," in *Topics in Cryptology–CT-RSA 2017: The Cryptographers' Track at the RSA Conference 2017*, Springer: San Francisco, CA, February 14–17, 2017, pp. 3–18.

[101] Bellare, M., and P. Rogaway, *Introduction to Modern Cryptography*, University of California Davis Department of Computer Science, 2005, p. 283.

[102] Goldreich, O., *Foundations of Cryptography: Volume 2, Basic Applications*, Cambridge, UK: Cambridge University Press, 2009.

[103] Oppliger, R., *Cryptography 101: From Theory to Practice*, Norwood, MA: Artech House, 2021, https://rolf.esecurity.ch/?page_id=465.

[104] Menezes, A. J., P. C. van Oorschot, and S. A. Vanstone, *Handbook of Applied Cryptography*, Fifth Edition, Boca Raton, FL: CRC Press, 2001, https://cacr.uwaterloo.ca/hac/.

[105] Singh, S., *The Code Book: The Science of Secrecy from Ancient Egypt to Quantum Cryptography*, New York: Anchor Books, 1999.

[106] Paar, C., and J. Pelzl, *Understanding Cryptography–A Textbook for Students and Practioners*, Berlin: Springer Verlag, 2009, https://www.crypto-textbook.com/.

[107] Wong, D., *Real-World Cryptography*, Shelter Island, NY: Manning Publications, 2021, https://www.manning.com/books/real-world-cryptography.

[108] Aumasson, J.-P., *Serious Cryptography: A Practical Introduction to Modern Encryption*, San Francisco, CA: NoStarch Press, 2017, https://books.google.de/books?id=hLcrD-wAAQBAJ.

[109] Stamp, M., and R. M. Low, *Applied Cryptanalysis: Breaking Ciphers in the Real World*, Hoboken, NJ: Wiley-IEEE Press, 2007, https://www.cs.sjsu.edu/~stamp/crypto/.

[110] Stinson, D. R., *Cryptography: Theory and Practice*, Third Edition, Boca Raton, FL: Chapman & Hall/CRC Press, 2006.

[111] Raphael Chung-Wei Phan. "Mini Advanced Encryption Standard (Mini-AES): A Testbed for Cryptanalysis Students," *Cryptologia* 26.4 (2002), pp. 283–306.

[112] Nguyen, M. V., *Number Theory and the RSA Public Key Cryptosystem – An Introductory Tutorial on Using SageMath to Study Elementary Number Theory and Public Key Cryptography*, 2009, https://faculty.washington.edu/moishe/hanoiex/Number%20Theory%20Applications/numtheory-crypto.pdf.

[113] Nguyen, M.V., *Exploring Cryptography Using the Sage Computer Algebra System*, Bachelor of Science Thesis, Victoria University, Australia, 2009, www.sagemath.org/files/thesis/nguyen-thesis-2009.pdf url2: https://www.sagemath.org/library-publications.html.

[114] Schneier, B., "A Self-Study Course in Block-Cipher Cryptanalysis," *Cryptologia*, Vol. 24, 2000, pp. 18–34, https://www.schneier.com/wp-content/uploads/2015/01/paper-self-study.pdf.

[115] Schaefer, E. F., *Cryptography Research: Devising a Better Way to Teach and Learn the Advanced Encryption Standard*, Santa Clara University, 2011, https://web.archive.org/web/20110829213229/http://www.scu.edu/cas/research/cryptography.cfm.

[116] Schaefer, E. F., "A Simplified Data Encryption Standard Algorithm," *Cryptologia*, Vol. 20, No. 1, 1996, pp. 77–84.

[117] Chung-Wei Phan, R., "Impossible Differential Cryptanalysis of Mini-AES," *Cryptologia*, Vol. 27, No. 4, 2003, pp. 361–374, https://www.tandfonline.com/doi/abs/10.1080/0161-110391891964.

[118] Musa, M. A., E. F. Schaefer, and S. Wedig. "A Simplified AES Algorithm and Its Linear and Differential Cryptanalyses," *Cryptologia*, Vol. 17, No. 2, April 2003, pp. 148–177, https://www.rose-hulman.edu/~holden/Preprints/s-aes.pdf.

[119] Hoffman, N., "A Simplified IDEA Algorithm," 2006, https://www.nku.edu/~christensen/simplified%20IDEA%20algorithm.pdf.

[120] Davod Mansoori, S., and H. Khaleghei Bizaki, "On the Vulnerability of Simplified AES Algorithm Against Linear Cryptanalysis," *IJCSNS International Journal of Computer Science and Network Security*, Vol. 7, No. 7, 2007, pp. 257–263, http://paper.ijcsns.org/07_book/200707/20070735.pdf.

CHAPTER 2

Paper-and-Pencil and Precomputer Ciphers

The term paper-and-pencil methods (P&P methods, also called hand ciphers or pen'n'paper ciphers) embraces all techniques that people can apply manually to encrypt and decrypt (encipher or decipher) a message. This includes all classic encryption methods (as opposed to those that require machines or computers), but also some newer ones that were deliberately developed with the aim of achieving a very high level of security by hand (which was not always successful), such as ElsieFour, Solitaire, Hutton, or Handycipher.

P&P methods were also popular with secret services, as a writing pad and a pencil are unsuspicious.

This chapter provides a broad overview (encyclopedic approach) of many of these P&P methods, each with an example and references to deeper information.[1]

Section 2.5 in this chapter presents Hagelin rotor machines as an example of electromechanical cipher machines that were in use until the 1970s.

At the end of this chapter (Section 2.8) you find sample code (for ciphers like Caesar, Atbash, monoalphabetic substitution, Vigenè, Hill, columnar transposition) written for the computer-algebra system SageMath.

While this chapter describes the procedures from the point of view of modern cryptography (and so that the ciphertext is already available correctly), Chapter 3 takes the view of a historian who first has to transcribe a found document before he can cryptographically examine it.

The first paper-and-pencil methods already arose about 3,500 years ago in Mesopotamia. All paper-and-pencil methods are a matter of symmetric methods. Even the earliest encryption algorithms use the basic principles such as transposition, substitution, block construction, and their combinations. Hence, it is worthwhile to closely consider these ancient methods especially under didactic aspects.

Ciphers to be successful and widespread had to fulfill the same characteristics required for modern algorithms:

1. The footnotes of this chapter describe how the cryptographic methods can be performed using the offline programs CrypTool 1 (CT1), CrypTool 2 (CT2), and JCrypTool (JCT). See Sections A.2 and A.3. Many of the methods can also be performed within a browser (e.g., on the website CrypTool-Online (CTO)). See Section A.4.

   While the CrypTool websites and programs offer both classic and modern ciphers, there are several sites related to the American Cryptogram Association (ACA) [1] that provide very detailed focus only on classic ciphers: for instance the sites of Bion [2] and Pilcrow [3].

- Exhaustive description, almost standardization (including special cases, padding, etc.);
- Good balance between security and usability (because methods being too complicated were error-prone or unacceptably slow).

Convention: If the alphabet uses only 26 letters, we write from now onward the plaintext in lowercase letters and the ciphertext in capital letters.

The letters of the ciphertext are—as used historically—grouped within blocks of five letters. It does not matter if a different (constant) block length is used for the output or if there is no separation by blanks.

2.1 Transposition Ciphers

Encrypting a message by means of transposition does not change the original characters of this message, only their order in the plaintext is modified (transposition = exchange). There is no ciphertext alphabet.

Sometimes, the name *permutation* is used to describe how characters, groups of characters, numbers, or columns of the plaintext are exchanged; for example, $(1, 2, 3, 4, 5) \Leftrightarrow (3, 4, 2, 1, 5)$.

2.1.1 Introductory Samples of Different Transposition Ciphers

- *Rail fence cipher*[2] [4]: The characters of a message are alternately written in two (or more) lines, creating a zigzag pattern. The resulting ciphertext is read out line by line. This is more a children's method.

  For an example, see Table 2.1.

  Plaintext: an example of transposition

Table 2.1 Rail Fence Cipher

	n		x		m		l		o		t		a		s		o		i		i		n
a		e		a		p		e		f		r		n		p		s		t		o	

Ciphertext: NXMLO TASOI INAEA PEFRN PSTO

2. - Using CTO, this cipher can be seen in the browser in the plugin "Rail fence": `https://www.cryptool.org/en/cto/railfence`. Here, not only the result but also the graphical zigzag display is output.
- This method can also be found using `CT1 Encrypt/Decrypt` ▷ `Symmetric (classic)` ▷ `Scytale / Rail Fence`.
- A rail fence with 2 lines and offset 1 can be simulated as simple columnar transposition; that is, using `CT1 Encrypt/Decrypt` ▷ `Symmetric (classic)` ▷ `Permutation`, using as key "(B,A)" and accepting the default settings (only one permutation, where your input is done line-by-line and the output is taken column-by-column). This was done in Table 2.1.

Using the key "(A,B)" would start the zigzag pattern in Table 2.1 in such a way that the first letter is written into the first line instead of the second line. Normally, the rail fence cipher cannot be simulated via a simple columnar transposition.

- *Scytale*[3] [4]: This method has probably been used since 600 B.C.—a description of how it operated is not known from before Plutarch (50–120 B.C.). A long strip of paper is wrapped around a wooden cylinder and then the message is written along the length of this strip. After unwinding, the strip contains the ciphertext. For decryption the recipient needs to have a previously agreed cylinder of the same diameter.
- *Grille cipher* [5]: Both parties use identical stencils. Line by line, their holes are filled with plaintext that is read out column by column to produce the ciphertext. If there is plaintext left, the procedure is repeated.
- *Turning grille* [6]: The German army used turning grilles during WW1. It was invented in 1881 by Eduard Fleissner von Wostrowitz.[4] A square grille serves as a stencil, a quarter of its fields being holes. The first part of the message is written on a piece of paper through these holes, then the grille is rotated by 90 degrees and the user can write down the second part of the message, and so forth. But this method only works if the holes are chosen carefully: Every field has to be used, and no field may be used twice. The ciphertext is read out line by line.

  In the example for a turning grille in Table 2.2 you can write 4 times 16 characters of the plaintext on a piece of paper (each character into the circles which should symbolize a punched hole).

Table 2.2 8 × 8 Turning Grille

O	-	-	-	-	O	-	-
-	-	-	O	O	-	-	O
-	-	-	O	-	-	O	-
-	-	O	-	-	-	-	-
-	-	-	-	O	-	-	-
O	-	O	-	-	-	O	-
-	O	-	-	-	-	-	O
-	-	-	O	O	-	-	-

3. - Using CTO, this cipher can be seen in the plugin "Scytale": `https://www.cryptool.org/en/cto/scytale`. Here, also the graphical arrangement is displayed.
   - This method can also be found using `CT1 Encrypt/Decrypt` ▷ `Symmetric (classic)` ▷ `Scytale / Rail Fence`. As this method is a special case of a simple columnar transposition, you also can simulate Scytale via `CT1 Encrypt/Decrypt` ▷ `Symmetric (classic)` ▷ `Permutation`: Use within the dialog box only the first permutation. If the wood has, for example, four angles use as key "1,2,3,4." This is equivalent to writing the text horizontally in blocks of four letters in a matrix and to read it out vertically. Because the key is in an ascending order, the Scytale is denoted as an identical permutation. And because writing and read-out is done only once it is a simple (and not double) permutation.
   - The Scytale can also be found in `CT2 Startcenter` ▷ `Templates` ▷ `Cryptography` ▷ `Classical`.
4. Using `JCT Default Perspective` ▷ `Analysis` ▷ `Fleissner-Grille-Analysis` you can encrypt and decrypt texts, but you can also attack ciphertexts encrypted with this turning grille.

   Another nice visualization can be found archived under `https://web.archive.org/web/20050922123752/http://www.turning-grille.com:80/`.

2.1.2 Column and Row Transposition[5]

Simple columnar transposition [6]: First, a keyword is chosen that is written above the columns of a table. This table is filled with the text to be encrypted line by line. Then the columns are rearranged by sorting the letters of the keyword alphabetically. Afterward the columns are read out from left to right to build the ciphertext.[6]

See Table 2.3.

Plaintext: an example of transposition

Table 2.3 Simple Columnar Transposition

K	*E*	*Y*
a	n	e
x	a	m
p	l	e
o	f	t
r	a	n
s	p	o
s	i	t
i	o	n

Transposition key: K=2; E=1; Y=3.

Ciphertext: NALFA PIOAX PORSS IEMET NOTN

AMSCO cipher [1]: The characters of the plaintext are written in alternating groups of one respectively two letters into a grille. Then the columns are swapped and the text can be read out.

Double column transposition (DCT) [6]: Double columnar transposition was used in WW2 and during the Cold War. Two simple columnar transpositions with different keys are executed successively.[7]

If the keys are different and long enough (at least each 20 characters), then this is even for today's computer a real challenge.[8]

5. Most of the following methods can be simulated in CT:
- In CTO in the plugin "Simple Column Transposition": `https://www.cryptool.org/en/cto/transposition`.
- Using `CT1 Encrypt/Decrypt ⊳ Symmetric (classic) ⊳ Permutation`.
- Using `CT2 Templates ⊳ Cryptography ⊳ Classical ⊳ Transposition Cipher`. In addition, CT2 contains several analyzers for this cipher.
- Using `JCT Default Perspective ⊳ Algorithms ⊳ Classic ⊳ Transposition`.
6. Using CT1: Choose a key for the first permutation, input line by line, permute and output column by column.
The component in CT2 also visualizes how the text is put into and taken off the matrix and how the columns are permuted.
7. Using CT1 or CT2: Choose a key for the 1st permutation, input line by line, permute and output column by column. Then choose a (different) key for the second permutation, input line by line, permute and output column by column.
8. MTC3 offers corresponding challenges (the ones called "reloaded" also explain the conditions for the keys in more detail), for instance `https://mysterytwister.org/challenges/level-x/double-column-transposition` and `https://mysterytwister.org/challenges/level-3/double-column-transposition-reloaded-part-1`.

Column transposition, General Luigi Sacco [6]: The columns of a table are numbered according to the letters of the keyword. The plaintext is entered line by line, in the first line up to column number one, in the second line up to column number two, and so forth. Again, the ciphertext is read out in columns.

See Table 2.4.

Plaintext: an example of transposition

Table 2.4 Columnar Transposition (General Luigi Sacco)

C 1	O 5	L 2	U 6	M 3	N 4
a					
n	e	x			
a	m	p	l	e	
o	f	t	r	a	n
s	p				
o	s	i	t		
i	o	n			

Ciphertext: ANAOS OIEMF PSOXP TINLR TEAN

Column transposition, French army in WW1 [6]: After executing a simple columnar transposition, diagonal rows are read out.

Row transposition [6]: The plaintext is divided into blocks of equal length and a keyword is chosen. Now the letters of the keyword are numbered and permutation is done only within each block according to this numbering.[9]

2.1.3 Further Transposition Algorithm Ciphers

Geometric figures [5]: Write the message into a grille following one pattern and read it out using another.

Union Route Cipher [5]: The Union Route Cipher derives from the Civil War. This method does not rearrange letters of a given plaintext, but whole words. Particularly sensitive names and terms are substituted by codewords that are recorded in codebooks together with the existing routes. A route determines the size of a grille and the pattern that is used to read out the ciphertext. In addition, a number of filler words are defined.

Nihilist transposition [1]: Insert the plaintext into a square grille and write the same keyword above the columns and next to the lines. As this keyword is sorted alphabetically, the contents of the grille are rearranged, too. Read out the ciphertext line by line.

See Table 2.5: You get the left matrix of the table by filling it with the plaintext. After switching rows and columns you get the right matrix of the table.

Plaintext: `an example of transposition`

Ciphertext: `SPOIS EPLOM ATRNF NIOTX NEAA`

9. Using CT1: Choose a key for first permutation, input line by line, permute column by column and output line by line. The component in CT2 also visualizes the row-wise transposition.

Table 2.5 Nihilist Transposition

	W	O	R	D	S			D	O	R	S	W
W	a	n	e	x	a		D	s	p	o	i	s
O	m	p	l	e	o		O	e	p	l	o	m
R	f	t	r	a	n		R	a	t	r	n	f
D	s	p	o	s	i		S	n	i	o	-	t
S	t	i	o	n	-		W	x	n	e	a	a

Cadenus cipher [1]: Cadenus is a form of a columnar transposition that uses two keywords. Cadenus combines a complete columnar transposition with a keyed column rotation. The procedure—both for encryption and decryption—is rather error-prone.

Let's say the first keyword (*KEY*) has a length of $n = 3$. The second keyword has a maximal length of the alphabet. It can be shorter, but it must contain the letters of the first keyword. The second keyword is a permutation of the used elements.

We have a table with a first column (*stub column*), and three blocks each consisting of n columns. The first keyword is written in the header row of each block and used to swap the columns of the second and third block later.

The second keyword is used to build the stub column and so define the initial letter of each column of the third block.

The plaintext is filled line by line in the first block. Then each column of the first block is moved to the second block, transposed in the order of the first keyword. The columns of the third block are created by copying each column from block 2 to block 3 and moving the elements within a column wrap-around such that it begins with the letter, which is in the same line as the corresponding key letter of the first keyword within the second keyword. The ciphertext is then read out line by line from the third block.

See Table 2.6: The letters of the first keyword are marked **bold** in the stub column. Within the second block, those letters are printed blue, which will be at

Table 2.6 Cadenus Cipher

	K	*E*	*Y*	*E*	*K*	*Y*	*E*	*K*	*Y*
A	c	a	d	a	c	d	s	a	a
D	e	n	u	n	e	u	s	r	p
X	s	i	s	i	s	s	i	f	i
K	a	f	o	f	a	o	u	l	o
C	r	m	o	m	r	o	n	n	s
W	f	c	o	c	f	o	k	t	g
N	l	u	m	u	l	m	w	n	e
S	n	a	r	a	n	r	d	o	o
Y	t	r	a	r	t	a	a	t	x
E	n	s	p	s	n	p	n	n	d
D	o	s	i	s	o	i	i	i	u
T	t	i	o	i	t	o	f	a	s
U	n	u	s	u	n	s	m	y	o
B	i	n	g	n	i	g	c	r	o
R	a	k	e	k	a	e	u	c	o
G	y	w	o	w	y	o	a	e	m
H	r	d	x	d	r	x	r	s	r

the top of the columns in block 3 (after wrapping around the selected columns). The sample here is a slightly enhanced version of Cadenus, as the length of the second keyword must not be exactly 25. But still the plaintext length (here 51 with the filler "x") has to be a multiple of the length of the second keyword (here 17).

Plaintext: `cadenus is a form of columnar transposition using a keywordx.`

First keyword: `KEY`
Second keyword: `ADXKCWNSYEDTUBRGH`

Ciphertext: `SAASR PIFIU LONNS KTGWN EDOOA TXNND IIUFA SMYOC ROUCO AEMRS R.`

2.2 Substitution Ciphers

Substitution methods assign to each plaintext object (i.e., to each element of the plaintext alphabet or to each element of the nomenclator) a ciphertext object (a ciphertext character or the corresponding nomenclator code).

Monoalphabetic substitution ciphers keep the same assignment for the whole encryption process once the assignment is decided—in contrast to *polyalphabetic* substitution. Treating more than one letter as one object is called *polygraphic* substitution. If an object in the plaintext alphabet can be mapped to more than one object in the ciphertext alphabet, it is called *homophonic*. If an object in the ciphertext alphabet can be mapped to more than one object in the plaintext alphabet, it is called *polyphonic*—this has been rarely used in the wider world because the decryption is no longer unique, but it is a common cipher type in the ACA where it's known as a key phrase cipher.

2.2.1 Monoalphabetic Substitution

MASC assigns each character of the plaintext alphabet to one character of the ciphertext alphabet. This mapping remains unchanged during the whole process of encryption. In the literature, this is differentiated even more finely: Simple MASC operate on single letters while MASC can also operate on groups.

General monoalphabetic substitution/random letter pairs[10] [10]: The substitution occurs by a given assignment of single letters. This is the usual description: The key is then a permutation of the plaintext alphabet. This permutation is used as the ciphertext alphabet. With a size of 26!, this general case has a significantly larger key space than Caesar or Atbash.

10. The MASC can be simulated in CT via:
- CTO plugin "Monoalphabetic Substitution": `https://www.cryptool.org/en/cto/monoalpha`.
- `CT1 Encrypt/Decrypt ▷ Symmetric (classic) ▷ Substitution / Atbash`.
- `CT2 Templates ▷ Cryptography ▷ Classical`. In CT2 you also can assign more than one character to an item (homophones). Several analyzers can be found via `CT2 Templates ▷ Cryptanalysis ▷ Classical`.

Atbash cipher[11] [4]: Replace the first letter of the alphabet by the last letter of the alphabet, the second one by the second-to-last, and so forth. So the key is an inversion of the given alphabet.

Shift cipher or Caesar cipher[12] [4]: The ciphertext alphabet is created by shifting the whole plaintext alphabet by a determined number of letters. The shift value is less than the length of the alphabet. Remark: Emperor Caesar always shifted by 3 positions.

Plaintext: `three positions to the right`

Ciphertext: `WKUHH SRVLWLRQV WR WKH ULJKW`

Affine cipher[13]: This is a generalization of the shift cipher. A plaintext character is first substituted by another character and then the result is encrypted using the shift cipher. The name "affine cipher" was chosen because its encryption and decryption can be described as affine or linear function.

Substitution with symbols [4], for instance the so-called freemason cipher: Each letter is replaced with a symbol.

Variants: Fill characters, intentional mistakes [4].

Nihilist substitution[14] [1]: The alphabet is entered into a 5×5 matrix and during processing each plaintext letter is assigned the number formed from the row and column number. The alphabet is built from a first keyword (`SUBSTITUTION ==> SUBTION`) and the remaining letters of the alphabet in the normal order. Then a second keyword (`KEY`) is chosen and placed above the columns of a second table. The plaintext is entered line by line (row by row) into this table. The ciphertext's numerical values are written below each plaintext letter—each of these values is the sum of the number of the plaintext letter and the number of the keyword letter. Numbers between 100 and 110 are transformed to numbers between 00 and 10, so that each letter is represented by a two-digit number. Here is the example calculation for the first letter in the plaintext: Since $A = 23$ and $K = 35$, a becomes $23 + 35 = 58$.

11. Atbash can be simulated in CT via:
 - CTO plugin "ATBASH": `https://www.cryptool.org/en/cto/atbash.`
 - `CT1 Encrypt/Decrypt ▷ Symmetric (classic) ▷ Substitution / Atbash.`
12. Caesar and the special case Rot13 (sometimes hyphenated ROT-13) can be simulated in CT via:
 - CTO plugin "Caesar / Rot13": `https://www.cryptool.org/en/cto/caesar.`
 - `CT1 Encrypt/Decrypt ▷ Symmetric (classic) ▷ Caesar / ROT13.`
 - `CT1 Analysis ▷ Symmetric Encryption (classic) ▷ Ciphertext only ▷ Caesar.`
 - `CT1 Indiv. Procedures ▷ Visualization of Algorithms ▷ Caesar.`
 - `CT2 Templates ▷ Cryptography ▷ Classical ▷ Caesar Cipher.`
 Several analyzers can be found via `CT2 Templates ▷ Cryptanalysis ▷ Classical.`
13. - CTO plugin "Affine / Multiplicative": `https://www.cryptool.org/en/cto/multiplicative.`
 - Further details about the affine cipher and an implementation in SageMath can be found in Section 2.8.2.4.
 - MTC3 offers a corresponding challenge for kids: `https://mysterytwister.org/challenges/level-1/affine-codes--modulo-arithmetic-with-n--extended-euclid.`
14. - An animation of this Nihilist method: `CT1 Indiv. Procedures ▷ Visualization of Algorithms ▷ Nihilist.`
 - `CT2 Templates ▷ Cryptography ▷ Classical ▷ Nihilist Cipher.`

See Table 2.7.
Plaintext: an example of substitution

Table 2.7 Nihilist Substitution

	1	2	3	4	5
1	S	U	B	T	I
2	O	N	A	C	D
3	E	F	G	H	K
4	L	M	P	Q	R
5	V	W	X	Y	Z

(a) Matrix

K (35)	E (31)	Y (54)
a (58)	n (53)	e (85)
x (88)	a (54)	m (96)
p (78)	l (72)	e (85)
o (56)	f (63)	s (65)
u (47)	b (44)	s (65)
t (49)	i (46)	t (68)
u (47)	t (55)	i (69)
o (56)	n (53)	

(b) Table

Ciphertext: 58 53 85 88 54 96 78 72 85 56 63 65 47 44 65 49 46 68 47 55 69 56 53

Codes [4]: In the course of time, codebooks were used again and again. A codebook assigns a codeword, a symbol, or a number to every possible word of a message. Only if both parties hold identical codebooks and if the assignment of codewords to plaintext words is not revealed, a successful and secret communication can take place.

Nomenclator [4]: A nomenclator refers to techniques that combine the use of a cipher algorithm with a codebook. Often the encryption system is based upon a ciphertext alphabet. This alphabet is used to encrypt (via substitution) the bigger part of the message. Particularly frequent or top-secret words are replaced by a limited number of codewords existing besides the ciphertext alphabet.

Map cipher [7]: This method constitutes a combination of substitution and steganography.[15] Plaintext characters are replaced by symbols that are arranged in a map following certain rules.

15. Pure steganography tries to conceal the existence of the message instead of encrypting it.
Different steganographic techniques can be found in CT2: For example, image steganography (with BPCS and with LSB), text steganography [with capital letters (alone or in binary mode), with letter marking, with zero width spaces], or watermarking (invisible, robust, or visible). See `CT2 Templates ▷ Steganography ▷ Image Steganography with BPCS`.

Straddling checkerboard[16] [5]: A 3×10 matrix is filled with the 26 letters of the used alphabet and two arbitrary digits (or special characters) as follows: The different letters of a keyword and the remaining characters are written into the grille. The columns are numbered 0 to 9, the second and the third line are numbered 1 and 2. Each plaintext character is replaced by the corresponding digit, respectively the corresponding pair of digits. As "1" and "2" are the first digits of the possible two-digit numbers, they are not used as single digits. The special thing is that the ciphertext characters sometimes have 1 digit and sometimes 2 digits: The letter "K" becomes "0," "E" becomes "3," but "B" becomes "10."

See Table 2.8.

Plaintext: an example of substitution

Table 2.8 Straddling Checkerboard with Password "Keyword"

	0	1	2	3	4	5	6	7	8	9
	K	-	-	E	Y	W	O	R	D	A
1	B	C	F	G	H	I	J	L	M	N
2	P	Q	S	T	U	V	X	Z	.	/

Ciphertext: 91932 69182 01736 12222 41022 23152 32423 15619

Here it is obvious, that "1" and "2" are the most frequently used digits in the ciphertext, but this will be fixed with the following version. Note that nonalphabet characters like spaces are filtered out before encryption.

Straddling checkerboard, variant[17] [5]: This variant of the straddling checkerboard was developed by Soviet spies during World War II. Ernesto (Ché) Guevara and Fidel Castro allegedly used this cipher for their secret communication. A grille is filled with the alphabet (number of columns = length of keyword). This additional grille is used to shuffle the alphabetic order when the letters are moved to the matrix. Then two arbitrary digits are chosen as "reserved" to indicate the second and third line of a 3 × 10-matrix (in our example 3 and 7). For a faster encryption, the eight most common letters (ETAONRIS for the English language) are assigned the digits from 0 to 9, the reserved 2 digits are not assigned. The remaining letters are

16. `CT2 Templates ▷ Cryptography ▷ Classical ▷ Straddling Checkerboard.`
To run the sample Table 2.8 in CT2, the input components of the template need the following (key and alphabet define the matrix of the checkerboard):
- Plaintext: anexampleofsubstitution
- Rows and columns: 12 0123456789
- Alphabet: abcdefghijklmnopqrstuvwxyz
- Key: keyword: The "Alphabet Permutator" component creates the order for the matrix from alphabet and key, as shown in Table 2.8.

17. `CT2 Templates ▷ Cryptography ▷ Classical ▷ Straddling Checkerboard.` CT2 supports all three straddling-checkerboard variants. To run the sample Table 2.9 in CT2, the input components of the template need the following (where the keyword "keyword" was used to build up the grid; the matrix alphabet comes from reading out the grid):
- Plaintext: anexampleofsubstitution
- Rows and columns: 37 0123456789
- Matrix alphabet: atenoriskjbluycmvwfxgpzhq.d/ [Matrix alphabet directly as input instead of alphabet and key as in the example before].

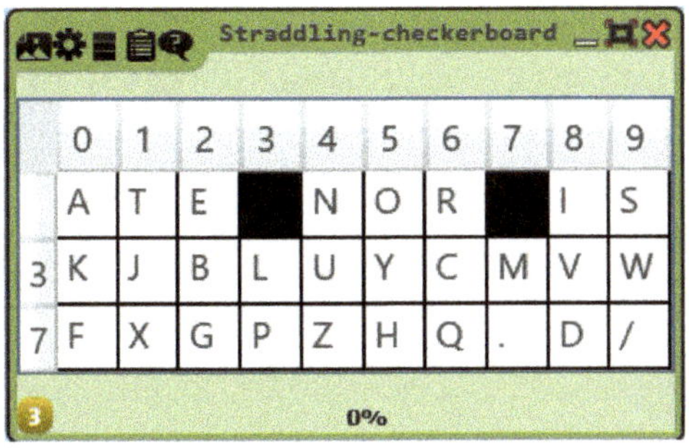

Figure 2.1 CT2 component for the straddling checkerboard.

traversed column by column from the grille and transferred row by row into the matrix.

See Table 2.9.

Plaintext: an example of substitution

Table 2.9 Variant of the Straddling Checkerboard

Grille							
	K	**E**	Y	W	**O**	**R**	D
	A	B	C	F	G	H	**I**
	J	L	M	**N**	P	Q	**S**
	T	U	V	X	Z	.	/

Matrix		0	1	2	3	4	5	6	7	8	9
		A	**T**	**E**	-	**N**	**O**	**R**	-	**I**	**S**
	3	K	J	B	L	U	Y	C	M	V	W
	7	F	X	G	P	Z	H	Q	.	D	/

Ciphertext: 04271 03773 33257 09343 29181 34185 4

Ché Guevara cipher[18]:
With a high probability, Ché Guevara used a special case of the checkerboard variant above (this case has an additional substitution step and a slightly changed checkerboard):

- The seven most frequent letters in Spanish are distributed in the first row;
- Four instead of three rows are used;
- So one could encrypt $10 \cdot 4 - 3 = 37$ different characters.

Tridigital cipher [1]: A keyword with 10 letters is used to create a numeric key by numbering its letters corresponding to their alphabetical order. This key is written above the columns of the 3 × 10-matrix. This matrix is filled line by line with the alphabet as follows: The different letters of a keyword are inserted first, followed by the remaining letters. The last column is left out. Plaintext characters are substituted with numbers, and the number of the last column is used to separate words.

18. `CT2 Templates ▷ Cryptography ▷ Classical ▷ Ché Guevara Cipher`

Baconian cipher[19] [1]: This is actually a code and not a cipher (there is no key). Assign a five-digit binary code to every letter and to six numbers or special characters (00000 = A, 00001 = B, etc.), and replace the plaintext characters with this binary code. Now use a second, unsuspicious message to hide the ciphertext inside of it. This may happen by upper and lower case or italicized letters: For example, all letters of the unsuspicious message below a binary "1" are capitalized. Overall this is obtrusive.

See Table 2.10.

Table 2.10 Baconian Cipher

Plaintext / message	*F*	*I*	*G*	*H*	*T*
Intermediate ciphertext	`00101`	`01000`	`00110`	`00111`	`10011`
Unsuspicious message	`itisw`	`arman`	`thesu`	`nissh`	`ining`
Ciphertext	`itIsW`	`aRman`	`thESu`	`niSSH`	`IniNG`

2.2.2 Homophonic Substitution

Homophonic methods constitute a special form of monoalphabetic substitution. Several or all characters of the plaintext alphabet are assigned to more than one ciphertext character. This methodology, in many variations, was widespread in historical reality, for example, in the Copiale cipher, an encrypted manuscript from the 18th century (see Figure 2.2).

Homophonic monoalphabetic substitution[20] [4]: Each language has a typical frequency distribution of the letters of the alphabet. To conceal this distribution, each plaintext letter is assigned to several ciphertext characters. The number of ciphertext characters assigned depends on the frequency of the letter to be encrypted.

Book cipher: The words of a message are substituted by triples "page-line-position." This method requires a detailed agreement of which book to use, especially regarding the edition (layout, error correction, etc.).

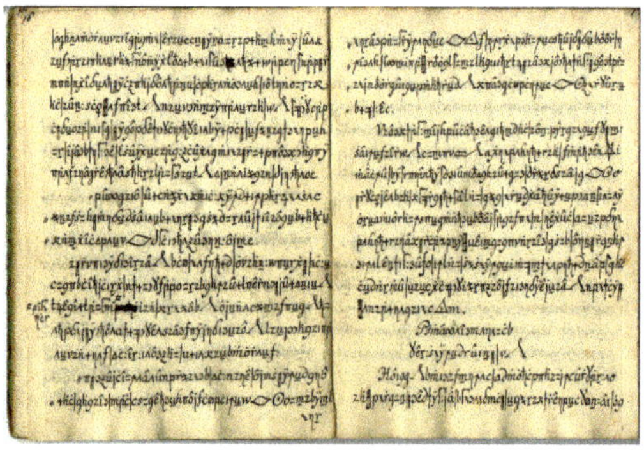

Figure 2.2 Copiale cipher, scaled page 16/17. (Source: [8].)

19. `CT2 Templates ▷ Cryptography ▷ Classical ▷ Baconian Cipher`
20. `- CT1 Encrypt/Decrypt ▷ Symmetric (classic) ▷ Homophone`
    `- CT2 Templates ▷ Cryptography ▷ Classical ▷ Homophonic Substitution Cipher and Nomenclature`. In addition, CT2 contains powerful analyzers for this cipher type.

Beale cipher [4]: The Beale cipher is a book cipher that numbers the words of a key text. These numbers replace the plaintext letters by the words' initial letters.

Grandpré cipher [6]: A square grille with 10 columns (other layouts are possible, too) is filled with 10 words. The initial letters should result in an eleventh word. As columns and rows are numbered from 0 to 9, letters can be replaced by two-digit numbers. It is obvious that with the table having 100 fields, most letters can be represented by more than one pair of numbers. You should keep in mind that those 10 words have to contain all letters of the plaintext alphabet.

Spanish strip cipher (SSC) [9]: This homophonic substitution cipher was the official method of ciphering by the Spanish ministries in the late 19th century, and the method mostly used by both sides, the Republicans and the Nationalists, in the Spanish Civil War (1936–1939).[21]

2.2.3 Polygraphic Substitution

Polygraphic techniques do not work solely by replacing single characters; they (also) replace whole groups of characters. In most cases, these groups are digrams, trigrams, or syllables. For these letter sequences of defined length, different alternative designations are known that are also used in the frequency analysis: 2-gram, digram, bigram, and digraph all mean the same thing. Because of the better readability and the practical comparability of different lengths, the notation n-gram slowly wins through.

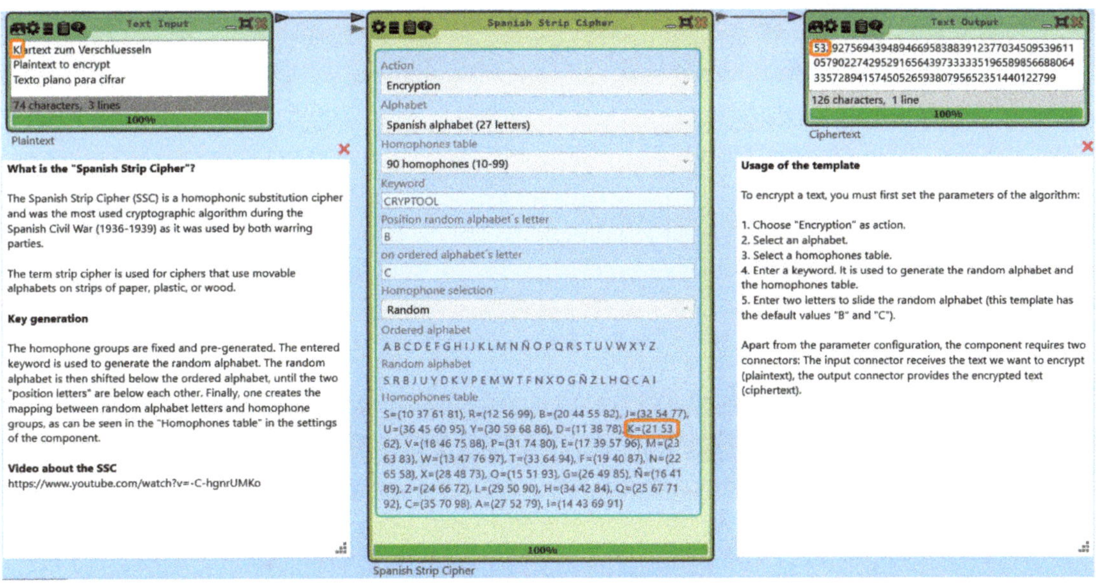

Figure 2.3 CT2 template for the Spanish strip cipher.

21. - `CT2 Templates` ▷ `Cryptography` ▷ `Classical` ▷ `Spanish Strip Cipher`. See Figure 2.3. CT2 can break it with the Homophonic Analyzer.
- MTC3: For example, `https://mysterytwister.org/challenges/level-2/spanish-strip-cipher-part-1`.
- The Kopal video explains in detail how the keys are generated and how to break this cipher with CT2: `https://www.youtube.com/watch?v=-C-hgnrUMKo`.

Grand Chiffre [4]: This cipher was used by Louis XIV in the 17th century and was not solved until the end of the 19th century. Cryptograms consisted of 587 different numbers, every number representing a syllable. The inventors of the Grand Chiffre (Rossignol, father and son) constructed additional traps to increase security. For example, a number could assign a different meaning to or delete the preceding one. Compare the "operational code elements" in Chapter 3.

Playfair cipher[22] [4]: A 5 × 5-matrix is filled with the plaintext alphabet (e.g., the Latin alphabet without the "J") in the following way: The different letters of a keyword are inserted first, followed by the remaining letters (like in the straddling checkerboard). The plaintext (PT) is then divided into pairs; these digraphs are encrypted using the following rules:

1. If both letters can be found in the same column, they are replaced by the letters underneath.
2. If both letters can be found in the same row, take the letters to their right.
3. If both letters of the digraph are in different columns and rows, the replacement letters are obtained by scanning along the row of the first letter up to the column where the other letter occurs and vice versa.
4. Double letters are treated by special rules, if they appear in one digraph. They can be separated by a filler, for example.

See Table 2.11.
Unformatted PT: `plaintext letters are encrypted in pairs`
1. Formatted PT: `pl ai nt ex tl et te rs ar ee ncrypted in pairs`
2. Formatted PT: `pl ai nt ex tl et te rs ar ex en cr yp te di np ai rs`

Table 2.11 5 × 5 Playfair Matrix with Password "Keyword"

K	E	Y	W	O
R	D	A	B	C
F	G	H	I	L
M	N	P	Q	S
T	U	V	X	Z

Ciphertext: SHBHM UWUZF KUUKC MBDWU DURDA VUKBG PQBHC M

Trigraphic Playfair: A 5 × 5-matrix is filled with the alphabet and the plaintext is divided into trigraphs. Trigraphs are encrypted according to the following rules:

1. Three equal letters are substituted by three equal letters, which is the letter on the right underneath the original letter (example from Table 2.11: `BBB` ⇒ LLL).
2. A trigraph with two different letters is encrypted like a digraph in Playfair.

22. - `CT1 Encrypt/Decrypt ▷ Symmetric (classic) ▷ Playfair`
- `CT2 Templates ▷ Cryptography ▷ Classical ▷ Playfair`. In addition, CT2 contains powerful analyzers for this cipher.
- `JCT Default Perspective ▷ Algorithms ▷ Classic ▷ Playfair`

3. If a trigraph contains three different characters, very complex rules come into effect. See [6].

Substituting digraphs by symbols [6]: Giovanni Battista della Porta in the 15th century created a 20 × 20-matrix that contained one symbol for every possible combination of letters (his alphabet did not comprise more than 20 letters).

Four-square cipher [6]: This method is similar to Playfair, because it is based on a system of coordinates whose four quadrants are each filled with the alphabet. The layout of letters can differ from quadrant to quadrant. To encipher a message, act in the following way: Look up the first plaintext letter in the first quadrant and the second one in the third quadrant. These two letters are opposite corners of a rectangle and the ciphertext letters can be found in quadrant numbers two and four.

See Table 2.12.

Plaintext: plaintext letters are encrypted in pairs

Table 2.12 Four-square Cipher

d	w	x	y	m	E	P	T	O	L
r	q	e	k	i	C	V	I	Q	Z
u	v	h	p	s	R	M	A	G	U
a	l	b	z	n	F	W	Y	H	S
g	c	o	f	t	B	N	D	X	K
Q	T	B	L	E	v	q	i	p	g
Z	H	N	D	X	s	t	u	o	h
P	M	I	Y	C	n	r	d	x	y
V	S	K	W	O	b	l	w	m	f
U	A	F	R	G	c	z	k	a	e

Ciphertext: MWYQW XQINO VNKGC ZWPZF FGZPM DIICC GRVCS

Two-square cipher [6]: The two-square cipher resembles the four-square cipher, but the matrix is reduced to two quadrants. Both letters of the digraph part of the same row, they are just exchanged. Otherwise, the plaintext letters are considered as opposite corners of a rectangle and substituted by the other vertices. Quadrants can be arranged horizontal and vertical.

Tri-square cipher [1]: Three quadrants are filled with the same alphabet. The first plaintext letter is looked up in the first quadrant and can be encrypted with every letter of that column. The second plaintext letter is looked up in the second quadrant (diagonally across) and can be encrypted with every letter of that row. Between these two ciphertext characters, the letter at the intersection point is set.

Dockyard cipher [6]: Used by the German Navy during World War II.

2.2.4 Polyalphabetic Substitution

Concerning polyalphabetic substitution, the assignment of ciphertext characters to plaintext characters is not static, but changes during the process of encryption (depending on the key). Polyalphabetic ciphers are historically far more important than polygraphic ciphers.

If the key is generated like a key stream from existing plaintext or ciphertext parts, this is called an autokey method. An incorrect letter at any point renders the entire text that follows unusable. Such errors can occur at the encryptor, in transmission, in reception, or occur at the decryptor (i.e., also at the parties who know the correct key).

- **Vigenère**[23] [4]: Each plaintext character is encrypted with a different ciphertext alphabet that is determined by the characters of a keyword (the so-called Vigenère tableau serves auxiliary means). If the plaintext is longer than the key, the key is (cyclically) repeated.
  See Table 2.13.

Table 2.13 Vigenère Tableau

Plaintext	the	alphabet	is	changing
Key	KEY	KEYKEYKE	YK	EYKEYKEY
Ciphertext	DLC	KPNREZOX	GC	GFKRESRE

-	A	B	C	D	E	F	G	H	I	J	K	L	M	N	O	P	Q	R	S	T	U	V	W	X	Y	Z
A	A	B	C	D	E	F	G	H	I	J	K	L	M	N	O	P	Q	R	S	T	U	V	W	X	Y	Z
B	B	C	D	E	F	G	H	I	J	K	L	M	N	O	P	Q	R	S	T	U	V	W	X	Y	Z	A
C	C	D	E	F	G	H	I	J	K	L	M	N	O	P	Q	R	S	T	U	V	W	X	Y	Z	A	B
D	D	E	F	G	H	I	J	K	L	M	N	O	P	Q	R	S	T	U	V	W	X	Y	Z	A	B	C
E	E	F	G	H	I	J	K	L	M	N	O	P	Q	R	S	T	U	V	W	X	Y	Z	A	B	C	D
F	F	G	H	I	J	K	L	M	N	O	P	Q	R	S	T	U	V	W	X	Y	Z	A	B	C	D	E
G	G	H	I	J	K	L	M	N	O	P	Q	R	S	T	U	V	W	X	Y	Z	A	B	C	D	E	F
H	H	I	J	K	L	M	N	O	P	Q	R	S	T	U	V	W	X	Y	Z	A	B	C	D	E	F	G
I	I	J	K	L	M	N	O	P	Q	R	S	T	U	V	W	X	Y	Z	A	B	C	D	E	F	G	H
J	J	K	L	M	N	O	P	Q	R	S	T	U	V	W	X	Y	Z	A	B	C	D	E	F	G	H	I
K	K	L	M	N	O	P	Q	R	S	T	U	V	W	X	Y	Z	A	B	C	D	E	F	G	H	I	J
...	... ...																									

Vigenère variants:

- *Interrupted key:* The key is not repeated continuously, but starts again with every new word of the message.
- *Autokey*[24] [6]: After using the agreed key, use the message itself as a key. See Table 2.14.

Table 2.14 Autokey Variant of Vigenère

Plaintext	the	alphabet	is	changing
Key	KEY	THEALPHA	BE	TISCHANG
Ciphertext	DLC	TSTHLQLT	JW	VPSPNIAM

23. - `CTO` in the plugins "Vigenère," "Autokey," "Beaufort," "Porta," and "Trithemius." In addition, CTO contains an interactive analyzer to determine the key length and the smallest value of the autocorrelation: `https://www.cryptool.org/en/cto/vigenerebreak`. For an automated analysis the autocorrelation is used: `https://www.cryptool.org/en/cto/autocorrelation`.
   - `CT1 Encrypt/Decrypt ▷ Symmetric (classic) ▷ Vigenère`
   Visualized in `CT1 Indiv. Procedures ▷ Visualization of Algorithms ▷ Vigenère...`
   - `CT2 Templates ▷ Cryptography ▷ Classical ▷ Vigenère`. In addition, CT2 contains powerful analyzers for this cipher.
   - `JCT Default Perspective ▷ Algorithms ▷ Classic ▷ Vigenère`
24. Autokey can be found in the CT variants mainly the same way as Vigenère.

- *Progressive key* [6]: The key changes during the process of encryption. With every repetition, the characters of the keyword are shifted about one position. KEY becomes LFZ.
- *Gronsfeld* [6]: Variant of Vigenère that uses a numeric key.
- *Beaufort*[25] [6]: Variant of Vigenère, the key is subtracted, not added. The ciphertext alphabets may be written backwards.
- *Porta* [1]: Variant of Vigenère with only 13 alphabets. As a consequence, two letters of the keyword are assigned the same ciphertext alphabet and the first and the second half of the alphabet are reciprocal.
- *Slidefair* [1]: This method can be used as a variant of Vigenère, Gronsfeld, or Beaufort. Slidefair does encrypt digraphs according to the following rules: Look up the first letter in the plaintext alphabet above the tableau. Then look up the second one in the row belonging to the corresponding keyword letter. These two letters make up opposite corners of an imaginary rectangle. The letters at the two remaining corners substitute the digraph.

One-time pad (OTP)[26,27]: This is a major concept: A sequence of key bytes (pad) is XORed byte-by-byte to the plaintext, or a sequence of key digits is added. OTP was the first information theoretically secure scheme (see Section 1.8.3). This can be considered as a generalization of Vigenère's cipher. Vernam applied for a patent for this process in 1918.

To fulfill this claim the pad must be random, and it must be used only once (to eliminate any semblance of pattern from the ciphertext).

Reason to use the key pad only once: Given ciphertext C, plaintext P, pad K, and two plaintexts encrypted with the same key: $C1 = P1 \oplus K$; $C2 = P2 \oplus K$; thus, $C1 \oplus C2 = (P1 \oplus K) \oplus (P2 \oplus K) = P1 \oplus P2$; which effectively could leak the plaintexts.[28]

Superposition (some variants of the OTP)

- *Running-key cipher*: A key text (for example out of a book) is added to the plaintext.
- *Superposition with numbers*: A sequence like Fibonacci or a number of sufficient length (for example pi) is added.

25. - In CTO in the plugin "Beaufort": `https://www.cryptool.org/en/cto/beaufort`
    - In CT2 part of the Vigenère component and template (including the autokey variant).
26. On a big scale OTPs have been successfully broken by Americans and British during the "Venona" project because of wrong usage by the Soviet spies.
27. Implementations of the OTP can be found in:
    - CTO in the plugins "Vernam" and "XOR": `https://www.cryptool.org/en/cto/vernam`, `https://www.cryptool.org/en/cto/xor`
    - `CT1 Encrypt/Decrypt ▷ Symmetric (classic) ▷ Vernam / OTP`
    `CT1 Encrypt/Decrypt ▷ Symmetric (classic) ▷ XOR`
    - Two templates, one for Vernam and one for XOR, can be found at `CT2 ▷ Cryptography ▷ Classical`
    - `JCT Default Perspective ▷ Algorithms ▷ Classic ▷ XOR`
    - Section 9.3.1 contains a detailed description of the OTP as bitstream cipher and its implementation in SageMath.
28. Via `JCT Default Perspective ▷ Analysis ▷ Viterbi Analysis` you can play with an automatic cryptanalysis of running-key ciphertexts. You can see how astonishing it is, if you get little by little—from XORed ciphertexts or from XORed plaintexts—both original plaintexts. See Figure 2.4.

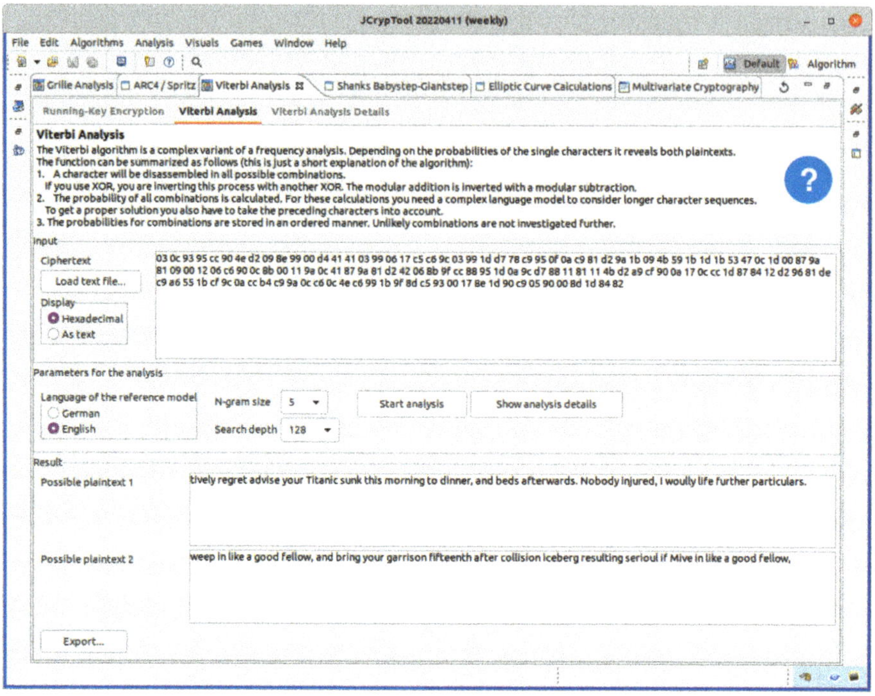

Figure 2.4 Viterbi analysis of an XOR cipher (like in the OTP) in JCT.

Phillips cipher [1]: The alphabet is filled into a square table with five columns. Seven more tables are generated by first shifting the first row one position toward the bottom, then shifting the second row toward the bottom. The plaintext is divided into blocks of five that are encrypted with one matrix each. Letters are substituted by the ones on their right and underneath.

Ragbaby cipher [1]: Construct an alphabet with 24 characters. Then number the plaintext characters, starting the numeration of the first word with "1," the numeration of the second one with "2," and so forth. Number 25 corresponds to number 1. Each letter of the message is encrypted by shifting it the corresponding positions to the right.

See Table 2.15.

Alphabet: KEYWORDABCFGHILMNPSTUVXZ

Table 2.15 Ragbaby Cipher

Plaintext	t h e	a l p h a b e t	i s	c h a n g i n g
Numbering	1 2 3	2 3 4 5 6 7 8 9	3 4	4 5 6 7 8 9 10 11
Ciphertext	U L O	C P V P I M C O	N X	I P I Z T X Y X

2.3 Combining Substitution and Transposition

In the history of cryptography one often comes across combinations of the previously mentioned methods. These have become—on average—proven to be safer than procedures that are based only on one of the principles of transposition or substitution. However, cascades[29] do not have to be safer (e.g., the sequential execution of two Caesar ciphers with different shift values is simpler than the unique executing of Caesar with a third shift value), so that's not at all a security gain. On the other hand, executing two column transpositions one after the other can lead to a significant gain in security; it is equivalent to a column transposition in which the key is as long as the product of the two individual key lengths, provided the keys are constructed correctly. See `https://mysterytwister.org/challenges/level-3/double-column-transposition-reloaded-part-1`.

ADFG(V)X[30] [4]: ADFG(V)X-encryption was developed in Germany during World War I. The alphabet is filled into a 5 × 5 or 6 × 6 matrix, and columns and rows are marked with the letters ADFGX and V, depending on the size of the grille. Each plaintext character is substituted by the corresponding pair of letters. Finally, a (row-) transposition cipher is performed on the resulting text.

Fractionation [6]: Generic term for all kinds of methods that encrypt one plaintext character by several ciphertext characters and then apply a transposition cipher to this intermediate ciphertext so that ciphertext characters originally belonging to each other are separated.

- *Bifid/Polybius square/checkerboard* [5]: Bifid encryption is the basic form of fractionation. A 5 × 5 matrix is filled with the plaintext alphabet (see Playfair encryption), rows and columns are numbered so that each plaintext character can be substituted by a pair of digits. Mostly the plaintext is divided into blocks of equal length. The length of blocks (here 5) is another configuration parameter of this cipher. Block-by-block all line numbers are read out first, followed by all numbers naming the columns. To obtain the ciphertext, the digits are pairwise transformed into letters again. The numbers can be any permutation of (1,2,3,4,5): in our case (1,4,2,3,5) for the rows and (2,4,5,1,3) for the columns. This is one key or configuration parameter of this cipher. Instead of numbering rows and columns, a keyword can also be used. See Table 2.16.

Table 2.16 Bifid Cipher

	2	4	5	1	3
1	K	E	Y	W	O
4	R	D	A	B	C
2	F	G	H	I	L
3	M	N	P	Q	S
5	T	U	V	X	Z

Plaintext	combi	nings	ubsti	tutio	nandt	ransp	ositi
Rows	41342	32323	54352	55521	34345	44333	13252
Columns	33211	41443	41321	24213	45442	25435	33121

29. Running the same cipher twice in a row is also called a cascade.
30. - CTO plugin "ADFG(V)X": `https://www.cryptool.org/en/cto/adfg-v-x`
- `CT1 Encrypt/Decrypt ▷ Symmetric (classic) ▷ ADFGVX`
- `CT2 Templates ▷ Cryptography ▷ Classical ▷ ADFGVX`. In addition, CT2 contains analyzers for this cipher.
- `JCT Default Perspective ▷ Algorithms ▷ Classic ▷ ADFGVX`

Plaintext: combining substitution and transposition
Read out intermediate step:

41342 32323 54352 55521 34345 44333 13252 33211 41443 41321 24213 45442 25435 33121

Ciphertext: BNLLL UPHVI NNUCS OHLMW BDNOI GINUR HCZQI

- *Trifid* [6]: Twenty-seven characters (alphabet + 1 special character) may be represented by a triple consisting of the digits 1 to 3. The message to be encrypted is divided into blocks of three and the relevant triple is written underneath each plaintext character as a column. The resulting numbers below the plaintext blocks are read out line by line and are substituted with the corresponding characters.

Bazeries cipher [1]: The plaintext alphabet is filled into a 5 × 5-matrix column by column, a second matrix is filled line by line with a keyword (a number smaller than a million) followed by the remaining letters of the alphabet. Then the message is divided into blocks of arbitrary length and their characters' order is inverted. Finally, each letter is substituted—according to its position in the original matrix—by its counterpart in the second matrix.

See Table 2.17.

Table 2.17 Bazeries Cipher

Alphabet					Keyword Alphabet				
a	f	l	q	v	N	I	E	H	U
b	g	m	r	w	D	R	T	O	S
c	h	n	s	x	A	F	B	C	G
d	i	o	t	y	K	L	M	P	Q
e	k	p	u	z	V	W	X	Y	Z

com	bini	ngs	ub	stitu	tiona	ndt	ran	sposi	ti	on
moc	inib	sgn	bu	utits	anoit	tdn	nar	isops	it	no
TMA	LBLD	CRB	DY	YPLPC	NBMLP	PKB	BNO	LCMXC	LP	BM

Plaintext: combining substitution and transposition
Keyword: 900004 (nine hundred thousand and four)

The second table in Table 2.17 has three lines: The first shows the plaintext split into blocks, the second shows the inverted blocks, and the third the ciphertext. The encryption is clear, but the decryption is only unique if the lengths of the blocks is exchanged as an additional key parameter.

Digrafid cipher [1]: To explain the Digrafid substitution, Table 2.18 is used (to simplify matters, the alphabet was left in its original order). Look up the first letter (here **c**) of the first digraph in the horizontal alphabet and write down the column number **3**. Then look up the second letter **o** in the vertical alphabet and write down the corresponding line number **6**. Between these two numbers, the number **2** at the intersection point is set. Afterward, the triples are written vertically underneath the digraphs that are arranged in groups of three. The three-digit numbers **3,4,9** arising horizontally are transformed back into digraphs **LI**.

Note that for a complete description, it is necessary to explain how sender and receiver handle texts that fill in the last block only 1–5 characters. The possibilities range from ignoring a last and incomplete block to padding it with random characters or with characters predefined in advance.

See Table 2.18.

Table 2.18 Digrafid Cipher

1	2	3	4	5	6	7	8	9				
A	B	C	D	E	F	G	H	I	1	2	3	
J	K	L	M	N	O	P	Q	R	4	5	6	
S	T	U	V	W	X	Y	Z	.	7	8	9	
									A	J	S	1
									B	K	T	2
									C	L	U	3
									D	M	V	4
									E	N	W	5
									F	O	X	6
									G	P	Y	7
									H	Q	Z	8
									I	R	.	9

co mb in	in gs ub	st it ut	io na nd	tr an sp	os it io
3 4 9	9 7 3	1 9 3	9 5 5	2 1 1	6 9 9
2 4 2	2 3 7	9 3 9	2 4 4	8 2 8	6 3 2
6 2 5	5 1 2	2 2 2	6 1 4	9 5 7	1 2 6
LI KB FN	.C BY EB	SU I. BK	RN KD FD	BA HQ RP	X. FT AO

Nicodemus cipher [1]: First, a simple columnar transposition is carried out. Before reading out the columns, the message is encrypted additionally by Vigenère (all letters of a column are enciphered with the corresponding keyword letter). The ciphertext is read out in vertical blocks.

See Table 2.19.

Table 2.19 Nicodemus Cipher

K	*E*	*Y*	*E*	*K*	*Y*	*E*	*K*	*Y*
c	o	m	o	c	m	S	M	K
b	i	n	i	b	n	M	L	L
i	n	g	n	i	g	R	S	E
s	u	b	u	s	b	Y	C	Z
s	t	i	t	s	i	X	C	G
t	u	t	u	z	t	Y	J	R
i	o	n	o	i	n	S	S	L
a	n	d	n	a	d	R	K	B
t	r	a	r	t	a	V	D	Y
n	s	p	s	n	p	W	X	N
o	s	i	s	o	i	W	Y	G
t	i	o	i	t	o	M	D	N

Plaintext: combining substitution and transposition

Ciphertext: SMRYX MLSCC KLEZG YSRVW JSKDX RLBYN WMYDG N

Double column transposition (DCT) / Granit E160 [10]: Granit is a two-step cipher. The second step is the double column transposition. In the first step, the plaintext is substituted by a sequence of digits using a codebook and a matrix (a variant of the Polybius square).

The Granit cipher was used for instance by the spy Guenter Guillaume for his communication with the Ministry of State Security of the former German Democratic Republic (GDR) until about 1960.[31] If the passwords are chosen correctly, this is a very secure procedure.

2.4 Further P&P Methods

The P&P procedures include both classic procedures and newer ones—which should always be executable by hand and operate on a few printable characters. The new methods include, for example, LC4 (2017) and Handycipher (2014).

Pinprick encryption [4]: For centuries, this simple encryption method has been put into practice for different reasons (actually steganography). During the Victorian Age, for example, small holes underneath letters in newspaper articles marked the characters of a plaintext, as sending a newspaper was much cheaper than the postage on a letter.

Stencil: Stencils (cardboard with holes) are also known as a Cardinal-Richelieu-Key. Sender and receiver have to agree upon this stencil. The stencil is laid above a text and the letters that remain visible make up the ciphertext. This is not to be confused with the Grille of Section 2.1.1, because here not only the scrambled plaintext letters are transmitted, but the entire text, most of whose letters are meaningless (called nulls).

Card games [6]: The key is created by means of a pack of cards and rules that are agreed upon in advance. All methods mentioned in this paragraph are designed as paper-and-pencil methods (i.e., they are applicable without electronic aid). A pack of cards is unsuspicious to outsiders, shuffling the deck provides a certain amount of coincidence, and cards can be transformed into numbers easily to be used in a substitution cipher without any further aid.

Solitaire cipher (Bruce Schneier)[32] [11]: Sender and receiver have to own a deck of cards shuffled in the same manner. A key stream is generated that has to consist of as many characters as the message to be encrypted.

The algorithm to generate the key is based on a shuffled deck of 54 cards (ace, 2–10, jack, queen, king in four suits, and two jokers). The pack of cards is held face up:

1. Swap the first joker with the card beneath it.
2. Move the second joker two cards down.

31. MTC3 offers a number of related challenges: If you enter in your browser at `https://mysterytwister.org/challenges/level-2/` the search item "Granit," you'll find six challenges about it.
A detailed 20-page description about the Granit cipher can be found at:
`https://mysterytwister.org/media/challenges/pdf/mtc3-drobick-01-doppelwuerfel-01-en.pdf`

32. - `CT1 Encrypt/Decrypt` ▷ `Symmetric (classic)` ▷ `Solitaire` (detailed GUI window);
- `CT2 Templates` ▷ `Cryptography` ▷ `Classical` ▷ `Solitaire`. In addition, CT2 contains a brute-force analyzer for this cipher.

3. Now swap the cards above the first joker with those below the second one.
4. Look at the bottom card and convert it into a number from 1 to 53 (bridge order of suits: clubs, diamonds, hearts, spades; joker = 53). Write down this number and count down as many cards starting with the top card. These cards are swapped with the remaining cards; only the bottom card remains untouched.
5. Look at the top card and convert it into a number as well. Count down as many cards starting with the top card.
6. Write down the number of the following card. This card is converted into your first keystream character. As we need numbers from 1 to 26 to match the letters of our alphabet, clubs and hearts correspond to the numbers 1 to 13, diamonds and spades to 14 to 26. If your output card is a joker, start again.

For each keystream character you would like to generate, these six steps have to be carried out. This procedure is—manually—very lengthy (4 hours for 300 characters, dependent on your exercise) and requires high concentration.

Encryption takes place by addition modulo 26. Encryption is relatively fast compared to the key stream generation.

This P&P cipher creates a key stream that is so good, even today it is hard to crack the cipher if you don't know the originally sorted card deck (ciphertext-only attack).

Mirdek cipher (Paul Crowley) [12]: Even though this method is quite complicated, the author provides a good example to illustrate the procedure.

Playing card cipher (John Savard) [6]: This algorithm uses a shuffled deck of 52 cards (no joker). Separate rules describe how to shuffle the deck. A keystream is created via the following steps:

1. The pack of cards lies in front of the user, top down. Cards are turned up and dealt out in a row until the total of the cards is eight or more.
2. If the last card dealt out is a jack, queen, or king, write down its value; otherwise write down the sum of the cards dealt out (a number between 8 and 17). In a second row, deal out that number of cards.
3. The remaining cards are dealt out in rows under the second row. The first one ends under the lowest card of the top row, the second one under the next lowest card, and so on. If there are two identical cards, red is lower than black.
4. The cards dealt out under step 3 are collected column by column, starting with the column under the lowest card. The first card that is picked up becomes the bottom card (face up).
5. The cards dealt out in step 1 and 2 are picked up, beginning with the last card.
6. The deck is turned over, the top card is now the bottom card (face down). Afterward, steps 1 to 6 are repeated twice.

To generate a keystream character, write down the first card not being a jack, queen, or king. Count down that number of cards. The card selected has to be between 1 and 10. Now repeat these steps beginning with the last card. These two numbers are added and the last digit of the sum is your keystream character.

Josse's cipher[33] [13]: Josse's cipher was invented by the French major H. D. Josse (1852–1929). This method has only recently been rediscovered and was described by Rémi Géraud-Stewart and David Naccache. It is a relatively simple substitution cipher, but additional security features have been added. That includes that the first letter is encrypted differently than the rest of the letters, and that there is some kind of error propagation (autokey), so that an incorrect letter at any point makes the entire following text unusable.

VIC cipher[34] [6]: This is a highly complicated but relatively secure paper-and-pencil method. It was developed and applied by Soviet spies. Among other things, the user had to create 10 pseudorandom numbers out of a date, the first words of a sentence, and any five-digit number. A straddling checkerboard is part of the encryption, too. A detailed description can be found in [6].

Handycipher[35] [14]: Handycipher is a homophonic substitution cipher, which inserts randomly chosen null characters. The cipher was improved several times using the feedback of the solvers in MTC3.

ElsieFour cipher[36] [15]: ElsieFour (shortened LC4) combines ideas of modern RC4 stream cipher, historical Playfair cipher, and plaintext-dependent keystreams (autokey). It can be computed manually.

Hutton cipher[37]: Hutton is a polyalphabetic substitution using a password, building a keyed alphabet from a keyword, and then dynamically mix the alphabets during the encryption/decryption process. Hutton was invented in 2018, and

33. `CT2 Templates ▷ Cryptography ▷ Classical ▷ Josse cipher` with visualization.
34. `CT2 Templates ▷ Cryptography ▷ Classical ▷ VIC cipher` with extensive explanation.
35. In addition to the original paper, the 24 challenges clustered in four series in MTC3 (by the same author) deliver additional details:
`https://mysterytwister.org/challenges/level-2/weakened-handycipher-part-1`;
`https://mysterytwister.org/challenges/level-2/handycipher-part-1`;
`https://mysterytwister.org/challenges/level-2/extended-handycipher-part-1`;
`https://mysterytwister.org/challenges/level-2/handycipher-made-in-love-part-1`.
In `https://mysterytwister.org/challenges/level-3/handycipher-part-9` you can find a simulator by clicking on "Download additional files for this challenge": `https://mysterytwister.org/media/challenges/add/mtc3-kallick-23-HC9-add.zip`.
36. Complementing the original paper, the 4 challenges clustered in two series in MTC3 deliver additional details:
`https://mysterytwister.org/challenges/level-2/weakened-elsiefour-part-1`;
`https://mysterytwister.org/challenges/level-3/elsiefour-part-1`;
In `https://mysterytwister.org/media/challenges/add/mtc3-rotthaler-01-elsiefour-01-add.zip` you can find Java and Python source code of LC4.
Details and Python source code for the enhancement LS47 can be found at: `https://gitea.blesmrt.net/exa/ls47`.
37. For v2 there was a prize for a challenge with a ciphertext of 169081 letters: `https://old.reddit.com/r/codes/comments/ar1lbd/hutton_cipher_a_10000_challenge/`. Hutton can be performed directly at `https://hutton-cipher.netlify.app/`.

refined with version 2 in 2019. It is similar to the Quagmire III cipher defined by ACA.

2.5 Hagelin Machines as Models for Precomputer Ciphers

So far, in this chapter, we discussed the encryption methods that can be easily used by anyone who has just a pencil and a piece of paper. However, such methods are very slow when larger amounts of text have to be encrypted. They also are not secure enough, especially in the scenarios when someone who wants to break the encryption can collect a relatively large amount of ciphertext that was the result of encryption with the same key.

At the beginning of the 20th century, with the development of radio the volume of information that was transmitted vastly increased. As before, the military and diplomatic information was often sensitive and needed to be confidential. This situation caused a need for fast, more secure, and more accurate encryption (and decryption) methods. The problem was solved by rotor cipher machines, which began to appear right after World War I [16]. It is believed that they were invented by two Dutch navy officers, Theo A. van Hengel and R. P. C. Spengler [17], who built the first machine for the Dutch Navy already in 1915 [18]. The rotor machines were the state-of-the-art tools used for achieving confidentiality of data before they were replaced by computers in the 1970s. Probably the most famous examples of rotor machines are Enigma, Sigaba, Hebern, and Hagelin M-209 also known as CSP-1500, C-38, and AM-1.

In the next sections, we discuss several Hagelin machines and show their usage in CrypTool 2 (note that CrypTool 2 also includes the rotor machines Enigma, Typex, SZ42, Sigaba, and Fialka).

2.5.1 Overview of Early Hagelin Cipher Machines

Crypto AG, also known as Hagelin Cryptos, Hagelin Crypto Company, and CAG, was a Swiss manufacturer of cryptographic equipment that was founded by Boris Hagelin, and was the successor to A.B. Cryptoteknik [19]. Hagelin developed several rotor cipher machines that were sold to more than 60 different countries worldwide. They are called Hagelin machines.

The first cipher machine developed by Boris Hagelin was B-21. This was a battery-powered electromechanical machine with an indicator panel of 25 electric lamps (Figure 2.5).

In 1932, France announced a competition to develop a cipher machine for their army. The French military leaders became interested in the B-21, but they insisted on two important improvements: The cipher machine must be portable and capable of printing ciphertext. As a result, Hagelin developed another electromechanical cipher machine called B-211 [20]. B-21 and B-211 do not feature polyalphabetic substitution by means of wired electrical rotors, but a scrambling matrix of 25 keys (5 × 5) [19].

In 1934, following the request of the French army who wanted to have a machine that would fit a trousers pocket, Boris Hagelin developed the C-35 cipher machine. C-35 was significantly smaller than others in this class (Figure 2.6). This

Figure 2.5 Hagelin B-21 [19].

Figure 2.6 Hagelin C-35 [19].

was the first fully mechanical pin-and-lug cipher machine and became the basic design on which all Hagelin's later machines would be based (see Section 2.5.2.2 for more information about the Hagelin machine). C-35 consisted of a drum with 25 bars, five pinwheels, and a type-wheel [21].

C-36 and C-362 were the next models of Hagelin machines, which had only slight modifications as compared to C-35. These included protective casing, different arrangements of the bar lugs, and later on, movable lugs on the drum (in C-336). In the pre-war years, C-36 and C-362 models were purchased by France, Great Britain, Italy, Germany, and some other European countries. C-37 is an extremely rare successor of C-36 that was used by the French Navy and also for French-British liaisons.

The next device developed by Hagelin was the C-38 machine, which had six pinwheels as compared to five wheels in the previous models. Another change was that the drum lugs of C-38 could slide in one of five active or in an inactive position. These improvements considerably increased the key space and thus increased the security of the machine. M-209 (Figure 2.7) was the U.S. Navy and the U.S. military variant of the C-38 that served for the American Army during the period

Figure 2.7 Front view of the Hagelin M-209 [19].

of time starting shortly before World War II and until after the Vietnam War [21]. About 140,000 units of this model were produced. This model and its implementation in CrypTool 2 are nicely presented on the YouTube channel *Cryptography for Everybody* [22].

The BC-38 electromechanical cipher machine, as the name suggests, was an improved model of the mechanical C-38. "B" was added as a prefix to the model number to indicate the presence of a keyboard.

The latest Hagelin machines we discuss are the ones that refer to the model range C-52/CX-52, which were developed in 1952. Information about the machines that appeared later can be found in [19] and [21]. The C-52 had great success on the market. This machine had six removable cipher wheels—the pinwheels—that could be configured outside of the machine and installed in any order. A more advanced and secure version was the CX-52 where the rotation of the pinwheels became irregular. It is considered to be one of the most successful Hagelin mechanical cipher machines, which made the encryption much more secure.

Before the CX-machines were invented, the United States and United Kingdom were able to break the high-level cipher systems of most of the countries. The development of the CX-52 model changed the situation, which resulted in the Rubicon operation (see Section 2.5.2.4).

In the next section, the architecture of these latter machines is explained.

2.5.2 Hagelin C-52/CX-52 Models

The C-52 and CX-52 are all-mechanical pin-and-lug type cipher machines that do not require any power source. All the models have a relatively similar architecture. However, before the architecture of the machine is provided, to enable an easier understanding of their construction, we first explain the general encryption method they use.

2.5.2.1 Encryption Principle

In fact, these machines (as well as most other rotor cipher machines) are mechanical implementations of stream ciphers (see Chapter 9). This means that the heart of this machine is a pseudorandom keystream generator. In other words, the Hagelin machine produces a sequence of pseudorandom numbers that are used to encrypt the plaintext. The encryption is applied to each plaintext character one by one using a Beaufort encoding, which is in principle a Caesar cipher with an inverted alphabet. That is, each plaintext character is substituted by a different one from the reversed alphabet.

Let's use the following notations:

- $P = (p_1, p_2, ..., p_L)$ is the plaintext, which is the original message that has to be encrypted. It is a sequence of Latin characters $p_i, 1 \leq i \leq L$, where L is the length of the message.
- $C = (c_1, c_2, ..., c_L)$ is the ciphertext received after encrypting the plaintext with a Hagelin machine also represented as a sequence of Latin characters.
- $\text{Ind}(a)$ is the index of the character a in the alphabet, for example, $\text{Ind}(\texttt{'A'}) = 0; \text{Ind}(\texttt{'E'}) = 4$.
- $\text{Char}(n)$ is the character located at the nth position in the alphabet, for example, $\text{Char}(0) = \texttt{'A'}, \text{Char}(4) = \texttt{'E'}$.
- $D = \{d_1, d_2, ..., d_L\}$ is the sequence of pseudorandom integer numbers generated by the Hagelin machine during the encryption process. This sequence is also called *displacement* sequence.

The encryption takes place in accordance with (2.1).

$$c_i = \text{Char}(\text{Ind}(\texttt{'Z'}) - \text{Ind}(p_i) + d_i \bmod 26) \qquad (2.1)$$

In this equation $\text{Ind}(\texttt{'Z'})$ is the constant value that is the index of the character 'Z' in the alphabet and is equal to 25.

Let's consider the following example. Suppose that we want to encrypt plaintext $P = (\texttt{'K'}, \texttt{'C'})$ and the 2 displacement values generated by a given Hagelin machine are $D = \{d_1, d_2\} = \{4, 1\}$. Then the ciphertext is computed as follows:

$$c_1 = \text{Char}(\text{Ind}(\texttt{'Z'}) - \text{Ind}(\texttt{'K'}) + 4) \bmod 26) = \text{Char}(25 - 10 + 4) = \text{Char}(19) = \texttt{'T'}$$
$$c_2 = \text{Char}(\text{Ind}(\texttt{'Z'}) - \text{Ind}(\texttt{'C'}) + 1) \bmod 26) = \text{Char}(25 - 2 + 1) = \text{Char}(24) = \texttt{'Y'}$$

Hence, the received ciphertext is $C = (\texttt{'T'}, \texttt{'Y'})$.

To understand the key idea of the Beaufort encoding take a look at the positions of the plaintext characters in the original alphabet and at the positions of the ciphertext characters in its reverse representation. It's easy to see that the distance between the plaintext and the corresponding ciphertext characters is equal to the displacement values $d_1 = 4$ ('K' to 'T') and $d_2 = 1$ ('C' to 'Y'):

```
A B C D E F G H I J K L M N O P Q R S T U V W X Y Z
Z Y X W V U T S R Q P O N M L K J I H G F E D C B A
```

This encryption principle is the same for all the Hagelin models discussed here. However, the way these pseudorandom sequences of displacement values are computed is different for each of the machines and defines how secure the given machine is.

2.5.2.2 Architecture of the Machines

To keep the presentation in this section compact, we focus only on the most essential parts of the machines that are important from a cryptographic point of view (for a description of the other parts and their purpose, see [23]).

Pseudorandom displacement generator. The pseudorandom displacement generator of these machines compromises two main parts (Figure 2.8), which are:

- The set of 6 pinwheels (also known as keywheels, pin disks, rotors, cipher wheels or just wheels);
- Rotating cage (also known as drum) with 32 bars.

The pinwheels for each of the machine versions can be selected from a set of 12 with 25, 26, 29, 31, 34, 37, 38, 41, 42, 43, 46, and 47 pins. The wheels are labeled with both letters and numbers. Next to each label there is a pin that can be set to an active or nonactive state.

A printer located at the left of the machine (Figure 2.8) acts as the input/output device. It has a double print head, where one letter ring contains the alphabet in the regular order, and the rightmost ring—the alphabet in reversed order (Figure 2.9). We call the former the input ring and the latter the output ring.

To encrypt each character p_i the operator sets the input ring of the printwheel to the position of this character. Then he turns the handle, and the drum makes a complete revolution. Each of the 32 bars located in the drum can be affixed with the small lugs at one or more of the six positions against each of the wheels [21]. During the revolution of the drum all the lugs that are affixed to the bars will contact the active guide arm of the wheel if the pin currently located against the drum is active. All bars whose lugs had this contact will slide to the left. Each bar that is slid to the left becomes an additional tooth on the drum, which will displace the output ring by one extra position. That is, the total displacement value d_i is equal to the

Figure 2.8 Hagelin C-52 [19].

Figure 2.9 Double print head with two letter rings; in regular-alphabet and in reversed-alphabet order [19].

number of bars that were slid to the left during the revolution of the drum. After the displacement value d_i is computed, the encryption happens following (2.1).

To make sure that decryption is possible, both sender and receiver must set their machines in the same way. The wheel pins and the lugs that form an internal secret key were normally changed on a daily basis, according to the key lists that were distributed periodically [24]. To keep the communication secure, the initial position of the six wheels, which is called an external key, has to be changed for each new message.

Advancement of the wheels. In the case of the C-52 model, all the pin wheels regularly step by one position after each character encryption. However, in some of the CX-52 versions, this advancement of the wheels happens irregularly, which makes the encryption considerably more secure. This is realized by the special drum bars which have cams that can advance the pinwheels, so that the next cycle starts with a different pinwheel setting and thus another set of six active and inactive pins [23]. Those bars, which can influence the stepping of the wheels, are called advance bars. If a bar is forced to the left by an active guide arm that contacts a lug, the wheel that is commanded by that bar will move.

Key space size. Assuming that the pin wheels and bars are known and are not part of the key, the whole key space of the machine is composed of the wheel settings key space and the lug settings key space. The wheel settings key space depends on the set of the selected pin wheels. Suppose that the selected machine pin wheels have the sizes that form a set: $L = \{l_1, l_2, ..., l_6\}$. Then the number of all possible initial positions of the wheels is given by:

$$\mathsf{K_i} = l_1 \cdot l_2 \cdot l_3 \cdot l_4 \cdot l_5 \cdot l_6$$

Moreover, each of the wheel pins can be active or not active. In total there are $l_1 + l_2 + l_3 + l_4 + l_5 + l_6$ pins. Hence, the total number of possible pin settings is:

$$\mathsf{K_p} = 2^{l_1+l_2+l_3+l_4+l_5+l_6}$$

In addition, the machine has 32 bars, where each bar has six positions where lugs can be placed (against each of the pin wheels). Some of the bars cannot be lugged. If such bars are used in the machine, they should not be included in the key space size computation.

The total number of lugs is $32 \cdot 6 = 192$. Therefore the maximal number of all possible lug settings is:

$$\mathsf{K_l} = 2^{192}$$

This gives the total key space size of:

$$\mathsf{K} = \mathsf{K_i} \cdot \mathsf{K_p} \cdot \mathsf{K_l} = l_1 \cdot l_2 \cdot l_3 \cdot l_4 \cdot l_5 \cdot l_6 \cdot 2^{l_1+l_2+l_3+l_4+l_5+l_6} \cdot 2^{192}$$

For example, if the selected wheel sizes are $L = \{l_1, l_2, ..., l_6\} = \{29, 31, 37, 41, 43, 47\}$ and all the bars can be lugged then the total key space of such machine is:

$$\mathsf{K} = 29 \cdot 31 \cdot 37 \cdot 41 \cdot 43 \cdot 47 \cdot 2^{29+31+37+41+43+47} \cdot 2^{192} \approx 2^{31.3} \cdot 2^{228} \cdot 2^{192} \approx 2^{451.3}$$

We note, however, that the cryptographic-relevant key space is usually considerably smaller for the concrete machines due to the existence of equivalent keys. That is, different settings may lead to the same sequence of displacements D, which means that for an attacker it is sufficient to know only one of such equivalent keys. Moreover, settings may result in only slightly different displacement sequences that could also be sufficient for an attack. This can reduce the security of the concrete machine even further.

Printer offset and the fixed-variable feature (F-V feature). Some of the models allow applying an offset on the printer between the input ring and the output ring. If this is done, when encrypting a given character p_i with the offset o, the machine will encipher the letter $\text{Char}(\text{Ind}(p_i) - o) \bmod 26$. For example, if the offset $o = 2$ and the plaintext character $p_i =$ `'M'`, the machine will encipher character `'K'`.

When a given model is equipped with a fixed-variable feature, this offset can be increased after encrypting every character. When this feature is installed and activated, the relative position of the input ring and output ring grows by the displacement value d_i in each encrypting cycle i.

2.5.2.3 Differences Between C-52/CX-52 Models

The differences between each of the C-52/CX-52 models known to us are presented next.

The main differences between the machine versions are the supported pin wheel types and the way the pin wheels advance during the encryption process. The latter is achieved by various types of the bars used in the machines. We provide an overview of the main principles, and for concrete specifications of the wheels and bars used in C-52/CX-52 models refer to [23]. To better understand the machine mechanism and how the models differ, we also encourage the readers to use the CrypTool 2 component "Hagelin Machine," which is discussed in Section 2.5.3.

C-52. In Hagelin C-52, all pin wheels always step by one. This is similar to the previous Hagelin models, however, the important improvement is that the pin wheels are interchangeable. It is also possible to configure the machine to be compatible with earlier Hagelin models (e.g., M-209).

CX-52a. In this model all the six pin wheels are advanced independently of each other, from one to five times. The advancement of any of the wheels can be affected by the active pin of any other wheel.

CX-52b. All the pin wheels of CX-52b step the same number of times (from 1 to 32). The number of steps is $32 - n$, where n is the total displacement applied in this encryption cycle d_i.

CX-52c. In this machine the first two pin wheels step once always, and all the other wheels always do it twice. This gives the false impression of irregular stepping. It is particularly suited to mimic the older machines such as M-209.

CX-M. The first wheel of the model CX-M always steps, and the lugs on the stepping-affecting bars are configured so that the other five wheels step if any of the wheels with the lower number has its active pin on. For example, the fourth wheel will step if the active pin of either the first, or second, or the third wheel is on.

CX-52 French version. In the French version of the CX-52 machine, all wheels always step the same number of steps (from 1 to 5). Similar to CX-52a, the active pin of any wheel may affect the stepping of any other wheel, including itself.

CX-52 EIRE. In the EIRE version of the machine the first two wheels step independently from 1 to 3 times. The next two pairs of wheels step together from 0 to 2 times. That is, the third wheel steps the same number of times as the fourth wheel, whereas fifth wheel steps together with the sixth wheel. It is also possible to configure the machine to be compatible with M-209.

2.5.2.4 Operation RUBICON

As can be seen from the description of the CX-52 models in Section 2.5.2.3, the key difference between the versions of C(X)-52 machines is the way how an advancement of the pin wheels is arranged. In practice, the more irregular the movement of the wheels, the higher the security of encryption. Given this, it is clear that some of the machines are weaker than others.

The recent disclosures by the German television station *ZDF* [25] and the American newspaper *The Washington Post* [26] about Operation Rubicon shed light on the reasons for such a difference in the security of different machines.

In the years following World War II, Hagelin provided their cipher machines to many countries. All the models of the machines earlier than CX types were subject to successful cryptanalysis by the U.S. and U.K. intelligence services. Indeed, they were able to get access to the encrypted communications between businesses and diplomatic missions of almost every other country (with a very few exceptions).

Alarmed by the capabilities of the new CX-52 machines, U.S. officials made a deal with Boris Hagelin to restrict sales of his most sophisticated models only to the approved countries. This deal was never put in writing and is known as a gentleman's agreement, which was in effect until 1957. At this point in time Hagelin started selling the secure machines to denied countries, keeping this fact in secret. In 1960, the U.S. Central Intelligence Agency (CIA) and Hagelin signed a licensing agreement in which sales to certain listed countries were now prohibited legally. Later, the company Crypto AG was bought by the German Bundesnachrichtendienst (BND) and CIA, who used a complex scheme of companies and fiduciaries in order to hide the actual ownership from public view [19].

2.5.3 Hagelin Component in CT2

The Hagelin C-52/CX52 models were implemented as a component in CrypTool 2. The name of the component is Hagelin Machine, which allows a user to select one of nine Hagelin machines used in practice or to build a virtual machine with an architecture that never existed in real life. When designing a virtual machine, the user can combine any known parts that were ever used in each model. That is, the user can freely select the wheels and the bars as well as define by themselves the number of bars. The number of wheels cannot be adjusted as this would need usage of the bars that never existed.

2.5.3.1 CT2 Template of Hagelin

To start learning the models with the help of the component it is recommended to use the prepared template that demonstrates the encryption and decryption operations of Hagelin. The screenshot in Figure 2.10 shows the CT2 template.

In this screenshot, it is demonstrated how the text "TO BE OR NOT TO BE" is encrypted and decrypted using the Hagelin model CX-52a. The component includes the following inputs:

- Input text—plaintext that has to be encrypted or ciphertext that has to be decrypted.

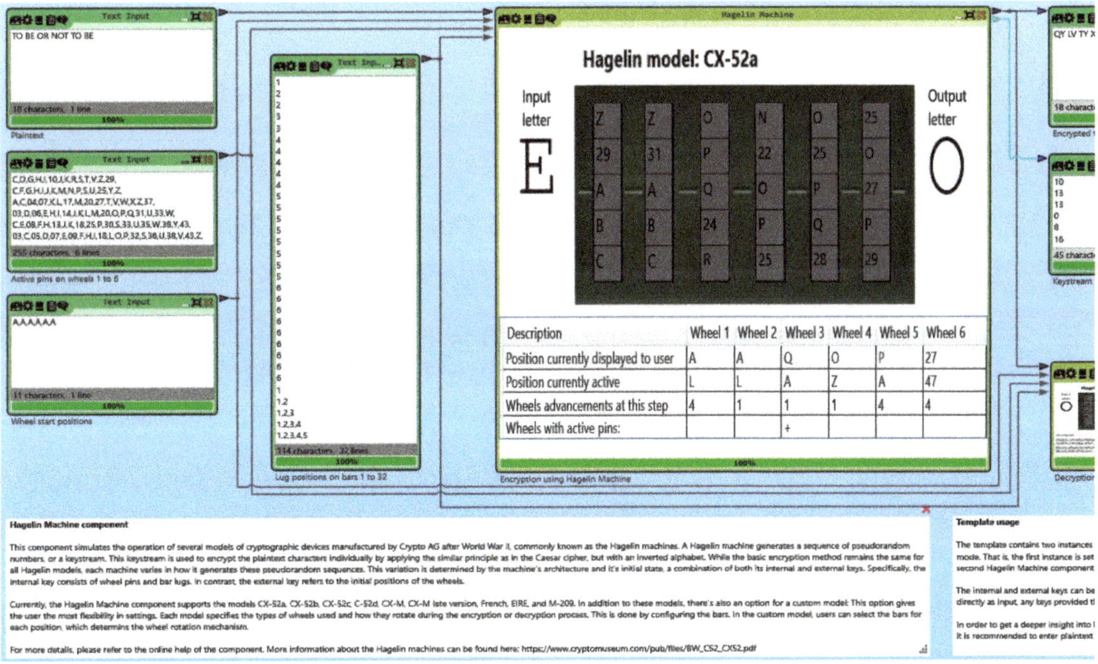

Figure 2.10 Hagelin template in CrypTool 2.

- Wheels state—indexes (letters or numbers) of the initial positions of the six wheels (see input box "Wheel start positions" in the template in Figure 2.10).
- Wheel pins—indexes (letters or numbers) of the pins that are activated for each wheel (see input box "Active pins on wheels 1 to 6" in the template).
- Bar lugs—numbers of the wheels for which lugs on each of the bars are active (see input box "Lug positions on bars 1 to 32" in the template).

and the following outputs:

- Output text—the result of encryption or decryption;
- Keystream—the array of the generated displacement values D;
- Report—report explaining intermediate steps.

The template contains two instances of the Hagelin Machine component. They share the same settings except for the operation mode. That is, the first instance is set to encryption mode and the second one to decryption mode. After the encrypted text is processed by the second Hagelin Machine component in the template, the received output text is equal to the original input text (see output box "Decrypted text" in the template in Figure 2.10).

Figure 2.10 also shows that the presentation tab of the Hagelin component has two main parts. The graphics demonstrate the current state of the six wheels after the last input character was processed. The table below the graphics presents the essential values of the internal state of the machine.

2.5.3.2 Hagelin Machine Settings

Using the necessary settings, the component can be adjusted in such a way that it implements the machine with the needed functionality.

As the very first step the user has to select the model of the machine (Figure 2.11). By default, the model CX-52a is selected. The models define the sets of the available pin wheels and which bars are used in the machine. The configuration of the bars defines how pin wheels are advanced during the encryption process.

Besides the predefined models, the user has an option to select the "Custom" model, which allows for the most flexible settings.

2.5.4 Recap on C(X)-52: Evolution and Influence

The C(X)-52 is a purely mechanical rotor cipher machine developed by the Swedish inventor and entrepreneur Boris Hagelin in 1952 and manufactured by his company Crypto AG. The machine was very flexible and easily customized according to the requirements of the customers and the desired level of security. Within a few years, different variations of the machine were exported to more than 60 countries [27], which made C(X)-52 the most used rotor machine of its time.

The successors of C(X)-52 are the HX-63 rotor machine developed in 1963 and the fully electronic H-460 cipher machine. Despite the emergence of new models, many countries, including France, continued to order the CX-52 by the end of the 1960s. Due to its purely mechanical architecture, CX-52 had the advantage over into successors of being immune to the so-called electromagnetic pulse (EMP)

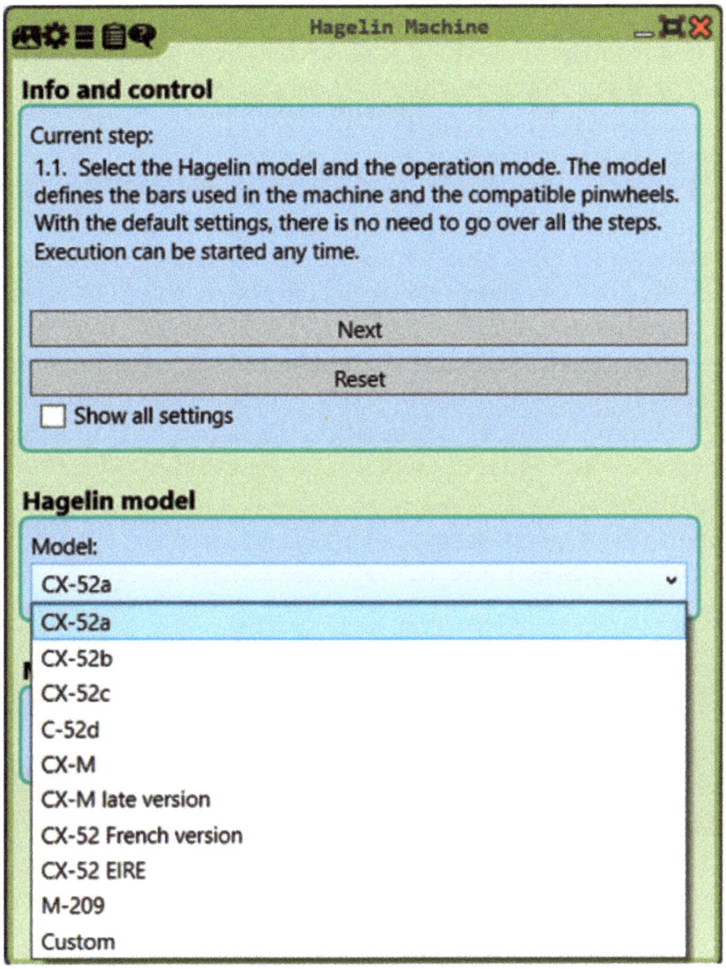

Figure 2.11 Hagelin settings with model selection.

attack. Because of this property, CX-52 was used as a backup cipher machine during the Cold War. In some countries, such as Belgium, this was the case up to the 1990s [19].

In this section, we showed how the machine was constructed and demonstrated how various C(X)-52 models work by implementing them in CT2.

2.6 Ciphers Defined by the American Cryptogram Association

There are several sites related to the American Cryptogram Association (ACA) (https://www.cryptogram.org/) that focus only on classic ciphers: for instance, `https://sites.google.com/site/bionspot/` from William Mason (Bion).

The ACA was founded in 1930 and is a nonprofit organization devoted to the hobby of classical cryptography. They try to understand these methods and how they can be solved without knowing the key and mainly by pencil and paper.

They classified and described a total of 55 ciphers. Among them are variants of well-known methods that they invented and gave them their own names, such as Gromark. Variants can arise, for example, from how the ciphertext alphabet is generated from the keyword in a monoalphabetic substitution or how the characters are arranged in a Polybius square. These puzzles are good training for cryptanalytic skills. In contrast to this, in the DECRYPT project or in the journal *Cryptologia*, real methods actually used in history are described and examined.

All 55 ACA procedures can also be found in CT2 and in CTO,[38] at Bion [2], and many at Phil Pilcrow [3] and at Oliver Kuhlemann [28].

In addition, CTO has Neural Cipher Identifier[39] (NCID), a plugin that can identify via different neural networks the type of the ACA cipher using only a short piece of ciphertext: With ciphertexts of a length of 100 characters, on average a correct detection rate over 80% was achieved [29]. A similar implementation called neural net ID test, but without a description of the internals, is also provided by Bion [2]. From Q1/2024, the classifiers developed by Dalton/Stamp to differentiate between five rotor machines will also be included in NCID [30].

2.7 Examples of Open-Access Publications on Cracking Classical Ciphers

Here are some good examples of open-access publications on cracking historical/classical ciphers:

- Homophones and nomenclatures used in diplomatic letters of Emperor Maximilian II in the 16th century [31];
- Simple monoalphabetic substitution with seven-code words used by a Dutch global player in the 17th century [32];
- Simple monoalphabetic substitution, homophones, polyphones, and nomenklator used in papal ciphers in the 16th to the 18th century [33];
- "Deciphering Mary Stuart's lost letters from 1578–1584" [34];
- Whitepaper about historical manuscripts investigated at the DECRYPT project [35];
- Josse, analysis of a French cipher from the late 19th century [36];
- SIGABA, broken with modern methods[40] [37];
- Madness' book on cryptanalysis of classical ciphers [38].

2.8 Examples Using SageMath

This section can be read as a kind of chapter appendix, implementing several classic ciphers by using the open-source computer-algebra system `SageMath`.[41] All the following ciphers have been explained previously in this chapter.

38. - CTO plugin "ACA Ciphers": `https://www.cryptool.org/en/cto/ACA-ciphers`;
    - `CT2 Templates ▷ Cryptography ▷ Classical ▷ ACA ciphers`.
39. `https://www.cryptool.org/de/cto/ncid`.
40. SIGABA is no P&P method but a US cipher rotor machine from WW2.
41. An introduction to the CAS SageMath can be found e.g. at `https://www.cryptool.org/en/documentation/ctbook/`.

To make the sample code[42] of the ciphers easier to understand, we used the structure and processes shown in Figure 2.12 and the following naming conventions: For practical reasons, most "plaintext" and "message" are not distinguished. The input to the enciphering function is just called plaintext. (Using the term "cleartext" instead of "plaintext" is wrong. Cleartext is used if an encrypted historical document has some parts in natural language in between. See Section 3.1.)

- Encryption consists of the two steps of encoding and enciphering. (This is a special term definition for the program examples, which does not apply in general.)
  - Encoding adapts the letters in the given plaintext P to the upper/lower case defined in the given alphabet, and all nonalphabet characters are filtered out. The plaintext is transferred to the message. In modern ciphers, the encoding to binary data is meant.
  - Enciphering creates the ciphertext C.

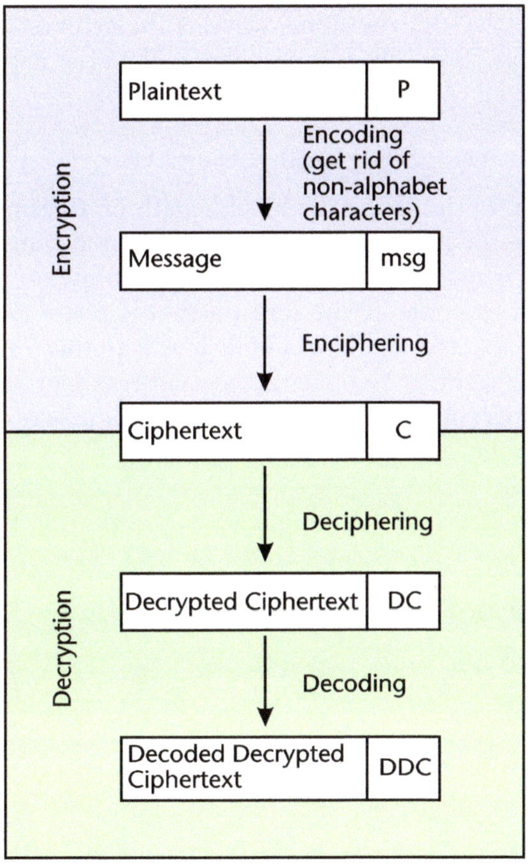

Figure 2.12 Structure and naming convention of the SageMath code examples of this chapter.

42. Further examples with SageMath concerning classic crypto methods can be found, for example:
- As a PDF in the 240-page document (SageMath 10.0) `https://doc.sagemath.org/pdf/en/reference/cryptography/cryptography.pdf`
- In the thesis of Minh Van Nguyen [39].

- Decryption also consists of two steps: deciphering and decoding. (This is a special term definition for the program examples.)
  - A final decoding step is only necessary if the symbols in the alphabet are not ASCII characters.

The SageMath code samples (SageMath scripts) first contain the code, and then as a comment the generated output at the end of each script. Note that the printed listings mostly don't show the comments from the end of the scripts. All SageMath examples of this book can be found on the CrypTool website: https://www.cryptool.org/en/documentation/ctbook/sagemath.

2.8.1 Transposition Ciphers

Transposition ciphers are implemented in the SageMath class

`sage.crypto.classical.TranspositionCryptosystem`

To construct and work with a transposition cipher in SageMath, we first need to determine the alphabet that contains the symbols used to build our plaintext and ciphertext. Typically, this alphabet will be the uppercase letters of the English alphabet, which can be accessed via the function

`sage.monoids.string_monoid.AlphabeticStrings`

We then need to decide on the block length of a permutation, which is the length of the row vector to be used in the simple column transposition. This row vector is our key, and it specifies a permutation of a plaintext.

The first example of the transposition ciphers (SageMath Example 2.1) has block length 14, and the key is built in a way that every letter in the plaintext is shifted to the right by two characters, with wraparound at the end of the block. That is the encryption process. The decryption process is shifting each letter of the ciphertext to the left by $14 - 2 = 12$.

SageMath Example 2.1: Simple Transposition by Shifting the Letters in the Message (Key is Constructed with "Range")

```
print("\n# CHAP02 -- Sage-Script-SAMPLE 010: =========")

# transposition cipher using a block length of 14 and upper-case ▸
  ▸alphabet
blocklen = keylen = 14; shift = 2
T = TranspositionCryptosystem(AlphabeticStrings(), blocklen)

# create encryption key (shifts each letter in the block by 2 positions▸
  ▸ to the right)
# key = [3, 4, 5, 6, 7, 8, 9, 10, 11, 12, 13, 14, 1, 2]
key = [(i+shift).mod(keylen) + 1 for i in range(keylen)]
print("key:  ", key, "  keylen: ", len(key), sep="")

# given plaintext
P   = "a b c d e f g h i j k l m n"; print("P:    ", P)
```

SageMath Example 2.1 (continued)

```
# encode plaintext (get rid of non-alphabet chars, convert lower-case ▸
  ▸into upper-case)
msg = T.encoding(P); print("msg: ", msg)

# encrypt plaintext message
C   = T.enciphering(key, msg); print("C:   ", C)

# decrypt using built-in deciphering method. Requires to change type of▸
  ▸ key
DC  = T.deciphering(T(key).key(), C); print("DC:  ", DC)

# Test correctness of decryption
print("msg == DC:", msg == DC)

#------------------------------------
# CHAP02 -- Sage-Script-SAMPLE 010: =========
# key:   [3, 4, 5, 6, 7, 8, 9, 10, 11, 12, 13, 14, 1, 2]  keylen: 14
# P:     a b c d e f g h i j k l m n
# msg:   ABCDEFGHIJKLMN
# C:     CDEFGHIJKLMNAB
# DC:    ABCDEFGHIJKLMN
# msg == DC: True
```

The second example of transposition ciphers (SageMath Example 2.2) is also a simple shifting column transposition. But now the generation of the plaintext is a bit automated: The sample plaintext is generated from the alphabet. Contrary to the first sample, the inverse key is generated explicitly—once from the shift parameter and once by using the transposition function from SageMath.

SageMath Example 2.2: Simple Transposition by Shifting (Key and Inverse Key are Constructed)

```
print("\n# CHAP02 -- Sage-Script-SAMPLE 020: =========")

# transposition cipher using a block length of 14, code more flexible
blocklen = keylen = 14; shift = 2
A = AlphabeticStrings()
T = TranspositionCryptosystem(A, blocklen)

# encryption key
# key = [3, 4, 5, 6, 7, 8, 9, 10, 11, 12, 13, 14, 1, 2]
key = [(i+shift).mod(keylen) + 1 for i in range(keylen)]
print("key:     ", key, "  keylen: ", len(key), sep="")

# construct the plaintext from the first 14 letters of the alphabet ▸
  ▸plus blanks
# P = "A B C D E F G H I J K L M N"
print("A.gens():\n", A.gens(), sep="")
P=''
for i in range(keylen): P = P + " " + str(A.gen(i))
print("P:   ", P)

# encode plaintext (get rid of non-alphabet chars)
msg = T.encoding(P); print("msg: ", msg)
```

SageMath Example 2.2 (continued)

```
# encrypt plaintext by shifting to the left by 2 letters (do it in one ▸
  ▸step)
C   = T.enciphering(key, msg); print("C:    ", C)

# decryption key [shifting to the left by 12 letters (= 2 to the right)▸
  ▸]
# keyInv = [13, 14, 1, 2, 3, 4, 5, 6, 7, 8, 9, 10, 11, 12]
shiftInv=keylen-shift;
keyInv = [(i+shiftInv).mod(keylen) + 1 for i in range(keylen)]; print("▸
  ▸keyInv: ", keyInv, sep="")

# decrypt by using keyInv and enciphering
DC   = T.enciphering(keyInv, C); print("1-DC:", DC)

# decrypt by using the "deciphering method with key" (without the need ▸
  ▸to calculate keyInv)
# Remark: Strangely, using the deciphering method requires to change ▸
  ▸the type of the variable key
#         The following does not work: DC  = T.deciphering(key, C); ▸
  ▸print("2-DC:", DC)
DC  = T.deciphering(T(key).key(), C); print("2-DC:", DC)

# Remarks about different representations of key and its inverse key
print("\n---Remark 1: TranspositionCryptosystem describing itself and ▸
  ▸the used cyclic groups")
print("T:", T)
print("T(key).key():    ", T(key).key())
print("T.inverse_key():", T.inverse_key(T(key).key()))
```

The third example of transposition ciphers (SageMath Example 2.3) is only a small modification of the second example. Here the order of the characters within a block are just inverted.

SageMath Example 2.3: Simple Column Transposition Just Reverting Each Plaintext Block

```
print("\n# CHAP02 -- Sage-Script-SAMPLE 030: =========")

blocklen = keylen = 14
T = TranspositionCryptosystem(AlphabeticStrings(), keylen);

key = [ keylen-i for i in range(keylen) ]; print("key: ", key, "   ▸
  ▸keylen: ", keylen, sep="")

P   = "THECATINTHEHAT"
msg = T.encoding(P); print("msg:", msg, "    msglen:", len(msg))
C   = T.enciphering(key, msg); print("C:   ", C)
DC  = T.deciphering(T(key).key(), C); print("DC: ", DC)

#------------------------------------
# CHAP02 -- Sage-Script-SAMPLE 030: =========
# key: [14, 13, 12, 11, 10, 9, 8, 7, 6, 5, 4, 3, 2, 1]  keylen: 14
# msg: THECATINTHEHAT    msglen: 14
# C:   TAHEHTNITACEHT
# DC:  THECATINTHEHAT
```

In the fourth example of transposition ciphers (SageMath Example 2.4) we use an arbitrary permutation as key in the encryption and decryption processes in order to scramble the characters within each block (permutation block length = number of columns in a simple column transposition). In addition, it shows the key space of the simple column transposition.

If the block length is n, then the key must be a permutation on n symbols. This example uses the method `random_key()` of the class `TranspositionCryptosystem` to create a random key. Each call to `random_key()` most likely produces a different key. Note that therefore your results (key and ciphertext) most likely will be different from the results shown in the following example.

SageMath Example 2.4: Simple Column Transposition with Randomly Generated (Permutation) Key and Showing the Size of the Key Space

```
print("\n# CHAP02 -- Sage-Script-SAMPLE 040: =========")

blocklen = keylen = 14
A = AlphabeticStrings()
T = TranspositionCryptosystem(A, blocklen);

# key_space() delivers generic info; order() of keyspace delivers the ▸
  ▸number of keys
print("T.key_space():", T.key_space())
print("T.key_space().order():", T.key_space().order())
print('                        {0:.2e}'.format( T.key_space().order() ))

P = "a b c d e f g h i j k l m n o p q r s t u v w x y z a b"
msg = T.encoding(P); print("msg:", msg)
key = T.random_key(); print("key: ", key)
Pkey = Permutation(key); print("Pkey:", Pkey, "Pkeylen:", len(Pkey))

# enciphering in one and in two steps
C   = T.enciphering(key, msg); print("1-C: ", C)

E = T(key)
C = E(msg); print("2-C: ", C)

# deciphering
DC   = T.deciphering(key, C); print("DC: ", DC)

# Just another way of decryption: Using "enciphering" with the inverse ▸
  ▸key
keyInv = T.inverse_key(key); print("keyInv: ", keyInv)
PkeyInv = Permutation(keyInv); print("PkeyInv:", PkeyInv, "PkeyInvlen▸
  ▸:", len(PkeyInv))
DDC     = T.enciphering(keyInv, C)

# Test correctness of both ways of decryption
print("msg == DC == DDC:", msg == DC == DDC)  # Expect True
```

The fifth example of transposition ciphers (SageMath Example 2.5) demonstrates the built-in encoding and ensures the correct message length (after encoding it must be a multiple of the block length). Here we don't apply padding.

SageMath Example 2.5: Simple Column Transposition (Demonstrates Encoding and Ensures Correct Message Length)

```
print("\n# CHAP02 -- Sage-Script-SAMPLE 050: =========")

keylen = 14
T = TranspositionCryptosystem(AlphabeticStrings(), keylen)
key = T.random_key(); print("key: ", key, "  keylen: ", keylen, sep="")

# P = "The cat in the hat.--The cat in the hat.XYZ" or P = "▸
  ▸ABCDEFGHIJUUUU"
P = "The cat in the hat.--The cat in the hat."  # plaintext
print("P:  ", P, "   Plen:", len(P))

# Encoding takes only letters from the alphabet (and potentially ▸
  ▸converts it to uppercase).
msg = T.encoding(P); print("msg:", msg, "        msglen:", len(msg))

# Ensure that the length of msg is a multiple of keylen
# if (len(msg)%keylen!=0): print("Error: msg length isn't a multiple of▸
  ▸ blocklength. len(msg) =",len(msg))
rest = len(msg) % keylen
if (rest != 0):
  print("Error: msg length isn't a multiple of blocklength. len(msg) ▸
    ▸=", len(msg))
  chunk_size = len(msg) - rest
  msg = msg[0:chunk_size]
  print("Truncated message =", msg)
assert len(msg) % keylen == 0

# Enciphering only works, if the number of symbols in msg is a multiple▸
  ▸ of keylen
C  = T.enciphering(key, msg); print("C:  ", C)

# Decryption using method deciphering
DC = T.deciphering(key, C); print("DC: ", DC)

print("msg == DC:", msg == DC)  # Expect True
```

2.8.2 Substitution Ciphers

Substitution cryptosystems are implemented in SageMath in the class

`sage.crypto.classical.SubstitutionCryptosystem`

SageMath Example 2.6 constructs a substitution cipher with a random key. A random key can be generated using the method `random_key()` of the class `SubstitutionCryptosystem`. Different keys determine different substitution ciphers: With each call to `random_key()` a different result is expected to be returned.

SageMath Example 2.6: Monoalphabetic Substitution with Randomly Generated Key

```
print("\n# CHAP02 -- Sage-Script-SAMPLE 060: =========")

S = SubstitutionCryptosystem(AlphabeticStrings())
```

SageMath Example 2.6 (continued)

```
key = S.random_key(); print("key: ", key, "  keylen: ", len(key), sep▸
 ▸="")

P   = "MASC: Substitute this with something else using a random key"
print("P:  ", P, "   Plen:", len(P))

msg = S.encoding(P); print("msg:", msg, "       msglen:", len(msg))
C   = S.enciphering(key, msg); print("C:  ", C)
DC  = S.deciphering(key, C); print("DC: ", DC)

print("msg == DC:", msg == DC)  # Expect True

#------------------------------------
# CHAP02 -- Sage-Script-SAMPLE 060: =========
# key: ZMUPXCHBVGTIYLKDQOSENFWARJ  keylen: 26
# P:   MASC: Substitute this with something else using a random key    ▸
 ▸Plen: 60
# msg: MASCSUBSTITUTETHISWITHSOMETHINGELSEUSINGARANDOMKEY        msglen▸
 ▸: 50
# C:   YZSUSNMSEVENEXEBVSWVEBSKYXEBVLHXISXNSVLHZOZLPKYTXR
# DC:  MASCSUBSTITUTETHISWITHSOMETHINGELSEUSINGARANDOMKEY
# msg == DC: True
```

2.8.2.1 Atbash Cipher

SageMath Example 2.7 constructs an Atbash cipher. Strictly speaking, Atbash like Rot13 is a "code" not a "cipher," because the key is fixed. See Section 3.1.

SageMath Example 2.7: Atbash (Substitution by Reverting the Alphabet)

```
print("\n# CHAP02 -- Sage-Script-SAMPLE 065: =========")

A = AlphabeticStrings(); print(A)

# key = A([25-i for i in range(26)]); print("key: ", key, sep="")
lenA = len(A.alphabet())  # lenA = 26
key = A([lenA-1-i for i in range(lenA)]); print("key: ", key, sep="")

# Atbash: using the Substitution cipher and reverted alphabet as key
S   = SubstitutionCryptosystem(A); print(S)

msg = S.encoding("Substitute this with something else.")
print("msg:", msg, "       msglen:", len(msg))

# shift the plaintext
C = S.enciphering(key, msg); print("C:  ", C)

# decrypt the ciphertext and ensure that it is the original plaintext
DC = S.deciphering(key, C); print("DC: ", DC)

print("msg == DC:", msg == DC)  # Expect True

#------------------------------------
# CHAP02 -- Sage-Script-SAMPLE 065: =========
# Free alphabetic string monoid on A-Z
# key: ZYXWVUTSRQPONMLKJIHGFEDCBA
# Substitution cryptosystem on Free alphabetic string monoid on A-Z
# msg: SUBSTITUTETHISWITHSOMETHINGELSE        msglen: 31
# C:   HFYHGRGFGVGSRHDRGSHLNVGSRMTVOHV
# DC:  SUBSTITUTETHISWITHSOMETHINGELSE
# msg == DC: True
```

2.8.2.2 Caesar Cipher

SageMath Example 2.8 constructs a Caesar cipher. The key is a number describing how far the alphabet is shifted.

SageMath Example 2.8: Caesar (Substitution by Shifting the Alphabet; Key Explicitly Given, Step-by-Step Approach)

```
print("\n# CHAP02 -- Sage-Script-SAMPLE 070: =========")

A = AlphabeticStrings()  # plaintext/ciphertext alphabet
keylen = len(A.alphabet()); # Alternative: keylen = len(A.gens())
shift = 3; shiftInv=keylen-shift;
print("keylen: ", keylen, "   shift: ", shift, "   shiftInv: ", shiftInv, sep="")

S = SubstitutionCryptosystem(A)

# construct key for Caesar cipher
numkey = [(i+shift).mod(keylen) for i in range(keylen)]  ### Now without +1 (after mod)
print("numkey", numkey)
key = A(numkey); print("key: ", key, "  keylen: ", len(key), sep="")

P = "Shift the alphabet three positions to the right."
print("P:   ", P, "   Plen:", len(P))

# encrypt message (method encoding can be called from A or from S)
msg = S.encoding(P); print("msg:", msg, "        msglen:", len(msg))
C   = S.enciphering(key, msg); print("C:   ", C)

# decrypt message (using inverse key and enciphering instead of deciphering)
numkeyInv = [(i+shiftInv).mod(keylen) for i in range(keylen)]; print("numkeyInv", numkeyInv)
keyInv = A(numkeyInv); print("keyInv: ", keyInv, sep="")

DC = S.enciphering(keyInv, C); print("DC: ", DC)

print("msg == DC:", msg == DC)  # Expect True
```

2.8.2.3 Shift Cipher

The shift cipher can also be thought of as a generalization of the Caesar cipher. While the Caesar cipher restricts us to shift exactly three positions along an alphabet, the shift cipher allows us to shift any number of positions along the alphabet.

In the above samples we applied the `SubstitutionCryptosystem` and built Caesar as a special kind of substitution. In contrast, here Caesar can be built as a special kind of shift cipher.

The shift cipher is implemented directly in the SageMath class

`sage.crypto.classical.ShiftCryptosystem`

In SageMath Example 2.9, we construct a shift cipher over the capital letters of the English alphabet. We then encrypt a plaintext P by shifting it `key` positions along the alphabet. Finally, we decrypt the ciphertext C and check whether the result (DC) is indeed the original plaintext.

Shifting is a special way of substitution. The original Caesar cipher is just simply a shift cipher whose shifting key is 3.

SageMath Example 2.9: Constructing the Caesar Cipher Using the Shift Cipher

```
print("\n# CHAP02 -- Sage-Script-SAMPLE 080: =========")

key = 3 # shift can be any integer number. Ancient Caesar used 3.
print("shift key: ", key, sep="")

# Caesar cipher using the Shift cipher instead of the more general Substitution cipher
S   = ShiftCryptosystem(AlphabeticStrings()); print(S)
msg = S.encoding("Shift me any number of positions."); print("msg:", msg, "      msglen:", len▸
 ▸(msg))

# shift the plaintext
C = S.enciphering(key, msg); print("C:  ", C)

# decrypt the ciphertext and ensure that it is the original plaintext
DC = S.deciphering(key, C); print("DC: ", DC)

print("msg == DC:", msg == DC)  # Expect True
```

2.8.2.4 Affine Cipher

The affine cipher is implemented in the SageMath class

`sage.crypto.classical.AffineCryptosystem`

In SageMath Example 2.11, we construct an affine cipher with the key $(a, b) = (3, 13)$. A given plaintext P (strictly speaking the encoded message msg) is then encrypted which results in the ciphertext C. The ciphertext is then decrypted and the result DC is compared to the original plaintext.

Mathematically, an affine cipher can be described as follows: Let n be the number of letters in the alphabet (in the example $n = 26$, as the alphabet consists only of the 26 capital letters). The individual letters of the plaintext $P = (p_1, p_2, ..., p_k)$ are encrypted as follows: $c_i = a \cdot p_i + b \pmod{n}$.

The following applies:

- To the factor a: $0 < a < n$ and a relatively prime to n (coprime with 26).
- To the shift value b: $0 \leq b \leq n - 1$.

As a is coprime to n, here a is one of the 12 numbers calculated in the SageMath Example 2.10.

SageMath Example 2.10: n.coprime_integers()

```
sage: n=26
sage: n.coprime_integers(n)
[1, 3, 5, 7, 9, 11, 15, 17, 19, 21, 23, 25]
```

If a is not relatively prime to n, that is, $gcd(a, n) > 1$, then the mapping cannot be inverted, and the ciphertext cannot be uniquely deciphered. That would be the case here with an even a and with $a = 13$.

In SageMath Example 2.11 we catch the case where a is not coprime to n. Not doing so, the command AffineCryptosystem(AlphabeticStrings())! catches this case and shows: `ValueError: (a,b) = (4,13) is outside the range of acceptable values for a key of this affine cryptosystem.`

The key space is $12 \cdot 26 = 312$ (all possible combinations (a, b) result from the 12 values for a and the 26 different values for b).

SageMath Example 2.11: An Affine Cipher with Key (3, 13)

```
print("\n# CHAP02 -- Sage-Script-SAMPLE 090: =========")

A = AlphabeticStrings()
# int conversion needed: A.ngens() creates type <class 'int'>, but we need <class 'sage.rings.
  integer.Integer'>
# via ngens() no hard coded "26" is needed.
n=Integer(A.ngens())  # n = 26 = number of free alphabetic string monoid on A-Z.

# a and b must be < n (this is checked by AffineCryptosystem(AlphabeticStrings()) too)
key = a, b = (3%n, 13%n); print("affine key: ", key, sep="")

cop_list=n.coprime_integers(n) # Here it's necessary that n is of type Sage integer, not of
  Python int
print("coprimes of n=%d:" % n, cop_list)
if a not in cop_list: # a must be coprime to n
  print("Exit, because a is no coprime to 26.")
  sys.exit(); # exit sage script

# create an affine cipher
AS  = AffineCryptosystem(A); print(AS)
msg = AS.encoding("The affine cryptosystem."); print("msg:", msg, "       msglen:", len(msg))

# encrypt the plaintext using the affine key
C = AS.enciphering(a, b, msg); print("C:   ", C)

# decrypt the ciphertext and make sure that it is equivalent to the original plaintext
DC = AS.deciphering(a, b, C); print("DC: ", DC)

print("msg == DC:", msg == DC)  # Expect True
```

We can also construct a shift cipher using the affine cipher. To do so, we need to restrict keys of the affine cipher to be of the form $(a = 1, b)$ where b is any nonnegative integer.

To create the Caesar cipher using the affine cipher, the encryption/decryption key must be $(1, 3)$. SageMath Example 2.9 works analogously with the affine cipher in SageMath Example 2.12.

SageMath Example 2.12: Constructing the Caesar Cipher Using the Affine Cipher

```
print("\n# CHAP02 -- Sage-Script-SAMPLE 100: =========")

key = a, b = (1, 3); print("affine key: ", key, sep="")

# construct a shift cipher using an affine cipher with a=1
AS  = AffineCryptosystem(AlphabeticStrings())
msg = AS.encoding("Shift the alphabet by three positions to the right.")
print("msg:", msg, "       msglen:", len(msg))

# shift the plaintext to get the ciphertext
C = AS.enciphering(a, b, msg); print("C:   ", C)

# decrypt the ciphertext and ensure that it is the original plaintext
DC = AS.deciphering(a, b, C); print("DC: ", DC)
```

SageMath Example 2.12 (continued)

```
print("msg == DC:", msg == DC)  # Expect True

#------------------------------------
# CHAP02 -- Sage-Script-SAMPLE 100: =========
# affine key: (1, 3)
# msg: SHIFTTHEALPHABETBYTHREEPOSITIONSTOTHERIGHT         msglen: 42
# C:   VKLIWWKHDOSKDEHWEBWKUHHSRVLWLRQVWRWKHULJKW
# DC:  SHIFTTHEALPHABETBYTHREEPOSITIONSTOTHERIGHT
# msg == DC: True
```

2.8.2.5 Vigenère Cipher

The Vigenère cipher is implemented in the SageMath class

`sage.crypto.classical.VigenereCryptosystem`

For our ciphertext/plaintext space, we can work with the upper-case letters of the English alphabet, the binary number system, the octal number system, or the hexadecimal number system. SageMath Example 2.13 uses the class `AlphabeticStrings`, which implements the English capital letters.

SageMath Example 2.13: Vigenère Cipher

```
print("\n# CHAP02 -- Sage-Script-SAMPLE 110: =========")

# construct Vigenere cipher
keylen = 14
A = AlphabeticStrings()
V = VigenereCryptosystem(A, keylen)

# Here, a random key of length keylen is generated.
# Alternatively, a key could be given explicitly like key = A('ABCDEFGHIJKLMN')
key = V.random_key(); print("key: ", key, "  keylen: ", len(key), sep="")

P = "The Vigenere cipher is polyalphabetic."; print("P:   ", P, "   Plen:", len(P))
msg = V.encoding(P); print("msg:", msg, "   msglen:", len(msg))

C   = V.enciphering(key, msg); print("C:   ", C, "  Clen:   ", len(C))

DC  = V.deciphering(key, C); print("DC: ", DC)

print("msg == DC:", msg == DC)  # Expect True

#------------------------------------
# CHAP02 -- Sage-Script-SAMPLE 110: =========
# key: OHZZMJTRCFOWKN  keylen: 14
# P:   The Vigenere cipher is polyalphabetic.   Plen: 38
# msg: THEVIGENERECIPHERISPOLYALPHABETIC   msglen: 33
# C:   HODUUPXEGWSYSCVLQHEYHCAFZLRNPLSHO   Clen:   33
# DC:  THEVIGENERECIPHERISPOLYALPHABETIC
# msg == DC: True
```

2.8.2.6 Hill Cipher

The Hill [40, 41] or matrix cipher[43] is based on linear algebra and was invented by Lester S. Hill in 1929. It was the first polygraphic cipher in which it was practical to

43. - CT1 `Encrypt/Decrypt ▷ Symmetric (classic) ▷ Hill.`
- CT2 `Templates ▷ Cryptography ▷ Classical` and CT2 `Templates ▷ Cryptanalysis ▷ Classical.`

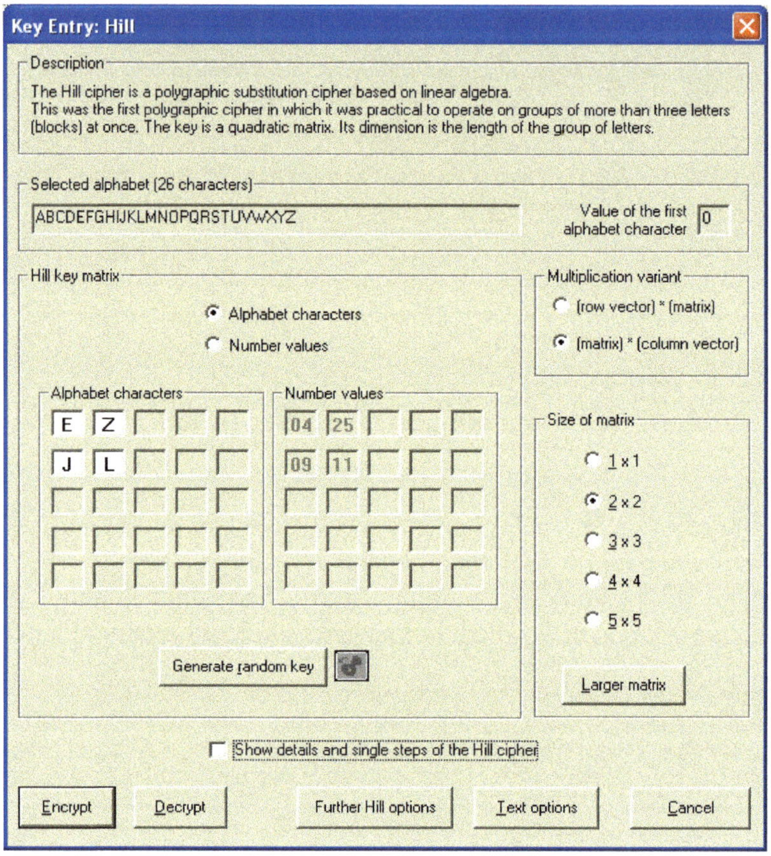

Figure 2.13 Hill dialog in CT1 showing the operations and options available.

operate on more than three symbols at once. The Hill cipher is not important from a security point of view, but because it was the first cipher trying to apply mathematics to cryptography. The encryption key of this cipher is an invertible square matrix (here called *key*) whose determinant is relatively prime to 26. Originally, plaintext and ciphertext are vectors (P and C). The encryption and decryption processes use matrix operations modulo 26: $C = P \cdot \text{key} \pmod{26}$.

The Hill cipher is implemented in the SageMath class

`sage.crypto.classical.HillCryptosystem`

In SageMath Example 2.14, our plaintext/ciphertext space is the capital letters of the English alphabet. The Hill cipher assigns each letter of this alphabet a unique integer modulo 26. The size of the key matrix (also called its dimension) is not restricted by the cipher.

Comparing the Hill implementation in CrypTool v1.4.42 and in SageMath version 9.3:

- SageMath offers fast command-line operations; CT1 offers its functionality within a GUI.
- SageMath offers for the key matrix any dimension; CT1 is restricted to a matrix size between 1 and 10.

- SageMath allows negative numbers in the key matrix, and converts them automatically into appropriate nonnegative numbers; CT1 doesn't allow negative numbers in the key matrix.
- SageMath always sets the first alphabet character to 0, only allows the 26 capital letters as an alphabet, and it uses only the multiplication variant plaintext row vector · key matrix: $C = P \cdot \text{key}$.
- CT1 offers to choose also 1 as value for the first alphabet character, you can customize your alphabet within the text options dialog, and it also offers to use a reverse multiplication variant: $C = \text{key} \cdot P$.

SageMath only provides the function for encryption and decryption for the classic ciphers and usually with a very restrictive alphabet. You have to implement methods for cryptanalysis yourself. A KPA against the Hill cipher is introduced in Section 2.8.3.2.

While SageMath Example 2.14 calculates for the entered characters always with their ASCII numerical values, SageMath Example 2.19 cannot only carry out a KPA analysis, but also put the key matrix *in front of* the plaintext (order of the multiplication) and the characters in the alphabet start from 0 or 1.[44]

Reference [43] is a very good article developing the formulas for how many invertible Hill matrices there are for a given dimension (compared to the total number of all matrices and to the number of involutory matrices).

SageMath Example 2.14: Hill Cipher with Given Key Matrix

```
print("\n# CHAP02 -- Sage-Script-SAMPLE 120: =========")

keylen = 3    # Alternative key length: keylen=13 --- ensure msg length is a multiple of keylen
A = AlphabeticStrings()
H = HillCryptosystem(A, keylen)

# Alternative 1: Non-random key creation (needs HKS; even H.key_space() is not enough)
HKS = H.key_space()
key = HKS([[1,0,1],[0,1,1],[2,2,3]]); print("key: \n", key, sep="")

# Alternative 2: Random key creation
# key = H.random_key(); print("key: \n", key, sep="")

# the key object has no method len(), but block_length()
print("block_length(): ", H.block_length(), "    key.det(): ", key.det(), sep="")

# encoding (Length of msg is a multiple of matrix dimension (block_length))
P = "HHill or matrix cipher uses matrix operations."
print("P:   ", P, "  Plen:", len(P))
msg = H.encoding(P); print("msg:", msg, "  msglen:", len(msg))

# encryption
C  = H.enciphering(key, msg); print("C:   ", C, "  Clen:   ", len(C))

# decryption
DC   = H.deciphering(key, C); print("DC: ", DC)
print("msg == DC:", msg == DC)  # Expect True

# alternative way to decrypt using inverse matrix
keyInv = key.inverse(); keyInv
keyInv = H.inverse_key(key); print("\nkeyInv: \n", keyInv, sep="")
DC      = H.enciphering(keyInv, C); print("DC: ", DC)
print("msg == DC:", msg == DC)  # Expect True
```

44. \$ sage chap02_hill_enc_dec_kpa.sage -enc -dim 3 -pt "ACTCAT" -A [[6,24,1],[13,16,10],[20,17,15]]" encrypts the example "ACTCAT" to "POHFIN" [42].

SageMath Example 2.14 (continued)

```
print("\n---Remark: Output C as a sequence of ASCII characters and their according numbers")
# print("type(C):", type(C)) # 'sage.monoids.string_monoid_element.StringMonoidElement'
# 'StringMonoidElement' object has no attribute to directly convert to integer sequence
from sage.crypto.util import ascii_to_bin, ascii_integer
# print("a_to_b: ", ascii_to_bin(str(C)))
print("C[i]:", [C[i] for i in range(len(C))])
print("binary C[i]:", [ascii_to_bin(str(C[i])) for i in range(len(C))])
print("integer C[i]:", [ascii_integer(ascii_to_bin(str(C[i]))) for i in range(len(C))])
```

2.8.2.7 Substitution with Symbols Using Not Only Capital Letters

Up to now, we used the capital letters for the classical ciphers. This set $\{A, B, ..., Z\}$ is the standard alphabet used for the classical ciphers in SageMath.

A substitution cipher can be considered as a stream cipher that acts on the plaintext by making a substitution of the characters with elements of a new ciphertext alphabet or by a permutation of the characters in the plaintext alphabet. Besides the capital letters, the predefined functions for classical ciphers in SageMath only offer the hexadecimal and binary systems as alphabets.

These alphabets can be called via the three functions that implement the free string monoids (i.e., sets whose elements can be concatenated to any finite length):

```
S = AlphabeticStrings()
H = HexadecimalStrings()
B = BinaryStrings()
```

The following samples demonstrate that one can vary the alphabet. The first two samples use the hexadecimal and the binary system. The last sample shows how to define your own alphabet. This currently requires you to also write your own cipher algorithm. We do this by defining an own MASC with a longer alphabet.

MASC with a hexadecimal alphabet. In SageMath Example 2.15, the hexadecimal number system is used as a substitution alphabet for plaintext/ciphertext.

SageMath Example 2.15: Monoalphabetic Substitution with a Hexadecimal Alphabet (and Decoding in Both SageMath and Python)

```
print("\n# CHAP02 -- Sage-Script-SAMPLE 130: =========")

A = HexadecimalStrings()
S = SubstitutionCryptosystem(A)

key = S.random_key(); print("key: ", key, "  keylen: ", len(key), sep="")
print("Number of possible keys: ", len(key), "! = ", factorial(len(key)), sep="")

P   = "Working with a larger alphabet."; print("P:   ", P, "  Plen:", len(P))
msg = A.encoding(P); print("msg:", msg, "  msglen:", len(msg))

C   = S.enciphering(key, msg); print("C:   ", C, "  Clen:  ", len(C))
DC  = S.deciphering(key, C); print("DC: ", DC)

print("msg == DC:", msg == DC)  # Expect True

# Conversion hex in DC back to ASCII:
DDC = DC.decoding(); # print("DDC:", DDC)
print("P == DDC:", P == DDC)  # Expect True
```

SageMath Example 2.15 (continued)

```
## Remark: Other ways for the decoding transformation using sage.crypto.util
# - AlphabeticStrings() and HexadecimalStrings() don't have an according method.
# - http://doc.sagemath.org/html/en/reference/cryptography/sage/crypto/util.html
# from sage.crypto.util import ascii_integer
# print("ascii_integer: ", ascii_integer("01000100"))
# from sage.crypto.util import bin_to_ascii
# print("bin_to_ascii:", bin_to_ascii("01000100"))
#
## Remark: Alternative conversion hex back to ASCII, using native Python
# import binascii
# DDC = binascii.a2b_hex(repr(DC));

#-----------------------------------
# CHAP02 -- Sage-Script-SAMPLE 130: =========
# key: c7834de0f1a65b29  keylen: 16
# Number of possible keys: 16! = 20922789888000
# P:   Working with a larger alphabet.   Plen: 31
# msg: 576f726b696e6720776974682061206c617267657220616c7068616265742e   msglen: 62
# C:   d0e908e6e1e2e08c00e104ef8ce78ce5e708e0ed088ce7e50cefe7e8ed0482   Clen:   62
# DC:  576f726b696e6720776974682061206c617267657220616c7068616265742e
# msg == DC: True
# P == DDC: True
```

MASC with a binary alphabet. In SageMath Example 2.16 the three cipher types, monoalphabetic substitution, shift, and Vigenère are used. The alphabet symbols here are in each case from the binary number system.

Except for the Vigenère cipher, which can be enhanced as a one-time pad, these variants are very unsecure. Because the plaintext/ciphertext alphabet has only the two elements 0 and 1, there are—for example, with monoalphabetic substitution—then only two possible keys: (0 1) and (1 0). The key of a monoalphabetic substitution cipher must contain all symbols of the alphabet exactly once.

SageMath Example 2.16: Different Substitution Ciphers on the Binary Alphabet

```
print("\n# CHAP02 -- Sage-Script-SAMPLE 140: =========")

# the plaintext/ciphertext alphabet is a binary sequence
B = BinaryStrings(); # print("B", B); print("B.alphabet()", B.alphabet())

# substitution cipher over the alphabet B; no keylen argument possible
S = SubstitutionCryptosystem(B)
print("1. Substitution: alphabet_size:", S.alphabet_size())

# Alternative: To get a substitute for each symbol, key has always the length of the alphabet
# key = S.random_key()
key = B("10") # "10" inverts all bits, "01" leaves them unchanged; "0", "1", "00" and "11" ▸
  ▸cause exception!
print("key: ", key, "  keylen: ", len(key))
# print("##type key: ", type(key))

P = "MA-Substitution on binary alphabet is very unsecure (flip bit or not)."; print("P:  ", P, ▸
  ▸"       Plen:", len(P))
msg = B.encoding(P); print("msg:", msg, "  msglen:", len(msg))

C  = S.enciphering(key, msg); print("C:  ", C, "  Clen: ", len(C))
DC = S.deciphering(key, C); # print("DC: ", DC)

print("msg == DC:", msg == DC)  # Expect True


S = ShiftCryptosystem(B) # Shift in the binary alphabet B which has only two elements.
print("\n2. Shift: alphabet_size:", S.alphabet_size())
```

SageMath Example 2.16 (continued)

```
# Alternative: key = S.random_key(); print("randK:", key)
# print("##type key: ", type(key))
key   = 1; print("key:", key)   # 1 inverts each bit; 0 leaves each bit unchanged.

P = "Shift on binary alphabet offers only 2 possibilities: very unsecure."; print("P:  ", P, " ▸
  ▸      Plen:", len(P))
msg = B.encoding(P); print("msg:", msg, "  msglen:", len(msg))

C    = S.enciphering(key, msg); print("C:   ", C, "  Clen: ", len(C))
DC   = S.deciphering(key, C); # print("DC:  ", DC)

print("msg == DC:", msg == DC)  # Expect True


keylen = 14;
# Alternative settings for alphabet and using a given key
# B = AlphabeticStrings(); key = B('ABCDEFGHIJKLMN'); print("key:", key, "  keylen: ", len(key)▸
  ▸)
# B = BinaryStrings(); key = B('11111111111111'); print("key:", key, "  keylen: ", len(key))
V = VigenereCryptosystem(B, keylen)
print("\n3. Vigenere: alphabet_size:", V.alphabet_size())
key = V.random_key(); print("randkey:", key, "  keylen: ", len(key))

msg = V.encoding("Vigenere on binary alphabet with long key is close to OTP."); print("msg:", ▸
  ▸msg, "  msglen:", len(msg))
C    = V.enciphering(key, msg); print("C:   ", C, "  Clen: ", len(C))
DC   = V.deciphering(key, C); # print("DC:  ", DC)

print("msg == DC:", msg == DC)  # Expect True
```

MASC with a self-defined alphabet. SageMath Example 2.17 uses an augmented alphabet—one that contains the alphabetic letters (upper and lowercase) plus numbers and/or some extra symbols.

SageMath Example 2.17: MASC Over Own Alphabet

```
print("\n# CHAP02 -- Sage-Script-SAMPLE 150: =========")

# Using own definitions for a classical cipher instead of the r2r Sage commands (r2r=ready-to-▸
  ▸run)
# An arbitrary alphabet made from A..Z, a..z, 0..9, and some symbols which are arranged in a ▸
  ▸wished order.

import string
import random

alph1Lower_string = string.ascii_lowercase
# print(alph1Lower_string)
alph2Upper_string = string.ascii_uppercase
# print(alph2Upper_string)
alph3Digits_string = string.digits
# print(alph3Digits_string)
alph4Punctuation_string = string.punctuation
# print(alph4Punctuation_string)

# Determine the order of the elements of the alphabet by ordering the 4 parts
alphabet = alph1Lower_string + alph2Upper_string + alph4Punctuation_string + alph3Digits_string
print(alphabet)

print('Length alphabet:', len(alphabet))
print('Check: Value of letter B in this alphabet:', alphabet.index('B'))

# Shuffle rearranges the given object. As strings and tuples are immutable,
# we have to use random.sample() instead of random.shuffle().
random.seed(int(15)) # argument not necessary. Initialized the PRNG just to have always the ▸
  ▸same to compare with.
        # argument 15 without casting throws TypeError: The only supported seed types are: None,▸
          ▸ int, float, str, bytes, and bytearray.
r_alphabet = ''.join(random.sample(alphabet, len(alphabet)))
```

SageMath Example 2.17 (continued)

```
print('1. shuffled alphabet:', r_alphabet)
r_alphabet = ''.join(random.sample(alphabet, len(alphabet)))
print('2. shuffled alphabet:', r_alphabet)

# Use these two alphabets to build a MASC
PA = alph1Lower_string + alph2Upper_string + alph4Punctuation_string + alph3Digits_string
print('\nPlaintext alphabet PA :', PA, ' Length of PA', len(PA))

random.seed(int(0)) # Initialized the seed to generate a fixed permutation
CA = ''.join(random.sample(PA, len(PA)))
print('Ciphertext alphabet CA:', CA, ' Length of CA', len(CA))

codetableC2P = str.maketrans(CA,PA) # requires the 2 strings CA, PA to have the same len
codetableP2C = str.maketrans(PA,CA) # requires the 2 strings CA, PA to have the same len

P1 = '''ATESTZtestTEST1234'''
P2 = '''DWHVWCCNBCWHVWNOPQ'''

C1 = P1.translate(codetableP2C);
C2 = P2.translate(codetableP2C);

P1_revealed = C1.translate(codetableC2P);
P2_revealed = C2.translate(codetableC2P);
```

2.8.3 Cryptanalysis of Classical Ciphers with SageMath

Of course, you can also do the cryptanalysis of classical methods with Python or SageMath. Good examples can be found in Kohel's book [44] on page 19 ff and page 110 ff. You can also find cryptanalysis examples for affine ciphers and shift ciphers at `https://doc.sagemath.org /pdf/en/reference/cryptography/cryptography.pdf`. Very sophisticated cryptanalysis methods for classical ciphers are part of CT2.

Two simple analyses are presented here as examples. For a shift cipher, the brute-force method built-in SageMath is used to match a ciphertext to the correct plaintext and key. A self-written program is presented for the Hill cipher: The attack determines the correct key from a given plaintext/ciphertext pair.

2.8.3.1 Cryptanalysis with SageMath: Ciphertext-Only Attack Against Shift Cipher

The Caesar method built into SageMath has only 26 possible keys for an alphabet consisting of 26 capital letters. So a brute-force approach needs to try very few possibilities. This is why brute force is already integrated as a method in the cryptographic procedure `ShiftCryptosystem`.[45]

It is clear that one of the 26 possible keys again generates the plaintext. Therefore, some authors subtract this key from the key space. But this is a matter of definition and the majority of the authors like [45] count it to the key space. In [46, page 27], these keys are referred to as "trivial." They form together with the "non-trivial" keys the key space. So Caesar has 25 nontrivial keys and one trivial key, the key set has 26 elements, and the key space has the value 26.

45. In SageMath 9.7, only the two classic cryptosystems `ShiftCryptosystem` and `AffineCryptosystem` have this method built in. For example, `SubstitutionCryptosystem` does not have it because the search space (26!) is too large. See `https://doc.sagemath.org/html/en/reference/cryptography/sage/crypto/classical.html`.

In SageMath Example 2.18, first the built-in method `brute_force` is executed and then a statistical test (chi-square or squared differences) is used, which finds with a high probability the right plaintext (in human language)—depending on the length of the given ciphertext.

The plaintext found does not contain any spaces, since these do not belong to the alphabet. Thus, the words are not clearly separated from each other.[46]

SageMath Example 2.18: Ciphertext-Only Attack Against Shift Cipher (Caesar)

```
print("\n# CHAP02 -- Sage-Script-SAMPLE 160: =========")

# Find the most likely plaintext of a ciphertext encrypted via a shift cipher
# - Instead of explicitely looping over all revealed plaintexts,
#   use the built-in brute-force method and apply a statistical measure
# - automated ciphertext-only attack against Caesar

# pt = "Hello this is a test. Please enter your text here."
# ct = "Pmttw BpqA qA i BmAB. XtmiAm mvBmz GwCz BmFB pmzm."  # via k=8
ct = "PmttwBpqAqAiBmABXtmiAmmvBmzGwCzBmFBpmzm"
print(f"Given ciphertext ct:\n{ct}")

S = AlphabeticStrings()
E = ShiftCryptosystem(S)

# both, ct and ctstr have the value PMTTWBPQAQAIBMABXTMIAMMVBMZGWCZBMFBPMZM
ct = S.encoding(ct)  # type: sage.monoids.string_monoid_element.StringMonoidElement'
ctstr = str(ct)      # type: str

print("\n---------- brute-force, No ranking")
dict = E.brute_force(ct)  # type(dict)=dict
print( *sorted(dict.items())[:26], sep="\n" ) # output one element per line

print("\n---------- chi_square ranking")
L = E.brute_force(ct, ranking="chisquare")  # type(L))=list
print(*L[:5], sep="\n") # display only the top 5 candidate keys and plaintexts

print("\n---------- squared_differences ranking")
L = E.brute_force(ct, ranking="squared_differences")  # type(L))=list
print(*L[:5], sep="\n") # display only the top 5 candidate keys and plaintexts

print("\n---------- Probably correct values")
print("Probable correct key: ", L[0][0])
print("Probable correct pt:  ", L[0][1])
```

2.8.3.2 Cryptanalysis with SageMath: KPA Against Hill Cipher

The Hill cipher is very difficult to break if only ciphertext is given, but it is vulnerable to KPAs. Given the corresponding plaintext and ciphertext, it is very likely that the key (matrix) can be determined.

Now the SageMath program hill_enc_dec_kpa.sage will be presented, which can execute a known-plaintext attack against the Hill cipher. While SageMath

46. Within `CT2 Startcenter` ▷ `Templates` ▷ `Tools` you can find the template "Split a Text" which recognizes words of different languages and restores the separators in the revealed plaintext almost automatically – With only little manual reworking (see Figure 2.14):
   HELLOTHISISATESTPLEASEENTERYOURTEXTHERE.
→ HELLO THIS IS ATE S T PLEASE ENTER YOUR TEXT HERE.
→ HELLO THIS IS A TEST PLEASE ENTER YOUR TEXT HERE.
Instead of using CT2, you can also do the splitting very well with AI tools like ChatGPT [47] or YouChat [48].

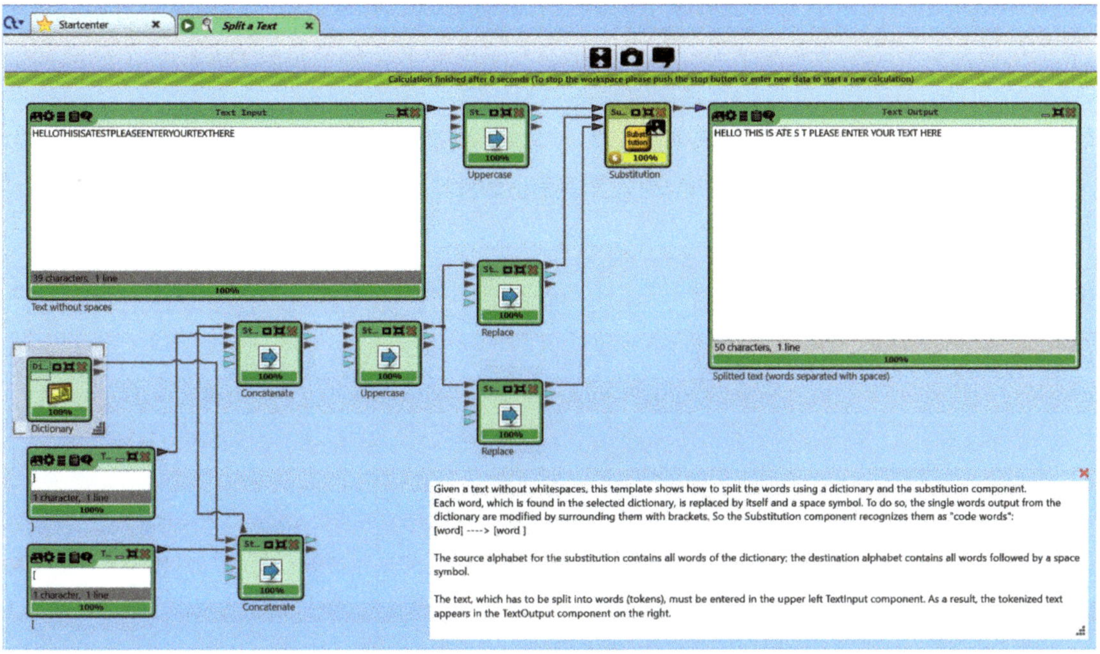

Figure 2.14 Increasing the readability of the decrypted text using the "Split a Text" template in CT2.

Example 2.14 in Section 2.8.2.6 could only encrypt and decrypt the data that was hard-encoded in the program, SageMath Example 2.19 is more professional: It contains different test data in a dictionary and can also read all necessary data and options from the command line.

To do the KPA, the key matrix A is calculated from the ciphertext C and the inverse of the plaintext P. The order of matrix multiplication in the analysis depends on whether the key is multiplied with the plaintext from the left or from the right during encryption:

$$\begin{aligned} C &= A \cdot P \pmod{26} \\ A \cdot P &= C \pmod{26} \\ A &= C \cdot P^{-1} \pmod{26} \end{aligned}$$

or

$$\begin{aligned} C &= P \cdot A \pmod{26} \\ P \cdot A &= C \pmod{26} \\ A &= P^{-1} \cdot C \pmod{26} \end{aligned}$$

It should be noted that the program first looks for the correct sections from the given plaintext so that the matrix P is invertible. This happens in the function PerformKPA in the for loop that determines the correct slices from P.

Since the entire SageMath Example 2.19 is over 700 lines long, only the file header is listed here. The entire file is available on the CT server: see `https://www.cryptool.org/en/documentation/ctbook/sagemath`.

SageMath Example 2.19: KPA Against the Hill Cipher

```
# Apply Hill cipher from SageMath in 3 modes: encrypt, decrypt, or do ▸
▸known-plaintext analysis.
# - these modes can be set via command line option. Internally they ▸
▸call the
#   functions PerformEncDev("Enc"), PerformEncDev("Dec"), or PerformKPA▸
▸()
# - the data used can be set
#   - either via -s SelectedExampleNo (to select one of the predefined ▸
▸examples stored in JSON syntax)
#   - or by entering all arguments on the command line [they overwrite ▸
▸values from a stored sample, if both are given]
#
# Usage: $ sage chap02_hill_enc_dec_kpa.sage [-h] [-V]  (-enc | -dec | ▸
▸-kpa)  [-v] [-dim DIM]
#                          [-pt STRING] [-ct STRING] [-A STRING] [-i ▸
▸INDEX] [-kl] [-s SAMPLE]
```

References

[1] ACA, *Length and Standards for All ACA Ciphers*, 2021, https://www.cryptogram.org/resourc e-area/cipher-types/.

[2] Bion, *Recreational Cryptography Programs*, https://williammason.github.io/rec-crypt/.

[3] Pilcrow, P., *CryptoPrograms*, http://www.cryptoprograms.com.

[4] Singh, S., *The Code Book: The Science of Secrecy from Ancient Egypt to Quantum Cryptography*, New York: Anchor Books, 1999.

[5] Goebel, G., *Codes, Ciphers and Codebreaking*, Version 2.3.2, 2014, http://web.archive.org/web/20151027000247/http://www.vectorsite.net/ttcode.html.

[6] Savard, J. J. G., *A Cryptographic Compendium*, 1999, http://www.quadibloc.com/crypto/jscrypt.htm.

[7] ThinkQuest Team 27158, *Data Encryption*, 1999.

[8] Knight, K., B. Megyesi, and C. Schaefer, *Copiale Cipher; Scaled Page 16/17*, Wikimedia Commons, 2011, https://commons.wikimedia.org/wiki/File:Copial e-cipher09s.png.

[9] Sanguino, L. A. B., et al., "Analyzing the Spanish Strip Cipher by Combining Combinatorial and Statistical Methods," in *Cryptologia*, Vol. 40, No. 3, 2016, pp. 261–284, https://www.semanticscholar.org/paper/Analyzing-the-Spanish-strip-cipher-by-combining-and-Sanguino-Leander/b4278e62c804ec0bf349a1e5c74a1b35bb276d83.

[10] Drobick, J., *Abriss DDR-Chiffriergeschichte: SAS- und Chiffrierdienst*, 2015, http://scz.bpla ced.net/m.html#dwa.

[11] Schneier, B., *The Solitaire Encryption Algorithm*, v. 1.2, 1999, https://www.schneier.com/ac ademic/solitaire/.

[12] Crowley, P., *Mirdek: A Card Cipher Inspired by "Solitaire,"* 2000, http://www.ciphergoth.org/crypto/mirdek/.

[13] Géraud-Stewart, R., and D. Naccache, "A French Cipher from the Late 19th Century," *Cryptologia*, 2020, pp. 1–29, https://doi.org/10.1080/01611194.2020.1753265.

[14] Kallick, B., *Handycipher: A Low-Tech, Randomized, Symmetric-key Cryptosystem*, Cryptology ePrint Archive, Report 2014/257, 2014, https://eprint.iacr.org/2014/257.

[15] Kaminsky, A., *ElsieFour: A Low-Tech Authenticated Encryption Algorithm for Human-to-Human Communication*, Cryptology ePrint Archive, Report 2017/339, 2017, https://eprint.iacr.org/2017/339.

[16] Dooley, J. F., *History of Cryptography and Cryptanalysis: Codes, Ciphers, and Their Algorithms*, Cham, Switzerland: Springer, 2018.

[17] De Leeuw, K., "The Dutch Invention of the Rotor Machine, 1915–1923," *Cryptologia*, Vol. 27, No. 1, 2003, pp. 73–94.

[18] Pommerening, K., *Cryptology. Lecture Notes*, Johannes Gutenberg Universität Mainz, 2021, https://www.staff.uni-mainz.de/pommeren/Cryptology/Classic/5_Rotor/HistRot.html.

[19] Crypto Museum Official Website, https://www.cryptomuseum.com/manuf/crypto/index.htm.

[20] Fridrih, T., *Hagelin–A Genius of Scientific and Technological Thought*, web blog post, 2019, https://habr.com/ru/company/ua-hosting/blog/271387/.

[21] Rijmenants, D., *Cipher Machines and Cryptology. Technical and Historical Information about Cipher Machines and the Fascinating World of Cryptology*, 2022, https://www.ciphermachinesandcryptology.com/.

[22] Kopal, N., "How Does the M-209 Cipher Machine Work? – A Brilliant Non-Electrical Encryption Device," 2020, YouTube channel *Cryptography for Everybody*, https://www.youtube.com/watch?v=Nhf6kHGujQ4&t=56s.

[23] Wessel, B., *The Hagelin Cryptographers C-52 and CX-52*, February 2021, https://www.cryptomuseum.com/pub/files/BW_C52_CX52.pdf.

[24] Lasry, G., N. Kopal, and A. Wacker, "Ciphertext-Only Cryptanalysis of Hagelin M-209 Pins and Lugs," *Cryptologia*, Vol. 40, No. 2, 2016, pp. 141–176.

[25] Theveßen, E., P. F. Müller, and U. Stoll, *Operation Rubikon. Wie BND und CIA die Welt belauschten*, German television station ZDF, February 2020, https://www.zdf.de/politik/frontal/operation-rubi kon-100.html.

[26] Miller, G., "The Intelligence Coup of the Century," *The Washington Post*, February 2020, https://www.washingtonpost.com/graphics/2020/world/national-security/cia-crypto-encryption-machines-espionage/.

[27] *CX-52*, Wikipedia, German version, 2022, https://de.wikipedia.org/wiki/CX-52.

[28] Kuhlemann, O., *Kryptografie.de.*, https://kryptografie.de.

[29] Leierzopf, E., et al., "Detection of Classical Cipher Types with Feature-Learning Approaches," in *Data Mining: 19th Australian Conference on Data Mining, AusDM 2021*, Brisbane, Australia, December 14–15, 2021, Springer Singapore, https://doi.org/10.1007/978-981-16-8531-6_11.

[30] Dalton, B., and M. Stamp, "Classifying World War II Era Ciphers with Machine Learning," Cryptology ePrint Archive, 2023, https://arxiv.org/abs/2307.00501.

[31] Kopal, N., and M. Waldispühl, "Two Encrypted Diplomatic Letters Sent by Jan Chodkiewicz to Emperor Maximilian II in 1574–1575," in *Proceedings of the 4th International Conference on Historical Cryptology*, 2021, pp. 80–89, doi: https://doi.org/10.3384/ecp188409.

[32] Dinnissen, J., and N. Kopal, "Island Ramanacoil a Bridge too Far. A Dutch Ciphertext from 1674" in *Proceedings of the 4th International Conference on Historical Cryptology*, 2021, pp. 48–57, https://ecp.ep.liu.se/index.php/histocrypt/article/view/156.

[33] Lasry, G., B. Megyesi, and N. Kopal, "Deciphering Papal Ciphers from the 16th to the 18th Century," *Cryptologia*, Vol. 45, No. 6, 2021, pp. 479–540, https://www.tandfonline.com/doi/full/10.1080/01611194.2020.1755915.

[34] Lasry, G., N. Biermann, and S. Tomokiyo. "Deciphering Mary Stuart's Lost Letters from 1578–1584," *Cryptologia*, 2023, doi: 10.1080/01611194.2022.2160677.

[35] Megyesi, B., et al., "Decryption of Historical Manuscripts: The DECRYPT Project," *Cryptologia*, Vol. 44, No. 6, 2020, pp. 545–559, https://doi.org/10.1080/01611194.2020.1716410.

[36] Lasry, G., "Analysis of a Late 19th Century French Cipher Created by Major Josse," *Cryptologia*, 2021, pp. 1–15, https://www.tandfonline.com/doi/full/10.1080/01611194.2021.1996484.

[37] Lasry, G., "Cracking SIGABA in Less than 24 Hours on a Consumer PC," *Cryptologia*, 2021, pp. 1–37, https://www.tandfonline.com/doi/full/10.1080/01611194.2021.1989522.

[38] Madness, *A Book on Classical Cryptography*, https://github.com/themaddoctor/classical_crypto_book.

[39] Van Nguyen, M., *Exploring Cryptography Using the Sage Computer Algebra System*, 2009, https://www.sagemath.org/files/thesis/nguyen-thesis-2009.pdf, and https://www.sagemath.org/library-publications.html.

[40] Hill, L. S., "Cryptography in an Algebraic Alphabet," *The American Mathematical Monthly*, Vol. 36, No. 6, 1929, pp. 306–312.

[41] Hill, L. S., "Concerning Certain Linear Transformation Apparatus of Cryptography," *The American Mathematical Monthly*, Vol. 38, No. 3, 1931, pp. 135–154.

[42] Wikipedia, *Hill Cipher*, https://en.wikipedia.org/wiki/Hill_cipher.

[43] Overbey, J. L., W. Traves, and J. Wojdylo, "On the Keyspace of the Hill Cipher," *Cryptologia*, Vol. 29, No. 1, 2005, pp. 59–72, doi: 10.1080/0161-110591893771, and https://www.tandfonline.com/doi/abs/10.1080/0161-110591893771.

[44] Kohel, D. R., *Cryptography*, Creative Commons, 2008, https://www.sagemath.org/files/kohel-book-2008.pdf.

[45] Stinson, D. R., *Cryptography—Theory and Practice*, 3rd ed., Chapman & Hall/CRC, 2006.

[46] Freiermuth, K., et al., *Einführung in die Kryptologie*, 1st ed., Vieweg+Teubner, 2010.

[47] OpenAI, *ChatGPT*, https://chat.openai.com.

[48] YOU.com, *AI Chatbot to Search the Web*, https://you.com.

CHAPTER 3

Historical Cryptology

Historical cryptology studies (original) encrypted manuscripts, often handwritten sources, produced in our history. These historical sources can be found in archives, often hidden without any indexing and therefore hard to locate. Once found they need to be digitized and turned into a machine-readable text format before they can be deciphered with computational methods. The focus of historical cryptology is not primarily the development of sophisticated algorithms for decipherment, but rather the entire process of analysis of the encrypted source from collection and digitization to transcription and decryption. The process also includes the interpretation and contextualization of the message set in its historical context. There are many challenges on the way, such as mistakes made by the scribe, errors made by the transcriber, damaged pages, handwriting styles that are difficult to interpret, historical languages from various time periods, and hidden underlying language of the message. Ciphertexts vary greatly in terms of their code system and symbol sets used with more or less distinguishable symbols. Ciphertexts can be embedded in clearly written text, or shorter or longer sequences of cleartext can be embedded in the ciphertext. The ciphers used mostly in historical times are substitutions (simple, homophonic, or polyphonic), with or without nomenclatures, encoded as digits or symbol sequences, with or without spaces.

So the circumstances are different from those in modern cryptography which focuses on methods (algorithms) and their strengths and assumes that the algorithm is applied correctly. For both historical and modern cryptology, attack vectors outside the algorithm are applied like implementation flaws and side-channel attacks.

In this chapter, we give an introduction to the field of historical cryptology and present an overview of how researchers today process historical encrypted sources.

3.1 Introduction

Historical cryptology deals with the encryption and decryption of historical, manually constructed ciphers. An encrypted source usually counts as historical if it has been produced no later than the mid-20th century. There is no exact break-even point; however, the development of telegraphy (from the 1830s) led to more sophisticated and complex mathematical methods applied to encryption requiring more advanced cryptanalysis.

Historical cryptology involves the field of cryptography (the art and science of code making and the encryption of messages), and the field of cryptanalysis (the art and science of code breaking [1], i.e., the decipherment of messages without the key). In everyday language, the terms "cryptography" and "cryptology" are often used interchangeably.

As in all scientific fields, historical cryptology has its own terminology. We start the chapter by introducing the most important terms and give a brief overview of historical ciphers and keys before we move on to the components needed to process and decipher historical ciphers. Given that historical cryptology as a scientific field of study is rather new, the terminology standards and the usage of the terms are still to be established and under discussion in the historical cryptology community (see for example [2] and [3]). We summarize the important terms in Figure 3.6 and Table 3.1, as well as illustrate the crypto process in Figure 3.7.

A cipher (sometimes also "cypher," which is simply the old spelling) refers to an algorithm that describes the procedure of encryption or decryption. The encrypted source itself is called ciphertext, though the terms "code" and "cipher" are often not distinguished from ciphertext in everyday language. A ciphertext consists of a sequence of symbols from a ciphertext alphabet. The ciphertext alphabet can be the same as the plaintext alphabet (e.g., the Latin letters), but often it consists of different symbol systems and alphabets, such as Greek letters, digits, graphic signs (e.g., alchemical or zodiac signs), or Chinese hieroglyphs. Figure 3.1 illustrates the variation of the symbol systems from ciphertext alphabets in three ciphertexts. The ciphertexts are extracts taken from the Borg cipher [4], a digit-based cipher from the National Archives of Sweden [5], and the Copiale cipher [6].

In ciphertexts, we can find regular usage of space marking word boundaries as in the Borg cipher (see Figure 3.1) even though most of the ciphertexts from the past use continuous script (scriptio continua) without any spaces, as shown in the examples from the Swedish National Archives and the Copiale cipher. Word boundaries were often removed in historical ciphers to make codebreaking more difficult. A ciphertext might also contain additional information such as accents and other diacritics, or punctuation marks appearing more or less systematically in connection to symbols, as in the example of the Copiale cipher. We can also find overwritings for corrections, underlined sequences, and unintentional ink spots in the manuscripts.

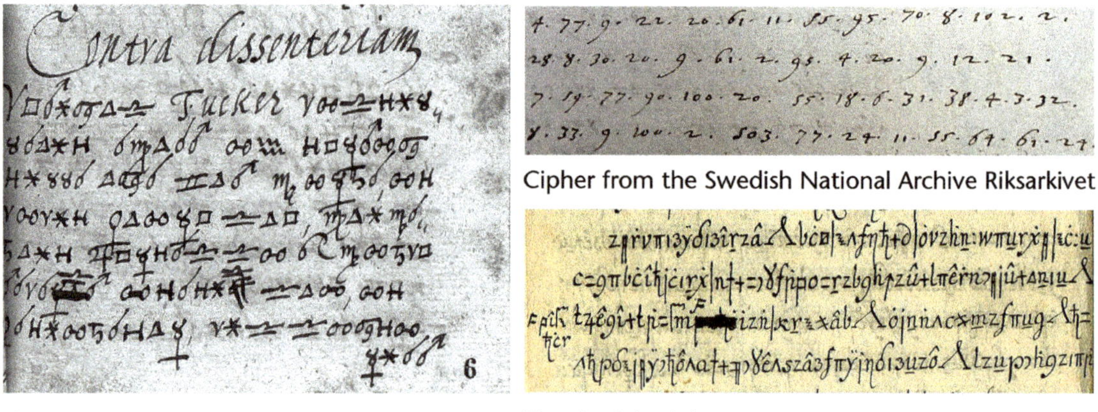

The Borg cipher

The Copiale cipher

Figure 3.1 Three examples of ciphertexts. (*From*: [4–6].)

Apart from the ciphertext in an encrypted message, nonencrypted sequences of texts that we call cleartext are also common. For example, the first line of the Borg cipher in Figure 3.1 contains a cleartext "Contra dissenteriam" in Latin. Cleartext passages bear important information about the possible underlying language(s) and the topic of the encrypted source.

The encrypted source might also include decrypted plaintext, written on the same page, often found above the ciphertext lines, as illustrated in Figure 3.2.

A cipher key (Figure 3.3) using a given cipher defines how to encrypt a plaintext and how to decrypt a ciphertext. Historical cipher keys usually contain a list of plaintext elements (letters, syllables, words, names, phrases) and the corresponding symbol or combination of symbols taken from the ciphertext alphabet, henceforth the code elements. Two examples of cipher keys are shown in Figures 3.3 and 3.4.

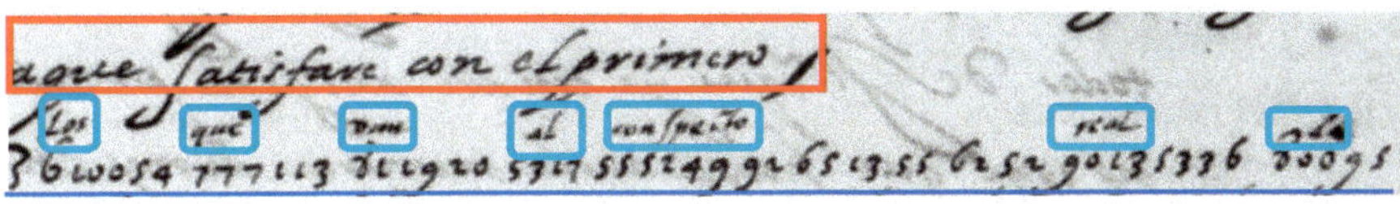

Figure 3.2 Ciphertext (underlined), cleartext (in red), and plaintext (in blue) in an encrypted manuscript.

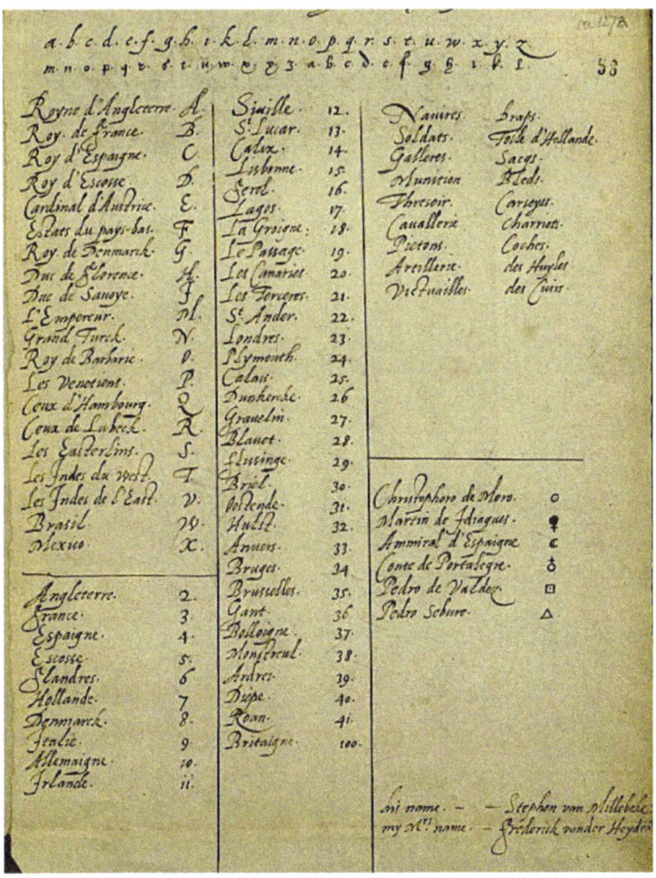

Figure 3.3 Cipher key: simple substitution. (Flanders, 1596 [7].)

In both keys, the letters of the plaintext alphabet (A–Z) are listed horizontally in the first line of the key tables. Moreover, underneath each plaintext letter, we can find either one (Figure 3.3) or several ciphertext symbols (Figure 3.4), henceforth alphabet-code elements, assigned to each plaintext letter. In Figure 3.3 these single ciphertext letters are taken from the plaintext alphabet but in a different position. In Figure 3.4, on the other hand, the lengths of the alphabet-code elements vary; two-digit code elements to encode the plaintext alphabet and three-digit code elements to encode the words. Note that the most frequently occurring plaintext alphabet letters have four alphabet-code elements, whereas the least frequent ones received three code elements. Adding several code elements to the frequently occurring plaintext elements leads to an increased difficulty of decipherment and renders a cipher homophonic.

In the columns of both keys we find a shorter or longer list of plaintext elements (names, content, and function words) with code elements assigned to each. Such a list as part of the key is called nomenclature, sometimes also spelled nomenclator. Sometimes the entire key that contains a nomenclature (i.e., a list of plaintext elements) is called a nomenclator. Here, we make a distinction between the various parts of the key. The nomenclature shown in Figure 3.3 consists of roughly 100 items in which we can see code elements using a single ciphertext symbol, for example "A" for "Royne d'Angleterre" and others with multiple ciphertext symbols, such as "12" for "Siuille." Here, the various types of nomenclature elements receive

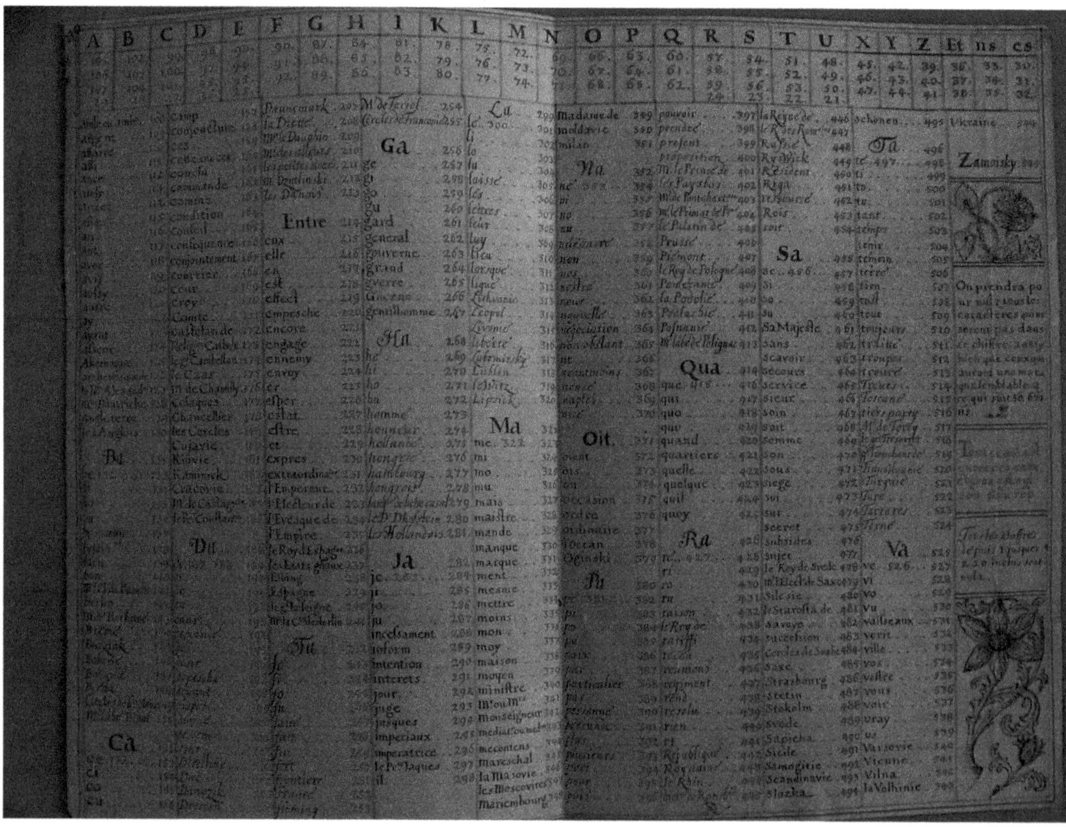

Figure 3.4 Cipher key: simple and homophonic substitution. (Hungary, 1703–1711 [8].)

different cipher symbol types: Personal names are encoded by capital letters, place names by numerals, military titles by other words, and dignitaries by graphic signs. However, this assignment is not fully consistent: "Brazil" and "Mexico" are listed among the personal names. Such inconsistencies are not uncommon in historical encrypted sources. In the other key in Figure 3.4, the nomenclature is larger, consisting of over 400 entities. Here, we can find syllables shown as section headings ("Ba," "Ca," "Da," ...), function and content words, and names and phrases, all in French. The last column contains additional information about the key to give instructions or details about the cipher.

Historical cipher keys were typically structured as tables, in which the alphabet elements and the nomenclature elements were graphically clearly separated; the former horizontally as lines and the latter vertically as columns. Content-wise, however, the boundary is not as clear-cut; double letters, syllables, or function words might be listed as part of the alphabet line. It is also noteworthy that the nomenclature tables usually have a certain structure in which plaintext elements can be ordered alphabetically (see the key in Figure 3.4) or thematically (as shown in Figure 3.3), or in a combination where the words in the themes can be alphabetically ordered. In turn, the code elements can be grouped thematically depending on the type of plaintext element they encode (as in Figure 3.3), and/or numerically when the code elements are represented by digits. The key creators often assigned code elements to the alphabetically or thematically listed plaintext elements in some structure. Code elements of the nomenclature list were typically numbered consecutively in increasing or decreasing order, either vertically following the order of the columns or horizontally, following the lines across the columns. The construction of the nomenclature list has an impact on the cryptanalysis (decipherment)—alphabetical order of the plaintext elements with increasing order of numbers can ease cryptanalysis as higher code numbers represent words starting with letters at the end of the alphabet.

To make cryptanalysis more difficult, operational code elements (i.e., code elements that operate either on the plaintext or on other code elements) have been used. A commonly occurring type are nulls, which can also be named in historical cipher keys as nullities and called by the public as "blenders"—fake code elements that encode an empty string in the plaintext. Note that keys might also contain code elements without any given plaintext in the nomenclature table treated as placeholders to be filled in later, which are not defined as nulls but empty code elements. Other types of operational code elements with special function on the plaintext include cancellation signs (also called nullifiers or deleters) that mark the removal of a certain sequence of ciphertext, and repetition signs that repeat the preceding symbol used for the reduplication of a plaintext letter.

Historical cipher keys changed and developed over time leading to the emergence of new ciphers. In fact, all the historical ciphers discussed in this chapter are variations of the substitution cipher. The specific substitution method was entirely determined by the key type used with it. Therefore, when we discuss the development of the keys, we also speak about the evolution of the ciphers. The earliest keys in Europe were based on simple substitution, in which each plaintext element is assigned to exactly one code element represented as a ciphertext symbol. An example of a simple substitution cipher is shown in Figure 3.3. The top two lines of this document illustrate a nice example of the Caesar cipher (see Section 2.2.1), in which the plaintext alphabet is also used for encryption but shifted (here by 11 positions). To complicate the cryptanalysis, a nomenclature table was added, which became the norm in Europe in the 15th century [9]. Simple substitution ciphers were then further developed into homophonic substitution ciphers, where the same plaintext entities—often the most frequently occurring ones, such as vowels and some consonants—could be encrypted with different code elements, as illustrated in Figure 3.4. The nomenclature list evolved from the 17th century and onward from several hundred elements to thick codebooks, in which not only content words but also grammatical categories (e.g., singular, plural; grammatical cases) or inflected word forms (e.g., "see, sees, saw, seen" for the verb "to see") were listed with their own code elements [9]. In some keys, different plaintext entities could also be assigned to the same code element, intentionally or unintentionally. Ciphers with one code element assigned to several plaintext symbols are called polyphonic substitution ciphers. Figure 3.5 illustrates such a cipher key. Here, the ciphertext symbol "3" can be decrypted as either "A" or "s," and the symbol "6" as either "t" or "r."

The three types of encryption methods—simple, homophonic, and polyphonic—are the most frequently occurring types in European history [9]. The interested reader can find more details about the structure and evolution of cipher keys throughout the centuries in Europe in [9].

In addition, not only monoalphabetic substitution ciphers have been used throughout history. After the early modern time, polyalphabetic substitution ciphers became common, such as the Vigenère cipher (see Section 2.2.4). In these ciphers, the plaintext alphabet is mapped to different ciphertext alphabets—see Section 2.2.4. Transposition ciphers (Section 2.1) are another type, in which the letters of the plaintext are switched around in some systematic way to form the ciphertext. In later centuries, we can also find ciphers that are actually cascades of different ciphers that we call composed ciphers. An example of such a cipher is the ADFGVX cipher [10], which is a combination of substitution (using a Polybius square—see Section 2.3) and (columnar) transposition.

In recent years, by far the greatest attention worldwide for historical cryptology has been given to the successful cryptanalysis of over 50 newly discovered letters written by Mary Stuart between 1578 and 1584. George Lasry, Norbert Biermann, and Satoshi Tomokiyo worked for over one year to transcribe, decipher, and place these letters containing over 150,000 symbols in their proper historical context [11]. Mary Stuart's letters were classified under Italian letters in the French National Library, without telling sender or recipient or the actual language used (French). The procedure used by Mary Stuart was a difficult cipher because she used a nomenclature with 191 different characters, which included well over 100 words in addition to the 26 letters of the alphabet, but also homophones (several symbols representing the same letter), symbols without meaning (nulls or blenders),

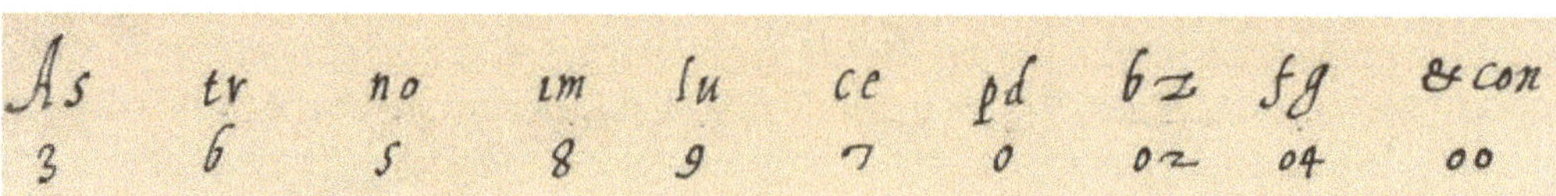

Figure 3.5 Cipher key example: polyphonic substitution from the 16th century.

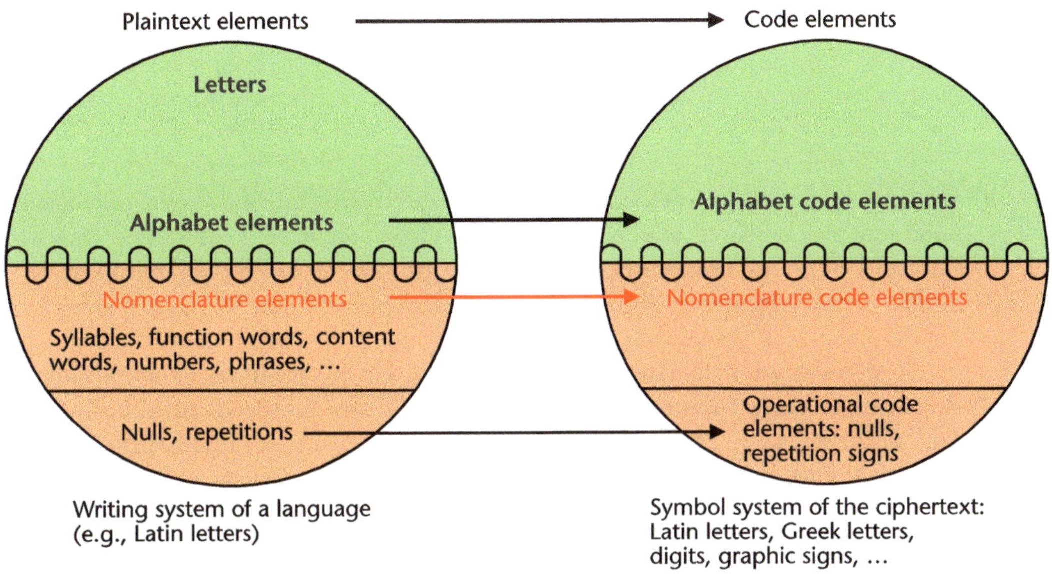

Figure 3.6 Terminology: Mapping of important terms.

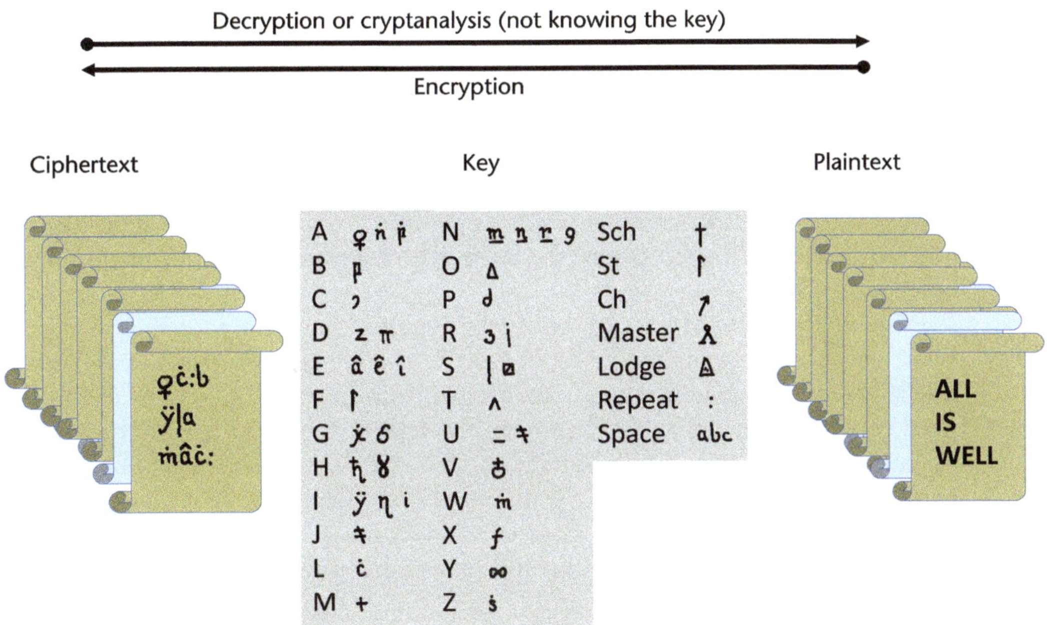

Figure 3.7 The crypto process: Components of encryption and decryption of historical sources.

symbols that cancel the previous symbol (nullifier), and symbols that repeat the previous symbol.

3.2 Analyzing Historical Ciphers: From Collection to Interpretation

Next, we describe the components involved in the processing and analysis of historical encrypted sources.

Historical ciphertexts are handwritten or printed manuscripts buried in archives, libraries, museums, or private collections. They might be difficult to find as they are hardly indexed as ciphers in archive or library catalogs. Only a small but increasing percentage of the historical encrypted sources are digitized and made available online, and even fewer are turned into a computer-readable text format. Finding, analyzing, and deciphering encrypted manuscripts are challenging and need various kinds of expertise. In this section, we give a bird's-eye view on the different steps and components involved in processing encrypted manuscripts from collection through transcription to decipherment, as illustrated in Figure 3.9. Then we describe each step of the process in detail in the subsequent sections.

Collecting encrypted sources requires knowledge about the whereabouts of the documents. Once found, the documents need to be digitized, turned into images, and described with a set of metadata according to some standard. Information can include the sender and receiver of the documents, the time and place when the encrypted source was produced or sent, and a description of its content. Describing historical sources in terms of metadata is as important as the content of the document itself.

Before we can cryptanalyze a ciphertext, we usually need to transcribe it (i.e., turning the ciphertext image(s) into a computer-readable text format). By doing so, we look closely at the symbol set and group the similar ciphertext symbols into types, which helps us in the identification of the entire ciphertext alphabet. A transcription is a text representing the ciphertext symbols from the image(s) symbol by symbol and line by line. This requires interpreting the handwriting style and motion educated guesses about the intentions of the scribe; in other words to interpret the handwriting. The transcription needs to be thorough; all symbols, diacritics, punctuation marks, and spaces must be transcribed to avoid error propagation during decipherment.

Given a (couple of lines of) transcription we can go on with the cryptanalysis. First, we need to segment the ciphertext into code elements and analyze the frequencies and co-occurrences of the various symbol types and code elements. We need to make educated guesses about the cipher type and about the underlying language. Dictionaries and language models for various time periods might be of help on the way when guessing the plaintext underneath. Once we have a decrypted text, we interpret the plaintext, correct wrongly transcribed symbols, and adjust the assumed key to get an appropriate and reasonable plaintext output. We might then translate the text to one or several languages, and set the plaintext in a historical context; what was written, by whom, to whom, and why.

Deciphering a ciphertext—albeit lots of fun—is often challenging. In the past, many historians and people worked individually in an uncoordinated fashion on the identification and deciphering of secret writings. Without access to automatic methods that can accelerate the decipherment, it's a time-consuming process. At the same time, cryptanalysts, computer scientists, and computational linguists develop automatic cryptanalysis algorithms to identify cipher types and to break various ciphers without having access to real historical ciphertexts.

To coordinate the efforts of various expertise and build research infrastructure in terms of resources and tools for historical cryptology, an international research program was created in 2018: the DECRYPT project [12]. The aim of the project was to establish a new cross-disciplinary scientific field of historical cryptology by bringing the expertise of the different disciplines together to digitize encrypted historical documents, build a database of historical ciphers, and develop software tools for transcription and cryptanalysis. We are not aware of any other cross-disciplinary project in the field that takes a holistic approach from collection through transcription to decipherment by developing open-source resources and tools for historical cryptology in large scale. Therefore, we base this chapter largely on the experiences and findings of the results of the cross-disciplinary cooperation in the DECRYPT project. However, there are many relevant high-quality studies on various aspects of historical cryptology and we will refer to the most prominent ones in the relevant parts of the subsequent sections.

To be able to study the characteristics of historical ciphers with the ultimate goal to decipher all cipher types from historical times, we need a large set of historical sources to be collected and stored from various places and time periods. The DECODE database [13] was created to store images of ciphertexts, encryption keys, and information about their provenance, transcriptions, and possible decryptions. The process of (semi-) automatic decryption involves, as mentioned before, transcription by applying image recognition to automatically convert the images to machine-readable format and a mapping of symbols to a transcription scheme. The detection of the underlying plaintext language of the ciphertext on the basis of historical text sources, the automatic identification of the cipher type, the cryptanalysis of the ciphertext, and finally its decryption are taken care of in the cryptanalysis step. The methods developed are based on a wide range of algorithms: from classical cryptanalysis to advanced deep-learning architectures taken from artificial intelligence. Various (neural network) models for transcription are released in the TranscriptTool [14] (see Figure 3.8), while the algorithms for cryptanalysis have been implemented in CrypTool 2 (CT2) [15]. CT2 is called in a command line version on the webserver of the DECRYPT pipeline. Both tools (TranscriptTool

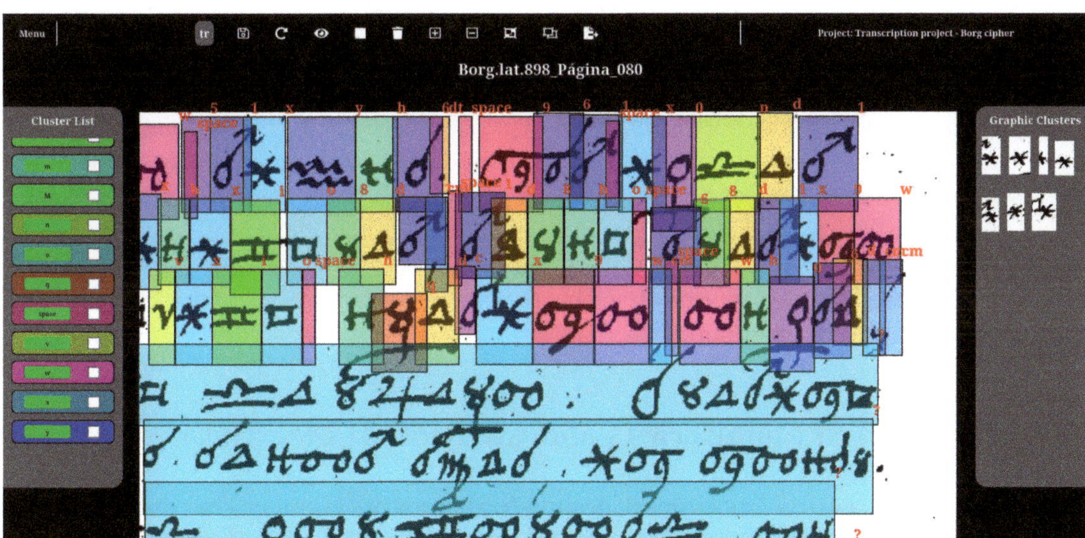

Figure 3.8 TranscriptTool for creating transcriptions of scanned historical manuscripts, offered as part of the DECRYPT pipeline.

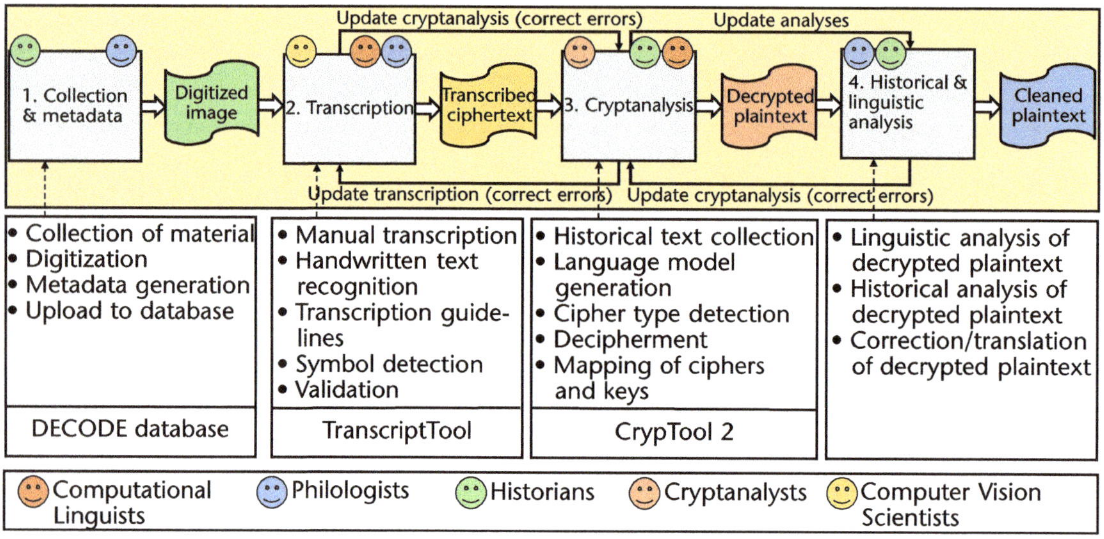

Figure 3.9 Overview of the DECRYPT pipeline (see `https://de-crypt.org/`).

and CT2) are released as open-source and are under continued development (as of 2023). The DECODE database and the two tools are included into a framework as a pipeline for processing the historical encrypted manuscripts to allow feedback loops and error reduction between the various steps in the pipeline. In addition to the TranscriptTool in the pipeline on the web, there is a standalone offline tool called CTTS. See Section 3.4.2. For ciphers that do not consist of numbers, CTTS or TranscriptTool are currently the best choice. For numeric ciphers, Transkribus.ai can be an alternative.

The steps for breaking a cipher need careful combination and cooperation of experts from different fields. Computational linguists provide the database with keys and ciphers, define transcription schemes for various symbol sets, and build and evaluate historical language models generated from historical texts. Historical linguists and philologists collect and analyze historical texts to develop models for language variation and language change. Cryptanalysts develop efficient algorithms for the cryptanalysis of various cipher types, and computer vision scientists provide a typology of symbol transcription and models to turn images into a machine-readable format. Historians contribute to the collection, contextualization, and interpretation of the hidden sources. By doing so the encrypted sources can be systematically handled, studied in large scale, and made available to the public.

The following sections describe the main parts shown in the pipeline and highlight the challenges in each step.

3.3 Collection of Manuscripts and Creation of Metadata

A general experience of experts looking for handwritten cipher keys and encrypted documents is that they are easy to recognize but hard to find (see Section 3.2). It is easy to recognize the keys because they have a typical structure: A plaintext alphabet and a ciphertext alphabet are written next to each other, often followed by a nomenclature table where words and corresponding code elements are listed. A typical historical key usually looks like a short note on a piece of paper (if it is a monoalphabetic cipher) or a large table on one or two approximately A4-measure pages. They are either separate sheets or part of an extensive collection, with pages in a book entirely dedicated to cipher keys. The encrypted documents are usually easy to *recognize* because they are text-like documents partially or entirely composed of numbers, letters, or graphic signs, often separated by dots. Even though sometimes inventories are mistaken for encrypted documents, and there might be some uncertainty about whether a text is encrypted or written in an unknown writing system or language, most of the time these documents are recognized without any problem. They might be only a few words, a paragraph-long ciphertext in an otherwise readable message, or a several-page (even a book-length) entirely encrypted document.

However, it is not easy to *find* the encrypted sources. Cipher keys and encrypted documents are found in two different places: in the archives and the manuscript collections of libraries. Imagine that a crypto-history expert pays a visit in a foreign country wishing to study that area's cryptology. Such a research trip should be thoroughly prepared because entering an archive and asking for cipher keys without any preparation rarely leads to success. This preparation includes consulting the secondary literature using that specific manuscript collection and writing directly to the archivists/librarians. Asking for advice from historians dealing with the period (but not necessarily with encrypted documents) might also be of considerable help. The importance of personal contacts is not to be underestimated. Finally, precious input can also arrive from blog authors, including the portal about the Voynich manuscript by René Zandbergen [16], Nick Pelling's *Cipher Mysteries* [17], or Klaus Schmeh's science blog [18] with a wide range of encrypted sources.

Manuscript collections in libraries usually have proper catalogs, but the reference materials of archives do not always specify that a given source is encrypted. Even when thoroughly cataloged, their description is rarely on document-level; they remain more frequently on a higher collection level, and thus individual documents remain invisible. Archives usually have boxes with a lot of documents in them. Often, the box is described (e.g., political documents from this or that war), but the individual letters, or documents, are not described one by one. However, even in those rare cases when the indexes list each individual record, a further problem arises: which search word to look for? "encrypted," "cipher," "in cifra" (or ciffra), "enchiffré," "crypté," and "chiffriert" are certainly good choices, but following the results of "en chiffre" in the Bibliothèque Nationale de Paris might be problematic, because one gets thousands of documents, the description of which involves "number" (chiffre).

Usually, it is easier to find the keys because they are often stored together in thematic collections. The two most frequent cases are (1) a whole handwritten book (either in a library or an archive) in which cipher keys are copied, contains one key per page, and (2) a folder (usually in an archive) stores separate sheets of various sizes, one key being on each sheet. Catalogs and reference books usually mention such collections. However, when an individual key occurs somewhere alone, it is hardly mentioned and can only be found by chance.

Encrypted documents are harder to find because the catalogs (of the libraries) and the reference books (of the archives) often do not specify in the indices that they are entirely or partially encrypted. In such cases, the crypto historian can ask for diplomatic or military correspondences of a specific period in general. Diplomatic letters (particularly ambassadors' letters and intelligence reports) and military messages will include encrypted messages with high probability. Even family collections (the kind of sources that make up a large portion of the totality of archival collections) might also contain encrypted documents, not to mention personal diaries and scientific and religious books. There is no systematic way to find them; one has to ask for whole folders and leaf them through. According to the conjecture of a crypto historian, *one percent* of the archival material is partly or entirely encrypted [19].

There is also a problem of matching the encrypted document with the corresponding key. Even if the collectors found both, it is not evident that they recognize the relationship between the two. This task gets harder as the collections grow. It is tough to index the records in a way that corresponding sources become identifiable.

Once crypto historians find cipher keys and encrypted documents, they face several further difficulties. First, the attached metadata might not be correct. The collections are dated, and the origins of the sources are also indicated in the archival folders; however, this information is usually too broad, and the documents and the keys are not dated separately. Some of the records contain dates and names, and in those cases when these are not later additions (by 19th-century archivists and librarians, for example) but historical data, they are reliable. In other cases, they are not always trustworthy, or just contain information that is too unspecific.

Describing a manuscript in terms of its location, structure, origin, and content is invaluable for research. Such descriptions are called metadata, which help us to interpret the manuscript. The more robust and detailed the description is the more accurate analysis we can carry out. Metadata of historical encrypted sources might include—albeit not limited to—information about:

1. The current location of the manuscript (index number in the archive/library, place, city, country).
2. The origin of the document including information about the place and dating, the sender and the receiver of the source, or the creator and/or the user of the cipher key.
3. The content of the document including its type (e.g., a ciphertext, a cipher key, or a manual about cryptology), and the language(s) involved.
4. Additional information might describe the symbol set of the ciphertext alphabet (e.g., digits, alphabets, graphic signs), the cipher type (simple, homophonic, or polyphonic substitution), the nature of nomenclature elements, or instructions.

Unfortunately, such metadata for encrypted sources is difficult to find in the archives and libraries, as they are hardly indexed and only a few know about their whereabouts. As a result of this—hardly operationalizable—process several online collections are available that also offer digital scans. Besides the blog authors already mentioned, Satoshi Tomokiyo's private homepage *Cryptiana* [20] contains original ciphers and keys from the 15th to the 20th centuries and also helpful material on the cryptanalysis of historical ciphers. Eugen Antal and Pavol Zajac's *Portal of Historical Ciphers* [21] hosts a yet small but growing database of original historical ciphers from the 17th up to the 20th century focusing on Central-European encrypted sources released with a nice graphical interface. And finally, being part of the DECRYPT project, the DECODE database [13] is the largest source for historical ciphers and keys today. At the time of writing (November 2023), the database contains over 7,000 historical encrypted sources, all stored with their original image(s) and annotated with metadata along with related documents such as transcriptions.

All collections of encrypted sources face two difficulties, one legal and one technical. First, the owner of the given records (let them be archives or libraries) usually does not allow making public high-resolution images in the online collection for copyright reasons. Thus, often only a low-resolution reproduction can be shared with the public. Second, visual recognition software requires good quality high-resolution (at least 300 DPI) copies. However, there has been considerable improvement in this second field, and thus sufficiently readable documents can be offered to the transcription tool, the next phase of the pipeline.

3.4 Transcription

Once collected, the images of the encrypted source must be turned into some computer-readable text format needed for the cryptanalysis part of the process. The digitization involves the conversion of the ciphertext as well as cleartext and/or plaintext passages appearing in the manuscript into a text representation. This means in particular that the symbols of the ciphertext in the images are replaced by machine-readable symbols and the cleartext and plaintext sequences are interpreted and transcribed. There are different methods and approaches how this can be done. In the following, we focus on the transcription of ciphertext and describe two methods: a manual option and a semiautomatic option. While the manual option relies entirely on human effort, the semiautomatic option uses computer-vision technology based on artificial intelligence (AI) methods followed by manual postcorrection of the AI output. We show the challenges with both methods and discuss their advantages and disadvantages in the last section.

3.4.1 Manual Transcription

Transcribing a historical source, especially those that are handwritten in a foreign language, is far from easy and needs trained eyes and hands. Here, the main challenges, standards, and current practices are summarized when transcribing encrypted sources.

The aim of the transcription is to convert the text appearing in the image into a text representation. The transcription of the historical document should be as accurate as possible. This concerns of course the delimitation of the distinct ciphertext symbols and the identification of the symbol types that appear in the manuscript. Sometimes it is an easy task if the ciphertext alphabet consists of a limited, known set of symbols such as digits. Oftentimes, the encrypted sources also contain other symbols such as dots, punctuation marks, accents and other diacritic signs, or underlined sequences.

Handwriting styles vary across individuals, and some writing is more clear than others. But it also changed across time periods and geographic areas. However, for these script types scholar descriptions can be found in handbooks of paleography. Script models in tables can serve as support. Also, abbreviations commonly used in historical texts changed over time.

Manual transcription of historical texts in general and probably historical ciphertexts in particular is laborious and time-consuming. It requires a high level of concentration and despite all efforts it is prone to inconsistencies and mistakes. In addition, the personnel needed causes expenses.

Even if the transcription should be as accurate as possible, the transcriber has to make decisions with regard to how detailed a transcription should be. In general, we can differentiate between two different levels of granularity. Either we transcribe very close to the historical writing and represent all word boundaries, all punctuation, all line and page breaks, and give spelling and abbreviations exactly as they appear in the original text (diplomatic transcription), or we modernize for instance punctuation and spelling, correct obvious mistakes, and dissolve abbreviations to help the modern reader (normalized transcription).

For historical ciphertext, we apply a high degree of granularity and aim to capture as many details as possible, for instance spacing, diacritics, and punctuation marks (i.e., everything that might be of relevance to be able to recover the plaintext). In the DECRYPT project, diplomatic transcription is applied.

One of the first tasks of the transcription process is to identify and segment each symbol in the ciphertext. Sometimes it is straightforward, as in the case of the clearly segmented digit-based cipher or the eclectic collection of symbols in the Copiale cipher, shown in Figure 3.1. Sometimes symbol segmentation is rather difficult, especially when the scribe used connected handwriting style with touching symbols, as in the case of the Borg cipher in Figure 3.1. To segment symbols correctly, it is helpful to look at highly similar symbols as they occur in the manuscript, especially in connection to other symbols to see where the symbol boundaries should be drawn. Spaces as shown in the original should not be left out from the observation. Spaces in ciphertexts can be intentional, often marking symbol boundaries and also word boundaries from the plaintext. However, spaces are sometimes just added to make decipherment harder. Spaces can also be unintentional where the scribe happened to put a space during writing that actually can reveal an actual word boundary in the plaintext. Therefore, spaces should be carefully observed and transcribed.

At the same time or as a next step, it is natural to group the similar symbols into a type and assign a unique letter or symbol to each symbol type to be used for transcription. The main difficulty at this step lies in the definition of a group. How similar shall the symbols be in order to be clustered into one group? Should a, a., á, à, å, and ä be one or several groups? How many? Investigating what types of symbols the ciphertext alphabet consists of and how frequent specific symbols are and in what context of other symbols (n-grams) they appear in can be of help. For example, if we can find some digits (1–3), then it is probable that we can find all digits (0–9). Similarly, if we can find some zodiac symbols, we can expect to find more of them, or even all 12. If a symbol with a dot appears only in one or a few cases, the dot could be an ink spot; but if it appears and is used systematically, it should be treated as a symbol type.

A big challenge for the transcription of ciphertexts is with eclectic symbol sets using a large variation of graphic signs; see examples of the Borg and Copiale ciphers in Figure 3.1. Many symbols look similar making it unclear whether we have to do with two distinct cipher symbols or the same symbol with some graphic variation due to the handwriting. For example, the zodiac signs ♍ and ♏ (UTF-8 char: U+264D and U+264F, respectively), look similar at the first sight but if we are familiar with zodiac signs, we can easily distinguish between the two. Human creativity many times invented their own signs with tiny differences between some symbol types, representing different plaintext entities. The challenge of identifying the unique ciphertext alphabet can often be only solved together with the following decipherment process.

To be able to study ciphers and compare them over time and across geographic areas, it is an advantage to have a transcription standard for encrypted sources so that the same symbol types are transcribed similarly across ciphertexts as well as cipher keys. A standardized transcription of all encrypted sources allows matching of ciphertexts with their corresponding key, which makes both decryption and historical contextualization more straightforward.

Within the DECRYPT project, transcription guidelines were developed; see [22] and [23]. The guidelines deal with the systematic transcription of ciphertext images, cipher-key images, and cleartext images.

The basic principle of the transcription is to transcribe the manuscript as close to the original as possible with a special attention directed on the ciphertext itself. Each line is transcribed symbol by symbol with line breaks, spaces, punctuation marks (periods, commas, question marks), diacritics, and underlined sequences marked. Symbols are represented in Unicode using the UTF-8 encoding scheme [24]. Uncertain symbols are transcribed with the guessed symbol followed by a question mark. Unknown letters are marked with an asterisk (*). Figure 3.10 shows

Unicode names

R <SPACE> gemini conjunction taurus conjunction ironore arsenic cancer H sextile <SPACE> H virgo sextile ironore sextile

Copperore fire taurus arsenic delta ? <SPACE> gemini conjunction taurus aries sextile <SPACE> 5 arsenic cancer arsenic aries sextile gemini

Unicode codes

R <SPACE> 264A 260C 2649 260C 2642 29df 264b H 26b9 <SPACE> H 264d 26b9 2642 26b9

2640 25b3 2649 29df 03b4 ? <SPACE> 264a 260C 2649 2648 25b3 26b9 <SPACE> 5 29df 264b 29df 2648 26b9 264a

Unicode characters

R <SPACE> ♊ ☌ ♉ ☌ ♂ ⧟ ♋ H ⚹ <SPACE> H ♍ ⚹ ♂ ⚹
♀ △ ♉ ⧟ δ x? <SPACE> ♊ ☌ ♉ ♈ △ ⚹ <SPACE> 5 ⧟ ♋ ⧟ ♈ ⚹ ♊

Figure 3.10 Transcription of the Borg cipher [4] represented as Unicode names, converted to Unicode codes, and visualized as original symbols.

a transcription of the Borg cipher with its eclectic symbol set using Unicode names that can be automatically converted to the actual Unicode codes, and finally represented graphically as icons. It is up to the transcriber's preference to use the Unicode names, which are easier to memorize, or to transcribe graphic signs directly as Unicode codes. Either way, using the keyboard for digits, punctuation marks, and the Latin letters is always preferable for faster progress.

To make the process of decipherment easier, transcription does not always keep to the original image. Instead, the transcription in some cases needs to reflect the intention of the encoder. This means that corrections in the manuscript are transcribed as was presumably intended by the scribe. For example, notes in the margin denoting corrections are transcribed and added to the place as indicated by the given mark in the original, as illustrated in Figure 3.11. Crossed-off symbols in the original are not transcribed but should be added as a comment in the metadata of the transcription file.

Like ciphertexts, cipher keys are transcribed using UTF-8 encoding. However, since cipher keys can be structured in many ways, we do some generalization in the representation of the layout. We separate the plaintext and the code elements onto two sides (different columns), showing this by adding "code" or "plaintext." Each pair is written in a separate line. In cases where several code elements (in the case of homophonic ciphers) or plaintext elements (in the case of polyphonic ciphers) are listed, the alternative elements are transcribed sequentially separated by a bar ("|"), followed by " – " and the plaintext unit(s), regardless of whether the alternatives are written on several lines in the original or not. Special functions in keys (called "operational code elements" in Table 3.1) are also transcribed. A transcription of the cipher key in Figure 3.5 is illustrated in Table 3.2.

The transcription of cleartexts and plaintexts also should represent the original text shown in the image. To be able to distinguish between ciphertext and cleartext sequences, the latter is marked in brackets with a description of the language, as ⟨ `CLEARTEXT LANG-ID Letter_sequence` ⟩. The language ID is a two-letter code defined by ISO 639-1. In addition, catchwords (i.e., a sequence of symbols anticipated as the first symbol(s) of the following page, served to mark page order), are written in brackets. These are marked as ⟨ `CATCHWORD Symbol_sequence` ⟩.

Some documents are damaged and the readability of cipher symbols and other text passages are therefore limited. In these cases, a transcriber marks insecurities in the transcription with a question mark or an asterisk for missing elements. The type of material damage causing the insecurity is described in the metadata, which should be part of the transcription file, and/or as a comment in the transcription. A similar problem might occur when the image quality provided by the archive or library is too poor. Problems caused by low resolution can to some extent be solved

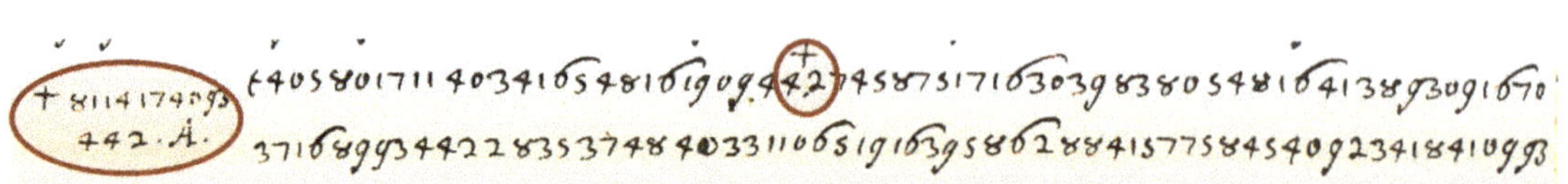

1 4 0 5 8 0 1 7 1 1 4 0 3 4 1 6 5 4 8 1 6 1 9 0 9 4 8 1 1 4 1 7 4 0 9 3 4 4 2 .A.? 2 7 4 5 8 7 5 1 7 1 6 3 0 3 9 8 3 8 0 5 4 8 1 6 4 1 3 8 9 3 0 9 1 6 7 0
3 7 1 6 8 9 9 3 4 4 2 2 8 3 5 3 7 4 8 4 0 3 3 1 1 0 6 5 1 9 1 6 3 9 5 8 6 2 8 8 4 1 5 7 7 5 8 4 5 4 0 9 2 3 4 1 8 4 1 0 9 9 3

Figure 3.11 Transcribing margin notes.

Table 3.1 Important Terms and Definitions in This Book

Plaintext	The text (or message) intended for encryption and/or the decrypted text.
Cleartext	Intentionally unencrypted text in an encrypted document.
Ciphertext	The encrypted text.
Encryption	The process of transforming a plaintext into a ciphertext using a given key.
Decryption	The process of transforming a ciphertext into a plaintext using a given key.
Cipher	A set of rules (algorithm) describing the process of encryption/decryption.
Key	A piece of information needed for encryption and decryption. A key has to be kept secret for security.
Nomenclature	A part of the key with a list of linguistic entities, such as syllables, words, phrases, or sentences, with their corresponding code elements. Thus, it contains both the nomenclature elements and nomenclature-code elements.
Cryptanalysis (decipherment/ code-breaking)	The process of analyzing a ciphertext without knowing or only partially knowing a key to reveal the original plaintext (and maybe also the key). Some authors emphasize with decipherment that the cryptanalysis process was successful.
Plaintext alphabet	Set of elements used in the plaintext, for example, letters, digits, punctuation marks, spaces.
Ciphertext alphabet	The set of symbols used in the ciphertext (e.g., digits, Latin and Greek letters, alchemical, or zodiac signs). We find these symbols not only in the ciphertext but also in the manuscript containing the key.
Plaintext elements	All types of plaintext entities that have corresponding code elements assigned to them. They usually represent letters, syllables, names, function (e.g., prepositions) and content (e.g., nouns, verbs) words, as well as phrases. The plaintext elements include the alphabet elements and the nomenclature elements.
Alphabet elements	Constitute a subset of plaintext elements. All letters in the alphabet of the writing system that have corresponding code elements assigned to them.
Nomenclature elements	Constitute a subset of plaintext elements. These are above the alphabet level. It may include syllables, names, function and content words, as well as phrases.
Code elements	A symbol or a concatenation of symbols of the ciphertext alphabet used for substitution of the plaintext elements or to indicate that an operation on the revealed plaintext is needed. We distinguish between three types of code elements: alphabet-code elements, nomenclature-code elements, and operational code elements.
Alphabet-code elements	Code elements used for encryption of the alphabet elements.
Nomenclature-code elements	Code elements used for encryption of the nomenclature elements. Nomenclature elements are often encrypted using a different symbol type or of a different length than used for the alphabet-code elements.
Operational code elements	Elements with a special function to carry out an operation on the revealed plaintext. Examples are repetition signs to repeat the preceding letter and cancellation signs (i.e., special code elements that mark the removal of a certain sequence of ciphertext).
Nulls/nullities	A subset of the operational code elements that represent an empty string in the plaintext. Their purpose is to confuse the codebreaker or to mark the start and/or the end of the nomenclature elements.
Code separator/ token separator	A symbol or a concatenation of symbols that separates code elements or groups of code elements from each other. The main intention is to help the receiver to tokenize the ciphertext. In the case of cryptanalysis, it can help to break the cipher more easily.

thanks to methods developed in computer vision science to increase the image quality.

Automatic methods for transcription developed within image processing in general and handwritten text recognition in particular, as parts of one of the scientific fields of artificial intelligence called computer vision, will be the topic of Section 3.4.3.

Table 3.2 Transcription of the Key in Figure 3.5

Code	–	*Plaintexts*
3	–	A | s
6	–	t | r
5	–	n | o
8	–	*ı* | m
9	–	l | u
7	–	c | e
0	–	p | d
02	–	b | z
04	–	f | g
00	–	& | con

3.4.2 CTTS: Offline Tool for Manual Transcription

To support the time-consuming human labor of manual transcription, George Lasry developed a transcription tool called CrypTool Transcriber and Solver (CTTS). The tool can be executed on Windows, macOS, and Linux, and be downloaded through CrypTool.[1]

CTTS is designed for efficient manual transcription of historical ciphertexts. It also includes a solver for homophonic substitution ciphers. CTTS encourages a cyclic process of review and iteratively editing of transcriptions and decryptions. It provides multidocument support so that users can work on several documents using the same symbol sets simultaneously. CTTS allows to store and load transcription projects and export both the transcribed ciphertexts as well as the decrypted plaintexts.

The nonpublic predecessor version of CTTS was successfully used to crack several real manuscripts (like the Mary Stuart ciphers [11] and the Armand de Bourbon cipher [25]), leading to several publications in *Cryptologia* or at HistoCrypt.

Ciphertexts in historical documents often contain graphic symbols, letters, or digits. The manual process of transcribing such a document with CTTS is as follows:

Step 1: The user loads an image file containing the ciphertext.

Step 2: The user uses the mouse to frame each ciphertext symbol with a box and associates the ciphertext symbols with each other. This is what is described above as grouping the similar symbols.

Step 3: The program generates a transcribed text. In a scenario for a 26-letter alphabet and a simple substitution cipher, it consists of a maximum of 26 clusters of ciphertext letters. Clearly, homophonic substitution ciphers will have many more than 26 clusters, plus additional clusters for punctuation marks, spaces, and other types of delimiters.

Step 4: The user may optionally apply a built-in cryptanalysis algorithm (simulated annealing) on the (so-far) transcribed text to cryptanalyze the cipher and reveal the plaintext.

1. https://www.cryptool.org/en/ctts.

Steps 2 through 4 are performed iteratively in a loop to improve transcription and decryption.

Figure 3.12 shows a screenshot of the tool. In the upper right section of the application, a historical encrypted document has been loaded and manually transcribed. Each of the graphical ciphertext symbols is enclosed by a user-drawn box. Boxes of the same color are used to mark ciphertext symbols belonging to the same cluster of symbols. The left side of the application displays a list of all the symbols transcribed so far. Additionally, transcription assignments can be seen; for instance, the first symbol of the list, a 90-degree-rotated letter T, is transcribed as "02." Next to the "02" there is a letter "E," which is the assigned plaintext symbol. Users can manually assign plaintext symbols or an automatic cryptanalysis algorithm can be executed to try and find the best assignments using simulated annealing.

At the bottom of the application, all symbols of the currently selected cluster are visible. Here, all ciphertext symbols transcribed as "02" are grouped in this cluster. This allows users to see which symbols share the same transcription symbol and identify transcription errors. Users can easily correct errors by dragging and dropping incorrectly assigned symbols into a different cluster.

Figure 3.13 shows how the result of step 4 (cryptanalysis) is included into the CTTS GUI again.

3.4.3 Automatic Transcription

Computer vision is the discipline of computer science that makes machines see. In artificial vision, the eyes are the cameras, formed by a matrix of light sensors. These sensors convert the intensity of the light that reaches them into numerical values, generating digital images. But these matrices of points (i.e., pixels), need their brains: computer programs that can associate the sets of pixels with concepts, according to their shape, color, layout, and so forth. In particular, document

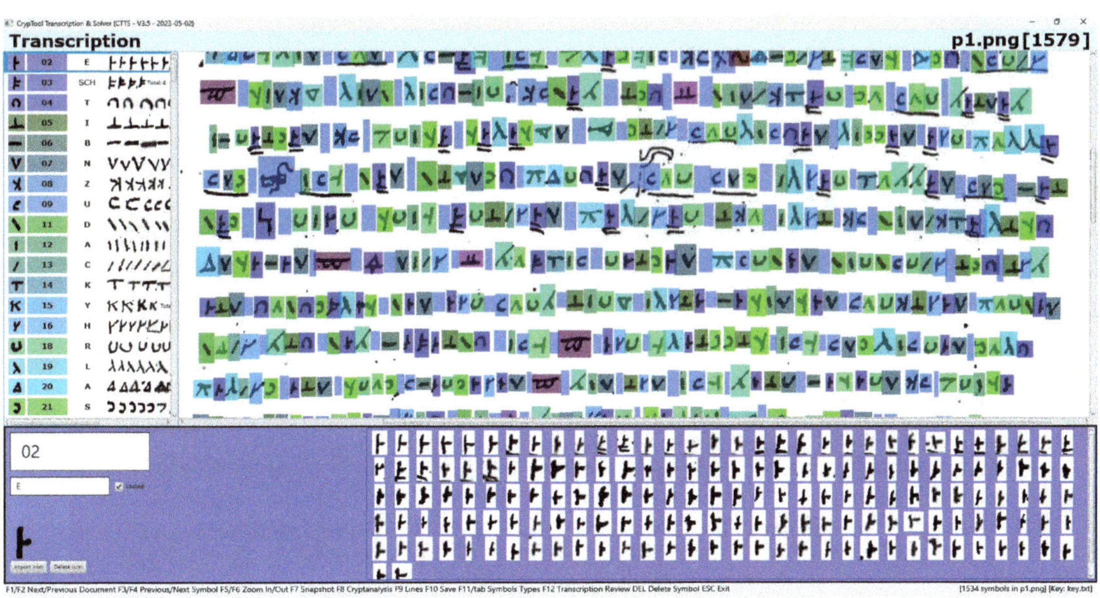

Figure 3.12 Ciphertext transcribed with the program CrypTool Transcriber and Solver.

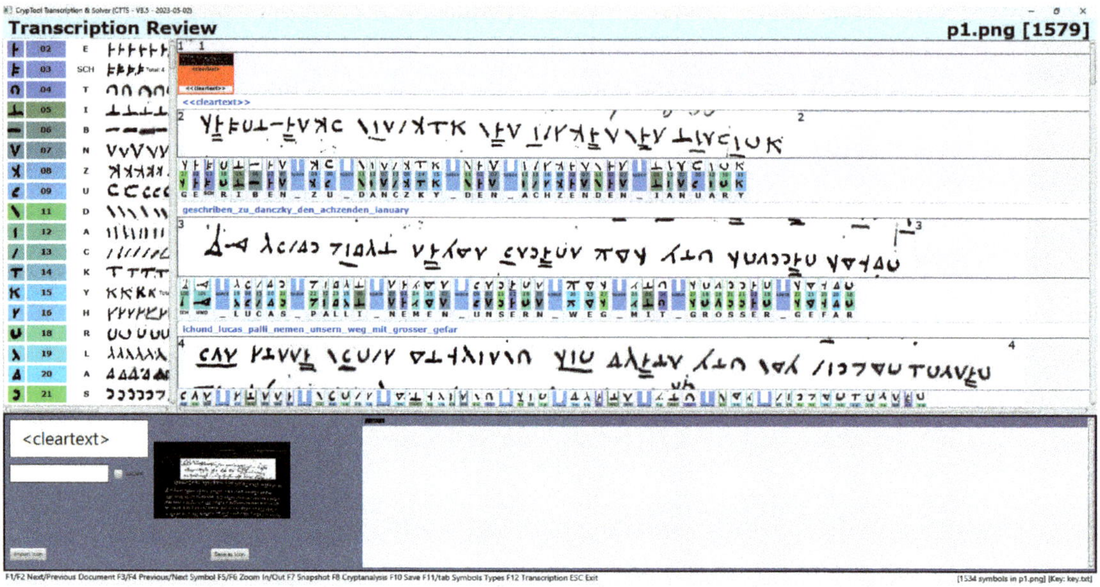

Figure 3.13 Ciphertext cryptanalyzed with the program CrypTool Transcriber and Solver.

analysis addresses the problem of automatically recognizing document content being it printed text, handwritten text, or graphic elements. Traditionally, optical character recognition (OCR) programs recognize clusters of pixels as letters and, at a higher level, validate joint interpretations to end up transforming a digital image into an editable text file. Despite advances during the last decades, reading systems still have limitations, and document analysis research must advance to offer large-scale solutions. In the case of historical handwritten documents, the different handwriting styles, the paper degradation, or the use of ancient languages makes the recognition difficult. Moreover, the use of unknown alphabets, which is commonly the case in such encrypted sources, makes its automatic transcription even more challenging. For this reason, recognition methods must be guided by human experts, and, once the transcription is provided, it must be validated to correct any transcription errors.

Typically, the stages when recognizing text include preprocessing, layout segmentation, and transcription. Given that labeled data (transcribed data) is often not available, the recognition methods are divided into learning-free and learning-based techniques. Next, the main stages of automatic transcription are described.

3.4.3.1 Document Preprocessing

The processing of the image includes those techniques that are usually applied after the digitization of the document. These techniques are essentially applied for improving the quality of the images to make the document more readable, both for people and also for automatic reading systems. In the case of very old and poorly preserved documents, it is necessary to apply document enhancement techniques for minimizing show-through or bleed-through effects, paper discoloration, or loss of ink intensity. Although many document enhancement methods can be directly applied to any input document image, recent deep learning-based methods, such as generative adversarial networks and transformer networks [26] have demonstrated a superior performance. They need similar (labeled) data to train such systems for a good performance. An example of a document enhancement, which includes binarization, is shown in Figure 3.14. Binarization here means converting a color or grayscale image into a binary image with only black and white pixels.

3.4.3.2 Layout Segmentation

Once the document has been preprocessed and enhanced, the central area of the page must be identified within the image. Layout analysis methods aim to identify the structure and nature of the regions within the document. Many historical documents contain heterogeneous contents, such as text, drawings, or music scores. In the case of ciphertexts, this stage is usually focused on detecting the blocks of text and separating them into lines, words, and ideally, into characters/symbols [27]. However, in many manuscripts, symbols are touching or even overlapping, which makes the segmentation at symbol level difficult, as shown in Figure 3.15 (see the bounding boxes in red color). In such cases, it is preferable to opt for learning-based models.

3.4.3.3 Text/Cipher Recognition

Once the structure of the document has been analyzed and the text regions, lines, and/or symbols have been extracted, these are processed to obtain the final

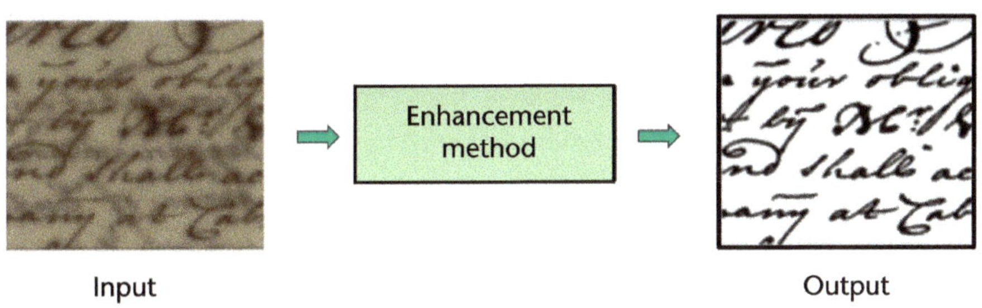

Figure 3.14 Example of a document enhancement method (binarization).

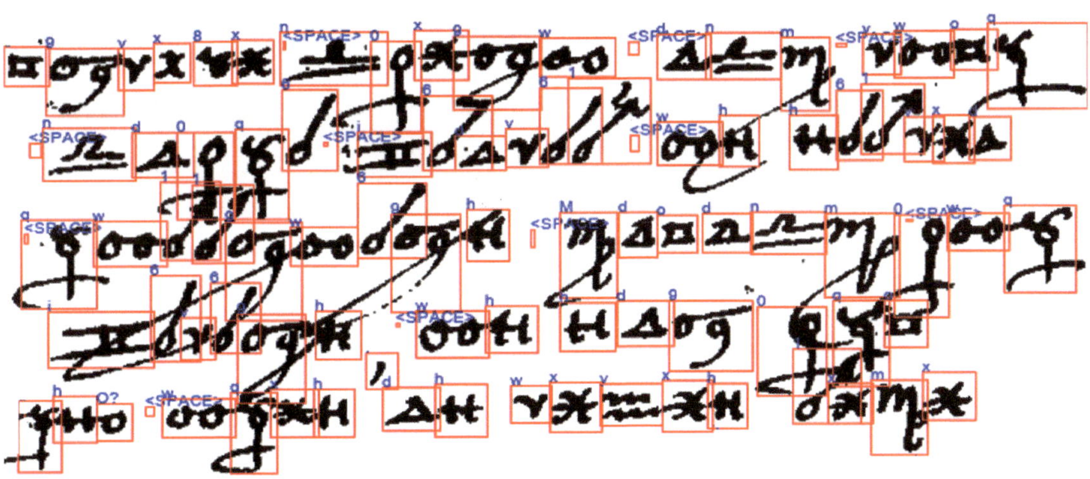

Figure 3.15 Example of symbol segmentation and transcription; segmentation shown in red, transcription in blue.

transcription. Most commercial OCR software only focuses on typewritten text, which means that these programs expect the same visual appearance for each character in the alphabet (e.g., every 'a' looks exactly the same, at pixel level). However, in the case of handwritten documents, the high variability of handwriting styles requires more sophisticated and flexible techniques.

Handwritten text recognition (HTR) methods [28] have been designed for this purpose, which tend to transcribe at line level, avoiding the segmentation into characters that is so prone to errors. Current HTR methods use deep *learning-based* architectures, such as long short-term memory recurrent neural networks (LMRNN), convolutional neural networks (CNN), sequence-to-sequence models (S2S), and transformer networks (TN) [29]. In these systems, the input is usually a text line and the output is the transcribed text. These deep learning-based methods have very good performance, but they require a lot of labeled data to train (more than 100 pages) to learn the shape or visual appearance of each character. But this need for providing examples of text images with their corresponding transcriptions can be a problem in the case of uncommon or unknown alphabets, such as the ones used in many historical encrypted documents. When there is few annotated data to train, the performance of deep learning models dramatically decreases.

For this reason, some researchers opt for learning-free transcription methods, such as learning-free spotting for cuneiform[2] [30] or unsupervised clustering for cipher alphabets like in [31], where the system segments symbols in the document and then groups them according to their visual appearance, using, for example, k-means clustering and label propagation. K-means clustering is an unsupervised method used in machine learning for grouping data into clusters (or groups). It consists in partitioning the elements into k clusters (or groups) so that each element belongs to the cluster with the nearest mean (cluster centers, or prototype of the cluster). Label propagation iteratively propagates the label of each cluster center or prototype through the rest of the nearest elements. The process finishes when all elements are assigned to a cluster, with a label. Then, each cluster corresponds to a particular symbol in the alphabet. Learning-free methods are very flexible and can be applied to any alphabet, but their performance is moderate compared to learning-based approaches, especially when alphabets contain very similar symbols or when characters are difficult to segment, as shown in Figure 3.15.

Lately, different strategies have been explored to deal with the lack of labeled data to train, including few-shot learning, semisupervised and self-supervised learning, transfer learning, and domain adaptation. Few-shot learning aims to mimic how humans learn novel concepts and adapt to unseen data. Concretely, few-shot learning can learn with limited data and the classes (i.e., alphabet symbols) for training and testing can differ. This is especially useful for recognizing manuscripts with rare scripts, unknown alphabets, or very different handwriting styles without retraining the whole model. Rare scripts are those alphabets that are not commonly used today (like Egyptian hieroglyphs, cuneiform, runes, or cipher alphabets). For example, a transcription method based on few-shot learning could

2. Cuneiform is a logosyllabic script used to write several languages of the ancient Near East (from around 3500 BC).

learn how to transcribe symbols from one alphabet, and then use this knowledge when transcribing symbols from an unseen new alphabet. Secondly, semi- and self-supervised learning aim to learn representations from few or no labeled data, which can transfer well to recognition tasks. These types of methods can also be combined with few-shot learning. For example, in [32] a few-shot learning method incrementally transcribes the symbols with a higher confidence rate (namely pseudolabels), assuming that their labels are correct, and uses these pseudolabels as training data for the next iterations, as shown in Figure 3.16. It must be noted that all these types of approaches require only a few annotated examples compared to standard deep learning methods, while reaching a performance only slightly below the typical deep learning-based ones.

3.4.4 The Future of Automatic Transcription

When comparing the manual transcription versus the automatic transcription, it is obvious that, in general, the use of automatic transcription methods are preferable because they minimize the human effort (see Section 3.4.1).

Automatic transcription decreases time-consumption significantly, especially for larger documents. However, for an automatic transcription, the user is required at the beginning to provide labeled data for learning-based methods, and at the end to validate the transcriptions and correct any possible errors. Besides, even though this manual postcorrection can be facilitated since the mistakes by automatic transcriptions are systematic, it requires time. For this reason, a manual transcription can be preferable for transcribing short manuscripts (a few pages). For anything else, the automatic transcription plus manual postcorrection is preferable: In this scenario, semi-interactive software tools are desired, so that the user can guide the automatic transcription (following the idea of *AI in the loop*), and benefit from intuitive graphical user interfaces for the postcorrection. The reader can find a deeper discussion about manual versus automatic transcription in [33].

The field of computer vision develops quickly, as do other branches of AI, and sooner or later we will have access to tools that not only can produce a reliable transcription but also decipher the encrypted manuscript in one step.

Next, we will turn to methods to analyze and decipher encrypted sources.

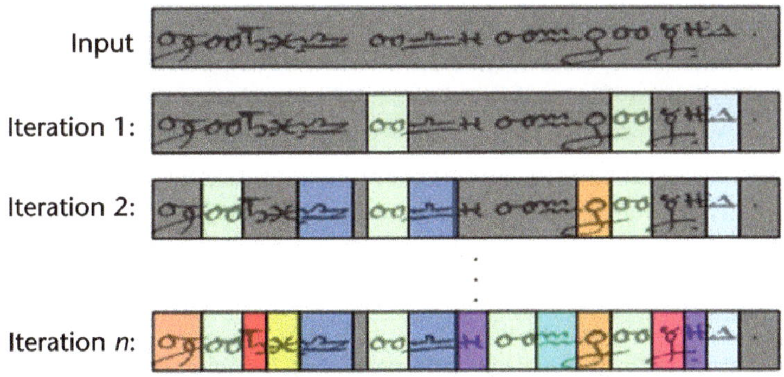

Figure 3.16 Example of incremental transcription by pseudolabeling. At each iteration, the method transcribes the symbols with higher confidence. Each color corresponds to one label. (*From*: [32]. Reprinted with permission.)

3.5 Cryptanalysis

Historical ciphers can be attacked automatically using a computer with heuristic methods like hill climbing. In the previous sections, we presented the different ways in which historical ciphers were built based on alphabet-code elements and nomenclature-code elements. While the alphabet-code elements can be recovered using properties of the original plaintext by methods such as counting frequencies of unigrams, bigrams, and trigrams, as these still show through the encryption, the nomenclature elements cannot really be recovered by automatic cryptanalysis. This is because nomenclature-code elements do not appear as regularly or as frequently as alphabet-code elements do. Nomenclature-code elements can be deciphered either by having access to the original key showing the corresponding plaintext element, or by linguistic and/or historical analysis through contextual interpretation. Contextual analysis (see Section 3.6) might involve the investigation of the surrounding words to reveal the linguistic type in terms of part-of-speech of the plaintext element (e.g., preposition, proper noun, common noun, verb), and/or historical analysis of the entire text to make educated guesses about probable certain persons or places mentioned in the underlying plaintext.

For cryptanalysis, the cipher type and the cipher alphabet used to encrypt the plaintext have to be determined. In the previous sections we showed that, for example, letters, graphic signs, digits, or a combination of them were used as alphabet-code elements. One recognizes only after the decipherment whether, for example, two symbols transcribed together into the same cluster (e.g., A and Ä, whereby one overlooked the points of the Ä) are actually two different symbols, that should have been transcribed differently.

While individual alphabet-code elements with graphic symbols and alphabet symbols are easily distinguishable (mostly, one symbol corresponds to one alphabet-code element), digit-based ciphers are often challenging to segment. Only a few digit-based ciphertexts have visible separations of the code elements (e.g., spaces, a comma, or a dot). Many ciphertexts use scriptio continua with a consecutive sequence of digits without any separation between them (see Section 3.1). Here, tokenization needs to be applied to cut the digit sequences into code elements and identify them, which is far from straightforward as the length of codes can vary within a single ciphertext (e.g., two-digit and three-digit code, or a combination of them).

In the subsequent sections, first we describe the tokenization of ciphertexts. Then, we present two algorithms using heuristics—namely hill climbing and simulated annealing—for the automatic recovery of alphabet-code elements from the transcribed text. Finally, we discuss cost and fitness functions as well as language models used during cryptanalysis.

3.5.1 Tokenization

Tokenization in the context of historical ciphers is defined as the separation of ciphertext into single code elements, be it alphabet or nomenclature codes. Tokenization can be straightforward if the code elements are clearly segmented from each other by separators like a space. Tokenizing a ciphertext that consists of graphic symbols (e.g., alchemical or zodiac symbols) is often also easy as each symbol being regarded as one token (i.e., one alphabet-code element). However, the tokenization of graphic ciphers sometimes has to be refined or corrected during the cryptanalysis because the creator of a transcription of a ciphertext falsely regarded two symbols as one token.

In contrast, tokenizing digit-based ciphers that are written in a continuous script (scriptio continua), without segmentation between the code elements, is challenging. So far, no solution has been found that allows the generally automated tokenization of such ciphertexts. At the time of writing, tokenizers need to be developed and adapted to individual ciphertexts.

Before attempting to develop a new tokenizer, we can start by applying the most trivial one—tokenizing the ciphertext into two-digit alphabet-code elements, which occur commonly in early modern ciphers. We can also apply already existing tokenizers developed for particular sets of ciphers originating from the same source to new ciphertext of the same collection, such as the papal ciphers from the Vatican or diplomatic correspondence between two sources. If the abovementioned alternatives do not lead to a correctly tokenized ciphertext, a new tokenizer has to be developed. To do so, the ciphers and the corresponding ciphertexts have to be statistically analyzed to find a set of rules the tokenizer is based on. Counting and analyzing unigram, bigram, and trigram frequencies of single digits, two-digit codes, three-digit-codes, and so forth are normally performed. Analysis contains to discover various structures in the code system. For example, if we see that the digit "2" is always in front of an odd digit, it may indicate that the combinations "21," "23," "25," "27," and "29" are valid tokens and may represent alphabet-code elements. In the end, one has to manually look for such peculiarities in the frequencies. The tokenizer can then be applied to the ciphertext and its output be run by the cryptanalysis algorithm(s) of choice (e.g., CT2) to recover the key. If cryptanalysis fails, the tokenizer is probably incorrect and needs adjustment. In the end, the process of tokenization of the ciphertext and the development of a valid tokenizer is a trial-and-error but inevitable process for successful cryptanalysis.

3.5.2 Heuristic Algorithms for Cryptanalysis

A basic flaw (and our advantage) of all simple and homophonic substitution ciphers is the fact that a partially correct key may already allow us to read the content of an encrypted text. Also, text frequencies of the original plaintext may be still visible in the encrypted text. For example the most frequent ciphertext letter in a simple substitution cipher or the most frequent homophone in a homophonic substitution cipher most likely encrypts the most frequent letter. In case of the English language, this would be the letter "E." Both these properties—the ability to have partial correct keys and the appearance of plaintext frequencies in the ciphertext—allow using heuristic algorithms to incrementally solve such ciphertexts. In the following, the two most used and most successful algorithms to break these ciphers are presented. Even though we focus here on the aforementioned two types of substitution ciphers, the algorithms shown can be applied to many other pen-and-paper ciphers as well as to rotor encryption machines. The heuristic algorithms have to be adapted specifically for each cipher.

3.5.2.1 Hill Climbing

The main goal of a hill-climbing algorithm is to find a solution of a search problem that cannot be solved by means of an exhaustive search (i.e., brute force). For example, a simple substitution cipher with 26 ciphertext letters has a total key space (search space) with 26! elements, which are about $2^{88.4} \approx 4 * 10^{26}$ = four hundred octillion different keys. Finding the correct key by testing all possible keys to decrypt the ciphertext is impossible in practice. With hill climbing, the search is possible in practice, but in some cases, for example, very short ciphertexts or poorly transcribed ciphertexts, it might not find the correct key. However, luckily the vast majority of simple monoalphabetically encrypted ciphertexts can be deciphered easily.

The basic hill-climbing algorithm for finding the correct key k_c of a ciphertext ct encrypted with the simple substitution cipher consists of five steps:

1. Select a randomly chosen start key k
2. Decrypt the ciphertext ct to get $pt := \text{decrypt}(ct, k)$
3. Compute the cost value of pt with $f := \text{cost}(pt)$
4. Loop while a defined termination criteria is not met:
   a. Generate a new key k' which is a slightly modified k
   b. Decrypt the ciphertext ct to get $pt' := \text{decrypt}(ct, k')$
   c. Compute the cost value of pt' with $f' := \text{cost}(pt')$
   d. if $f' > f$ then assign $f := f'$ and assign $k := k'$
5. Output the key k (which most likely is the correct key k_c)

The five steps of the hill-climbing algorithm can be clustered into two parts: The first part is the initialization, which is steps (1) to (3). It first generates a random start key and rates its "cost" using a cost (or fitness) function. The higher the cost value, the closer the decrypted plaintext is to real text. In the second part, the algorithm incrementally improves the key. To do so, it generates in step (4a) a slightly modified key, which it then rates in step (4c) using the same cost function as in the initialization part. When the cost value is higher than the previous one it keeps the new cost value as well as the new key. The algorithm loops as long as a defined termination criterion is not met. Finally, in step (5) it outputs the key k, which is with high probability the correct key k_c.

The algorithm can be visualized in a two-dimensional graph as shown in Figure 3.17. Here, the keys are drawn at the x-axis, and the corresponding cost values at the y-axis. The hill-climbing algorithm follows the cost function to find the global maximum (= the key k_c). The figure shows a potential problem of the hill-climbing algorithm, namely local maxima where the algorithm might get stuck (sitting stick figure). Later in this section, we will discuss how to mitigate the effects of local maxima on the success rate of cryptanalysis. Also, keep in mind that while the algorithm can be nicely drawn in a two-dimensional manner, the real problem is a multidimensional problem with, for example, 26 dimensions in the case of the simple substitution cipher with a 26-letter alphabet.

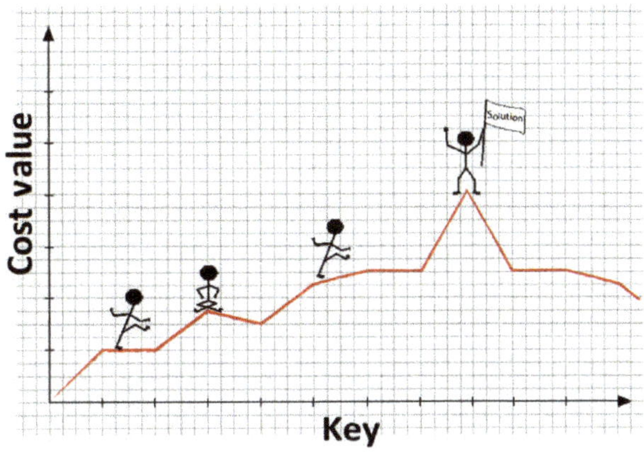

Figure 3.17 A visualization of the hill-climbing algorithm.

In the following, we discuss different aspects and design ideas of the hill-climbing algorithm to break simple substitution ciphers.

Decrypt function and key representation. For the simple substitution cipher, our decryption function requires both the ciphertext and a key as input. The key is represented by a string or array of characters with the same length as the plaintext alphabet. For example, the key "WDNBZCJHOKQRPEISFTUGVXYALM" means that the "W" is decrypted to "A," the "D" is decrypted to "B," ..., and the "M" is decrypted to "Z." The actual decryption is performed by walking letter by letter through the ciphertext and replacing the ciphertext letters with plaintext letters as described before.

Start key. The generation of the start key can be crucial for the success of a hill-climbing algorithm. For some ciphers, a "good" start key is needed to allow the algorithm to converge to the correct solution. With the simple substitution cipher, the start key can just be chosen at random. To do so, we take the alphabet of the assumed plaintext language (e.g., the Latin 26-letter alphabet for the English language) and create a key by shuffling it:

ABCDEFGHIJKLMNOPQRSTUVWXYZ → WDNBZCJHOKQRPEISFTUGVXYALM

With historical encrypted manuscripts, the used alphabet can differ from the alphabet we use today. Some letters may be represented by the same single letter (e.g., "I"="J" and "U"="V"). This depends on the plaintext language and the time of the creation of the manuscript. Sometimes, letters may be intentionally omitted for security purposes, such as by writing a single "L" instead of "LL" or "VV" instead of "W." Sometimes, the alphabets are extended, for example, by adding a symbol for double letters ("LL"), "SCH," or letters with diacritics ("á"). This all has to be taken into account when generating an alphabet and keys with an automated heuristic-based analyzer.

Cost function. The cost or fitness function evaluates the quality (cost or fitness value) of a supposedly decrypted plaintext. Depending on the problem (the cipher), a special cost function may have to be implemented. For the simple substitution cipher, a language model (n-gram statistics; n being between 3 and 5) is used. In Section 3.5.3 we discuss cost and fitness functions in more detail.

Key modification. The next important part of the hill-climbing algorithm is how to modify the key k to obtain a new key k' during a single iteration. Only a small change in the key allows the algorithm to smoothly follow the curvature of the graph of the cost function to potentially reach its global maximum.

Figure 3.18 shows how the key k is modified to create a new key k' during hill climbing by swapping only two letters (here "C" and "F") at the same time. There are different strategies how to choose which two letters should be swapped:

1. *Perform a single random swap:* In every iteration of the hill-climbing algorithm, use two random indices i and j with $i \neq j$. The two letters at position i and j are swapped. Clearly, only "good" swaps are kept and "bad" swaps are discarded.
2. *Take only the "best" swap:* In every iteration of the hill-climbing algorithm, all indices i and j with $i \neq j$ are tested. The "best" swap of all possible letter swaps is kept. The "best" swap of all possible swaps is the "good" swap, that increases the cost value the most.
3. *Take all "good" swaps:* In every iteration of the hill-climbing algorithm, all indices i and j with $i \neq j$ are tested. Every time a "good" swap occurs, the swap is kept. This means, that during a test of all indices i and j in an iteration, multiple consecutive "good" swaps may occur.

The classical hill-climbing algorithm as described in the literature uses random swaps of two letters—the strategy (1) above. While this works well in most cases, the two other strategies may improve the success rate as well and reduce the computational time needed by the cryptanalysis algorithm. With strategy (2), we test all possible swaps and only take the "best" possible "good" swap. A "good" swap increases the current best cost value while a "bad" swap leads to the same or even a worse cost value. With a 26-letter alphabet, there are $\frac{26 \cdot 25}{2} = 325$ different swaps that need to be tested in every iteration. Clearly, this slows down the algorithm and increases the needed computation time in the worst case by a factor of 325. To mitigate the effect of testing all possible two-letter-swaps, there is strategy (3) that allows already to keep a "good" swap while all remaining swaps still need to be tested. For example, the Vigenère analyzer component of CT2 uses strategy (3), which allows solving really short Vigenère ciphertexts with high success rate in very short times [34].

Termination criteria. In theory, a hill-climbing algorithm should terminate, when it reaches the global maximum (success) or it got stuck in a local maximum (fail).

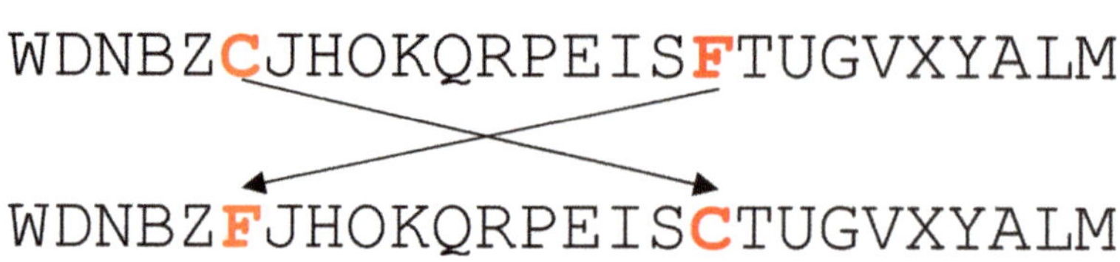

Figure 3.18 Swapping two letters of key k to obtain a modified key k'.

Depending on the selected key-modification strategy, it is possible to detect if the algorithm got stuck or not. For example, with random swaps, it is possible that it by chance never selects a new swap that allows us to increase the cost value, despite there exists another "good" swap. Thus, a suitable termination criterion for random swaps is to count the number of consecutive randomly chosen "bad" swaps and then terminate when a specific number of "bad" consecutively chosen swaps is met. With the two other strategies, (2) and (3), we can actually find out if the algorithm got stuck because in every iteration all possible swaps are tested. If all of these swaps are "bad" swaps, the algorithm terminates.

Strategies to counter getting stuck. There are different strategies to counter getting stuck with hill climbing in a local maximum:

1. *Better start keys.* With some ciphers, it is possible to already generate "good" start keys that are close to the global maximum. In the case of the simple substitution cipher, this is not needed, since any randomly created start key can be used and will lead to the correct solution in nearly all cases. In contrast, with homophonic substitution ciphers, a good start key improves the success rate and performance of the algorithm. We describe this later in Section 3.5.2.2.
2. *Better key modification(s).* For example, instead of swapping only two elements of the key at the same time, one could perform a triple swap, where element i becomes j, j becomes k, and k becomes i while $i \neq j \neq k$. With the simple substitution cipher and with the homophonic substitution cipher, swapping only two letters at the same time is good enough.
3. *Better cost function.* When hill climbing does not find the correct key, it is probably a good idea to change the cost function. For example, instead of using n-gram models with $n = 2$, we could increase the dimension of the language model to $n = 3$. With simple and homophonic substitution ciphers, $n = 5$ works very well. Sometimes, it can also be useful to change to a lower n, especially with bad transcriptions or many errors in the ciphertext. See Section 3.5.3.
4. *Shotgun hill climbing/random restarts.* Another idea of improving the algorithm is to restart it several times (e.g., 100 times) with different randomly chosen start keys. This is also referred to as shotgun hill climbing, since the start keys are distributed over the key space like shotgun shrapnels. With the simple substitution cipher, this strategy is very effective.
5. *Use of simulated annealing.* This algorithm is an alternative to hill climbing. See Section 3.5.2.2.

When working on a historical ciphertext, all the aforementioned improvements have usually to be tested individually. For example, evaluations with different key modifications and cost functions have to be performed to test the impact of the changes on the cryptanalysis success rates. For CT2, the implemented cryptanalysis algorithms were tested and tweaked with millions of artificially generated test records until sufficient success rates were achieved. Additionally, all CT2 cryptanalysis components allow exchanging the language model or set different parameters in the corresponding components' settings. A few examples of such CT2 components are the substitution analyzer, the Vigenère analyzer, the homophonic substitution analyzer [35], and the Enigma analyzer.

Figure 3.19 shows a screenshot of the CT2 homophonic substitution analyzer[3] solving an encrypted letter written by Holy Roman Emperor Maximilian II and sent to Polish delegates in 1575. The upper part of the analyzer has some helpful information about the currently analyzed ciphertext, such as the number of used homophones. The large middle part shows the analyzed ciphertext. The lower part shows the currently revealed plaintext. Green marked symbols are already locked, meaning they won't change any more during the ongoing cryptanalysis. Blue marked symbols show German words found in a predefined dictionary. A CT2 user can stop the automatic analysis process at any time and manually change and improve plaintext-ciphertext symbol-mappings on his own.

3.5.2.2 Simulated Annealing

Simulated annealing is a generalization of hill climbing: The basic idea is that with a defined probability modifications of the key are also chosen, which lead to a bad key, which means the cost value may decrease in an iteration. Over time, the probability for selecting a bad key is reduced until it reaches zero. Then, simulated annealing behaves exactly the same way hill climbing does.

The simulated annealing heuristic is inspired by the physical annealing in metallurgy. Here, annealing is a slow process of heat treatment of metals to alter the

Figure 3.19 The CT2 homophonic substitution analyzer solving an encrypted letter from Maximilian II.

3. In `CT2 Startcenter` ▷ `Templates` ▷ `Cryptanalysis` ▷ `Classical` ▷ `Homophonic Substitution Analysis`. In CTO, a similar homophonic analyzer can be found.

physical properties of the material. While in physical annealing, the real temperature is slowly decreased; with simulated annealing a virtual temperature value is used. The basic simulated-annealing algorithm consists of six steps:

1. Select a randomly chosen start key k
2. Set the temperature to a start value $t := t_{\text{start}}$
3. Decrypt the ciphertext ct to get $pt := \text{decrypt}(ct, k)$
4. Compute the cost value of pt with $f := \text{cost}(pt)$
5. Loop `while` $t > 0$
    a. Generate a new key k', which is a slightly modified k
    b. Decrypt the ciphertext ct to get $pt' := \text{decrypt}(ct, k')$
    c. Compute the cost value of pt' with $f' := \text{cost}(pt')$
    d. If $f' \geq f$ then assign $f := f'$ and assign $k := k'$ `else`
        - Compute a degradation value $d := -\text{abs}(f - f')$
        - Compute an acceptance probability $p = e^{\frac{d}{t}}$
        - Choose a random value r in the interval $]0;1[$
        - If $p > p_{\text{min}}$ and $r < p$ then assign $f := f'$ and assign $k := k'$
    e. Decrease temperature, for example, by using a defined step size ss to get $t := t - ss$
6. Output the key k (which most likely is the correct key k_c)

In step (2) a start temperature is set. The start temperature, among other new values needed for simulated annealing, has to be tweaked for each type of cipher and often also for each individual ciphertext, which you want to cryptanalyze. The termination criterion in step (5) now checks if the temperature t is still higher than 0. Inside the main loop of the algorithm, when a key k' is not accepted in step (5d), a probability p based on the degradation value is computed and a random value r is chosen. If r is smaller than the computed probability and the computed probability is greater than a minimum probability p_{min}, the bad key is kept. In practice, we set the minimum probability to $p_{min} = 0.85\%$, which gave us good results. This allows the simulated-annealing algorithm to jump away from local maxima. While the algorithm is being executed, the temperature value t is reduced by a step size ss. The value of s is predefined and can be determined, for example, by dividing the start temperature by the number of wanted steps s, and then the algorithm should perform. So $ss := \frac{t_{\text{start}}}{s}$. Other temperature reduction strategies are also possible. For example, instead of reducing the temperature by the same value ss all the time, it could also be reduced by a percentage value of t with $t := t - 0.01 \cdot t$. The different strategies have to be evaluated to find the best one for the specific case. Figure 3.20 shows a simulation of the key acceptance probability of simulated annealing over time with a fixed temperature step size and Figure 3.21 shows a simulation of simulated annealing with a percentage-based temperature step size.

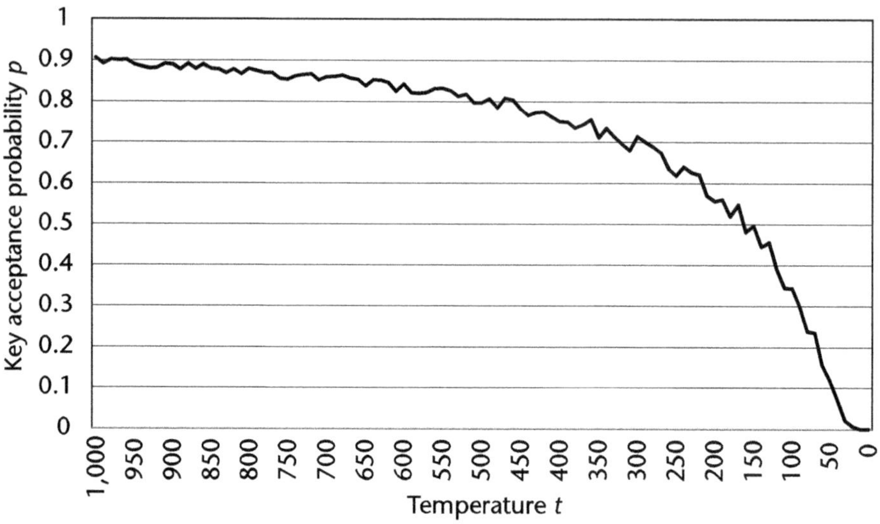

Figure 3.20 Key acceptance probability of simulated annealing with linear decreased temperature over time.

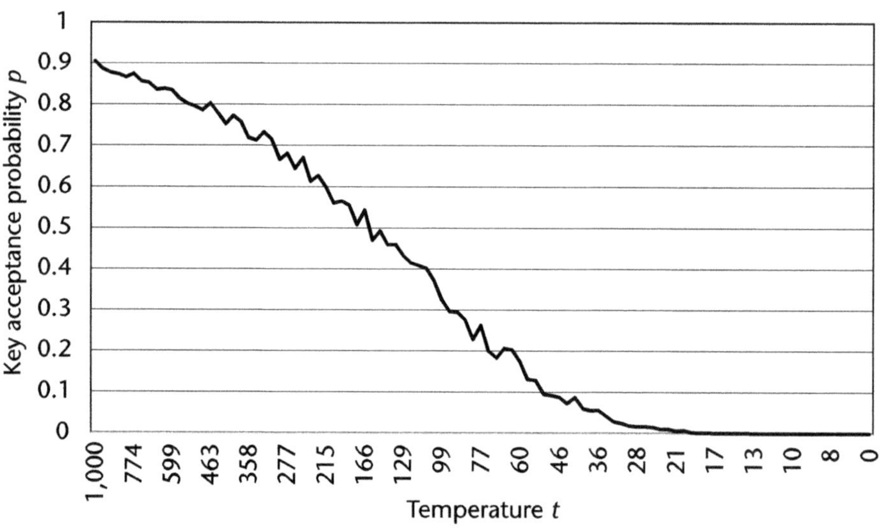

Figure 3.21 Key acceptance probability of simulated annealing with percentage decreased temperature over time.

Improving simulated annealing for homophonic substitution ciphers. During the cryptanalysis of the homophonic substitution cipher, plaintext letters from the plaintext alphabet are assigned to all homophones and the ciphertext is decrypted for testing. During a single iteration of the simulated-annealing algorithm, we swap the assignments of two plaintext letters. As with the simple substitution, we test all possible two-letter swaps of all homophones.

In the following, we present some adaptions and strategies to be applied to the simulated-annealing algorithm to improve its performance, especially for the cryptanalysis of homophonic substitution ciphers.

1. *Good start keys.* With the homophonic substitution cipher, it is helpful when the start keys for the cryptanalysis algorithm are already chosen in a

way that reflects the distribution of letter frequencies of the language. For example, it is better to assign more homophones to more frequent plaintext letters (e.g., the "E" with English) than to less frequent letters (e.g., the "X" with English). Therefore, the Homophonic Substitution Analyzer of CT2 allows distributing the letters among the homophones based on probabilities that are based on the original text frequencies of the language.

2. *Homophone locking (manual).* When analyzing homophonic substitution ciphers, it may improve the cryptanalysis if already correctly assigned letters can be fixed by the user. The Homophonic Substitution Analyzer of CT2 allows this in the semiautomatic mode. Here, the user may pause the analysis and lock homophones, meaning the corresponding assignment of plaintext letters to the homophones cannot be changed anymore by the cryptanalysis algorithm during the further iterations. Also, the user may change and correct the already made assignments.
3. *Homophone locking (automatic with a dictionary).* Besides manually locking homophones as described above, it is possible to automatically lock homophones based on words found in a dictionary. Therefore, the Homophonic Substitution Analyzer of CT2 provides a dictionary to the cryptanalysis algorithm. Every time a new global best value (best key) is found, the analyzer searches for words with a minimum and maximum length. If it finds more words than a specified threshold value, it automatically locks all corresponding homophones to their corresponding plaintext letters. This can also be combined with the manual method for homophone locking described in the second adaption.

3.5.3 Cost Functions

While optimizing a key k with hill climbing or simulated annealing, the algorithm needs a way to decide if a modified key k' is better or worse than the original key k. To rate a key, we use cost or fitness functions on the text previously decrypted with the key k'.

The basic idea of a cost function $cost(t)$ is that it calculates a number that reflects how natural a given text t is. The closer the text is to a real text, the higher the cost value should be. The more random (not natural) a text is, the lower the cost value should be. In the best case, the cost function returns the highest value when we enter the original plaintext. Between the lowest and the highest value, there should be a smooth curve that the cryptanalysis algorithm can follow during the optimization of the key.

A common practice is to use a language model built from a large text corpus. For historical ciphers, it also turned out that the cryptanalysis algorithm can benefit from using a language model based on a historical text corpus [36]. A language model returns the probability of a given text being a text of the language it was built for.

The language models used in our cost functions are n-gram models. Such an n-gram model provides a value (probability) for a given n-gram. Clearly, frequent n-grams of the language, such as "ING" in English, return a higher n-gram value than less frequent n-grams, such as "XYZ." An overview of English and German language frequencies can be found in CrypTool-Online[4] and a set including different other language n-grams can be found on Practical Cryptography.[5]

We created different language models by using large corpora of text. To create a model, we first count the number of occurrences of all individual n-grams (e.g., from "AAA" to "ZZZ" for a 3-gram model) of the set. Also, we count the total number of all n-grams of the corpus. Then, for each individual n-gram, we divide its number by the number of all n-grams to obtain its probability. To compute the cost value of a given text, we could multiply all the computed values of all n-grams of that particular text to obtain a probability (the cost value) of the text. Here, we have two problems: (1) the probability values of each n-gram are very small numbers, which will result in many precision errors when multiplying these numbers on a computer, and (2) multiplications can be costly, so the performance of the computation may be poor. On modern PCs, problem 2 is negligible, but problem 1 is a huge problem. A common way to get rid of both problems is the usage of logarithmic values. Instead of multiplying all small values of the language model, we add the logarithms of each value. This is possible due to the logarithm law $\log_b(x \cdot y) = \log_b(x) + \log_b(y)$. In the end, to obtain the final value, we could raise the used base b to the power of the sum c, meaning b^c. But this is not needed since the optimization algorithm can also run on the logarithmic values. In CT2, the cost values are normalized to double precision floating point values in the interval of $[0 : 10000000]$. By doing so, the CT2 language models are comparable to each other.

A final note on the data format of language models: During cryptanalysis, the letters are mapped into an integer number space based on the used alphabet. For example, with the 26-letter Latin alphabet, the letter "A" is represented by 0, the letter "B" by 1, ..., and the letter "Z" is represented by 25. A language model is an n-dimensional array. To look up, for example, the 3-gram "ABC," which is encoded as integers $0, 1, 2$, we can just look up the language model array using the integers as indices. Doing the encoding of letters this way is easy and fast.

The CT2 language model files have a specific binary file format:

```
Header:
    "CTLM"            4 ASCII characters           (magic number)
    LanguageCode      0-terminated UTF-8 string    (language code)
    GramLength        4 byte integer               (length of n-grams)
    Alphabet          0-terminated UTF-8 string    (alphabet)
Data:
    (Alphabet.Length ^ GramLength) * 4 bytes       (model data)
```

A language model file starts with the four ASCII characters "CTLM" (CrypTool Language Model) to identify the file type. The "LanguageCode" string identifies the language model. The "GramLength" defines the size of the n-gram model. The "Alphabet" defines the used alphabet. In the data section, the actual language model data is stored as 4-byte float values containing the logarithmic values computed using a text corpus. The sizes of the n-gram models increase quickly with n, so the models are compressed using the gzip algorithm. For the English language with 26 characters the file

4. See `https://www.cryptool.org/en/cto/frequency-analysis`.

5. See `http://practicalcryptography.com/cryptanalysis/letter-frequencies-various-languages/`.

sizes on disc are (rounded): 1-gram: 1 kB, 2-gram: 3 kB, 3-gram: 50 kB, 4-gram: 800 kB, 5-gram: 8500 kB. Decompressed in RAM (rounded on 1 kB): 1-gram: 1 kB, 2-gram: 3 kB, 3-gram: 71 kB, 4-gram: 1828 kB, 5-gram: 47526 kB. One observation here is that the more data (texts) are used to create these language models, the smaller the amount of file size reduction achieved by compressing the models. The reason for this is that the increase in entropy (aka amount of information) of the data used leads to lower compressibility.

3.6 Contextualization and Interpretation: Historical and Philological Analysis

Once we have managed to reveal (parts of) the plaintext, we aim to set the manuscript in a historical context to recover what was written, by whom, to whom, and why. Such a contextualization concerns historical and philological interpretation, which will be the topic of this section. These approaches involve a broader type of analysis than cryptanalysis described above, because they do not primarily restitute the message, but rather investigate the linguistic and historical context in which the message was written, encrypted, and sent. *Linguistic* analysis involves the contextualization of the given ciphertext into the contemporary language usage, which presupposes that we have sufficient knowledge about how languages were used in the given time period and geographical area. *Historical* analyses do not only involves the identification of the sender and receiver (and perhaps the code-breaker) of the ciphertext, and the political context, but also the transfer of knowledge in the field of cryptology, as well as the social history of those who applied this technology of secrecy.

3.6.1 Analysis of Historical Languages (Linguistic Analysis)

Historical languages pose some specific challenges to the cryptanalyst. One important aspect is that most languages show a great deal of variation before they were standardized sometime in the eighteenth century. This means, for instance, that one and the same word could be written in many different ways (i.e., orthography was not normalized and even the same scribe could use various spellings for the same word in one text [37]). Moreover, in languages such as English, German, or Italian, we find a lot of different dialectal forms in the same language. Languages also change over time, certain words or word forms disappear, new ones emerge. The pilot study [36] on the decipherment of German and English historical homophonic substitution ciphertexts showed that using 4-gram models derived from century-specific texts leads to significantly better performance than language models built on more modern, contemporary texts for ciphertexts produced in the 17th century or earlier. A corpus of historical texts such as a digital library of online texts like the Project Gutenberg or the collection of historical texts with diplomatic transcriptions for 16 European languages available within the HistCorp collection [38] can serve well as a basis for the creation of language models.

Another general aspect to bear in mind in the use of algorithms for cryptanalysis is that in the plaintext alphabet a historical cipher is based on might differ from modern alphabets in specific languages: In many cases, only one letter is used for both *u* and *v*, for instance, and usually, letters with diacritics (such as *ä, ö, ü* in German; or *á, é, í, ő*, etc. in Hungarian) do not form part of plaintext alphabets. At the same time, plaintext alphabets also might merge commonly co-occurring alphabet letters and treat these as one plaintext element, such as *ss* or *sch* in the Copiale [6] cipher with German as its plaintext language.

In historical ciphertexts, especially in the domain of diplomacy and military correspondence, often more than one language was used [12]. Several languages, such as German and Latin, could be combined in one and the same sentence, as was the case in a letter written by a Lithuanian nobleman to the Habsburg Emperor Maximilian II in 1574 [39]. Initially, this fact caused problems in the decipherment process because the analysis was based on a monolingual German language model and the switching was not detected. Only afterwards, in a closer linguistic analysis, the change of language was identified.

It is also possible that different languages were used for passages in cleartext and passages in ciphertext [40–42], or that the plaintext language used in the key and the language of the plaintext of the encrypted letter are not the same. For example, a Swedish envoy based in Germany during the Thirty Years' War used a German key in his correspondence with the Swedish Lord High Chancellor. However, the underlying plaintext in his letters is in Swedish and Latin [43]. Hence, even when the language of cleartext passages or of a key is identified, other languages may still be encountered in the ciphertext.

These examples show that the linguistic analysis of ciphertexts can form part of the process of cryptanalysis and functions as an auxiliary method to solve a cipher and to reveal information about the underlying language, the provenance, and the dating of a ciphertext. In fact, already in the Middle Ages, codebreakers used linguistic analysis in cryptanalysis: Arabic scholars realized that there is a certain frequency distribution of letters in different languages—a tool they used to decipher monoalphabetic substitution ciphers [44, 45]. Linguistic knowledge also helps to detect transcription errors and to resolve certain decipherment problems. Finally, knowledge in historical languages is often needed to fully understand the content of the deciphered documents.

On the other hand, linguistic analysis can serve its own purpose and be aimed at understanding linguistic patterns and language practices in historical cryptographic texts. Examples for this research path are, for instance, studies on what languages were chosen in ciphers in different geographical areas and different times or which and how different languages were combined in documents [42]. Further, the linguistic analysis of a recovered plaintext can complement the historical analysis and contribute to the understanding of scribal practices and language usage at chanceries and black chambers. Historical ciphertexts can also serve as sources for the analysis of written dialects and languages, and language change.

The linguistic analysis can be fully or partly automatized by algorithms developed within computational linguistics and natural language processing. Spelling variation in historical texts can be automatically discovered and normalized to a modern version, cleartext sequences can be detected and its language(s) identified by applying automatic language identification. The computational analysis of language heavily relies on language models derived from large samples of diplomatic transcriptions of historical texts from various time periods and genres. Such collections are not easy to find and their creation requires linguistic and philological expertise.

3.6.2 Historical Analysis and Different Research Approaches

Similar to the linguistic analysis, historical analysis in historical cryptology plays a double role: It might be the *goal* of the whole procedure described above, or alternatively, it might also be a *tool* used in the process. It is the goal when the historian aims to reconstruct certain past events and study a particular historical context. Solving the ciphers, pairing the keys and the messages, and exploring the ways cryptography was used help her in this task. In other cases, however, it is rather a tool: Most homophonic cipher keys consist of an alphabet part and a nomenclature table. One needs mathematical and linguistic knowledge to analyze the alphabet, but reconstructing the nomenclature table requires a deep knowledge of the historical context. In this second type of case, history is an auxiliary science of the crypto-historian.

In the following, we provide a—by no means exhaustive—typology of the different (sometimes contradictory, sometimes complementary) approaches when historical analysis comes to the picture, and we exemplify each approach with a corresponding publication.

1. One typical research path aims at getting new, previously unknown knowledge by *solving a given encrypted source*. This approach enriches our picture of a particular historical period and becomes useful for traditional history writing, but the emphasis is more on cryptanalysis, the solution of a riddle [46].

2. A second typical research path follows the agenda of *political history*. Ciphers were primarily used in diplomacy. The analysis of the correspondences of political centers with their ambassadors, messengers, and spies can provide new insight into the history of a given era even if the exchanged letters had always been readable because the historical addressee wrote the solution above the ciphertext characters. Examples for this category include studies on diplomatic history [47–49], analyses on the earliest black chambers, such as codebreaking offices [50], and the reconstruction of particular encryption practices (polyphonic and fixed length ciphers) used in the 16th century in the Vatican [51].

3. It is not the aim but the scope of the *microhistory approach* that makes it different from the previous ones. In this case, a temporarily limited series of events (a few years or a few exchanged letters) is analyzed with a variety of tools in order to have better insight into one particular historical event, such as the study on encrypted letters sent by and to the Habsburg Emperor Maximilian II in 1574–1575 [39, 52, 53].

4. The previous approach might be enriched with a *linguistic analysis* of the sources, as described in the previous section. The two fields have always been close: study of languages and cryptology have walked hand in hand from the earliest times.

5. An opposite approach is followed by those who perform *large-scale statistical analyses* of cipher keys and/or encrypted documents. The emphasis is not on particular sources but on conclusions, tendencies, and correlations that can be pointed out on the basis of relatively big data. An example for this approach are the studies on the typology and change of early modern cipher key documents [9, 54, 55].

6. Cryptology is both a technology and a scientific endeavor neighboring mathematics; thus, it is a genuine topic for a *history of science* approach. Basic issues include knowledge transfer (the ways this secretive knowledge is transferred from one generation to another, from one political center to another), the relations of cryptology to other scientific fields (statistics, algebra, poetics, etc.), its technology use, and the evolution of encrypting and codebreaking practices over time [19].

7. A separate category is populated by articles and book-length studies on specific *famous ciphers*, solved or unsolved, such as the Voynich manuscript [56, 57], the Copiale manuscript [6], the Borg cipher [4], or the Beale ciphers [58].

8. Sometimes it is not the ciphers and keys but the social background of the users that is under study. A *social history of cryptology* relies on the same sources but attempts to answer different questions: Who are the human actors of crypto-history, what are their attitudes to the technology they are using, what do they wish to keep as a secret, and so forth [19].

9. And, finally, further approaches are conceivable and can be exemplified by the continuously growing number of publications, including studies on personal diaries, private ciphers, and so forth.

3.7 Conclusion

Historical cryptology is a cross-disciplinary scientific field aiming at the systematic study of historical encrypted sources: ciphertexts, cipher keys, and related documents. The aim is not only to shed light on the content behind the encrypted sources by breaking their code, but also to study the evolution of cryptography and cryptanalysis over time periods and geographic areas.

As with all scientific disciplines, historical cryptology is in need of research infrastructure including resources and tools for the automatic processing of the encrypted documents. In this chapter, we presented several databases containing smaller or larger collections of historical ciphertexts and cipher keys, with the largest—at the time of writing—being the DECODE database [59]. The collections make it possible to study the evolution of cipher keys over time and to identify the most commonly occurring cipher types. We presented the structure and the peculiarities of three commonly occurring cipher types in early modern times in Europe: simple, homophonic, and polyphonic substitution ciphers, all monoalphabetic with or without nomenclatures. Surprisingly, transposition and polyalphabetic ciphers were used very rarely in Europe in these centuries, even though the cryptographic techniques were known. In contrast, in the U.S. Civil War from 1861 to 1865 the Vigenère cipher was used by the Confederates [60].

To break the historical ciphertexts, we introduced a set of tools for both transcription (to turn the images into a machine-readable text format) and for cryptanalysis (to decrypt the ciphertext). We presented transcription guidelines for the consistent transcription of symbol sets across ciphertexts and described the challenges and pitfalls of manual transcription. We then introduced how current handwritten text recognition techniques developed in computer vision are applied to ease the time-consuming and expensive transcription process. Given the ciphertext in text format, we described algorithms for cipher-type identification, cryptanalysis, and decipherment for the most commonly occurring European historical ciphers. We pointed out the importance of language models and various heuristics for the generation of cipher keys. Lastly, we gave an overview of the linguistic and historical interpretation of encrypted sources and the great challenge of their contextualization.

The latest and rapid development in AI provides us with efficient algorithms and models. It's challenging how AI can be efficiently used to produce an error-free and complete transcription to minimize error propagation to the subsequent step of codebreaking, to identify the cipher type used for producing a given ciphertext, and even to get the original message by breaking the cipher. Another future extension could be the selection and analysis of non-European ciphertexts, especially with languages not using a Latin-based alphabet.

The field of historical cryptology requires expertise from various scientific disciplines in order to collect, describe, transcribe, break, and analyze historical encrypted manuscripts. Historians contribute to the contextualization and interpretation of the hidden sources and linguists analyze the historical plaintext by acquiring models for language variation and language change. Cryptanalysts develop efficient algorithms for breaking of various cipher types, and image processing specialists provide models to process images to a machine-readable format. Computational linguists build and evaluate historical language models generated from historical texts. By close cooperation a hidden class of sources, encrypted to hide the content of importance in the past, can be systematically handled and made available to the public.

The interested reader can find scientific articles on the topic in publication channels of various disciplines from history, linguistics, natural language processing, and digital humanities to image processing and cryptology. The most well-known scientific publication sources for historical cryptology are, however, the proceedings of the annual International Conference on Historical Cryptology (HistoCrypt) [61] and the journal *Cryptologia* [62]. The community of historical cryptology has also a network called HICRYPT that can be reached through the email address hicrypt@ling.su.se.

The work of this chapter was supported by the Swedish Research Council, grant 2018-06074, DECRYPT – Decryption of Historical Manuscripts https://de-crypt.org/.

References

[1] Friedman, W. F. D., and L. Callimahos, "Military Cryptanalytics, Part I," National Security Agency, United States Government, Washington, DC, 1959 (available through Aegean Park Press, Laguna Hills, CA).

[2] Schmeh, K., *Revisited: A Terminology for Codes and Nomenclators*, 2018, https://scienceblogs.de/klausis-krypto-kolumne/2018/10/07/revisited-a-terminology-for-codes-and-nomenclators/.

[3] Mikhalev, V., et al., "What is the Code for the Code? Historical Cryptology Terminology," in *Proceedings of the 6th International Conference on Historical Cryptology*, 2023, pp. 130–138, https://ecp.ep.liu.se/index.php/histocrypt/article/view/702.

[4] Aldarrab, N., Kevin Knight, and Beáta Megyesi, *The Borg Cipher*, https://cl.lingfil.uu.se/~bea/borg.

[5] Cipher ID-3816,reproduced image from the Swedish National Archive Riksarkivet 1637, https://de-crypt.org/decrypt-web/RecordsView/189.

[6] Knight, K., B. Megyesi, and C. Schaefer, "The Copiale Cipher," *invited talk at ACL Workshop on Building and Using Comparable Corpora (BUCC)*, Association for Computational Linguistics, 2011.

[7] Key ID-345, Reproduced image from the National Archives in Kew, State Papers. *TNA_SP106/2_ElizabethI_f58(0069)*. 1596. url: https://de-crypt.org/decrypt-web/RecordsView/345.

[8] Key ID-633, Reproduced image from the National Archives in Hungary, G15 Caps. C. Fasc. 44. 01, DECODE ID 633, 1703–1711, https://de-crypt.org/decrypt-web/RecordsView/633.

[9] Megyesi, B., et al. "Keys with Nomenclatures in the Early Modern Europe," *Cryptologia*, 2022, doi: 10.1080/01611194.2022.2113185.

[10] Lasry, G., et al., "Deciphering ADFGVX Messages from the Eastern Front of World War I," *Cryptologia*, Vol. 41, No. 2, 2017, pp. 101–136.

[11] Lasry, G., N. Biermann, and S. Tomokiyo, "Deciphering Mary Stuart's Lost Letters from 1578–1584," *Cryptologia*, 2023, doi: 10.1080/01611194.2022.2160677.

[12] Megyesi, B., et al., "Decryption of Historical Manuscripts: The DECRYPT Project," *Cryptologia*, Vol. 44, No. 6, 2020, pp. 545–559, https://doi.org/10.1080/01611194.2020.1716410.

[13] Megyesi, B., N. Blomqvist, and E. Pettersson, "The DECODE Database: Collection of Ciphers and Keys," in *Proceedings of the 2nd International Conference on Historical Cryptology*, 2019.

[14] Szigeti, F., and M. Héder, "The TRANSCRIPT Tool for Historical Ciphers by the DECRYPT Project," in *Proceedings of the 5th International Conference on Historical Cryptology*, 2022, pp. 208–211, https://ecp.ep.liu.se/index.php/histocrypt/article/view/409/367.

[15] Kopal, N., and B. Esslinger, "New Ciphers and Cryptanalysis Components in CrypTool 2," in *Proceedings of the 5th International Conference on Historical Cryptology*, 2022, pp. 127–136.

[16] Zandbergen, R., *The Voynich Manuscript*, http://www.voynich.nu/.

[17] Pelling, N., *The Cipher Mysteries Blog*, www.ciphermysteries.com.

[18] Schmeh, K., *Cipherbrain*, https://scienceblogs.de/klausis-krypto-kolumne/ (updates on this website stopped end of 2022).

[19] Láng, B., *Real Life Cryptology: Ciphers and Secrets in Early Modern Hungary*, Amsterdam: Atlantis Press, Amsterdam University Press, 2018.

[20] Tomokiyo, S., *Cryptiana: Articles on Historical Cryptography*, http://cryptiana.web.fc2.com/code/crypto.htm.

[21] Antal, E., and P. Zajac, "HCPortal Overview," in *Proceedings of the 3rd International Conference on Historical Cryptology*, 2020, pp. 18–20, doi: 10.3384/ecp2020171003, https://hcportal.eu.

[22] Megyesi, B., "Transcription of Historical Ciphers and Keys," in *Proceedings of the 3rd International Conference on Historical Cryptology*, 2020.

[23] Megyesi, B., and C. Tudor, *Transcription of Historical Ciphers and Keys: Guidelines, version 2.0*, https://cl.lingfil.uu.se/~bea/publ/transcription-guidelines-v2.pdf.

[24] Unicode, The Unicode® Standard Version 12.0–Core Specification, 2019, https://unicode.org/standard/standard.html.

[25] Lasry, G., "Armand de Bourbon's Poly-Homophonic Cipher–1649," in *Proceedings of the 6th International Conference on Historical Cryptology*, 2023, pp. 105–112, https://ecp.ep.liu.se/index.php/histocrypt/article/view/699.

[26] Souibgui, M. A., et al. "DocEnTr: An End-to-End document Image Enhancement Transformer," in *26th International Conference on Pattern Recognition (ICPR)*, 2022.

[27] Axler, G., and L. Wolf, "Toward a Dataset-Agnostic Word Segmentation Method," in *25th IEEE International Conference on Image Processing (ICIP)*, IEEE, 2018, pp. 2635–2639.

[28] Frinken, V., and H. Bunke, "Continuous Handwritten Script Recognition," in *Handbook of Document Image Processing and Recognition* (D. Doermann and K. Tombre, eds.), Springer, 2014, pp. 391–425.

[29] Kang, L., et al., "Pay Attention to What You Read: Non-Recurrent Handwritten Text-Line Recognition," *Pattern Recognition*, Vol. 129, 2022, p. 108766.

[30] Bogacz, B., N. Howe, and H. Mara, "Segmentation Free Spotting of Cuneiform Using Part Structured Models," in *15th International Conference on Frontiers in Handwriting Recognition (ICFHR)*, IEEE, 2016, pp. 301–306.

[31] Baró, A., et al., "Towards a Generic Unsupervised Method for Transcription of Encoded Manuscripts," in *Proceedings of the 3rd International Conference on Digital Access to Textual Cultural Heritage*, 2019, pp. 73–78.

[32] Souibgui, M. A., et al., "Few Shots Are All You Need: A Progressive Learning Approach for Low Resource Handwritten Text Recognition," *Pattern Recognition Letters*, Vol. 160, 2022, pp. 43–49, https://doi.org/10.1016/j.patrec.2022.06.003.

[33] Souibgui, M. A., et al., "A User Perspective on HTR Methods for the Automatic Transcription of Rare Scripts: The Case of Codex Runicus," *Journal on Computing and Cultural Heritage*, 2022.

[34] Kopal, N., "Solving Classical Ciphers with CrypTool 2," in *Proceedings of the 1st International Conference on Historical Cryptology*, 2018, pp. 29–38.

[35] Kopal, N., "Cryptanalysis of Homophonic Substitution Ciphers Using Simulated Annealing with Fixed Temperature," in *Proceedings of the 2nd International Conference on Historical Cryptology*, 2019, pp. 107–116.

[36] Megyesi, B., et al., "Historical Language Models in Cryptanalysis: Case Studies on English and German," in *Proceedings of the 6th International Conference on Historical Cryptology*, 2023, pp. 120–129. url: https://ecp.ep.liu.se/index.php/histocrypt/article/view/701.

[37] Waldispühl, M., "Variation and Change," in *The Cambridge Handbook of Historical Orthography* (M. Condorelli and H. Rutkowska, eds.), Cambridge University Press, 2023, pp. 245–264.

[38] Pettersson, E., and B. Megyesi, "The HistCorp Collection of Historical Corpora and Resources," in *Proceedings of the Digital Humanities in the Nordic Countries 3rd Conference*, March 2018.

[39] Kopal, N., and M. Waldispühl, "Two Encrypted Diplomatic Letters Sent by Jan Chodkiewicz to Emperor Maximilian II in 1574–1575," in *Proceedings of the 4th International Conference on Historical Cryptology*, 2021, pp. 80–89, doi: https://doi.org/10.3384/ecp188409.

[40] Pettersson, E., and B. Megyesi, "Matching Keys and Encrypted Manuscript," in *Proceedings of the 22nd Nordic Conference on Computational Linguistics*, October 2019, pp. 253–261.

[41] Gambardella, M.-E., B. Megyesi, and E. Pettersson. "Identifying Cleartext in Historical Ciphers," in *Proceedings of the Workshop on Language Technologies for Historical and Ancient Languages, LT4HALA 2022*, 2022.

[42] Waldispühl, M., and B. Megyesi, "Language Choice in Eighteenth-Century Diplomatic Ciphers from Europe," in *Languages of Diplomacy in the Eighteenth Century* (V. Rjéoutski and G. Kazakov, eds.), Amsterdam University Press, 2023.

[43] Waldispühl, M., "Verschlüsselte Briefe: Mehrsprachigkeit und Geheimschrift im Schwedischen Reich," in *Praktiken der Mehrsprachigkeit im Schwedischen Reich (1611–1721)* (M. Prinz and D. Stoeva-Holm, eds.), Harrassowitz, 2023.

[44] Kahn, D., "The Future of the Past—Questions in Cryptologic History," *Cryptologia,* Vol. 32, 2008, pp. 56–61.

[45] Mrayati, M., Y. MeerAlam, and M. Hassan at-Tayyan, eds., *The Arabic Origins of Cryptology*, Volumes 1–6, KFCRIS & KACST, 2003–2006.

[46] Lasry, G., "Deciphering a Letter from the French Wars of Religion," in *Proceedings of the 5th International Conference on Historical Cryptology*, 2022, pp. 147–152.

[47] Braun, G., and S. Lachenicht, eds, *Spies, Espionage and Secret Diplomacy in the Early Modern Period*, Kohlhammer, 2021.

[48] Bullard, M. M., "Secrecy, Diplomacy and Language in the Renaissance," in *Das Geheimnis am Beginn der europäischen Moderne*, G. Engel, et al. (eds.), Klostermann, 2002, pp. 77–97.

[49] Desenclos, C., "Unsealing the Secret: Rebuilding the Renaissance French Cryptographic Sources (1530–1630)," in *Proceedings of the 1st International Conference on Historical Cryptology*, 2018, pp. 9–17.

[50] De Leeuw, K., "The Black Chamber in the Dutch Republic During the War of the Spanish Succession and Its Aftermath, 1707–1715," *The Historical Journal*, Vol. 42, No. 1, 1999, pp. 133–156.

[51] Lasry, G., B. Megyesi, and N. Kopal. "Deciphering Papal Ciphers from the 16th to the 18th Century," *Cryptologia*, 2020, pp. 479–540, https://www.tandfonline.com/doi/full/10.1080/01611194.2020.1755915.

[52] Kopal, N., and M. Waldispühl, "Deciphering Three Diplomatic Letters sent by Maximilian II in1575," *Cryptologia*, Vol. 46, No. 2, 2022, pp. 103–127, doi: 10.1080/01611194.2020.1858370.

[53] Dinnissen, J., and N. Kopal, "Island Ramanacoil a Bridge too Far. A Dutch Ciphertext from 1674," in *Proceedings of the 4th International Conference on Historical Cryptology*, 2021, pp. 48–57, https://ecp.ep.liu.se/index.php/histocrypt/article/view/156.

[54] Megyesi, B., et al. "Key Design in the Early Modern Era in Europe," in *Proceedings of the 4th International Conference on Historical Cryptology*, 2021.

[55] Megyesi, B., et al. "What Was Encoded in Historical Cipher Keys in the Early Modern Era?" in *Proceedings of the 5th International Conference on Historical Cryptology*, 2022.

[56] Pelling, N., *The Curse of the Voynich: The Secret History of the World's Most Mysterious Manuscript; The Intriguing Story of the People, Places, and Politics Behind the Enigmatic "Voynich Manuscript,"* Compelling Press, 2006.

[57] Kennedy, G., and R. Churchill, *The Voynich Manuscript: The Mysterious Code that Has Defied Interpretation for Centuries*, Rochester, VT: Inner Traditions, 2006.

[58] Kruh, L., "A Basic Probe of the Beale Cipher as a Bamboozlement," *Cryptologia*, Vol. 6, No. 4, 1982, pp. 378–382.

[59] *DECODE Records*, https://de-crypt.org/decrypt-web.

[60] Tomokiyo, S., *Confederate Ciphers During the Civil War: Various Vigenère Keywords*, 2022, http://cryptiana.web.fc2.com/code/civilwar4.htm.

[61] *HistoCrypt–International Conference on Historical Cryptology*, https://histocrypt.org/.

[62] *Cryptologia*, https://www.tandfonline.com/journals/ucry20.

CHAPTER 4

Prime Numbers

This chapter introduces prime numbers (primes) and corresponding results from number theory by asking and answering questions. It provides lots of examples and is not as rigorous as mathematical textbooks usually are. At the end, you will have a good understanding what primes are, what their distribution is like, and why they are very useful in cryptography.

4.1 What Are Prime Numbers?

Prime numbers are integers greater than or equal to 2 that can only be divided by 1 and themselves. All other natural numbers that are not primes and greater or equal to 4 are composite, they can be created by multiplying prime numbers.

The *natural* numbers $\mathbb{N} = \{1, 2, 3, 4, \ldots\}$ thus comprise

- The number 1 (the unit value);
- The primes;
- The composite numbers.

Prime numbers are particularly important for three reasons:

- In number theory, they are considered to be the basic components of natural numbers, upon which numerous brilliant mathematical ideas are based.
- They are of extreme practical importance in modern cryptography (public-key cryptography). The most common public-key procedure, invented at the end of the 1970s, is the RSA encryption. Using large prime numbers is required for particular parameters to guarantee that the RSA procedure is secure, and also further modern procedures (e.g., elliptic curves).
- The search for the largest known prime numbers does not have any practical usage known to date, but it is an excellent benchmark (e.g., for the possibility of determining the performance of computers) [1].

Many people have been fascinated by prime numbers over the past two millennia. Ambition to make new discoveries about prime numbers has often resulted in brilliant ideas and conclusions. The following section provides an easily comprehensible introduction to the basics of prime numbers. We will also explain what is known about the distribution of prime numbers (i.e., density, number of prime numbers; intervals in particular intervals), and how prime number tests work.

4.2 Prime Numbers in Mathematics

Each integer number has a factor. Only the number 1 has one factor, itself, whereas the number 12 has the six factors $1, 2, 3, 4, 6, 12$. Many numbers can only be divided by themselves and by 1. With respect to multiplication, these are the atoms in the area of numbers. Such numbers are called prime numbers.

In mathematics, a slightly different (but equivalent) definition is used.

Definition 4.1 An integer $p \in \mathbb{N}$ is called prime if $p > 1$ and p only possesses the trivial factors ± 1 and $\pm p$.

By definition, the number 1 is not a prime number. In the following sections, p will always denote a prime number (notation).

The sequence of prime numbers starts with:

$$2,\ 3,\ 5,\ 7,\ 11,\ 13,\ 17,\ 19,\ 23,\ 29,\ 31,\ 37,\ 41,\ 43,\ 47,\ 53,\ 59,\ 61,\ 67,\ 71,\\ 73,\ 79,\ 83,\ 89,\ 97, \ldots$$

The first 100 numbers include precisely 25 prime numbers. After this, the percentage of primes constantly decreases. Prime numbers can be factorized in a uniquely *trivial* way: $p = 1 \cdot p$

$$5 = 1 \cdot 5, \quad 17 = 1 \cdot 17, \quad 1013 = 1 \cdot 1013, \quad 1296409 = 1 \cdot 1296409$$

All numbers that have 2 or more factors not equal to 1 are called *composite* numbers. These include

$$4 = 2 \cdot 2, \quad 6 = 2 \cdot 3$$

as well as numbers that *look like primes*, but are in fact composite:

$$91 = 7 \cdot 13, \quad 161 = 7 \cdot 23, \quad 767 = 13 \cdot 59$$

Figure 4.1 gives a first impression of how primes are distributed between natural numbers. There are many graphical forms of representation (the most well-known is the Ulam spiral; see Figures 4.2 and 4.3). However, until now, these graphical forms gained no new insights, but for some people they created the impression that there are at least local patterns within the random distribution.

Theorem 4.1 Each integer m greater than 1 possesses a lowest factor greater than 1. This is a prime number p. Unless m is a prime number itself, then: p is less than or equal to the square root of m.

Sample: $6 = 2 \cdot 3$ and $2 < \sqrt{6} = 2.45$

All integers greater than 1 can be expressed as a product of prime numbers—uniquely. This is the claim of the *first fundamental theorem of number theory* (= fundamental theorem of arithmetic = fundamental building block of all positive integers).

1	**2**	**3**	4	**5**	6	**7**	8	9	10	**11**	12	**13**	14	15	16	**17**	18	**19**	20	21	22	**23**	24	25	26	27	28	**29**	30
31	32	33	34	35	36	**37**	38	39	40	**41**	42	**43**	44	45	46	**47**	48	49	50	51	52	**53**	54	55	56	57	58	**59**	60
61	62	63	64	65	66	**67**	68	69	70	**71**	72	**73**	74	75	76	77	78	**79**	80	81	82	**83**	84	85	86	87	88	**89**	90
91	92	93	94	95	96	**97**	98	99	100	**101**	102	**103**	104	105	106	**107**	108	**109**	110	111	112	**113**	114	115	116	117	118	119	120
121	122	123	124	125	126	**127**	128	129	130	**131**	132	133	134	135	136	**137**	138	**139**	140	141	142	143	144	145	146	147	148	**149**	150
151	152	153	154	155	156	**157**	158	159	160	161	162	**163**	164	165	166	**167**	168	169	170	171	172	**173**	174	175	176	177	178	**179**	180
181	182	183	184	185	186	187	188	189	190	**191**	192	**193**	194	195	196	**197**	198	**199**	200	201	202	203	204	205	206	207	208	209	210
211	212	213	214	215	216	217	218	219	220	221	222	**223**	224	225	226	**227**	228	**229**	230	231	232	**233**	234	235	236	237	238	**239**	240
241	242	243	244	245	246	247	248	249	250	**251**	252	253	254	255	256	**257**	258	259	260	261	262	**263**	264	265	266	267	268	**269**	270
271	272	273	274	275	276	**277**	278	279	280	**281**	282	**283**	284	285	286	287	288	289	290	291	292	**293**	294	295	296	297	298	299	300
301	302	303	304	305	306	**307**	308	309	310	**311**	312	**313**	314	315	316	**317**	318	319	320	321	322	323	324	325	326	327	328	329	330
331	332	333	334	335	336	**337**	338	339	340	341	342	343	344	345	346	**347**	348	**349**	350	351	352	**353**	354	355	356	357	358	**359**	360
361	362	363	364	365	366	**367**	368	369	370	371	372	**373**	374	375	376	377	378	**379**	380	381	382	**383**	384	385	386	387	388	**389**	390

Figure 4.1 Primes within the first 390 integers in a $(30 \cdot 13)$ rectangle—marked with color.

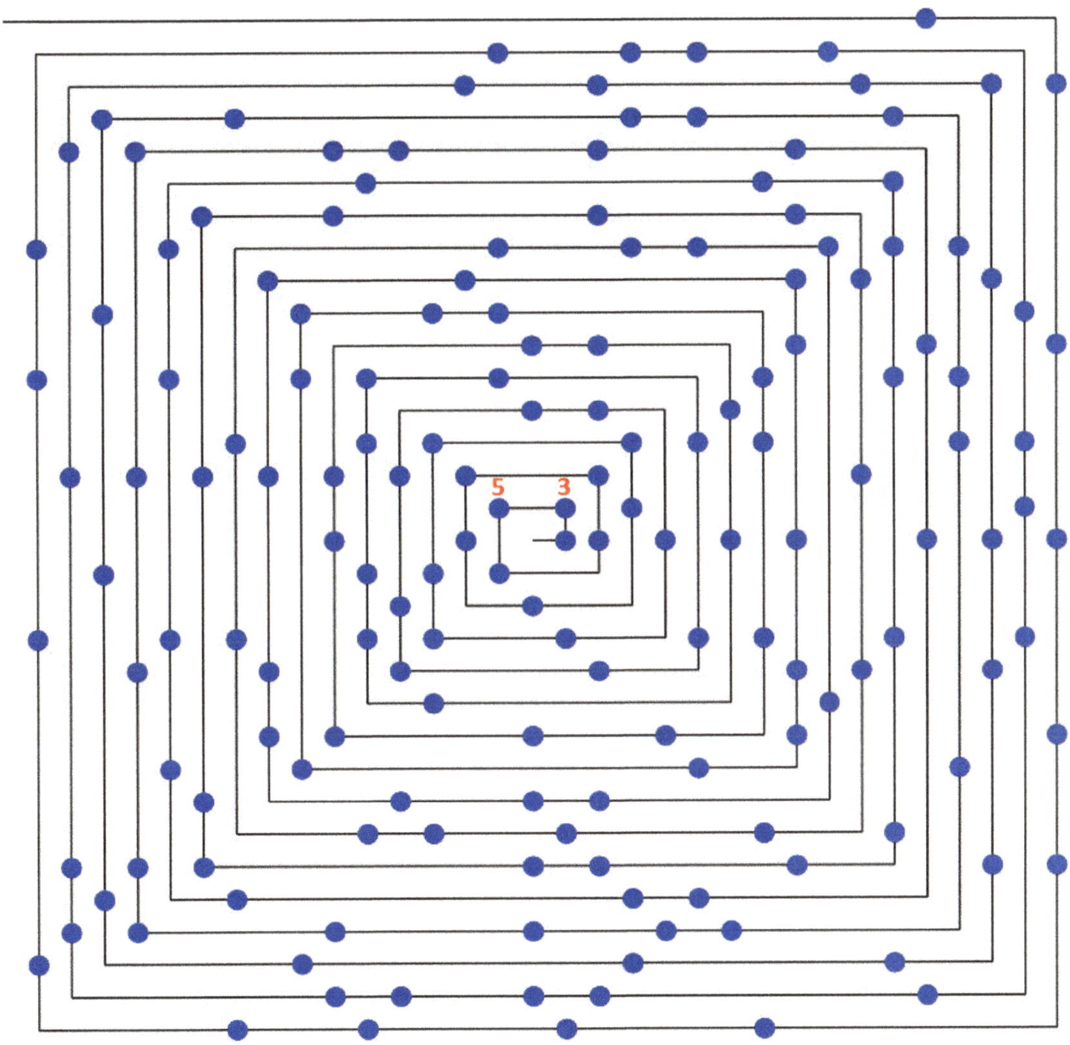

Figure 4.2 Primes within the first 999 integers in a $(33 \cdot 33)$ rectangle as Ulam spiral (graphics from CT2 `Crypto Tutorials ▷ World of Primes ▷ Distribution of primes ▷ Ulam's spiral`).

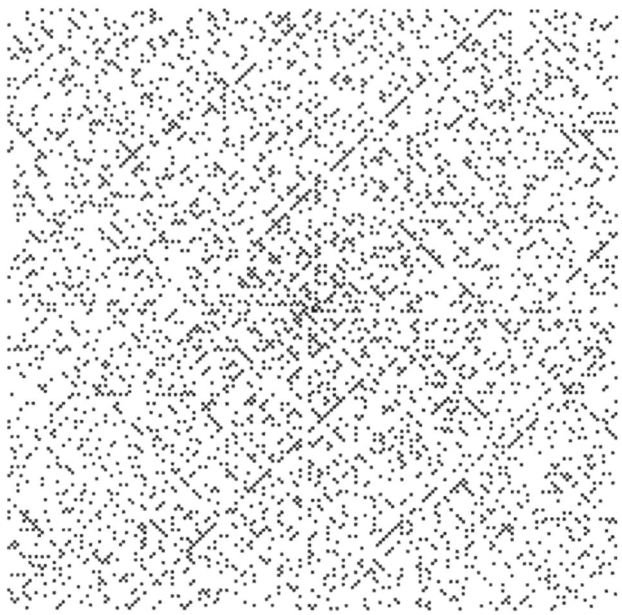

Figure 4.3 Primes within the first 40,000 integers in a $(200 \cdot 200)$ rectangle as Ulam spiral.

Theorem 4.2 Each element n of the natural numbers greater than 1 can be written as the product $n = p_1 \cdot p_2 \ldots p_m$ of prime numbers ($p_1, p_2, \ldots, p_m$ are called the prime factors of n). If two such factorizations

$$n = p_1 \cdot p_2 \cdots p_m = p'_1 \cdot p'_2 \cdots p'_{m'}$$

are given, then they can be reordered such that $m = m'$ and for all i: $p_i = p'_i$.

In other words, each natural number other than 1 can be written as a product of prime numbers in precisely one way, if we ignore the order of the factors. The factors are therefore unique (or you can say: the *expression as a product of factors* is unique). For example,

$$60 = 2 \cdot 2 \cdot 3 \cdot 5 = 2^2 \cdot 3^1 \cdot 5^1.$$

And this—other than changing the order of the factors—is the only way in which the number 60 can be factorized. If you allow numbers other than primes as factors, there are several ways of factorizing integers and the uniqueness is lost:

$$60 = 1 \cdot 60 = 2 \cdot 30 = 4 \cdot 15 = 5 \cdot 12 = 6 \cdot 10 = 2 \cdot 3 \cdot 10 = 2 \cdot 5 \cdot 6 = 3 \cdot 4 \cdot 5 = \cdots .$$

This paragraph is for those familiar with mathematical logic: The first fundamental theorem may appear to be obvious, but we can construct numerous other sets of numbers (i.e., other than positive integers greater than 1), for which numbers in the set cannot be expressed uniquely as a product of the prime numbers of the set: In the set $M = \{1, 5, 10, 15, 20, \ldots\}$ there is no equivalent to the fundamental theorem under multiplication. The first five prime numbers of this sequence are

5, 10, 15, 20, 30 (note: 10 is prime, because 5 is not a factor of 10 in this set—the result is not an element of the given basic set M). As the following applies in M:

$$100 = 5 \cdot 20 = 10 \cdot 10$$

and 5, 10, 20 are all prime numbers in this set, the expression as a product of prime factors is not unique here.

4.3 How Many Prime Numbers Are There?

For the natural numbers, the primes can be compared to elements in chemistry or the elementary particles in physics (see [2, p. 22]) as their building blocks. Although there are only 92 natural chemical elements, the number of prime numbers is unlimited.

Even the Greeks knew this in the third century B.C.; the theorem of the infiniteness of the primes had already been distinguished and proven in *Euclid's Elements* (Book IX, theorem 20). Euclid was a Greek mathematician in fourth and third century B.C. who worked at the Egyptian academy of Alexandria and wrote *The Elements*, the most well known systematic textbook of the Greek mathematics.

The following theorem of Euclid does not denote Euclid as the inventor of the theorem; rather the true inventor is not known. The phraseology in the Greek original is remarkable due to the fact that the word infinite is not used. The text reads as follows:

Οί πρῶτοι ἀριθμοὶ πλείους εἰσὶ παντὸς τοῦ προτεθέντος πλήθους πρώτων ἀριθμῶν

The English translation is: The prime numbers are more than any previously existing amount of prime numbers.

Or in a less literal translation:

Theorem 4.3 (Euclid). The sequence of prime numbers does not discontinue. Therefore, the quantity of prime numbers is infinite.

His proof that there is an infinite number of primes is still considered to be a brilliant mathematical consideration and conclusion today (*proof by contradiction*). He assumed that there is only a finite number of primes and therefore there exists a largest prime number. Based on this assumption, he drew logical conclusions until he obtained an obvious contradiction. This meant that something must be wrong. As there were no mistakes in the chain of conclusions, it could only be the assumption that was wrong. Therefore, there must be an infinite number of primes!

According to Euclid (Proof by Contradiction)

Assumption: There is a *finite* number of primes.

Conclusion: Then these can be listed $p_1 < p_2 < p_3 < \cdots < p_n$, where n is the (finite) number of prime numbers. p_n is therefore the largest prime. Euclid now looks at the number $a = p_1 \cdot p_2 \cdots p_n + 1$. This number cannot be a prime number because it is not included in our list of primes. It must therefore be divisible by a prime; that is, there is a natural number i between 1 and n, such that p_i divides the number a. Of course, p_i also divides the product $a - 1 = p_1 \cdot p_2 \cdots p_n$ because p_i is a factor of $a - 1$. Since p_i divides the numbers a and $a - 1$, it also divides the difference of these numbers. Thus: p_i divides $a - (a - 1) = 1$. p_i must therefore divide 1, which is impossible.

Contradiction: Our assumption was false. Thus, there is an *infinite* number of primes.

(Cross-reference: See the overview in Section 4.10 of the number of prime numbers in various intervals.)

4.4 The Search for Extremely Large Primes

The largest prime numbers known today have several million digits.[1] This is too big for us to imagine. The number of elementary particles in the universe is estimated to be "only" a 80-digit decimal number (see the overview in Section 4.12 about various orders of magnitude / dimensions).

4.4.1 The 20+ Largest Known Primes

Table 4.1 contains the largest currently known primes as of April 2022 and a description of its particular number type.

Note the terms used in the column "Description" of Table 4.1: Pure Mersenne numbers have a base of 2 and an exponent n; pure Fermat numbers have a base of 2 and an exponent, which itself is a power of 2. Generalizations add a factor k to the power or change the base b.

Mersenne: $f(n) = 2^n - 1$; generalized Mersenne: $f(k, b, n) = k \cdot b^n \pm 1$ (with $b \neq 2, k \neq 1, k, b \in \mathbb{N}$)

Fermat: $f(n) = 2^{2^n} + 1$; generalized Fermat: $f(b, n) = b^{2^n} + 1$ (with $b > 1$, $b \in \mathbb{N}$)

Note that there are rarely major changes in the top 10. Therefore, we intentionally show the table from April 2022. Until July 2023, the first 12 ranks are still the same, then three new entries have appeared in the top 20 since April 2022.

The development over time is shown in Figure 4.4. Note the logarithmic vertical scale.

The largest currently known prime is a Mersenne prime (see Section 4.4.2), found by the GIMPS project. Within the largest known primes there are also numbers of the type generalized Mersenne number (see Section 4.6.2) and generalized Fermat numbers (see Section 4.6.5).

4.4.2 Special Number Types: Mersenne Numbers and Mersenne Primes

Mersenne numbers have the form $f(n) = 2^n - 1$ with $n \in \mathbb{N}$. These numbers are often abbreviated as $M(n)$.

1. Using `CT1 Indiv. Procedures ⊳ Number Theory Interactive ⊳ Compute Mersenne Numbers` you can calculate all digits of such a big number very quickly.

Table 4.1 The 20+ Largest Known Primes and Their Particular Number Types*

	Definition	*Decimal Digits*	*Year*	*Description*
1	$2^{82589933}-1$	24862048	2018	Mersenne, 51st known
2	$2^{77232917}-1$	23249425	2018	Mersenne, 50th known
3	$2^{74207281}-1$	22338618	2016	Mersenne, 49th known
4	$2^{57885161}-1$	17425170	2013	Mersenne, M-48
5	$2^{43112609}-1$	12978189	2008	Mersenne, M-47
6	$2^{42643801}-1$	12837064	2009	Mersenne, M-46
7	$2^{37156667}-1$	11185272	2008	Mersenne, M-45
8	$2^{32582657}-1$	9808358	2006	Mersenne, M-44
9	$10223 \cdot 2^{31172165}+1$	9383761	2016	Generalized Mersenne
10	$2^{30402457}-1$	9152052	2005	Mersenne, M-43
11	$2^{25964951}-1$	7816230	2005	Mersenne, M-42
12	$2^{24036583}-1$	7235733	2004	Mersenne, M-41
13	$202705 \cdot 2^{21320516}+1$	6418121	2021	Generalized Mersenne
14	$2^{20996011}-1$	6320430	2003	Mersenne, M-40
15	$1059094^{1048576}+1$	6317602	2018	Generalized Fermat[1]
16	$919444^{1048576}+1$	6253210	2017	Generalized Fermat
17	$168451 \cdot 2^{19375200}+1$	5832522	2017	Generalized Mersenne
18	$3 \cdot 2^{18924988}+1$	5696990	2022	Generalized Mersenne
19	$69 \cdot 2^{18831865}-1$	5668959	2021	Generalized Mersenne
20	$7 \cdot 2^{18233956}+1$	5488969	2020	Generalized Mersenne[2]
...				
42	$2^{13466917}-1$	4053946	2001	Mersenne, M-39
...				
46	$19249 \cdot 2^{13018586}+1$	3918990	2007	Generalized Mersenne
...				
152	$2^{6972593}-1$	2098960	1999	Mersenne, M-38
...				
2115	$1372930^{131072}+1$	804474	2003	Generalized Fermat[3]
...				
2152	$342673 \cdot 2^{2639439}-1$	794556	2007	Generalized Mersenne

* As of April 2022.

[1] Generalized Fermat number: $1059094^{1048576}+1 = 1059094^{(2^{20})}+1$.

[2] Generalized Mersenne number: As 18233956 is no power of 2 it is no Fermat number.

[3] Generalized Fermat number: $1372930^{131072}+1 = 1372930^{(2^{17})}+1$.

Written out in binary form, a Mersenne number consists only of 1s. $M(2) = 3$, or in binary digits 11. $M(3) = 7$, or 111. $M(4) = 15$, or 1111. $M(5) = 31$, or 11111.

Almost all of the biggest known prime numbers are special candidates of the form $2^p - 1$, where the exponent p is a prime. Not all Mersenne numbers of this form are prime:

$$\begin{array}{rcl} M(2): & 2^2 - 1 = 3 & \Rightarrow \text{prime} \\ M(3): & 2^3 - 1 = 7 & \Rightarrow \text{prime} \\ M(5): & 2^5 - 1 = 31 & \Rightarrow \text{prime} \\ M(7): & 2^7 - 1 = 127 & \Rightarrow \text{prime} \\ M(11): & 2^{11} - 1 = 2047 = 23 \cdot 89 & \Rightarrow \text{NOT prime !} \end{array}$$

Even Mersenne knew already that not all Mersenne numbers are prime (see exponent $p = 11$). A prime Mersenne number is called *Mersenne prime number*.

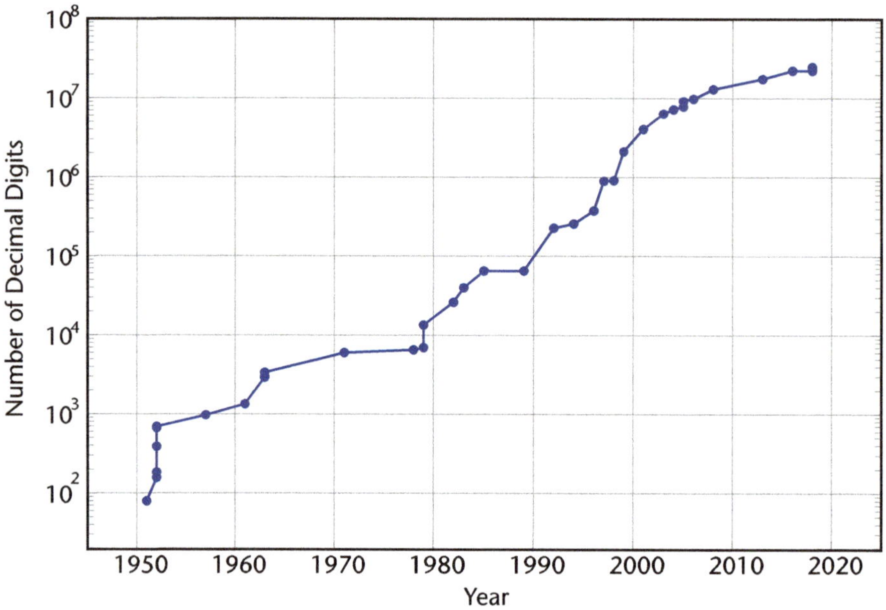

Figure 4.4 Number of digits of largest known prime by year (as of April 2022) (own plot, thanks to data from Chris Caldwell [3]).

However, it is with thanks to Mersenne for the interesting conclusion that a number of the form $2^n - 1$ cannot be a prime number if n is a composite number:

Theorem 4.4 (Mersenne) If $2^n - 1$ is a prime number, then n is also a prime number (or to put it another way: $2^n - 1$ is prime, only if n is prime).

Proof
The theorem of Mersenne can be proved by contradiction. We therefore assume that there exists a composite natural number n (with real factorization) $n = n_1 \cdot n_2$, with the property that $2^n - 1$ is a prime number.

Abbreviated, the theorem is:

$$[\ M(n) \text{ is prime } \Rightarrow n \text{ is prime}\ ]$$

Consequently, our assumption for the proof by contradiction then is:

$$[\ n \text{ is composite } \textbf{and } M(n) \text{ is a prime number}\ ]$$

From

$$\begin{aligned}(x^r - 1)((x^r)^{s-1} + (x^r)^{s-2} + \cdots + x^r + 1) &= ((x^r)^s + (x^r)^{s-1} + (x^r)^{s-2} + \cdots + x^r) \\ &\quad - ((x^r)^{s-1} + (x^r)^{s-2} + \cdots + x^r + 1) \\ &= (x^r)^s - 1 = x^{rs} - 1,\end{aligned}$$

we conclude

$$2^{n_1 n_2} - 1 = (2^{n_1} - 1)((2^{n_1})^{n_2-1} + (2^{n_1})^{n_2-2} + \cdots + 2^{n_1} + 1).$$

Because $2^n - 1$ is a prime number, one of the above two factors on the right-hand side must be equal to 1. This is the case if and only if $n_1 = 1$ or $n_2 = 1$. But this contradicts our assumption. Therefore, the assumption is false. This means that there is no composite number n, such that $2^n - 1$ is a prime. □

Notes:
The following two statements are equivalent because from $A \Rightarrow B$ follows $\neg B \Rightarrow \neg A$:

$$[M(n) \text{ is a prime number} \Rightarrow n \text{ is prime}] \equiv [n \text{ is composite} \\ \Rightarrow M(n) \text{ is not a prime number}]$$

Unfortunately, Theorem 4.4 only applies in one direction (the inverse statement does not apply, no equivalence): That means that there are prime exponents for which the Mersenne number is *not* prime (see the above example $2^{11} - 1$, where 11 is prime, but $2^{11} - 1$ is not).

Mersenne claimed that $2^{67} - 1$ is a prime number. There is also a mathematical history behind this claim: It took over 200 years before Edouard Lucas (1842–1891) proved that this number is composite. However, he argued indirectly and did not name any of the factors. In 1903, Frank Nelson Cole showed which factors make up this composite number:

$$2^{67} - 1 = 147573952589676412927 = 193707721 \cdot 761838257287$$

Cole admitted to having worked for 20 years on the *factorization* (dissection as a product of prime factors)[2] of this 21-digit decimal number.

Due to the fact that the exponents of the Mersenne numbers do not use all natural numbers, but only the primes, the *experimental space* is considerably limited. The *exponents* of the currently known 51 Mersenne prime numbers are listed in Table 4.2.

The 19th number with the exponent 4253 was the first with at least 1000 digits in decimal system (the mathematician Samual Yates coined the expression *titanic* prime for this; it was discovered by Hurwitz in 1961); the 27th number with the exponent 44497 was the first with at least 10000 digits in the decimal system—Yates coined the expression *gigantic* prime for this. Today, these expressions are long outdated.

For the first 48 Mersenne prime numbers we know that this list is complete. The exponents up to the 51st Mersenne prime number have not yet been checked completely [4]. See Section 4.5 for hints on checking the primality of a number.

2. Factorization algorithms can be found in CTO, CT2, and CT1.
- In CTO in the plugin "Msieve Factorizer": `https://www.cryptool.org/en/cto/msieve`.
- In CT2 `Startcenter ▷ Templates ▷ Mathematics ▷ Factorization with Quadratic Sieve (QS)`.
- Using `CT1 Indiv. Procedures ▷ RSA Cryptosystem ▷ Factorization of a Number` you can factorize numbers. With the quadratic sieve (QS), CT1 factorizes numbers up to 250 bit in a reasonable time (on a single PC).
- The current factorization records are listed in Section 5.12.4. Single factoring algorithms like Bill Hart's quadratic sieve and Paul Zimmermann's GMP-ECM are also available in SageMath: see `https://doc.sagemath.org/html/en/thematic_tutorials/explicit_methods_in_number_theory/integer_factorization.html`.

Table 4.2 Exponents of Currently Known Mersenne Prime Numbers

1:	2	14:	607	27:	44497	40:	20996011
2:	3	15:	1279	28:	86243	41:	24036583
3:	5	16:	2203	29:	110503	42:	25964951
4:	7	17:	2281	30:	132049	43:	30402457
5:	13	18:	3217	31:	216091	44:	32582657
6:	17	19:	4253	32:	756839	45:	37156667
7:	19	20:	4423	33:	859433	46:	42643801
8:	31	21:	9689	34:	1257787	47:	43112609
9:	61	22:	9941	35:	1398269	48:	57885161
10:	89	23:	11213	36:	2976221	49:	74207281
11:	107	24:	19937	37:	3021377	50:	77232917
12:	127	25:	21701	38:	6972593	51:	82589933
13:	521	26:	23207	39:	13466917		

As of May 1, 2023 all prime exponents smaller than 63,589,987 have been tested and double-checked. So we can be certain that M-48 is really the 48th Mersenne prime number, and that there are no smaller undiscovered Mersenne primes (it is common not to use the notation M-nn until it is proven that the nn-th "known" Mersenne prime is really the nn-th Mersenne prime).

Here are some examples in more detail:

M-37 – January 1998

The 37th Mersenne prime, called M-37,

$$2^{3021377} - 1$$

has 909,526 digits in the decimal system, which are equivalent to 33 pages of a newspaper.

M-38 – June 1999

The 38th Mersenne prime, called M-38,

$$2^{6972593} - 1$$

has 2,098,960 digits in the decimal system (that are equivalent to around 77 pages of a newspaper).

M-39 – December 2001

The 39th Mersenne prime, called M-39,

$$2^{13466917} - 1,$$

was published on December 6, 2001—more exactly, the verification of this number, found on November 14, 2001, by Canadian student Michael Cameron, was successfully completed. This number has about 4 million decimal digits (exactly 4,053,946 digits). Trying only to print this number

$$(924947738006701322247758 \cdots 1130073855470256259071)$$

would require around 200 pages in the *Financial Times*.

GIMPS

The GIMPS project (Great Internet Mersenne Prime Search) was founded in 1996 by George Woltman to search for new largest Mersenne primes (https://www.mersenne.org). Further explanations about this number type can be found in the Sections Mersenne numbers and Mersenne primes.

So far, the GIMPS project has discovered 17 largest Mersenne primes, including the largest known prime number ever. Table 4.3 contains these Mersenne record primes.

Richard Crandall discovered the advanced transform algorithm used by the GIMPS program. George Woltman implemented Crandall's algorithm in machine language, thereby producing a prime-search program that has unprecedented efficiency.

On June 1st, 2003 a possible Mersenne prime was reported to the GIMPS server that was checked afterward as usual, before it was to be published. Unfortunately, mid-June the initiator and GIMPS project leader George Woltman had to announce that two independent verification runs proved the number was composite. This was the first false-positive report of a client in 7 years. Since the end of 2020 new proofs are used that eliminate the need for double checks.

As of May 2023, more than 250,000 volunteers, amateurs and experts have participated in the GIMPS project. They connect their computers into the PrimeNet, originally organized by the company Entropia.

Table 4.3 The Largest 17 Primes Found by the GIMPS Project*

	Definition	*Decimal Digits*	*Date*	*Who*
1	$2^{82589933}-1$	24862048	Dec 7, 2018	Patrick Laroche
2	$2^{77232917}-1$	23249425	Dec 26, 2017	Jonathan Pace
3	$2^{74207281}-1$	22338618	Jan 7, 2016	Curtis Cooper
4	$2^{57885161}-1$	17425170	Jan 25, 2013	Curtis Cooper
5	$2^{43112609}-1$	12978189	Aug 23, 2008	Edson Smith
6	$2^{42643801}-1$	12837064	Apr 12, 2009	Odd Magnar Strindmo
7	$2^{37156667}-1$	11185272	Sep 6, 2008	Hans-Michael Elvenich
8	$2^{32582657}-1$	9808358	Sep 4, 2006	Curtis Cooper/Steven Boone
9	$2^{30402457}-1$	9152052	Dec 15, 2005	Curtis Cooper/Steven Boone
10	$2^{25964951}-1$	7816230	Feb 18, 2005	Martin Nowak
11	$2^{24036583}-1$	7235733	May 15, 2004	Josh Findley
12	$2^{20996011}-1$	6320430	Nov 17, 2003	Michael Shafer
13	$2^{13466917}-1$	4053946	Nov 14, 2001	Michael Cameron
14	$2^{6972593}-1$	2098960	Jun 1, 1999	Nayan Hajratwala
15	$2^{3021377}-1$	909526	Jan 27 1998	Roland Clarkson
16	$2^{2976221}-1$	895932	Aug 24, 1997	Gordon Spence
17	$2^{1398269}-1$	420921	Nov 13, 1996	Joel Armengaud

* As of April 2022.

4.4.3 Challenge of the Electronic Frontier Foundation

This search is also spurred on by a competition started by the nonprofit organization EFF using the means of an unknown donor. The participants are rewarded with a total of $500,000 USD if they find the longest prime number. In promoting this project, the unknown donor is not looking for the quickest computer, but rather wants to draw people's attention to the opportunities offered by *cooperative networking*: https://www.eff.org/awards/coop.

The discoverer of M-38 received $50,000 USD from the EFF for discovering the first prime with more than 1 million decimal digits.

The next prize of $100,000 USD for a proven prime with more than 10 million decimal digits was awarded to Edson Smith, who found the number $2^{43112609} - 1$ within the GIMPS project.

According to the EFF rules for their prizes, in the next stage $150,000 USD is being offered for a proven prime with more than 100 million decimal digits.

Edouard Lucas (1842–1891) held the record for the longest prime number for over 70 years by proving that $2^{127} - 1$ is prime. No new record is likely to last that long.

4.5 Prime Number Tests[3]

In order to implement secure encryption procedures we need extremely large prime numbers (but still much smaller than the prime records). These numbers in the region of 2^{2048} have more than 600 digits in the decimal system.

In order to do this, random numbers are considered and then examined whether they are prime or not. If even the smallest prime factor is huge, it would take far too long to factor the candidates. Factorizing numbers using systematic division (brute-force) or the Eratosthenes' sieve can be used with today's computers for numbers with up to about 20 digits in the decimal system. The largest number that has so far been factorized into its two approximately equal prime factors in a multimonth undertaking with sophisticated methods has 250 decimal digits (see Section 5.12.4).

Instead of factoring, there are very fast, heuristic methods that can be used to test the primality of a number. Such fast algorithms that can very reliably state whether a number is prime or not are the Fermat primality test, Lucas test, Solovay-Strassen test, Miller-Rabin test, and Baillie-PSW test. Because these algorithms are imperfect (and probabilistic), tests are also passed by a few numbers that are not prime, called pseudoprimes.

3. - Using `CT1 Indiv. Procedures ▷ RSA Cryptosystem ▷ Prime Number Test` the following tests can be performed: Miller-Rabin, Fermat, Solovay-Strassen und AKS. The first three are probabilistic tests.
- With the educational tool for number theory **NT** you can apply the tests of Fermat and of Miller-Rabin: See in there the NT learning units 3.2 and 3.3, pages 3-11/11. NT can be called via `CT1 Indiv. Procedures ▷ Number Theory Interactive ▷ Learning Tool for Number Theory`.
- Using `CT2 Templates ▷ Mathematics ▷ Primes Test` a brute-force test with small factors and the Miller-Rabin test is performed.
- In `CT2 Crypto Tutorials ▷ World of Primes ▷ Primality test` the following methods are visualized and their single steps can be reproduced: Sieve of Eratosthenes, Miller-Rabin test and Sieve of Atkin.

We will not go into the various test methods here, but show the most important property of prime numbers and which numbers are false witnesses in this context (i.e., they fulfill the primality property even though they are composite). Primality tests only answer the primality property of a number, but cannot determine the prime factors of composite numbers. Small implementations for this can also be found in the supplied Python and SageMath programs.

Obviously, determining primality cannot be harder than factoring: If we know how to factor, we have a test for primality. But it is surprising and fundamental for some asymmetric algorithms that primality testing is so much easier than factoring.

4.5.1 Special Properties of Primes for Tests

This section is a good example of applying mathematical logic.

Fermat put forward in 1640 an important theorem: Many rapid prime number tests are based on the (little) Fermat theorem (see also Section 5.8.3).

Theorem 4.5 ("little" Fermat). Let p be a prime number and a be any integer, then for all a

$$a^p \equiv a \bmod p$$

This could also be formulated as follows:
Let p be a prime number and a be any integer that is not a multiple of p (so $a \not\equiv 0 \bmod p$ or $gcd(a, p) \neq p$), then $a^{p-1} \equiv 1 \bmod p$.

Since numbers that satisfy this congruence are rare, much rarer than prime numbers, and because modular exponentiation (i.e., the computation of $a^{n-1} \bmod n$), is efficient even for relatively large numbers, Fermat's little theorem is in principle suitable as a primality test. This congruence (satisfying an equation modulo a number) can be checked much faster than doing factorization.

Unfortunately, the converse to Fermat's theorem does not hold—otherwise we would have a simple proof of the prime number property (or to put it in other words, we would have a simple prime number criterion).

If you are not used to calculating with remainders (modulo), please simply accept the theorem or first read Chapter 5 "Introduction to Elementary Number Theory with Examples." It is important here to realize this theorem implies that if this equation is not met for any integer a, then p is not a prime. The tests (e.g., for the first formulation) can easily be performed using the *test basis* $a = 2$.

This gives us a criterion for nonprime numbers; that is, a *negative test* (criterion for exclusion), but no proof that the number p is prime:

$$[a^p \not\equiv a \bmod p \;\Rightarrow\; p \text{ not prime}]$$

or

$$[(\text{a not divisible by p} \;\wedge\; a^{p-1} \not\equiv 1 \bmod p) \Rightarrow p \text{ not prime}]$$

If an a exists where the congruence from Theorem 4.5 is *not* met, we say a is a "Fermat witness" to the composite nature of p. So witnesses can very quickly inform us that a number p is *not* prime.

One can also say that Fermat's prime property is a necessary but not a sufficient condition for the number p to be prime.

4.5.2 Pseudoprime Numbers

A pseudoprime is a probable prime (PRP). So a PRP is an integer that shares a property common to all prime numbers, but is actually not prime. Pseudoprimes are classified according to which property of primes they satisfy. They are called pseudoprime in reference to this property.

The following three types of numbers successfully pass such negative tests even though they are not prime numbers:

1. *Fermat pseudoprime numbers.* Numbers n that have the property

$$2^n \equiv 2 \mod n$$

but are not prime are called *pseudoprime numbers* for the basis 2 (i.e., the exponent and the modulus n are not prime). The first pseudoprime number for the basis 2 is

$$341 = 11 \cdot 31$$

In SageMath[4] you can easily verify that 341 satisfies the negative test $2^{341} \equiv 2 \mod 341$, which is derived from Theorem 4.5:

SageMath Example 4.1: Find on the Sage Command Line Nonprimes That Pass the Fermat Test

```
# Calculate pseudoprimes for base a = 2 (which pass the Fermat ▸
  ▸primality test)
# Note: 2.powermod(n,n) is circa 3 times faster than power_mod(2,n,n) ▸
  ▸cause usage of gmp
sage: a=2; count=0
....: for n in range(1,4000):
....:    if  n not in Primes()  and  a.powermod(n,n) == 2:
....:       print("%
....: print("How many numbers found with this property:", count)
....:
2^n mod n == n for n = 341
2^n mod n == n for n = 561
2^n mod n == n for n = 645
2^n mod n == n for n = 1105
2^n mod n == n for n = 1387
2^n mod n == n for n = 1729
2^n mod n == n for n = 1905
2^n mod n == n for n = 2047
2^n mod n == n for n = 2465
2^n mod n == n for n = 2701
2^n mod n == n for n = 2821
2^n mod n == n for n = 3277
How many numbers found with this property: 12
```

There are infinitely many Fermat pseudoprimes for each basis.

4. SageMath is a free computer-algebra system (CAS). See the introduction at `https://www.cryptool.org/en/documentation/ctbook/`.

2. Carmichael Numbers. There are pseudoprime numbers n that pass the Fermat test

$$a^{n-1} \equiv 1 \mod n$$

with all bases a that are relatively prime to n [$gcd(a, n) = 1$], even though the numbers n tested are not prime. These numbers are called *Carmichael numbers.*

So the set of powers to be tested is restricted to those where a and n are relatively prime. For an n now it is not enough to test just an arbitrarily chosen a, but all $a < n$ that are coprime to n. The first of these is

$$561 = 3 \cdot 11 \cdot 17$$

Example: The number to be tested is $n = 561$.

Because $561 = 3 \cdot 11 \cdot 17$, the test condition is $a^{560} \mod 561 = 1$. This congruence

- Is satisfied for $a = 2, 4, 5, 7, 8, 10, \cdots$,
- But not for $a = 3, 6, 9, 11, 12, 15, 17, 18, 21, 22, \cdots$.

The test condition does "not" have to be fulfilled either for multiples of the prime factors 3, 11, or 17: For instance, the test applied for
- $a = 3$ results in: $3^{560} \mod 561 = 375$,
- $a = 5$ results in: $5^{560} \mod 561 = 1$.

SageMath Example 4.2 can find all Carmichael numbers up to a specified limit. There are very fast methods to construct single large Carmichael numbers, but the methods we know that generate the list completely, are relatively slow. SageMath Example 4.3 (chap04_sample080.sage) is significantly faster than SageMath Example 4.2 (chap04_sample070.sage), but still very slow with larger numbers.

SageMath Example 4.2: Find Carmichael Numbers up to n = 10000 (Plus Some More Information)

```
print("\n# CHAP04 -- Sage-Script-SAMPLE 070: =========")

# Calculate the Carmichael numbers n (composite numbers with a prime property) up to upper=1000▸
▸0
# Like primes they fulfill the Fermat congruence a^n = a mod n,
# plus: for them the congruence is valid for all a with gcd(a,n)=1
# Remarks:
# - With "for a in range(2, 4000):" a's type is <class 'int'>. Using then
#   "a.powermod(n,n)" causes AttributeError: 'int' object has no attribute 'powermod'
#   Therefore, the range over Sage integers (class Integer) is used.
# - It's good enough to test "a" up to the given "n" (must not always go to "upper")
# - This is a very straightforward implementation. It could be made quicker :-)

verbose = False # if True the coprime bases up to n are listed too
count1=0; count2=0;
upper=10000  # 20,4000,341,561 (up to 4000 are 3449 nonprimes; up to 10000: 8770)
Carm_list=[]
for n in range(3, upper+1, 2):  # a Carmichael number is an odd composite number.
  if n not in Primes():
    count2 += 1 # counts how many non-primes n will be considered
    count_gcd = 0; count_pmod = 0
    a_list=[]
    for a in [2..n]:  # for a in [2..upper]:  # for a in [2..7]:
      if gcd(a,n) == 1:
        count_gcd += 1
        if a.powermod(n,n) == a:
          count_pmod += 1
```

SageMath Example 4.2 (continued)

```
            a_list.append(a)
    if count_gcd>0  and  len(a_list)==count_gcd  and  count_gcd==count_pmod:
      if verbose:
        print("a^n mod n == n for n = %d and all %d a (%s) with gcd(a,n)==1" % (n, count_gcd, ▸
          ▸a_list));
      else:
        print("a^n mod n == n for n = %d and all %d bases a with gcd(a,n)==1" % (n, count_gcd))▸
          ▸;
      count1 += 1
      Carm_list.append(n)
print("How many numbers found with this property:", count1, "[how many nonprimes n considered: ▸
  ▸%d]" % count2)
print("List of Carmichael numbers found up to %d: %s" % (upper, Carm_list))
```

SageMath Example 4.3: Find All Carmichael Numbers up to n = 100,000

```
print("\n# CHAP04 -- Sage-Script-SAMPLE 080: =========")

# Calculate the Carmichael numbers n (composite numbers with a prime property)
# Remarks:
# - This script (sample 08) is much faster, but less explicit than sample 07.
# - Slightly modified [in Primes()" instead of "is_prime(n)"] the pure Python 3 script from
#   https://stackoverflow.com/questions/58944035/printing-carmichael-numbers-in-a-given-limit

from math import gcd
upper = 100_000  # 30000

def is_carmichael(n):  # now expects only odd numbers >= 3 !
    if n in Primes():
        return False
    for a in range(3, n, 2):  # Why not start with a=2 and why handle only odd bases?
        if gcd(a, n) == 1:
            if pow(a, n-1, n) != 1:
                return False
    return True

def print_carmichael(maximum):
    for n in range(3, maximum+1, 2):  # consider only odd numbers >= 3
        if is_carmichael(n):
            print(n)

print_carmichael(upper)
```

For some composite numbers it's hard to find a witness with the Fermat test. With Carmichael numbers the Fermat test actually always fails. They are "liars" for all coprime bases.

The first 16 Carmichael numbers are 561, 1105, 1729, 2465, 2821, 6601, 8911, 10585, 15841, 29341, 41041, 46657, 52633, 62745, 63973, and 75361. They start to become increasingly rare after that.

There are 2,0138,200 Carmichael numbers between 1 and 10^{21}. This is approximately one in 50 trillion ($50 \cdot 10^{12}$) numbers [5].

The largest known Carmichael number has almost $300 \cdot 10^9$ digits [6]. This number is much greater than the largest known prime (see Table 4.1).

In 1992, Carl Pomerance proved that there are infinitely many Carmichael numbers. Carmichael numbers have at least three prime factors, none of which are duplicates. So they are square free.

The Carmichael numbers are sequence A002997 at OEIS. This list contains all Carmichael numbers up to $1,713,045,574,801 \approx 1.7 \cdot 10^{12}$ (these are the first 10,000 ones).

Feitsma/Galway prepared a very good website (http://www.cecm.sfu.ca/Pseudo primes/index-2-to-64.html) that contains gzipped lists of the Carmichael numbers and their factors up to $2^{64} = 1.84 \cdot 10^{19}$.

3. *Strong pseudoprime numbers.* A stronger test is provided by Miller/Rabin [7]: It is only passed by prime numbers and *strong pseudoprime numbers.*

Let $n \in \mathbb{N}$ be of the form $n = 1 + 2^s \cdot m$ for an odd natural number m and $s \in \mathbb{N}$. Then n is called a *strong pseudoprime with base b* if n itself is not a prime number and

$$\begin{array}{lrcl} \text{either} & b^m & \equiv & 1 \mod n \\ \text{or} \quad \exists\, i \in \{0, 1, \ldots, s-1\}: & b^{2^i m} & \equiv & -1 \mod n \end{array}$$

holds. A number n is called a *strong pseudoprime* if a base $b \geq 2$ exists, so that n is a strong pseudoprime with base b.

Again, these strong pseudoprime numbers are not primes, but they are rare as compared to simple pseudoprime numbers or to Carmichael numbers. The smallest strong pseudoprime number base 2 is

$$15841 = 7 \cdot 31 \cdot 73$$

If you test all four bases, 2, 3, 5, and 7, you will find only one strong pseudoprime number up to $25 \cdot 10^9$ (i.e., a number that passes the test and yet is not a prime number).

More extensive mathematics behind the Rabin test delivers the probability that the number examined is nonprime (such probabilities are currently around 10^{-60}).

Detailed descriptions of tests for finding out whether a number is prime can be found at [8, 9].

4.6 Special Types of Numbers and the Search for a Formula for Primes

There are currently no useful, open, nonrecursive polynomial-like formulae known that *only* deliver prime numbers (recursive means that in order to calculate the function the same function is used with a smaller variable). Mathematicians would be happy if they could find a formula that leaves gaps (i.e., does not deliver all prime numbers) but does not deliver any composite (nonprime) numbers.

Ideally, we would like for the number n to immediately be able to obtain the nth prime number; that is, for $f(8) = 19$ or for $f(52) = 239$.

Ideas for that are seriously discussed in [10].

Table 4.12 in Section 4.11 contains the precise values for the nth prime numbers for selected n.

For prime number formulae usually very special types of numbers are used. The following enumeration contains the most common ideas for prime number formulae and what our current knowledge is about very big elements of the number series: Is their primality proven? If they are compound numbers, could their prime factors be determined?

4.6.1 Mersenne Numbers $f(n) = 2^n - 1$ for n Prime

As seen in Section 4.4.2, this formula yields a relatively large number of large prime numbers, but—like for $n = 11$ with $f(11) = 2^{11} - 1 = 2047$—the result is not always prime even for prime exponents. Currently (as of May 2023) we know all the Mersenne prime numbers up to 17,000,000 decimal digits (M-48).

4.6.2 Generalized Mersenne Numbers $f(k, n) = k \cdot 2^n \pm 1$ for n Prime and k Small Prime/Proth Numbers

This first generalization of the Mersenne numbers creates the Proth numbers. There are (for small k) extremely quick prime number tests (see [11]). These can be performed in practice using software such as Proth 20 [12].

4.6.3 Generalized Mersenne Numbers $f(b, n) = b^n \pm 1$ / The Cunningham Project

This is another possible generalization of the Mersenne numbers. The *Cunningham Project* determines the factors of all composite numbers that are formed as follows:

$$f(b, n) = b^n \pm 1 \quad \text{for} \quad b = 2, 3, 5, 6, 7, 10, 11, 12$$

(b is not equal to powers of bases already used, such as 4, 8, 9).

Details of this can be found at [13].

4.6.4 Fermat Numbers $F_n = f(n) = 2^{2^n} + 1$

In the seventeenth century, Pierre de Fermat wrote to Mersenne that he presumed that all numbers of the form

$$f(n) = 2^{2^n} + 1$$

are prime for all integers $n \geq 0$.

The first 5 numbers $F_0 = 3$, $F_1 = 5$, $F_2 = 17$, $F_3 = 257$, $F_4 = 65537$ are all prime (see Table 4.4). As mentioned, Fermat wrote to Mersenne regarding his assumption that all numbers of this type are primes. This assumption was already disproved by Euler in 1732. The prime number 641 divides $F_5 = f(5)$.

And as early as in the 19th century, it was discovered that the 39-digit number

$$f(7) = 2^{2^7} + 1 \qquad (2^7 = 128)$$

is not prime. However, it was not until 1970 that Morrison/Billhart managed to factorize it.

$$\begin{aligned} f(7) &= 340282366920938463463374607431768211457 \\ &= 59649589127497217 \cdot 5704689200685129054721 \end{aligned}$$

The project Distributed Search for Fermat Number Dividers [14] finds both new compound Fermat numbers and new monster primes.

Example: On February 22, 2003, John Cosgrave discovered:

- The largest composite Fermat number up to that time;

Table 4.4 List of the First Eight Fermat Numbers and Their Factorization

f(n)	*Value*	*Prime?*
$f(0) = 2^{2^0} + 1 = 2^1 + 1$	= 3	Prime
$f(1) = 2^{2^1} + 1 = 2^2 + 1$	= 5	Prime
$f(2) = 2^{2^2} + 1 = 2^4 + 1$	= 17	Prime
$f(3) = 2^{2^3} + 1 = 2^8 + 1$	= 257	Prime
$f(4) = 2^{2^4} + 1 = 2^{16} + 1$	= 65537	Prime
$f(5) = 2^{2^5} + 1 = 2^{32} + 1$	= 4294967297 = 641 · 6700417	Not prime !
$f(6) = 2^{2^6} + 1 = 2^{64} + 1$	= 18446744073709551617	
	= 274177 · 67280421310721	Not prime !
$f(7) = 2^{2^7} + 1 = 2^{128} + 1$	= (see Section 4.6.4)	Not prime !

- The largest prime nonsimple Mersenne number so far with 645,817 decimal digits.

This Fermat number

$$f(2145351) = 2^{(2^{2145351})} + 1$$

is divisible by the prime

$$p = 3 \cdot 2^{2145353} + 1$$

At that time this prime p was the largest known prime generalized Mersenne number and the fifth largest known prime number of all.

$f(18233954)$ is the biggest Fermat number of which a factor is known (as of July 2023).

It is assumed that $f(4) = 65,537$ is the last (and thus also the largest) Fermat prime.

4.6.5 Generalized Fermat Numbers $f(b, n) = b^{2^n} + 1$

With generalized Fermat numbers, the base of the power is no longer restricted to 2. Generalized Fermat numbers are more numerous than Mersenne numbers of an equal size and many of them are waiting to be discovered to fill the big gaps between the Mersenne primes already found or still undiscovered. Progress in number theory made it possible that numbers, where the representation is not limited to the base 2, can be tested at almost the same speed as Mersenne numbers.

The program Proth.exe was widely used to investigate generalized Fermat numbers. Proth.exe was created by Yves Gallot in 1998 as a single-threaded CPU program that found many prime number records more than 20 years ago. The successor *genefer* is a highly optimized GPU application, created in 2022 [12].

Using the original program, on February 16, 2003, Michael Angel discovered the largest of them with 628,808 digits, which at that time became the fifth largest known prime number:

$$b^{2^{17}} + 1 = 62722^{131072} + 1$$

4.6.6 Idea Based on Euclid's Proof: $p_1 \cdot p_2 \cdot \ldots \cdot p_n + 1$

This idea is based on Euclid's proof (see Section 4.3) that there are infinite prime numbers:

$$
\begin{array}{lll}
2{\cdot}3 + 1 & = 7 & \mapsto \text{Prime} \\
2{\cdot}3{\cdot}5 + 1 & = 31 & \mapsto \text{Prime} \\
2{\cdot}3{\cdot}5{\cdot}7 + 1 & = 211 & \mapsto \text{Prime} \\
2{\cdot}3{\cdot}\ldots{\cdot}11 + 1 & = 2311 & \mapsto \text{Prime} \\
2 \cdot 3 \cdot \ldots \cdot 13 + 1 & = 59 \cdot 509 & \mapsto \text{Not prime!} \\
2 \cdot 3 \cdot \ldots \cdot 17 + 1 & = 19 \cdot 97 \cdot 277 & \mapsto \text{Not prime!}
\end{array}
$$

4.6.7 As Above but -1 except $+1$: $p_1 \cdot p_2 \cdot \ldots \cdot p_n - 1$

$$
\begin{array}{lll}
2 \cdot 3 - 1 & = 5 & \mapsto \text{Prime} \\
2 \cdot 3 \cdot 5 - 1 & = 29 & \mapsto \text{Prime} \\
2 \cdot 3 \cdot \ldots \cdot 7 - 1 & = 11 \cdot 19 & \mapsto \text{Not prime!} \\
2 \cdot 3 \cdot \ldots \cdot 11 - 1 & = 2309 & \mapsto \text{Prime} \\
2 \cdot 3 \cdot \ldots \cdot 13 - 1 & = 30029 & \mapsto \text{Prime} \\
2 \cdot 3 \cdot \ldots \cdot 17 - 1 & = 61 \cdot 8369 & \mapsto \text{Not prime!}
\end{array}
$$

4.6.8 Euclid Numbers $e_n = e_0 \cdot e_1 \cdot \ldots \cdot e_{n-1} + 1$ with $n \geq 1$ and $e_0 := 1$

The number e_{n-1} is not the $(n-1)$th prime number, but the number previously found here. Unfortunately this formula is not open but recursive. The sequence starts with

$$
\begin{array}{lll}
e_1 = 1 + 1 & = 2 & \mapsto \text{Prime} \\
e_2 = e_1 + 1 & = 3 & \mapsto \text{Prime} \\
e_3 = e_1 \cdot e_2 + 1 & = 7 & \mapsto \text{Prime} \\
e_4 = e_1 \cdot e_2 \cdot e_3 + 1 & = 43 & \mapsto \text{Prime} \\
e_5 = e_1 \cdot \ldots \cdot e_4 + 1 & = 13 \cdot 139 & \mapsto \text{Not prime!} \\
e_6 = e_1 \cdot \ldots \cdot e_5 + 1 & = 3263443 & \mapsto \text{Prime} \\
e_7 = e_1 \cdot \ldots \cdot e_6 + 1 & = 547 \cdot 607 \cdot 1033 \cdot 31051 & \mapsto \text{Not prime!} \\
e_8 = e_1 \cdot \ldots \cdot e_7 + 1 & = 29881 \cdot 67003 \cdot 9119521 \cdot 6212157481 & \mapsto \text{Not prime!}
\end{array}
$$

Also, $e_9, \ldots, e_{17}$ are composite, which means that this formula is not particularly useful.

Comment:
However, it is very particular that no pair of any of these numbers does have a common factor other than 1. Therefore, they are *relatively prime*.

SageMath Example 4.4 calculates the Euclid numbers with $+1$ and -1.

SageMath Example 4.4: Euclid Numbers with -1

```
print("\n# CHAP04 -- Sage-Script-SAMPLE 015: =========")
print("# Euclid numbers with +1 (and a variant with -1)")

def euclidnumbers(beg, end, variant, startProd):
```

SageMath Example 4.4 (continued)

```
    EProdn = startProd
    for n in (beg..end):
      En = EProdn + variant
      EProdn = EProdn*En
      B = is_prime(En)
      print(n,En,B);

# Initialization
e1 = 2; En=e1; print(1, En, is_prime(En))
e2 = 3; En=e2; print(2, En, is_prime(En))
Prod = e1 * e2

print("--Variant with +1:")
euclidnumbers(3,9, +1, Prod) # default calculation of next Euclid number
print("--Variant with -1:")
euclidnumbers(3,9, -1, Prod) # modified calculation with -1

#------------------------------------
# CHAP04 -- Sage-Script-SAMPLE 015: =========
# Euclid numbers with +1 (and a variant with -1
# 1 2 True
# 2 3 True
# --Variant with +1:
# 3 7 True
# 4 43 True
# 5 1807 False
# 6 3263443 True
# 7 10650056950807 False
# 8 113423713055421844361000443 False
# 9 12864938683278671740537145998360961546653259485195807 False
# --Variant with -1:
# 3 5 True
# 4 29 True
# 5 869 False
# 6 756029 False
# 7 571580604869 False
# 8 326704387862983487112029 False
# 9 106735757048926752040856495274871386126283608869 False
```

4.6.9 $f(n) = n^2 + n + 41$

This sequence starts off *promisingly*, but that is by no means proof that things will continue like this:

$$
\begin{array}{lcl}
f(0) = 41 & \mapsto & \text{Prime} \\
f(1) = 43 & \mapsto & \text{Prime} \\
f(2) = 47 & \mapsto & \text{Prime} \\
f(3) = 53 & \mapsto & \text{Prime} \\
f(4) = 61 & \mapsto & \text{Prime} \\
f(5) = 71 & \mapsto & \text{Prime} \\
f(6) = 83 & \mapsto & \text{Prime} \\
f(7) = 97 & \mapsto & \text{Prime} \\
\vdots & & \\
f(33) = 1163 & \mapsto & \text{Prime} \\
f(34) = 1231 & \mapsto & \text{Prime}
\end{array}
$$

$$\begin{aligned}
f(35) &= 1301 && \mapsto \text{Prime}\\
f(36) &= 1373 && \mapsto \text{Prime}\\
f(37) &= 1447 && \mapsto \text{Prime}\\
f(38) &= 1523 && \mapsto \text{Prime}\\
f(39) &= 1601 && \mapsto \text{Prime}\\
f(40) &= 1681 = 41 \cdot 41 && \mapsto \text{Not prime!}\\
f(41) &= 1763 = 41 \cdot 43 && \mapsto \text{Not prime!}
\end{aligned}$$

The first 40 sequence values are different prime numbers (which have the obvious regularity that their difference starts with 2 and increases by 2 each time). Of the 240 possible prime numbers p with $41 \le p \le 1601$, there are 40 that occur in the sequence.[5]

But the 41st and 42nd values are not prime numbers. It is easy to recognize that $f(41)$ cannot be a prime number:

$$f(41) = 41^2 + 41 + 41 = 41(41 + 1 + 1) = 41 \cdot 43$$

The Euler polynomial $f(n) = n^2+n+41$ can be made visible in the Ulam spiral (antidiagonal) by setting 41 as the starting value in the center. See Figure 4.5.[6]

4.6.10 $f(n) = n^2 - 79n + 1601$ and Heegner Numbers

This function delivers prime numbers for all function values from $n = 0$ to $n = 79$. See Table 4.5. The source to compute this table can be found in SageMath Example 4.12.

Unfortunately, $f(80) = 1681 = 41 \cdot 41$ is not a prime number. To this date, no function has been found that delivers more prime numbers in a row. On the other hand, each prime occurs twice (first in the decreasing then in the increasing sequence), which means that the algorithm delivers a total of 40 different prime values. These are the same ones as delivered by the function in Section 4.6.9.

But there are polynomials that deliver more than 40 distinct prime numbers in a row. In this context, the Heegner numbers play a role. Longer polynomials have been found since 2000, for example, in Al Zimmermann's Programming Contests [15]. For example, $f(n) = n^4 - 97 \cdot n^3 + 3294 \cdot n^2 - 45458 \cdot n + 213589$ returns 49 different, consecutive prime numbers—but only if you count the 9 negative ones as prime.

The polynomial A121887 from OEIS [16] $(n^5-133{\cdot}n^4+6729{\cdot}n^3-158379{\cdot}n^2+1720294 \cdot n - 6823316)/4$ even yields 57 different primes in a row (14 of which are negative), but the coefficients of the polynomial are not integers. You can simulate this with SageMath Example 4.14.

5. How many prime numbers are really in a range can be easily determined with SageMath. Here, for example, with `len(list(primes(41,1602)))` or with `pari(1601).primepi() - pari(40).primepi() = 252-12 = 240`.

6. Graphics from `CT2 Crypto Tutorials ▷ World of Primes ▷ Distribution of primes ▷ Ulam's spiral`.

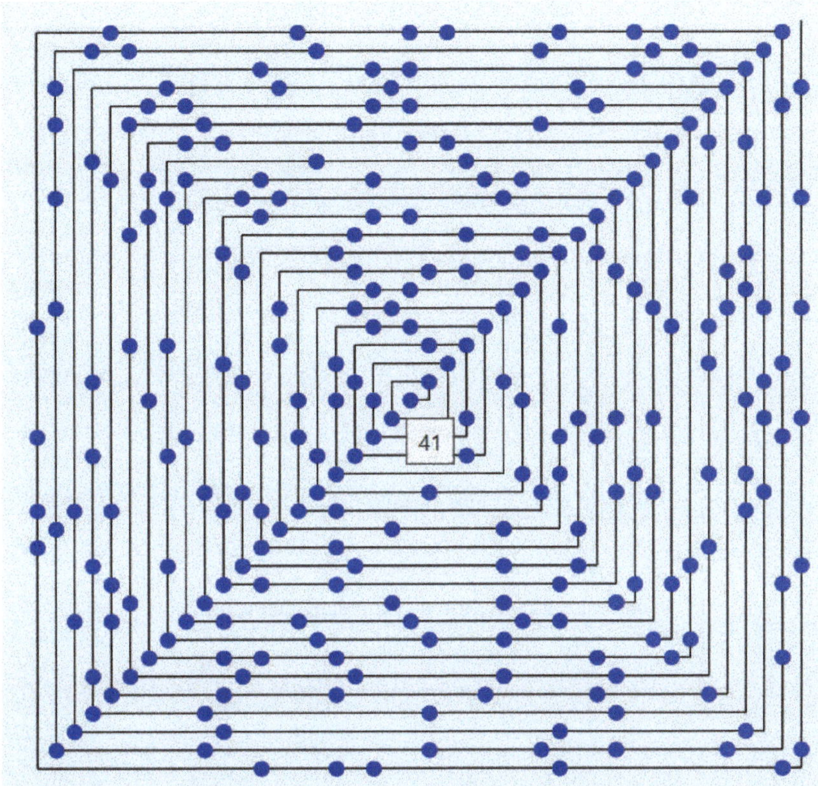

Figure 4.5 Ulam's prime spiral for the Euler polynomial $f(n) = n^2 + n + 41$, starting with 41 in the center.

4.6.11 Polynomial Functions $f(x) = a_n x^n + a_{n-1} x^{n-1} + \cdots + a_1 x^1 + a_0$ ($a_i \in \mathbb{Z}, n \geq 1$)

There is no polynomial that delivers prime values only for all x in $\mathbb{Z}$. For a proof of this, please refer to [17, p. 83 f.], where you will also find further details about prime number formulae.

This means there is no hope in looking for further formulae (functions) similar to that in Section 4.6.9 or Section 4.6.10, if one expects that these produce only primes for all n.

4.6.12 Catalan's Mersenne Conjecture

Eugene Charles Catalan conjectured that C_4 and any further term in this sequence is a prime:

$$
\begin{aligned}
C_0 &= 2, \\
C_1 &= 2^{C_0} - 1, \\
C_2 &= 2^{C_1} - 1, \\
C_3 &= 2^{C_2} - 1, \\
C_4 &= 2^{C_3} - 1, \quad \ldots
\end{aligned}
$$

Table 4.5 Values of the Prime Number Function $f(n) = n^2 - 79 \cdot n + 1601$

$f(0) = 1601$	$\mapsto$ prime	$f(26) = 223$	$\mapsto$ prime
$f(1) = 1523$	$\mapsto$ prime	$f(27) = 197$	$\mapsto$ prime
$f(2) = 1447$	$\mapsto$ prime	$f(28) = 173$	$\mapsto$ prime
$f(3) = 1373$	$\mapsto$ prime	$f(29) = 151$	$\mapsto$ prime
$f(4) = 1301$	$\mapsto$ prime	$f(30) = 131$	$\mapsto$ prime
$f(5) = 1231,$	$\mapsto$ prime	$f(31) = 113$	$\mapsto$ prime
$f(6) = 1163$	$\mapsto$ prime	$f(32) = 97$	$\mapsto$ prime
$f(7) = 1097$	$\mapsto$ prime	$f(33) = 83$	$\mapsto$ prime
$f(8) = 1033$	$\mapsto$ prime	$f(34) = 71$	$\mapsto$ prime
$f(9) = 971$	$\mapsto$ prime	$f(35) = 61$	$\mapsto$ prime
$f(10) = 911$	$\mapsto$ prime	$f(36) = 53$	$\mapsto$ prime
$f(11) = 853$	$\mapsto$ prime	$f(37) = 47$	$\mapsto$ prime
$f(12) = 797$	$\mapsto$ prime	$f(38) = 43$	$\mapsto$ prime
$f(13) = 743$	$\mapsto$ prime	$f(39) = 41$	$\mapsto$ prime
$f(14) = 691$	$\mapsto$ prime	$f(40) = 41$	$\mapsto$ prime
$f(15) = 641$	$\mapsto$ prime	$f(41) = 43$	$\mapsto$ prime
$f(16) = 593$	$\mapsto$ prime	$f(42) = 47$	$\mapsto$ prime
$f(17) = 547$	$\mapsto$ prime	$f(43) = 53$	$\mapsto$ prime
$f(18) = 503$	$\mapsto$ prime	...	
$f(19) = 461$	$\mapsto$ prime	$f(77) = 1447$	$\mapsto$ prime
$f(20) = 421$	$\mapsto$ prime	$f(78) = 1523$	$\mapsto$ prime
$f(21) = 383$	$\mapsto$ prime	$f(79) = 1601$	$\mapsto$ prime
$f(22) = 347$	$\mapsto$ prime	$f(80) = 41 \cdot 41$	$\mapsto$ NOT prime!
$f(23) = 313$	$\mapsto$ prime	$f(81) = 41 \cdot 43$	$\mapsto$ NOT prime!
$f(24) = 281$	$\mapsto$ prime	$f(82) = 1847$	$\mapsto$ prime
$f(25) = 251$	$\mapsto$ prime	$f(83) = 1933$	$\mapsto$ prime
		$f(84) = 43 \cdot 47$	$\mapsto$ NOT prime!

This sequence is defined recursively and increases extremely fast (much quicker than the Mersenne prime numbers). Does this sequence consist only of primes?

$$
\begin{array}{lll}
C_0 = 2 & & \mapsto \text{Prime} \\
C_1 = 2^2 - 1 & = 3 & \mapsto \text{Prime} \\
C_2 = 2^3 - 1 & = 7 & \mapsto \text{Prime} \\
C_3 = 2^7 - 1 & = 127 & \mapsto \text{Prime} \\
C_4 = 2^{127} - 1 & = 170141183460469231731687303715884105727 & \mapsto \text{Prime}
\end{array}
$$

It is not (yet) known whether $C_5 = 2^{C_4} - 1$ and all higher elements are prime. In any case, it has not been proved that this formula delivers only primes.

It seems very unlikely that C_5 (or many of the larger terms) would be prime. So this could be another example of Guy's law of small numbers.

4.6.13 Double Mersenne Primes

From C_2 onwards, the above Catalan-Mersenne numbers are a subset of the double Mersenne primes. A double Mersenne prime is a Mersenne prime of the form

$$M_{M_p} = 2^{2^p - 1} - 1$$

where p is a Mersenne prime exponent and M_p is a prime Mersenne number.

The first values of p for which M_p is prime are $p = 2, 3, 5, 7, 13, 17, 19, 31, 61, 89, 107, 127, 521, \ldots$ (see above).

M_{M_p} is known to be prime for $p = 2, 3, 5, 7$, and has the appropriate values:

$$7, 127, 2147483647, 170141183460469231731687303715884105727.$$

SageMath Example 4.5 calculates these values.

SageMath Example 4.5: Double Mersenne Primes

```
print("\n# CHAP04 -- Sage-Script-SAMPLE 010: =========")
print("# Double Mersenne primes")

for p in (2,3,5,7):
    Mp=2^p-1
    MMp=2^Mp-1
    B=is_prime(MMp)
    print(p,Mp,MMp,B);

#------------------------------------
# CHAP04 -- Sage-Script-SAMPLE 010: =========
# 2 3 7 True
# 3 7 127 True
# 5 31 2147483647 True
# 7 127 170141183460469231731687303715884105727 True
```

For $p = 11, 13, 17, 19$, and 31, the corresponding double Mersenne numbers are not prime. The next candidate for the next double Mersenne prime is $M_{M_{61}} = 2^{2305843009213693951} - 1$.

Being approximately $1695 \cdot 10^{694127911065419641}$ this number—like C_5 (see Section 4.6.12)—is far too large for any currently known primality test to be successfully applied.

4.7 Density and Distribution of the Primes

As Euclid proved, there is an infinite number of primes. However, some infinite sets are *denser* than others.

Within the set of natural numbers, there is an infinite number of even, uneven, and square numbers. How to compare the density of two infinite sets is shown with the sets of even and square numbers.

Whereas in colloquial language you can often hear that "there are more" even numbers than square ones, mathematicians say that from both there are infinitely many that their sets are equivalent to $\mathbb{N}$ (so both are infinite and countable; i.e., one can assign an integer to each even number and to each square number). However, the set of even numbers is denser than the set of square numbers. The following proves that the even numbers are distributed more densely than square ones:

- The size of the nth element:

  The nth element of the even numbers is $2n$; the nth element of the square numbers is n^2. Because for all $n > 2$: $2n < n^2$, the nth even number occurs much earlier than the nth square number.

- The numbers of values that are less than or equal to a certain *maximum value* x in $\mathbb{R}$ are:

  There are $\lfloor x/2 \rfloor$ such even numbers and $\lfloor \sqrt{x} \rfloor$ square numbers. Because for all $x > 6$ the value $\lfloor x/2 \rfloor$ is greater than the largest integer smaller or equal to the square root of x, the even numbers are distributed more densely.

The Value of the nth Prime $P(n)$

Theorem 4.6 For large n: The value of the nth prime $P(n)$ is asymptotic to $n \cdot \ln(n)$; that is, the limit of the relation $P(n)/(n \cdot \ln n)$ is equal to 1 if n tends to infinity.

For $n > 5$, $P(n)$ lies between $2n$ and n^2. This means that prime number are less dense than natural numbers, but denser than square numbers. See Section 4.10.

The number of prime numbers $PI(x)$. The definition for the number $PI(x)$—also conventionally written as $\Pi(x)$, is similar: It is the number of all primes that does not exceed the maximum value x.

Theorem 4.7 $PI(x)$ is asymptotic to $x/\ln(x)$.

This is the famous *prime number theorem*. It was put forward by Adrien-Marie Legendre and Carl Friedrich Gauss but not proved until over 100 years later.

Alternative ways of expressing this are:

$$PI \to x/\ln(x)$$

$$\lim_{x\to\infty}\left(PI \cdot \frac{\ln(x)}{x}\right) = 1$$

$$\lim_{x\to\infty}\left(\ln(x) = \frac{x}{PI}\right)$$

The distribution is graphically presented in Figure 4.10 in Section 4.14.

The formulae for the prime number theorem only apply when n tends to infinity. The formula of Gauss can be replaced by more precise formulae. For $x \geq 67$:

$$\ln(x) - 1,5 < x/PI(x) < \ln(x) - 0,5$$

Given that we know $PI(x) = x/\ln x$ only for very large x (x tending towards infinity), we can create the following overview:

x	$\ln(x)$	$x/\ln(x)$	$PI(x)$(counted)	$PI(x)/(x/\ln(x))$
10^3	6.908	144	168	1.160
10^6	13.816	72386	78498	1.085
10^9	20.723	48254942	50847534	1.054

For a binary number (these consist only of the digits 0 and 1) x of length of 250 bits (2^{250} is approximately $1.81 \cdot 10^{75}$) and because the quotient $PI(x)/(x/\ln(x))$

is moving closer to 1, PI(x) can be estimated very well this way:

$$PI(x) = x/\ln x = 2^{250}/(250 \cdot \ln 2) \approx 2^{250}/173.28677 \approx 1.05 \cdot 10^{73}$$

We can therefore expect that the set of numbers with a bit length of less than 250 contains approximately 10^{73} primes—a reassuring result!

We can also express this as follows: Let us consider a *random* natural number n. Then the probability that this number is prime is around $1/\ln(n)$. For example, let us take numbers in the range of 10^{16}. Then we must consider $16 \cdot \ln 10 = 36.8$ numbers (on average) until we find a prime. A precise count shows that there are 10 prime numbers between $10^{16} - 370$ and $10^{16} - 1$.

Another way to express this is: The *average gap* between two consecutive primes near the number n is close to the natural logarithm of n. For example, for a number n close to 100, $\ln(n) \approx 4.6$, so roughly every fifth number in this range should be prime. Further details about prime gaps can be found on [18] and in Section 4.9.5.

Under the heading *How Many Primes Are There* in [19], you can find numerous other details. Using the website in [20] you can easily determine $P(n)$ and $PI(x)$.

The *distribution* of primes displays several irregularities for which no system has been found yet. On the one hand, many occur closely together, like 2 and 3, 11 and 13, 809 and 811, on the other hand large gaps containing no primes also occur. For example, no primes lie between 113 and 127, 293 and 307, 317 and 331, 523 and 541, 773 and 787, 839 and 853 as well as between 887 and 907. Discovering the secrets of these irregularities is precisely part of what motivates mathematicians. Some visualizations (plots) of the quantity of primes in different number dimensions can be found in Section 4.14.

Sieve of Eratosthenes. An easy way of calculating all $PI(x)$ primes less than or equal to x is to use the sieve of Eratosthenes. In the third century B.C., he found an extremely easy, automatic way of discovering this. To begin with, you write down all numbers from 2 to x, circle 2, then cross out all multiples of 2. Next, you circle the lowest number that hasn't been circled or crossed out (now 3) and again cross out all multiples of this number, and so on.

You have to keep crossing out only until you reach the largest number whose square is less than or equal to x (here up to 10, as 11^2 is already >120).[7]

Prime numbers never end in the digit 0. Apart from 2, prime numbers are never even. Apart from 2 and 5, prime numbers never end in 2 or 5. So you only need to consider numbers ending in 1, 3, 7, or 9 anyway (there are infinite primes ending in each of these digits; see [22, Vol. 1, p. 137]).

4.8 Outlook

Currently, you can find large *databases* that contain either many primes or the factorization of numerous composite numbers. The fastest *factorizers* on the internet are FactorDB by Markus Tervooren [23] and Alpertron, the integer factorization

7. `CT2 Crypto Tutorials` ▷ `World of Primes` also contains a visualization of this method.

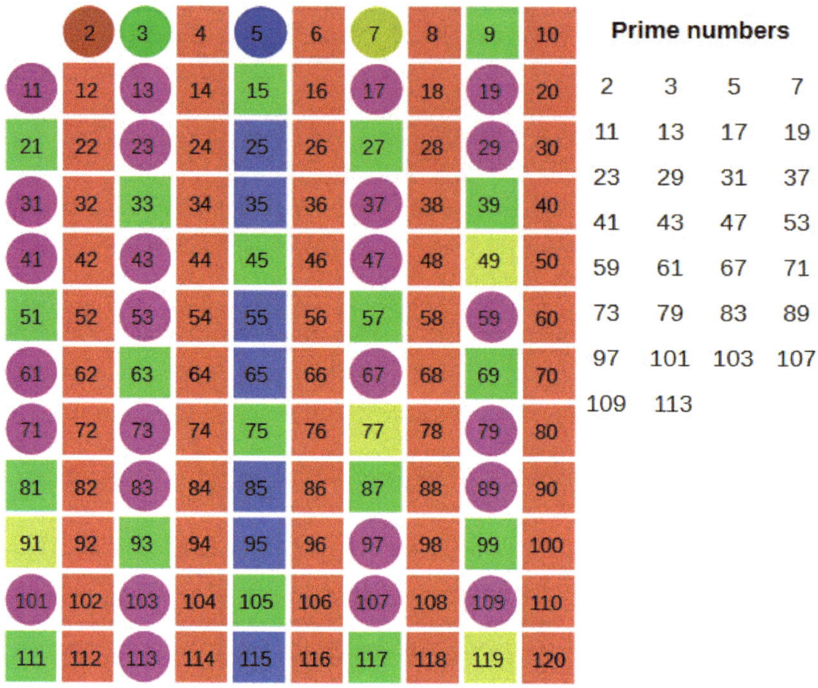

Figure 4.6 The sieve of Eratosthenes, applied to the first 120 numbers [21].

calculator by Dario Alpern [24]. Sometimes they just look in their databases to see if they already know how to factor the entered number; then it is particularly quick. Alpertron has implemented the two algorithms elliptic curve method (ECM) and self-initializing quadratic sieve (SIQS) with WebAssembly so that they can run purely locally in the browser. Another factorizer that runs purely locally in the browser uses the Msieve library (https://www.cryptool.org/en/cto/msieve).

Further Interesting Topics Regarding Prime Numbers

This chapter didn't consider other number theory topics such as divisibility rules, modulus calculation, modular inverses, modular powers, modular roots, Chinese remainder theorem, Euler Phi function, or perfect numbers. Some of these topics are considered in the *next chapter*.

4.9 Notes about Primes

The following lists some interesting theorems, conjectures, and open questions about primes, as well as some peculiar things and overviews.

4.9.1 Proven Statements and Theorems about Primes

- For each number n in $\mathbb{N}$ there are n consecutive natural numbers that are not primes (prime gaps). A proof of this can be found in [17, p. 79]. See Section 4.9.6.

- Paul Erdös proved that between each random number not equal to 1 and its double, there is at least one prime. He was not the first to prove this theorem, but he proved it in a much simpler manner than those before him.
- There is a real number a such that the function $f : \mathbb{N} \to \mathbb{Z}$ where $n \mapsto \lfloor a^{3^n} \rfloor$ only delivers primes for all n (see [17, p. 82]). The Gauss bracket $\lfloor x \rfloor$ of a real number x is defined via: $\lfloor x \rfloor$ is the next integer less or equal x. Unfortunately, problems arise when we try to determine a (see Section 4.9.3).

4.9.2 Arithmetic Prime Sequences

There are arithmetic prime sequences of arbitrary length. An *arithmetic progression* or arithmetic sequence is a sequence of numbers such that the difference between the consecutive terms is constant. Arithmetic sequences, consisting only of primes, are called *prime arithmetic progressions*. If such a sequence has exactly k elements it's abbreviated with PAP-k or AP-k. In 1923 the famous British mathematician Godfrey Harold Hardy made the conjecture that there are arithmetic sequences of arbitrary length that consist of primes only. This conjecture was proven in 2004 by two young American mathematicians.

In school, children normally learn about *arithmetic number sequences*. These are sequences of numbers for which the difference between any two consecutive numbers is equal or constant (an arithmetic sequence must have at least three elements but can also have indefinitely many). In the sample sequence 5, 8, 11, 14, 17, 20 the difference between the sequence's elements is 3 and the length of the sequence is 6.

Arithmetic sequences have been known for millennia and one would think they have no more secrets. They become more interesting again if we impose additional constraints on the sequence's elements, as the prime example shows.

For example, 5, 17, 29, 41, 53 is an *arithmetic prime sequence* that consists of five elements and the difference between the elements is always 12. These sequences are abbreviated to PAP-k. Note that the prime numbers here do not necessarily have to be consecutive. See the stronger requirement about CPAP-k in Section 4.9.2.1.

The sequence 5, 17, 29, 41, 53 is not extendable—the next element would be 65, but 65 is not prime (65 is the product of 5 and 13). Therefore, this sequence has the maximal length of $k = 5$ and belongs to PAP-5, described by $f(n) = d \cdot n + a$ with $d = 12$ and $a = 5$ (a is the first or start element). $f(n)$ is usually expressed with $n = 0$ to $k - 1$.

A further sample is the PAP-10 sequence 199, 409, 619, 829, 1039, 1249, 1459, 1669, 1879, and 2089, where there is a difference of 210 between the consecutive primes.

How many elements are possible within an arithmetic prime number sequence? Around 1770 the French Joseph-Louis Lagrange and the British Edward Waring investigated this question. In 1923 Godfrey Harold Hardy and his colleague John Littlewood theorized that there is no upper limit for the number of elements. But they could not prove this. In 1939 more progress was achieved: The Dutch mathematician Johannes van der Corput was able to prove that there are infinitely many different arithmetic prime number sequences with exactly three elements. Two examples are 3, 5, 7 and 47, 53, 59. Within the first 5,000 prime numbers we counted 244 such triples.

The longest known arithmetic prime number sequence contains 27 elements (as of January 2023). Table 4.6 lists the longest currently known arithmetic prime number sequences with minimal difference. In Table 4.6, n is the number of elements of the sequence and $Digits$ is the number of digits of the difference d of the sequence elements. How to read table Table 4.6 (Table 4.7 shows the values of $k\#$)?

For $n = 3$: $d = 2 = 2\#$ (so $k = 2$). The sequence is: $3, 5, 7$.

For $n = 4$: $d = 6 = 3\#$ (so $k = 3$). The sequence is: $5, 11, 17, 23$.

For $n = 5$: $d = 6 = 4\# = 3\#$ (so $k = 3$). The sequence is: $5, 11, 17, 23, 29$.

For $n = 6$: $d = 30 = 5\#$ (so $k = 5$). The sequence is: $7, 37, 67, 97, 127, 157$.

For $n = 7$: $d = 150 = 5 * 5\#$ (so multiple of k# with $k = 5$). The sequence is: $7, 157, 307, 457, 607, 757, 907$.

Table 4.6 was built using [25] and lists the sequences that have the smallest known difference for a given length. In contrast, the "largest known AP-k" listed in [26] contain as the last sequence element a prime as large as possible.

As a team, the two mathematicians Ben Green and Terence Tao were able in 2004 to prove Hardy's conjecture, which had puzzled mathematicians for over 80 years. It states that for any arbitrary length there exists an arithmetic prime number sequence (PAP). Additionally, they managed to prove that for any given length there are infinitely many different sequences.

Table 4.6 Arithmetic Prime Number Sequences with Minimal Distance*,**

n	*First Element*	*Distance d*	*Year* / *Digits*	*Discovered By*
3	3	2 $= 2\#$	- / 1	
4	5	6 $= 3\#$	- / 1	
5	5	6 $= 3\#$	- / 1	
6	7	30 $= 5\#$	1909 / 2	G. Lenaire
7	7	150 $= 5 \cdot 5\#$	1909 / 3	G. Lenaire
...				
...				
21	28112131522731197609	9699690 $= 19\#$	2008 / 7	Jaroslaw Wroblewski
22	166537312120867	96599212710 $= 9959 \cdot 19\#$	2006 / 11	Markus Frind
23	403185216600637	2124513401010 $= 9523 \cdot 23\#$	2006 / 13	Markus Frind
24	158209144596158501	14517322329510 $= 65073 \cdot 23\#$	2014 / 14	Bryan Little
25	6171054912832631	81737658082080 $= 366384 \cdot 23\#$	2008 / 14	Raanan Chermoni, Jaroslaw Wroblewski
26	3486107472997423	371891575525470 $= 1666981 \cdot 23\#$	2012 / 15	James Fry
27	224584605939537911	18135696597948930 $= 81292139 \cdot 23\#$	2019 / 18	Rob Gahan

* Smallest known AP-n.

** As of April 2022.

Table 4.7 Products of the First Primes <= k, Called k Primorial or $k\#$

i	k	$k\#$
1	2	2
2	3	6
3	5	30
4	7	210
5	11	2310
6	13	30030
7	17	510510
8	19	9699690
9	23	223092870
10	29	6469693230
11	31	200560490130

Green and Tao intended to prove that there are infinitely many arithmetic sequences of length four. For this purpose they considered sets of numbers consisting of primes and near primes. These are numbers with a small set of divisors like numbers that are the product of exactly two primes—these numbers are called half primes. Thus, they managed to considerably simplify their work because there were already a lot of useful theorems about near primes. Finally, they discovered that the results of their theorem were far more reaching than they had assumed. Therefore, they were able to prove Hardy's conjecture.

Anyone who believes that it is easy to use Green's and Tao's 49-page proof to compute arithmetic prime number sequences of arbitrary length will soon become disappointed, because the proof is nonconstructive. It is a proof of existence. This means that these mathematicians have shown "only" that these sequences exist, but not how to find them in practice.

This indicates that in the set of the natural numbers there is, for example, a sequence of one billion primes that all have the same distance; and there are infinitely many of them. However, these sequences lie extremely far beyond the numbers we usually use (*far outside*).

The length of a sequence determines the minimal distance between the single primes of the sequence. Given a sequence with $n = 6$ elements the distance between them has to be 30 or a multiple of 30. More precisely: If a PAP-k does *not* begin with the prime k, then the common difference is a multiple of the primorial k#. The number 30 results from the product of all primes smaller than the length of the sequence (here 6): $6\# = 5\# = 2 \cdot 3 \cdot 5 = 30$. Another example: $10\# = 7\# = 2 \cdot 3 \cdot 5 \cdot 7 = 210$. If you look for a sequence with 15 elements, then the common distance is at least $15\# = 13\# = 2 \cdot 3 \cdot 5 \cdot 7 \cdot 11 \cdot 13 = 30030$. The primorial notation starts with $k = 2$: $2\# = 2,\ 3\# = 2 \cdot 3 = 6,\ 5\# = 2 \cdot 3 \cdot 5 = 30$.

This signals that the length of an arithmetic prime sequence can be arbitrary big, but the distance between the elements cannot be any arbitrary number. For example, there is no arithmetic prime sequence with the distance 100 because 100 cannot be divided by 3.

The minimal distances for sequences of length n are multiples of $k\#$ with $k = n - 1$. The k-primorials (prime factorials) are shown in Table 4.7—the values were calculated by SageMath Example 4.6.

4.9.2.1 Further Restriction for the Arithmetic Progression

If you look at arithmetic prime sequences that fulfill the *additional* requirement that all primes are *consecutive*, also called *consecutive prime arithmetic progressions* (or CPAP-k or CAP-k) then the longest such sequence (as of May 2022), has a length of 10. The first CPAP-10 was found in 1998; it has the common distance $10\# = 7\# = 210$ and starts with the 93-digit prime: 100 9969724697 1424763778 6655587969 8403295093 2468919004 1803603417 7589043417 0334888215 9067229719.

SageMath Example 4.6: Calculate the k Primorials (k#) for Table 4.7

```
print("\n# CHAP04 -- Sage-Script-SAMPLE 060: =========")

# calculate the k primorials (k#) for tables in chapter 4
# k# = product of first k primes. Note, function defines k#=1 for k=1 (besides k=1 is no prime)
# k# = defined in Sage as a Sloane sequence
# https://doc.sagemath.org/html/en/reference/combinat/sage/combinat/sloane_functions.html

print('\nGet k# = primorial numbers from OEIS Integer Sequence for Table 4.8:')
n=33; print(' Upper border for prime k (k<n): n =', n)

primlist = [i for i in primes(1,n)]
lenprimlist = len(primlist)
print(' lenprimlist =', lenprimlist, '; pList =', primlist)
# P=Primes(); print(' Alternative to get prime k using unrank: ', [P.unrank(i-1) for i in range▸
 ▸(1,lenprimlist+1)])

a = sloane.A002110  # Begins with 1 instead of 2
aa = a.list(lenprimlist+1)[1:] # get lenprimlist elements and discard 1st element
print(' lenaa =', len(aa), '; aaList =', aa)

L = [(i,primlist[i-1],aa[i-1]) for i in range(1, lenprimlist+1)]
print(' List of tuples (i, k, k#):')
print(' ', L)

print('\nSome distances for Table 4.7:')
print(' 19# =', a[8])
print(' 9959·19# =', 9959*a[8])
print(' 9523·23# =', 9523*a[9])
print(' 65 073·23# =', 65073*a[9])
print(' 366 384·23# =', 366384*a[9])
print(' 1 666 981·23# =', 1666981*a[9])
print(' 81 292 139·23# =', 81292139*a[9])
```

If one compares the maximum length of prime number sequences, depending on how they were created, it turns out that the more restrictions one has, the shorter the maximum length found so far. Given the following definitions, arithmetic prime number sequences (PAP) are a special case of polynomial functions (see Section 4.6.11); CPAP are a special case of PAP. The abbreviations are: PAP = prime in arithmetic progression; CPAP = consecutive PAP; PbP = prime by polynomial:

max length CPAP < max length PAP < max length PbP. So currently $10 < 27 < 57$.

4.9.3 Unproven Statements, Conjectures, and Open Questions about Primes

- Christian Goldbach speculated that every even natural number greater than 2 can be represented as the sum of two prime numbers (see Section 4.9.4). Sample: $12 = 7 + 5$.

- Bernhard Riemann put forward an important but still unproved hypothesis about the location of the nontrivial zeros of the Riemann zeta function. A consequence is an improved estimation within the prime number theorem (distribution of primes).
- Benford's law does not apply to primes. According to Benford's law, also called the first-digit law, the single digits in lists of numbers from many (but not all) real-life sources of data are distributed in a nonuniform way. In particular, the leading digit is much more often 1 than any other digit.

  Which empirical data applies to this law is not completely clear yet. Timo Eckhardt extensively analyzed attributes of prime numbers in his thesis in 2008. For example, all primes up to 7,052,046,499 were described with different bases of the positional notation.

  Comparing the bases 3 to 10 the deviation from Benford's law was lowest with base 3. Comparing the first digit for base 10 he found all digits are almost equally distributed. Analyzing bigger bases showed strong differences.
- The proof (mentioned in Section 4.9.1) of the function $f : N \to Z$ with $n \mapsto \lfloor a^{3^n} \rfloor$ only guarantees the existence of such a number a. How can we determine this number a and will it have a value, also making the function of some practical interest?
- Is there an infinite number of Mersenne prime numbers?
- Is there an infinite number of Fermat prime numbers?
- Does a polynomial time algorithm exist for calculating the prime factors of a number (see [27, p. 167])? This question can be divided into the three following questions:

  - Does a polynomial time algorithm exist that decides whether a number is prime?

    This question has been answered by the AKS algorithm (see Section 5.12.5.3: Primality testing is polynomial).
  - Does a polynomial time algorithm exist that determines for a composite number how many prime factors it is made up of (without calculating these factors)?
  - Does a polynomial time algorithm exist that calculates for a composite number n a nontrivial factor (nontrivial means other than 1 and n itself)? See Sections 5.12.4 and 5.12.5.1.

  Table 5.13 shows the different orders of magnitude up to which the current algorithms for primality testing and for factorization deliver good results.

4.9.4 The Goldbach Conjecture

Here we will have a closer look at the *Goldbach conjecture.* There are two forms: The "strong" one is considered the actual Goldbach conjecture. The "weak" one is also known as the *odd* or *ternary* Goldbach conjecture.

4.9.4.1 The Weak Goldbach Conjecture

In a letter to the mathematician Euler, Goldbach conjectured in 1742:

Every *odd* integer greater than 5 can be
represented as the sum of *exactly three* prime numbers.

Samples: $7 = 3 + 2 + 2$ or $27 = 19 + 5 + 3$ or $27 = 17 + 5 + 5$.

The weak Goldbach conjecture was proven by the Peruvian mathematician Harald Helfgott—more than 250 years after the original conjecture was stated. Previous results had proved that the weak Goldbach conjecture is true for all odd integers larger $e^{3100} \approx 2 \cdot 10^{1346}$. In another preliminary work [28], Terence Tao from the University of California proved that every *odd* natural number greater than 1 can be represented as the sum of *at most five* prime numbers.

Before Helfgott's proof (rigorous result), people tried to find out with the assistance of computers for as many numbers as possible, whether the conjecture is true or not. Computers verified the *weak Goldbach conjecture* for all odd natural numbers up to: $4 \cdot 10^{18}$ (simple check April 2012) and $4 \cdot 10^{17}$ (double check May 2013).

The lower limit set by Helfgott's proof in 2013 was 10^{30}. This was small enough to handle the remaining cases by computer (which he did with David Platt, and thus checked the validity for all numbers below $8.875 \cdot 10^{30}$ with computer help) [29].

4.9.4.2 The Strong Goldbach Conjecture

Goldbach's strong prime number hypothesis, also known as the *even* or *binary* Goldbach conjecture, was formulated by Euler after a mail exchange with Goldbach. This is now called the Goldbach conjecture:

Every *even* integer greater than 2 can be represented as the sum of
exactly two prime numbers.

Samples of Goldbach partitions: $8 = 5 + 3$ or $28 = 23 + 5$.

It is generally accepted today that the Goldbach conjecture is true (i.e., valid for all even natural numbers greater than 2). With computers the Goldbach conjecture was verified for all even numbers up to $4 \cdot 10^{18}$ (May 2013), but no general proof has been found yet. The theorem that has come closest so far to Goldbach's conjecture was proved by Chen Jingrun in 1966 [30] in a way that is somewhat hard to understand: Each even integer greater than 2 is the sum of one prime and of the product of two primes. For example, $20 = 5 + 3 \cdot 5$.

This conjecture specifically makes it clear that even today we do not have a complete understanding of the deeper connections between addition and multiplication of natural numbers.

The bigger an even number is, the more such binary Goldbach partitions can be found—on average. For four there is only one partition $2 + 2$; for 16 there are two, $3 + 13$ and $5 + 11$. With 100 there are six such partitions: $3 + 97$, $11 + 89$, $17 + 83$, $29 + 71$, $41 + 59$, $47 + 53$.

Alon Amit classified the meaning of the strong Goldbach conjecture this way: The conjecture itself is not useful for solving other things. Contrary to the Riemann hypothesis, Goldbach's conjecture isn't a lemma, a stepping stone, or useful tool.

4.9.4.3 Interconnection Between the Two Goldbach Conjectures

If the strong Goldbach conjecture holds, then the weak one is also true (so the strong implies the weak conjecture). The proof for this is relatively simple:

Prerequisite: u is an odd number bigger than 5.

Each such odd number u can be written as sum $u = (u - 3) + 3$. The first summand then is even and ≥ 4, so it fulfills the prerequisite of the strong Goldbach conjecture, and can be written as sum of two primes p_1 and p_2 (p_1 and p_2 are not necessarily different). So we found a partition of u into the three primes p_1, p_2, and 3. This means it is always possible to find such a sum, where one of the primes is the number 3.

Similarly, it can be shown that the weak Goldbach conjecture implies the above-mentioned conjecture from Terence Tao (both hold for odd numbers) and a third statement following from the proof of the weak Goldbach conjecture by Helfgott:

- For odd numbers $u > 5$, the weak Goldbach conjecture directly implies that the sum consists of five primes at most.
- For the remaining odd numbers 3 and 5 you can directly check it:
  $3 = 3$ (the sum has only one and therefore at most five prime summands);
  $5 = 2 + 3$ (the sum has two and therefore at most five prime summands).
- Every even number $n \geq 4$ is the sum of at most 4 primes.

As with many famous conjectures in mathematics, there are also a number of purported proofs of the Goldbach conjecture, but none have been accepted by the mathematical community yet.

4.9.5 Open Questions about Twin Primes

Twin primes are prime numbers whose difference is exactly 2. Examples include 5 and 7, or 101 and 103, or $1693965 \cdot 2^{66443} \pm 1$, or $318032361 \cdot 2^{107001} \pm 1$. *Cousin* primes are prime numbers that differ, for example, by four, like 13 and 17.

The conjecture that there are infinite many twin primes is not obvious. It's known that for large numbers in average the expected gap between primes is constantly growing at around 2.3 times the number of decimal digits. For example, among 100-digit decimal numbers the expected gap between primes is in average 230. But this statement is true just on average—often the gap is much bigger, or much smaller.

Note: There is only one *triplet* prime: 3, 5, 7. For all other sets of three consecutive odd numbers, one of them is always divisible by 3 and thus not a prime.

The biggest known twin pairs (as of April 2023) are:

$$2,996,863,034,895 \cdot 2^{1290000} \pm 1 \text{ with 388342 decimal digits}$$

$$3,756,801,695,685 \cdot 2^{666669} \pm 1 \text{ with 200700 decimal digits}$$

They were found in 2011 and 2016.

Open questions about twin primes are:

- What is the number of twin primes: Are there infinitely many or only a limited number?
- Does a formula exist for calculating the number of twin primes per interval?

In order to approach such questions, one can take different paths and ask different leading questions.

One was asked by the Norwegian mathematician Viggo Brun (1885–1978), who looked at the sum of the reciprocals of successive twin primes. In 1919, he proved that this sum converges to a specific numerical value (≈ 1.90216), which is now called *Brun's constant*. The fact that this sum converges shows that twin primes are relatively rare, even though there might be infinitely many of them. In contrast, the sum of the reciprocals of all primes diverges.

It's interesting that the *Pentium FDIV bug* was found in 1994 by Thomas Nicely when he used massive computing power to calculate Brun's constant. The flaw in the Pentium microprocessor caused only certain types of arithmetic errors. However, his discovery forced the chip's manufacturer to replace about one million faulty processors (costing Intel about half a billion dollars).

In the following subsections, two major milestones are explained that may allow us to come closer to the riddle of the number of twin primes.

4.9.5.1 GPY 2003, Proof Correction 2005

A big step toward the solution of this problem was made by Dan Goldston, János Pintz, and Cem Yildirim in 2003. The three mathematicians were investigating the distribution of prime numbers. They could prove that

$$\liminf_{n\to\infty} \frac{p_{n+1} - p_n}{\log p_n} = 0,$$

where p_n denotes the nth prime number.

This means that the smallest limit point (lim inf) of the sequence $\frac{p_{n+1}-p_n}{\log p_n}$ equals zero. A point is called limit point of a sequence, if—in any arbitrary small neighborhood of that point—there lie infinitely many elements of the sequence. $\log p_n$ is about the average distance between the prime p_n and the next prime p_{n+1}. Hence, the term above implies that there are infinitely many consecutive primes with a gap between them, which is arbitrarily small compared to the expected average gap.

Moreover, it was proved that

$$p_{n+1} - p_n < (\log p_n)^{8/9}$$

holds true for infinitely many primes.

4.9.5.2 Zhang 2013

In May 2013, the results of Yitang Zhang became known. Zhang proved that there are infinitely many cousin primes, or more explicitely, that there is some number H smaller than 70 million such that there are infinitely many pairs of primes that differ by H. Whereas the gap between the primes of a twin prime is exactly 2, cousin primes do denote two primes that have a gap between them, which has a value of a bigger, even, but finite number H. In the meantime this minimal gap H of 70 millions was improved in further work. The corresponding progress is documented in the Polymath8 project Bounded Gaps between Primes. The best known value of H was 4680 (as of August 2013) and is until now (as of April 2022) 2460—this is good progress compared to 70 million, but far away from 2. H then has been reduced to 6 (but only on the assumption of the Elliott-Halberstam conjecture).

Those results could be the basis for a final proof that infinitely many twin primes exist.

4.9.6 Prime Gaps

A prime gap is the difference between two consecutive primes: $g_m = p_{m+1} - p_m$. The smallest prime number gap is the first one: $g_1 = 3 - 2 = 1$. All other prime number gaps are even, since 2 is the first and only even prime number and thus the difference of all others is formed between two odd numbers. The second prime number gap is $g_2 = 5 - 3 = 2$. Some authors use *prime number gap* differentiating this as the number of composite numbers between two prime numbers, which is one less than according to the definition used here. So their g_2 would be $(5 - 3) - 1 = 1$. Gaps between consecutive prime numbers were already discussed briefly in Section 4.7.

Whether there are infinitely many twin primes (i.e., gaps of length 2) is one of the great unsolved problems in mathematics. See Section 4.9.5.

Here we mention a fact that initially is somewhat surprising: In the sequence of *all* prime numbers $p_1, p_2, p_3, \ldots$, there are gaps between prime numbers of an arbitrary long length n. In other words, there exists a sequence of $n - 1$ consecutive *composite* integers ≥ 0 for any given value of n. That is, for any positive integer n, there is an integer m with $g_m \geq n$, where m is the index of the gap.

It is easy to argue that such gaps of length n exist. Let N be a natural number that is *not* coprime to any of the numbers $2, 3, 4, \ldots, n$. Then the numbers $N + 2, N + 3, N + 4, \ldots, N + n$ are also not coprime to N, and consequently they are not prime numbers. So the largest prime number before this sequence is at most $N + 1$; the smallest afterward is at least $N + n + 1$, so that the length of this prime number gap is at least n.

Such an N can be constructed in at least three different ways:

1. Factorial $N = n!$

   This is technically the easiest way to prove it. Then the considered $N + k$ in the sequence $N + 2, N + 3, N + 4, \ldots, N + n$ are each divisible by the k.

2. $N = lcm(2, \ldots, n)$

   You can also choose the least common multiple of the numbers from 2 to n.

3. Primorial $N = n\#$

   The smallest possible candidate for N is found through the primorial. If p_m is the smallest prime number greater than n, then $n\# = (p_{m-1})\#$ applies.

   This method of constructing such n-gaps (prime number gaps of length n) uses primorials: $n\#$ = product of all primes $\leq n$. See the introduction to primorials in Section 4.9.2.

Although N was chosen as small as possible in the last case, it is still not guaranteed that the gap found is always the first gap of the required length n. In this respect, all of these methods provide solid evidence and a specific gap. However, they are only of limited use when searching for the *first* occurrence of large gaps.

The first prime number gap of length n usually occurs well before $n!+2$, respectively, $N+2$.

Examples:

1. For example, to find a gap of at least length 4 (4-gap); that is, having at least a triple of composite numbers in the gap, you set $n = 4$ and get the sequence $4!+2, 4!+3, 4!+4$, where $4! = 24$. So a triple of composite numbers inside the gap is $(26, 27, 28)$. The first occurrence of such a triple is already before $4!$ at $(8, 9, 10)$.
2. With $n = 6$ one finds a prime number gap of at least length 6 between the following prime number candidates:
   - Via factorial: $N = 6! = 720 \Rightarrow N+2 = 722; N+6 = 726$.
     Since 721 is not prime, the gap is even larger. It is framed by the prime numbers [719 and 727] and thus has the length 8.
   - Via lcm: $N = lcm(1, \ldots, 6) = 60 \Rightarrow N+2 = 62; N+6 = 66$.
     Since both borders in [61,67] happen to be prime, the length is also exactly 6.
   - Via primorial: $N = 6\# = 2 \cdot 3 \cdot 5 = 30 \Rightarrow N+2 = 32; N+6 = 36$.
     The gap found [31, 37] has exactly the length 6 since both are prime numbers.
   - First gap of length 6: [23, 29] with $g_9 = 6$. The index 9 means that it is the gap between the 9th and the 10th prime number.
3. The factorial is the fastest growing function among the three functions considered.
   - For $n = 6$ this was: $n! = 720$, $lcm(2, \ldots, 6) = 60$, and $n\# = 30$.
     The first gap of exactly length 6 is [23,29] with $g_9 = 6$.
     You can calculate this with SageMath:
     ```
     sage: n=6; factorial(n); lcm(2..n);
     primorial=sloane.A002110; primorial(int(pari(n).primepi()))
     ```
   - For $n = 10$ the following applies: $n! = 3628800$, $lcm(2, \ldots, 10) = 2520$, and $n\# = 210$.
     The first gap of exactly length 10 is [139,149] with $g_{34} = 10$.
   - For $n = 14$ the following applies:
     $n! = 87178291200$, $lcm(2, \ldots, 14) = 360360$, and $n\# = 30030$.
     The first gap of exactly length 14 is [113,127] with $g_{30} = 14$.
     Gaps of lengths 10 and 12 follow *after* the first occurrence of the gap of length 14.
   - For $n = 20$ the following applies:
     $n! = 2432902008176640000$, $lcm(2, \ldots, 20) = 232792560$, and $n\# = 9699690$.
     The first gap of exactly length 20 is [887,907] with $g_{154} = 20$.
     You can get the gap index via SageMath: `pari(887).primepi() -->` `154`.

Table 4.8 shows the prime number gaps for the first prime numbers, whether the gap is maximal and the values of the functions the factorial, lcm, and primorial for the respective gap length.

4.9.6.1 Maximal Prime Gaps and Their Distribution

We call a prime gap *maximal* if it is bigger than all earlier gaps. Or more formally: Considering the mth gap with $g_m = p_{m+1} - p_m$: g_m is a maximal gap, if $g_m > g_i \quad \forall i < m$.

According to this, there is no *maximal* gap of length 10 because $g_{34} = 10$ and $g_{30} = 14$ (i.e., the earliest gap of length 14 occurs earlier than the earliest gap of length 10). So a maximal gap is always the first gap of this length, but being the first gap of a given length alone does not qualify to be maximal.

No general method is known to be more efficient than an exhaustive search for the determination of first occurrences and of maximal prime gaps [31].

By the prime number theorem we know there are approximately $n/\ln n$ primes less than n with ln = natural log. So the average gap between primes up to n is about $\ln n$.

The $[887, 907]$ gap we found above is the first one of size 20 and it is also a maximal prime gap. With numbers of this size, the average gap length is $\ln 887 \approx 6.78$. A metrics for how outstanding a gap is, is the *merit* that is the actual gap divided by the average gap. For this gap, the merit is $20/\ln 887 \approx 2.95$. The higher the merit, the more interesting the gap is.

As of October 2020, the *highest known merit* is ≈ 41.9 for $g_m = 8350$ starting at an 87-digit prime p_m. This gap was found by the Gapcoin network (Jonnie Frey) in 2017.

As of June 2022, the *largest known maximal prime gap* has a length $G = 1550$, found 2014 by Bertil Nyman. It is the 80th maximal gap, and it occurs after the prime 18,361,375,334,787,046,697 (20 digits). The merit of this record maximal prime gap is $M = 34.94$.

Largest known prime gap: Martin Raab found in 2017 a new first (and largest) known prime gap of length $G = 6582144$, following the 216841-digit prime $499973\#/30030 - 4509212$. The gap has merit $M = 13.18$. With today's technology it cannot be claimed whether this largest known gap is also a maximal one.

There are many conjectures about lower and upper bounds of prime gaps, first occurrences, maximal gaps, and largest gaps in the literature. The most interesting overview is the 2020 paper by Kourbatov and Wolf [32].

For example, on page 17 they conjecture about the distribution of maximal gaps: The number of maximal prime gaps up to a prime x is $\approx 2 \cdot \ln x$. So this count has—according to the Bachmann-Landau notation—the order $\mathcal{O}(\ln x)$ for $x \to \infty$.

Table 4.8 shows the prime number gaps for the first prime numbers. The g_i column contains the length $= n$ of the $i - th$ gap. If the gap is maximum, new_{gmax} is "True." The last three columns then show the values N of the functions factorial, lcm, and primorial named in Section 4.9.6. After that value, at the latest, a gap of this length is to be expected or can be constructed.

The sequence of prime gaps from the column g_i can also be found at A001223 from OEIS [16]: "Prime Gaps: Differences between Consecutive Primes."

Table 4.8 Gaps Between the First Primes: Gap Length $g_i = n$

i	p	p_{next}	g_i	new_{gmax}	idx_{gmax}	$gmax$	$n!$	lcm	$n\#$
1	2	3	1	True	1	1	1	1	1
2	3	5	2	True	2	2	2	2	2
3	5	7	2	False	–	–	–	–	–
4	7	11	4	True	3	4	24	12	6
5	11	13	2	False	–	–	–	–	–
6	13	17	4	False	–	–	–	–	–
7	17	19	2	False	–	–	–	–	–
8	19	23	4	False	–	–	–	–	–
9	23	29	6	True	4	6	720	60	30
10	29	31	2	False	–	–	–	–	–
...									
23	83	89	6	False	–	–	–	–	–
24	89	97	8	True	5	8	40320	840	210
25	97	101	4	False	–	–	–	–	–
26	101	103	2	False	–	–	–	–	–
27	103	107	4	False	–	–	–	–	–
28	107	109	2	False	–	–	–	–	–
29	109	113	4	False	–	–	–	–	–
30	113	127	14	True	6	14	87178291200	360360	30030
31	127	131	4	False	–	–	–	–	–
32	131	137	6	False	–	–	–	–	–
33	137	139	2	False	–	–	–	–	–
34	139	149	10	False	–	–	3628800	2520	210
35	149	151	2	False	–	–	–	–	–

Table 4.8 was completely created with the SageMath script `chap04_sample100.sage` (this script is not printed here, but can be downloaded from the CT website).

Table 4.9 shows the maximal prime number gaps for the first prime numbers. The first six maximum gaps occur after one of the first 30 prime numbers. Column g_i again contains the length n of the gap. The last three columns again show the values N of the functions factorial, lcm, and primorial mentioned in Section 4.9.6. After these candidates, at the latest, a gap of this length is to be expected or can be constructed in this way. The specific gap, its actual length, and its merit value are output for each candidate.

Table 4.9 has the same content as the website of Jens Kruse Andersen [33]. Table 4.9 was completely created with the SageMath Example 4.7.

Since the entire SageMath Example 4.7 is over 100 lines long, only the file header is listed here. The entire file is available on the CT server: see https://www.cryptool.org/en/documentation/ctbook/sagemath.

SageMath Example 4.7: List of First Maximal Prime Gaps with Merits

```
print("\n# CHAP04 -- Sage-Script-SAMPLE 110: =========")

# Calculate table with details for the maximal prime gaps (starting with p_1 = 2)
# - plus some candidate gaps starting with N+2 according to three formulas
# - plus the latex code for the table used in the CTB (currently commented out)
```

Table 4.9 List of First Maximal Prime Gaps with Merits Plus Additional Information

No	*i* *i*th prime	g_i	*Gap[]*	*Merit*	*n! candidate* real gap real len/merit	*lcm candidate* real gap real len/merit	*n# candidate* real gap real len/merit
1	1	1	[2, 3]	1.443	1 [2, 3] 1 // 1.443	1 [2, 3] 1 // 1.443	1 [2, 3] 1 // 1.443
2	2	2	[3, 5]	1.820	2 [3, 5] 2 // 1.820	2 [3, 5] 2 // 1.820	2 [3, 5] 2 // 1.820
3	4	4	[7, 11]	2.056	24 [23, 29] 6 // 1.914	12 [13, 17] 4 // 1.559	6 [7, 11] 4 // 2.056
4	9	6	[23, 29]	1.914	720 [719, 727] 8 // 1.216	60 [61, 67] 6 // 1.460	30 [31, 37] 6 // 1.747
5	24	8	[89, 97]	1.782	40320 [40289, 40343] 54 // 5.092	840 [839, 853] 14 // 2.080	210 [211, 223] 12 // 2.242
6	30	14	[113, 127]	2.961	87178291200 [87178291199, 87178291219] 20 // 0.7939	360360 [360337, 360391] 54 // 4.220	30030 [30029, 30047] 18 // 1.746

4.9.7 Peculiar and Interesting Things about Primes

Primes are not only a very active and serious research area in mathematics. Many people enjoy working with them in their free time and outside the scientific research.

4.9.7.1 Recruitment at Google in 2004

In summer 2004, Google used the number e to attract potential employees. The base of the natural logarithm e is approximately 2.718281828459.

On a prominent billboard in California's Silicon Valley on July 12, the following mysterious puzzle appeared:

(first 10 digit prime in consecutive digits of e).com

Finding the first 10-digit prime in the decimal expansion of e is not easy, but with various software tools, one can determine that the answer is

$$7427466391$$

Then, if you visited the website www.7427466391.com, you were presented with an even more difficult puzzle. Having accomplished the second puzzle, you were taken to a web page that asked you to submit your CV to Google. This ad campaign got high attention.

Presumably Google's ulterior motive was that if you're smart enough to solve the puzzles, you're smart enough to work for them. Of course some days after the launch, anyone who really wanted to discover the answers without incurring a headache could merely do a Google search for them, since many solvers immediately posted their solutions online. The second level of the puzzle, which involved finding the 5th term of a given number sequence had nothing to do with primes anymore.

4.9.7.2 Primes Helping to Contact Aliens: The 1997 Movie *Contact*

The movie *Contact*, directed by Robert Zemeckis, originated from Carl Sagan's book of the same title.

The plot of the movie is as follows. After years of unavailing search, the radio astronomer Dr. Ellie Arroway (Jodie Foster) discovers signals from the solar system Vega, 26 light years away. These signals contain the primes in the right order and without a gap. This makes the hero confident that this message is different from the radio signals that permanently hit earth. These are random and of cosmic origin (radio galaxies, pulsars). In an unmasking scene a politician then asks her why these intelligent aliens didn't just speak English ....

Doing communication with absolute strange and unknown beings from deep space is very hard for two reasons: First, the great distance and therefore the long transfer time make it impossible to exchange more than one message in each direction within an average lifetime. Second, the first contact must give the receiver of the radio signals a good chance to notice the message and to categorize it as something from intelligent beings. Therefore, the aliens send numbers at the beginning of their message, which can be considered as the easiest part of any higher language, and which are not too trivial. So they chose the sequence of primes. These special numbers play such a fundamental role in mathematics that one can assume that they are well known to each species who has the technical know-how to receive radio waves.

The aliens then send a plan to build a mysterious machine ....

4.9.7.3 Listen to Primes

At the end of the last century, Chris Caldwell of the University of Tennessee developed a scheme for listening to prime sequences. Maybe you can hear both simple patterns and perplexing irregularities.

The following information is mostly from Ivars Peterson's MathTrek editorial from June 22, 1998, and from Caldwell's page [34], where you find descriptions, some sample audio files, and the "primal sounds program" to create your own music by using prime sequences.

The Musical Instrument Digital Interface (MIDI) specification assigns a number to each note: the middle C is 60, C-sharp is 61, D is 62, and so on. In total there are 128 notes assigned to numbers.

As there are infinitely many primes, they have to be shrunk to 128 possible values. So Caldwell plays just the remainder modulo a given number. For example, if the modulus is 7, then for the primes $2, 3, 5, 7, 11, 13, 17, 19, 23, \ldots$ it plays $2, 3, 5, 0, 4, 6, 3, 5, 2, \ldots$. As the notes 0 to 6 would be too low in frequency to be audible, a constant such as 56 is added. Hence, the first prime, 2, is played as the note A. The website creates the midi files on the server. The source code for the "primal sounds program" is not offered.

4.10 Number of Prime Numbers in Various Intervals

Tables 4.10 and 4.11 show the number of primes within different intervals.

A visualization of the number of primes in higher intervals of powers of 10 can be found in Section 4.14.

Table 4.10 How Many Primes Exist within the First Intervals of Tens, Hundreds, and Thousands?

Ten-Sized Intervals		*Hundred-Sized Intervals*		*Thousand-Sized Intervals*	
Interval	*Number*	*Interval*	*Number*	*Interval*	*Number*
1-10	4	1-100	25	1-1000	168
11-20	4	101-200	21	1001-2000	135
21-30	2	201-300	16	2001-3000	127
31-40	2	301-400	16	3001-4000	120
41-50	3	401-500	17	4001-5000	119
51-60	2	501-600	14	5001-6000	114
61-70	2	601-700	16	6001-7000	117
71-80	3	701-800	14	7001-8000	107
81-90	2	801-900	15	8001-9000	110
91-100	1	901-1000	14	9001-10000	112

Table 4.11 How Many Primes Exist within the First Intervals of Dimensions?

Dimension	*Interval*	*Number*	*Average Number per 1000*
4	1 - 10000	1229	122.90
5	1 - 100000	9592	95.92
6	1 - 1000000	78498	78.50
7	1 - 10000000	664579	66.46
8	1 - 100000000	5761455	57.62
9	1 - 1000000000	50847534	50.85
10	1 - 10000000000	455052512	45.51

4.11 Indexing Prime Numbers: *n*th Prime Number

Table 4.12 shows the index for a few selected prime numbers. The index in the first column starts with 1.

It is very easy to calculate the nth prime if the given n is not too big. For example, SageMath responds almost instantaneously (30 μsec) to get the billionth prime with the `unrank` function. As this function starts indexing from 0 (so the index of the first prime 2 is 0), we have to reduce the index in SageMath Example 4.8 by 1. However, to find the trillionth prime number, SageMath did not come back even after 2 days.

SageMath Example 4.8: Get the *n*th Prime Number with SageMath

```
sage: P=Primes(); P.unrank(10^9-1)
22801763489
```

Does the opposite work too, given a prime p get its index or position? Above, with `Primes().unrank(n)` we got a prime p. For example, `Primes().unrank( 999)` delivers 7919. Now we want something like `Primes().ununrank(7919)` to get $n = 999$, but `Primes()` doesn't have such a method. However, there is the prime counting function `prime_pi` to determine the number of primes up to a given number (and this upper number may also be a prime): This n is then the position of a prime p; that is, to find the n when p is the nth prime. So $prime\_pi(7919)$ delivers 1000. See SageMath Example 4.9.

Table 4.12 List of Specific *n*th Prime Numbers P(n)

Index n	*Precise value P(n)*	*Rounded value*	*Comment*
1	2		
2	3		
3	5		
4	7		
5	11		
6	13		
7	17		
8	19		
9	23		
10	29		
100	541		
1000	7919		
664579	9999991	9,99999E+06	All prime numbers up to 1E+07 were known at the beginning of the 20th century.
1E+06	15485863	1,54859E+07	
6E+06	104395301	1,04395E+08	This prime was discovered in 1959.
1E+07	179424673	1,79425E+08	
1E+09	22801763489	2,28018E+10	
1E+12	29996224275833	2,99962E+13	

SageMath Example 4.9: Get the Position of a Prime Number

```
sage: P=Primes(); P.unrank(4)
11
sage: prime_pi(11)
5
```

Note that with gaps, extremely large prime numbers were discovered at an early stage. However, for the biggest ones like the Mersenne primes we don't know their concrete index number [3, 20].

4.12 Orders of Magnitude and Dimensions in Reality

In the description of cryptographic protocols and algorithms, numbers occur that are so large or so small that they are inaccessible to our intuitive understanding. It may therefore be useful to provide comparative numbers from the real world around us so that we can develop a feeling for the security of cryptographic algorithms. Some of the numbers in Table 4.13 originate from [35] and [36, p.18].

4.13 Special Values of the Binary and Decimal Systems

Special values of the binary and decimal systems like in Table 4.14 can be used to conclude from a key length in bits to the corresponding decimal number of possible

Table 4.13 Likelihoods and Dimensions from Physics and Everyday Life

Probability that you will be hijacked on your next flight	$5.5 \cdot 10^{-6}$
Annual probability of being hit by lightning	10^{-7}
Probability of 6 correct numbers in the lottery	$7.1 \cdot 10^{-8}$
Risk of being hit by a meteorite	$1.6 \cdot 10^{-12}$
Time until the next ice age (in years)	$14000 = (2^{14})$
Time until the sun dies (in years)	$10^9 = (2^{30})$
Age of the earth (in years)	$10^9 = (2^{30})$
Age of the universe (in years)	$10^{10} = (2^{34})$
Number of molecules within one water drop	$10^{20} = (2^{63})$
Number of bacteria living on earth	$10^{30.7} = (2^{102})$
Number of the earth's atoms	$10^{51} = (2^{170})$
Number of the sun's atoms	$10^{57} = (2^{190})$
Number of atoms in the universe (without dark material)	$10^{77} = (2^{265})$
Volume of the universe (in cm^3)	$10^{84} = (2^{280})$

Table 4.14 Corresponding Special Values of the Binary and Decimal Systems

Binary System	*Decimal System*
2^{10}	1024
2^{40}	$1.09951 \cdot 10^{12}$
2^{56}	$7.20576 \cdot 10^{16}$
2^{64}	$1.84467 \cdot 10^{19}$
2^{80}	$1.20893 \cdot 10^{24}$
2^{90}	$1.23794 \cdot 10^{27}$
2^{112}	$5.19230 \cdot 10^{33}$
2^{128}	$3.40282 \cdot 10^{38}$
2^{150}	$1.42725 \cdot 10^{45}$
2^{160}	$1.46150 \cdot 10^{48}$
2^{192}	$6.27710 \cdot 10^{57}$
2^{250}	$1.80925 \cdot 10^{75}$
2^{256}	$1.15792 \cdot 10^{77}$
2^{320}	$2.13599 \cdot 10^{96}$
2^{512}	$1.34078 \cdot 10^{154}$
2^{768}	$1.55252 \cdot 10^{231}$
2^{1024}	$1.79769 \cdot 10^{308}$
2^{2048}	$3.23170 \cdot 10^{616}$
2^{4096}	$1.04439 \cdot 10^{1233}$

keys and the search effort. This can be done provided that, for example, one million keys can be tested within one second.

Such tables can easily be generated using computer algebra systems (CAS) as here with SageMath Example 4.10.

SageMath Example 4.10: Special Values of the Binary and Decimal Systems

```
print("\n# CHAP04 -- Sage-Script-SAMPLE 020: =========")

E = [10,40,56,64,80,90,112,128,150,160,192,256,1024,2048,4096]
for e in E:
    print( '2^%4d  --- ' % e, RR(2^e).n(24) )
```

4.14 Visualization of the Quantity of Primes in Higher Ranges

4.14.1 The Distribution of Primes

There are four primes between 1 and 10. There are already 1061 primes between 10^3 and 10^4. In the interval $[10^9, 10^{10}]$ lie $404204977 \approx 4 \cdot 10^8$ primes, and in the interval from 10^{19} to 10^{20} there are

$$1986761935284574233 \approx 1.9 \cdot 10^{18}$$

primes.

4.14.1.1 The Prime Number Theorem

The number $PI(x)$ of primes up to a given number x can be approximately determined by a formula, derived from the prime number theorem (see Section 4.7). $PI(x)$ denotes the number of primes that are smaller or equal to x:

$$PI(x) \sim \frac{x}{\ln x}$$

Note that this formula only gives an approximation of the number of primes smaller or equal to x. It becomes more exact as the number x increases. In the following we are using the prime number theorem to examine the distribution of primes.

In order to understand why the number of primes is growing so rapidly, although the boundaries of the intervals only differ by the exponent 1. Let's have a closer look at both components of the right side of the formula: x and $\ln x$.

4.14.1.2 The Functions x and 10^x

The function x is a straight line. It is shown in Figure 4.7(a).

In the next step the function of the boundaries of the intervals are drawn in Figure 4.7(b). To get an idea of how the functions look, like the domain of definition was chosen to be from 0 to 10^{10} and from 0 to 10, respectively. You can see that with increasing exponent x the numbers grow stronger.

4.14.1.3 The Function $\ln x$

In comparison to functions x and 10^x, we now consider the function $\ln x$. Figure 4.8(a) shows the graph with the domain of definition from 1 to 100. Figure 4.8(b) the domain of definition was chosen between 1 and 10^{10}.

One can see that the values of the function $\ln x$ grow slowly compared to the growth of the function x. This is visualized by the graph of both functions in one picture shown in Figure 4.9. In addition to that, the graph of the function $\frac{x}{\ln x}$ was drawn in the same figure.

4.14.1.4 The Function $PI(x) = \frac{x}{\ln x}$

The function $\frac{x}{\ln x}$ consists of the function x as the numerator and the function $\ln x$ in the denominator which, in comparison to x, increases very slowly. Compared to

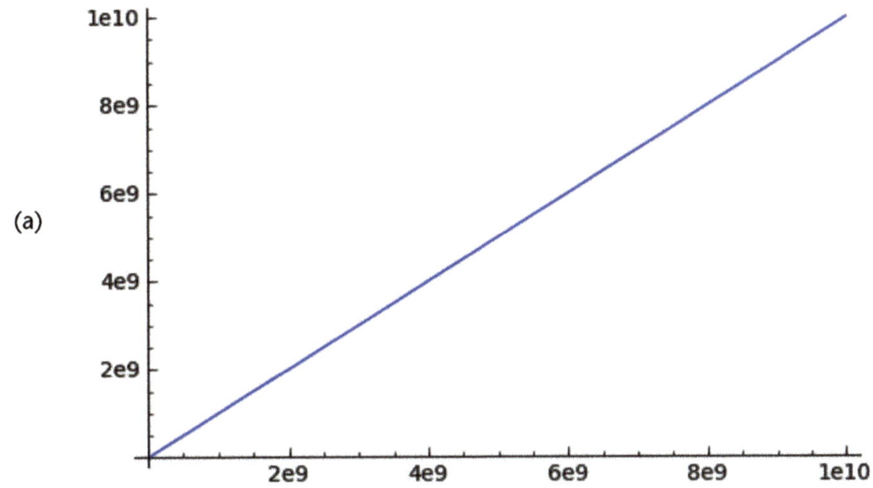

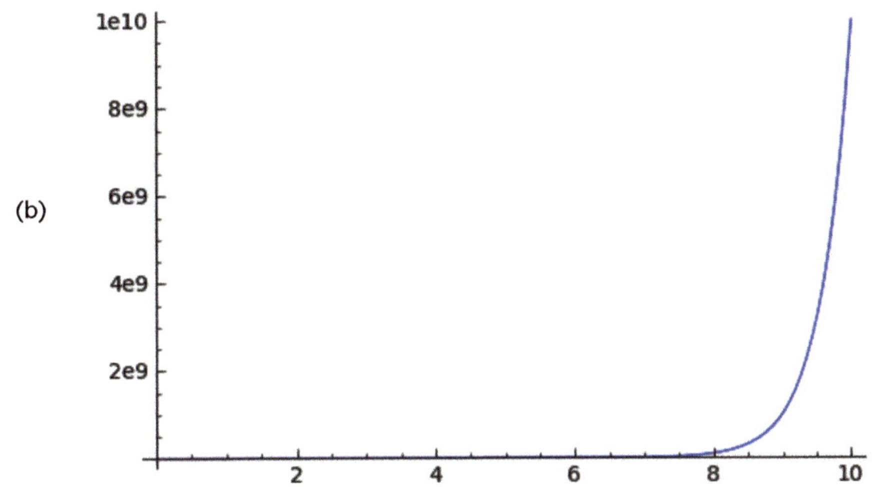

Figure 4.7 Graph of the functions (a) x and (b) 10^x.

the number x itself, the number of primes less or equal to x is small. But still, $\frac{x}{\ln x}$ is an increasing function as you can see in Figure 4.9.

4.14.1.5 The Number of Primes in the Different Intervals

Figure 4.10 visualizes how the number of primes behaves in the intervals $[1, 10^x]$ and $[10^{x-1}, 10^x]$. The result of the approximation function is used to calculate it faster (not the exact numbers like in Tables 4.10 and 4.11).

Here for each base 10 exponent two bars are drawn: $\frac{10^x}{\ln 10^x}$ and $\frac{10^x}{\ln 10^x} - \frac{10^{x-1}}{\ln 10^{x-1}}$: The top chart in Figure 4.10 shows the values for the exponents x from 1 to 5, and the right one for x from 1 to 10, where x is the base 10 exponent.

The blue bars represent the overall number of primes up to 10^x. The red bars show how many primes accrue in the interval $[10^{x-1}, 10^x]$, respectively. This makes

(a)

(b)

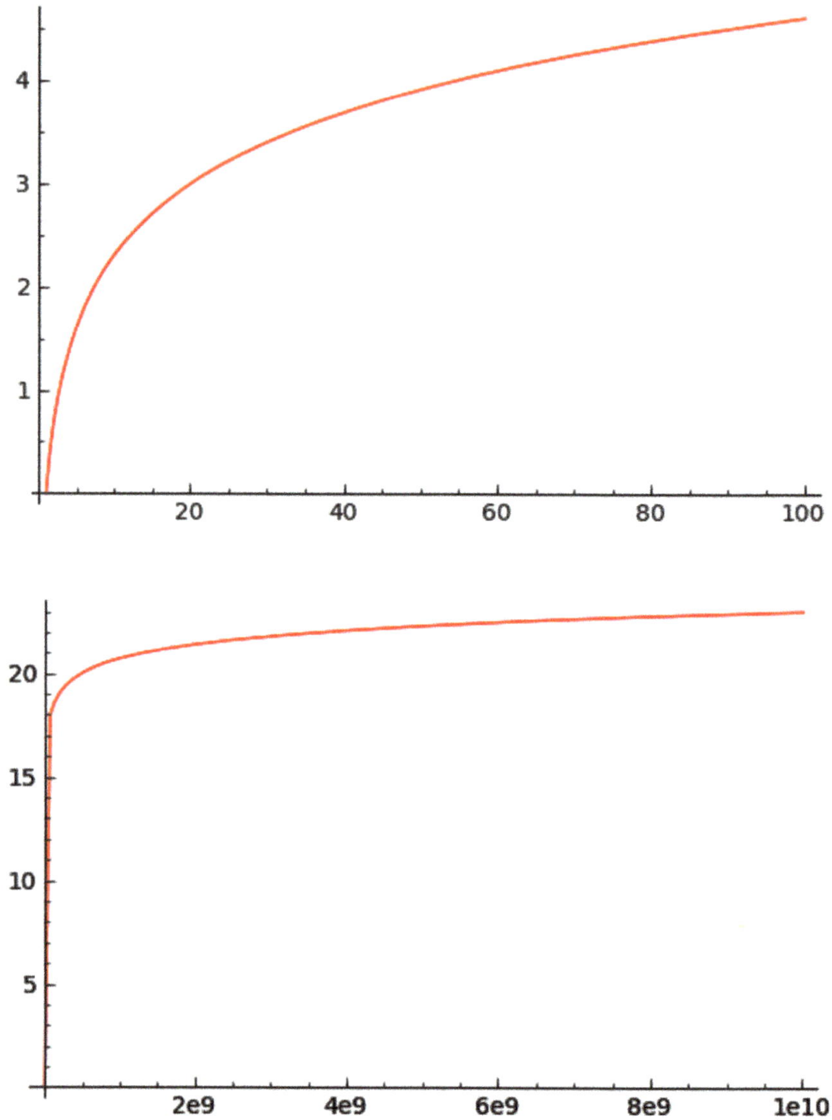

Figure 4.8 Graph of the function $\ln x$ (a) up to 100 and (b) up to 10^{10}.

clear that the number of primes in intervals of higher exponents keeps growing quite fast.

A table containing the number of primes in some dedicated intervals can be found in Section 4.10. For example, within the interval $[1, 10^4]$ there are 1229 primes; thereof are in the interval $[10^3, 10^4]$ 1229 - 168 = 1061 primes.

More theory about the prime number theorem and the function PI(x) can be found in Section 4.7.

SageMath Example 4.11 creates the graphs for the three functions x, log(x), and x/log(x), shown in Figures 4.7 to 4.9. It also calculates values for the bars in Figure 4.10.

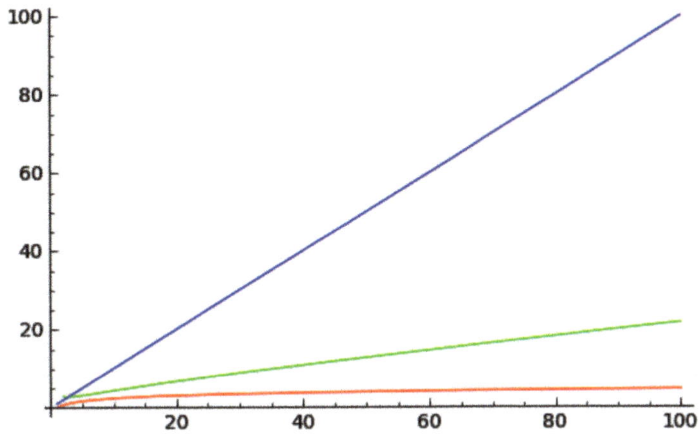

Figure 4.9 The functions x (blue), $\ln x$ (red), and $\frac{x}{\ln x}$ (green).

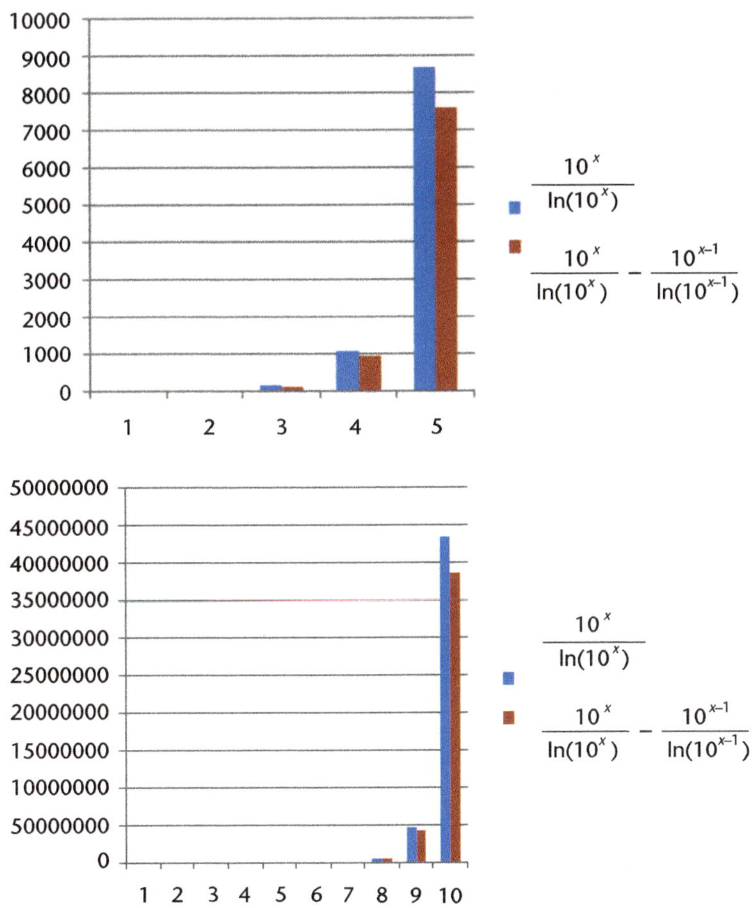

Figure 4.10 Numbers of primes in the interval $[1, 10^x]$ (blue) and in the interval $[10^{x-1}, 10^x]$ (red) for different exponents x.

SageMath Example 4.11: Generation of the Graphs of the Three Functions x, log(x), and x/log(x)

```
print("\n# CHAP04 -- Sage-Script-SAMPLE 030: =========")

def Display(F,fname):
    # parameter 'fname' should contain no blanks

    ### The following commands work in Sage CLI for all OS, but not when called from script
    # F.show()  # Alternative, also working in Sage CLI: F.plot()
      # CLI outputs: 'Launched png viewer for Graphics object consisting of 1 graphics ▸
        ▸primitive'
      # This automatically comes with annotated axes / Achsen sind automatisch beschriftet

    ### This works from a Sage script under Ubuntu (adapt path and viewer name for other OS)
    pngfile='/tmp/'+fname+'.png'; # print("pngfile=", pngfile)
    F.save(pngfile,axes=True)
    imv = 'feh '  # 'okular '  # 'gwenview '  # 'eog '   ## image viewer to start from the ▸
      ▸terminal under Linux
    # imv = 'open -a preview '   ## image viewer to start from the terminal under macOS
    oscommand = imv + pngfile + ' &'; # print("oscommand=", oscommand)
    os.system(oscommand)  # With Ubuntu: eog = Eye of GNOME file viewer
    ## os.system('display /tmp/F.png &')  # Alternative: The display command needs 'imagemagick▸
      ▸' to be installed.
    return


# Definition of function f(x)=x and plots for the domains 0 to 100 and 0 to 10^10
def f(x):return x
F1=plot(f,(0,100)); Display(F1, "F1")  # it doesn't matter whether range starts from 0 or 1
F2=plot(f,(0,10^10)); Display(F2, "F2")

# Definition of function g(x)=10^x and plots for the domain 0 to 10
def g(x): return 10^x
G=plot(g,(0,10)); Display(G, "G")

# Definition of function h(x)=log(x) and plots for the domains 1 to 100 and 1 to 10^10
def h(x): return log(x)
H1=plot(h,(1,100),color="red"); Display(H1, "H1")
H2=plot(h,(1,10^10),color="red"); Display(H2, "H2")

# Definition of function k(x)=x/log(x) and plots for the domain 2 to 100
def k(x): return x/log(x)
K1=plot(k,(1,100),color="green"); Display(K1, "K1")
# K=plot(k,(2,100),color="green"); Display(K, "K")

# Plots of the functions f, k and h for the domain up to 100
Display(F1+K1+H1, "F1+K1+H1")


# Generation of the data for the bar charts ..........................
# Determination of the number of primes in the interval [1,10]
print("#p in %13s:" % "[1,10]", pari(10).primepi() - pari(1).primepi() )

# Determination of the number of primes in the interval [1,100]
print("#p in %13s:" % "[1,100]", pari(100).primepi() - pari(1).primepi() )

# Determination of the number of primes in the interval [10^3,10^4]
print("#p in %13s:" % "[10^3,10^4]", pari(10**4).primepi()-pari(10**3).primepi() )

# Determination of the number of primes in the interval [10^8,10^9]
print("#p in %13s:" % "[10^8,10^9]", pari(10**9).primepi()-pari(10**8).primepi() )

# Determination of the number of primes in the interval [10^9,10^10]
print("#p in %13s:" % "[10^9,10^10]", pari(10**10).primepi()-pari(10**9).primepi() )

# Determination of the number of primes in the interval [10^10,10^11]
print("#p in %13s:" % "[10^10,10^11]", pari(10**11).primepi()-pari(10**10).primepi() )

# Determination of the number of primes in the interval [1,10^11]
print("#p in %13s:" % "[1,10^11]", pari(10**11).primepi()-pari(1).primepi() )
```

4.15 Examples Using SageMath

Below is SageMath source code related to the contents of this chapter.

4.15.1 Some Basic Functions about Primes Using SageMath

SageMath Example 4.12 shows some calls to answer very simple questions about primes.

SageMath Example 4.12: Some Basic Functions about Primes

```
# Methods of the class of the set of prime numbers
sage: P=Primes(); P
Set of all prime numbers: 2, 3, 5, 7, ...

sage: P.next(5)
7

# unrank(n): Return the n-th prime number
sage: P.unrank(0); P.unrank(5)
2
13

sage: P[5]
13

# Function to return the next prime number
sage: next_prime(5)
7

# Returns how many primes <=x are there
sage: pari(10).primepi()
4

# Returns the first x primes
sage: primes_first_n(6)
[2, 3, 5, 7, 11, 13]

# Returns the primes in an interval
sage: list(primes(4)); list(primes(2,10))
[2, 3]
[2, 3, 5, 7]
```

4.15.2 Check Primality of Integers Generated by Quadratic Functions

SageMath Example 4.13 verifies the primality of integers generated by the function $f(n) = n^2 - 9n + 61$. The code defines a function called `quadratic_prime_formula()` that takes three arguments:

- `start` — An integer that is the lower bound for integers in the sequence $\texttt{start}, \texttt{start} + 1, \texttt{start} + 2, \ldots, \texttt{end} - 1, \texttt{end}$.
- `end` — An integer that is the upper bound for the integers in this sequence.

- `verbose` — (default: `True`) a flag to signify whether to print a message indicating the primality of an integer generated by $f(n)$. If `False`, only the statistics are printed.

Since the output of SageMath Example 4.13 has many lines, only the beginning and the end of the output are listed here.

SageMath Example 4.13: Verify the Primality of Integers Generated by a Quadratic Function

```
print("\n# CHAP04 -- Sage-Script-SAMPLE 040: =========")

def quadratic_prime_formula(start, end, verbose=True):
    print("N -- N^2 - 9*N + 61")
    P = 0  # the number of primes between start and end
    for n in range(start, end + 1):
        X = n^2 - 9*n + 61
        if is_prime(X):
            P += 1
            if verbose:
                print(str(n) + " -- " + str(X) + " is prime")
        else:
            if verbose:
                print(str(n) + " -- " + str(X) + " is NOT prime")
    print("Number of primes: " + str(P) + " in range (%d, %d)" % (start,end))
    print("Percentage of primes: " + str(float((P * 100) / (end - start + 1))))

# Compute the values of f(n) = n^2 - 9n + 61 for n = 0, ..., 50
#  and verify the primality of the generated integers
quadratic_prime_formula(0, 50)

#------------------------------------
# CHAP04 -- Sage-Script-SAMPLE 040: =========
# N -- N^2 - 9*N + 61
# 0 -- 61 is prime
# 1 -- 53 is prime
# 2 -- 47 is prime
# 3 -- 43 is prime
# 4 -- 41 is prime
# 5 -- 41 is prime
# 6 -- 43 is prime

# 46 -- 1763 is NOT prime
# 47 -- 1847 is prime
# 48 -- 1933 is prime
# 49 -- 2021 is NOT prime
# 50 -- 2111 is prime
# Number of primes: 48 in range (0, 50)
# Percentage of primes: 94.11764705882354
```

With the function call at the end of the code sample we compute the values of $f(n) = n^2 - 9n + 61$ for $n = 0, 1, 2, \ldots, 50$ and verify the primality of the generated integers.

The last two lines of the output contain a small statistics. You can see that $f(n)$ generates 48 primes when $0 \leq n \leq 50$, which is approximately 94% of the values generated by $f(n)$.

A modification of this code is in SageMath Example 4.14: There, the function values of other functions $f(n)$ are checked for their primality.

For larger sequences, it is impractical to print all single messages indicating the primality of integers. Therefore, SageMath Example 4.14 sets the verbose parameter of the function to `False`. So only the statistics at the end are printed: the overall number of primes and the percentage of primes, generated by $f(n)$ where $0 \leq n \leq 1000$.

Here the function `quadratic_prime_formula()` is slightly expanded: It can also handle negative prime numbers and noninteger coefficients: The 4th argument `countNegativePrimeToo` of the function has the default `False`. This is a flag that determines whether a function value $f(n)$ that is a negative prime should also be considered prime.

SageMath Example 4.14: Primality of Results of a Quadratic Function (Variable n Runs Until 1000)

```
print("\n# CHAP04 -- Sage-Script-SAMPLE 050: =========")

def quadratic_prime_formula(start, end, verbose=True, countNegativePrimeToo=False):
  a_list = []
  print("N -- N^2 - 79*N + 1601 .........") # all Y = f(n) are positive; 80 consecutive primes;▸
  ▸ 40 unique ones.
  #a# print("N -- N^2 - 9*N + 61 .........") # all Y = f(n) are positive; 45 consecutive primes▸
  ▸; 40 unique ones.
  #b# print("n -- 36*n^2 - 810*n + 2753 .........") # 14 Y = f(n) are negative; 45 consecutive ▸
  ▸primes; 45 unique ones.
  #c# print("n -- n^4 - 97*n^3 + 3294*n^2 - 45458*n + 213589 .........") # 9 neg / 50 con / 49 ▸
  ▸uni.
  #d# print("n -- (n^5 - 133*n^4 + 6729*n^3 - 158379*n^2 + 1720294*n - 6823316) / 4 .........")▸
  ▸ # 14 neg / 57 con / 57 uni.
  P = 0  # the number of calculated primes Y between start and end
  Pneg = 0 # how many negative primes found als function value of the polynomial
  for n in range(start, end + 1):
    Y = n^2 - 79*n + 1601    # [0,79] from 41 to 1601 and back
    #a# Y = n^2 - 9*n + 61    # [0,44] from 61 to 1601
    #b# Y = 36*n^2 - 810*n + 2753    # [0,44] from 2753 to 36809
    #c# Y = n^4 - 97*n^3 + 3294*n^2 - 45458*n + 213589    # [0,49] from 213589 to 247889
    #d# Y = (n^5 - 133*n^4 + 6729*n^3 - 158379*n^2 + 1720294*n - 6823316) / 4    # [0,56] from -▸
    ▸1705829 to 4325119
    if is_prime(int(Y)): # Y within int() is necessary if non-integer coeff in f(n)
      P += 1
      a_list.append(Y)
      if verbose:
        print(str(n) + " -- " + str(Y) + " is prime")
    elif is_prime(abs(int(Y))): # Y within int() is necessary if non-integer coeff
      Pneg += 1
      if verbose:
        print(str(n) + " -- " + str(Y) + " is NOT prime, but -n is prime")
      if countNegativePrimeToo:
        P += 1
        a_list.append(-Y)
    else:
      if verbose:
        print(str(n) + " -- " + str(Y) + " is NOT prime")

  Punique = len(set(a_list))
  a_range = end - start + 1
  print("Number of primes in f(n): %d in n-range (%d, %d) [n takes %d diff. values as both ▸
  ▸borders are included]" % (P, start,end, end-start+1))
  print("Number of unique primes: %d in the list of found primes (%d)" % (Punique, len(a_list))▸
  ▸)
  print("Percentage of primes: %.2f" % float((P * 100) / a_range))
  print("Percentage of unique primes: %.2f" % float((Punique * 100) / a_range))
  if Pneg > 0:
    if countNegativePrimeToo:
      print("Number of negative primes found: %d. These are counted as primes." % Pneg)
    else:
      print("Number of negative primes found: %d. These are not counted as primes." % Pneg)

# quadratic_prime_formula(0, 50) # Get same output as in chap04_sample040.sage, if #a# is ▸
 ▸activated
# quadratic_prime_formula(0, 79, countNegativePrimeToo=True) # (0,60)
quadratic_prime_formula(0, 1000, verbose=False) # If verbose==False, only the statistics are ▸
 ▸printed.
```

SageMath Example 4.14 (continued)

```
#------------------------------------
# CHAP04 -- Sage-Script-SAMPLE 050: =========
# N -- N^2 - 79*N + 1601 .........
# Number of primes in f(n): 602 in n-range (0, 1000) [n takes 1001 diff. values as both borders▸
 ▸ are included]
# Number of unique primes: 562 in the list of found primes (602)
# Percentage of primes: 60.14
# Percentage of unique primes: 56.14
```

References

[1] Great Internet Mersenne Prime Search (GIMPS), *GIMPS*, https://www.mersenne.org/primes/.

[2] Blum, W., *Die Grammatik der Logik*, dtv, 1999.

[3] Caldwell, C., *The Largest Known Prime by Year*, https://t5k.org/notes/by_year.html.

[4] Great Internet Mersenne Prime Search (GIMPS), *GIMPS PrimeNet Activity Summary*, https://www.mersenne.org/primenet/.

[5] Pinch, R., "The Carmichael Numbers Up to 10^{21}," *Proceedings of Conference on Algorithmic Number Theory*, Vol. 46, 2007, pp. 129–131, https://tucs.fi/publications/attachment.php?fname=G46.pdf.

[6] Alford, W.R., et al., "Constructing Carmichael Numbers Through Improved Subset-Product Algorithms," *Mathematics of Computation*, Vol. 83, No. 286, 2014, pp. 899–915, https://arxiv.org/abs/1203.6664.

[7] Witten, H., and R.-H. Schulz, "RSA & Co. in der Schule: Moderne Kryptologie, alte Mathematik, raffinierte Protokolle. NF Teil 5: Der Miller-Rabin-Primzahltest oder: Falltüren für RSA mit Primzahlen aus Monte Carlo," *LOG IN*, Vol. 166/167, 2010, pp. 92–106, https://informatik.schule.de/krypto/.

[8] PrimePages, *Mersenne Primes: History, Theorems and Lists*, https://t5k.org/mersenne/index.html.

[9] PrimePages, *Finding Primes & Proving Primality*, https://t5k.org/prove/index.html.

[10] Caldwell, C., *FAQ: Is There a Formula for the nth Prime?* https://t5k.org/notes/faq/p_n.html.

[11] Knuth, D. E., *The Art of Computer Programming, Volume 2: Seminumerical Algorithms*, Third Edition, Addison Wesley, 1998.

[12] Gallot, Y., *proth20: An OpenCL Implementation of Proth's Theorem*, https://github.com/-galloty/genefer22.

[13] Wagstaff, S., *The Cunningham Project*, https://homes.cerias.purdue.edu/~ssw/cun/.

[14] Morelli, L., *Distributed Search for Fermat Number Divisors*, http://www.fermatsearch.org.

[15] Zimmermann, A., *Al Zimmermann's Programming Contests*, http://azspcs.com/.

[16] Sloane, N. J. A., et al., *The On-Line Encyclopedia of Integer Sequences (OEIS)*, https://oeis.org/.

[17] Padberg, F., *Elementare Zahlentheorie*, Second Edition, Spektrum Akademischer Verlag, 1996.

[18] Caldwell, C., *The Gaps Between Primes*, https://t5k.org/notes/gaps.html.

[19] Caldwell, C., *How Many Primes Are There?*, https://t5k.org/howmany.html.

[20] Booker, A., *The Nth Prime Page*, https://t5k.org/nthprime/.

[21] Koppehel, S., *Animation of the Sieve of Eratosthenes*, https://upload.wikimedia.org/wikipedia/commons/0/0b/Sieve_of_Eratosthenes_animation.svg.

[22] Tietze, H., *Gelöste und ungelöste mathematische Probleme*, Sixth Edition, C.H. Beck, 1973.

[23] Tervooren, M., *FactorDB*, http://factordb.com/.

[24] Alpern, D., *Alpertron, the Integer Factorization Calculator*, https://www.alpertron.com.ar/ecm.htm.

[25] Andersen, J. K., *Primes in Arithmetic Progression Records*, http://primerecords.dk/aprecords.htm.

[26] Andersen, J. K., and N. Luh, *Primes in Arithmetic Progression Records*, https://www.pzk-tupel.de/JensKruseAndersen/aprecords.php.

[27] Klee, V., and S. Wagon, *Ungelöste Probleme in der Zahlentheorie und der Geometrie der Ebene*, Birkhäuser Verlag, 1997.

[28] Tao, T., *Every Odd Number Greater Than 1 is the Sum of at Most Five Primes*, 2012, https://arxiv.org/abs/1201.6656.

[29] Helfgott, H. A., and D. J. Platt, *Numerical Verification of the Ternary Goldbach Conjecture Up to 8.875e30*, 2014, https://arxiv.org/abs/1305.3062.

[30] Chen, J., On the Representation of a Larger Even Integer as the Sum of a Prime and the Product of at Most Two Primes, in *The Goldbach Conjecture* (Y. Wang, ed.), Singapore: World Scientific, 2002.

[31] Nicely, T. R., "New Maximal Prime Gaps and First Occurrences," *Mathematics of Computation*, Vol. 68, No. 227, 1999, pp. 1311–1315, https://www.ams.org/journals/mcom/1999-68-227/S0025-5718-99-01065-0/S0025-5718-99-01065-0.pdf, and https://faculty.lynchburg.edu/~nicely/gaps/gaps.html.

[32] Kourbatov, A., and M. Wolf, "On the First Occurrences of Gaps Between Primes in a Residue Class," *Journal of Integer Sequences*, Vol. 23, 2020, https://arxiv.org/abs/2002.02115.

[33] Kruse Andersen, J., *Maximal Prime Gaps*, http://primerecords.dk/primegaps/maximal.htm.

[34] Caldwell, C., *Prime Number Listening Guide*, https://t5k.org/programs/music/listen/.

[35] Schwenk, J., "Conditional Access," in *taschenbuch der telekom praxis*, B. Seiler (ed.), 1996.

[36] Schneier, B., *Applied Cryptography, Protocols, Algorithms, and Source Code in C*, Second Edition, Wiley, 1996.

CHAPTER 5

Introduction to Elementary Number Theory with Examples

This introduction is for people with a mathematical interest. No more previous knowledge is required than that taught in secondary or high school.

We intentionally had beginners in mind; we did not take the approach of mathematical textbooks, where the "introduction," of enabling cannot be understood at the first reading further than page 3 and which have the purpose to enable the reader to understand monographs. For this reason, requirements and ideas are explained in a comprehensible way and often illustrated with concrete numerical examples and sample programs.

5.1 Mathematics and Cryptography

A large proportion of modern, asymmetric cryptography is based on mathematical knowledge—on the properties of integers, which are investigated in elementary number theory. Here, the word "elementary" means that questions raised in number theory are essentially rooted in the set of natural and whole numbers (integers).

Further mathematical disciplines currently used in cryptography include (see [1, p. 2; 2, p. 3]):

- Group theory;
- Combination theory;
- Complexity theory;
- Stochastic (ergodic theory);
- Information theory.

Number theory or arithmetic (the emphasis here is more on the aspect of performing calculations with numbers) was established by Carl Friedrich Gauss as a special mathematical discipline. Its elementary features include the greatest common divisor (gcd), congruences (remainder classes), factorization, the Euler-Fermat theorem, and primitive roots. However, the most important aspect is prime numbers and their multiplicative operation.

For a long time, number theory was considered to be the epitome of pure research, the ideal example of research in the ivory tower. It delved into the mysterious laws of the realm of numbers, giving rise to philosophical considerations whether it described elements that exist everywhere in nature or whether it artificially constructed elements (numbers, operators, and properties).

With the number-theoretical applications of modern cryptography, it became clear that a discipline that had been regarded as purely theoretical for centuries was now being applied in practice. Today, experts in the field are sought after on the job market.

Applications in (computer) security now use cryptography because this mathematical discipline is simply better and easier to prove than all other creative substitution procedures that have been developed over the course of time and better than all sophisticated physical methods such as those used to print banknotes [3, p. 4].

This chapter explains the basics of elementary number theory in a way that you can easily understand. It provides numerous examples and very rarely goes into any proofs (these can be found in mathematical textbooks).

The goal is not to exhaustively explain the number theory findings, but to show the essential procedures. The scope of the material is oriented towards being able and apply the RSA method in more detail.

For this purpose we will use both theory and examples to explain how to perform calculations in finite sets and describe how these techniques are applied in cryptography. Particular attention will be paid to the traditional Diffie-Hellman (DH) and RSA public-key procedures.

It was important to me to make verifiable statements about the security of the RSA algorithm, and to add runnable Python or SageMath code for as many examples as possible. SageMath is an open-source Python-based computer-algebra system (CAS); see [4].

5.2 Introduction to Number Theory

Number theory studies positive integers $1, 2, 3, 4, \cdots$, also referred to as the set of *natural numbers* $\mathbb{N}$. These are the first mathematical constructs used by human civilization. According to Leopold Kronecker, they are a creation of God. In Julius Dedekind's opinion, they are a creation of the human intellect. Dependent upon one's ideology, this is an unsolvable contradiction or one and the same thing.

In ancient times, no distinction was made between number theory and numerology, which attributed a mystical significance to specific numbers. In the same way as astronomy and chemistry gradually detached themselves from astrology and alchemy during the Renaissance (from the 14th century), number theory also separated itself from numerology.

Number theory has always been a source of fascination for both amateurs and professional mathematicians. In contrast to other areas of mathematics, many of the problems and theorems in number theory can be understood by nonexperts. On the other hand, the solutions to these problems or the proof to the theorems often resisted to the mathematicians for a very long time. It is therefore one thing to pose good questions but quite another matter to find the answer. One example of this is what is known as Fermat's last theorem. One of the things we learn in mathematics at school is Pythagoras' theorem, which states for a right-angle triangle: $a^2 + b^2 = c^2$, where a and b are the real-valued lengths of the sides next to the right angle and c is the length of the hypotenuse. Fermat famously proposed that $a^n + b^n \neq c^n$ for $a, b, c \in \mathbb{N}$ and integer exponents $n > 2$. Unfortunately, the border of his book from Diophant where he made the claim did not have enough space for him to prove it. The theorem was not proven until over 300 years later [5, pp. 433–551]. The name "last" got attached to it because it has been the last conjecture Fermat had made that remained open. See Section 8.2.

Up until the mid-20th century, number theory was considered to be the purest area of mathematics, an area that had no practical use in the real world. This changed with the development of computers and digital communication, as number theory was able to provide several unexpected solutions to real-life tasks. At the same time, advances in information technology allowed specialists in number theory to make huge progress in factorizing large numbers, finding new prime numbers, testing (old) conjectures, and solving numerical problems that were previously impossible to solve. Modern number theory is made up of areas such as:

- Elementary number theory
- Algebraic number theory
- Analytic number theory
- Geometric number theory
- Combinatorial number theory
- Numeric number theory
- Probability theory

All the different areas are concerned with questions regarding integers (both positive and negative whole numbers plus zero). However, they each have different methods to deal with them.

This chapter mainly deals with the area of elementary number theory.

5.2.1 Convention and Notation

Unless stated otherwise:

- The letters $a, b, c, d, e, k, n, m, q$ are used to represent integers (whole numbers).
- The letters i and j represent natural numbers.
- The letter p always represents a prime number.
- The sets $\mathbb{N} = \{1, 2, 3, \cdots\}$ and $\mathbb{Z} = \{\cdots, -3, -2, -1, 0, 1, 2, 3, \cdots\}$ are the *natural numbers* and *integers*, respectively.
- $\mathbb{Z}_n = \{0, 1, 2, \cdots, n-2, n-1\}$ and $\mathbb{Z}_p^*$ ($= \mathbb{Z}_p \setminus \{0\}$, where p is prime) are *finite sets* with n or $p - 1$ elements.

Often instead of $\mathbb{Z}_n$ the notation $\mathbb{Z}/n\mathbb{Z}$ is used. However, we use the first notation here, as it is easier to write and as there is no danger of confusion with the so-called p-adic numbers $\mathbb{Z}_p$ for $n = p$. The * representation also exists for integers (that is, compound n and not only prime p), see Definition 5.9.

The following list shows the SageMath commands to call the three important structures $\mathbb{Z}, \mathbb{Z}_n$, and $\mathbb{Z}_n^*$ – these commands are applied in the SageMath Example 5.1:

- $\mathbb{Z}$: `IntegerRing()` or `ZZ`
- $\mathbb{Z}_n$: `Integers(n)` or `Zmod(n)` or `IntegerModRing(n)`
- $\mathbb{Z}_n^*$: `[a for a in Integers(n) if gcd(a,n) == 1]` or `Zmod(n).list_of_elements_of_multiplicative_group()` or `Zmod(n).unit_group()`

SageMath Example 5.1: $\mathbb{Z}, \mathbb{Z}_n$, and $\mathbb{Z}_n^*$ in SageMath

```
print("\n# CHAP05 -- Sage-Script-SAMPLE 017: =========")
print("Different ways in Sage to define Z, Z/nZ, and (Z/nZ)* :")

n = 10  # n = 17
print("- n: ", n, "  type(n): ", type(n))


print("1### Z  ==>  Sage: IntegerRing() = ZZ")
# Ring of all integers
#
R1 = IntegerRing();   e = R1(5)
print("- R1 = IntegerRing(): ", R1, type(R1))
print("  e: ", e, "  ", type(e))
print("  R1.range(n): ", R1.range(n))

print("- xx IntegerRing()==ZZ: ", IntegerRing()==ZZ)

# print("- ZZ.range(0,50,5): ", ZZ.range(0,50,5))  # [0, 5, 10, 15, 20,▸
  ▸ 25, 30, 35, 40, 45]
# a = ZZ('1234')  # Alternative arguments for ZZ: a = ZZ(1234)  or  ZZ▸
  ▸('0x4D2') because of
                  # i=1234; l=i.digits(base=16); l; j=ZZ(l,base=16); j; i▸
                    ▸==j
                  # [2, 13, 4]  //  1234  //  True
# print("-   a:", a, " //  type(a): ", type(a))   # <class 'sage.rings.▸
  ▸integer.Integer'>
# b = 1234
# print("  xx a==b:", a==b)


print("2### quotient ring Z/nZ  ==>  Sage: Integers(n) = Zmod(n) = ▸
  ▸IntegerModRing(n)")
# Ring of integers from 0 to n-1  or  ring of integers modulo n = ▸
  ▸additive group
#
R2 = Integers(n)
print("- R2 = Integers(n): ", R2)
print("  list(R2): ", list(R2))
print("  R2.order(): ", R2.order())

R3 = Zmod(n)
```

SageMath Example 5.1 (continued)

```
print("- R3 = Zmod(n): ", R3)
print("  list(R3): ", list(R3))  # Alternative:  L = [a for a in R3]; ▸
 ▸print("  L: ", L)
print("  R3.order(): ", R3.order())

R4 = IntegerModRing(n)
print("- R4 = IntegerModRing(n): ", R4)
print("  list(R4): ", list(R4))
print("  R4.order(): ", R4.order())

print("- xx Integers(n)==Zmod(n): ", Integers(n)==Zmod(n), "  //  ",
      "Zmod(n)==IntegerModRing(n): ", Zmod(n)==IntegerModRing(n))

# a = R4(5)  # No alternative is: R4(5**(10^62)
# print("-   a**(10^62): ", a**(10^62))  # calculating in the finite ▸
 ▸ring is very quick!


print("3### Multiplicative group (Z/nZ)*  ==>  Sage: further handle ▸
 ▸Zmod(n) or Integers(n)")
#
# Return a list of all invertible elements (type of each is Sage int)
L1 = [a for a in R2 if gcd(a,n) == 1]
print("- L1 (via Integers(n)): ", L1);  print("  type(L1[1]): ", type(L▸
 ▸1[1]))
# print(" R2.multiplicative_table(): ", R2.multiplication_table())  # ▸
 ▸this works well
# print(" R2.multiplicative_order(): ", R2.multiplicative_order())  # ▸
 ▸no such attribute
#   --> L1 has no attribute 'order'; L1, R2, R3, R3 have no attribute '▸
 ▸multiplicative_order'
m=7  #  m=4; multiplicative order of m is only defined if m is a unit ▸
 ▸modulo n !
print("  R4(%d):" % m, R4(m), "  R4(m).multiplicative_order(): ", R4(m)▸
 ▸.multiplicative_order())

G3 = R3.unit_group()
print("- G3 (via Zmod.unit_group): ", G3)
print("  list(G3): ", list(G3))
print("  G3.order(): ", G3.order())

# Return a list of all invertible elements (type of each is Python int)
L2 = R3.list_of_elements_of_multiplicative_group()
print("- L2 (via Zmod): ", L2);  print("  type(L2[1]): ", type(L2[1]))
```

5.3 Prime Numbers and the First Fundamental Theorem of Elementary Number Theory

Many of the problems in elementary number theory are concerned with prime numbers (see Chapter 4).

Every integer has divisors or factors. The number 1 has just one—itself, whereas the number 12 has the six factors 1, 2, 3, 4, 6, and 12.

The SageMath method `divisors()` gives a list of all divisors of a number n. For instance of $n = 12$ in the SageMath Example 5.2.

SageMath Example 5.2: Edit all Divisors of an Integer a and the Number of Divisors $\tau(a)$

```
sage: a=12; a.divisors(); number_of_divisors(a)
[1, 2, 3, 4, 6, 12]
6
```

Many numbers are only divisible by themselves and by 1. When it comes to multiplication, these can be regarded as the atoms in the realm of numbers.

Definition 5.1 Prime numbers are natural numbers greater than 1 that can only be divided by 1 and themselves.

By definition, 1 is not a prime number. Every integer is either prime, composite, or 1.

If we write down the prime numbers in ascending order (prime number sequence), then we get:

$$2,\ 3,\ 5,\ 7,\ 11,\ 13,\ 17,\ 19,\ 23,\ 29,\ 31,\ 37,\ 41,\ 43,\ 47,\ 53,\ 59,\ 61,\ 67,\ 71,\ 73,\ 79,\ 83,\ 89,\ 97, \cdots$$

The first 100 numbers include precisely 25 prime numbers. After this, the percentage of primes decreases, but never reaches zero. See Tables 4.10 and 4.11.

We come across integers that are prime fairly often. In the last decade of the 20th century, only three years were prime: 1993, 1997, and 1999. If they were rare, cryptography would not be able to work with them to the extent it does.

Prime numbers can be factorized in a unique (*trivial*) way:

$$\begin{aligned} 5 &= 1 \cdot 5 \\ 17 &= 1 \cdot 17 \\ 1013 &= 1 \cdot 1013 \\ 1296409 &= 1 \cdot 1296409 \end{aligned}$$

Definition 5.2 Natural numbers greater than 1 that are not prime are called *composite numbers*. These have at least two factors other than 1.

The dissection of a number into its prime factors is called *(complete) factorization*. Examples of the unique decomposition of composite numbers into prime factors:

$$\begin{aligned} 4 &= 2 \cdot 2 \\ 6 &= 2 \cdot 3 \\ 91 &= 7 \cdot 13 \\ 161 &= 7 \cdot 23 \\ 767 &= 13 \cdot 59 \\ 1029 &= 3 \cdot 7^3 \\ 5324 &= 22 \cdot 11^3 \end{aligned}$$

Theorem 5.1 Each composite number a has a lowest factor greater than 1. This factor is a prime number p and is less than or equal to the square root of a.

All integers greater than 1 can be expressed as a product of prime numbers—in a *unique* way.

This is the claim of the first *fundamental theorem of number theory* (= fundamental theorem of arithmetic = fundamental building block of all positive integers). It was formulated precisely for the first time by Carl Friedrich Gauss in his *Disquisitiones Arithmeticae* (1801).

Theorem 5.2 Gauss 1801 Every even natural number greater than 1 can be written as the product of prime numbers. Given two such decompositions $a = p_1 \cdot p_2 \cdot \ldots \cdot p_n = q_1 \cdot q_2 \cdot \ldots \cdot q_m$, these can be resorted such that $n = m$ and for all i, $p_i = q_i$.

In other words, each natural number other than 1 can be written as a product of prime numbers in precisely one way (if we ignore the order of the factors). The factors are therefore unique (the expression as a product of factors is unique).

For example, $60 = 2 \cdot 2 \cdot 3 \cdot 5 = 2^2 \cdot 3 \cdot 5$. And this—other than changing the order of the factors—is the only way in which the number 60 can be factorized.

If you allow numbers other than primes as factors, there are several ways of factorizing integers and the *uniqueness* is lost:

$$60 = 1 \cdot 60 = 2 \cdot 30 = 4 \cdot 15 = 5 \cdot 12 = 6 \cdot 10 = 2 \cdot 3 \cdot 10 = 2 \cdot 5 \cdot 6 = 3 \cdot 4 \cdot 5 = \cdots$$

In mathematics one also studies sets of numbers where the factorization into primes (or objects that have prime properties inside those sets) is not unique. An example for this (see Theorem 4.2) and further details on prime numbers (e.g., how Fermat's little theorem can be used to test extremely large numbers to determine whether they are prime) can be found in Chapter 4 of this book.

5.4 Divisibility, Modulus and Remainder Classes

There is a close connection between divisibility and congruences, which we will explain here using several examples.

5.4.1 Divisibility

If integers are added, subtracted, or multiplied, the result is always another integer. However, the division of two integers does not always result in an integer. For example, if we divide 158 by 10 the result is the decimal number 15.8, which is not an integer.

If, however, we divide 158 by 2 the result 79 is an integer. In number theory we express this by saying that 158 is *divisible* by 2 but not by 10. In general, we say:

Definition 5.3 An integer n is divisible by another integer d if the quotient n/d is an integer c such that $n = c \cdot d$.

In other words: An integer d divides an integer n, if $n = c \cdot d$ for some $c \in \mathbb{Z}$. n is called a *multiple* of d, whereas d is called a *divisor* or *factor* of n.

The mathematical notation for this is $d|n$ (read "d divides n"). The notation $d \nmid n$ means that d does not divide the number n. It is:

$$[\, d \mid n \,] \Leftrightarrow [\, n \text{ is a multiple of } d \,]$$

In our example therefore: $10 \nmid 158$ but $2|158$.

We have several possibilities to check with SageMath if the integer d divides the integer n. The most directly one is the method `is_integer()`. See SageMath Example 5.3.

SageMath Example 5.3: Check to Find Out If a Variable or a Term Is Integer

```
sage: n=158; d1=10; d2=2
sage: n.is_integer(); n in ZZ
True
True
sage: d1.divides(n);   (n % d1) == 0; (n / d1).is_integer(); (n / d1) in▸
  ▸ ZZ
False
False
False
False
sage: d2.divides(n);   (n % d2) == 0; (n / d2).is_integer(); (n / d2) in▸
  ▸ ZZ
True
True
True
True
```

Two further important definitions are those of the greatest common divisor (gcd) and the least common multiple (lcm) of two integers.

Definition 5.4 The $gcd(a, b)$ is the largest integer dividing both a and b.

Definition 5.5 The $lcm(a, b)$ is the smallest positive integer divisible by both a and b.

Two numbers are called relatively prime or *coprime*, if their greatest common divisor equals 1. For example, $9 = 3 \cdot 3$ and $28 = 2 \cdot 2 \cdot 7$ are coprime. The following equivalence holds:

$$[\, \gcd(a, b) = 1 \,] \Leftrightarrow [\, a \text{ and } b \text{ are coprime.}]$$

For coprimes a, b the following is true: $lcm(a, b) = a \cdot b$.

The functions `gcd` and `lcm` are available in SageMath—see SageMath Example 5.4.

For describing divisor relations for a set $\{a_1, \ldots, a_n\}$ of more than two elements, one has to be careful:

- $a_1, a_2, \ldots, a_n$ are *relatively prime*, if $\gcd(a_1, \ldots, a_n) = 1$. Here the $\gcd(a_1, \ldots, a_n)$ has to be computed stepwise by, for example, computing $\gcd(a_1, a_2) = g_1$ first and then $\gcd(g_1, a_3) = g_2$, and so forth, and finally $\gcd(g_{n-2}, a_n)$.

- An even stronger request for more than two numbers is: $a_1, \ldots, a_n$ are *pairwise relatively prime*, if for all $i = 1, \ldots, n$ and $j = 1, \ldots, n$ with $i \neq j$: $\gcd(a_i, a_j) = 1$.

Example:
2, 3, 6 are relatively prime, because $\gcd(2, 3, 6) = 1$. They are not pairwise relatively prime because $\gcd(2, 6) = 2 > 1$.

SageMath Example 5.4: Calculate gcd and lcm

```
sage: gcd(30, 160)
10
sage: gcd(9, 28)
1
sage: lcm(30, 160)
480
sage: lcm(9, 28)
252
```

As with the divisors, the integers coprime to a given integer a can also be calculated. There are different ways to do so in SageMath. The direct way is to use the method `coprime_integers()`. It requires an argument up to that the coprimes are calculated. Giving a as argument ensures to look only within $\{1, \ldots, a-1\}$. See SageMath Example 5.5.

Another way is to build the ring of integers modulo a, then list the multiplicative group of that ring. As this returns Python integers and we want Sage integers, we can convert them to SageMath integers via ZZ.

SageMath Example 5.5: Calculate the Coprimes of an Integer a

```
sage: a=15; a.coprime_integers(19)
[1, 2, 4, 7, 8, 11, 13, 14, 16, 17]
sage: a=8; a.coprime_integers(a)
[1, 3, 5, 7]
#
sage: a=8; Zmod(a).list_of_elements_of_multiplicative_group()
[1, 3, 5, 7]
sage: a=8; [ZZ(k) for k in Zmod(a).▸
 ▸list_of_elements_of_multiplicative_group()]
[1, 3, 5, 7]
```

5.4.2 The Modulo Operation: Working with Congruences

When we investigate divisibility, it is only the remainder of the division that is important. When dividing a number n by m, we often use the following notation:

$$\frac{n}{m} = c + \frac{r}{m},$$

where c is an integer and r is a number with the values $0, 1, \cdots, m-1$. This notation is called division with remainder, whereby c is called the integer quotient and r is the remainder of the division.

Example:

$$\frac{19}{7} = 2 + \frac{5}{7} \quad (m = 7,\ c = 2,\ r = 5)$$

What do the numbers $5, 12, 19, 26, \cdots$ have in common for division by 7? The remainder is always $r = 5$.

Dividing arbitrary integers by 7, only the following remainders are possible:

$$r = 0, 1, 2, \cdots, 6$$

If $r = 0$, then: $m|n$ ("m divides n").

The numbers that result in the same remainder r when divided by 7 are combined to form the remainder class r modulo 7. Two numbers a and b belonging to the same remainder class modulo 7 are said to be congruent modulo 7. Or in general:

Definition 5.6 The remainder class r modulo m is the set of all integers a that have the same remainder r when divided by m.

Example of remainder classes RC:

RC 0 mod 4 $= \{x|x = 4{\cdot}n;\ n \in \mathbb{Z}\} = \{\ldots, -16, -12, -8, -4, 0, 4, 8, 12, 16, \ldots\}$
RC 3 mod 4 $= \{x|x = 4{\cdot}n{+}3;\ n \in \mathbb{Z}\} = \{\ldots, -13, -9, -5, -1, 3, 7, 11, 15, \ldots\}$

As only the finitely many remainders $0, 1, 2, \cdots, m-1$ are possible for division modulo m, modular arithmetic works with finite sets. For each modulus m there are precisely m remainder classes.

The result of the modulo operation can be formulated as: $a \bmod m = a - m \cdot \lfloor a/m \rfloor$.

Definition 5.7 Two numbers $a, b \in \mathbb{N}$ are said to be congruent modulo $m \in \mathbb{N}$ if and only if they have the same remainder when divided by m.

We write: $a \equiv b \pmod{m}$) (read *a is congruent b modulo m*), which means that a and b belong to the same remainder class. The modulus m is therefore the divisor. This notation was introduced by Gauss. Although the divisor is usually positive, a and b can be any integer. This equivalence relation modulo m is also called *congruence*:

$$a \equiv b \pmod{m}$$

Example:
$19 \equiv 12 \pmod{7}$, because the remainders are equal: $19/7 = 2$ remainder 5 and $12/7 = 1$ remainder 5.

$23103 \equiv 0 \pmod{453}$, because $23103/453 = 51$ remainder 0 and $0/453 = 0$ remainder 0.

Theorem 5.3 $a \equiv b \pmod{m}$ if and only if the difference $(a - b)$ is divisible by m; that is, if $q \in \mathbb{Z}$ exists with $(a - b) = q \cdot m$.

In other words: $a \equiv b \pmod{m} \iff m|(a - b) \iff (a - b) \equiv 0 \pmod{m}$.

Therefore: If m divides the difference, there exists an integer q such that: $a = b + q \cdot m$. As an alternative to the congruence notation, we can also use the divisibility notation: $m|(a - b)$.

Remark:
This equivalence does apply only to the *difference* $(a-b)$, but not to the sum $(a+b)$.

Example: $11 \equiv 2 \pmod 3$, therefore $11 - 2 = 9 \equiv 0 \pmod 3$; but $11 + 2 = 13$ is not divisible by 3. The statement in Theorem 5.3 does not even apply to sums in one direction. It is correct for sums only if the remainder is 0 and only in the following direction: If a divisor divides both summands with no remainder, it also divides the sum with no remainder.

Example of equivalent statements:
$35 \equiv 11 \pmod 3 \iff 35 - 11 \equiv 0 \pmod 3)$, where $35 - 11 = 24$ is divisible by 3 without remainder while $35/3$ and $11/3$ leave the remainder 2.

SageMath Example 5.6: Division With and Without Remainder

```
sage: 10/4
5/2
sage: 10//4   # for integer arguments, "//" returns the integer ▸
 ▸quotient
2
sage: 10 % 3 # for integer arguments, "%" means mod, i.e., remainder
1
#
sage: 11//3
3
sage: 11 % 3
2
#
sage: int(11/3)
3
sage: type(int(11/3))
<class 'int'>
#
sage: type(11/3)
<class 'sage.rings.rational.Rational'>
sage: type(11//3)
<class 'sage.rings.integer.Integer'>
sage: type(11%3)
<class 'sage.rings.integer.Integer'>
```

We can apply the equivalence in Theorem 5.3 if we need a quick and easy method for determining whether large numbers are divisible by a certain number.

Example: Is 69993 divisible by 7?
The number can be written in the form of a difference in which it is clear that each operand is divisible by 7: $69993 = 70000 - 7$. Therefore, the difference is also divisible by 7.

Although these considerations and definitions may seem to be rather theoretical, we are so familiar with them in everyday life that we no longer think about the formal procedure. For example, the 24 hours on a clock are represented by the numbers $1, 2, \cdots, 12$. We obtain the hours after 12 noon as the remainder of a division by 12 and know immediately that 2 o'clock in the afternoon is the same as 14:00.

The modular arithmetic (based on division remainders) forms the basis of asymmetric encryption procedures. Cryptographic calculations are therefore not based on real numbers like the calculations mostly performed at school, but rather on number sets with a limited length (finite sets), like on positive integers that cannot exceed a certain value.

So we choose a large number m and calculate modulo m. That is, we ignore integer multiples of m and, rather than working with a number, we only work with the remainder when this number is divided by m. The result is that all results are in the range 0 to $m - 1$.

Since m is really large in practice, the set is also significantly larger than in our examples and cannot be completely stored in the computer's memory. But it has the advantages and properties of modular computing.

5.5 Calculations with Finite Sets

Here we consider congruences, which are modulo relations between integers. Congruences form a special equivalence relation (i.e., the relation is reflexive, symmetric, and transitive). From algebra, it follows that essential parts of the conventional calculation rules are kept to when we proceed to modular calculations over a basic set $\mathbb{Z}$. For example, addition remains commutative. The same goes for multiplication modulo m.

5.5.1 Laws of Modular Calculations

The known laws apply:

1. Associative law
   $((a + b) + c) \pmod m \equiv (a + (b + c)) \pmod m$
   $((a \cdot b) \cdot c) \pmod m \equiv (a \cdot (b \cdot c)) \pmod m$
2. Commutative law
   $(a + b) \pmod m \equiv (b + a) \pmod m$
   $(a \cdot b) \pmod m \equiv (b \cdot a) \pmod m$

The associative law and the commutative law apply to both addition and multiplication.

3. Distributive law
   $(a \cdot (b + c)) \pmod m \equiv (a \cdot b + a \cdot c) \pmod m$
4. Reducibility
   $(a + b) \pmod m \equiv (a \pmod m + b \pmod m) \pmod m$
   $(a \cdot b) \pmod m \equiv (a \pmod m \cdot b \pmod m) \pmod m$
   When adding or multiplying the order in which the modulo operation is performed does not matter.

5. Existence of an identity (neutral element)
   $(a+0) \pmod{m} \equiv (0+a) \pmod{m} \equiv a \pmod{m}$
   $(a \cdot 1) \pmod{m} \equiv (1 \cdot a) \pmod{m} \equiv a \pmod{m}$
6. Existence of an inverse element
   - Additive inverse
     For all integers a and m there exists another integer $-a$ such that: $(a+(-a)) \pmod{m} \equiv 0 \pmod{m}$
   - Multiplicative inverse modulo a prime p
     For each integer a (with $a \not\equiv 0 \pmod{p}$) and p prime) there exists an integer a^{-1} such that: $(a \cdot a^{-1}) \pmod{p} \equiv 1 \pmod{p}$
   - Multiplicative inverse modulo a compound number m
     For all integers a and m (with $a \not\equiv 0 \pmod{m}$) and $gcd(a,m) = 1$) there exists an integer a^{-1} such that: $(a \cdot a^{-1}) \pmod{m} \equiv 1 \pmod{m}$
7. Closure
   $a, b \in G \Longrightarrow (a+b) \in G$
   $a, b \in G \Longrightarrow (a \cdot b) \in G$

   More on the topic of closure can be found in Section 5.7.
8. Transitivity
   $[a \equiv b \pmod{m} \wedge b \equiv c \pmod{m}] \Longrightarrow [a \equiv c \pmod{m}]$
9. Modular division

   For k, x in $\{1, \ldots, m\}$ with $\gcd(k, m) = 1$ and arbitrary x the division of x by k is just the multiplication of x with the (existing) multiplicative inverse of k:

$$x : k \pmod{m} = x \cdot k^{-1} \pmod{m}$$

   If k^{-1} does not exist, the term $x : k$ is not defined. See also Table 5.3.

5.5.2 Patterns and Structures (Part 1)

In general, mathematicians investigate structures. They ask, for example, at $a \cdot x \equiv b \bmod m$, which values x can take for given values of a, b, m.

In particular the case is investigated where the result b of this operation is the neutral element. Then x is the (multiplicative) inverse of a regarding this operation.

5.6 Examples of Modular Calculations

As we have already seen:

- For two natural numbers a and m, $a \bmod m$ denotes the remainder obtained when we divide a by m. This means that $a \pmod{m}$ is always a number between 0 and $m-1$.
- For example, $1 \equiv 6 \equiv 41 \pmod{5}$ because the remainder is always 1. Another example is: $2000 \equiv 0 \pmod{4}$ because 4 divides 2000 with no remainder.

- Modular arithmetic only works on a limited quantity of nonnegative numbers. The number of these is specified by the modulus m. If, for example, the modulus is $m = 5$, then only the 5 numbers in the set $\{0, 1, 2, 3, 4\}$ are used.
- A calculation result larger than 4 is then reduced modulo 5. In other words, it is the remainder when the result is divided by 5. For example, $2 \cdot 4 \equiv 8 \equiv 3 \pmod 5$ because 3 is the remainder when we divide 8 by 5.

5.6.1 Addition and Multiplication

The following shows two tables:

- The addition table for mod 5 (Table 5.1);
- The multiplication tables for mod 5 (Table 5.2) and mod 6 (Table 5.3).

Those tables were generated with SageMath; see SageMath Example 5.14 for the source code.

Example of an Addition Table.
The result when we add 3 and 4 $\pmod 5$ is determined as follows: Calculate $3+4 = 7$ and keep subtracting 5 from the result until the result is less than the modulo: $7 - 5 = 2$. Therefore: $3 + 4 \equiv 2 \pmod 5$.

Example of a Multiplication Table.
The result of the multiplication $4 \cdot 4 \pmod 5$ is determined as follows: Calculate $4 \cdot 4 = 16$ and subtract 5 until the result is less than the modulus.

$$16 - 5 = 11;\ 11 - 5 = 6;\ 6 - 5 = 1$$

Table 5.2 directly shows that $4 \cdot 4 \equiv 1 \pmod 5$ because $16/5 = 3$ remainder 1.

Note that the multiplication is defined on the set $\mathbb{Z}$ excluding 0 (as $0 \cdot x$ is always 0, and 0 has no inverse).

5.6.2 Additive and Multiplicative Inverses

You can use the tables to read the inverses for each number in relation to addition and multiplication.

The inverse of a number is the number that gives the result 0 when the two numbers are added, and 1 when they are multiplied (i.e., as a result the neutral element for the respective operation). Thus, the inverse of 4 for addition mod 5 is 1, and the inverse of 4 for multiplication mod 5 is 4 itself, because

$$\begin{aligned} 4 + 1 = \ \ 5 &\equiv 0 \pmod 5; \\ 4 \cdot 4 = 16 &\equiv 1 \pmod 5. \end{aligned}$$

Table 5.1 Addition Table Modulo 5

+	*0*	*1*	*2*	*3*	*4*
0	0	1	2	3	4
1	1	2	3	4	0
2	2	3	4	0	1
3	3	4	0	1	2
4	4	0	1	2	3

Table 5.2 Multiplication Table Modulo 5: Operations Table of $(\mathbb{Z}_5^*, \cdot)$

$\cdot$	*1*	*2*	*3*	*4*
1	1	2	3	4
2	2	4	1	3
3	3	1	4	2
4	4	3	2	1

Table 5.3 Multiplication Table Modulo 6: Operations Table of $(\mathbb{Z}_6 \setminus \{0\}, \cdot)$

*	*1*	*2*	*3*	*4*	*5*
1	1	2	3	4	5
2	2	4	0	2	4
3	3	0	3	0	3
4	4	2	0	4	2
5	5	4	3	2	1

The inverse of 1 for multiplication mod 5 is 1, while the inverse modulo 5 of 2 is 3 and, since multiplication is commutative, the inverse of 3 is again 2.

If we take a random number (here 2) and add or multiply another number (here 4) and then add or multiply the corresponding inverse of the other number (1 or 4) to the interim result (1 or 3), then the end result is the same as the initial value.

Example:

$$2 + 4 \equiv 6 \equiv 1 \pmod 5; \quad 1 + 1 \equiv 2 \equiv 2 \pmod 5$$
$$2 \cdot 4 \equiv 8 \equiv 3 \pmod 5; \quad 3 \cdot 4 \equiv 12 \equiv 2 \pmod 5$$

In the set $\mathbb{Z}_5 = \{0, 1, 2, 3, 4\}$ for the addition, and in the set $\mathbb{Z}_5^* = \mathbb{Z}_5 \setminus \{0\}$ for the multiplication, all numbers have a unique inverse modulo 5.

In the case of modular addition, this is true for every integer used as modulus (not just for 5).

However, this is not the case for modular multiplication (important theorem):

Theorem 5.4 A natural number a from the set $\{1, \cdots, m-1\}$ has one modular multiplicative inverse if and only if this number and the modulus m are coprime, in other words if a and m have no common prime factors.

Since $m = 5$ is prime, the numbers 1 to 4 are relatively prime to 5 and each of these numbers has a multiplicative inverse in mod 5.

Table 5.3 shows as a counter example the multiplication table for mod 6 (since the modulus $m = 6$ is not prime, not all elements from $\mathbb{Z}_6 \setminus \{0\}$ are relatively prime to 6).

In addition to 0, also for the numbers 2, 3, and 4 there exists no other factor, so that the product equals 1 mod 6. We can say these numbers have *no* inverse.

The numbers 2, 3, and 4 have the factor 2 or 3 in common with the modulus 6. Only the numbers 1 and 5, which are relatively prime to 6, have multiplicative inverses, namely themselves.

The number of numbers that are relatively prime to the modulus m is the same as the number of numbers that have a multiplicative inverse (see the Euler function $\phi(m)$ in Section 5.8.2).

For the two moduli 5 and 6 used in the multiplication tables, this means the modulus 5 is a prime number itself. In mod 5, therefore, there are exactly $\phi(5) = 5 - 1 = 4$ numbers that are relatively prime to the modulus, which is all numbers from 1 to 4.

Since 6 is not a prime number, we write it as a product of its factors: $6 = 2 \cdot 3$. In mod 6, therefore, there are exactly $\phi(6) = (2-1) \cdot (3-1) = 1 \cdot 2 = 2$ numbers that have a multiplicative inverse; that is 1 and 5.

Although it may seem difficult to calculate the table of multiplicative inverses for large moduli, we can use Fermat's little theorem to create a simple algorithm for this [6, p. 80]. Quicker algorithms are described, for instance, in [7].

Cryptographically not only the unique nature of the inverse is important, but also that the set of possible values has been exhausted.

Theorem 5.5 For $a, i \in \{1, \ldots, m-1\}$ with $\gcd(a, m) = 1$, the product $a \cdot i \pmod{m}$ takes for any number a all values from $\{1, \ldots, m-1\}$ (exhaustive permutation of the length $m-1$).

See also Theorem 5.15 in Section 5.9.

Note that this is different from RSA, where a is raised to a fixed number e, while here the a is multiplied by all i. The following three examples illustrate the properties of multiplicative inverses (here only the lines for the factors 5 and 6 are listed, not the complete multiplication table).

Table 5.4 (multiplication table mod 17) was calculated for $i = 1, 2, \ldots, 18$:

$$(5 \cdot i)/17 = a \text{ remainder } r \text{ and highlighted } 5 \cdot i \equiv 1 \pmod{17} \text{ for } i = 7,$$
$$(6 \cdot i)/17 = a \text{ remainder } r \text{ and highlighted } 6 \cdot i \equiv 1 \pmod{17} \text{ for } i = 3.$$

We need to find the i for which the product remainder $a \cdot i$ modulo 17 with $a = 5$ or $a = 6$ has the value 1 (i.e., i is the multiplicative inverse of $a \pmod{17}$).

Between $i = 1, \ldots, m$ all values between $0, \ldots, m-1$ occur for the remainders, because both 5 and 6 are also relatively prime to the modulus $m = 17$.

The multiplicative inverse of 5 (mod 17) is 7, while the inverse of 6 (mod 17) is 3.

Table 5.5 (multiplication table mod 13) calculates the remainders of the products $5 \cdot i$ and $6 \cdot i$.

Between $i = 1, \ldots, m$, all values between $0, \ldots, m-1$ occur for the remainders, because both 5 and 6 are relatively prime to the modulus $m = 13$.

Table 5.4 Multiplication Table Modulo 17 for $a = 5$ and $a = 6$ Generated with SageMath Example 5.15

$i \Rightarrow$	1	2	3	4	5	6	7	8	9	10	11	12	13	14	15	16	17	18
$5 \cdot i$	5	10	15	20	25	30	35	40	45	50	55	60	65	70	75	80	85	90
remainder	5	10	15	3	8	13	**1**	6	11	16	4	9	14	2	7	12	0	5
$6 \cdot i$	6	12	18	24	30	36	42	48	54	60	66	72	78	84	90	96	102	108
remainder	6	12	**1**	7	13	2	8	14	3	9	15	4	10	16	5	11	0	6

Table 5.5 Multiplication Table Modulo 13 for $a = 5$ and $a = 6$

$i \Rightarrow$	*1*	*2*	*3*	*4*	*5*	*6*	*7*	*8*	*9*	*10*	*11*	*12*	*13*	*14*	*15*	*16*	*17*	*18*
$5 \cdot i$	5	10	15	20	25	30	35	40	45	50	55	60	65	70	75	80	85	90
remainder	5	10	2	7	12	4	9	**1**	6	11	3	8	0	5	10	2	7	12
$6 \cdot i$	6	12	18	24	30	36	42	48	54	60	66	72	78	84	90	96	102	108
remainder	6	12	5	11	4	10	3	9	2	8	**1**	7	0	6	12	5	11	4

The multiplicative inverse of 5 (mod 13) is 8, while the inverse of 6 (mod 13) is 11.

Table 5.6 contains an example where the modulus m and the number $a = 6$ are *not* relatively prime.

We calculated $5 \cdot i \pmod{12}$ and $6 \cdot i \pmod{12}$. Between $i = 1, \ldots, m$, not all values between $0, \ldots, m-1$ occur and 6 does not have an inverse mod 12, because 6 and the modulus $m = 12$ are not coprime.

The multiplicative inverse of 5 (mod 12) is 5. The number 6 has no inverse (mod 12).

5.6.3 Raising to the Power

In modular arithmetic, raising to the power is defined as repeated multiplication—which is standard. With small exceptions we can even apply the usual rules, such as:

$$a^{b+c} = a^b \cdot a^c$$
$$(a^b)^c = a^{b \cdot c} = a^{c \cdot b} = (a^c)^b$$

Modular powers work in the same way as modular addition and modular multiplication:

$$3^2 = 9 \equiv 4 \pmod{5}$$

Even consecutive powers work in the same way:

Example 1:

$$(4^3)^2 = 64^2 \equiv 4096 \equiv 1 \pmod{5}$$

1. We can speed up the calculation by reducing the interim results modulo 5, but we need to take care because *not* everything will then work in the same way as in standard arithmetic.

$$\begin{aligned}(4^3)^2 &\equiv (4^3 \pmod{5})^2 \pmod{5} \\ &\equiv (64 \pmod{5})^2 \pmod{5}\end{aligned}$$

Table 5.6 Multiplication Table Modulo 12 for $a = 5$ and $a = 6$

$i \Rightarrow$	*1*	*2*	*3*	*4*	*5*	*6*	*7*	*8*	*9*	*10*	*11*	*12*	*13*	*14*	*15*	*16*	*17*	*18*
$5 \cdot i$	5	10	15	20	25	30	35	40	45	50	55	60	65	70	75	80	85	90
remainder	5	10	3	8	**1**	6	11	4	9	2	7	0	5	10	3	8	1	6
$6 \cdot i$	6	12	18	24	30	36	42	48	54	60	66	72	78	84	90	96	102	108
remainder	6	0	6	0	6	0	6	0	6	0	6	0	6	0	6	0	6	0

$$\equiv 4^2 \pmod{5}$$
$$\equiv 16 \equiv 1 \pmod{5}$$

Remark: The time required to calculate the multiplication of two numbers normally depends on the length of the numbers. We can observe this if we use the school method to calculate, for instance, $474 \cdot 228$. The time required increases in a quadratic square manner because we need to multiply $3 \cdot 3$ numbers. The numbers become considerably smaller if we reduce the interim result.

2. In standard arithmetic, consecutive powers can be reduced to a single power by multiplying the exponents:

$$(4^3)^2 = 4^{3 \cdot 2} = 4^6 = 4096.$$

In modular arithmetic let's try what happens if we substitute $3 \cdot 2 \pmod{5}$ for the product of the exponents $3 \cdot 2$:

$$(4^3)^2 \equiv 4^{3 \cdot 2 \pmod{5}} \equiv 4^{6 \pmod{5}} \equiv 4^1 \equiv 4 \pmod{5}$$

But as we saw above, the correct result is 1.

3. Therefore, the rule is slightly different for consecutive powers in modular arithmetic: We do not multiply the exponents in $(\text{mod } m)$ but rather in $(\text{mod } \phi(m))$.

Using $\phi(5) = 4$ gives:

$$(4^3)^2 \equiv 4^{3 \cdot 2 \pmod{\phi(5)}} \equiv 4^{6 \pmod{4}} \equiv 4^2 \equiv 16 \equiv 1 \pmod{5}$$

This delivers the correct result for $m = 5$, but there are cases where it can't be done like that. For example if $m = 12$ we have $\phi(m) = 4$. The element 2 divides m, and if we calculate the 9th power, we get $2^9 \pmod{12} = 8$, but

$$2^{9 \pmod{\phi(12)}} = 2^{9 \pmod{4}} = 2^1 = 2 \neq 8.$$

Theorem 5.6 *Reduction in the exponent mod $\phi(m)$*
Let $\gcd(a, m) = 1$. Then

$$(a^b)^c \equiv a^{b \cdot c \pmod{\phi(m)}} \pmod{m}.$$

This is a consequence of the theorem of Euler and Fermat (see Theorem 5.13). Assume $bc = r + k\phi(m)$ with $r < m$ and $r, k \in \mathbb{N}_0$, then

$$a^{bc} = a^r \cdot a^{k \cdot \phi(m)} = a^r \cdot (\underbrace{a^{\phi(m)}}_{\equiv 1 (\text{mod } m)})^k \equiv a^r \pmod{m}.$$

Example 2:

$$3^{28} = 3^{4 \cdot 7} \equiv 3^{4 \cdot 7 \pmod{10}} \equiv 3^8 \equiv 6561 \equiv 5 \pmod{11}$$

5.6.4 Fast Calculation of High Powers (Square and Multiply)

RSA encryption and decryption are famous examples where calculating high powers modulo m is needed (see Sections 5.10 and 5.14). For example, the calculation 100^5 (mod 3) exceeds the 32-bit long integer number range if we calculate a^n by actually multiplying a with itself n times.

Remark: A 32-bit computer architecture refers to computer systems where all major system components like processor and memory can operate on data in 32-bit (4 byte) units—within registers. A 32-bit register can store 2^{32} different values. If an integer is represented as unsigned binary number, the range is 0 through $4294967295 = 2^{32} - 1$. Modern operating systems support 64 bits. Such a register can hold any of 2^{64} (over 18 quintillion = $1.8 \cdot 10^{19}$) different values. Representing an integer as unsigned binary number, the range is 0 through $18446744073709551615 = 2^{64} - 1$.

In case of extremely large numbers, even a fast computer would take longer than the age of the universe to calculate a single exponentiation. Luckily, there is an extremely effective shortcut for calculating exponentiations (but not for calculating logarithms).

If the expression is divided differently using the rules of modular arithmetic, then the calculation does not even exceed the 16-bit short integer number range:

$$(a^5) \equiv (((a^2 \pmod m))^2 \pmod m)) \cdot a) \pmod m \quad \text{cause } 5_{10} = 101_2$$

We can generalize this by representing the exponent as a binary number. For example, the naive method would require 36 multiplications in order to calculate a^n for $n = 37$. However, if we write n in the binary representation as $100101 = 1 \cdot 2^5 + 1 \cdot 2^2 + 1 \cdot 2^0$, then we can rewrite the expression as: $a^{37} = a^{2^5+2^2+2^0} = a^{2^5} \cdot a^{2^2} \cdot a^1$.

Example 3: 87^{43} (mod 103)
Since $43 = 32 + 8 + 2 + 1$, 103 is prime, $43 < \phi(103)$, and the squares (mod 103) can be calculated beforehand

$$\begin{aligned}
87^2 &\equiv 50 \pmod{103}, \\
87^4 &\equiv 50^2 \equiv 28 \pmod{103}, \\
87^8 &\equiv 28^2 \equiv 63 \pmod{103}, \\
87^{16} &\equiv 63^2 \equiv 55 \pmod{103}, \\
87^{32} &\equiv 55^2 \equiv 38 \pmod{103},
\end{aligned}$$

we have:

$$\begin{aligned}
87^{43} &\equiv 87^{32+8+2+1} \pmod{103}, \\
&\equiv 87^{32} \cdot 87^8 \cdot 87^2 \cdot 87 \pmod{103}, \\
&\equiv 38 \cdot 63 \cdot 50 \cdot 87 \equiv 85 \pmod{103}.
\end{aligned}$$

The powers $(a^2)^k$ can be determined easily by means of repeated squaring. As long as a does not change, a computer can calculate them beforehand and—if enough memory is available—save them. In order to then find a^n in each individual case, it now only needs to multiply those $(a^2)^k$ for which there is a one in the kth position of the binary representation of n. The typical effort is then reduced from 2^{600} to $2 \cdot 600$ multiplications! This frequently used algorithm is called *square and multiply*.

SageMath Example 5.8 contains source code implementing the square-and-multiply method in SageMath manually. It outputs the intermediate results, so you can reproduce the calculations above. See also Section 5.17.2 for a sample using the function `power_mod` built in SageMath.

5.6.5 Roots and Logarithms

Instead of computing the value of a power for a given basis and a given exponent modulo m, we can try to find a fitting exponent for a given value and a given basis (logarithm) or try to find a fitting basis for a given value and a given exponent (nth root).

The roots and logarithms are again integers. Yet in contrast to the usual situation, they are not only difficult to calculate but, in the case of large numbers, cannot be calculated at all within a reasonable amount of time.

Let us take the equation: $a \equiv b^c \pmod{m}$.

a. Existence and Uniqueness:
   If we restrict the numbers a, b, c to be elements of the set $\{0, 1, \ldots, m-1\}$ and $m > 2$ a natural number, then

   - $x \equiv b^c \pmod{m}$ for b, c, m is always well-defined and has a unique solution for x (trivial),
   - $a \equiv x^c \pmod{m}$ for a, c, m is not always solvable, and if it is solvable, it is not always uniquely solvable, for example, $x^2 \equiv 2 \pmod{15}$ has no solution while $x^2 \equiv 4 \pmod{15}$ has four different solutions $2, 7, 8, 13$.
   - $a \equiv b^x \pmod{m}$ for a, b, m is not always solvable, and if it is solvable, it is not always uniquely solvable, for example, $2^x \equiv 5 \pmod{15}$ has no solution while $2^x \equiv 1 \pmod{15}$ has solutions $x = 4, 8, 12, \ldots$.

b. Taking the Logarithm (Determining c); Discrete Logarithm Problem:
   If we know a and b of the three numbers a, b, and c that meet this equation, then every known method of finding c (if it exists) is approximately just as time-consuming as trying out all m possible values one after the other. For a typical m of the order of magnitude of 10^{180} (600-digit binary number), this is a hopeless task. Further details about the discrete logarithm problem can be found in Section 6.4. More on the complexity of this problem can be found in Sections 5.12.1 and 12.1.

c. Calculating the Root; Determining b:
   The situation is similar if b is the unknown variable, and we know the values of a and c. Here we use the Euler function (see Section 5.8.2). If we know the value of the Euler function $\phi(m)$ and $\gcd(\phi(m), c) = 1$ is true, then there exists a unique root b: For a given c we can easily calculate d with $c \cdot d \equiv 1 \pmod{\phi(m)}$ and use Theorem 5.6 to obtain

$$a^d \equiv (b^c)^d \equiv b^{c \cdot d} \equiv b^{c \cdot d \ (mod\ \phi(m))} \equiv b^1 \equiv b \pmod{m}$$

where b is the *c-th root* of a.

If $\phi(m)$ cannot be determined in a reasonable amount of time it is difficult to calculate the cth root. According to the first fundamental theorem of number theory and Theorem 5.11, we can determine $\phi(m)$ if we know the prime factors of m, but in real world settings those prime factors cannot be found quickly. This forms the basis for the security assumption used by the RSA encryption system (see Sections 5.10, 5.12.1, and 6.3.1).

Both logarithm and nth root—if well-defined and unique—can be seen as inverse operations of exponentiation. Compared with inverting exponentiation, the time required for inverting addition and multiplication is simply proportional to $\log m$ or $(\log m)^2$. Power functions ($x \mapsto x^k$ with k fixed) and exponential functions ($x \mapsto k^x$ with k fixed) are therefore typical one-way functions (compare Sections 5.13.1 and 6.1).

5.7 Groups and Modular Arithmetic in $\mathbb{Z}_n$ and $\mathbb{Z}_n^*$

Mathematical *groups* play a decisive role in number theory and cryptography. We only talk of groups if, for a defined set and a defined relation (an operation such as addition or multiplication), the following properties are fulfilled:

- The set is closed;
- A neutral element exists;
- An inverse element exists for each element;
- The associative law applies.

The abbreviated mathematical notation is $(G, +)$ or $(G, *)$.

After this somewhat more formal definition, we work again in $\mathbb{Z}_n$ (compare Section 5.5).

Definition 5.8 $\mathbb{Z}_n$:

$$\mathbb{Z}_n \text{ comprises all numbers from 0 to } n-1 : \mathbb{Z}_n = \{0, 1, 2, \ldots, n-2, n-1\}.$$

$\mathbb{Z}_n$ is an often used finite group of the natural numbers. It is sometimes also called the *remainder set* R modulo n. The remainder set and the reduced remainder set (resulting from multiplication) must not be confused with the remainder class (see Definition 5.6).

For example, today's 64-bit computers directly work only with integers in a finite set, that is the value range $0, 1, 2, \ldots, 2^{64} - 1$. Since 2003, 64-bit processors have been introduced. A 64-bit register can represent $2^{64} \approx 1.8 \cdot 10^{19}$ different integer values.

This value range is equivalent to the set $\mathbb{Z}_{2^{64}}$.

5.7.1 Addition in a Group

If we define the operation mod+ on such a set where

$$a \bmod+\ b := (a + b) \pmod{n},$$

then the set $\mathbb{Z}_n$ together with the relation mod+ is a group because the following properties of a group are valid for all elements in $\mathbb{Z}_n$:

- $a \text{ mod} + \ b$ is an element of $\mathbb{Z}_n$. The set is closed.
- There is a neutral element (the 0).
- Each element $a \in \mathbb{Z}_n$ has an inverse for this operation, namely $n - a$ [because $a \text{ mod} + \ (n - a) \equiv a + (n - a) (\text{mod } n) \equiv n \equiv 0 (\text{mod } n)$].
- $(a \text{ mod} + \ b) \text{ mod} + \ c \equiv a \text{ mod} + \ (b \text{ mod} + \ c)$.
  mod+ is associative.

Since additionally, the operation is commutative; that is, $(a \text{ mod} + \ b) = (b \text{ mod} + \ a)$, this structure is actually a commutative group.

5.7.2 Multiplication in a Group

If we define the operation mod* on the set $\mathbb{Z}_n$ where

$$a \text{ mod}^* \ b := (a \cdot b) \pmod{n},$$

then $\mathbb{Z}_n$ together with this operation is usually not a group because not all properties are fulfilled for each n.

Example:

a. In $\mathbb{Z}_{15}$, for example, the element 5 does not have an inverse. That is to say, there is no a with $5 \cdot a \equiv 1 \pmod{15}$. Each modulo product with 5 on this set gives 5, 10, or 0.

b. In $\mathbb{Z}_{55} \setminus \{0\}$, for example, the elements 5 and 11 do not have multiplicative inverses. That is to say, there is no $a \in \mathbb{Z}_{55}$ such that $5 \cdot a \equiv 1 \pmod{55}$ and no a such that $11 \cdot a \equiv 1 \pmod{55}$. This is because 5 and 11 are not relatively prime to 55. Each modulo product with 5 on this set gives $5, 10, 15, \ldots, 50$ or 0. Each modulo product with 11 on this set gives 11, 22, 33, 44, or 0.

On the other hand, there are subsets of $\mathbb{Z}_n$ that form a group with the operation mod*. If we choose all elements in $\mathbb{Z}_n$ that are relatively prime to n, then this set forms a group with the operation mod*. We call this set $\mathbb{Z}_n^*$.

Definition 5.9 $\mathbb{Z}_n^*$:

$$\mathbb{Z}_n^* = \{a \in \mathbb{Z}_n \,|\, \gcd(a, n) = 1\}.$$

$\mathbb{Z}_n^*$ is sometimes also called the reduced remainder set R' modulo n.

Example: For $n = 10 = 2 \cdot 5$, the following applies:

- Full remainder set $R = \mathbb{Z}_n = \{0, 1, 2, 3, 4, 5, 6, 7, 8, 9\}$;
- Reduced remainder set $R' = \mathbb{Z}_n^* = \{1, 3, 7, 9\} \longrightarrow \phi(n) = 4$.

SageMath Example 5.7 calculates the residue set R' and the Euler ϕ function of $n = 10$.

SageMath Example 5.7: Calculate Residue System R' of 10 with $\phi(10)$

```
sage: n=10; R1=n.coprime_integers(n); [i for i in R1]
....: eu1=euler_phi(n); eu1
[1, 3, 7, 9]
4
```

Comment:
R' or $\mathbb{Z}_n^*$ is always a genuine subset of R or $\mathbb{Z}_n$ because 0 is always an element of R but never an element of R'. Since 1 and $n-1$ are always relatively prime to n, they are always elements of both sets.

If we select a random element in $\mathbb{Z}_n^*$ and multiply it by every other element in $\mathbb{Z}_n^*$, then the products are all in $\mathbb{Z}_n^*$. This is due to the fact that $\mathbb{Z}_n^*$ is closed with respect to the multiplication and due to the gcd property:

$[a, b \in \mathbb{Z}_n^*] \Rightarrow [((a \cdot b) \pmod{n})) \in \mathbb{Z}_n^*]$, more precisely:

$[a, b \in \mathbb{Z}_n^*] \Rightarrow [\gcd(a, n) = 1, \gcd(b, n) = 1] \Rightarrow [\gcd(a \cdot b, n) = 1] \Rightarrow [((a \cdot b) \pmod{n})) \in \mathbb{Z}_n^*]$.

Those products also induce a unique permutation on the elements in $\mathbb{Z}_n^*$. Since 1 is always an element of $\mathbb{Z}_n^*$, there is a unique partner in this set such that the product is 1. In other words:

Theorem 5.7 Each element in $\mathbb{Z}_n^*$ has a multiplicative inverse.

Example:
For $a = 3$ modulo n with $n = 10$ and $\mathbb{Z}_n^* = \{1, 3, 7, 9\}$, we have that $a^{-1} = 7$ and multiplying $a = 3$ by any other number in $\mathbb{Z}_n^*$ gives a permutation of the values in $\mathbb{Z}_n^*$:

$$\begin{aligned} 3 &\equiv 3 \cdot 1 \pmod{10} \\ 9 &\equiv 3 \cdot 3 \pmod{10} \\ 1 &\equiv 3 \cdot 7 \pmod{10} \\ 7 &\equiv 3 \cdot 9 \pmod{10} \end{aligned}$$

The unique invertibility is an essential condition for cryptography (see Section 5.10).

5.8 Euler Function, Fermat's Little Theorem, and Euler-Fermat

Euler's phi function is an important function in number theory. Likewise, the Euler-Fermat theorem is of great importance for RSA.

5.8.1 Patterns and Structures (Part 2)

As mathematicians investigate the structure $a \cdot x \equiv b \pmod{m}$ (see Section 5.5.2), they also look at the structure $x^a \equiv b \pmod{m}$.

Again here they are interested in the cases, if $b = 1$ (then x is the multiplicative inverse of a) and if $b = x$ (then the function $f(x) = x^a \pmod{m}$ has a fixpoint). Compare the multiplicative order in Section 5.9 and the RSA fixed points in Section 5.17.7.

5.8.2 The Euler Phi Function

Given n, then the number of numbers from the set $\{1, \ldots, n-1\}$ that are relatively prime to n is equal to the value of the Euler function $\phi(n)$.

Definition 5.10 The Euler phi function $\phi(n)$ specifies the number of elements in $\mathbb{Z}_n^*$: $\phi(n) = |\mathbb{Z}_n^*|$

Compare this definition with Definition 5.9 and eventually read the explanations about the Euler function $\phi(n)$ in Section 6.3.1.

The Euler phi function sometimes is also written as $\Phi(n)$ or phi(n).

The number of these elements in the group is also called its cardinality or the order of the group.

$\phi(n)$ can be calculated very easily if we know the prime factors of n.

Theorem 5.8 For each prime number p holds: $\phi(p) = p - 1$.

Theorem 5.9 For the product of two distinct primes p and q, the following is true:

$$\phi(p \cdot q) = (p-1) \cdot (q-1) \quad \text{or} \quad \phi(p \cdot q) = \phi(p) \cdot \phi(q).$$

This case is important for the RSA procedure.

Theorem 5.10 If $n = p_1 \cdot p_2 \cdot \ldots \cdot p_k$ where p_1 to p_k are distinct prime numbers (i.e., no factor occurs more than once), then the following is true (as a generalization of Theorem 5.9):

$$\phi(n) = (p_1 - 1) \cdot (p_2 - 1) \cdot \ldots \cdot (p_k - 1).$$

Theorem 5.11 In general, the following is true for every prime number p and every n in $\mathbb{N}$:

1. $\phi(p^n) = p^{n-1} \cdot (p-1)$.
2. If $n = p_1^{e_1} \cdot p_2^{e_2} \cdot \ldots \cdot p_k^{e_k}$, where p_1 to p_k are distinct prime numbers, then:

$$\phi(n) = [(p_1^{e_1-1}) \cdot (p_1-1)] \cdot \ldots \cdot [(p_k^{e_k-1}) \cdot (p_k-1)] = n \cdot ([(p_1-1)/p_1] \cdot \ldots \cdot [(p_k-1)/p_k]).$$

Example:

- $n = 70 = 2 \cdot 5 \cdot 7 \Rightarrow$ using Theorem 5.10: $\phi(n) = 1 \cdot 4 \cdot 6 = 24$.
- $n = 9 = 3^2 \Rightarrow$ using Theorem 5.11: $\phi(n) = 3^1 \cdot 2 = 6$, because $\mathbb{Z}_9^* = \{1, 2, 4, 5, 7, 8\}$.
- $n = 2701125 = 3^2 \cdot 5^3 \cdot 7^4 \Rightarrow$ using Theorem 5.11:

$$\phi(n) = [3^1 \cdot 2] \cdot [5^2 \cdot 4] \cdot [7^3 \cdot 6] = 1234800.$$

Remark: Number-Theoretic Functions in CT2

The Euler phi function is just one of several number-theoretic functions or statistics used. In CT2 you can get an overview and a quick comparison for different numbers. In Figure 5.1 the phi function is highlighted for the number 24. Navigate to there in CT2 from `CT2 Crypto Tutorials ▷ World of Primes ▷ Distribution of primes ▷ Number line`.

5.8.3 The Theorem of Euler-Fermat

In order to prove the RSA procedure, we need Fermat's little theorem and its generalization (Euler-Fermat theorem).

Theorem 5.12 Fermat's Little Theorem
Let p be a prime number and a be a random integer, then:

$$a^p \equiv a \pmod{p}$$

An alternative formulation of Fermat's little theorem is as follows: Let p be a prime number and a be a random integer that is relatively prime to p, then:

$$a^{p-1} \equiv 1 \pmod{p}$$

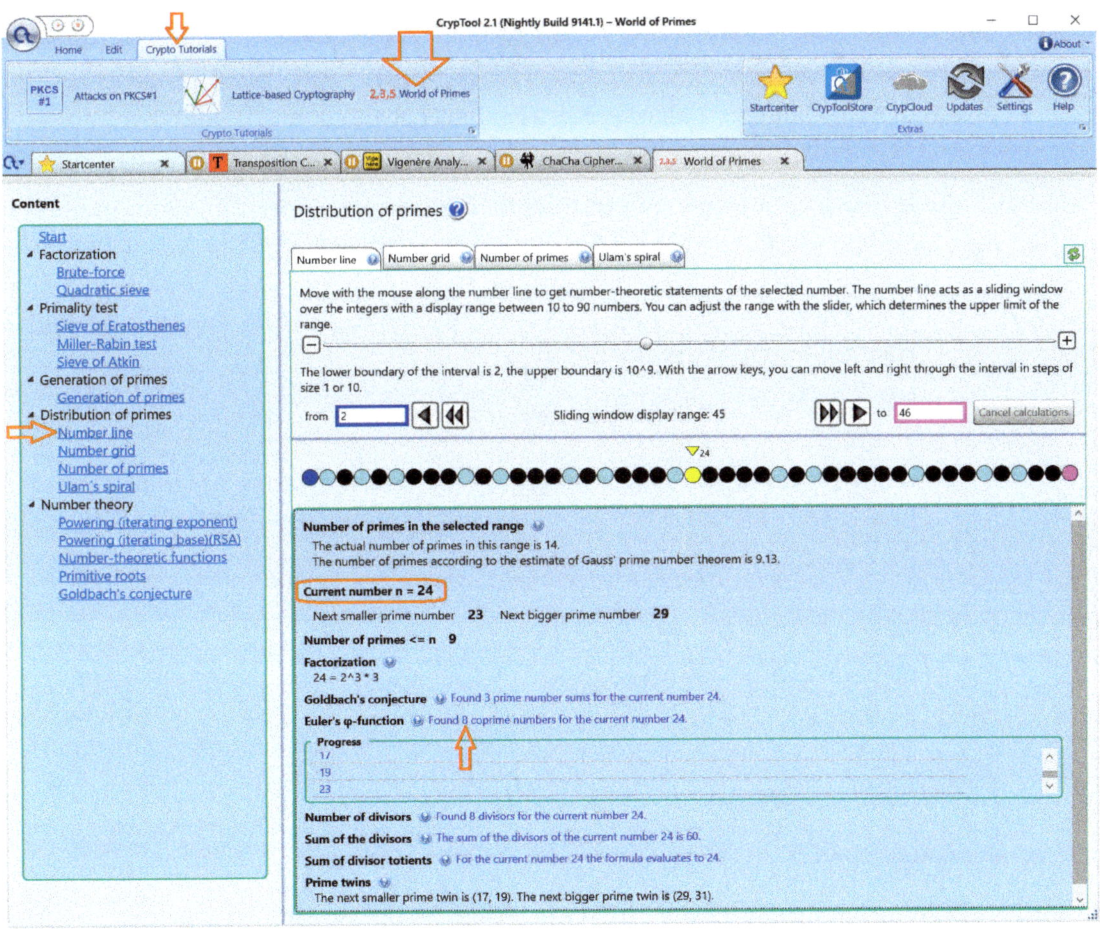

Figure 5.1 Number-theoretic functions in CT2.

Because if a and p are relatively prime (or coprime), an inverse $a^{-1} \pmod{p}$ always exists. Multiply the first congruence with $a^{-1} \pmod{p}$ from the left as well as from the right and the second congruence follows.

See also Section 4.5. Theorem 4.5 corresponds to Theorem 5.12 here.

Theorem 5.13 Euler-Fermat theorem (generalization of fermat's little theorem) For all elements a in the group $\mathbb{Z}_n^*$ (i.e., a and n are natural numbers that are coprime):

$$a^{\phi(n)} \equiv 1 \pmod{n}$$

This theorem states that if we raise a group element (here a) to the power of the order of the group (here $\phi(n)$), we always obtain the neutral element for multiplication (the number 1).

See for example [8, S. 94 ff] in the literature for a proof of this theorem.

The second formulation of Fermat's little theorem is derived directly from Euler's theorem if n is a prime number.

If n is not a prime number, then in most cases there do not exist primitive roots modulo n and the exponent $\phi(n)$ in Theorem 5.13 is not sharp; that is, can be replaced by a proper divisor of $\phi(n)$. The following formulation of the theorem is taken from an unpublished handout of Professor Geyer; see [9]. Alternatively this can be found in the famous classic of Hardy and Wright [10], on page 63 ff. There you can also find the proofs.

Theorem 5.14 Sharper Euler-Fermat theorem Let n not be divisible by 8 and not of the form $2u$ with $u \equiv 1 \bmod 2$.

1. If $n = p^r$ is a prime power, then there does exist a primitive root modulo n with order $\phi(n) = p^{r-1}(p-1)$ and the exponent $\phi(n)$ in 5.13 can not be replaced by a smaller one.
2. If n is not a prime power (and not a prime), then there exists no primitive root modulo n of order $\phi(n)$. If

$$n = p_1^{\alpha_1} \cdot p_2^{\alpha_2} \cdot \dots \cdot p_r^{\alpha_r} \qquad (r > 1)$$

is the prime factorization of n, then the multiplicative order of a residue modulo n is always a divisor of the least common multiple

$$\epsilon(n) = \text{lcm}(\phi(p_1^{\alpha_1}), \phi(p_2^{\alpha_2}), \dots, \phi(p_r^{\alpha_r}))$$

and $\epsilon(n)$ occurs as the order of some residue. Therefore, the Euler-Fermat theorem can be improved to

$$a^{\epsilon(n)} \equiv 1 \bmod n$$

for a, n coprime.

Remarks:

1. For numbers divisible by 8, $\epsilon(n)$ must be divided by 2 for getting the best exponent because $a^2 \equiv 1 \bmod 8$ for all odd a.

2. The condition that the number n is not of the form $2u$ is needed because for $n = 2u$ with odd u we have $\mathbb{Z}_{2u}^* \cong \mathbb{Z}_u^*$.
3. The SageMath Example 5.28 shows how such an element of maximal order can be constructed.

If n is the product of two prime numbers, we can—in certain cases—use Euler's theorem to calculate the result of a modular power very quickly. We have: $a^{(p-1)\cdot(q-1)} \equiv 1 \pmod{pq}$.

Examples for calculating a modular power:

- What is $5^2 \pmod 6$?
  With $2 = 1{\cdot}2$ and $6 = 2{\cdot}3$ where 2 and 3 are both prime; $\phi(6) = 2$ because only 1 and 5 are relatively prime to 6, we obtain the equation $5^2 \equiv 5^{\phi(6)} \equiv 1 \pmod 6$, without having to calculate the power.
- What is $31^{792} \pmod{851}$?
  With $792 = 22 \cdot 36$ and $23 \cdot 37 = 851$ where 23 and 37 are both prime, it follows for $31 \in \mathbb{Z}_{851}^*$ that $31^{792} \equiv 31^{\phi(23\cdot 37)} \equiv 31^{\phi(851)} \equiv 1 \pmod{851}$.

5.8.4 Calculation of the Multiplicative Inverse

Another interesting application is a special case of determining the multiplicative inverses using the Euler-Fermat theorem (multiplicative inverses are otherwise determined using the extended Euclidean algorithm).

Example:
Find the multiplicative inverse of 1579 modulo 7351.

According to Euler-Fermat: $a^{\phi(n)} = 1 \pmod n$ for all a in $\mathbb{Z}_n^*$. If we divide both sides by a, we get: $a^{\phi(n)-1} \equiv a^{-1} \pmod n$. For the special case that the modulus is prime, we have $\phi(n) = p - 1$. Therefore, the modular inverse is

$$a^{-1} = a^{\phi(n)-1} \equiv a^{(p-1)-1} \equiv a^{p-2} \pmod p.$$

For our example, this means:

Since the modulus 7351 is prime, $p - 2 = 7349$.
$1579^{-1} \equiv 1579^{7349} \pmod p$.

By cleverly breaking down the exponent, we can calculate this power relatively easy (see Section 5.6.4):

$$7349 = 4096 + 2048 + 1024 + 128 + 32 + 16 + 4 + 1$$
$$1579^{-1} \equiv 4716 \pmod{7351}$$

SageMath Example 5.8 contains source code implementing the square-and-multiply method in SageMath manually. It outputs the intermediate results, so you can reproduce the calculations above.

SageMath Example 5.8: Square and Multiply Done Manually in SageMath

```
print("\n# CHAP05 -- Sage-Script-SAMPLE 013: =========")

print("Square & multiply, applied to 2 samples:")
print("- Sample from 5.6.4 just quickly calculates the power 87^43 mod 103.")
print("- Sample from 5.8.4 calculates multiplicative inverse quickly in a special case, where ▸
 ▸exponent e = p-2.")
print("  a=1579; p = 7351; a^(-1) = a^(p-2) mod p")
print("  So a^(-1) = a^7349 mod 7351; and 7349=4096+2048+1024+128+32+16+4+1")

### Choose one of the two following examples:
### Sample from 5.6.4 ###
# p=103 ; R=Integers(p); a=R(87); e=43
# print("\nSample from 5.6.4: p, a, e:", p, a, e)

### Sample from 5.8.4 ###
p=7351; R=Integers(p); a=R(1579); e = p-2
print("\nSample from 5.8.4: p, a, e:", p, a, e)


print("\nFirst, calculate all squares:")
b = ZZ(e).bits(); blen = len(b); b = [1] * blen
print("b:", b, "  len(b):", len(b))

expo=0; sumall=0; prodall=1;
for pos, bit in enumerate(b):
  expo=2^pos;  sumall+=expo
  if expo == 1:
    z = a  # keep a
  else:
    z = power_mod(z, 2, p);  prodall*=z
  print("Pos: %2d" % pos, "  Expo:%5d" % expo, "  Sum:%5d" % sumall, "  z:%5d" % z, "  Prod:%5d▸
   ▸" % prodall)


print("\nSecond, calculate power a^e manually via square & multiply:")
b = ZZ(e).bits(); print("b:", b, "  len(b):", len(b), "  e:", e);  # b=b[::-1]; print(b);

expo=0; sumall=0; prodall=1; sum=0; prod=1
for pos, bit in enumerate(b):
  expo=2^pos;  sumall+=expo
  if expo == 1:
    z = a  # keep a
  else:
    z = power_mod(z, 2, p);  prodall*=z
  if bool(bit):
    sum+=expo; prod*=z
    print("Pos: %2d" % pos, "  Expo:%5d" % expo, "  Sum:%5d" % sum, "  z:%5d" % z, "  Prod:%5d"▸
     ▸ % prod)


# Using the build in Sage function
print("\nValidation:")
print("power_mod(87,43,103) = ", power_mod(87,43,103))  # 85
# print("power_mod(87,e,p) = ", power_mod(87,e,p))
# print("power_mod(1579,e,p) = ", power_mod(1579,e,p))
print("power_mod(1579,7349,7351) = ", power_mod(1579,7349,7351), "\n")  # 4716
```

5.8.5 How Many Private RSA Keys d Are There Modulo 26

This chapter addresses in detail some of the questions posed by thoughtful students, even if such no-gos are otherwise rather skipped in the literature.

According to Theorem 5.6, the arithmetic operations of modular expressions are performed in the exponents modulo $\phi(n)$ rather than modulo n. Note that here we adopt the usual practice for the RSA procedure to use "n" rather than "m" to denote the modulus.

In $a^{e \cdot d} \equiv a^1 \pmod{n}$, if we wish to determine the inverses for the factor e in the exponent, we need to calculate modulo $\phi(n)$.

Example: (with reference to the RSA algorithm)
If we calculate modulo 26, which set can e and d come from?

Solution: We have $e \cdot d \equiv 1 \pmod{\phi(26)}$.

> The reduced remainder set (reduced residue system) R' are the elements in $\mathbb{Z}_{26}$, which have a multiplicative inverse; that is, which are relatively prime to 26: $R' = \mathbb{Z}_{26}^* = \{1, 3, 5, 7, 9, 11, 15, 17, 19, 21, 23, 25\}$ (see Definition 5.9). R' has $\phi(26) = 12$ elements.
>
> The reduced remainder set R'' contains all numbers from 1 to $\phi(n)$ that are relatively prime to $\phi(n) = 12 : R'' = \{1, 5, 7, 11\}$. R'' has $\phi(\phi(26)) = 4$ elements. Note that in the general case not necessarily all elements of R'' have to also be contained in R'. For $n = 26$ however, this is the case: $R'' \subseteq R'$.
>
> For every e in R'', there exists a unique d in R'' such that $a \equiv (a^e)^d \pmod{n}$. So there are four values possible for key d mod(26).

For every e in R'', there exists precisely one element d such that $e \cdot d \equiv 1 \pmod{\phi(26)}$. This element d is not necessarily different from e: for example, $5 \cdot 5 \equiv 1 \pmod{12}$.

SageMath Example 5.9 calculates the two residue sets and the two ϕ functions.

SageMath Example 5.9: Calculate Residue System R' of 26 and Reduced Residue System R'' with $\phi(26)$

```
sage: n=26; R1=n.coprime_integers(n); [i for i in R1]
[1, 3, 5, 7, 9, 11, 15, 17, 19, 21, 23, 25]
sage: f=euler_phi(n); R2=f.coprime_integers(f); R2; f; euler_phi(f); ▸
 ▸len(R2)
[1, 5, 7, 11]
12
4
4
```

SageMath Example 5.9 considered the case for $n = 26$. With SageMath Example 5.34 you can consider the general case, where n can be any integer. See Section 5.17.6. The SageMath program delivers the number of all values d.

For all e that are coprime $\phi(n)$ we can calculate d as follows using the Euler-Fermat theorem:

$$d \equiv e^{-1} \pmod{\phi(n)}$$
$$\equiv e^{\phi(\phi(n))-1} \pmod{\phi(n)},$$

$$\text{because} \quad a^{\phi(n)} \equiv 1 \pmod{n} \quad \text{matches} \quad a^{\phi(n)-1} \equiv a^{-1} \pmod{n}$$

The problems of factorizing $n = pq$ with $q \neq p$ and of finding $\phi(n)$ have a similar degree of difficulty, and if we find a solution for one of the two problems, we also have a solution for the other one:

If we know the factors of $n = p \cdot q$ with $p \neq q$, then $\phi(n) = (p-1) \cdot (q-1) = n - (p+q) + 1$. Additionally, the values p and q are solutions of the quadratic equation $x^2 - (p+q)x + pq = 0$.

If only n and $\phi(n)$ are known, then mind that $pq = n$ and $p+q = n - \phi(n) + 1$. So you get p and q by solving the equation $x^2 + (\phi(n) - n - 1)x + n = 0$.

See also condition 3 in Section 5.10.1.

5.9 Multiplicative Order and Primitive Roots

The multiplicative order (see Definition 5.10) and the primitive root are two useful constructs (concepts) in elementary number theory.

Mathematicians often ask under which conditions the repeated application of an operation results in the neutral element (compare Section 5.8.1).

For the i-times successive modular multiplication of a number a by itself there is an i from $\{1, \ldots, m-1\}$ where the power $a^i \pmod m$ is the neutral element of the multiplication if and only if a and m are relatively prime.

Definition 5.11 The multiplicative order $ord_m(a)$ of an integer $a \pmod m$ (where a and m are coprime) is the smallest integer i for which $a^i \equiv 1 \pmod m$.

Example 1:
Table 5.7 shows the values a^i mod 11 for the exponents $i = 1, 2, \ldots, 10$, and for the bases $a = 1, 2, \ldots, 10$ as well as the resulting value $ord_{11}(a)$ for each a.

Table 5.7 also shows, for example, that the order of 3 modulo 11 has the value 5. SageMath Example 5.17 contains the source code to generate the entries. See Section 5.17.3.

In a multiplicative group (here $\mathbb{Z}_{11}^*$) not all numbers necessarily have the same order. The different orders in this case are 1, 2, 5, and 10, and we notice that:

1. The orders are all factors of 10.
2. The numbers $a = 2, 6, 7$, and 8 have the order 10, so we say that these numbers have the *maximum order* in $\mathbb{Z}_{11}^*$.

Definition 5.12 If a and m are coprime and if $ord_m(a) = \phi(m)$ (i.e., a has maximum order), then we say that a is a *primitive root* of m.

Table 5.7 Values of $a^i \pmod{11}$, $1 \leq a, i < 11$ and Corresponding Order of $a \pmod{11}$

	i=1	*i=2*	*i=3*	*i=4*	*i=5*	*i=6*	*i=7*	*i=8*	*i=9*	*i=10*	$ord_{11}(a)$
a=1	**1**	1	1	1	1	1	1	1	1	1	1
a=2	2	4	8	5	10	9	7	3	6	**1**	**10**
a=3	3	9	5	4	**1**	3	9	5	4	1	5
a=4	4	5	9	3	**1**	4	5	9	3	1	5
a=5	5	3	4	9	**1**	5	3	4	9	1	5
a=6	6	3	7	9	10	5	8	4	2	**1**	**10**
a=7	7	5	2	3	10	4	6	9	8	**1**	**10**
a=8	8	9	6	4	10	3	2	5	7	**1**	**10**
a=9	9	4	3	5	**1**	9	4	3	5	1	5
a=10	10	**1**	10	1	10	1	10	1	10	1	2

In Section 5.17.4, "Primitive Roots" there are SageMath examples to calculate primitive roots.

What is special with numbers a that are primitive roots, is that the powers $a^i \pmod{11}$ for running i with $1 \leq i < 11$ take on all values in $\mathbb{Z}_n^*$ (cf. Theorem 5.15).

As already mentioned in Theorem 5.14, not for every modulus m does not have a number a that is a primitive root. For example, $m = 45$ has no primitive roots a at all.

In Table 5.7, only $a = 2, 6, 7$, and 8 are a primitive root with respect to mod $m = 11$ ($ord_m(a) = \phi(11) = 10$). SageMath outputs only the first primitive root (2) via `primitive_root(11)`. The SageMath Example 5.22 can output *all* primitive roots a for a given m.

A few more comments based on Table 5.7:

- For your own experience it is good to be able to calculate values quickly. The values of a column of this table can be calculated with SageMath like this: `[power_mod(a,7,11) for a in [1..10]]`. This results in the seventh column: [1, 7, 9, 5, 3, 8, 6, 2, 4, 10]
- How would the table go on? What would the 11th column be? With `[power_mod(a,11,11) for a in [1..10]]` you can see that you get the first column again. The columns repeat with a cycle length of $\phi(m) = 10$. So $a^{11} = a^1 \pmod{11}$ for all $a = 1, 2, \ldots, 10$.
- It holds: $a^i = a^{i+k\cdot\phi(m)} \pmod{m}$, since the exponent is calculated modulo $\phi(m)$.

Using the primitive roots, we can clearly establish the conditions for which powers modulo m there is a unique inverse, and where the calculations in the exponents is manageable.

The two Tables 5.8 and 5.9 show the multiplicative orders and primitive roots modulo 45 and modulo 46.

Example 2:
Table 5.8 shows the values a^i mod 45 for the exponents $i = 1, 2, \cdots, 12$ and for the bases $a = 1, 2, \ldots, 12$ as well as the resulting value $ord_{45}(a)$ for each a.

SageMath Example 5.18 contains the source code to generate Table 5.8. See Section 5.17.3.

$\phi(45)$ is calculated using Theorem 5.11: $\phi(45) = \phi(3^2 \cdot 5) = 3^1 \cdot 2 \cdot 4 = 24$.

Since 45 is not a prime, there is no multiplicative order for all values of a (for all numbers that are not relatively prime to $45 : 3, 5, 6, 9, 10, 12, \cdots$, because $45 = 3^2 \cdot 5$).

Example 3:
Is 7 a primitive root modulo 45?

The necessary—but not sufficient—requirement/condition $\gcd(7, 45) = 1$ is fulfilled. Table 5.8 shows that the number $a = 7$ is not a primitive root of 45, because $ord_{45}(7) = 12 \neq 24 = \phi(45)$.

Example 4:
Table 5.9 answers the question as to whether the number $a = 7$ is a primitive root of 46.

Table 5.8 Values of $a^i \pmod{45}, 1 \le a, i < 13$ and Corresponding Order of $a \pmod{45}$

$a \setminus i$	1	2	3	4	5	6	7	8	9	10	11	12	$ord_{45}(a)$	$\phi(45)$
1	1	1	1	1	1	1	1	1	1	1	1	1	1	24
2	2	4	8	16	32	19	38	31	17	34	23	1	12	24
3	3	9	27	36	18	9	27	36	18	9	27	36	—	24
4	4	16	19	31	34	1	4	16	19	31	34	1	6	24
5	5	25	35	40	20	10	5	25	35	40	20	10	—	24
6	6	36	36	36	36	36	36	36	36	36	36	36	—	24
7	7	4	28	16	22	19	43	31	37	34	13	1	12	24
8	8	19	17	1	8	19	17	1	8	19	17	1	4	24
9	9	36	9	36	9	36	9	36	9	36	9	36	—	24
10	10	10	10	10	10	10	10	10	10	10	10	10	—	24
11	11	31	26	16	41	1	11	31	26	16	41	1	6	24
12	12	9	18	36	27	9	18	36	27	9	18	36	—	24

Table 5.9 Values of $a^i \pmod{46}, 1 \le a, i < 24$ and Corresponding Order of $a \pmod{46}$

$a \setminus i$	1	2	3	4	5	6	7	8	9	10	11	12	13	14	15	16	17	18	19	20	21	22	23	ord
1	1	1	1	1	1	1	1	1	1	1	1	1	1	1	1	1	1	1	1	1	1	1	1	1
2	2	4	8	16	32	18	36	26	6	12	24	2	4	8	16	32	18	36	26	6	12	24	2	–
3	3	9	27	35	13	39	25	29	41	31	1	3	9	27	35	13	39	25	29	41	31	1	3	11
4	4	16	18	26	12	2	8	32	36	6	24	4	16	18	26	12	2	8	32	36	6	24	4	–
5	5	25	33	27	43	31	17	39	11	9	45	41	21	13	19	3	15	29	7	35	37	1	5	22
6	6	36	32	8	2	12	26	18	16	4	24	6	36	32	8	2	12	26	18	16	4	24	6	–
7	7	3	21	9	17	27	5	35	15	13	45	39	43	25	37	29	19	41	11	31	33	**1**	7	**22**
8	8	18	6	2	16	36	12	4	32	26	24	8	18	6	2	16	36	12	4	32	26	24	8	–
9	9	35	39	29	31	3	27	13	25	41	1	9	35	39	29	31	3	27	13	25	41	1	9	11
10	10	8	34	18	42	6	14	2	20	16	22	36	38	12	28	4	40	32	44	26	30	24	10	–
11	11	29	43	13	5	9	7	31	19	25	45	35	17	3	33	41	37	39	15	27	21	1	11	22
12	12	6	26	36	18	32	16	8	4	2	24	12	6	26	36	18	32	16	8	4	2	24	12	–
13	13	31	35	41	27	29	9	25	3	39	1	13	31	35	41	27	29	9	25	3	39	1	13	11
14	14	12	30	6	38	26	42	36	44	18	22	32	34	16	40	8	20	4	10	2	28	24	14	–
15	15	41	17	25	7	13	11	27	37	3	45	31	5	29	21	39	33	35	19	9	43	1	15	22
16	16	26	2	32	6	4	18	12	8	36	24	16	26	2	32	6	4	18	12	8	36	24	16	–
17	17	13	37	31	21	35	43	41	7	27	45	29	33	9	15	25	11	3	5	39	19	1	17	22
18	18	2	36	4	26	8	6	16	12	32	24	18	2	36	4	26	8	6	16	12	32	24	18	–
19	19	39	5	3	11	25	15	9	33	29	45	27	7	41	43	35	21	31	37	13	17	1	19	22
20	20	32	42	12	10	16	44	6	28	8	22	26	14	4	34	36	30	2	40	18	38	24	20	–
21	21	27	15	39	37	41	33	3	17	35	45	25	19	31	7	9	5	13	43	29	11	1	21	22
22	22	24	22	24	22	24	22	24	22	24	22	24	22	24	22	24	22	24	22	24	22	24	22	–
23	23	23	23	23	23	23	23	23	23	23	23	23	23	23	23	23	23	23	23	23	23	23	23	–

The necessary but not sufficient requirement/condition $\gcd(7, 46) = 1$ is fulfilled.

$\phi(46)$ is calculated using Theorem 5.9: $\phi(46) = \phi(2 \cdot 23) = 1 \cdot 22 = 22$. The number 7 is a primitive root of 46, because $ord_{46}(7) = 22 = \phi(46)$.

SageMath Example 5.19 contains the source code to generate Table 5.9–see Section 5.17.3.

Theorem 5.15 Given a modulus m and a number a, relative prime to m, the following is true:

The set $\{a^i \pmod{m} | \ i = 1, \dots, \phi(m)\}$ equals the multiplicative group Z_m^* if and only if $ord_m(a) = \phi(m)$.

Even for prime moduli p and $0 < a < p$, not all a are of order $\phi(p) = p - 1$. Compare Table 5.7 as an example. But if $ord_m(a) = \phi(m)$, $a^i \pmod{p}$ goes through all the values $1, \dots, p-1$. Exhausting all possible values of the set is an important cryptographic proposition (compare Theorem 5.5). This determines a permutation $\pi(p-1)$.

Table 5.9 demonstrates that also for composite moduli m not all a are of maximal order $\phi(m)$. In this example only $5, 7, 11, 15, 17, 19,$ and 21 are of order 22.

The left-hand side of Theorem 5.15 holds exactly if a is a primitive root modulo m (see Definition 5.12).

The multiplicative group Z_m^* (see Definition 5.9) contains all values from 1 to $m - 1$ if and only if m is prime.

Example 5: Length of Cycles

Tables 5.10 and 5.11 serve as samples to introduce cycle lengths. This is a topic that goes beyond the multiplicative order.

Cycle here means a sequence of numbers $a^i \bmod n$ with $1 \leq i < n$ for a given a, and a repeating sequence. According to the generation method as modular power, here each number is unique within a cycle. The cycles here don't have to contain the 1 unless this cycle belongs to a multiplicative order: Then they have the 1 at the end of the cycle and at the position $a^{n-1} \bmod n$.

With l we now mean the cycle length. The maximum cycle length l_{max} is $\phi(n)$. For elements that do not belong to $\mathbb{Z}_n^*$ this can be explained with the Chinese remainder theorem (see e.g., [11, p. 167]):

Let π be the mapping that maps every $a \in \mathbb{Z}_n$ to the tuple of its remainders $(a_1, \dots, a_r)$ with $a_i \equiv a \bmod m_i$ for $i = 1, \dots, r$. If $a \neq 0$ is not invertible in $\mathbb{Z}_n$ and therefore has no group order, at least one of those a_i—but not all—are equal to 0.

Now imagine substituting 1's for every 0 component in this tuple. Then this element is invertible and has a well-defined order in $\mathbb{Z}_n^*$ which is by the theorem of Lagrange (see also [11]) a divisor of $\phi(n)$ and cannot be larger than $\phi(n)$ of course. See also Theorem 5.14.

In Tables 5.10 and 5.11, a runs through different values. For instance, for $\phi(a)$ we have (according to Theorem 5.11):

- $\phi(14) = \phi(2 \cdot 7) = 1 \cdot 6 = 6$
- $\phi(22) = \phi(2 \cdot 11) = 1 \cdot 10 = 10$

The values in the tables can be explained this way:

a. If the multiplicative order exists for a (independently whether a is prime), we have:

$$ord_n(a) = l \quad \text{and} \quad l | \phi(n)$$

The maximum length l_{max} is achieved, for example, for:

Table 5.10 Values of $a^i \pmod{14}, 1 \le a < 17, i < 14$

$a \setminus i$	1	2	3	4	5	6	7	8	9	10	11	12	13	$ord_{14}(a)$	$\phi(14)$	l
1	1	1	1	1	1	1	1	1	1	1	1	1	1	1	6	1
2	2	4	8	2	4	8	2	4	8	2	4	8	2	0	6	3
3	3	9	13	11	5	1	3	9	13	11	5	1	3	6	6	**6**
4	4	2	8	4	2	8	4	2	8	4	2	8	4	0	6	3
5	5	11	13	9	3	1	5	11	13	9	3	1	5	6	6	**6**
6	6	8	6	8	6	8	6	8	6	8	6	8	6	0	6	2
7	7	7	7	7	7	7	7	7	7	7	7	7	7	0	6	1
8	8	8	8	8	8	8	8	8	8	8	8	8	8	0	6	1
9	9	11	1	9	11	1	9	11	1	9	11	1	9	3	6	3
10	10	2	6	4	12	8	10	2	6	4	12	8	10	0	6	**6**
11	11	9	1	11	9	1	11	9	1	11	9	1	11	3	6	3
12	12	4	6	2	10	8	12	4	6	2	10	8	12	0	6	**6**
13	13	1	13	1	13	1	13	1	13	1	13	1	13	2	6	2
14	0	0	0	0	0	0	0	0	0	0	0	0	0	0	6	1
15	1	1	1	1	1	1	1	1	1	1	1	1	1	1	6	1
16	2	4	8	2	4	8	2	4	8	2	4	8	2	0	6	3

Table 5.11 Values of $a^i \pmod{22}, 1 \le a < 26, i < 22$

$a \setminus i$	1	2	3	4	5	6	7	8	9	10	11	12	13	14	15	16	17	18	19	20	21	$ord_{22}(a)$	l
1	1	1	1	1	1	1	1	1	1	1	1	1	1	1	1	1	1	1	1	1	1	1	1
2	2	4	8	16	10	20	18	14	6	12	2	4	8	16	10	20	18	14	6	12	2	0	**10**
3	3	9	5	15	1	3	9	5	15	1	3	9	5	15	1	3	9	5	15	1	3	5	5
4	4	16	20	14	12	4	16	20	14	12	4	16	20	14	12	4	16	20	14	12	4	0	5
5	5	3	15	9	1	5	3	15	9	1	5	3	15	9	1	5	3	15	9	1	5	5	5
6	6	14	18	20	10	16	8	4	2	12	6	14	18	20	10	16	8	4	2	12	6	0	**10**
7	7	5	13	3	21	15	17	9	19	1	7	5	13	3	21	15	17	9	19	1	7	10	**10**
8	8	20	6	4	10	14	2	16	18	12	8	20	6	4	10	14	2	16	18	12	8	0	**10**
9	9	15	3	5	1	9	15	3	5	1	9	15	3	5	1	9	15	3	5	1	9	5	5
10	10	12	10	12	10	12	10	12	10	12	10	12	10	12	10	12	10	12	10	12	10	0	2
11	11	11	11	11	11	11	11	11	11	11	11	11	11	11	11	11	11	11	11	11	11	0	1
12	12	12	12	12	12	12	12	12	12	12	12	12	12	12	12	12	12	12	12	12	12	0	1
13	13	15	19	5	21	9	7	3	17	1	13	15	19	5	21	9	7	3	17	1	13	10	**10**
14	14	20	16	4	12	14	20	16	4	12	14	20	16	4	12	14	20	16	4	12	14	0	5
15	15	5	9	3	1	15	5	9	3	1	15	5	9	3	1	15	5	9	3	1	15	5	5
16	16	14	4	20	12	16	14	4	20	12	16	14	4	20	12	16	14	4	20	12	16	0	5
17	17	3	7	9	21	5	19	15	13	1	17	3	7	9	21	5	19	15	13	1	17	10	**10**
18	18	16	2	14	10	4	6	20	8	12	18	16	2	14	10	4	6	20	8	12	18	0	**10**
19	19	9	17	15	21	3	13	5	7	1	19	9	17	15	21	3	13	5	7	1	19	10	**10**
20	20	4	14	16	12	20	4	14	16	12	20	4	14	16	12	20	4	14	16	12	20	0	5
21	21	1	21	1	21	1	21	1	21	1	21	1	21	1	21	1	21	1	21	1	21	2	2
22	0	0	0	0	0	0	0	0	0	0	0	0	0	0	0	0	0	0	0	0	0	0	1
23	1	1	1	1	1	1	1	1	1	1	1	1	1	1	1	1	1	1	1	1	1	1	1
24	2	4	8	16	10	20	18	14	6	12	2	4	8	16	10	20	18	14	6	12	2	0	10
25	3	9	5	15	1	3	9	5	15	1	3	9	5	15	1	3	9	5	15	1	3	5	5

- $a = 3, 5$ with $l_{max} = ord_{14}(a) = 6$ in Table 5.10 (cell highlighted in green)
- $a = 7, 13, 17, 19$ with $l_{max} = ord_{22}(a) = 10$ in Table 5.11

If an element of maximal order has to be computed, think of this element as a tuple with the help of the Chinese remainder theorem. Then for every single component of that tuple find a primitive root modulo the corresponding m_i, the modulus of that component. In this manner we get a tuple of primitive roots—with respect to the moduli m_i—for which by the Chinese remainder theorem there corresponds a unique number $a \in \{1, \ldots, n-1\}$. This number then generates a cycle of maximal length $\text{lcm}(\phi(m_1), \phi(m_2), \ldots, \phi(m_r))$ as already mentioned in Theorem 5.14. The listing 5.29 shows a SageMath example for computing such an element.

b. In some cases the maximum cycle length can be achieved for some values of a despite no multiplicative order exists for them (then in Tables 5.10 and 5.11 under the column header $ord_m(a)$ there is a "0" instead of a dash like in Table 5.8).
   Samples:
   - In Table 5.10: $l_{max} = \phi(14) = 6$ for $a = 10, 12$ (cell highlighted in red)
   - In Table 5.11: $l_{max} = \phi(22) = 10$ for $a = 2, 6, 8, 18$

   Both cases are special cases of Theorem 5.14 because 14 and 22 are of the form $2u$ with u not only odd but also prime.

   SageMath Example 5.20 contains the source code to generate Tables 5.10 and 5.11—see Section 5.17.3.

The topic of cycles and their lengths is also treated in detail in context with the RSA plane, where the notions *orbit* and *path* is used. See Section 6.5, especially Sections 6.5.8, 6.5.8.2, and 6.5.8.3.

5.10 Proof of the RSA Procedure with Euler-Fermat

Using the Euler-Fermat theorem (see Theorem 5.13) we can prove the RSA procedure in the group $\mathbb{Z}_n^*$.

The RSA procedure is the most common asymmetric cryptography procedure. Developed in 1978 by Ronald Rivest, Adi Shamir, and Leonard Adleman, it can be used both for signatures and for encryption.

5.10.1 Basic Idea of Public-Key Cryptography and Requirements for Encryption Systems

The basic idea behind public-key cryptography is that all participants possess a different pair of keys (P and S) and the public keys for all recipients are published. You can retrieve the public key P for a recipient from a directory just as you would look up someone's phone number in the phone book. Furthermore, each recipient has a secret key S that is needed in order to decrypt the message and that is not known to anyone else. If the sender wishes to send a message M, he encrypts it before sending using the public key P of the recipient.

The ciphertext C is determined as $C = E(P, M)$, where E (encryption) is the encryption rule. The recipient uses his private key S to decrypt the message with the decryption rule $M = D(S, C)$.

In order to ensure that this system works for every message M, the following four requirements must be met:

1. $D(S, E(P, M)) = M$ for every M (*invertibility*).
2. All (S, P) pairs are different for all participants.
3. The time required to derive S from P is at least as high as the time required to decrypt M with no knowledge of S.
4. Both C and M can be calculated relatively easily if the corresponding key is known.

The first requirement is a general condition for all cryptographic encryption algorithms.

The prerequisite of the second requirement can easily be met because there is a very large number of prime numbers. According to the prime number theorem (Theorem 4.7) of Legendre and Gauss, there are approximately $n/\ln(n)$ prime numbers up to the number n. This means, for example, that there are $6.5 \cdot 10^{74}$ prime numbers under $n = 2^{256}$ ($= 1.1 \cdot 10^{77}$) and $3.2 \cdot 10^{74}$ prime numbers under $n = 2^{255}$. Between 2^{255} and 2^{256} there are therefore $3.3 \cdot 10^{74}$ prime numbers with precisely 256 bits. Because of this large number of primes we cannot simply store them all—just because of physics (see the number of atoms in the universe in the overview under Section 4.12).

In addition, the second requirement can be ensured by a central office that issues certificates (see Section 5.12.5.4).

It is the last requirement that makes the procedure actually usable. This is because it is possible to calculate the powers in a linear amount of time (because there is a restriction on the length of the numbers).

Although Whitfield Diffie and Martin Hellman formulated the general method as early as 1976, the actual procedure that met all four requirements was publicly discovered later by Rivest, Shamir, and Adleman in 1978.

5.10.2 How the RSA Procedure Works

The RSA procedure including its prerequisites and secondary conditions is described here in detail. The seven individual steps for performing the RSA procedure can be clustered as follows. Steps 1 to 3 constitute key generation, steps 4 and 5 are the encryption, and steps 6 and 7 are the decryption:

1. Select two distinct random prime numbers p and q and calculate $n = p \cdot q$. The value n is called the RSA modulus.

   In CT1 and often in the literature, the RSA modulus is denoted with a capital "N."
2. Select an arbitrary $e \in \{3, \cdots, n-1\}$ such that e is relatively prime to $\phi(n) = (p-1) \cdot (q-1)$.

   We can then throw away p and q.
3. Calculate $d \in \{1, \cdots, n-1\}$ with $e \cdot d \equiv 1 \mod \phi(n))$; that is, d is the multiplicative inverse of e modulo $\phi(n)$. We can then throw away $\phi(n)$.

   → (n, e) is the public key P.

   → (n, d) is the private key S (only d must be kept secret).

4. For encryption, the message represented as a (binary) number is divided into parts such that each part of the number is less than n.
5. Encryption of the plaintext (or the parts of it) $M \in \{1, \cdots, n-1\}$:

$$C = E((n,e), M) = M^e \pmod{n}$$

6. For decryption, the ciphertext represented as a binary number is divided into parts such that each part of the number is less than n.
7. Decryption of the ciphertext (or the parts of it) $C \in \{1, \cdots, n-1\}$:

$$M = D((n,d), C) = C^d \pmod{n}$$

Remarks:

1. The numbers p, q, n chosen in step 1 are extremely large in practice (e.g., p and q have 1000 bit each, n 2000 bit length).
2. Further security aspects of the implementation and the algorithm itself are discussed in Sections 5.11 and 5.12.
3. In Section 6.5 the RSA algorithm is more deeply reasoned from number theory: The RSA plane is a model to illustrate the processes in this algorithm using pictures of rectangles.
4. Compaq introduced the multiprime method with high marketing effort in 2000. n was not the product of two primes, but of three: of two big ones and one relative small prime: $n = o \cdot p \cdot q$. With Theorem 5.10 we get: $\phi(n) = (o-1) \cdot (p-1) \cdot (q-1)$. This method did not assert itself.

   One reason probably was that Compaq claimed a patent on it. Generally there is less understanding in Europe and within the open-source community that one can claim patents on algorithms. But there is really no understanding outside the United States, that one can get a patent for a special case (3 instead of 2 factors) of an algorithm (RSA), although the patent for the general case was almost expired.[1]
5. If the two primes p and q are equal then $(m^e)^d \equiv m \bmod n$ is not true for all $m < n$ (although $e \cdot d \equiv 1 \bmod \phi(n)$ is fulfilled).

   Example: If $n = 5^2$ then according to Theorem 5.11, it is $\phi(n) = 5 \cdot 4 = 20$, $e = 3$, $d = 7$, $e \cdot d = 21 \equiv 1 \bmod \phi(n)$.

   But it is $(5^3)^7 \equiv 0 \bmod 25$. Therefore, p and q must be different.
6. The BSI (German Information Security Agency) recommends to choose the prime factors p and q almost the same, but not too close:

$$0.5 < |\log_2(p) - \log_2(q)| < 30$$

   They recommend generating the primes independently and check that the restriction is fulfilled (see [12]).

1. The multiprime RSA method is contained in `JCT Default Perspective` ▷ `Visuals` as well as in the JCT Algorithm Perspective.

7. For reasons of security, the selected e should not be **too** small. Since $\phi(n) = (p-1)\cdot(q-1)$ is even and e has to be relatively prime to $\phi(n)$, e cannot be 2. So the smallest value for e is 3 [13, Chap. 7.2.7].

   The BSI reference [12] recommends $2^{16}+1 \leq 2^{256}-1$. The procedure also allows us to select d freely and then calculate e. However, this has practical disadvantages. We usually want to be able to encrypt messages quickly, which is why we choose a public exponent e such that it has a short bit length compared to the modulus n and as few binary ones as possible (e.g., $2^{16}+1$). So a fast exponentiation is possible when encrypting. The prime numbers 3, 17, and 65537 have proved to be particularly practical for this purpose. The most often used number is $65537 = 2^{16}+1$, or in binary: $10\cdots00\cdots01$ (this number is prime and therefore relatively prime to many other numbers).

5.10.3 Proof that RSA Fulfills Requirement 1 (Invertibility)

Four requirements were set out in Section 5.10.1 that every practical asymmetric encryption method must meet. Requirement 1 was that the procedure must be able to be reversed unambiguously, such as a bijective mapping (this is fulfilled by RSA, but not by the default Rabin cryptosystem).

For pairs of keys (n, e) and (n, d) that possess the properties defined in steps 1 to 3 of the RSA procedure, the following must be true for all $M < n$:

$$M \equiv (M^e)^d \pmod n \quad \text{with} \quad (M^e)^d = M^{e\cdot d}$$

This means that the deciphering algorithm above works correctly.

We therefore need to show that:

$$M^{e\cdot d} \equiv M \pmod n$$

We will show this in three steps using Theorem 5.12 (Fermat's little theorem) (according to [3, p. 131ff]).

Step 1:
In the first step we show that: $M^{e\cdot d} \equiv M \pmod{\mathbf{p}}$.

Since $n = p\cdot q$ and $\phi(p\cdot q) = (p-1)\cdot(q-1)$ and since e and d are selected in such a way that $e\cdot d \equiv 1 \pmod{\phi(n)}$, there is an integer k such that: $e\cdot d = 1 + k\cdot(p-1)\cdot(q-1)$.

$$\begin{aligned}
M^{e\cdot d} &\equiv M^{1+k\cdot\phi(n)} \equiv M\cdot M^{k\cdot\phi(n)} \equiv M\cdot M^{k\cdot(p-1)\cdot(q-1)} \pmod p \\
&\equiv M\cdot(M^{p-1})^{k\cdot(q-1)} \pmod p \text{ based on little Fermat: } M^{p-1} \equiv 1 \pmod p \\
&\equiv M\cdot(1)^{k\cdot(q-1)} \pmod p \\
&\equiv M \pmod p
\end{aligned}$$

The requirement for using the simplified Euler-Fermat (Theorem 5.12) was that M and p are relatively prime.

Since this is not true in general, we need to consider the case when M and p are not relatively prime. Since p is a prime number, this implies that p is a factor of M. But this means:

$$M \equiv 0 \pmod{p}.$$

If p is a factor of M, then p is also a factor of $M^{e \cdot d}$. Therefore:

$$M^{e \cdot d} \equiv 0 \pmod{p}.$$

Since p is a factor of both M and $M^{e \cdot d}$, it is also a factor of their difference:

$$(M^{e \cdot d} - M) \equiv 0 \pmod{p}.$$

And therefore our conjecture is also true in this special case.

Step 2:
In exactly the same way we prove that: $M^{e \cdot d} \equiv M \pmod{\mathbf{q}}$.

Step 3:
We now combine the conjectures from step 1 and 2 for $n = p \cdot q$ to show that:

$$M^{e \cdot d} \equiv M \pmod{\mathbf{n}} \quad \text{for all } M < n.$$

From step 1 and 2 we have $(M^{e \cdot d} - M) \equiv 0 \pmod{p}$ and $(M^{e \cdot d} - M) \equiv 0 \pmod{q}$. Therefore, p and q are both factors of the same number $z = (M^{e \cdot d} - M)$. Since p and q are *distinct* prime numbers, their product must also be a factor of this number z. Thus:

$$(M^{e \cdot d} - M) \equiv 0 \pmod{p \cdot q} \quad \text{or} \quad M^{e \cdot d} \equiv M \pmod{p \cdot q} \quad \text{or}$$
$$M^{e \cdot d} \equiv M \pmod{n}.$$

□

Comment 1:
We can also condense the three steps if we use the Theorem 5.13 (Euler-Fermat), that is, not the simplified theorem where $n = p$ and which corresponds to Fermat's little theorem:

$$(M^e)^d \equiv M^{e \cdot d} \equiv M^{(p-1)(q-1) \cdot k+1} \equiv (\underbrace{M^{(p-1)(q-1)}}_{\equiv M^{\phi(n)} \equiv 1 \pmod{n}})^k \cdot M \equiv 1^k \cdot M \equiv M \pmod{n}.$$

Comment 2:
When it comes to signing messages, we perform the same operations but first use the secret key d (for signing), followed by the public key e (for validation). The RSA procedure can also be used to create digital signatures because:

$$M \equiv (M^d)^e \pmod{n}.$$

5.11 Regarding the Security of RSA Implementations

Section 5.12 deals with the security of the actual algorithm. On the other hand, this section is about practical security.

As we have presented RSA[2] so far in this chapter, it is also called textbook RSA; that is, the use of the algorithm itself. RSA is a kind of monoalphabetic substitution (see Section 2.2.1), except that the range of values does not only include 26 characters as in simple classical methods, but 2^n values (n is the modulus).

In general, you can avoid the practical problems and many (simple) attacks by always padding with additional and random data before encrypting the message with RSA. This reduces the range of values for the message. It is recommended to use, for example, the cryptographic padding method Optimal Asymmetric Encryption Padding (OAEP). OAEP is also defined in the standard PKCS#1 (version 2.2, 2012-10-27) and in RFC 8017. So a component of randomness is added to the deterministic RSA algorithm. One of the modern characteristics of a secure encryption system is that it is indistinguishable. There are two particular types of indistinguishability that are of importance: IND-CPA (indistinguishability under a chosen plaintext attack), and IND-CCA (indistinguishability under a chosen ciphertext attack). We will not go into the theoretical details here, but just state the results for RSA under reasonable assumptions: Textbook RSA cannot be IND-CPA secure (even less it is IND-CCA secure). RSA with OAEP on the other hand is CCA secure in the random oracle model. More details can be found in [14] and Section 1.8.2.

Figure 5.2 shows a screenshot from CTO: In the GUI, a selection is made to encrypt a text with RSA. The RSA public key comes from the file "my_rsa.pub" and by default the called OpenSSL implementation uses OAEP for padding. The internally used OpenSSL command (`openssl pkeyutl -encrypt -pubin -inkey my_rsa.pub -hexdump`) is displayed in the console window (below the "Execute" button).

5.12 Regarding the Security of the RSA Algorithm

The first part of this section follows the article "Vorzüge und Grenzen des RSA-Verfahrens" written by F. Bourseau, D. Fox, and C. Thiel [15].

2. The RSA cryptosystem can be executed with CT in many variations:
   - CTO has two broad plugins for RSA:
     Via "RSA (Step-by-step)" `https://www.cryptool.org/en/cto/rsa-step-by-step`.
     Via "RSA visual and more" `https://www.cryptool.org/en/cto/rsa-visual` you can see with graphics how RSA assigns its input values when encrypting, you can test textbook RSA with big numbers, and also use RSA with OAEP padding and certificates as it is used in practice.
   - The menu path of `CT1 Individual Procedures` ▷ `RSA Cryptosystem` ▷ `RSA Demonstration` offers variants for block size and alphabet of textbook RSA. Furthermore, under `CT1 Encrypt/Decrypt` ▷ `Asymmetric` messages can be encrypted and decrypted with RSA quickly.
   - Under `CT2 Templates` ▷ `Cryptography` ▷ `Modern` ▷ `Asymmetric` you can find asymmetric methods like RSA.
   - Both `JCT Default Perspective` ▷ `Visuals` and the `JCT Algorithm Perspective` offer asymmetric methods like RSA.

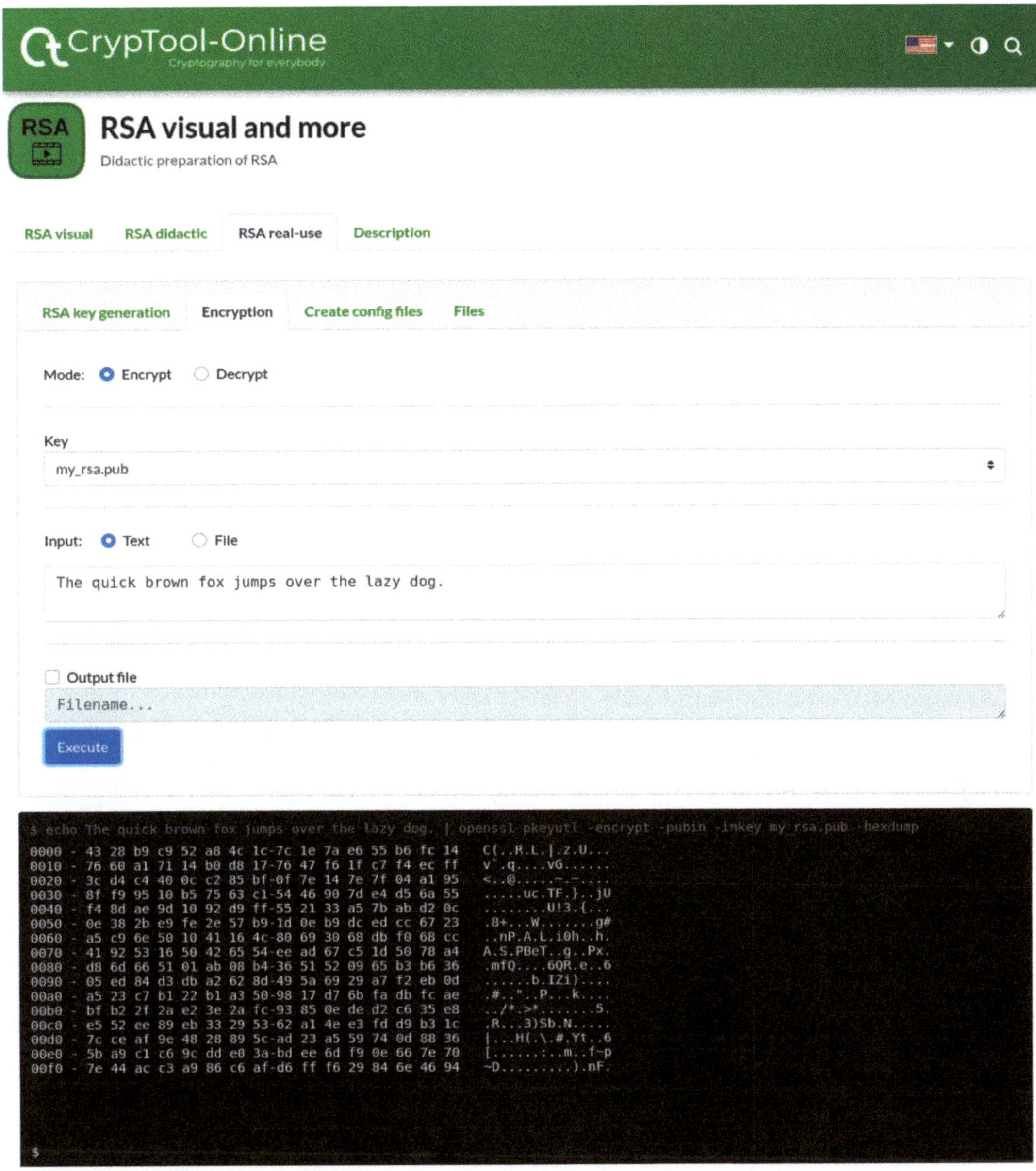

Figure 5.2 CrypTool-Online: Encryption with OpenSSL (using padding via OAEP).

When new breakthroughs in factorization are published, the discussion keeps coming up whether the RSA algorithm is still suitable for digital signatures and encryption. Nevertheless, the RSA algorithm is still the asymmetric de facto standard (compare Section 8.1).

The security of the RSA algorithm rests—as with all asymmetric cryptographic methods—on the following four central pillars:

- The complexity of the number theoretical problem on which the algorithm is based (here factorization of big numbers);
- The election of fitting parameters (here the length of the modulus n);

- The adequate usage of the algorithm and key generation (here the choice of p, q, e, d);
- The correct implementation of the algorithm.

Usage and key generation are well understood today. The actual implementation based on long integer arithmetic is very easy.

The following two sections examine the RSA algorithm with respect to the first two points.

5.12.1 Complexity

The security of the RSA algorithm depends—as with all public-key methods—on the difficulty to calculate the private key (here d) from the public key (n, e).

Especially for the RSA method this means:

1. It is hard to calculate $\phi(n)$ for big compounds n;
2. It is hard to calculate the prime factors of big compounds n (integer factorization problem IFP).

   There is no reason for the concern sometimes mentioned that there are not enough primes: Raising the dimension of the modulus always offers enough primes to consider. This is visualized in Section 4.14.

Successful decryption or forgery of a signature—without knowing the private key—therefore requires calculating the eth root mod n. The private key, which is the multiplicative inverse of e mod $\phi(n)$, then can be easily determined if $\phi(n)$ is known. $\phi(n)$ again can be calculated from the prime factors of n. Breaking of RSA therefore cannot be more difficult than factorization of the modulus n.

The inverse proposition that the RSA algorithm can be broken only by factorization of n is still not proven. Most number theorists consider the RSA problem and the factorization problem equivalent in terms of time complexity.

The best factorization method known today is a further development of the general number field sieve (GNFS), which was originally devised to factor only numbers of a special form (like Fermat numbers).

More details about GNFS and its complexity can be found in Section 12.4. The discussion there shows that the GNFS belongs to the class of problems with subexponential time complexity (i.e., time complexity grows asymptotically not as fast as exponential functions like e^l or 2^l, but strictly slower, like $e^{\sqrt{l}}$).

This classification is current knowledge; it does not preclude the possibility that the factorization problem can be solved in polynomial time (see Section 5.12.5.1 and Sections 5.10 and 6.3.1).

5.12.2 Security Parameters Because of New Algorithms

Factorization Algorithms[3]

The complexity of an attack is essentially determined by the length l of the modulus n. How large this essential parameter is chosen depends on the possibilities of the current factorization algorithms:

3. The quadratic sieve (QS) can be found in CT1, CT2, and CTO (see Msieve).

- In 1994 a 129-digit RSA modulus (428 bits), published in 1977, was factorized by a distributed implementation of the quadratic sieve algorithm (QS), developed 1982 by Pomerance. This effort took 8 months. Please refer to [16].
- In 1999 a 155-digit modulus (512 bits) was factorized with an implementation of the GNFS developed by Buhler, Lenstra, and Pomerance. The GNFS is more efficient than QS if n is longer than about 116 decimal digits. This effort took 5 months. Please refer to [17].
- Ten years later, at the end of 2009, a 232-digit modulus (768 bits) was factorized by Kleinjung after 2 ½ years. See [18].

This clearly demonstrates that a modulus length of 768 bits no longer provides sufficient protection against attackers.

For details about factorization progress since 1999, see Section 5.12.4.

A good website for online factorization is Dario Alpern's *Integer Factorization Calculator*; see [19].

RSA cannot only be attacked by factorization but by several—well known—poorly chosen settings. A tool that implemented almost all these attacks (mostly in Python, some in SageMath) is **RsaCtfTool.** See [20].

Lattice Base Reduction Algorithms

The modulus length l is not the only parameter relevant for security. Beneath requirements from implementation and engineering the sizes and the proportions of the parameters e, d, and n are relevant.

Corresponding attacks based on lattice reductions are a real threat for (too) simple implementations of RSA. These attacks can be structured into the following four categories:

- Attacks against very small public keys e (e.g., $e = 3$);
- Attacks against relatively small private exponents d (e.g., $d < n^{0.5}$);
- Factorization of the modulus n, if one of the factors p or q is *partly* known;
- Attacks requiring that a *part* of the private key d is known (the motivation for these partial key exposure attacks mainly arises from the study of side-channel attacks on RSA).

Sections 11.8.2 and 11.9 go into more detail on lattice-based attacks. A very good overview can be found on the website. *LatticeHacks* which is a joint work by Daniel J. Bernstein, Nadia Heninger, and Tanja Lange; see [21]. On this website you will find both a lecture at the CCC 2017 and the SageMath sources for this.

The four categories mentioned above are implemented in CTT; see Matthias Schneider's diploma thesis [22].

You can also find out which lattice-based methods and attacks are offered in CrypTool in Section 11.12.

5.12.3 Forecasts about Factorization of Large Integers

Since 1980 a lot of progress has been made regarding factorization of large integers. Estimations about the future development of the ability to factor RSA moduli vary and depend on some assumptions:

- Progression in computing performance (Moore's law: every 18 months computing power will double) and progression in grid computing;
- Development of new algorithms.

Within the last few years, the RSA modulus bit length feasible for factorization increased on average by 10 bits per year—even without new algorithms. Larger numbers require not only more time to be factorized, but also huge RAM storage for the solutions matrix being used by the best algorithms known today. This need for storage grows like the square root of the computation time (i.e., also subexponentially). Because RAM availability increased exponentially in the recent decades, it seems that this should not be the limiting factor.

A very well-founded estimation of the evolution of secure key lengths was done by Lenstra/Verheul in 1999 [23] (compare Figures 8.1 and 13.1). Another forecast can be found in Section 12.4.3.

In 2001, Dirk Fox et al. [15] predicted an almost linear factorization progression (see Figure 5.3): Each year the modulus length feasible for factorization increases by 20 bits on average. Their forecast then was below the more optimistic estimations of BSI and NIST. This forecast proved true by the factorization records of RSA-200 and RSA-768 (see Section 5.12.4). The estimation for the year 2005, to achieve a bit length of 660 bits, was almost a precision landing. Then the forecast became too optimistic as it expected the factorization of an RSA modulus of 1024 bits by 2020.

We disregard speculations about advances in quantum computers. To attack current RSA parameters, significantly more stable *and* interconnected qubits would have to be available than is currently the case.

5.12.4 Status Regarding Factorization of Specific Large Numbers

An exhaustive overview about the factorization records of composed integers using different methods can be found at *Wikipedia* (e.g., [24, 25]). Usually we recommend

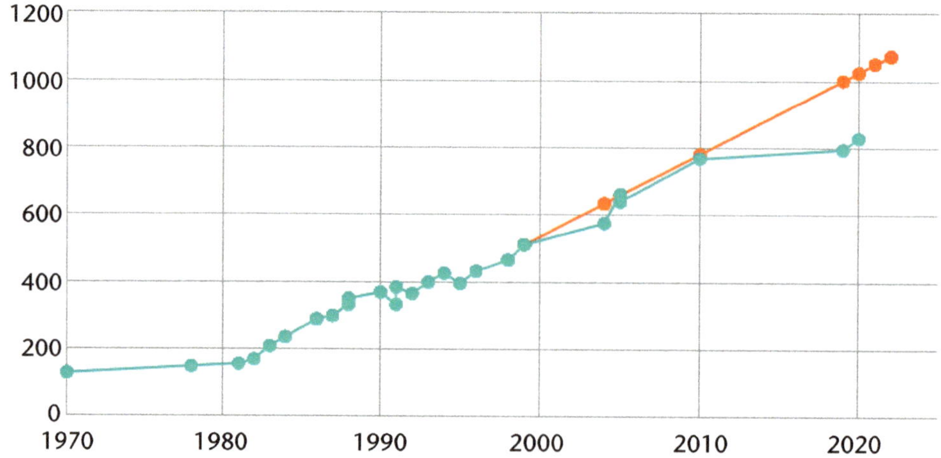

Figure 5.3 Comparison between the published, real factorization records (blue) and the predicted development (orange). [Forecast by Fox 2001; last real addition 2020 (see Table 5.12).]

to cite original sources, but for overviews websites are often more up-to-date and these two Wikipedia websites are frequently updated. Further websites are *primerecords* [26] (but its last edit was in 2018 and it still does not offer https) and *FactorDB* by Markus Tervooren, which provides over 2 billion fully factorized composite numbers; see [27].

The last records with factorization algorithms for composed numbers are listed in Table 5.12: The RSA numbers in the first column are certain large semiprime numbers (i.e., numbers with exactly two prime factors). The “C” numbers are compound and special numbers: They are either a Mersenne/Cunningham number (see Sections 4.4.2 and 4.6.3) themselves or factors of such a number.

The RSA numbers were generated and published by the company RSA Security. In the RSA Factoring Challenge the prime factors for these numbers are sought.

RSA Labs has offered its challenges since the beginning of the 1990s. The first challenge labeled the numbers, from RSA-100 to RSA-500, according to their number of decimal digits; the second RSA Factoring Challenge labeled the numbers after their number of binary digits. Within the second challenge cash prizes were offered for successful factorizations of RSA-576 to RSA-2048 (RSA-576, RSA-640, etc. using 64-bit steps upwards. An exception to this is RSA-617, which was created prior to the change in the numbering scheme). But the RSA challenges ended ahead of time in 2007, when RSA Inc. retracted the prize. All unsolved RSA challenges of RSA Labs can be found at the website of the cipher challenge “MysteryTwister” [28].

The C numbers originate from the Cunningham project [29], which seeks to factor Mersenne numbers. These have a very special form that makes it orders of magnitude easier to factor them compared to RSA moduli of the same length.

Table 5.12 shows for each number its length as a binary number and as a decimal number, then the length of the two last and largest prime factors: p123 means that the number is prime and has 123 decimal places. This is the notation as also used in the standard book [30]. How difficult it is to factorize the Mersenne numbers depends above all on the size of their last two (largest) factors; see also [29].

Table 5.12 The Current Factoring Records as of May 2023 (Compare with Figure 5.3)

	Binary Digits	*Decimal Digits*	*Last Prime Factors*		*Factorized On*	*By*
RSA-250	829	250	p125	p125	Feb 2020	F. Baudot et al.
RSA-240	795	240	p120	p120	Dec 2019	F. Baudot et al.
RSA-768	768	232	p116	p116	Dec 2009	T. Kleinjung et al.
RSA-200	663	200	p100	p100	May 2005	Jens Franke et al.
RSA-640	640	193	p97	p97	Nov 2005	Jens Franke et al.
RSA-576	576	174	p87	p87	Dec 2003	Jens Franke et al.
RSA-160	530	160	p80	p80	Apr 2003	Jens Franke et al.
RSA-155	512	155	p78	p78	Aug 1999	H. te Riele et al.
...						
C355 in $2^{1193}-1$	1177	355	p104	p251	Aug 2014	T. Kleinjung et al.
C320 = $2^{1061}-1$	1061	320	p143	p177	Aug 2012	G. Childers et al.
C307 in $2^{1039}-1$	1017	307	p80	p227	May 2007	K. Aoki et al.
C274 in $6^{353}-1$	911	274	p120	p155	Jan 2006	K. Aoki et al.
C176 in $11^{281}+1$	583	176	p87	p89	May 2005	K. Aoki et al.
C158 in $2^{953}-1$	523	158	p73	p86	Jan 2002	Jens Franke et al.

The last two columns of the table show when and by whom the number was fully factorized.

If you are looking for a challenge, you will find many incompletely factorized Mersenne numbers in the [29] database—these have the status code "CF" such as:

- C337: $2^{1207} - 1 = 131071 \cdot 228479 \cdot 48544121 \cdot 212885833 \cdot 7121450524 \ldots 71$
- C297: $2^{1213} - 1 = 327511 \cdot 7150798418 \ldots 71 \cdot 6022881435 \ldots 11$
- C284: $2^{1229} - 1 = 36871 \cdot 46703 \cdot 105431792806619161 21033 \cdot 9536289355 \ldots 57 \cdot 5339295584 \ldots 87$

Further tasks/challenges can be found in the "Wanted list" in [29].

The current record (as of May 2023) obtained using the GNFS method factorized a general 250 decimal-digit integer (829 bits) into its both prime factors.

Experiments about the elapsed time of factorization with the open-source software Pari-GP, SageMath, CrypTool 1, and CrypTool 2 can be found in [31].

Considerations by Martin Ziegler and Samuel S. Wagstaff Jr. (Cunningham table maintainer), which (non-)sense the factoring of such large numbers make, can be found in [32] and [33].

Some of the records listed in Table 5.12 are explained in more detail below.

RSA-155

On August 22, 1999, researchers from the Netherlands found the solution of the RSA-155 challenge. They factorized a 155-digit number into its both 78-digit primes (see Section 5.12.2).

This 512-bit RSA-155 meant to reach a kind of magic border.

C158

On January 18, 2002, researchers at the University of Bonn factorized a 158-digit decimal number into its both prime factors (these are built with 73 and 86 decimal digits) using the GNFS method.

This record got much less attention within the press than the solution of RSA-155. The task of the researchers was not initiated by a challenge, but they wanted to find the last prime factors of the integer $2^{953} - 1$ (see "Wanted List" in the Cunningham Project [29]). The six smaller prime factors, already found before have been:

$$\begin{gathered}3, 1907, 425796183929,\\ 1624700279478894385598779655842584377,\\ 3802306738549441324432139091271828121 \text{ and}\\ 128064886830166671444802576129115872060027\end{gathered}$$

The first three factors can be easily computed.[4] The next three prime factors were found by P. Zimmermann, T. Grandlund, and R. Harley during 1999 and 2000 using the elliptic curve factorization method.

4. For example, using `CT1 Indiv. Procedures ▷ RSA Cryptosystem ▷ Factorization of a Number`. CTO's Msieve shows errors. Alpertron's Calculator also finds the first three factors immediately. CT1 can factorize in a reasonable time numbers only not longer than 250 bits.

The last remaining factor, called C158, was known to be composite by then, but its factors were not known (the following three lines contain one number):

$$
\begin{gathered}
3950587458326514452641976780061448199602077646030493 6\\
45413937605157935562652945068360972784246821953509354\\
43058704902519956553357102097992264849779494429556 03
\end{gathered}
$$

The factorization of C158 resulted in the following two 73- and 86-digit prime factors:

$$
\begin{gathered}
3388495837466721394368393204672181522\\
815830368604993048084925840555281177
\end{gathered}
$$

and

$$
\begin{gathered}
1165882340667125990314837655838327081813101\\
2258146392600439520994131344334162924536139.
\end{gathered}
$$

So now all eight prime factors of $2^{953} - 1$ have been found.

RSA-160

On January 18, 2002, researchers at the University of Bonn factorized a 160-digit number into its both prime factors (each with 80 decimal digits) using the GNFS method. The computations for the factorization of RSA-160 also took place at the German Information Security Agency (BSI) in Bonn.

The 160-digit decimal number origins from the old challenge list of RSA Security. This number was retracted after RSA-155 had been factorized successfully. The prime factors of RSA-160 were still unknown. So this record of the team of Franke provides the solution of the old challenge, for which no prize is awarded anymore.

The composite number called RSA-160 is (the following three lines contain one number):

$$
\begin{gathered}
215274110271888970189601520131282542925777358884567598017049\\
767677813314521885913567301105977349105960249790711158521430\\
2079314665202840140619946994927570407753
\end{gathered}
$$

The factorization of RSA-160 resulted in the following two prime factors:

$$
\begin{gathered}
p = 4542789285848139407168619064973883 1\\
65613714577846979325095998470925000415733 5359
\end{gathered}
$$

and

$$
\begin{gathered}
q = 4738809060383201619663383230378895 1\\
97326892292104095794474135464881202849390 9367
\end{gathered}
$$

The calculations took place between December 2002 and April 2003.

RSA-200

On May 9, 2005, the research group of Jens Franke at the University of Bonn announced that they factorized a 200-digit number into its both prime factors (each with 100 decimal digits) using the GNFS method.

The composite number called RSA-200 is (the following three lines contain one number):

$$\begin{gathered}27997833911221327870829467638722601621070446786955428537560009929326128400107609345671052955360856061822351910951365788637105954482006 \\ 5767750985805576135790987349501441788631789462951872378692218239832 \\ 5767750985805576135790987349501441788631789462951872378692218239833\end{gathered}$$

The factorization of RSA-200 resulted in the following two prime factors:

$$\begin{aligned}p = {} & 3532461934402770121272604978198464368671197400197 6 \\ & 2502364930346877612125367942320005854795652808834 9\end{aligned}$$

and

$$\begin{aligned}q = {} & 7925869954478333033347085841480059687737975857364 2 \\ & 1996073433034145576787281815213538140930474018546 7\end{aligned}$$

The calculations took place between December 2003 and May 2005. The research group included Bahr, Böhm, Franke, Kleinjung, Montgomery, and te Riele. The operating expense of the calculations was about 120,000 MIPS-years. A MIPS-year (MY) is the quantity of operations a machine can perform in one year if the machine constantly achieves one million integer operations per second (MIPS). For context, an Intel Pentium processor then had about 10 MIPS.

To factorize a 2048-bit modulus it is estimated to need about $8.5 \cdot 10^{40}$ MY. A current processor (such as AMD Ryzen 5900) achieved around 10^5 MIPS at the end of 2021.

C307/M1039

In May 2007, Franke, Kleinjung (University of Bonn), the Japanese telecommunication company NTT, and Arjen Lenstra (Polytechnical University of Lausanne) announced that they managed to factorize a 307-digit decimal number into its both prime factors with the SNFS method (special number field sieve) within 11 months (the two factors have 80 and 227 decimal digits).

The task of the researchers was not initiated by a challenge, but they wanted to find the last prime factors of the Mersenne number $2^{1039} + 1$ from the "Wanted List" of the Cunningham Project [29].

The numbers in the Cunningham table have the following notation: "(2,n)-" means $2^n - 1$; "(2,n)+" means $2^n + 1$.

To describe the magnitude one writes "p<n>" or "c<n>": "n" is the number of decimal digits and "p" and "c" tell whether the number is prime or composite: So $2^{1039} - 1 = p7 \cdot c307 = p7 \cdot p80 \cdot p227$.

It is explained more precisely in [34]. "2,651+" means $2^{651} + 1$ and the size (c209 means 209 decimal digits) of the number that was factored. Then come the new factor(s), the discoverer, and the method used. Recently, only the multiple polynomial quadratic sieve (ppmpqs), the elliptic curve method (ecm), and the number field sieve (nfs) have been used. "hmpqs" stands for hypercube multiple polynomial quadratic sieve. Under "new factors," "p90" means a 90-digit prime and "c201" is a 201-digit composite number.

The number $2^{1039} - 1$ consists of three prime factors. The smallest one, $p7 = 5080711$, was already known.[5]

To complete this, the second factor (codivider) "C307" had to be factorized. Until then it was only known that the last remaining factor was composite, but it was unknown how many prime factors it had and what were the prime factors. The following five lines contain one number:

$$
\begin{aligned}
C307 = {} & 1159420574072573064369807148876894640753899791702017724986 \\
& 8683535388224838599667566080006095408005179472053993261230 20487 \\
& 4402860435302861914101440934535123347127396798885022630757 52809 \\
& 3791660285551055004258107711761776100941379707879738061870 08437 \\
& 7771868286808898447128220029352018060747554515413707110 23817
\end{aligned}
$$

The factorization of C307 resulted in the following two 80- and 227-digit prime factors:

$$
\begin{aligned}
p80 = {} & 55853666619936291260749204658315944968646527018 4 \\
& 8863764801005234631985328837475 3
\end{aligned}
$$

and

$$
\begin{aligned}
p227 = {} & 2075818194644238276457048137035946951629397080073952098812 08 \\
& 3870379272909032467938234314388414483488253405334476911222 30 \\
& 2815832769652537609141018910524199389933410971162435896206 59 \\
& 72167481161749004803659735573409253205425523689.
\end{aligned}
$$

So now the number $2^{1039} - 1$ is completely factorized in its three prime factors.

RSA-768

On December 12, 2009, the research group of Thorsten Kleinjung announced that they factorized a 232-digit number into its both prime factors (both factors have 116 decimal digits). They used the GNFS method in a way where they did oversieving on several hundred computers before starting the matrix step.

The composite number called "RSA-768" is (the following three lines contain one number):

$$
\begin{aligned}
& 123018668453011775513049495838496272077285356959533479219732245215 \\
& 172640050726365751874520219978646938995647494277406384592519255732 \\
& 630345373154826850791702612214291346167042921431160222124047927473 \\
& 7794080665351419597459856902143413
\end{aligned}
$$

5. This one can also be found using `CT1 Indiv. Procedures` ▷ `RSA Cryptosystem` ▷ `Factorization of a Number` with the algorithms of Brent, Williams, or Lenstra, which are good to separate relatively small factors.

The factorization of RSA-768 resulted in the following two prime factors (each with 384 bits):

$$\begin{aligned} p = {} & 334780716989568987860441698482126908177047949837137685689 1 \\ & 2431388982883793878002287614711652531743087737814467999489 \end{aligned}$$

and

$$\begin{aligned} q = {} & 3674604366679959042824463379962795263227915816434308764267 \\ & 6032283815739666511279233373417143396810270092798736308917 \end{aligned}$$

The calculations took about 2 ½ years. This was an academic effort—organizations with bigger resources could do it much faster.

Size of Factorized RSA Numbers Compared to Primality Proven Numbers

As you notice, the factorized compound numbers built of two prime factors are much smaller than the especially structured numbers, for which primality tests are able to decide whether these numbers are prime or not (see Sections 4.4 to 4.6).

Bit length of the current world records are in Table 5.13.

5.12.5 Further Research Results about Factorization and Prime Number Tests

Prime numbers are part of many topical research areas in number theory and computer science. Progress made with factorization is greater than was estimated in 2005—this is not only due to faster computers but also new mathematical knowledge. The current status of the corresponding research is discussed in Chapter 12.

The security of the RSA algorithm is based on the empirical observation that factoring large numbers is a hard problem. A modulus n (typically 2048 bits) can be easily constructed as the product of two large primes p, q (typically 1200 bits each), by calculating $n = pq$. However, it is a hard problem to (reversely) extract p, q from n. In order to calculate the private key from the public key, you either need to know p and q, or the value of the Euler phi function $\phi(n)$.

Thus, any progress in efficiency of factorizing large integers will affect the security of RSA. As a consequence, the underlying primes p, q and, thus the modulus n must be increased. In case of a quantum leap in factorization, the RSA algorithm would be compromised and has to be omitted.

Despite the following four publications dating from 2001 to 2012, in my opinion they have received the most attention among the corresponding research results due to their practical importance.

Table 5.13 Comparing the Record Sizes of Factorized RSA Numbers vs Primality Proven Numbers

[RSA-250 number]	$\longleftrightarrow$	[51st known Mersenne prime]
829 bit	$\longleftrightarrow$	82589933 bit
[see Table 5.12]	$\longleftrightarrow$	[see Table 4.1]

5.12.5.1 Bernstein's Paper and Its Implication on the Security of the RSA Algorithm

In his paper "Circuits for Integer Factorization: A Proposal," published November 2001, D. J. Bernstein [35] addresses the problem of factorizing large integers. As a main result Bernstein claims that the implementation of the GNFS algorithm can be improved to factorize integers with three times more digits—with the same effort as before.

Here the definition of *effort* is a crucial point: Bernstein claims that effort is the product of time and costs of the machine (including the memory used). The gist of the paper lies in the fact that he can reduce a big part of factorizing to sorting. Using Schimmler's scheme, sorting can be optimized by massive parallel computing. At the end of Section 3, Bernstein explains this effect: The costs of m^2 parallel computers with a constant amount of memory is a constant time m^2. The costs of a computer with a single processor and memory of size m^2 is also of the order of m^2, but with a different constant factor. With m^2 processors in parallel, sorting of m^2 numbers (with Schimmler's scheme) can be achieved in time m, while a m^2-memory computer needs time of the order of m^2. Decreasing memory and increasing the number of processors, the computing time can be reduced by a factor $1/m$ without additional effort in terms of total costs. In Section 5 it is said that massive parallel computing can also increase efficiency of factorizing using Lenstra's elliptic-curve-method (a search algorithm has costs that increase in a quadratic square manner instead of cubically).

All results achieved so far are asymptotic results. This means that they only hold in the limit n to infinity. Unfortunately, there is no upper limit for the residual error (i.e., the difference between the real and the asymptotic value) for finite n – a problem that has already been addressed by the author. As a consequence, one cannot conclude whether the costs (in the sense of Bernstein) for factorizing 1024–2048-bit RSA moduli can be significantly reduced.

There is no doubt that Bernstein's approach is innovative. However, the reduction of computing time under constant costs comes along with a massive use of parallel computing—a scenario that seems not to be realistic yet. For example, formally 1 sec computing time on one machine and 1/1,000,000 sec time parallel computing time on 1,000,000 machines might have same costs. In reality, it is much harder to realize the second situation. Although distributed computing over a large network might help to overcome this problem, realistic costs for data transfer have to be taken into account.

Arjen Lenstra, Adi Shamir, et al. analyzed the paper of Bernstein [36]. In summary, they expect a factorization improvement on how much longer the bit length of the keys could be with a factor of 1.17 (instead of factor 3 as proposed by Bernstein).

The abstract of their paper "Analysis of Bernstein's Factorization Circuit" says:

> Bernstein proposed a circuit-based implementation of the matrix step of the number field sieve factorization algorithm. We show that under the non-standard cost function used in [1], these circuits indeed offer an asymptotic improvement over other methods but to a lesser degree than previously claimed: For a given cost, the new method can factor integers that are 1.17 times larger (rather than 3.01). We also propose an improved circuit design based on a new mesh routing algorithm, and show that for factorization

> of 1024-bit integers the matrix step can, under an optimistic assumption about the matrix size, be completed within a day by a device that costs a few thousand dollars. We conclude that from a practical standpoint, the security of RSA relies exclusively on the hardness of the relation collection step of the number field sieve.

RSA Security concludes in its analysis of the Bernstein paper [37] from April 8, 2002, also—as expected—that RSA is still not compromised.

5.12.5.2 The TWIRL Device

In January 2003, Adi Shamir and Eran Tromer from the Weizmann Institute of Science published a preliminary draft called "Factoring Large Numbers with the TWIRL Device," raising concerns about the security of key sizes up to 1024 bits [38].

Their abstract summarizes their results very well:

> The security of the RSA cryptosystem depends on the difficulty of factoring large integers. The best current factoring algorithm is the number field sieve (NFS), and its most difficult part is the sieving step. In 1999 a large distributed computation involving thousands of workstations working for many months managed to factor a 512-bit RSA key, but 1024-bit keys were believed to be safe for the next 15–20 years. In this paper we describe a new hardware implementation of the NFS sieving step ...which is 3–4 orders of magnitude more cost-effective than the best previously published designs ...Based on a detailed analysis of all the critical components (but without an actual implementation), we believe that the NFS sieving step for 1024-bit RSA keys can be completed in less than a year by a $10M device, and that the NFS sieving step for 512-bit RSA keys can be completed in less than ten minutes by a $10K device. Coupled with recent results about the difficulty of the NFS matrix step ... this raises some concerns about the security of those key sizes.

A detailed explanation from these two authors also can be found in the RSA Laboratories CryptoBytes [39].

The three-page article in the DuD issue of June 2003 [40] contains a good explanation of how the attack using the GNFS works and what progress is made to factorize numbers. With GNFS we can distinguish two general steps: The sieve step (relation collecting) and the matrix reduction. Besides that the sieve step is highly parallelizable, it also dominates the overall calculation burden. Shamir and Tromer haven't built a TWIRL device yet, but the estimated costs of 10 to 50 million Euro (in order to factorize a 1024-bit number) is not prohibitive for secret agencies or big criminal organizations, as the costs for a single espionage satellite is estimated, for example, to be several billion USD. The authors therefore recommend getting rid of sensible RSA keys with a key length below 2048 bit as soon as possible.

This fits with recommendations like the BSI's annual technical guideline [41] to switch to longer RSA key lengths.

5.12.5.3 Primes in P: Primality Testing is Polynomial

In August 2002, the three Indian researchers M. Agrawal, N. Kayal, and N. Saxena published the paper "Primes in P" about a new primality testing algorithm called AKS [42]. They discovered a polynomial time deterministic algorithm for determining if a number is prime or not.

The importance of this discovery is that it provides number theorists with new insights and opportunities for further research. Lots of people over the centuries have been looking for a polynomial time test for primality, and this result is a major theoretic breakthrough. It shows that new results can be generated from already known facts.

But even its authors note that other known algorithms may be faster (for example ECPP). The new algorithm works on any integer. For example the GIMPS project uses the Lucas-Lehmer primality test which takes advantage of the special properties of Mersenne numbers. This makes the Lucas-Lehmer test much faster, allowing to test numbers with millions of digits, while general-purpose algorithms are limited to numbers with a few thousand digits.

5.12.5.4 Shared Primes: Moduli with Common Prime Factors

The RSA algorithm is based on the presumed difficulty of factorizing large biprime integers (moduli), the factorizing problem. However, as pointed out in Lenstra et al. [43] it is possible, given a set of moduli, to factorize quickly those that share prime factors. In this case, the factorization problem is bypassed using the—relatively easy—greatest common divisor (gcd) operation. On the other hand, it is no trivial task to extract common shared primes and to factorize the corresponding moduli efficiently for a very big number of given moduli (several millions).

Shared primes only occur if the RSA keys were not generated randomly. Taking into consideration the significance of strong cryptographic keys, it is important to verify that all keys were generated following the principle of true randomness [44].

When Lenstra et al. published their paper [43] in February 2012, they did not publish the source code. However, soon afterward the source code of a similar program was published at the CrypTool website [45] in Python and C++, and—again a bit later—at the page used by [46]. The fastest code known to me comes from [46].

These applications find all shared factors that may exist, given a finite set of moduli—even if this set includes millions of moduli. Such an application enables system administrators to test their own RSA keys.

The quite naive way to find all shared factors would be to compare each modulus with all other moduli, which has a complexity growing quadratically with the number of moduli.

A very efficient method using trees for finding shared prime factors is based on a publication of Dan Bernstein in 2005 [47]. Bernstein uses a precalculation that leads to the product of all moduli.

This is another example showing how helpful *precalculations* can be to break cryptographic systems. Another famous example are *rainbow tables* used to find the origin of a hash value [48].

Comparing the Execution Time to Calculate gcd and to Factorize

SageMath Example 5.10 shows the very different run times when calculating a gcd and a factorization. The following operations are very fast: multiplication of factors, dividing a modulus by a known factor, or calculating the gcd. However, factorizing moduli steeply increases with longer moduli. Even the relatively small moduli used in this example show this: For the smaller modulus $m1$ (69 decimal digits, 228 bits) 35 sec were needed; the larger $m2$ (72 decimal digits, 239 bits) took 99 sec. Furthermore, the operations (multiplication, division, and gcd) show big differences in execution time when the used operands are very different in size.

SageMath Example 5.10: Comparing the Execution Time for Calculating a gcd and Performing a Factorization

```
print("\n# CHAP05 -- Sage-Script-SAMPLE 010: =========")

import time # in scripts: measure time like in Python and calculate ▸
  ▸execution time

print("Multiplication: -----------------------")
cf=35938757044958237573881998942687731534 39
Start_Time = time.time()
m1=cf * 84115747449047881488635567801
print(" Time = %f sec \n  m1:" % (time.time()-Start_Time), m1)

Start_Time = time.time()
m2=cf * 162259276829213363391578010288127
print(" Time = %f sec \n  m2:" % (time.time()-Start_Time), m2)


print("Division: -----------------------")
Start_Time = time.time()
r=302301541122639745170382530168903859625492057067780948293331060817639▸
  ▸ / \
     35938757044958237573881998942687731534 39
print(" Time = %f sec \n  m1/cf:" % (time.time()-Start_Time), r)

Start_Time = time.time()
r=583139672825572068433667900695808357466165186436234672858047078770918▸
  ▸753 / \
     35938757044958237573881998942687731534 39
print(" Time = %f sec \n  m2/cf:" % (time.time()-Start_Time), r)


print("gcd: -----------------------")
Start_Time = time.time()
r=gcd(583139672825572068433667900695808357466165186436234672858047078 77▸
  ▸0918753, \
          302301541122639745170382530168903859625492057067780948293331 0▸
            ▸60817639)
print(" Time = %f sec \n  gcd(m2,m1):" % (time.time()-Start_Time), r)


print("Factorization: -----------------------")
Start_Time = time.time()
r=factor(583139672825572068433667900695808357466165186436234672858047 07▸
  ▸8770918753)
print(" Time = %f sec \n  m2 =" % (time.time()-Start_Time), r)
```

SageMath Example 5.10 (continued)

```
Start_Time = time.time()
r=factor(3023015411226397451703825301689038596254920570677809482933310 6▸
 ▸0817639)
print(" Time = %f sec \n  m1 =" % (time.time()-Start_Time), r)
```

Differences Between Sage Script: Sage Command Line

The calls from the script in the SageMath Example 5.11 can also be executed on the SageMath CLI (command line interface); see Section 5.12.5.4. For short sequences of commands the CLI is easier: Operations output their results immediately (without print()); and the execution time is obtained syntactically easier, namely simply by putting `%time` in front of it. The usual way to achieve this in a script is to get current time and calculate execution time as the difference for each statement: `ExecutionTime = time.time() - StartTime`.

Hint 1: In both cases, a terminating backslash is used to span a single line of code over multiple lines.

Hint 2: To stop and exit your script somewhere in between use the corresponding Python call: `sys.exit(1)`.

Remark: The factor() command can be used with the PARI defaults or you can specify the dedicated algorithm (here Bill Hart's quadratic sieve and Paul Zimmermann's GMP-ECM).

SageMath Example 5.11: Printing the Execution Time on the Sage Command Line / Factorize

```
sage: # Calculate gcd
sage: %
....:                  302301541122639745170382530168903859625492057067780 9▸
 ▸48293331060817639)
CPU times: user 15 µs, sys: 0 ns, total: 15 µs
Wall time: 16.2 µs
35938757044958237573881998942687731534 39

sage: # Factorize (using PARI at the time of writing)
sage: %
CPU times: user 1min 33s, sys: 129 ms, total: 1min 33s
Wall time: 1min 33s
1622592768292133633915780102881 27 * 35938757044958237573881998942687731▸
 ▸53439

sage: # Factorize (select dedicated algorithms)
sage: n=583139672825572068433667900695808357466165186436234672858047078▸
 ▸770918753

sage: %
/usr/lib/python3/dist-packages/IPython/core/interactiveshell.py:2364: ▸
 ▸RuntimeWarning:
the factorization returned by qsieve may be incomplete (the factors may▸
 ▸ not be prime)
or even wrong; see qsieve? for details result = fn(*args, **kwargs)
CPU times: user 1.92 ms, sys: 10 µs, total: 1.93 ms
```

SageMath Example 5.11 (continued)

```
Wall time: 48.8 s
162259276829213363391578010288127 * 3593875704495823757388199894268773▶
 ▶153439

sage: %
CPU times: user 74 ms, sys: 28 ms, total: 102 ms
Wall time: 3min 48s
162259276829213363391578010288127 * 3593875704495823757388199894268773▶
 ▶153439

sage: # Knowing the size of the number and its factors
sage: n=162259276829213363391578010288127; len(n.digits())
33
sage: n=3593875704495823757388199894268773153439; len(n.digits())
40
sage: n=583139672825572068433667900695808357466165186436234672858047078▶
 ▶770918753; len(n.digits())
72
```

Efficient Computing of All Shared Primes

The paper "Mining Your Ps and Qs: Detection of Widespread Weak Keys in Network Devices" [46] explains how the algorithm efficiently calculates the gcds (greatest common divisors) of every pair of RSA moduli (taken from a given but huge set of RSA moduli). More precisely, the gcds of all pairs of RSA moduli are not computed but the gcds with other arguments, which is faster and delivers the same result: shared primes.

This section explains the essential part of the method used in this paper: Using two trees greatly accelerates the calculation of the gcds.

First the product P of all moduli m_i for $i = 1, \ldots, k$ is calculated by using a product tree (see Figure 5.4):

$$P = \prod_{i=1}^{k} m_i$$

Then, using a remainder tree, for each i the remainder z_i of the division of P by m_i^2 is computed:

$$z_i \equiv P \bmod m_i^2, \quad z_i \in \{0, 1, \ldots, m_i^2 - 1\}$$

Now this remainder modulo the square of m_i is divided by m_i, yielding an integer value because m_i^2 as well as P are divisible by m_i and therefore also z_i is divisible by m_i:

$$r_i = \frac{z_i}{m_i}; \qquad r_i \leq m_i$$

To finish, we only have to compute the gcd:

$$\gcd(r_i, m_i)$$

This is visualized in Figure 5.4, which is taken from [46] with some minor changes.

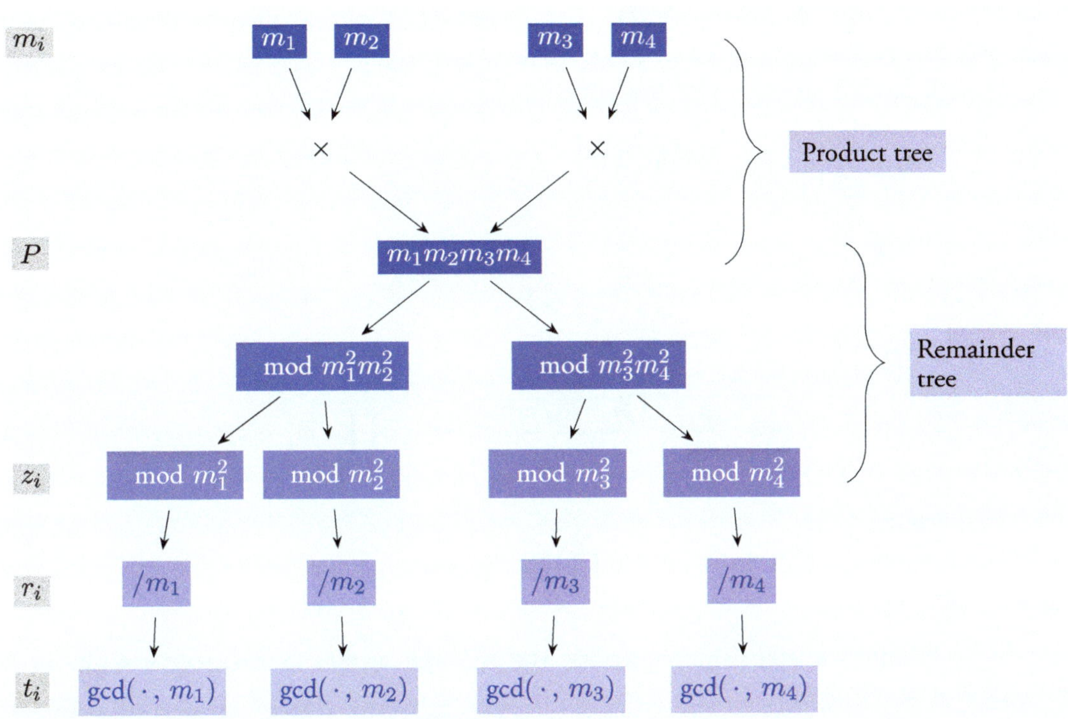

Figure 5.4 Efficient computation of shared primes (quasi-linear gcd finding). (*From*: [46].)

The paper we are referring to explains well *how* the algorithm works, but not as well *why*. The product P of all moduli is a very big number, even compared to a single modulus. Without the simplifications from the remainder tree you would go the following way: Calculate $g_i := \gcd_i = \gcd(P/m_i, m_i)$ for all i. Now for every i there are three possibilities:

- $g_i = \gcd_i = 1$
- $g_i = \gcd_i$ is a prime number
- $g_i = \gcd_i = m_i$

The third case is a special case that occurs if, $m_1 = p_1 \cdot q_1$ and p_1 divides m_2 and q_1 divides m_3. This case occurred only "in a handful of instances in our dataset" ([46, p. 5]) and was solved by computing the gcd pairwise.

In the second case, one has found a prime factor of m_i. In the first case, no information about m_i can be retrieved.

Here an example with very small moduli:

$$m_1 = 2 \cdot 3 = 6; \quad m_2 = 2 \cdot 7 = 14; \quad P = 6 \cdot 14 = 84$$
$$P \bmod m_1 = 84 \bmod 6 = 0; \quad P \bmod m_1^2 = 84 \bmod 36 = 12$$
$$P \bmod m_2 = 84 \bmod 14 = 0; \quad P \bmod m_2^2 = 84 \bmod 196 = 84$$
$$g_1 = \gcd_1 = \gcd(12/6, 6) = \gcd(2, 6) = 2$$
$$g_2 = \gcd_2 = \gcd(84/14, 14) = \gcd(6, 14) = 2$$

Why does $\gcd((P \bmod m_i^2)/m_i, m_i)$ deliver the same result as $\gcd(P/m_i, m_i)$? We have a closer look at why this identity is correct.

Let, as before, P be the product $\prod_{i=1}^{k} m_i$ and $z_i \equiv P \bmod m_i^2$ with $z_i \in \{0, 1, \ldots, m_i^2 - 1\}$ for $i = 1, 2, \ldots, k$.

Then, if we again denote with r_i the (integer) quotient $\frac{z_i}{m_i}$, we have

$$P = c_i m_i^2 + z_i \quad \text{or} \quad z_i = \prod_{j=1}^{k} m_j - c_i m_i^2 \quad \text{for some integer} \quad c_i$$

and therefore

$$r_i = \frac{z_i}{m_i} = \frac{\prod_{j=1}^{k} m_j - c_i m_i^2}{m_i} = \prod_{j=1, j\neq i}^{k} m_j - c_i m_i \tag{5.1}$$

as well as:

$$\prod_{j=1, j\neq i}^{k} m_j = r_i + c_i m_i \tag{5.2}$$

Finally, the algorithm computes the gcd of r_i and m_i, we denote it with

$$t_i = \gcd(r_i, m_i).$$

We also write $g_i = \gcd\left(\frac{P}{m_i}, m_i\right)$. Clearly in (5.1) one can always factor out g_i and so g_i divides t_i.

Conversely, t_i always divides g_i: Because of t_i dividing both m_i and r_i, it must because of (5.2) also divide $\prod_{j\neq i} m_j = \frac{P}{m_i}$ and therefore also g_i. It follows $t_i = g_i$ for all i.

The latter is only an alternative formulation of the statement we had made before:

$$\gcd((P \bmod m_i^2)/m_i,\ m_i) \ = \ \gcd(P/m_i,\ m_i).$$

5.13 Applications of Asymmetric Cryptography Using Numerical Examples

The results of modular arithmetic are used extensively in modern cryptography. Here we will provide a few examples from cryptography using small numbers. In the RSA procedure, we call numbers "small" if the bit lengths are much shorter than currently recommended. In practice, 2048 bits (which is about 600 decimal points) is currently considered the minimum length for a secure RSA modulus.

5.13.1 Problem Description for Nonmathematicians

To encrypt data, the data (which is given as text or as binary data) is converted into numbers. The encryption then consists in applying a function (mathematical operations) that produces another number from it.

Decrypting means reversing this function; that is, restoring the original domain from the distorted codomain that the function made from the plaintext. For example, the sender of a message could add a number to be kept secret (the key S) to the plaintext number M and thereby obtain the ciphertext number C:

$$C = M + S$$

By reversing this operation, that is, by subtracting S, the receiver can reconstruct the plaintext:

$$M = C - S$$

Adding S reliably obfuscates the plaintext. Nevertheless, this "encryption" is very weak: If an eavesdropper gets her hands on just one pair of plaintext and ciphertext numbers, she can calculate the key

$$S = C - M$$

and read all subsequent messages encrypted with S. A key reason for this is that subtraction is as simple an operation as addition.

One-Way Functions

If we want to make it impossible to determine the key even with the knowledge of both the plaintext and the ciphertext, we need a function that is, on the one hand, relatively easy to calculate. On the other hand, the inverse function should exist (otherwise information would be lost during encryption), but should be de facto incalculable.

What are possible candidates for such a one-way function? We could take multiplication rather than addition, but even primary school children know that the inverse function, division, is only slightly more difficult than multiplication itself. We need to go one step higher in the hierarchy of calculation methods. It is still relatively simple to calculate the power of a number, but the corresponding two reverse functions—*taking roots* (find b in the equation $a = b^c$ when a and c are known) and *calculating logarithms* (find c in the equation $a = b^c$ when a and b are known) are so complicated that students normally do not learn them at school.

Knowing a few values of the function then for addition and multiplication, a certain structure can still be recognized, but raising numbers to the power of another one or calculating exponentiations doesn't tell us much about the function parameters. Taking the logarithm becomes even more difficult if you don't work in infinite sets like $\mathbb{N}$ or $\mathbb{Z}$, but in large finite sets.

5.13.2 The Diffie-Hellman Key-Exchange Protocol

Before we get back to an encryption function, let's first consider a protocol that allows two parties to securely agree on a shared secret.

Whitfield Diffie, Martin E. Hellman, and Ralph Merkle developed this key-exchange protocol in Stanford in 1976.[6]

6. - In `CT1 Indiv. Procedures ▷ Protocols ▷ Diffie-Hellman Demonstration` this exchange protocol is visualized: You can execute the single steps with specific numbers.
   - You also find an enhanced version in `JCT Default Perspective ▷ Visuals ▷ Diffie-Hellman Key Exchange (EC)`.

For allocating the participants in the protocol, Bob and Alice are used, which are the default names for the two authorized participants (see [49, p. 23]).

Alice and Bob use a one-way function to obtain a secret key S, the session key, for subsequent correspondence; see Figure 5.5. The session key can be used, for example, as a key in a symmetrical procedure such as AES. This session key is only known to the two parties. How do the protocol works: Alice selects a random number a and keeps it secret. She applies a one-way function to a to calculate the number $A = g^a$ and sends it to Bob. He does the same, by selecting a secret random number b, calculating $B = g^b$ and sending it to Alice. The number g is random and can be publicly known. Alice applies the one-way function together with her secret number a to B, while Bob does the same with his secret number b and the received number A.

The result S is the same in each case because the one-way function is commutative: $(g^a)^b = (g^b)^a$. But even Bob cannot reconstruct Alice's secret number a from the data available to him, while Alice cannot determine Bob's secret number b. And an eavesdropper (Eve) who knows g and has intercepted both A and B cannot use this knowledge to determine a, b, or S.

Procedure:
Alice and Bob want to negotiate a secret session key S via a channel that may be intercepted.

1. They select a prime number p and a random number g and exchange this information openly.
2. Alice now selects a, a random number less than p and keeps it secret. Similarly, Bob selects b, a random number less than p and keeps it secret.
3. Alice now calculates $A \equiv g^a \pmod{p}$. Bob calculates $B \equiv g^b \pmod{p}$.

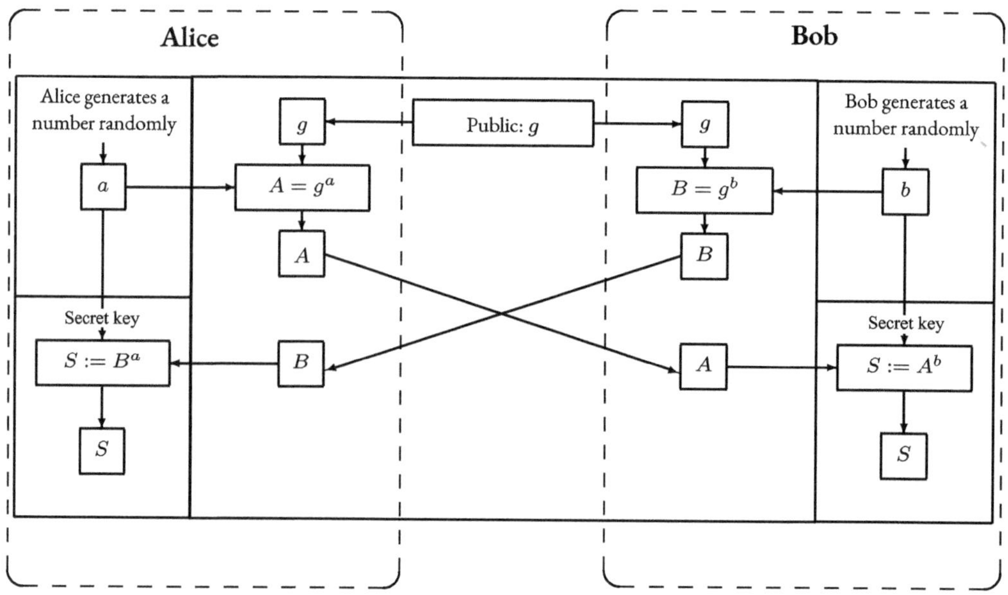

Figure 5.5 Process of the DH key-exchange protocol (all operations modulo p).

4. Alice sends the result A to Bob.
   Bob sends the result B to Alice.
5. In order to determine the session key to be used by both parties, they both separately raise the respective results they have received to the power of their secret random number modulo p. This means:
   - Alice calculates $S \equiv B^a \pmod{p}$;
   - Bob calculates $S \equiv A^b \pmod{p}$.

Even if a spy (Eve) intercepts g, p, and the interim results A and B, she cannot use these in order to determine the session key used due to the difficulty of calculating the discrete logarithm: $a = \log_g(A)$.

Further details about the discrete logarithm problem can be found in Section 6.4 and Chapter 12.

We will now use an example with (unrealistically) small numbers to illustrate this.

Example using small numbers:

1. Alice and Bob select $g = 11, p = 347$.
2. Alice selects $a = 240$, Bob selects $b = 39$; a and b are kept secret.
3. Alice calculates $A \equiv g^a \equiv 11^{240} \equiv 49 \pmod{347}$.
   Bob calculates $B \equiv g^b \equiv 11^{39} \equiv 285 \pmod{347}$.
4. Alice sends to Bob: $A \equiv 49$,
   Bob sends to Alice: $B \equiv 285$.
5. Alice calculates $B^a \equiv 285^{240} \equiv 268 \pmod{347}$,
   Bob calculates $A^b \equiv 49^{39} \equiv 268 \pmod{347}$.

Alice and Bob can now communicate securely using their shared session key S. Even if a spy can intercept everything transferred via the connection ($g = 11, p = 347, A = 49$, and $B = 285$) she would not be able to calculate the secret key S. However, this is only true for large numbers because then the discrete logarithm is extremely difficult to solve (see Chapter 12). After revealing a or b, S can be calculated in the same way as Alice or Bob do it.

To get the discrete logarithms x or y, here we need to calculate one of the following equations:

a from Alice: $11^x \equiv 49 \pmod{347}$, that means $\log_{11}(49) \pmod{347}$.

b from Bob: $11^y \equiv 285 \pmod{347}$, that means $\log_{11}(285) \pmod{347}$.

SageMath Example 5.12 determines the discrete logarithm (for both Alice and Bob).

SageMath Example 5.12: Sample with Small Numbers: Calculating the Discrete Logs a and b in Order to Attack DH

```
print("\n# CHAP05 -- Sage-Script-SAMPLE 020: =========")

print("Get the secret key of Alice (with g=11, p=347, A=49, a=240 or 67▸
 ▸) ---")
```

SageMath Example 5.12 (continued)

```
print("a) via 'normal' integer numbers")
print(" a:", discrete_log(mod(49,347),mod(11,347)))

print("b) via the ring of integers (better)")
R=Integers(347)
g=R(11)
A=R(49)
print(" a:", discrete_log(A,g))

print("Get the secret key of Bob: (with g=11, p=347, B=285, b=39) ---")
B=R(285)
print(" b:", discrete_log(B,g))

#------------------------------------
# CHAP05 -- Sage-Script-SAMPLE 020: =========
# Get the secret key of Alice (with g=11, p=347, A=49, a=240 or 67) ---
# a) via 'normal' integer numbers
#  a: 67
# b) via the ring of integers (better)
#  a: 67
# Get the secret key of Bob: (with g=11, p=347, B=285, b=39) ---
#  b: 39
```

As the SageMath function `discrete_log` expects as arguments only elements of a ring (integers between 0 and an upper limit), we can enforce this type by entering the numbers directly with the corresponding modulo operator:

```
discrete_log( mod(49, 347), mod(11, 347) )
```

A much better alternative is to let SageMath know from the very beginning that they are elements of a ring (as in SageMath Example 5.12). After this extra "burden" for the initialization, you can write the formulas as you are used to:

```
discrete_log(A, g)
```

Such number theoretic tasks can also be solved using other tools like PariGP, BC, or Mathematica. Here is the corresponding syntax to get the discrete log for Alice (all function calls deliver the result 67):

- Pari-GP: `znlog(Mod(49,347),Mod(11,347))`.
- Mathematica: `MultiplicativeOrder[11, 347, 49]`
  The general "Solve" function provides the "em tdep message": The equations appear to involve the variables to be solved for in an essentially nonalgebraic way.

Why did the functions deliver the value 67 for the discrete logarithm of Alice rather than 240, which Alice selected as exponent a?

The discrete logarithm is the smallest natural exponent that solves the equation $11^x \equiv 49 \pmod{347}$. Both $x = 67$ and $x = 240$ (the number selected in the example) satisfy the equation and can therefore be used to calculate the session key: $285^{240} \equiv 285^{67} \equiv 268 \pmod{347}$. If Alice and Bob had selected a primitive root modulo p as base g, then for every remainder from the set $\{1, 2, \ldots, p-1\}$ there is exactly one exponent from the set $\{0, 1, \ldots, p-2\}$.

As an aside, there are 172 different primitive roots modulo 347, 32 of which are prime (not necessary). Since the number 11 selected for g in the example is not a primitive root of 347, the remainders do not take all values from the set $\{1, 2, \ldots, 346\}$. Thus, for a particular remainder there may be more than one exponent or even no exponent at all in the set $\{0, 1, \ldots, 345\}$ that satisfies the equation.

With the relevant SageMath commands you find:

`is_prime(347)=True`, `euler_phi(347)=346`, `gcd(11,347)=1`, and `multiplicative_order(mod(11, 347))=173`.

i	$11^i \bmod 347$	
0	1	
1	11	
2	121	
3	290	
67	49	searched exponent
172	284	
173	1	= multiplicative order of $11^i \bmod 347$
174	11	
175	121	
176	290	
240	49	searched exponent

Further information can be found in Section 5.17.4.

5.14 The RSA Procedure with Specific Numbers

Having described in Section 5.10.2 how the RSA procedure works, we will now work through the steps using specific, but still small, numbers—and still only textbook RSA.

5.14.1 RSA with Small Prime Numbers and with a Number as Message

Before applying the RSA procedure to a text, we will first demonstrate it directly using a single number as message. In practice, RSA is not applied on texts, but only on big numbers.[7]

1. Let the selected prime numbers be $p = 5$ and $q = 11$.
   Thus, $n = 55$ and $\phi(n) = (p-1) \cdot (q-1) = 40$.
2. $e = 7$ (e must be relatively prime to 40).

7. - You can handle this, for example, using `CT1 Indiv. Procedures ▷ RSA Cryptosystem ▷ RSA Demonstration`.
- Or in CTO: either in the plugin “RSA (step-by-step)”: `https://www.cryptool.org/en/cto/rsa-step-by-step` or in the plugin “RSA visual and more”: `https://www.cryptool.org/en/cto/rsa-visual` in the tabs “RSA visual” or “RSA didactic.”
- Or using `CT2 Templates ▷ Mathematics ▷ RSA with big numbers`.

3. $d = 23$ (since $23 \cdot 7 \equiv 161 \equiv 1 \pmod{40}$).

   $\rightarrow$ Public key of the recipient: (55, 7).

   $\rightarrow$ Private key of the recipient: (55, 23).

4. Let the message be the number $M = 2$ (so no division into blocks is required).
5. Encryption: $C \equiv 2^7 \equiv 18 \pmod{55}$.
6. The ciphertext is simply the number $C = 18$.
7. Decryption: $M \equiv 18^{23} \equiv 18^{(1+2+4+16)} \equiv 18 \cdot 49 \cdot 36 \cdot 26 \equiv 2 \pmod{55}$.

We will now apply the RSA procedure to a text, first using the upper case alphabet (26 characters), then using the entire ASCII character set as the basis for the messages. Once again, only the numerical value of the individual character is used, but the individual characters can be combined into blocks.

5.14.2 RSA with Slightly Larger Primes and a Text of Uppercase Letters

We have the text "ATTACK AT DAWN," and the characters (including the blank) are coded according to Table 5.14.[8]

Key generation (steps 1 to 3):

1. $p = 47, q = 79$ ($n = 3713$; $\phi(n) = (p-1) \cdot (q-1) = 3588$).
2. $e = 37$ (e must be relatively prime to 3588).
3. $d = 97$ (since $e \cdot d = 1 \mod \phi(n); 37 \cdot 97 \equiv 3589 \equiv 1 \pmod{3588}$).

4. Encryption:

Text:	A	T	T	A	C	K		A	T		D	A	W	N
Number:	01	20	20	01	03	11	00	01	20	00	04	01	23	14

Table 5.14 Alphabet of Capital Letters Plus Blank

Character	*Numerical Value*	*Character*	*Numerical Value*
Blank	0	M	13
A	1	N	14
B	2	O	15
C	3	P	16
D	4	Q	17
E	5	R	18
F	6	S	19
G	7	T	20
H	8	U	21
I	9	V	22
J	10	W	23
K	11	X	24
L	12	Y	25
		Z	26

8. - You can handle this using `CT1 Indiv. Procedures` ▷ `RSA Cryptosystem` ▷ `RSA Demonstration`. This is also described in the tutorial/scenario in CT1's online help [Options: specify alphabet, number system, block length 2 and decimal representation].
- In CTO in the plugin "RSA visual and more": `https://www.cryptool.org/en/cto/rsa-visual`. See Figures 5.6 and 5.7.

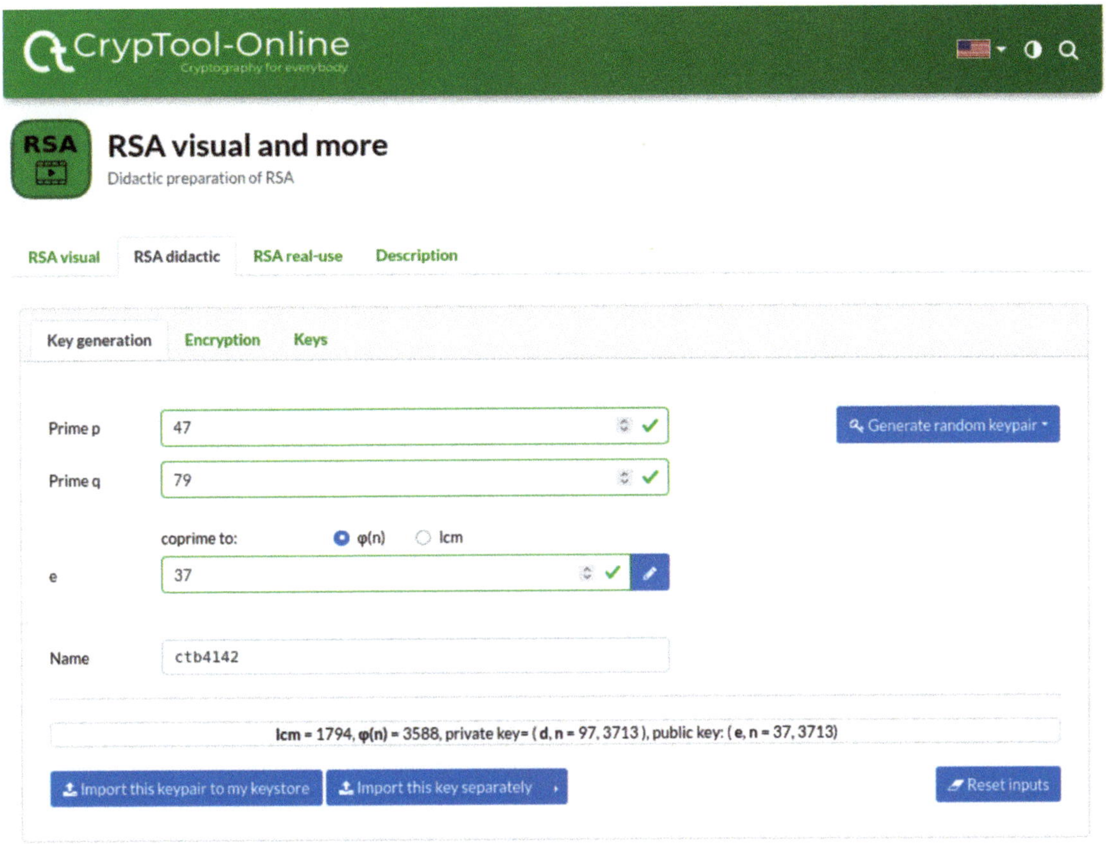

Figure 5.6 RSA in CTO: text encryption, own alphabet, decimal concatenation, block length 2; Part 1: generate and store key.

This 28-digit number is divided into four-digit parts (because 2626 is still smaller than $n = 3713$). This means that the block length is 2 (the numerical values of two characters are combined):

`0120 2001 0311 0001 2000 0401 2314`

All 7 parts are encrypted using: $C \equiv M^{37} \pmod{3713}$:

`1404 2932 3536 0001 3284 2280 2235`

See Section 5.17.5 for source code to do this RSA encryption using SageMath.

5. Decryption:

Ciphertext: `1404 2932 3536 0001 3284 2280 2235`

This 28-digit number is divided into four-digit parts.

All 7 parts are decrypted using: $M \equiv C^{97} \pmod{3713}$:

`0120 2001 0311 0001 2000 0401 2314`

The two-digit numbers are transformed into capital letters and blanks.

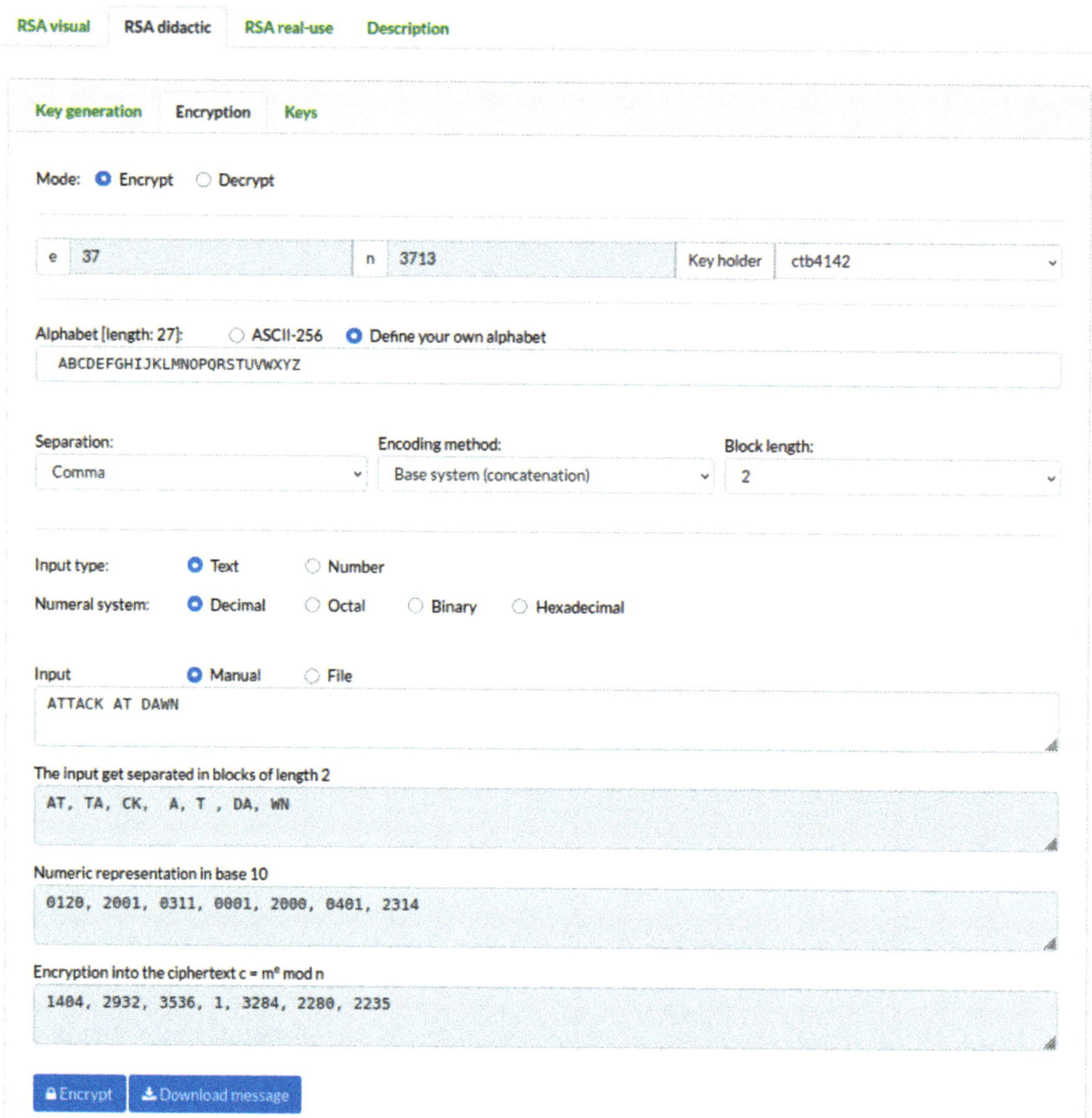

Figure 5.7 RSA in CTO: text encryption, own alphabet, decimal concatenation, block length 2; Part 2: encryption.

Using the selected values it is easy for a cryptanalyst to derive the secret values from the public parameters $n = 3713$ and $e = 37$ by factorizing n. However, if n is a 2048-bit number, there is, according to present knowledge, little chance to do a factorization. Nevertheless, this form of RSA is insecure even with large moduli (see Section 5.11).

5.14.3 RSA with Even Larger Primes and a Text Made up of ASCII Characters

In real life, the ASCII alphabet is used to code the individual characters of the message as 8-bit numbers.

The idea for this exercise[9] is taken from the example in [50, p. 271].
Coded in decimal notation, the text "RSA works!" is as follows:

Text:	R	S	A		w	o	r	k	s	!
Number:	82	83	65	32	119	111	114	107	115	33

We will work through the example in two variants. The steps 1 to 3 are common for both.

Key generation (steps 1 to 3):

1. $p = 503$, $q = 509$ $(n = 256027;\ \phi(n) = (p-1)(q-1) = 255016 = 2^3 \cdot 127 \cdot 251)$
2. $e = 65537$ (e must be relatively prime to 255016)
3. $d = 231953$ (since $e \equiv d^{-1} \pmod{\phi(n)}$: $65537 \cdot 231953 \equiv 15201503761 \equiv 1 \pmod{255016}$). Other possible combinations of (e, d) include: $(3, 170011), (5, 204013), (7, 36431)$.

Variant 1: All ASCII characters are en-/decrypted separately (no blocks are formed)
See Section 5.17.5 for the source code for RSA operations like modular exponentiation or the Euler function using SageMath.

4. Encryption:

Text:	R	S	A		w	o	r	k	s	!
Number:	82	83	65	32	119	111	114	107	115	33

The letters are not combined here. For secure procedures we need large numbers that accept—as far as possible—all values up to $n - 1$. If the possible value set for the numbers in the message is too small, even large prime numbers cannot make the procedure secure. An ASCII character is represented by 8 bits. If we want larger values we must combine several numbers. Two characters need 16 bits, whereby the maximum value that can be represented is 65536. The modulus n must then be greater than $2^{16} = 65536$. This is applied in variant 2. When the numbers are combined, the leading zeros are kept in binary notation (just as if we were to write all numbers with three digits in decimal notation above and were then to obtain the sequence `082 083, 065 032, 119 111, 114 107, 115 033`).

Each character is encrypted using: $C = M^{65537} \pmod{256027}$:

```
212984 025546 104529 031692 248407
100412 054196 100184 058179 227433
```

9. - You can handle this exercise using `CT1 Indiv. Procedures ▷ RSA Cryptosystem ▷ RSA Demonstration`.
- Using `CT2 Templates ▷ Mathematics ▷ RSA with big numbers` for single numbers.
- Using `JCT Default Perspective ▷ Visuals ▷ RSA Cryptosystem` you can handle this task too.
- In CTO in the plugin "RSA visual and more" both variants (with and without block formation) and both sequences of coding (concatenation of the binary representation of the individual characters, or their decimal representation is concatenated first and then converted to binary) can also be used: `https://www.cryptool.org/en/cto/rsa-visual`. See Figures 5.8 and 5.9 for variant 2.

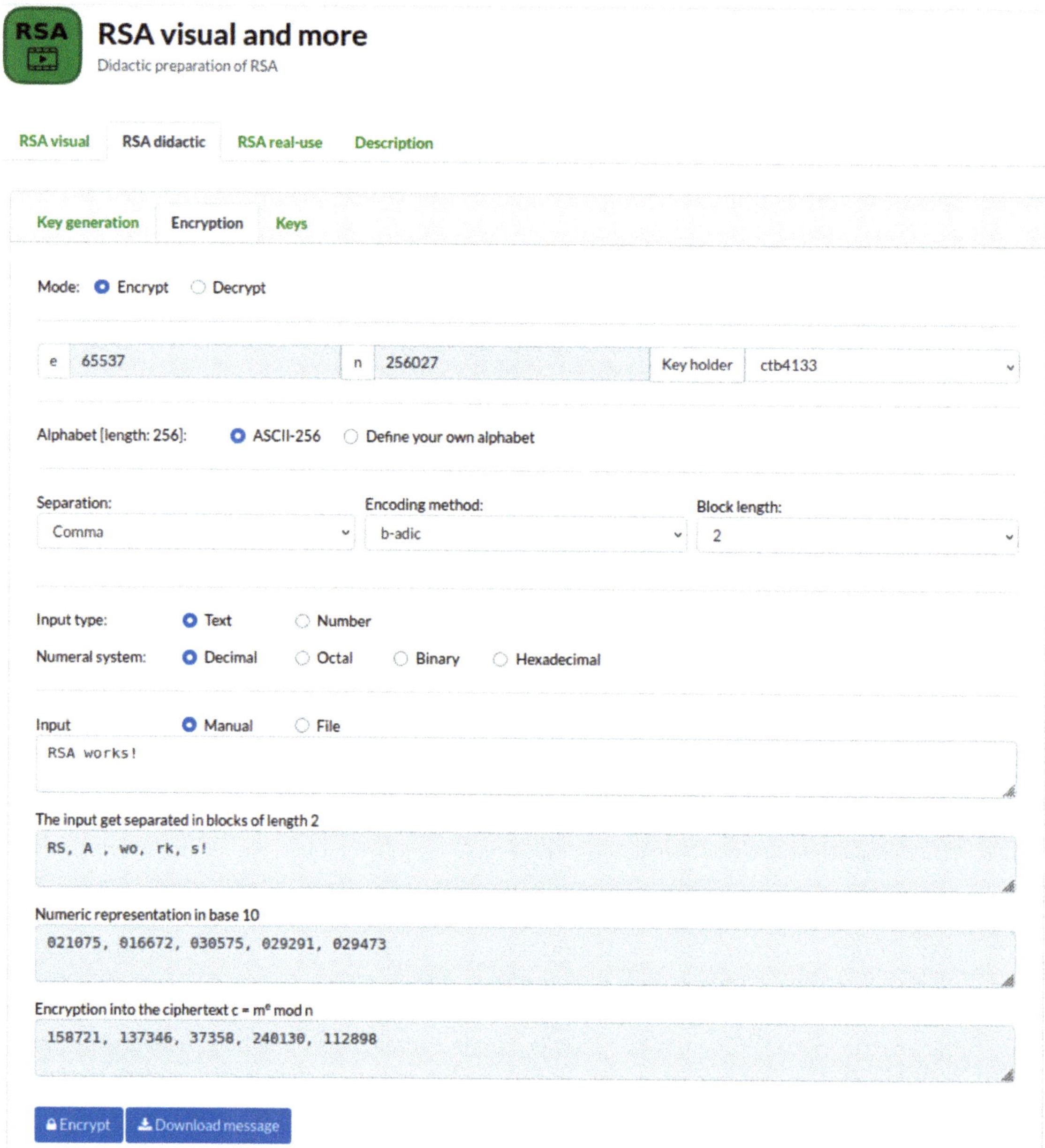

Figure 5.8 RSA in CTO: text encryption, ASCII alphabet, block length 2; Part 1: b-adic encoding.

5. Decryption:
Ciphertext:

212984 025546 104529 031692 248407
100412 054196 100184 058179 227433

Each character is decrypted using: $M \equiv C^{231953} \pmod{256027}$:
82 83 65 32 119 111 114 107 115 33

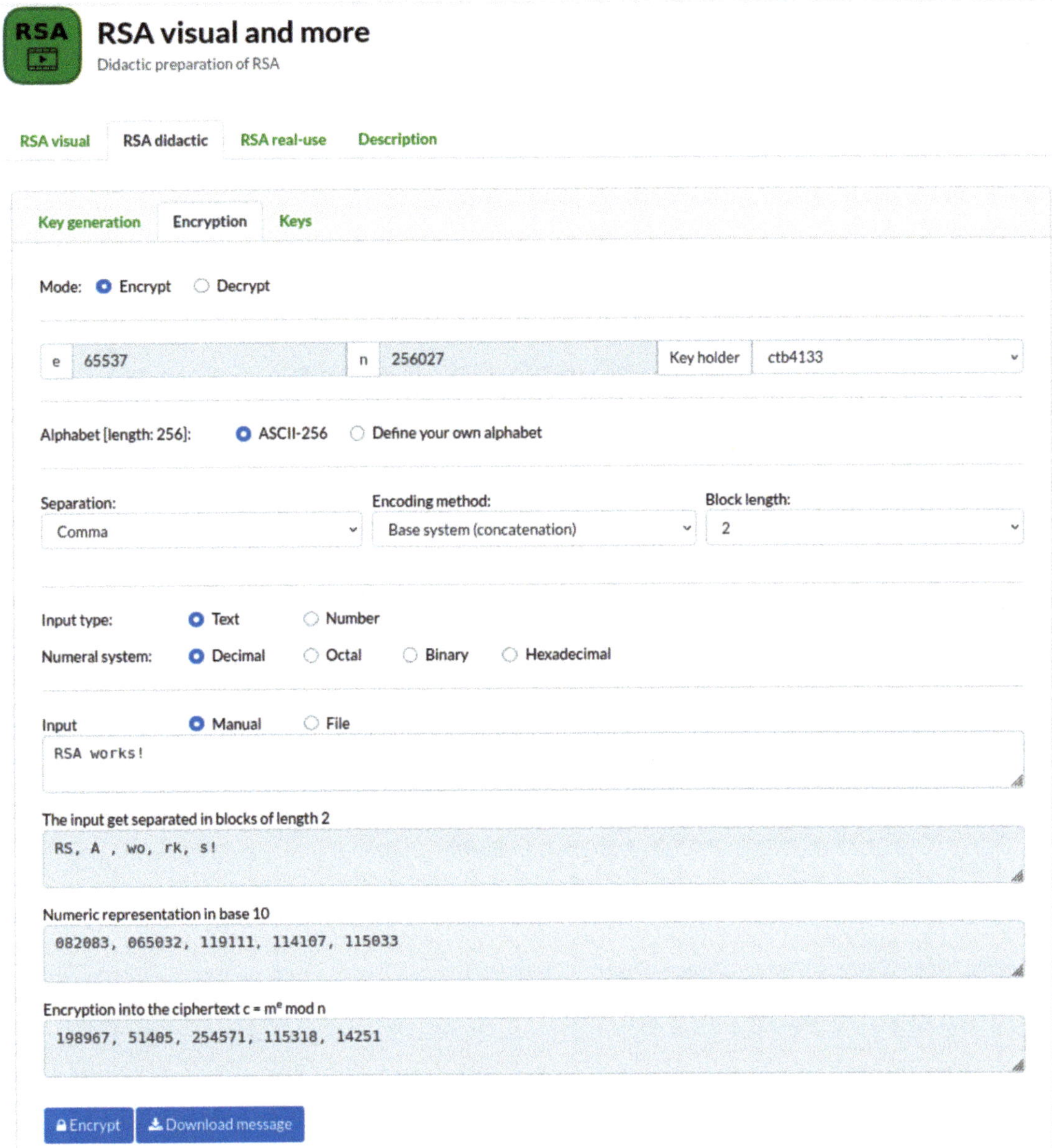

Figure 5.9 RSA in CTO: text encryption, ASCII alphabet, block length 2; Part 2: decimal conatenation.

Variant 2: The ASCII characters are en-/decrypted two at a time as blocks.[10]
In variant 2 the block formation is done in two different subvariants: (4./5. and 4'./5'.).

Text:	R	S	A		w	o	r	k	s	!
Number:	82	83	65	32	119	111	114	107	115	33

10. Also solvable with CTO in the plugin "RSA visual and more": `https://www.cryptool.org/en/cto/rsa-visual`. The adequate settings alphabet (ASCII or self-defined), b-adic or concatenation, and block length are set in Figures 5.8 and 5.9.

4. Encryption:
Blocks are formed by encoding each ASCII character into an 8-digit binary number and joining two binary numbers are:
Forming a block:

single character	binary representation	decimal representation
01010010, 82	01010010 01010011	= 21075
01010011, 83		
01000001, 65	01000001 00100000	= 16672
00100000, 32		
01110111, 119	01110111 01101111	= 30575
01101111, 111		
01110010, 114	01110010 01101011	= 29291
01101011, 107		
01110011, 115	01110011 00100001	= 29473
00100001, 33		

Altogether:[11]

```
21075 16672 30575 29291 29473
```

Each block is encrypted using: $C \equiv M^{65537} \pmod{256027}$:

```
158721 137346 37358 240130 112898
```

5. Decryption:
Ciphertext:

```
158721 137346 37358 240130 112898
```

Each block is decrypted using: $M \equiv C^{231953} \pmod{256027}$:

```
21075 16672 30575 29291 29473
```

4'. Encryption:
Blocks are formed (each block contains two ASCII characters, and the ASCII characters are written as two 3-digit decimal numbers one after the other):

```
82083 65032 119111 114107 115033
```

RSA encryption works correctly with the modulus $n = 256027$ because each ASCII block of two characters will be encoded into a number that is smaller or equal to the number 255,255.

11. You can solve this using `CT1 Indiv. Procedures` ▷ `RSA Cryptosystem` ▷ `RSA Demonstration` with the following options: all 256 ASCII characters, b-adic, block length 2, and decimal representation.

Each block is encrypted using: $C \equiv M^{65537} \pmod{256027}$:

```
198967 051405 254571 115318 014251
```

5'. Decryption:
Ciphertext:

```
198967 051405 254571 115318 014251
```

Each block is decrypted using: $M \equiv C^{2473} \pmod{67519}$:

```
82083 65032 119111 114107 115033
```

5.14.4 A Small RSA Cipher Challenge, Part 1

The following task is taken from [11, Exercise 4.6] and the pure solution has been published by Douglas Stinson. However, it is not the result that is important here but rather the individual steps of the solution; that is, the explanation of the cryptanalysis. The method of solving the problem is outlined in the scenario of the online help to CT1 and in the CT1 presentation on the CT website.

Here is the task in its original text:

Two samples of RSA ciphertext are presented in Tables 5.15 and 5.16. Your task is to decrypt them. The public parameters of the system are

$n = 18923$ and $e = 1261$ (for Table 5.15);
$n = 31313$ and $e = 4913$ (for Table 5.16).

The cryptanalysis can be accomplished as follows. First, factor n (which is easy because it is so small). Then compute the exponent d from $\phi(n)$, and, finally, decrypt the ciphertext. Use the square-and-multiply algorithm to exponentiate modulo n.

In order to translate the plaintext back into ordinary English text, you need to know how alphabetic characters are encoded as elements in $\mathbb{Z}_n$. Each element of $\mathbb{Z}_n$ represents three alphabetic characters as in the following examples (with $A = 0$):

$$\begin{aligned} \texttt{DOG} &\mapsto 3 \cdot 26^2 + 14 \cdot 26 + 6 = 2398 \\ \texttt{CAT} &\mapsto 2 \cdot 26^2 + 0 \cdot 26 + 19 = 1371 \\ \texttt{ZZZ} &\mapsto 25 \cdot 26^2 + 25 \cdot 26 + 25 = 17575 \end{aligned}$$

You will have to invert this process as the final step in your program.

The first plaintext was taken from *The Diary of Samuel Marchbanks* by Robertson Davies, 1947, and the second was taken from *Lake Wobegon Days* by Garrison Keillor, 1985.

5.14.5 A Small RSA Cipher Challenge, Part 2

The following task is a corrected version from the book written by Song Yan [51, Example 3.3.7, p. 318]. Like in Section 5.14.4, it is not the result that is important here but understanding the individual steps of the solution. The method of solving the problem is outlined in the scenario of the online help to CT1 and in

Table 5.15 RSA Ciphertext A

12423	11524	7243	7459	14303	6127	10964	16399
9792	13629	14407	18817	18830	13556	3159	16647
5300	13951	81	8986	8007	13167	10022	17213
2264	961	17459	4101	2999	14569	17183	15827
12693	9553	18194	3830	2664	13998	12501	18873
12161	13071	16900	7233	8270	17086	9792	14266
13236	5300	13951	8850	12129	6091	18110	3332
15061	12347	7817	7946	11675	13924	13892	18031
2620	6276	8500	201	8850	11178	16477	10161
3533	13842	7537	12259	18110	44	2364	15570
3460	9886	8687	4481	11231	7547	11383	17910
12867	13203	5102	4742	5053	15407	2976	9330
12192	56	2471	15334	841	13995	17592	13297
2430	9741	11675	424	6686	738	13874	8168
7913	6246	14301	1144	9056	15967	7328	13203
796	195	9872	16979	15404	14130	9105	2001
9792	14251	1498	11296	1105	4502	16979	1105
56	4118	11302	5988	3363	15827	6928	4191
4277	10617	874	13211	11821	3090	18110	44
2364	15570	3460	9886	9988	3798	1158	9872
16979	15404	6127	9872	3652	14838	7437	2540
1367	2512	14407	5053	1521	297	10935	17137
2186	9433	13293	7555	13618	13000	6490	5310
18676	4782	11374	446	4165	11634	3846	14611
2364	6789	11634	4493	4063	4576	17955	7965
11748	14616	11453	17666	925	56	4118	18031
9522	14838	7437	3880	11476	8305	5102	2999
18628	14326	9175	9061	650	18110	8720	15404
2951	722	15334	841	15610	2443	11056	2186

Table 5.16 RSA Ciphertext B

6340	8309	14010	8936	27358	25023	16481	25809
23614	7135	24996	30590	27570	26486	30388	9395
27584	14999	4517	12146	29421	26439	1606	17881
25774	7647	23901	7372	25774	18436	12056	13547
7908	8635	2149	1908	22076	7372	8686	1304
4082	11803	5314	107	7359	22470	7372	22827
15698	30317	4685	14696	30388	8671	29956	15705
1417	26905	25809	28347	26277	7897	20240	21519
12437	1108	27106	18743	24144	10685	25234	30155
23005	8267	9917	7994	9694	2149	10042	27705
15930	29748	8635	23645	11738	24591	20240	27212
27486	9741	2149	29329	2149	5501	14015	30155
18154	22319	27705	20321	23254	13624	3249	5443
2149	16975	16087	14600	27705	19386	7325	26277
19554	23614	7553	4734	8091	23973	14015	107
3183	17347	25234	4595	21498	6360	19837	8463
6000	31280	29413	2066	369	23204	8425	7792
25973	4477	30989					

the CrypTool 1 presentation (see https://www.cryptool.org/assets/ct1/presentations/CrypTool1-Presentation-en.pdf, pp. 52–57).

There are three tasks with completely different degrees of difficulty here. In each case we know the ciphertext and the public key (e, n):

a. Known-plaintext attack: find the secret key d using the additionally known original message.
b. Ciphertext-only attack: find d and the plaintext.
c. Calculate the RSA modulus; in other words, factorization (with no knowledge of the message).

$n = 63978486879527143858831415041$, $e = 17579$
Message:

```
1401202118011200,
1421130205181900,
0118050013010405,
0002250007150400
```

Cipher:

```
45411667895024938209259253423,
16597091621432020076311552201,
46468979279750354732637631044,
32870167545903741339819671379
```

Comment:
The original message consisted of a sentence containing 31 characters (coded with the capital letters' alphabet from Section 5.14.2). Each group of 16 decimal numbers is then combined to form one number (the last number is filled with zeros). These numbers are raised to the power of e.

When you decrypt the message you must fill the calculated numbers with leading zeros in order to obtain plaintext. This needs to be stressed because the type of padding is extremely important during implementation and standardization for interoperable algorithms.

5.15 Didactic Comments on Modulo Subtraction

Comment on subtraction modulo 5: $2 - 4 = -2 \equiv 3 \mod 5$.

It is therefore *not* true that $-2 \equiv 2 \mod 5$.

People often make the mistake of equating this. It is easy to see why this is not the same if you place the permutation $(0, 1, 2, 3, 4)$ in $\mathbb{Z}_5$, for example from -11 to $+11$, over the range of numbers in $\mathbb{Z}$, like in Figure 5.10. Moving then on the number line of integers from 3 to the left by 5, one ends up with the next element belonging mod 5 to the same residue class as 3, which is -2.

Range of numbers modulo 5

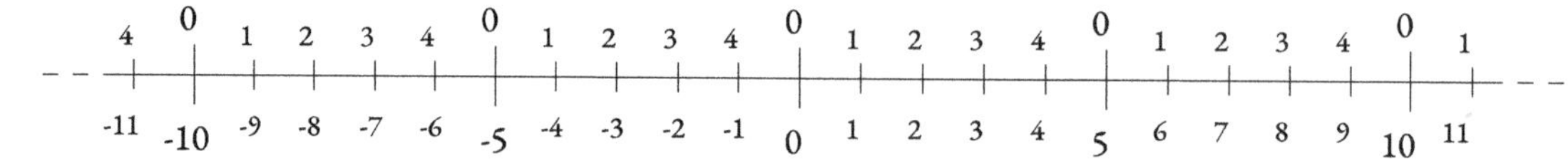

Range of numbers in $\mathbb{Z}$

Figure 5.10 Number line of integers compared to modulo 5 numbers.

Table 5.17 Schematic Representation of Which Integers Belong to the Same Residue Class Modulo 26

−26	−25	...	−2	−1
0	1	...	24	25
26	27	...	50	51

From time to time, some students ask how to deal with negative results (e.g., -1). For example, for affine ciphers with 26 characters, you calculate modulo 26. In $\mathbb{Z}_{26} = \{0, 1, 2, \cdots, 25\}$ it is $0 - 1 = -1 = 25 \pmod{26}$.

The modulo calculation and which numbers belong to the same residue class are illustrated in Table 5.17. You can see that numbers in the same column belong to the same residue class. Such didactic representations help certain types of learners more than formulas.

So here you can see that the following numbers are congruent: $51 \equiv 25 \equiv -1 \pmod{26}$. Being congruent means belonging to the same residue class. Two numbers are congruent if their difference can be divided by the modulus. So, $(25 - (-1)) = (51 - 25) = 26$ are all divisible by 26.

5.16 Base Representation and Base Transformation of Numbers and Estimation of Length of Digits

For a given number z one may ask how to represent such a number. In general we use representations like $z = 2374$ or $z = \sqrt{2}$. The second number consists of an infinite number of digits and therefore it can never be described precisely by the first representation. You can get around this problem by writing the number symbolically. But if you have to write it in digits, the number must be rounded.

We represent numbers usually in the decimal system (base 10). Computers are working with the binary representation of numbers—only for the display numbers are represented in decimal or sometimes hexadecimal (base 16) form.

This section describes how to generate arbitrary base representations of any positive integer and how to determine the number of required digits via the logarithm function.

5.16.1 b-adic Sum Representation of Positive Integers

Given base b, each positive integer z can be represented as a b-adic sum

$$z = a_n b^n + a_{n-1} b^{n-1} + \cdots + a_1 b + a_0,$$

where $a_i \in \{0, 1, \ldots, b-1\}$, $i = 0, 1, \ldots, n$ are called *digits*.
For this sum, we have:

1. For arbitrary digits $a_0, a_1, \ldots, a_n$ it holds: $b^{n+1} > a_n b^n + a_{n-1} b^{n-1} + \cdots + a_1 b + a_0$.
2. There exist digits $a_0, a_1, \ldots, a_n$ (namely $a_i = b-1$ for $i = 0, \ldots, n$), with $b^{n+1} - 1 \leq a_n b^n + a_{n-1} b^{n-1} + \cdots + a_1 b + a_0$.

(Using these inequalities it can be shown that each positive integer can be represented by a b-adic sum).

By writing the digits $a_n a_{n-1} \cdots a_1 a_0$ in a row directly after each other (without the b^i) the usual writing for numbers becomes available.

Example:

base $b = 10$: `10278` $= 1 \cdot 10^4 + 0 \cdot 10^3 + 2 \cdot 10^2 + 7 \cdot 10^1 + 8$.
base $b = 16$: `FE70A` $= 15 \cdot 16^4 + 14 \cdot 16^3 + 7 \cdot 16^2 + 0 \cdot 16^1 + 10$.

5.16.2 Number of Digits to Represent a Positive Integer

For a positive integer z the length of the b-adic representation can be determined via the following steps. Starting from the inequality $b^{n+1} > z \geq b^n$ we have—after applying the logarithm function on basis b—$n+1 > log_b z \geq n$. Therefore, we have $n = \lfloor log_b z \rfloor$ (the notion $\lfloor x \rfloor$ for a positive real number indicates to round down to the next positive integer or do nothing if x itself is an integer). We call $l_b(z)$ *the number of required digits to represent the number z on the base b*. We have

$$l_b(z) := \lfloor \log_b z \rfloor + 1$$

Applying the logarithm formula on base b and b' we have $\log_b z = \log_{b'} z / \log_{b'} b$. It is therefore easy using, for example, logarithm tables for the base $b' = 10$ to compute the logarithm of base $b = 2$. With SageMath it is even easier: The command `log(n,b)` returns the logarithm ob n to the base b, usually in an algebraic form: `log(101,10)` returns `log(101)/log(10)` where `log(n)` is the natural logarithm with base e. For numerical values use `log(101,10).n()`, then you get `2.0043...` and for rounding down use `floor(log(101,10))` then you get 2.

Example 1 (decimal→hex)
We compute for the decimal number $z = 234$ (EA in hex) the hexadecimal representation (number base $b = 16$)

$$l_{16}(z) = \lfloor \log_{16}(z) \rfloor + 1 = \lfloor \ln(z)/\ln(16) \rfloor + 1 = \lfloor 1.96\ldots \rfloor + 1 = 1 + 1 = 2.$$

Example 2 (decimal→binary)
We compute for the decimal number $z = 234$ (11101010 in binary) the binary representation (number base $b = 2$)

$$l_2(z) = \lfloor \log_2(z) \rfloor + 1 = \lfloor \ln(z)/\ln(2) \rfloor + 1 = \lfloor 7.87\ldots \rfloor + 1 = 7 + 1 = 8.$$

Example 3 (binary→decimal)
We compute for the binary number $z = 11101010$ (234 decimal) the decimal representation (number base $b = 10$)

$$l_{10}(z) = \lfloor \log_{10}(z) \rfloor + 1 = \lfloor \ln(z)/\ln(10) \rfloor + 1 = \lfloor 2,36\ldots \rfloor + 1 = 2 + 1 = 3.$$

5.16.3 Algorithm to Compute the Base Representation

Given the number z one can compute the base b representation of z using the following algorithm:

input: z, b
$n := 0, z' := z$
while $z' > 0$ **do**
  $a_n := z' \pmod{b}$
  $z' := \lfloor z'/b \rfloor$
  $n := n + 1$
end do
output: $a_n a_{n-1} \cdots a_1 a_0$ in base b representation.

Example 4 (decimal→hex)
The integer $z = 234$ on the number base 10 will be transformed into the hex representation via $a_0 = 234 \pmod{16} = 10 = A$; $234/16 = 14 = E$, $a_1 = 14 \pmod{16} = E$, and therefore we have EA.

Example 5 (binary→decimal):
The binary number $z = 1000100101110101$ is transformed into the decimal representation via the following steps:

$1000100101110101 = 1001 \pmod{1010} \Longrightarrow a_0 = 9, \quad 1000100101110101/1010 = 110110111110$

$110110111110 = 1000 \pmod{1010} \Longrightarrow a_1 = 8, 110110111110/1010 = 101011111$

$101011111 = 1 \pmod{1010} \Longrightarrow a_2 = 1, 10101111/1010 = 100011$

$100011 = 101 \pmod{1010} \Longrightarrow a_3 = 5, 100011/1010 = 1$

$11 = 11 \pmod{1010} \Longrightarrow a_4 = 3$

therefore $z = 35189$.

SageMath Example 5.13 contains code for the examples about digit length and converting the representation between different bases. Sage integers can be read as a decimal, octal, hexadecimal, or binary number: `Integer()` or `ZZ()` interpret strings that begin with "0o" as octal numbers, strings that begin with "0x" as hexadecimal numbers, and strings that begin with "0b" as binary numbers. We can omit `Integer()` and `ZZ()` when entering a number on the Sage command line, as the interpretation as a Sage integer is the default.

SageMath Example 5.13: Number of Digits Representing a Positive Integer and Transformation Between Different Bases

```
# Length of decimal number
z=234; z.ndigits()
3

# Sample 1 -- Length of hex: decimal --> hex
sage: z.ndigits(16)
2

# Sample 2 -- Length of binary: decimal --> binary
sage: z.ndigits(2)  # Alternative: 234.nbits()
8

# Sample 3 -- Length of decimal: binary --> decimal
sage: z=Integer('0b11101010'); z; z.ndigits()
234
3

# Enter a number of a given base as Sage integer
sage: Integer('0x12')
18
sage: Integer('-0o12')
-10
sage: Integer('+0b101010')
42

# Output a Sage integer as number of a required base
sage: Integer(2^10).str(2)
'10000000000'
sage: print(Integer(800).oct())
1440

# Sample 4 -- show representation of hex: decimal --> hex
sage: print(Integer(234).hex())
ea

# Sample 5 -- show representation of decimal: binary --> decimal
# a) via a conversion using Python int() and str()
sage: z=1000100101110101; z
1000100101110101
sage: type(z)
<class 'sage.rings.integer.Integer'>
sage: int(str(z), base=2)
35189
# b) more directly using Sage integer
# sage: z=ZZ('0b1000100101110101'); z
sage: z=0b1000100101110101; z
35189

sage: s = '12d'; zx=ZZ('0x'+s); zx; print(Integer(zx).hex())
301
12d

sage: a = ZZ(18986613671930846201827115924243716853 2); a.binary(); ▸
  ▸print(a.binary())
# '10001110110101101110001101000 1 ... 10100000000110110010100'
# 100011101101011011100011010001 ... 10100000000110110010100
```

5.17 Examples Using SageMath

Below you can find SageMath source code related to contents of this Chapter 5.

We also recommend the short article by Nguyen, which is didactically very clear and treats basic number theory and SageMath usage [52].

5.17.1 Addition and Multiplication Tables Modulo *m*

SageMath Example 5.14 calculates the addition and multiplication Tables 5.1, 5.2, and 5.3. The calculation is done once with self-written code, and once ready-made functions from SageMath are used.

SageMath Example 5.14: Creating Small Addition and Multiplication Tables with SageMath

```
print("\n# CHAP05 -- Sage-Script-SAMPLE 025: =========")

# Create tables with own code .....................................
m = 5; print("Addition table mod %d" % m)
for i in range(0,m):
  print( [mod(i+j, m) for j in range(0,m)] )

m = 5; print("Multiplication table mod %d" % m)
for i in range(1,m):
  print( [mod(i*j, m) for j in range(1,m)] )

m = 6; print("Multiplication table mod %d" % m)
for i in range(1,m):
  print( [mod(i*j, m) for j in range(1,m)] )

# Create tables with predefined Sage functions for magmas .....................................
# https://doc.sagemath.org/html/en/reference/categories/sage/categories/magmas.html#sage.
  categories.magmas.Magmas.ParentMethods.multiplication_table
m = 5; R=Zmod(m)
T = R.addition_table(names='digits'); print("\n", T, sep='')  # same result with (names='digits
  ')

m = 5; R=Zmod(m)
T = R.multiplication_table(names='digits'); print(T)
# print( latex(T) )  # get the code for LaTeX to print the according table

m = 5; R=Zmod(m); elem = [str(i) for i in range(1,m)]  #; print("elem: ", elem)
T = R.multiplication_table(names='elements', elements=elem); print(T)  # =elem instead of
  elements=('1','2','3','4')
# print(T.column_keys(), "\n")  #  (1, 2, 3, 4)

m = 6; R=Zmod(m)  # m = 15
T = R.multiplication_table(names='digits'); print(T)
```

SageMath Example 5.15 calculates the multiplication Table 5.4. for $a \cdot i \pmod{m}$, with $m = 17$, $a = 5$ and $a = 6$, and i from 0 to 16, respectively.

SageMath Example 5.15: Multiplication Tables for $a \cdot i \pmod{m}$ with $m = 17$, $a = 5$, and $a = 6$

```
print("\n# CHAP05 -- Sage-Script-SAMPLE 030: =========")

m = 17; a = 5
print( [mod(a * i, m) for i in range(m)] )
a = 6
print( [mod(a * i, m) for i in range(m)] )
```

SageMath Example 5.15 (continued)

```
#------------------------------------
# CHAP05 -- Sage-Script-SAMPLE 030: =========
# [0, 5, 10, 15, 3, 8, 13, 1, 6, 11, 16, 4, 9, 14, 2, 7, 12]
# [0, 6, 12, 1, 7, 13, 2, 8, 14, 3, 9, 15, 4, 10, 16, 5, 11]
```

The function `mod()` returns an object that represents integers modulo m (in our case $m = 17$).

The other multiplication table examples modulo 13 (Table 5.5) and modulo 12 (Table 5.6) can be computed similarly by replacing `m = 17` with `m = 13` and `m = 12`, respectively.

5.17.2 Fast Exponentiation

The fast exponentiation modulo m can be computed using the SageMath function `power_mod()`. The result of this function is an integer.

With the SageMath Example 5.16 you can reproduce the idea of the square-and-multiply method, as shown in the example in Section 5.6.4.

SageMath Example 5.16: Fast Exponentiation of $a^e \bmod m = 103$

```
print("\n# CHAP05 -- Sage-Script-SAMPLE 040: =========")

a = 87; m = 103
exp = [2, 4, 8, 16, 32, 43]
z = [power_mod(a, e, m) for e in exp]
print( type(z), "\n", z )

#------------------------------------
# CHAP05 -- Sage-Script-SAMPLE 040: =========
# <class 'list'>
#  [50, 28, 63, 55, 38, 85]
```

5.17.3 Multiplicative Order

The order $\text{ord}_m(a)$ of a number a in the multiplicative group $\mathbf{Z}_m^*$ is the smallest number $i \geq 1$ such that $a^i \equiv 1 \pmod m$ (see Section 5.9).

To create Table 5.7 we can print all exponentiation $a^i \pmod{11}$ like in SageMath Example 5.17.

SageMath Example 5.17: Table with All Powers $a^i \pmod m$ for $m = 11$, $a, i = 1, ..., 10$

```
print("\n# CHAP05 -- Sage-Script-SAMPLE 050: =========")

m = 11
for a in range(1, m):
    print( [power_mod(a, i, m) for i in range(1, m)] )

# E: adding a last column with the order of each 'a' mod (11)
# D: die letzte Spalte um die Ordnung des jeweiligen 'a' mod (11) ergänzen
print()  # add an empty line between the two tables
```

SageMath Example 5.17 (continued)

```
for a in range(1, m):
  lst = [power_mod(a, i, m) for i in range(1, m)]
  lst.append(multiplicative_order(mod(a,m)))

  for k in range(0,m-1): # some beautifier formatting
    print ("{:>4}".format(lst[k]), end = '')  # print first m-1 cells of current row
  print ("{:>6}".format(lst[m-1]))  # print last cell of current row
```

Table 5.8 gives examples for $\text{ord}_{45}(a)$ and the Euler number $\phi(45)$. SageMath Example 5.18 constructs a table similar to Table 5.8. In addition, this sample uses the method `table` in order to print the layout of the table.

SageMath Example 5.18: Table with All Powers $a^i \pmod{45}$ for $a, i = 1, \ldots, 12$ Plus the Order of a

```
print("\n# CHAP05 -- Sage-Script-SAMPLE 060: =========")

tbl = []
m = 45
noCols=m; noRows=m     # so whole table isprinted
noCols=13; noRows=13   # so smaller, more clear table isprinted

for a in range(1, noRows):
    lst = [power_mod(a, i, m) for i in range(1, noCols)]
    try:
        lst.append(multiplicative_order(mod(a, m)))
    except:
        lst.append("No mult. order")
    lst.append(euler_phi(m))
    # print(lst)
    tbl.append(lst)  # build up a table from the single rows (lst)

print(table(tbl, align='center'))  # print whole table
```

The number $\text{ord}_m(a)$ only exists if a is relatively prime to m, which can be checked with `gcd(a,m)`. For our example with $m = 45 = 3^2 \cdot 5$, the values $a = 3, 5, 6, 9, 10, 12, \ldots$ are not relatively prime to m and so have no multiplicative order.

Programming hint: In the code example 5.18, we put the calculation of the multiplicative order within a `try-except` block. So you can catch any exceptions or errors raised by the function `multiplicative_order()`. If an exception or error is raised in the `try` block, then we know that $\text{ord}_m(a)$ does not exist for that particular value of a. Hence in the `except` block we append the string `"No mult. order"` to the row represented by the object `lst`.

Table 5.9 displays exponentiation $a^i \pmod{46}$ as well as the order $\text{ord}_{46}(a)$. SageMath Example 5.19 creates such a table.

SageMath Example 5.19: Table with All Powers $a^i \pmod{46}$ for $a, i = 1, \ldots, 23$ Plus the Order of a

```
print("\n# CHAP05 -- Sage-Script-SAMPLE 070: =========")

m = 46
print( euler_phi(m) ); print()

for a in range(1, 24):
```

SageMath Example 5.19 (continued)

```
lst = [power_mod(a, i, m) for i in range(1, 24)]
try:
    lst.append(multiplicative_order(mod(a, m)))
except:
    lst.append("None")  # No multiplicative order exists for this 'a'
print(lst)
```

SageMath Example 5.20 generates Tables 5.10 and 5.11. It also delivers the result in a way that can be easily processed in LaTeX. The prerequisite is that all content is assigned to one SageMath object (here the matrix r).

A note about SageMath Example 5.20, especially about the SageMath indices:

- `for x in range(2, 5)` delivers 2,3,4.
- `m = matrix(ZZ, 2, 5)` has 2 rows and 5 columns. The cells are named m(0,0) to m(1,4).
- All elements of the matrix have to be numerical, so "0" is used instead of `None` as in the tables before.
- The output of matrices can be controlled in SageMath with:

```
sage: from sage.matrix.matrix import set_max_cols, set_max_rows
sage: set_max_cols(100)
sage: set_max_rows(100)
```

- The length of the cycle in the last column of the Tables 5.10 and 5.11 was added manually.

SageMath Example 5.20: Code for Tables with All Powers $a^i \pmod m$ for Variables a and i Plus Order of a and Eulerphi of m

```
print("\n# CHAP05 -- Sage-Script-SAMPLE 080: =========")

def power_mod_order_matrix(m, max_a, max_i):
    r = matrix(ZZ, max_a+1, max_i+3)
    for a in range(0, max_a+1):
        r[a, 0] = a
        for i in range(1, max_i+1):
            if a==0:
                r[a,i] = i
            else:
                r[a, i] = power_mod(a, i, m)
        try:
            r[a, max_i+1] = multiplicative_order(mod(a, m))
        except:
            r[a, max_i+1] = 0
        r[a, max_i+2] = euler_phi(m)
    return r

print("\n#1: m=45; max_i=13; max_a=13"); m=45; max_i=13; max_a=13
r = power_mod_order_matrix(m, max_a, max_i)
print(r); print( latex(r) )

print("\n#2: m=46; max_i=25; max_a=25"); m=46; max_i=25; max_a=25
r = power_mod_order_matrix(m, max_a, max_i)
print( r.str() ); print( latex(r) )

print("\n#3: m=14; max_i=13; max_a=16"); m=14; max_i=13; max_a=16
r = power_mod_order_matrix(m, max_a, max_i)
```

SageMath Example 5.20 (continued)

```
print(r); print( latex(r) )

print("\n#4: m=22; max_i=21; max_a=25"); m=22; max_i=21; max_a=25
r = power_mod_order_matrix(m, max_a, max_i)
print( r.str() ); print( latex(r) )
```

5.17.4 Primitive Roots

Section 5.9 explained what primitive roots are and why they are useful. Computing a primitive root in SageMath is very straightforward. If `n` is an integer, the command `primitive_root(n)` delivers *one* primitive root of the multiplicative group $\mathbb{Z}_n^*$, if such a primitive root exists. If n is prime then this is the same as calculating a primitive root of $\mathbb{Z}_n$. If the number under consideration is a prime number, `primitive_root(n)` returns the smallest primitive root. For nonprimes this is not always the case as you can see with $m = 10$, when comparing SageMath Example 5.21 and SageMath Example 5.22: The function returns 7 instead of 3.

a. SageMath Example 5.21.

   The example 5.21 calculates primitive roots of a few integers first from some nonprimes (see the special cases of Theorem 5.14 in the remarks there), then from the first 15 primes.

   SageMath Example 5.21: Calculating One Primitive Root for a Given Number

```
print("\n# CHAP05 -- Sage-Script-SAMPLE 090: =========")

print( " 4:", primitive_root(4) )
print( " 6:", primitive_root(6) )
print( " 9:", primitive_root(9) )  # Remark: 8 has no primitive ▸
  ▸root
print( "10:", primitive_root(10) )
print( "22:", primitive_root(22) )
for p in primes(1, 50):
    print( "%2d: %3d" % (p, primitive_root(p)) )
```

b. SageMath Example 5.22.

   If p is prime, then $\mathbb{Z}_p$ has at least one primitive root. But also for composite numbers there are cases where a primitive root exists; for example, if n is the product of 2 and some odd prime power (see the special cases in Theorem 5.14).

   Sometimes we want to compute for an integer n `all` the primitive roots of $\mathbb{Z}_n^*$ (if they do exist), not just any primitive root of it. The self-written function `enum_PrimitiveRoots_of_an_Integer` in the example 5.22 can do this. After some smaller test cases this Sage sample lists of all primitive roots of the prime number 541. The listing shows only the beginning of the Sage sample.

SageMath Example 5.22: Function `enum_PrimitiveRoots_of_an_Integer` to Calculate All Primitive Roots for a Given Number

```
# CHAP05 -- Sage-Script-SAMPLE 100: =========
# This file can be used both as script and imported as library,
# so printing its name is moved to the __main__ part at the end.

def enum_PrimitiveRoots_of_an_Integer(M):
    """
    Return all the primitive roots of the integer M (if possible).
    """
    try:
        g = primitive_root(M)
    except:
        return None
    targetOrder = euler_phi(M)
    L=[]
    # Stepping through all odd integers from 1 up to M, not including
    # M. So this loop only considers values of i where 1 <= i < M.
    for i in range(1,M,2):
            testGen = Mod(g^i,M)
            if testGen.multiplicative_order() == targetOrder:
                L.append(testGen.lift())
    # removing duplicates
    return Set(L)


if __name__ == "__main__":
    print("\n# CHAP05 -- Sage-Script-SAMPLE 100: =========")

    # AA_Start -- Testcases for enum_PrimitiveRoots_of_an_Integer(M)
    print("AA_Start -- Testcases for enum_PrimitiveRoots_of_an_Integer(M)")

    M=5; print( "-----------Testcase 1: M = %s" % M )
    LL = enum_PrimitiveRoots_of_an_Integer(M)
    if LL==None: print("None")
    else:        print(LL)

...
```

c. SageMath Example 5.23.

With a little bit of programming, we also can count how many primitive roots are there in a given range of integers. The example 5.23 does this: Depending on the third argument of the function `count_PrimitiveRoots_of_an_IntegerRange` this amount is calculated either for all numbers or only for the primes within this range.

Programming hint: Code sample 5.23 uses the function `enum_PrimitiveRoots_of_an_Integer`. This function was previously defined in the Sage script `chap05_sample100.sage` and therefore, it should not be rewritten here.

Sadly, the Sage command `load("./chap05_sample100.sage")` doesn't allow yet to just load selected functions (as of SageMath v9.3). Instead, `load` always performs all commands and functions of the script. Concerning `load` and `attach` see the introduction to SageMath at https://www.cryptool.org/en/documentation/ctbook/.

Instead of creating your own SageMath file for the function `enum_PrimitiveRoots_of_an_Integer` (like in `bitciphers.sage` for the generic functions of

Chapter 9), here Python is used to import just this function from another Sage script.

To do so, the importing script uses the function `my_import` which gets as argument the function name to be imported. `my_import` then uses Python mechanisms to load and execute the function from the argument from another Sage script. Note that the function is loaded from the Python version of the Sage script. This works because SageMath always transforms the Sage script into a Python script, which then is performed (sample: When calling `$sage myfile.sage` a Python file `myfile.sage.py` is created from `myfile.sage`). Currently, this `my_import` function is written in such a way that it requires a typical Unix shell (e.g., Bash), which is available by default on Linux and macOS. However, this also works on Windows, where SageMath for Windows is delivered together with the Bash shell (see https://wiki.sagemath.org/SageWindows).

The call within the own Sage script looks like this:

`my_import("chap05_sample100", "enum_PrimitiveRoots_of_an_Integer")`.

SageMath Example 5.23: Function `count_PrimitiveRoots_of_an_IntegerRange` to Calculate the Number of All Primitive Roots for a Given Range of Integers

```
print("\n# CHAP05 -- Sage-Script-SAMPLE 110: =========")

# function to include 'enum_PrimitiveRoots_of_an_Integer' from another sage script
def my_import(module_name, func_name='*'):
    import os
    command = 'sage --preparse ' + module_name + '.sage'
    os.system(command)  # this creates the *.sage.py file
    command = 'mv ' + module_name + '.sage.py ' + module_name + '.py'
    os.system(command)  # this renames *.sage.py to *.py
    from sage.misc.python import Python; python = Python()
    command = 'from ' + module_name + ' import ' + func_name
    python.eval(command, globals())


def count_PrimitiveRoots_of_an_IntegerRange(start, end, bPrimesOnly=True, ▸
  ▸bShowPrimRoots=False):
    """
    Compute all primitive roots of all numbers between start and end,
    inclusive, and count them.
    - If the flag bPrimesOnly is True (default), it performs primality tests and the ▸
      ▸calculation is
      done only for primes (printing the count of found primes from start to end, ▸
        ▸inclusive).
    - If the flag bPrimesOnly is False, it not only considers the primes; and it
      additionally prints the count of (even) numbers which have NO primitive root.
    - If the flag bShowPrimRoots is True (default is False), then the primitive roots ▸
      ▸are listed too.
    """
    nCheckedNumb = 0
    nCheckedNumb_WithoutPrimitivRoots = 0
    nPrimitiveRoots = 0
    for n in range(start, end+1):
        if bPrimesOnly:
            if is_prime(n):
                nCheckedNumb += 1
                L = enum_PrimitiveRoots_of_an_Integer(n)
                nPrimitiveRoots += len(L)
                if bShowPrimRoots==True: print("n=%d, lenL=%d, L=%s" % (n, len(L), L))
        else:
            nCheckedNumb += 1
```

SageMath Example 5.23 (continued)

```
            L = enum_PrimitiveRoots_of_an_Integer(n)
            if L==None:
                nCheckedNumb_WithoutPrimitivRoots += 1
            else:
                nPrimitiveRoots += len(L)
                if bShowPrimRoots==True: print("n=%d, lenL=%d, L=%s" % (n, len(L), L))

    if bPrimesOnly:
        print("Found all %s" % nPrimitiveRoots + \
              " primitive roots of %s prime(s)." % nCheckedNumb)
    else:
        if nCheckedNumb_WithoutPrimitivRoots == 0:
            print("Found all %s " % nPrimitiveRoots + \
                  "primitive roots of %s number(s)." % nCheckedNumb)
        else:
            print("Found all %s " % nPrimitiveRoots + \
                  "primitive roots of %s number(s)." % \
                     (nCheckedNumb - nCheckedNumb_WithoutPrimitivRoots))
            print("(Total of numbers checked: %s  " % nCheckedNumb + \
                  "Amount of numbers without primitive roots: %s)" % \
                     nCheckedNumb_WithoutPrimitivRoots)

import time # in sage scripts: measure time like in Python and calculate execution time

my_import("chap05_sample100",
"enum_PrimitiveRoots_of_an_Integer") # import from 'chap05_sample100.sage'

print("\nBB_Start -- Testcases for count_PrimitiveRoots_of_an_IntegerRange(start, end, ▸
  ▸bPrimesOnly, bShowPrimRoots)")

print("\n-----------Testcase 1: (5, 6)"); StartTime = time.time() # default: only prime▸
  ▸ moduli in range
count_PrimitiveRoots_of_an_IntegerRange(5, 6)
print(" Time = %f sec" % (time.time()-StartTime))
...
```

According to the test cases with the function `count_PrimitiveRoots_of_an_IntegerRange`, it becomes clear that the elapsed time for the calculation raises exponentially with the size of the numbers. In 2021, on a modern PC and with SageMath 9.3, the execution of the script in 5.23 lasted more than 5 minutes. Most of that time was spent on the last test case.

d. SageMath Example 5.24

The example 5.24 counts how many primitive roots are there in a given primes range and enumerates all these primitive roots. Function `enum_PrimitiveRoots_of_an_Integer` is used, which lists all primitive roots of a single prime number p. This function again is included from a Sage script already introduced before.

From this list of primitive roots, we can determine the smallest and largest primitive root for $\mathbb{Z}_p$, as well as count the number of primitive roots of $\mathbb{Z}_p$.

SageMath Example 5.24: Function `count_PrimitiveRoots_of_a_PrimesRange` to Calculate the Number of Primitive Roots for a Given Range of Primes

```
print("\n# CHAP05 -- Script-SAMPLE 120: =========")
...
def count_PrimitiveRoots_of_a_PrimesRange(start, end):
    """
```

SageMath Example 5.24 (continued)

```
    Compute all primitive roots of all primes between start and end, inclusive.
    """
    nPrimes = 0
    nPrimitiveRoots = 0
    for p in primes(start, end+1):
        L = enum_PrimitiveRoots_of_an_Integer(p)
        print(p, len(L))
        nPrimes += 1
        nPrimitiveRoots += len(L)
    print("Found all %s" % nPrimitiveRoots + " primitive roots of %s primes." % nPrimes▸
    ▸)


import time
my_import("chap05_sample100", "enum_PrimitiveRoots_of_an_Integer")

print("\nCC_Start -- Testcases for count_PrimitiveRoots_of_a_PrimesRange(start, end)")
print("-----------Testcase: (1, 1500)"); StartTime = time.time()
count_PrimitiveRoots_of_a_PrimesRange(1, 1500)
print(" Time = %f sec" % (time.time()-StartTime))
```

e. SageMath Example 5.25

A variant of the function `count_PrimitiveRoots_of_a_PrimesRange` from SageMath Example 5.25 (slightly modified by Minh Van Nguyen for the example 5.25) was used to generate a database of all primitive roots of all primes between 1 and 100,000.

SageMath Example 5.25: Code to Generate the Database with All Primitive Roots for All Primes Between 1 and 100000

```
print("\n# CHAP05 -- Sage-Script-SAMPLE 130: =========")
...
my_import("chap05_sample100", "enum_PrimitiveRoots_of_an_Integer")
start = 1
end = 100  # 10^5 needs ca. an hour. For testing choose end=100
fileName = "./primroots.dat"
print("...Creating file %s with end=%d" % (fileName, end))
file = open(fileName, "w")
for p in primes(start, end+1):
    L = enum_PrimitiveRoots_of_an_Integer(p)
    # print(p, len(L)) # just temporarily active
    # Output to a file. The format is:
    # (1) the prime number p under consideration
    # (2) the number of primitive roots of Z/pZ
    # (3) all the primitive roots of Z/pZ
    file.write(str(p) + " " + str(len(L)) + " " + str(L) + "\n")
    file.flush()
file.close()
```

This code and the function `enum_PrimitiveRoots_of_an_Integer` were also executed in a Sage script noninteractively. In 2016, it took about 6 hours on a then-modern PC with SageMath 7.2. In 2021, it took less than 1 hour with SageMath 9.3.

Between 1 and 100,000 there are 9,592 primes. For them, more than 169 million primitive roots have been calculated. For each prime $p > 3$ it holds that between 20% and almost 50% of all integers between 1 and p are a corresponding primitive root. So, prime numbers have relatively many primitive roots.

The resulting file `primroots.dat` is a database of all primitive roots of all primes between 1 and 100,000 inclusive. It is a large file (about 1.1 GB uncompressed, and 153 MB compressed with 7Zip). You can find the compressed file at `https://www.cryptool.org/download/ctb/primroots_1-100000.7z`. Its content looks like this (prime number, count, and set of primitive roots):

```
2       1       {1}
3       1       {2}
5       2       {2, 3}
7       2       {3, 5}
11      4       {8, 2, 6, 7}
...
89      40      {3, 6,  7, 13, 14, 15, 19, ..., 66, 70, 74, 75, 76, 82, ▸
 ▸83, 86}
97      32      {5, 7, 10, 13, 14, 15, 17, ..., 76, 80, 82, 83, 84, 87, ▸
 ▸90, 92}
...
99989  42840  {2, 3, 8, 10, 11, 13, 14, ..., 99978, 99979, 99981, 9998▸
 ▸6, 99987}
99991  24000  {65539, 6, 65546, 11, 12, ..., 65518, 65520, 87379, 6552▸
 ▸6, 65528}
```

f. SageMath Example 5.26

The example 5.26 calculates all primitive roots for all primes up to one million, and outputs for each prime number four values: the prime number, the number of different primitive roots, and its smallest and its biggest primitive root.

SageMath Example 5.26: Code to Generate the Database with the Smallest Primitive Root for All Primes Between 1 and 1000000

```
print("\n# CHAP05 -- Sage-Script-SAMPLE 140: =========")
...
my_import("chap05_sample100", "enum_PrimitiveRoots_of_an_Integer")

import time
StartTime = time.time()
start = 1    # normal value: 1 // test value: 10^5+3
end = start+200  #  normal value: 10^6 // start+200 test value
fileName = "./primroot-smallest_up-to-one-million.dat"
print("...Creating file %s with start=%d, end=%d" % (fileName, start, end))
file = open(fileName, "w")
file.write("Used parameters:  start = " + str(start) + ",  end = " + str(end) + "\n")
file.write("---- StartTime:     " + str(StartTime) + " sec ----\n")
file.flush()
for p in primes(start, end+1):
    L = enum_PrimitiveRoots_of_an_Integer(p)
    # - To commandline, output only p and number of prim roots of Z_p
    # print(p, len(L)) # just temporarily active to see where you are while testing
    # - To file, output much more in the following format:
    # (1) the prime number p
    # (2) the number of primitive roots of Z_p
    # (3) the smallest primitive root of Z_p
    # (4) the largest  primitive root of Z_p
    LL = sorted(L) # sort necessary as the smallest primroot is
                   # not always found first (see L of p=43)
    file.write(str(p) + " " + str(len(L)) + " " + str(LL[0]) + " " + str(LL[-1]) + "\n▸
     ▸")
    file.flush()

EndTime = time.time(); EllapsedTime = EndTime-StartTime
file.write("---- EndTime:       " + str(EndTime) + " sec ----\n")
```

SageMath Example 5.26 (continued)

```
file.write("---- EllapsedTime: " + str(EllapsedTime) + " sec ----\n")
file.flush()
file.close()
```

SageMath Example 5.26 was stopped after several weeks (running on a modern PC with SageMath 7.2) after investigating all primes up to half a million. The result was stored in the file `primroot_number-of-and-smallest_up-to-prime-500107.dat`, which is 617 kB uncompressed, and 178 kB compressed with 7Zip. You can find the compressed file at `https://www.cryptool.org/download/ctb/primroot_number-of-and-smallest_up-to-prime-500107.7z`.

This file contains all primes p between 1 and 500,107 together with the corresponding number of primitive roots and the corresponding smallest prime root mod p. It holds that the number of primitive roots (for $p > 3$) is always an odd number. The number of primitive roots modulo a prime p is always equal to $\phi(\phi(p)) = \phi(p-1)$ because the set of in $\mathbb{Z}_p$ invertible elements forms the cyclic group $\mathbb{Z}_p^*$ and this group has $p-1$ elements and $\phi(p-1)$ generators, the latter being exactly the primitive roots modulo p. A cyclic group of order n has always $\phi(n)$ generators. A proof for this can be found in [53, p. 36].

So this file may be interesting to some number theorists. Its content looks like this:

```
2       1       1
3       1       2
5       2       2
7       2       3
11      4       2
13      4       2
17      8       3
...
99989   42840   2
99991   24000   6
100003  28560   2
...
500069  250032  2
500083  151520  2
500107  156864  2
```

If you are looking `only` for the smallest primitive root, then this script could be accelerated dramatically by applying mathematical theory and searching more directly for possible candidates (instead of first generating all primitive roots with `enum_PrimitiveRoots_of_an_Integer`).

g. SageMath Example 5.27

The database file `primroots_1-100000.dat` from the SageMath Example 5.27 then was used as input to create three graphics using the example 5.27). For a change, here the Sage code is from the CLI instead of from a Sage script.

SageMath Example 5.27: Code to Generate the Graphics about the Primitive Roots (for Big Datasets)

```
sage: # open a database file on primitive roots for p between 1 ▸
 ▸and 100,000
sage: file = open("./primroots.dat", "r")
sage: plist = []     # list of all primes between 1 and 100,000
sage: nlist = []     # number of primitive roots modulo prime p
sage: minlist = []   # smallest primitive root modulo prime p
sage: maxlist = []   # largest primitive root modulo prime p
sage: for line in file:
....:     # get a line from the database file and tokenize it for▸
 ▸ processing
....:     line = line.strip().split(" ", 2)
....:     # extract the prime number p in question
....:     plist.append(Integer(line[0]))
....:     # extract the number of primitive roots modulo p
....:     nlist.append(Integer(line[1]))
....:     # extract the list of all primitive roots modulo p
....:     line = line[-1]
....:     line = line.replace("{", "")
....:     line = line.replace("}", "")
....:     line = line.split(", ")
....:     # sort the list in non-decreasing order
....:     line = [Integer(s) for s in line]
....:     line.sort()
....:     # get the smallest primitive root modulo p
....:     minlist.append(line[0])
....:     # get the largest primitive root modulo p
....:     maxlist.append(line[-1])
....:
sage: file.close()  # close the database file
sage: # plot of number of primitive roots modulo p
sage: nplot = point2d(zip(plist, nlist), pointsize=1)
sage: nplot.axes_labels(["x", "y"])
sage: nplot
sage: # plot of smallest primitive root modulo prime p
sage: minplot = point2d(zip(plist, minlist), pointsize=1)
sage: minplot.axes_labels(["x", "y"])
sage: minplot
sage: # plot of largest primitive root modulo prime p
sage: maxplot = point2d(zip(plist, maxlist), pointsize=1)
sage: maxplot.axes_labels(["x", "y"])
sage: maxplot
```

Figure 5.11 graphs the number of primitive roots for each prime between 1 and 100,000. The x-axis represents primes between 1 and 100,000, while the y-axis counts the number of primitive roots for each prime.

Figure 5.12 graphs the *smallest* primitive roots of all primes between 1 and 100,000. The x-axis again represents primes between 1 and 100,000. The y-axis represents the smallest primitive root of each prime.

Figure 5.13 shows a corresponding graph for the *largest* primitive root of each prime within the same interval between 1 and 100,000.

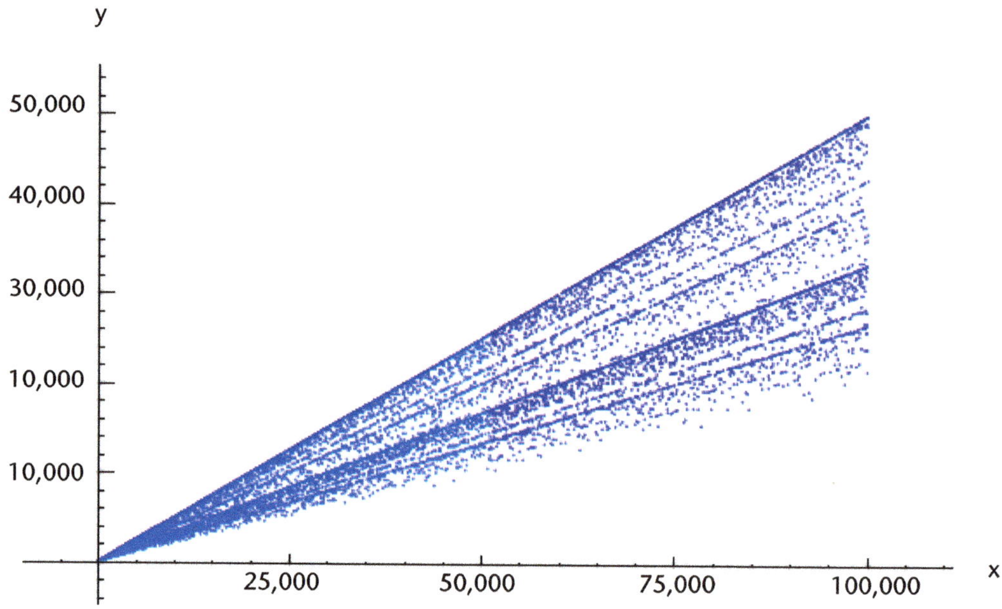

Figure 5.11 The number of primitive roots of all primes between 1 and 100,000.

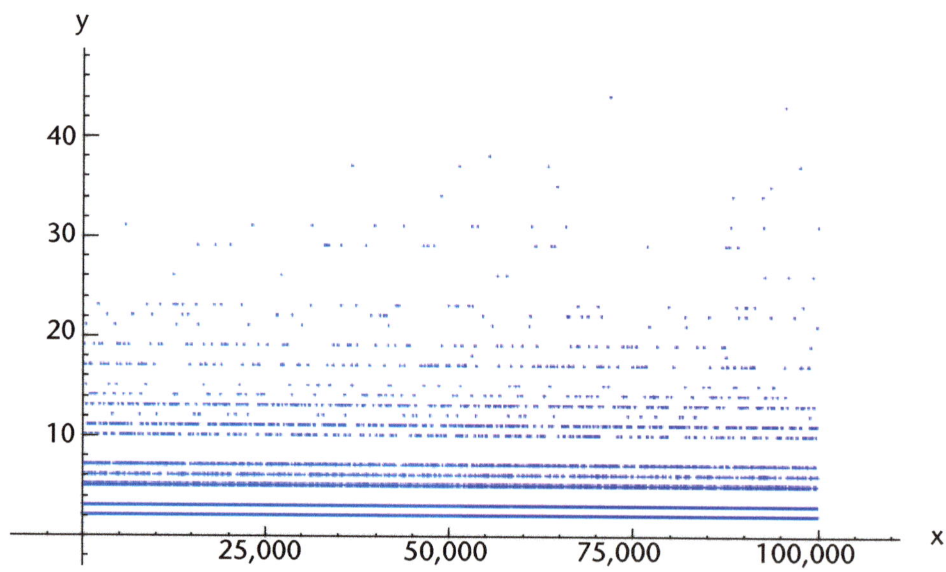

Figure 5.12 The smallest primitive roots of all primes between 1 and 100,000.

h. SageMath Example 5.28

In order to do some experiments, another much smaller dataset was used, which considered only the primitive roots for primes in the range between 1 and 100. The database file `primroots.dat` from SageMath Example 5.24 then was used as input to create three graphics using the SageMath Example 5.28 (Figure 5.14).

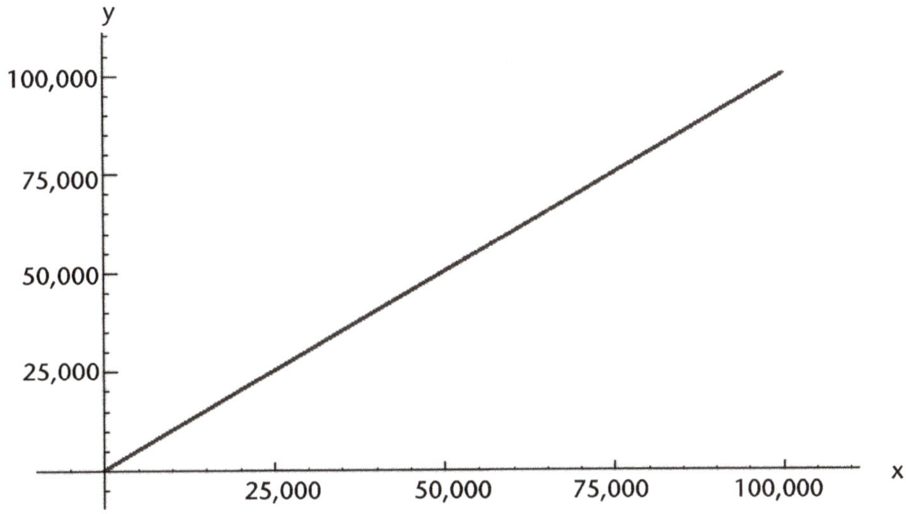

Figure 5.13 The largest primitive roots of all primes between 1 and 100,000.

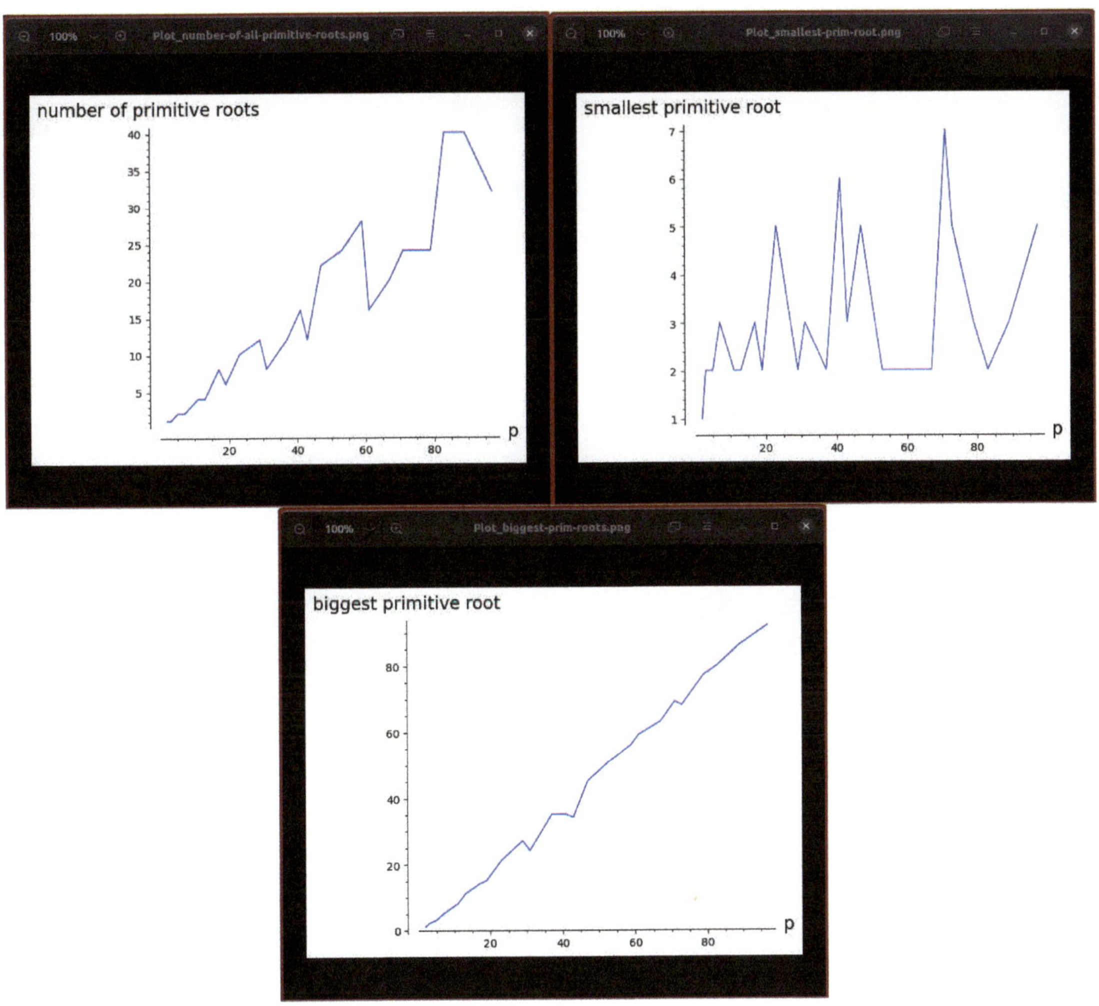

Figure 5.14 Number and smallest and biggest primitive root for all primes up to 100.

SageMath Example 5.28: Code to Generate the Graphics about the Primitive Roots

```
print("\n# CHAP05 -- Sage-Script-SAMPLE 150: =========")

def Display(F,fname):
...

# Open a database file on primitive roots (the file "primroots.dat" used
# here, was created with the script chap05_sample130.sage where end=100)
file = open("./primroots.dat", "r")
plist = []     # list of all primes between 1 and 100
nlist = []     # number of primitive roots modulo prime p
minlist = []  # smallest primitive root modulo prime p
maxlist = []  # largest primitive root modulo prime p
for line in file:
    # get a line from the database file and tokenize it for processing
    line = line.strip().split(" ", 2)
    # extract the prime number p in question
    plist.append(Integer(line[0]))
    # extract the number of primitive roots modulo p
    nlist.append(Integer(line[1]))
    # extract the list of all primitive roots modulo p
    line = line[-1]
    line = line.replace("{", "")
    line = line.replace("}", "")
    line = line.split(", ")
    # sort the list in non-decreasing order
    line = [Integer(s) for s in line]
    line.sort()
    # get the smallest primitive root modulo p
    minlist.append(line[0])
    # get the largest primitive root modulo p
    maxlist.append(line[-1])

file.close()  # close the database file
print("length :", len(nlist))    # print just for test purposes
print("plist  :", plist)    # print just for test purposes
print("nlist  :", nlist)    # print just for test purposes
print("minlist:", minlist)  # print just for test purposes
print("maxlist:", maxlist)  # print just for test purposes
print("zipped :", list(zip(plist, nlist))) # just for test purposes

# Generate 3 graphics:
# 1) Plot of number of primitive roots modulo p
# a) Either plot with the 2D plotting class "point2d()" built in sage
#    Remark 1: If you have a smaller primes range, use bigger
#              pointsize values or get rid of this parameter
#              For huge sets, "pointsize=1" is fine.
#    Remark 2: point2d() has no option "plotjoined=True"
#    nplot = point2d(zip(plist, nlist))
#    nplot = point2d(zip(plist, nlist), pointsize=1)
# b) or plot with "list_plot()"
#    Remark 3: "list_plot(list(zip(plist, nlist))" needs cast with list()
#              and has same result as "point2d(zip(plist, nlist))"
#    Remark 4: list_plot() has option "plotjoined=True" to connect the
#              points. This gives a better impression for smaller sets.
nplot = list_plot(list(zip(plist, nlist)), plotjoined=True)
nplot.axes_labels(["p", "number of primitive roots"])
Display(nplot, "Plot_number-of-all-primitive-roots")

# 2) Plot of smallest primitive root modulo prime p
## minplot = point2d(zip(plist, minlist), pointsize=1)
minplot = list_plot(list(zip(plist, minlist)), plotjoined=True)
minplot.axes_labels(["p", "smallest primitive root"])
Display(minplot, "Plot_smallest-prim-root")

# 3) Plot of largest primitive root modulo prime p
## maxplot = point2d(zip(plist, maxlist), pointsize=1)
maxplot = list_plot(list(zip(plist, maxlist)), plotjoined=True)
maxplot.axes_labels(["p", "biggest primitive root"])
Display(maxplot, "Plot_biggest-prim-roots")
```

Figure 5.14 contains three graphs. The x-axis always represents the primes between 1 and 100. In the top-left graph the y values are the number of primitive roots for each prime. In the graph in the top-right, the y value represents the smallest primitive root for the corresponding prime number. In the bottom graph, the y value represents the biggest primitive root for the corresponding prime number.

Compared to SageMath Example 5.27, here `list_plot()` is used which offers the option `plotjoined`. This option connects the single points with lines, which only makes sense for small sets of points.

i. SageMath Example 5.29

SageMath Example 5.29: Code to Generate an in $\mathbb{Z}_n$ Invertible Element of Maximal Order

```
sage: n=45 #change n as desired
sage: l=list(factor(n))
sage: ms=[l[i][0]^l[i][1] for i in range(len(l))]
sage: dim=len(ms)
sage: m=[primitive_root(ms[i]) for i in range(dim)]
sage: maxelt=crt(m,ms)
sage: maxelt
2
```

5.17.5 RSA Examples with SageMath

Below is SageMath source code for the simple RSA examples in Section 5.14.

Example in Section 5.14.2:

SageMath Example 5.30 executes the RSA exponentiation M^{37} (mod 3713) on message $M = 120$.

SageMath Example 5.30: RSA Exponentiation

```
sage: power_mod(120, 37, 3713)
1404
```

Example in Section 5.14.3:

SageMath Example 5.31 executes the factorization of $\phi(256027) = 255016 = 2^3 \cdot 127 \cdot 251$.

SageMath Example 5.31: Factoring a Number

```
sage: factor(255016)
2^3 * 127 * 251
```

Example in Section 5.14.3:

SageMath Example 5.32 encrypts the integer values of a sequence of ASCII characters. SageMath can do RSA encryption as follows:

SageMath Example 5.32: RSA Encryption by Modular Exponentiation of the Number Values of the Characters of a Message

```
sage: A = [82, 83, 65, 32, 119, 111, 114, 107, 115, 33]
sage: e = 65537; m = 256027
sage: [power_mod(a, e, m) for a in A]
[212984, 25546, 104529, 31692, 248407, 100412, 54196, 100184, 58179, 22▸
▸7433]
```

Example in Section 5.14.3:

SageMath Example 5.33: RSA Encryption Using SageMath

```
sage: A = [21075, 16672, 30575, 29291, 29473]
sage: e = 65537; m = 256027
sage: [power_mod(a, e, m) for a in A]
[158721, 137346, 37358, 240130, 112898]
```

Example in Section 5.14.3:

SageMath Example 5.34: RSA Encryption Using SageMath

```
sage: A = [82083, 65032, 119111, 114107, 115033]
sage: e = 65537; m = 256027
sage: [power_mod(a, e, m) for a in A]
[198967, 51405, 254571, 115318, 14251]
```

5.17.6 How Many Private RSA Keys d Exist within a Given Modulo Range?

The RSA encryption procedure was described in Section 5.10.2. Steps 1 to 3 constitute key generation, steps 4 and 5 are the encryption:

1. Select two distinct random prime numbers p and q and calculate $n = p \cdot q$. The value n is called the RSA modulus.
2. Select an arbitrary $e \in \{2, \cdots, n-1\}$ such that:
   e is relatively prime to $\phi(n) = (p-1) \cdot (q-1)$.
   We can then throw away p and q.
3. Select $d \in \{1, \cdots, n-1\}$ with $e \cdot d \equiv 1 \pmod{\phi(n)}$.
   That is, d is the multiplicative inverse of e modulo $\phi(n)$. We can then throw away $\phi(n)$.

   $\rightarrow$ (n, e) is the public key P.

   $\rightarrow$ (n, d) is the private key S (only d must be kept secret).

4. For encryption, the message represented as a (binary) number is divided into parts such that each part of the number represents a number less than n.
5. Encryption of the plaintext (or the parts of it) $M \in \{1, \cdots, n-1\}$:

$$C = E((n,e), M) = M^e \pmod{n}.$$

To crack a given RSA ciphertext C, the default way would be to use the public key of the recipient and to try to factorize n. Then you can go through the steps 2 and 3 and generate the private key e, which is normally used to decrypt a ciphertext.

According to the prime number theorem (see Theorem 4.7), the number of prime numbers $PI(x)$ is asymptotic to $x/ln(x)$. Between 1 and a given n there are about $n/ln(n)$ different primes.

If you don't want to use factorization you may ask a question like in classic encryption: Does an exhaustive search over all keys make sense? Therefore, you may want to find out how many possible private keys (n, d) are there for a given n or for a given range $n \in [a, b]$? Section 5.8.5 deals with the special case $n = 26$.

The general question is answered by the function `count_Number_of_RSA_Keys` (if the modulus is not too big). This function is defined in SageMath Example 5.35.

Some remarks to the code of sample 5.35:

a. Calling `sage: count_Number_of_RSA_Keys(100,1000)` means to consider the interval $[100, 1000]$ for n. We define n by the two primes p and q as: $n = p \cdot q$. So here one prime can have the maximal value 500 because $2 \cdot 500 = 1000$ (while then the other prime will have the smallest possible prime value 2).
   - The number of primes in the given range is 143 (`prime_pi(1000) - prime_pi(100) = 168 - 25`).
   - The number of possible combinations of primes is $comb = 258$.
   - The number of private keys is 34816.

b. Calling `sage: count_Number_of_RSA_Keys(100, 100, True)` has the following output:
   - Number of private keys for modulus in a given range: 0.
   - Number of primes in a given range: 0.

   The reason for that is with this call only $n = 100$ is considered, and the function investigates only semiprime n: 100 is not semiprime, which means 100 is not the product of only two primes.

c. The output of `count_Number_of_RSA_Keys` with either the parameters `(26,26,True)` or `(713,713,True)` shows, that if there is only one possible combination of primes (as only one semiprime n is given), then the number of elements in $\{R2\} \setminus \{1\}$ is the number of possible keys: With $n = 26 = 2 \cdot 13$ there are three possible private keys (there are three elements > 1 in $R2$, so d can be 5, 7, or 11); with $n = 713 = 23 \cdot 31$ there are 159 possible private RSA keys.

The program also calculates the reduced residue systems: The identifiers R' and R'' defined in Section 5.8.5 are referred to as $R1$ and $R2$ in SageMath Example 5.35.

SageMath Example 5.35: How Many Private RSA Keys d Are There If You Know a Range for the Public Key n?

```
print("\n# CHAP05 -- Sage-Script-SAMPLE 160: =========")

def count_Number_of_RSA_Keys(start, end, Verbose=False):
  """
  How many private RSA keys (n,d) exist, if only modulus n is given, and start <= n <= end?
  - prime_range(u,o) delivers all primes >=u und < o
  - verbose = True adds p,q,n, phi, residue systems,
    verbose = 2 in addition prints the primelist and the given ranges.
  """
  a = start; b = end
  s = 0;      comb = 0

  for p in prime_range(1, b/2+1):
    for q in prime_range(p+1, b/2+1):
      n=p*q
      if Verbose>=3:
        print("     Considered primes: p=%d, q=%d, n=%d" % (p,q,n))
      if a <= n and n <= b:
        comb = comb+1
        eu1 = euler_phi(n)
        eu2 = euler_phi(eu1)-1 # Not only determine the count, but also the values of d.
        s = s + eu2
        if Verbose:
          print("p=%s, q=%s, n=%s, eu1=%s, eu2=%s, s=%s" % (p, q, n, eu1, eu2, s))
          R1=n.coprime_integers(n)
          print("R1: ", [i for i in R1])
          R2=[i for i in R1 if gcd(i,eu1)==1 and i <= eu1]
          print("R2: ", [i for i in R2])

  print("Number of private keys d for moduli in given range: %s (#comb=%s), " % (s, comb))

  # Just for comparison: How many primes are in the given moduli range [a,b+1]?
  if Verbose:
    s = 0
    primeslist=[]
    for p in prime_range(a, b+1):
      if Verbose>=2:
        primeslist.append(p)
      s = s + 1
    if Verbose>=2:
      print("   a=%s, " % a + "b=%s, " % b + "\n   primeslist=%s" % primeslist)
    print("   Number of primes in given range: %s" % s)

import time

print("\nDD_Start -- Testcases for count_Number_of_RSA_Keys(start, end)")

print("\n----------Testcase: Modulus n=26, verbose=True (just 1 semi prime number)")
count_Number_of_RSA_Keys(26, 26, True)

print("\n----------Testcase: Modulus n=713, verbose=True (just 1 semi prime number)")
count_Number_of_RSA_Keys(713, 713, True)

print("\n----------Testcase: Moduli range = (100, 107, True)[this range has 3 primes]")
StartTime = time.time()
count_Number_of_RSA_Keys(100, 107, True)
print("Time = %f sec" % (time.time()-StartTime))
...
```

As there are so many private keys (n, d) within a bigger range of values for n, even brute-force factoring would be more efficient than brute-force trying all possible private keys.

5.17.7 RSA Fixed Points $m \in \{1, ..., n-1\}$ with $m^e = m \bmod n$

Encryption methods can also have fixed plaintext messages where the corresponding ciphertext matches the original. In mathematics, variables mapped by the algorithm (function) onto themselves are called fixed points. In cryptography the corresponding messages are called unconcealed messages.

Generally speaking, fixed points are undesirable and there should be only very few compared to the overall range of possible messages. The more fixed points an encryption algorithm contains, the easier it is to break it. To deal with this and with rainbow attacks, in practice random padding like in OAEP is added (instead of using textbook RSA).

With the RSA procedure, $n = pq$ is the product of two different prime numbers, and there exists e where $gcd(e, (p-1)(q-1)) = 1$. The encryption is then $c = m^e \mod n$. A fixed point in the RSA procedure is a message m, where: $m = m^e \mod n$. The result of the encryption is the given message.

When the size of n is sufficiently *big*, the probability of the occurrence of fixed points in RSA is very small, as illustrated in Figure 5.16. Experimentally, we found on average not more than 40 fixed points for a single given RSA parameter set.

Students often presume the occurrence of fixed points is high, because they encounter a relatively large number of fixed point examples when experimenting with *small* prime numbers, as $m = 0, 1, n-1$ are always fixed points with RSA.

In practice, where large prime numbers are chosen, fixed points have no significance for the security of RSA. Therefore, here we refer more to the mathematical questions.

5.17.7.1 The Number of RSA Fixed Points

In this section we show how many RSA fixed points $m \in \{1, \ldots, n-1\}$ exist with $m^e = m \mod n$ for a fixed key e. Theorem 5.17 gives an answer to this question. For the proof of Theorem 5.17 we first need Theorem 5.16.

This theorem characterizes *roots of unity*, these are numbers x that equal 1 when raised to some integer power n. A priori they can be complex numbers but in our context they are integers $\bmod n$: An nth root of unity x is *primitive* if there is no kth root of unity for all integers k smaller than n:

$$x^n = 1 \text{ and } x^k \neq 1 \quad (k = 1, 2, 3, ..., n-1)$$

If F is a finite field and n is a positive integer, then an n-th root of unity in F is a solution of the equation

$$x^n - 1 = 0 \quad \text{in } F$$

Keep in mind that $\mathbb{Z}_n$ is a field if n is prime.

Theorem 5.16 from [54, p. 69] characterizes all $(e-1)$th roots of unity in $\mathbb{Z}_p^*$:

Theorem 5.16 g^α is a $(e-1)$th root of unity in $\mathbb{Z}_p^*$ if and only if $(e-1)\alpha = 0 \mod p-1$. There are $gcd(p-1, e-1)$ of these.

Proof

The first theorem results directly from the little theorem from Fermat (Theorem 5.12):

$$g^{\alpha(e-1)} = 1 \mod p \;\Rightarrow\; \alpha(e-1) = 0 \mod p-1$$

Let $\delta = gcd(p-1, e-1)$. Then $\alpha(e-1) = 0 \mod p-1$ implies $\frac{\alpha(e-1)}{\delta} = 0 \mod \frac{p-1}{\delta}$. Since $\frac{e-1}{\delta}$ and $\frac{p-1}{\delta}$ are coprime (each was reduced by the gcd of their corresponding numerator), α must be a multiple of $\frac{p-1}{\delta}$.

$$\alpha = k \cdot \frac{p-1}{\delta} \text{ for some } k \in \mathbb{Z}$$

For $k = 1, \ldots, \delta$ we get δ different possible values for α. For larger k write $k = r \cdot \delta + k'$ with $k' < \delta$, then we have

$$g^{\alpha} = g^{k \cdot \frac{p-1}{\delta}} = g^{(r \cdot \delta + k') \cdot \frac{p-1}{\delta}} = g^{r \cdot (p-1) + k' \cdot \frac{p-1}{\delta}} = (g^r)^{p-1} \cdot g^{k' \cdot \frac{p-1}{\delta}} = g^{k' \cdot \frac{p-1}{\delta}}$$

and therefore k was already counted by k', so we can assume $k \leq \delta$. The negative case k can be omitted. Indeed for the power g^{α} of the generator g we did not require $\alpha \geq 0$, but we could, because for every natural number $\beta > 0$ we have $g^{-\beta} = (g^{-1})^{\beta}$, this is a positive power of the multiplikative inverse of g. The span of a group element g and its inverse g^{-1} is always the same. So if $\alpha = -\beta$ for some $\beta > 0$, it was already counted.

So then these δ different powers ($k = 1, \ldots, \delta$) correspond to the $(e-1)$th roots of unity $g^{k \cdot \frac{p-1}{\delta}} \bmod p$ in $\mathbb{Z}_p^*$. □

Analog for q: For $m^{e-1} = 1 \bmod q$ we then have $gcd(q-1, e-1)$ many of $(e-1)$th roots of unity.

Theorem 5.17 The number of RSA fixed points $m \in \{1, \ldots, n-1\}$ for an e with $\gcd(e, \varphi(n)) = 1$ is

$$\gcd(p-1, e-1) \cdot \gcd(q-1, e-1).$$

Proof

Given $m^e = m \bmod n$. According to the Chinese remainder theorem (CRT), the following statements are equivalent:

$$[m^e = m \bmod n] \Leftrightarrow [m^e = m \bmod p \text{ and } m^e = m \bmod q]$$

Furthermore, the decomposition on the right side is equivalent to:

$$m^{e-1} = 1 \bmod p \text{ and } m^{e-1} = 1 \bmod q.$$

We consider $m^{e-1} = 1 \bmod p$ and search all $(e-1)$th roots of unity in $\mathbb{Z}_p^*$.

We know that $\mathbb{Z}_p^*$ for a prime p is cyclic. $\Rightarrow$ A generator g exists which produces $\mathbb{Z}_p^*$: $\mathbb{Z}_p^* =< g >$.

Theorem 5.16 delivers the remaining part. □

The number of combinations of the $(e-1)$th root of unity in $\mathbb{Z}_p^*$ and $\mathbb{Z}_q^*$ gives the total quantity of RSA fixed points: $m^e = m \bmod n$ with $m \in \{1, ..., n-1\}$:

$$\gcd(p-1, e-1) \cdot \gcd(q-1, e-1)$$

Adding $m = 0$ to the above, results in the Theorem 5.18:

Theorem 5.18 If $m \in \{0, ..., n-1\}$, then the quantity of the RSA fixed points is:

$$(gcd(p-1, e-1) + 1) \cdot (gcd(q-1, e-1) + 1)$$

5.17.7.2 Lower Bound for the Quantity of RSA Fixed Points

In the following section, we show that there is a lower bound for the quantity of RSA fixed points. This lower bound 6 exists when the two different RSA prime numbers are the smallest possible values (2 and 3).

Theorem 5.19 Given: $p = 2, q = 3$
The quantity of RSA fixed points is $(\underbrace{gcd(p-1, e-1)}_{=1} + 1) \cdot (\underbrace{gcd(q-1, e-1)}_{=2} + 1)$
$= 2 \cdot 3 = 6$

Theorem 5.20 Given: $p \neq q; p > 2, q > 2$
The quantity of RSA fixed points is ≥ 9.

Proof
Each value $(p-1)$ and $(q-1)$ is even for primes $p, q > 2$.
The RSA algorithm requires choosing e so that $1 < e < \phi(n) = (p-1)(q-1)$ and $gcd(e, (p-1)(q-1)) = 1$.
Since $(p-1)$ and $(q-1)$ are even, e is odd $\Rightarrow e - 1$ is even.
Since $(p-1)$ and $(e-1)$ are even, then: $gcd(p-1, e-1) \geq 2$.
$\Rightarrow (gcd(p-1, e-1) + 1) \geq 3$ and $(gcd(q-1, e-1) + 1) \geq 3$.
$\Rightarrow (gcd(p-1, e-1) + 1) \cdot (gcd(q-1, e-1) + 1) \geq 9$. □

Samples with $m \in \{0, ..., n-1\}$:

- For $(e, n) = (17, 6)$ with $6 = 2 \cdot 3$, all six possible messages $\{0, 1, 2, 3, 4, 5\}$ are fixed points (for $n = 6$, it is independent of the value of e).
- For $(e, n) = (17, 10)$ with $10 = 2 \cdot 5$, all 10 possible messages are fixed points.
- For $(e, n) = (19, 10)$, only 6 of the 10 possible messages are fixed points.
- For $(e, n) = (7, 55)$ with $66 = 5 * 11$, 9 of the 55 messages $(\{0, 1, 10, 11, 21, 34, 44, 45, 54\})$ are fixed points. Figure 5.15 visualizes this.

Figure 5.15 was taken from "RSA Visual and More," `https://www.cryptool.org/en/cto/rsa-visual`. In this plugin you can either see with graphics how RSA assigns its input values when encrypting, or you can test textbook RSA with big numbers (and subject keys in a key store to simulate certificate exchange in a pupils' lab), or use RSA with OAEP, padding, and certificates as it is used in practice.

5.17.7.3 Unfortunate Choice of e (Weak or Unsuitable e)

In this section, we show that with $e = 1 + lcm(p-1, q-1)$ each encryption results in a fixed point (independently of the size of p, q, or n, each m is mapped on itself); and then we generalize this to all unfortunate choices of e.

If $e = 1$, then for all m: $c = m^e = m$. This is the trivial case.

Theorem 5.21 Given: $p, q > 2$
If $e = 1 + lcm(p-1, q-1)$, then for all $m \in \{1, ..., n-1\}$: $m^e = m \bmod n$.

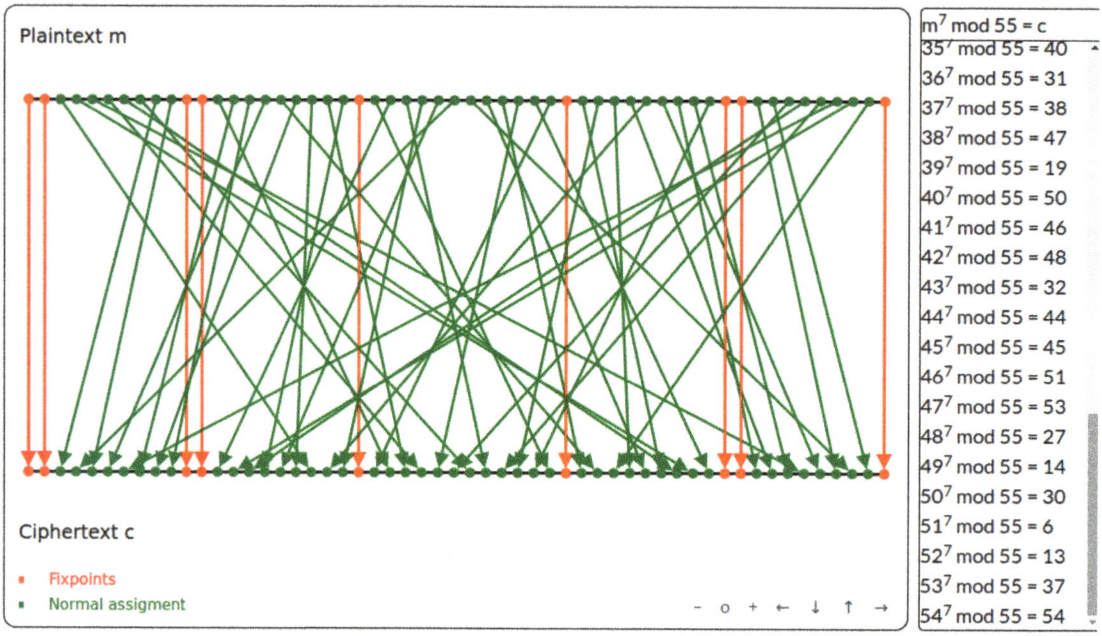

Figure 5.15 CTO: Nine fixed points (in red) with RSA using $(e, n) = (7, 55)$ with $m \in \{0, \ldots, 54\}$.

Proof Given:

- $e \cdot d = 1 \mod \phi(n)$ or $e \cdot d = 1 \mod lcm(p-1, q-1)$
- $m^x \mod n = m^{x \mod \phi(n)} \mod n$

Encryption of messages:

$c = m^e \mod n$, where c is the ciphertext and m is the plaintext.

Decryption of messages:

$m' = c^d \mod n$, where d is the multiplicative inverse of e.

We will show: $c = m \mod n$ for the chosen e.

$$c = m^e \mod n$$
$$c = m^{1+lcm(p-1,q-1)} \mod n$$
$$c = m^1 \cdot m^{k \cdot (p-1) \cdot (q-1)} \mod n$$
$$c = m^1 \cdot m^{[k \cdot \phi(n)] \mod \phi(n)} \mod n$$
$$c = m^1 \cdot m^0 = m \mod n$$

□

Example 1: Fixed Point Property for All m

Given $n = p \cdot q = 13 \cdot 37 = 481$

$\Rightarrow \phi(n) = (p-1)(q-1) = 12 \cdot 36 = 432$

$\Rightarrow e = lcm(p-1, q-1) + 1 = lcm(12, 36) + 1 = 36 + 1 = 37$

With $m \in \{4, 6, 7, 480\}$ we get in $m^e \mod n$ as:

$4^{37} \mod 481 = 4$

$6^{37} \mod 481 = 6$

$7^{37} \mod 481 = 7$

$480^{37} \mod 481 = 480$

There is not just one single e, so that all $m \in \{1, \ldots, n-1\}$ have the fixed point property $m^e = m \bmod n$. Sometimes these e, which make any message to a fixed point, are called weak keys (e, n) of the RSA algorithm. This interpretation is different to the weak keys k in DES, where every message m relates to itself if the encryption is done twice. To my knowledge, for larger n the RSA procedure does not have weaknesses in this meaning: $(m^e)^e = m$. A double operation with the public key also makes no sense with asymmetric procedures.

With JCT `Default Perspective` ▹ `Visuals` ▹ `Inner States of the Data Encryption Standard (DES)` you can find weak DES keys.

Theorem 5.22 The complete fixed point property of all m is valid for every $e = j \cdot lcm(p-1, q-1) + 1$, where $j = 0, 1, 2, 3, 4, \ldots$ to $e \leq \phi(n)$.

Example 2: Further Values for e with Fixed Point Properties
Given $n = p \cdot q = 13 \cdot 37 = 481$ with $lcm(p-1, q-1) = lcm(12, 36) = 36$.
Then, e can have the following values: $e = j \cdot lcm(p-1, q-1) + 1$ for $j = 0, 1, 2, \ldots, 11$:
$\Rightarrow e \in \{1, 37, 73, 109, 145, 181, 217, 253, 289, 325, 361, 397\}$.

Starting with $j = 12$, it holds: $e = 12 \cdot lcm(12, 36) + 1 = 432 + 1 = 433 > 432 = \phi(n)$.

Checking the same four values for m as in the example 1 above, but now with $e = 217$, the results are:

$$\begin{aligned}
4^{217} &\mod 481 = 4 \\
6^{217} &\mod 481 = 6 \\
7^{217} &\mod 481 = 7 \\
480^{217} &\mod 481 = 480
\end{aligned}$$

Theorem 5.23 The number of possible values for e with $m^e = m \bmod n$ can be computed by:

$$[\text{Quantity } e] = \left\lfloor \frac{\phi(n)}{lcm(p-1, q-1) + 1} \right\rfloor + 1 = \frac{\phi(n)}{lcm(p-1, q-1)}$$

In our example, this results in $\frac{432}{lcm(12,36)} = 12$ different values for e, where $m^e = m \bmod n$ for all m in $\mathbb{Z}_{481}$.

5.17.7.4 An Empirical Estimate of the Quantity of Fixed Points for Growing Moduli

In this section, we make an empirical estimate of the quantity of fixed points for growing moduli (here we did *not* use weak e as in Section 5.17.7.3).

For this, we randomly choose p and q from the six following ranges each characterized by its lower and upper bound: $(2^2, 2^{10})$, $(2^{10}, 2^{20})$, $(2^{20}, 2^{40})$, $(2^{40}, 2^{80})$, $(2^{80}, 2^{160})$, $(2^{160}, 2^{320})$.

For each range, 10 attempts were made. For the exponent e, the standard value $e = 2^{16} + 1$ was always chosen. The quantity of fixed points for all 60 attempts was computed with SageMath Example 5.36.

Figure 5.16 shows that the average number of fixed points was not greater than 40 in any of the six size ranges.

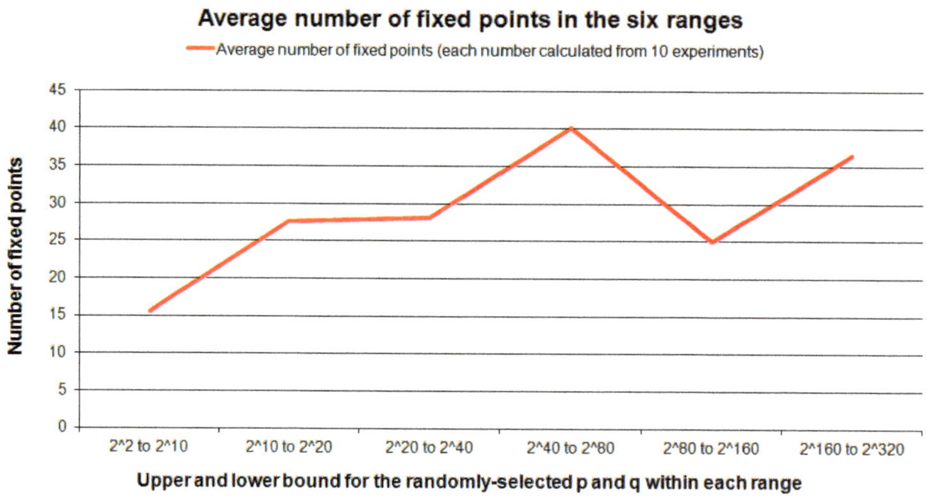

Figure 5.16 An empirical estimate of the quantity of fixed points for growing moduli.

Table 5.18 15 RSA Fixed Points for the Public Key $(n, e) = (866959, 17)$

0	1	23518	23519	47037
188964	212482	236000	630959	654477
677995	819922	843440	843441	866958

5.17.7.5 Example: Determine All Fixed Points for a Specific Public RSA Key

The exercise is to determine all fixed points for $(n, e) = (866959, 17)$.

Solution:
We start by factoring n: $866959 = 811 \cdot 1069$.
The quantity of RSA fixed points results from Theorem 5.18:

$(gcd(p-1, e-1)+1) \cdot (gcd(q-1, e-1)+1) = (gcd(811-1, 17-1)+1) \cdot (gcd(1069-1, 17-1)+1) = (2+1) \cdot (4+1) = 15$

SageMath Example 5.36 generates the 15 fixed points for $(n, e) = (866959, 17)$ shown in Table 5.18.

Sample validating the fixed point property for $m = 843441$: $843441^{17} \bmod 866959 = 843441$.

So $m = 843441$ is actually a fixed point for the given public key (n, e).

SageMath Example 5.36: Determine All Fixed Points for a Specific Public RSA Key

```
print("\n# CHAP05 -- Sage-Script-SAMPLE 170: =========")
print("--- Search for fixpoints in textbook RSA given p, q, e ---")

import numpy

### EDIT BEGIN ### --- Edit e,p,q here
e=17; p=811; q=1069
```

SageMath Example 5.36 (continued)

```
# e=7;  p=5;   q=11    # 4th sample from chapter 5.20.7.2
### EDIT END ###

fp=numpy.array([0])
fq=numpy.array([0])

n=p*q
print("Prime p:", p)
print("Prime q:", q)
print("Modul n:", n)
print("Public exponent e:", e)

r=Integers(p)
gen_f_p = r.multiplicative_generator(); print("\nGenerator of f_p:", gen_f_p)
s=Integers(q)
gen_f_q = s.multiplicative_generator(); print("Generator of f_q:", gen_f_q)

gcd_p = gcd(e-1,p-1)
gcd_q = gcd(e-1,q-1)
print("\ngcd(e-1,p-1):", gcd_p)
print("gcd(e-1,q-1):", gcd_q)
print("Number of fixpoints:", (gcd_p+1)*(gcd_q+1))

print("\nFixpoints modulo p:")
print("0 (this trivial fixpoint added manually)")
i=0;
for i in range(gcd_p):
    fix_p = power_mod(gen_f_p,Integer(i*(p-1)/gcd_p),p); print(fix_p)
    fp = numpy.append(fp,fix_p)

print("\nFixpoints modulo q:")
print("0 (this trivial fixpoint added manually)")
j=0;
for j in range(gcd_q):
    fix_q = power_mod(gen_f_q,Integer(j*(q-1)/gcd_q),q); print(fix_q)
    fq = numpy.append(fq,fix_q)

print("\nFixpoints for the public RSA key (n,e) = (%d, %d):" % (n, e) )
for r in fp:
    for s in fq:
        print(crt(Integer(r),Integer(s),Integer(p),Integer(q)))

print("\nRemark: You can verify each fixpoint with power_mod(m,e,n).")
```

Meaning of the Variables in SageMath Example 5.36

- `gen_f_p = r.multiplicative_generator()`
  r is a residue class ring modulo p, and `multiplicative_generator()` returns a generator element that was created by the ring modulo p.
- `power_mod(gen_f_p,Integer(i*(p-1)/gcd_p),p)`
  The `power_mod` function raises a number m to the power of e, and returns the results modulo n.
  Sample: `power_mod(m,e,n) := m^e modulo n`
- `numpy.append(fp,power_mod(gen_f_p,Integer(i*(p-1)/gcd_p),p))`
  Die append function extends an array fp by an additional element.
- `crt(Integer(r),Integer(s),Integer(p),Integer(q))`
  CRT is the acronym for the Chinese remainder theorem. `crt(r,s,p,q)` solves the congruences $x \equiv r \mod p$ and $x \equiv s \mod q$ with the help of the CRT.

References

[1] Bauer, F. L., *Entzifferte Geheimnisse*, Berlin: Springer, 1995.

[2] Bauer, F. L., *Decrypted Secrets*, Second Edition, Berlin: Springer, 2000.

[3] Beutelspacher, A., *Kryptologie*, Fifth Edition, Vieweg, 1996.

[4] SageMath Contributors,SageMath Landing Page, https://www.sagemath.org.

[5] Wiles, A., "Modular Elliptic Curves and Fermat's Last Theorem," *Annals of Mathematics*, Vol. 141, 1995.

[6] Pfleeger, C. P., *Security in Computing*, Second Edition, Prentice-Hall, 1997.

[7] Knuthm, D. E., *The Art of Computer Programming, Volume 2: Seminumerical Algorithms*, Third Edition, Boston: Addison-Wesley, 1998.

[8] Lang, S., *Algebra*, Third Edition, Reading, MA: Addison-Wesley, 1993.

[9] Geyer, W.-D., "Wie erkennt man Primzahlen? (How to Identify Prime Numbers?)," handout, Vortrag zur Lehrerfortbildung, Erlangen, 1982.

[10] Hardy, G. H., and E. M. Wright, *An Introduction to the Theory of Numbers*, Second Edition, Oxford, UK: Oxford University Press, Clarendon Press, 1945.

[11] Stinson, D. R., *Cryptography–Theory and Practice*, Third Edition, Boca Raton, FL: Chapman & Hall/CRC, 2006.

[12] BSI,*Technical Guideline TR-02102-1, Cryptographic Mechanisms: Recommendations and Key Lengths* (Version 2022-01), Tech. Rep, 2022, https://www.bsi.bund.de/SharedDocs/-Downloads/EN/BSI/Publications/TechGuidelines/TG02102/BSI-TR-02102-1.pdf.

[13] Buchmann, J., *Einführung in die Kryptographie*, Sixth Edition, Springer, 2016.

[14] Barthe, G., et al., "Beyond provable Security Verifiable IND-CCA Security of OAEP," in *Cryptographers' Track at the RSA Conference*, 2011, Springer, pp. 180–196.

[15] Bourseau, F., D. Fox, and C. Thiel. "Vorzüge und Grenzen des RSA-Verfahrens," *Datenschutz und Datensicherheit (DuD)*, Vol. 26, 2002, pp. 84–89, https://www.secorvo.de/publikationen/rsa-grenzen-fox-2002.pdf.

[16] Pomerance, C., "The Quadratic Sieve Factoring Algorithm," in *Proceedings of Crypto '84, LNCS 196*, (G.R. Blakley and D. Chaum, eds.), Springer, 1984, pp. 169–182.

[17] J. P. Buhler, H. W. Lenstra, and C. Pomerance. "Factoring Integers with the Number Field Sieve," in *The Development of the Number Field Sieve, Lecture Notes in Mathematics*, Vol. 1554 (K. Lenstra and H.W. Lenstra, eds.), Springer, 1993, pp. 50–94.

[18] Kleinjung, T., et al., *Factorization of a 768-Bit RSA Modulus, Version 1.4,* 2010, http://eprint.iacr.org/2010/006.pdf.

[19] Alpern, D., *Integer Factorization Calculator*, https://www.alpertron.com.ar/ecm.htm.

[20] Clavijo, D., *daedalus RsaCtfTool*, https://github.com/RsaCtfTool/RsaCtfTool.

[21] Bernstein, D. J., N. Heninger, and T. Lange. *LatticeHacks*, https://latticehacks.cr.yp.to/rsa.html.

[22] Schneider, M., "Analyse der Sicherheit des RSA-Algorithmus. Mögliche Angriffe, deren Einfluss auf sichere Implementierungen und ökonomische Konsequenzen," MA thesis,Universität Siegen, 2004, https://www.cryptool.org/assets/ctp/documents/Diplomarbeit_Schneider.pdf.

[23] Lenstra, A. K., and E. R. Verheul, *Selecting Cryptographic Key Sizes (1999 + 2001), Journal of Cryptology*, Vol. 14, 2001, pp. 255–293, https://www.cs.ru.nl/E.Verheul/papers/Joc2001/joc2001.pdf.

[24] *Integer Factorization Records*, https://en.wikipedia.org/wiki/Integer_factorization_records.

[25] *RSA Factoring Challenge*, https://en.wikipedia.org/wiki/RSA_Factoring_Challenge.

[26] Kruse Andersen, J., *Largest Consecutive Factorizations*, http://primerecords.dk/consecutive_factorizations.htm.

[27] Tervooren. M., factordb.com, http://factordb.com/index.php?query=2%5C%5E1193-1.

[28] *MysteryTwister (MTC3)*, https://www.mysterytwister.org.

[29] Wagstaff, S., *The Cunningham Project*, https://homes.cerias.purdue.edu/_ssw/cun/.

[30] Brillhart, J., et al., *Factorizations of bn ±1, b = 2, 3, 5, 6, 7, 10, 11, 12 Up to High Powers*, AmericanMathematical Society, 2002, http://citeseerx.ist.psu.edu/viewdoc/download?doi=10.1.1.121.4467&rep=rep1&type=pdf.

[31] Schulz. R.-H., and H. Witten, "Zeitexperimente zur Faktorisierung. Ein Beitrag zur Didaktik der Kryptographie," *LOG IN*, Vols. 166/167, 2010, pp. 113–120, https://informatik.schule.de/krypto/.

[32] Ziegler, G. M., *The Great Prime Number Record Races,Notices of the AMS*, Vol. 51, 2004, pp. 414–416, http://www.ams.org/notices/200404/comm-ziegler.pdf, and https://www.mi.fu-berlin.de/math/groups/discgeom/ziegler/Preprintfiles/095e-PREPRINT.pdf.

[33] Wagstaff, S. S. Jr., "The Cunningham Project," *Fields Institute Communications*, 2000, pp. 1–12, https://homes.cerias.purdue.edu/~ssw/cun1.pdf.

[34] Wagstaff, S., *The Cunningham Project*, https://homes.cerias.purdue.edu/~ssw/cun/notat.txt.

[35] Bernstein, D. J., *Circuits for Integer Factorization: A Proposal*, 2001, https://cr.yp.to/papers/nfscircuit.ps url2: https://cr.yp.to/djb.html.

[36] Lenstra, A. J., et al., *Analysis of Bernstein's Factorization Circuit*, 2002, https://tau.ac.il/~tromer/papers/meshc.pdf.

[37] RSA Security, *Has the RSA Algorithm Been Compromised as a Result of Bernstein's Paper?* Tech. Rep., RSASecurity, April 2002, http://www.networkdls.com/Articles/bernstein.pdf.

[38] Shamir, A., and E. Tromer, *Factoring Large Numbers with the TWIRL Device*, 2003, https://cs.uwec.edu/~tan/priv/www-docs/cs376/Readings/twirl2003.pdf.

[39] Shamir, A., and E. Tromer, "On the Cost of Factoring RSA-1024," *RSA Laboratories CryptoBytes*, Vol. 6, No. 2, 2003, pp. 11–20, https://www.tau.ac.il/~tromer/papers/cbtwirl.pdf.

[40] Weis, R., S. Lucks, and A. Bogk, "Sicherheit von 1024 bit RSA-Schlüsseln gefährdet," *Datenschutz und Datensicherheit (DuD)*, Vol. 27, No. 6, 2003, pp. 360–362.

[41] BSI, *Technische Richtlinie TR-02102-1, Kryptographische Verfahren: Empfehlungen und Schlüssellängen (Version 2022-01)*, Tech. Rep., 2022, https://www.bsi.bund.de/SharedDocs/Downloads/DE/BSI/Publikationen/TechnischeRichtlinien/TR02102/BSI-TR-02102.pdf.

[42] Agrawal, M., N. Kayal, and N. Saxena, *PRIMES in P*, corrected version, August 2002, https://www.cse.iitk.ac.in/users/manindra/algebra/primality_v6.pdf/, and http://fatphil.org/maths /AKS/.

[43] Lenstra, A. K., et al., "Ron Was Wrong, Whit Is Right, A Sanity Check of Public Keys Collected on the Web," *Cryptology ePrint Archive*, February 2012, https://eprint.iacr.org/2012/064.pdf.

[44] Esslinger, B., J. Schneider, and V. Simon, "RSA–Sicherheit in der Praxis," in *KES Zeitschrift für Informationssicherheit*, 2012.2, April 2012, pp. 22–27, https://www.cryptool.org/assets/ctp /documents/kes_2012_RSA_Sicherheit.pdf.

[45] CrypTool Contributors, *Testing RSA Moduli for Shared Prime Factors*, https://www.cryptool.org/en/posts/2012-05-24/rsa-sanity-check.

[46] Heninger, N., et al., "Mining Your Ps and Qs: Detection of Widespread Weak Keys in Network Devices," in *Proceedings of the 21st USENIX Security Symposium*, August 2012, https://factorable.net/paper.html.

[47] Bernstein, D. J., "Factoring into Coprimes in Essentially Linear Time," *Journal of Algorithms*, Vol. 54, 2005, https://cr.yp.to/lineartime/dcba-20040404.pdf.

[48] Oechslin, P., *Making a Faster Cryptanalytic Time-Memory Trade-Off*, Tech. Rep., Crypto 2003, https://lasecwww.epfl.ch/pub/lasec/doc/Oech03.pdf.

[49] Schneier, B., *Applied Cryptography, Protocols, Algorithms, and Source Code in C*, Second Edition, Indianapolis, IN: John Wiley & Sons, 1996.

[50] Eckert, C.,*IT-Sicherheit: Konzepte-Verfahren-Protokolle*, Ninth Edition, De Gruyter Oldenbourg, 2014.

[51] Yan, S. Y., *Number Theory for Computing*, Berlin: Springer, 2000.

[52] Nguyen, M. V., *Number Theory and the RSA Public Key Cryptosystem–An Introductory Tutorial on Using SageMath to Study Elementary Number Theory and Public Key Cryptography*, 2009, https://faculty.washington.edu/moishe/hanoiex/Number%20Theory%20Applications/numtheory-crypto.pdf.

[53] Hungerford, T. W., *Algebra*, New York: Springer, 1974.

[54] Katzenbeisser, S., *Recent Advances in RSA Cryptography*, New York: Springer, 2001.

[55] Giry, D., *BlueKrypt: Cryptographic Key Length Recommendation*, Version 32.3, May 2020, https://www.keylength.com/.

CHAPTER 6

The Mathematical Ideas Behind Modern Asymmetric Cryptography

This chapter provides an introduction to the algorithms of modern cryptography (mostly asymmetric cryptography as it appeared before the usage of post-quantum cryptography and of multiparty communication). Individual parts of this introduction are deepened or extended in further chapters, like elliptic curves in Chapter 8.

You can find corresponding functions within the programs CT1, CT2, JCT, and CTO: see the list of the included functions within Sections A.1, A.2, A.3, and A.4, and at the website `https://www.cryptool.org/en/documentation/functionvolume`.

Section 6.5 introduces the RSA plane. In principle, this is a two-dimensional graphic interpretation of the Chinese remainder theorem in the case of RSA. Many statements about RSA can be understood intuitively in the RSA plane without having to have previous knowledge of algebra, for example the Pollard $p-1$ algorithm for factoring natural numbers. Numerous graphics and examples lead step-by-step to an understanding of the RSA encryption process and associated mathematics, especially group theory. And although the meaning of the Chinese remainder theorem for RSA really is nothing new it seems that an explicit graphical model for this doesn't exist in the literature to date.

6.1 One-Way Functions with Trapdoor and Complexity Classes

A *one-way function* is a function that can be calculated efficiently, but whose inverse is extremely complicated and practically impossible to calculate.

To put it more precisely: A one-way function is a mapping f from a set X to a set Y, such that $f(x)$ can be calculated easily for each element x of X, whereas for (almost) every y from Y it is practically impossible to find an inverse image x (i.e., an x where $f(x) = y$).

An everyday example of a one-way function is a telephone book: the function to be performed is to assign a name to the corresponding telephone number. This can be done easily due to the fact that the names are sorted alphabetically. However, the inverse function—assigning a name to a given number—is obviously difficult if you only have a telephone book available.

One-way functions play a decisive role in cryptography. Almost all cryptographic terms can be rephrased using the term one-way function. Let's take for example public-key encryption (asymmetric cryptography):

Each subscriber T to the system is assigned a key pair: a private key d_T and a public key e_T. These keys must have the following property (public-key property):

For an opponent who knows the public key e_T, it is practically impossible to determine the private key d_T.

In order to construct useful public-key procedures, therefore, we look for a one-way function that is easy to calculate in one direction but is difficult (practically impossible) to calculate in the other direction, provided that a particular piece of additional information (trapdoor) is not available. This additional piece of information allows the inverse to be found efficiently. Such functions are called *trapdoor one-way functions*. In the above case, the one-way function is the encryption via exponentiation with the public key e_T as exponent. The private key d_T is the trapdoor information.

In this process, we describe a problem as "easy" if it can be solved in polynomial time as a function of the length of the input. If the length of the input is n bit, then the time for calculating the function is proportional to n^a, where a is a constant. We say that the complexity of such problems is $O(n^a)$ (Landau or big-O notation).

If you compare two functions 2^n and n^a, where a is a constant, then there always exists a value for n, from which for all further n it applies: $n^a < 2^n$. The function n^a has a lower complexity. So for $a = 5$ the following applies: From the length $n = 23$, 2^n is greater than n^5; for further n, 2^n clearly increases more quickly $[(2^{22} = 4194304,\ 22^5 = 5153632),\ (2^{23} = 8388608,\ 23^5 = 6436343),\ (2^{24} = 16777216,\ 24^5 = 7962624)]$.

The term "practically impossible" is slightly less precise. In general, we can say that a problem cannot be solved efficiently if the time required to solve it increases more quickly than the polynomial time as a function of the size of the input. If, for example, the length of the input is n bits and the time required for calculating the function is proportional to 2^n, then the following currently applies: The function cannot practically be calculated for $n > 80$.

In order to develop a public-key procedure that can be implemented in practice, it is therefore necessary to discover a suitable trapdoor one-way function.

In order to tidy things up among this confusing multitude of possible problems and their complexities, we group problems with similar complexities into classes.

The most important complexity classes are the classes **P** and **NP**:

- The class **P**: This class contains those problems that can be solved in a polynomial amount of time.
- The class **NP**: The definition of this class doesn't look at the time required to solve a problem, but rather at the time required to verify a given solution. The class **NP** consists of those problems for which a given solution can be verified in a polynomial amount of time. Hereby, the term **NP** "nondeterministic" means polynomial and is based on a calculation model (i.e., on a computer that only exists in theory and can guess correct solutions nondeterministically then verify them in polynomial time).

The class **P** is contained in the class **NP**. A well-known unsolved problem is whether or not **P** $\neq$ **NP** is true (i.e., whether or not **P** is a true subset). An important property of the class **NP** is that it also contains what are known as **NP**-complete problems. These are problems that represent the class **NP** as follows: If a "good" algorithm for such a problem exists, then "good" algorithms exist for all problems from **NP**. In particular, if **P** only contained one complete problem (i.e., if a polynomial solution algorithm existed for this problem), then **P** would be equal to **NP**. In this sense, the **NP**-complete problems are the most difficult problems in **NP**.

Many cryptographic protocols are formed in such a way that the "good" subscribers only have to solve problems from **P**, whereas a perpetrator is faced with problems from **NP**.

Unfortunately, we do not yet know whether one-way functions actually exist. However, we can prove that one-way functions exist if and only if $\mathbf{P} \neq \mathbf{NP}$ [1, p. 63].

Some mathematicians have again and again claimed to have proven this equivalence, but so far the claims have always turned out to be false [2].

A number of algorithms have been suggested for public-key procedures. In many cases—although they at first appeared promising—it was discovered that they could be solved in polynomial time. The most famous failed applicant is the knapsack with trapdoor suggested by Ralph Merkle [3]. Also see Section 11.8.1.

6.2 Knapsack Problem as a Basis for Public-Key Procedures

An exhaustive treatment of knapsacks can be found in *Knapsack Problems* by Kellerer, Pferschy, and Pisinger [4]. In Section 11.8.1 you can find further information and also a challenge with knapsacks.

6.2.1 Knapsack Problem

You are given n objects $G_1, \ldots, G_n$ with the weights $g_1, \ldots g_n$ and the values $w_1, \cdots, w_n$. The aim is to carry away as much as possible in terms of value while restricted to an upper weight limit g. You therefore need to find a subset of $\{G_1, \cdots, G_n\}$, that is $\{G_{i_1}, \ldots, G_{i_k}\}$, so that $w_{i_1} + \cdots + w_{i_k}$ is maximized under the condition $g_{i_1} + \cdots + g_{i_k} \leq g$.

Such questions belong to the NP-complete problems (not deterministically polynomial) that are difficult to calculate.

A special case of the knapsack problem is:

Given the natural numbers $a_1, \ldots, a_n$ and g. Find $x_1, \ldots, x_n \in \{0, 1\}$ where $g = \sum_{i=1}^{n} x_i a_i$ (i.e., where $g_i = a_i = w_i$ is selected). This problem is also called a *0-1 knapsack problem* and is identified with $K(a_1, \ldots, a_n; g)$.

Two 0-1 knapsack problems $K(a_1, \ldots, a_n; g)$ and $K(a'_1, \ldots, a'_n; g')$ are called congruent if two coprime numbers w and m exist in such a way that

(1) $m > \max\{\sum_{i=1}^{n} a_i, \sum_{i=1}^{n} a'_i\}$;

(2) $g \equiv wg' \mod m$;

(3) $a_i \equiv wa'_i \mod m$ for all $i = 1, \ldots, n$.

Comment:
Congruent 0-1 knapsack problems have the same solutions. No quick algorithm is known for clarifying whether two 0-1 knapsack problems are congruent.

A 0-1 knapsack problem can be solved by testing the 2^n possibilities for $x_1, \ldots, x_n$. The best method requires $O(2^{n/2})$ operations, which for $n = 100$ with $2^{100} \approx 1.27 \cdot 10^{30}$ and $2^{n/2} \approx 1.13 \cdot 10^{15}$ represents an insurmountable hurdle for computers. However, for special $a_1, \ldots, a_n$ the solution is quite easy to find; for example, for $a_i = 2^{i-1}$. The binary representation of g immediately delivers $x_1, \ldots, x_n$. In general, the 0-1 knapsack problem can be solved easily if a permutation π of $1, \ldots, n$ exists with $a_{\pi(j)} > \sum_{i=1}^{j-1} a_{\pi(i)}$ with $j = 1, \ldots, n$. If, in addition, π is the identity; that is, $\pi(i) = i$ for $i = 1, 2, \ldots, n$, then the sequence $a_1, \ldots, a_n$ is said to be super-increasing. Crypto Procedure 6.1 solves the knapsack problem with a super-increasing sequence in the time of $O(n)$.

Crypto Procedure 6.1: Solving Knapsack Problems with Super-Increasing Weights

for $i = n$ **to** 1 **do**
    if $T \geq a_i$ **then**
        $T := T - s_i$
        $x_i := 1$
    else
        $x_i := 0$
if $T = 0$ **then**
    $X := (x_1, \ldots, x_n)$ is the solution.
else
    No solution exists.

6.2.2 Merkle-Hellman Knapsack Encryption

In 1978, Merkle and Hellman [3] specified a public-key encryption procedure that is based on defamiliarizing the easy 0-1 knapsack problem with a super-increasing sequence into a congruent one with a super-increasing sequence. It is a block ciphering that ciphers an n-bit plaintext each time it runs; see Crypto Procedure 6.2 for the details.

Crypto Procedure 6.2: Merkle-Hellman (Based On Knapsack Problems)

Let $(a_1, \ldots, a_n)$ be super-increasing. Let m and w be two coprime numbers with $m > \sum_{i=1}^{n} a_i$ and $1 \leq w \leq m - 1$. Select $\bar{w}$ with $w\bar{w} \equiv 1 \bmod m$ the modular inverse of w and set $b_i := wa_i \bmod m$, $0 \leq b_i < m$ for $i = 1, \ldots, n$, and verify whether the sequence $b_1, \ldots b_n$ is not super-increasing. A permutation $b_{\pi(1)}, \ldots, b_{\pi(n)}$ of $b_1, \ldots, b_n$ is then published and the inverse permutation μ to π is defined secretly. A sender writes his/her message in

Crypto Procedure 6.2 (continued)

blocks $(x_1^{(j)}, \ldots, x_n^{(j)})$ of binary numbers n in length, calculates

$$g^{(j)} := \sum_{i=1}^{n} x_i^{(j)} b_{\pi(i)}$$

and sends $g^{(j)}, (j = 1, 2, \ldots)$.
The owner of the key calculates

$$G^{(j)} := \bar{w} g^{(j)} \mod m, \quad 0 \leq G^{(j)} < m$$

and obtains the $x_{\mu(i)}^{(j)} \in \{0, 1\}$ (and thus also the $x_i^{(j)}$) from

$$\begin{aligned} G^{(j)} &\equiv \bar{w} g^{(j)} = \sum_{i=1}^{n} x_i^{(j)} b_{\pi(i)} \bar{w} \equiv \sum_{i=1}^{n} x_i^{(j)} a_{\pi(i)} \mod m \\ &= \sum_{i=1}^{n} x_{\mu(i)}^{(j)} a_{\pi(\mu(i))} = \sum_{i=1}^{n} x_{\mu(i)}^{(j)} a_i \mod m \end{aligned}$$

by solving the easier 0-1 knapsack problems $K(a_1, \ldots, a_n; G^{(j)})$ with superincreasing sequence $a_1, \ldots, a_n$.

In 1982, Shamir [5] specified an algorithm for breaking the system in polynomial time without solving the general knapsack problem. Len Adleman [6] and Jeff Lagarias [7] specified an algorithm for breaking the twice iterated Merkle-Hellman knapsack encryption procedure in polynomial time. Ernst Brickell [8] then specified an algorithm for breaking multiply iterated Merkle-Hellman knapsack encryption procedures in polynomial time. This made this procedure unsuitable as an encryption procedure. It therefore delivers a one-way function whose trapdoor information (defamiliarization of the 0-1 knapsack problem) could be discovered by an eavesdropper.

6.3 Decomposition into Prime Factors as a Basis for Public-Key Procedures

Primes form the basis for numerous algorithms for public-key procedures.

6.3.1 The RSA Procedure

As early as 1978, R. Rivest, A. Shamir, and L. Adleman [9] introduced the most important asymmetric cryptography procedure to date (see algorithm in Crypto Procedure 6.3).

RSA is also exposed in this book, for example, Section 5.10 ff. or RSA2D in Section 6.5.

You can gain practical experience with the RSA procedure in CT under:

- CTO has two broad plugins for RSA:
  - Via "RSA (Step-by-step)" `https://www.cryptool.org/en/cto/rsa-step-by-step`.
  - Via "RSA visual and more" `kb=https://www.cryptool.org/en/cto/rsa-visual` you can see with graphics how RSA assigns its input values when encrypting, you can test textbook RSA with big numbers, and also use RSA with OAEP, padding, and certificates as it is used in practice.
- Using `CT1 Indiv. Procedures ▷ RSA Cryptosystem ▷ RSA Demonstration`.
- Using `CT2 Templates ▷ Mathematics ▷ RSA with big numbers` and further RSA templates in CT2.
- Using `JCT Default Perspective ▷ Visuals` and `JCT Algorithm Perspective`.

Crypto Procedure 6.3: RSA (Based On the Factorization Problem)

Key generation:
Let p and q be two different prime numbers and $N = pq$. Let e be any number relative prime to $\phi(N)$; that is, $\gcd(e, \phi(N)) = 1$. Using the Euclidean algorithm, we calculate the positive integer $d < \phi(N)$ such that

$$ed \equiv 1 \mod \phi(N)$$

whereby ϕ is the **Euler phi function.**
The output text is divided into blocks and encrypted, whereby each block has a binary value $x^{(j)} \leq N$.

Public key: N, e

Private key: N, d

Encryption: $y = \text{enc}(x) = x^e \mod N$

Decryption: $x = \text{dec}(y) = y^d \mod N$

Comment: Euler Phi Function

The Euler phi function is defined as:

$\phi(n)$ is the number of integers $1 \leq x < n$
that do not have a common factor with n.

"No common factor" is defined as: Two integers a and b are coprime if $\gcd(a, b) = 1$.

The first values of the Euler phi function are:

$$\phi(1) = 1, \;\; \phi(2) = 1, \;\; \phi(3) = 2, \;\; \phi(4) = 2, \;\; \phi(6) = 2, \;\; \phi(10) = 4, \;\; \phi(15) = 8.$$

For example, $\phi(24) = 8$, because

$$|\{x < 24 : \gcd(x, 24) = 1\}| = |\{1, 5, 7, 11, 13, 17, 19, 23\}|.$$

SageMath Example 6.1 shows how to get these values via SageMath and how to create the graphics for the Euler phi function (Figure 6.1).

SageMath Example 6.1: Phi and the List of Coprime Numbers via SageMath

```
sage: n=24; philist=[i for i in range(n) if gcd(n,i) == 1];
      print('n =', n, '; Len =', euler_phi(n), '; List =', philist)
n = 24 ; Len = 8 ; List = [1, 5, 7, 11, 13, 17, 19, 23]

sage: n=25; philist=[i for i in range(n) if gcd(n,i) == 1];
      print('n =', n, '; Len =', euler_phi(n), '; List =', philist)
n = 25 ; Len = 20 ;
List = [1, 2, 3, 4, 6, 7, 8, 9, 11, 12, 13, 14, 16, 17, 18, 19, 21, 22,▸
 ▸ 23, 24]

sage: n=26; philist=[i for i in range(n) if gcd(n,i) == 1];
      print('n =', n, '; Len =', euler_phi(n), '; List =', philist)
n = 26 ; Len = 12 ; List = [1, 3, 5, 7, 9, 11, 15, 17, 19, 21, 23, 25]

P = plot(euler_phi, -3, n); P
```

Table 6.1 shows values of $\phi(n)$ up to $n = 25$.
If p is a prime number, then $\phi(p) = p - 1$.
In the case of $N = pq$:

$$\phi(N) = pq(1 - 1/p)(1 - 1/q) = p(1 - 1/p)q(1 - 1/q) = (p - 1)(q - 1).$$

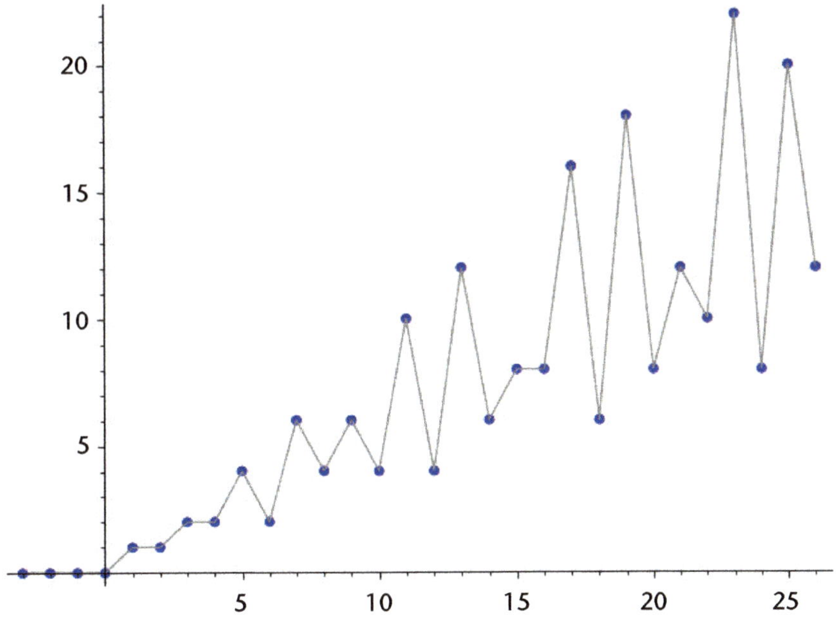

Figure 6.1 Values of the phi function up to $n = 26$.

Table 6.1 Euler Phi Function

n	$\phi(n)$	*The Natural Numbers That Are Coprime to n and Less Than n*
1	1	1
2	1	1
3	2	1, 2
4	2	1, 3
5	4	1, 2, 3, 4
6	2	1, 5
7	6	1, 2, 3, 4, 5, 6
8	4	1, 3, 5, 7
9	6	1, 2, 4, 5, 7, 8
10	4	1, 3, 7, 9
15	8	1, 2, 4, 7, 8, 11, 13, 14
20	8	1, 3, 7, 9, 11, 13, 17, 19
25	20	1, 2, 3, 4, 6, 7, 8, 9, 11, 12, 13, 14, 16, 17, 18, 19, 21, 22, 23, 24

Table 6.2 $L(N)$ Value Table*

N	10^{50}	10^{100}	10^{150}	10^{200}	10^{250}	10^{300}
$L(N)$	$1.42 \cdot 10^{10}$	$2.34 \cdot 10^{15}$	$3.26 \cdot 10^{19}$	$1.20 \cdot 10^{23}$	$1.86 \cdot 10^{26}$	$1.53 \cdot 10^{29}$

* Factorization effort related to the length of the modulus.

If we know the various prime factors $p_1, \ldots, p_k$ of n, then

$$\phi(n) = n \cdot \left(1 - \frac{1}{p_1}\right) \cdot \ \ldots \ \cdot \left(1 - \frac{1}{p_k}\right).$$

Further formulas for the Euler phi function are in Section 5.8.2.

The function enc is a one-way function whose trapdoor information is the decomposition of N into primes.

At the moment, no algorithm is known that can factorize two prime numbers sufficiently quickly for extremely large values (e.g., for several hundred decimal places). The quickest algorithms known today [10] factorize a compound integer N in a time period proportional to $L(N) = e^{\sqrt{\ln(N)\ln(\ln(N))}}$. Some example values can be found in Table 6.2.

To this date, it has not been proved that the problem of breaking RSA is equivalent to the factorization problem. Nevertheless, it is clear that the RSA procedure will no longer be safe if the factorization problem is solved.

6.3.2 Rabin Public-Key Procedure 1979

"Rabin's scheme [...] gets its security from the difficulty of finding square roots modulo a composit number. This problem is equivalent to factoring" [11]. Unfortunately, this procedure is susceptible to chosen-ciphertext attacks. See [11] for more details.

This algorithm can be performed step-by-step in `JCT Default Perspective ▷ Visuals ▷ Rabin Cryptosystem`.

Crypto Procedure 6.4: Rabin (Based On the Factorization Problem)

Let p and q be two different prime numbers with $p, q \equiv 3 \mod 4$ and $n = pq$. Let $0 \le B \le n - 1$.

Public key:	(n, B)
Private key:	(p, q)
Encryption:	$y = \text{enc}(x) = x(x + B) \mod n$
Decryption:	$\text{dec}(y) = \sqrt{y + B^2/4} - B/2 \mod n$

Note that the encryption function is not injective: If doing the decryption, for each ciphertext value you get exactly four different plaintext values that have $e_T(x)$ as inverse image: $x, -x - B, \omega(x + B/2) - B/2, -\omega(x + B/2) - B/2$ where ω is one of the four roots of unity.

Backdoor information is the decomposition into prime numbers of $n = pq$.

6.4 The Discrete Logarithm as a Basis for Public-Key Procedures

Discrete logarithms form the basis for many algorithms for public-key procedures.

6.4.1 The Discrete Logarithm in $\mathbb{Z}_p$

Let p be a prime number and let $g \in \mathbb{Z}_p^* = \{0, 1, \dots, p - 1\}$. Then the discrete exponential function with base g is defined as

$$e_g : k \longrightarrow y := g^k \mod p, \quad 1 \le k \le p - 1.$$

The inverse function is called a discrete logarithm function $\log_g$; the following holds:

$$\log_g(g^k) = k.$$

The problem of the discrete logarithm (in $\mathbb{Z}_p^*$) is understood to be as follows:

$$\text{Given } p, g \text{ and } y, \text{ determine } k \text{ such that } y = g^k \mod p$$

It is much more difficult to calculate the discrete logarithm than to evaluate the discrete exponential function (see Section 5.9). Table 6.3 lists several procedures for calculating the discrete logarithm and their complexity [10].

From the CrypTool Variants, you can try `CT2 Templates ▷ Mathematics ▷ Discrete Logarithm` or `JCT Default Perspective ▷ Visuals ▷ Shanks Babystep-Giantstep`. The baby-step giant-step algorithm computes the discrete logarithm or order of an element in a finite abelian group. The algorithm is based on a space-time trade-off. Although the algorithm is superior to nonely trying out all possibilities concerning the runtime, it is still not practicable for very large groups.

Table 6.3 Procedures for Calculating the Discrete Logarithm over $\mathbb{Z}_p^*$

Name	*Complexity*
Baby-step-giant-step	$O(\sqrt{p})$
Silver-Pohlig-Hellman	Polynomial in q, the greatest prime factor of $p-1$
Index-Calculus	$O\left(e^{(1+o(1))\sqrt{\ln(p)\ln(\ln(p))}}\right)$

In February 2007, the group Kleinjung, Franke, and Bahr at the University of Bonn set the then-record for calculating discrete logarithms. They calculated the discrete logarithm modulo a 160-digit (530-bit) prime number p and a generator g:

$$\begin{aligned}
p &= \lfloor 10^{159}\pi \rfloor + 119849 \\
&= 314159265358979323846264338327950288419716939937510582097494 \\
&\quad\ 459230781640628620899862803482534211706798214808651328230664 \\
&\quad\ 7093844609550582231725359408128481237299 \\
g &= 2
\end{aligned}$$

Note that this is indeed a prime number:

```
sage: is_prime(floor(10^159*pi)+119849)
True
```

More precisely, the discrete logarithm k of the following integer y—the first 159 digits of the Euler number e—was determined:

$$\begin{aligned}
y &= \lfloor 10^{159}e \rfloor \\
&= 271828182845904523536028747135266249775724709369995957496696 \\
&\quad\ 762772407663035354759457138217852516642742746639193200305992 \\
&\quad\ 1817413596629043572900334295260595630738 \\
k &= \log_g(y) \mod p \\
&= 829897164650348970518646802640757844024961469323126472198531 \\
&\quad\ 845186895984026448342666252850466126881437617381653942624307 \\
&\quad\ 537679319636711561053526082423513665596
\end{aligned}$$

The search was performed with GNFS method (index-calculus) and took about 17 CPU years on 3.2 GHz Xeon machines.

The current record (as state of November 2021) considering integers modulo p as the finite cyclic group G was on December 2, 2019 by Fabrice Boudot, Pierrick Gaudry, Aurore Guillevic, Nadia Heninger, Emmanuel Thomé, and Paul Zimmermann [12]. They computed the discrete logarithm modulo the 240-digit (795-bit) prime RSA-240 + 49204 (the first safe prime above RSA-240), using the number field sieve algorithm and the open-source software CADO-NFS.

6.4.2 Diffie-Hellman Key Agreement

The mechanisms and algorithms of classical cryptography only take effect when the subscribers have already exchanged the secret key. In classical cryptography you cannot avoid exchanging secrets without encrypting them. Transmission safety here must be achieved using noncryptographic methods. We say that we need a secret channel for exchanging secrets. This channel can be realized either physically or organizationally.

What is revolutionary about modern cryptography is, among other things, that you no longer need secret channels: You can agree on secret keys using nonsecret (i.e., public) channels.

One protocol that solves this problem is that of Diffie and Hellman (Crypto Procedure 6.5).

Crypto Procedure 6.5: Diffie-Hellman Key Agreement

Two subscribers A and B want to agree on a joint secret key.

Let p be a prime number and g a natural number. These two numbers do not need to be secret.

The two subscribers then select a secret number a and b from which they calculate the values $\alpha = g^a \mod p$ and $\beta = g^b \mod p$. They then exchange the numbers α and β. To end with, the two subscribers calculate the received value to the power of their secret value to get $\beta^a \mod p$ and $\alpha^b \mod p$.

Thus

$$\beta^a \equiv (g^b)^a \equiv g^{ba} \equiv g^{ab} \equiv (g^a)^b \equiv \alpha^b \mod p$$

This exchange protocol has been visualized, and you can execute the single steps with specific numbers in:

- `CT1 Indiv. Procedures ▷ Protocols ▷ Diffie-Hellman Demonstration`
- `JCT Default Perspective ▷ Visuals ▷ Diffie-Hellman Key Exchange (EC)`

The safety of the *Diffie-Hellman protocol* is closely connected to calculating the discrete logarithm mod p. It is even thought that these problems are equivalent.

6.4.3 ElGamal Public-Key Encryption Procedure in $\mathbb{Z}_p^*$

By varying the Diffie-Hellman key agreement protocol slightly, you can obtain an asymmetric encryption algorithm, Crypto Procedure 6.6. This observation was made by Taher ElGamal.

Crypto Procedure 6.6: ElGamal (Based On the Discrete Logarithm Problem)

Let p be a prime number such that the discrete logarithm in $\mathbb{Z}_p$ is difficult to compute. Let $\alpha \in \mathbb{Z}_p^*$ be a primitive element. Let $a \in \mathbb{N}$ and $\beta = \alpha^a \mod p$.

Public key:	p, α, β
Private key:	a

Let $k \in \mathbb{Z}_{p-1}$ be a random number and $x \in \mathbb{Z}_p^*$ the plaintext.

Crypto Procedure 6.6 (continued)

Encryption:	$\text{enc}(x, k) = (y_1, y_2)$
where	$y_1 = \alpha^k \mod p$
and	$y_2 = x\beta^k \mod p$
Decryption:	$\text{dec}(y_1, y_2) = y_2(y_1^a)^{-1} \mod p$

In Java CrypTool this can be explored with `JCT Default Perspective` ▷ `Visuals` ▷ `ElGamal`.

6.4.4 Generalized ElGamal Public-Key Encryption Procedure

The discrete logarithm can be generalized in any number of finite groups $(G, \circ)$. The following provides several properties of G that make the discrete logarithm problem difficult. Instead of $g \circ h$ we often write only gh.

Calculating the Discrete Exponential Function

Let G be a group with the operation $\circ$ and $g \in G$. The (discrete) exponential function base g is defined as

$$e_g : k \longmapsto g^k, \quad \text{for all } k \in \mathbb{N}.$$

with

$$g^k := \underbrace{g \circ \ldots \circ g}_{k \text{ times}}.$$

The exponential function is easy to calculate.

Lemma
The power g^k can be calculated in at most $2\log_2 k$ group operations.

Proof
Let $k = 2^n + k_{n-1}2^{n-1} + \cdots + k_1 2 + k_0$ be the binary representation of k. Then $n \leq \log_2(k)$, because $2^n \leq k < 2^{n+1}$. k can be written in the form $k = 2k' + k_0$ with $k' = 2^{n-1} + k_{n-1}2^{n-2} + \cdots + k_1$. Thus,

$$g^k = g^{2k'+k_0} = (g^{k'})^2 g^{k_0}.$$

We therefore obtain g^k from $g^{k'}$ by squaring and then multiplying by g. The claim is thus proved by induction to n. □

Problem of the Discrete Logarithm

> Let G by a finite group with the operation $\circ$. Let $\alpha \in G$ and $\beta \in H = \{\alpha^i : i \geq 0\}$.
> We need to find a unique $a \in \mathbb{N}$ with $0 \leq a \leq |H| - 1$ and $\beta = \alpha^a$.
> We define a as $\log_\alpha(\beta)$.

Calculating the Discrete Logarithm

A simple procedure for calculating the discrete logarithm of a group element, which is considerably more efficient than simply trying all possible values for k, is the baby-step-giant-step algorithm.

Theorem 6.1 (baby-step-giant-step algorithm). Let G be a group and $g \in G$. Let n be the smallest natural number with $|G| \leq n^2$. Then the discrete logarithm of an element $h \in G$ can be calculated base g by generating the following two lists each containing n elements and comparing these lists:

$$\text{Giant-step list: } \{1, g^n, g^{2n}, \ldots, g^{n \cdot n}\},$$
$$\text{Baby-step list: } \{hg^{-1}, hg^{-2}, \ldots, hg^{-n}\}.$$

After detecting a common element the calculation can be stopped. In order to calculate these lists, we need $2n$ group operations.

Proof
If $g^{jn} = hg^{-i}$, that is $h = g^{i+jn}$, then the problem is solved. If the lists are disjoint, then h cannot be represented as g^{i+jn}, $i, j \leq n$. As all powers of g are thus recorded, the logarithm problem does not have a solution. □

You can use the baby-step-giant-step algorithm to demonstrate that it is much more difficult to calculate the discrete logarithm than to calculate the discrete exponential function. If the numbers that occur have approximately 1,000 bits in length, then you only need around 2,000 multiplications (see Theorem 6.1) to calculate all g^k, but around $2^{500} \approx 10^{150}$ operations to calculate the discrete logarithm using the baby-step-giant-step algorithm.

In addition to the baby-step-giant-step algorithm, there are also numerous other procedures for calculating the discrete logarithm [10].

The Theorem from Silver-Pohlig-Hellman

In finite abelian groups, the discrete logarithm problem can be reduced to groups of a lower order.

Theorem 6.2 (Silver-Pohlig-Hellman) Let G be a finite abelian group with $|G| = p_1^{a_1} p_2^{a_2} \cdot \ldots \cdot p_s^{a_s}$. The discrete logarithm in G can then be reduced to solving logarithm problems in groups of the order $p_1, \ldots, p_s$.

If $|G|$ contains a dominant prime factor p, then the complexity of the logarithm problem is approximately

$$O(\sqrt{p}).$$

Therefore, if the logarithm problem is to be made difficult, the order of the group used G should have a large prime factor. In particular, if the discrete exponential function in the group $\mathbb{Z}_p^*$ is to be a one-way function, then $p - 1$ must be a large prime factor. In this case a generalized ElGamal procedure can be defined (see Crypto Procedure 6.7).

Crypto Procedure 6.7: Generalized ElGamal (Based On the Factorization Problem)

Let G be a finite group with operation $\circ$, and let $\alpha \in G$, so that the discrete logarithm in $H = \{\alpha^i : i \geq 0\}$ is difficult to calculate. Let a with $0 \leq a \leq |H| - 1$ and let $\beta = \alpha^a$.

Public key: α, β

Private key: a

Let $k \in \mathbb{Z}_{|H|}$ be a random number, $k \neq 0$, and $x \in G$ be a plaintext.

Encryption: $\text{enc}(x, k) = (y_1, y_2)$

where $y_1 = \alpha^k$

and $y_2 = x \circ \beta^k$

Decryption: $\text{dec}(y_1, y_2) = y_2 \circ (y_1^a)^{-1}$

6.5 The RSA Plane

Earlier in Sections 5.10 to 5.12 and also in Section 6.3.1 the RSA algorithm was discussed. We will now introduce the RSA plane.

The RSA plane is a model for illustrating the math behind the RSA algorithm by geometrical means: If the RSA modulus equals $n = pq$ we use rectangles with sides p and q, and inside these rectangles we obtain a two-dimensional arrangement of the integers from 0 to $n - 1$, the numbers becoming points or little squares as in the figures on the following pages.

Although RSA has been known for a long time and the mathematical context on which RSA is based on is not very sophisticated, there is no description of RSA in our two-dimensional model in the literature, at least as far as we know.

6.5.1 Definition of the RSA Plane

Let $n \in \mathbb{N}, n = p \cdot q$ for two different (usually large) primes p and q. Then the Chinese remainder theorem states that the ring $\mathbb{Z}_n$ is *isomorphic* to the ring $\mathbb{Z}_p \times \mathbb{Z}_q$.

We use the abbreviation CRT see; for example, [13, p. 94 ff]. People without a deeper algebraic background usually know the CRT only in the context of giving a procedure for solving systems of linear congruences over the natural numbers.

In shorter mathematical notation the isomorphism statet by the CRT is written like this:

$$\mathbb{Z}_n \cong \mathbb{Z}_p \times \mathbb{Z}_q$$

You don't have to know what a ring is; it suffices to know that computing "modulo n" takes place in such a ring, usually denoted by $\mathbb{Z}_n$.

In algebra, an isomorphism (sometimes also isomorphy) is a 1:1 map that also carries over the algebraic structure from the domain (or preimage) to the codomain (or image). If the domain and the codomain are the same set, the term *automorphism* is used instead of isomorphism. For example, if we have addition inducing an algebraic structure on a set, an isomorphism f satisfies $f(a+b) = f(a)+f(b)$. Therefore, the order is not important: We can first add two elements of the set and then apply f to the sum, or first apply f to every summand separately and then add the mapped elements. The same holds for multiplication. The CRT isomorphism $\mathbb{Z}_n \cong \mathbb{Z}_p \times \mathbb{Z}_q$ specifically gives a 1:1 mapping from the numbers from 0 to $n-1$ onto the pairs of numbers from $(0, 0)$ to $(p-1, q-1)$, together with the possibility to choose whether we do computations with *one* element, mod n, or with a pair of *two* elements, the first mod p, the second mod q.

The isomorphism $\mathbb{Z}_n \cong \mathbb{Z}_p \times \mathbb{Z}_q$ is not an automorphism.

In the first case the elements of $\mathbb{Z}_n$ are usually identified with $\{0, 1, \ldots, n-1\}$ as a subset of a line; that is, we think of a linear arrangement (see Figure 6.2).

In the second case where we have pairs of numbers, we identify $\mathbb{Z}_p \times \mathbb{Z}_q$ with all pairs

$$\{(x, y) : x = 0, \ldots, p-1, \ y = 0, \ldots, q-1\}.$$

This then corresponds to a set of points of the plane enclosed by a rectangle, which results in a two-dimensional arrangement of the elements. This is why we use the terms "RSA-2D" and "RSA plane." We mention that the RSA plane is not a plane in a strict mathematical sense. More on that in Section 6.5.2.

The isomorphism π from Z_n to $Z_p \times Z_q$ (with $p \neq q$ as before) is defined by:

$$\pi : \left\{ \begin{array}{ccc} \mathbb{Z}_n & \to & \mathbb{Z}_p \times \mathbb{Z}_q \\ z & \mapsto & (x, y) := (z \bmod p, z \bmod q) \end{array} \right\}$$

For $p = 11$ and $q = 7$ the image of; for example, 23 is $\pi(23) = (1, 2)$, because when dividing 23 by 11 it has remainder 1 and when dividing 23 by 7 it has remainder 2.

Figure 6.3 shows this example for $p = 11$ and $q = 7$: Following the blue line that starts at the point $(0, 0)$ and then first goes to the top right, it's easy to see how π maps the numbers from 0 to $n-1$ (in our case $n-1 = 76$) step by step onto the points (x, y) of the rectangle.

6.5.2 Finite Planes

In spite of RSA planes not being real planes in a mathematical sense, they do have a lot in common with finite planes. It is therefore appropriate to briefly introduce the concept of the finite plane at this point. Finite geometry also forms the framework on which cryptographic procedures that are using elliptical curves are based (see Chapter 8).

Figure 6.2 Linear, one-dimensional arrangement of $\mathbb{Z}_n$.

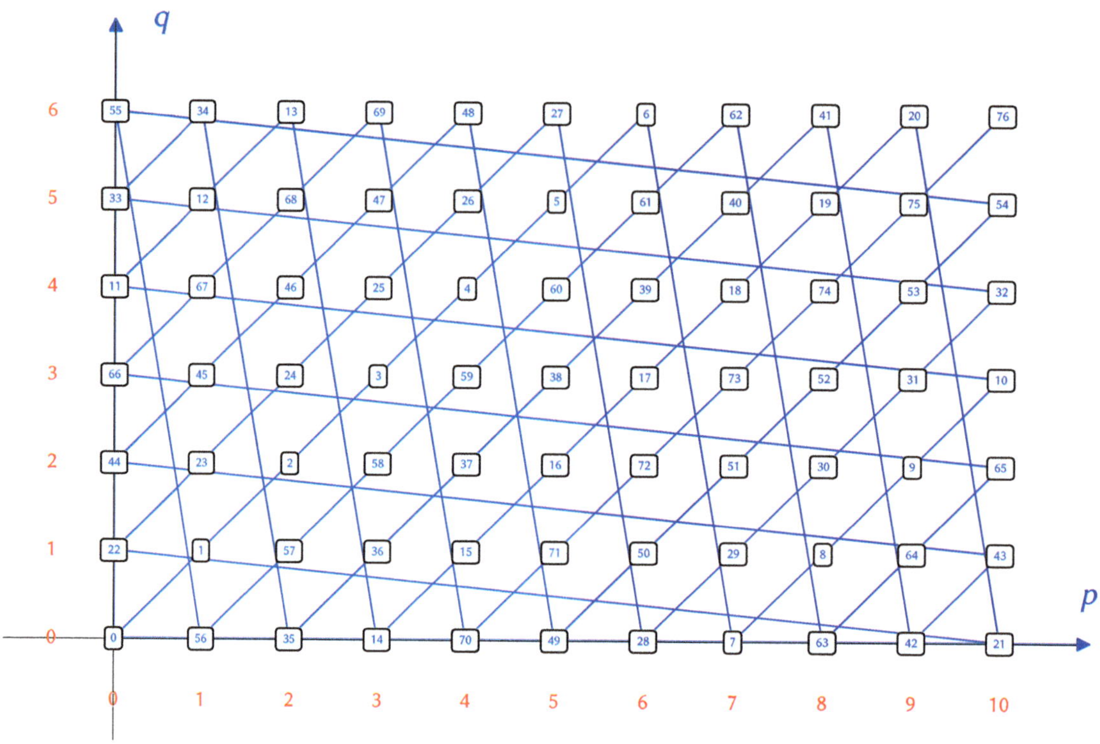

Figure 6.3 Rectangular, two-dimensional arrangement of the numbers from 0 to 76 ($p = 11$, $q = 7$, $n = 77$).

We define the finite plane as the set of all points (x, y) whose coordinates are elements of a finite field K. As usual, we associate a two-dimensional vector space with this plane. That means that we use vectors for describing, for example, the way from the origin $(0, 0)$ to the point $(2, 3)$. In vector notation this would be $\begin{pmatrix} 2 \\ 3 \end{pmatrix}$ using a column style, whereas points are written in row style. In geometry, points and vectors are not the same objects, which is why we use the term *associated*. Vectors are used because one cannot do computations with points; for example, it isn't possible to add them (that's actually not always true: It is indeed possible to define an addition on points of an elliptic curve, see Section 8.5).

Vectors, on the contrary, can be added in the way it is known from school mathematics. The main difference of addition in finite vector spaces, compared with infinite ones, is that when adding the same vector repeatedly one has to end up with the zero vector at some point because of the finite field from which the coordinates of the vector are taken from. It is known from algebra that finite fields are either equal to $\mathbb{Z}_p$ for a prime p or that they contain $\mathbb{Z}_p$ as a subset.

Remark: To be exact, instead of using the term "equal" to $\mathbb{Z}_p$ we should use the term "isomorphic" because the elements of a finite field are not always denoted by the "symbols" $\{0, 1, 2, \ldots\}$. Sometimes variables are used; for example, e instead of 1 or ϵ for a multiplicative generator. The same is true for the subset: To be more exact, we should say: They contain a subset that is isomorphic to $\mathbb{Z}_p$.

The number of elements in such a field is then always equal to a prime power p^N for some $N \in \mathbb{N}$. The prime p controls the behavior of the addition: For an arbitrary element x from such a field the sum $\underbrace{x + x + \cdots + x}_{p\text{-times}}$ must be equal to 0. This is a property of all finite fields with p^N elements. The number p is called the *characteristic*. For more about the characteristic, see Section 8.3.2.

For comparing finite planes with RSA planes restricting N to $N = 1$ is sufficient. In this case we can think of the finite plane as a square, similar to the RSA plane. The difference is that the model for the RSA plane is a rectangle but not a square with two sides of equal length. The two-dimensional vector space that corresponds to a finite plane is written as $\mathbb{Z}_p^2$ or $\mathbb{Z}_p \times \mathbb{Z}_p$, the same way as $\mathbb{R}^2 = \mathbb{R} \times \mathbb{R}$. Remember that the notation: $\mathbb{Z}_p^2$ is the two-dimensional vector space over the field $\mathbb{Z}_p$ (with p elements), whereas $\mathbb{Z}_{p^2}$ is the field with p^2 elements. Unfortunately, the notations used in mathematics in this context are somewhat inconsistent. For instance, $\mathbb{Z}_N$ is usually used for the additive group of integers modulo N but also for the ring if not only addition but also multiplication is under consideration. In the special case of $N = p$ for a prime p this ring is also a field and then one uses $\mathbb{F}_p$ or $\mathrm{GF}(p)$ rather than $\mathbb{Z}_p$. Finite fields are also called Galois fields, hence GF. The notion $\mathbb{Z}_p$ is also used for the p-adic numbers, which contain the integers $\mathbb{Z}$ as a subset and don't form a finite set. The p-adic numbers are not of interest in this book.

We are using $\mathbb{Z}_p$ and not $\mathbb{F}_p$ for the fields with p elements.

6.5.3 Lines in a Finite Plane

Every oblique line $g = \{(x, mx + t) \in \mathbb{Z}_p \times \mathbb{Z}_p : x \in \mathbb{Z}_p\}$ for $m, t \in \mathbb{Z}_p$, as well as every vertical line, has exactly p points.

Figure 6.4 shows the line through the origin with the equation $y = x$ in Z_{11} (lines through the origin have parameters $m = 1, t = 0$). We chose to represent the points of a finite plane as little squares, but dots are also often used in plots of finite planes. Our choice was mainly, but not only made out of aesthetic reasons. For our needs points would have been sufficient. Some short remarks on why we chose squares can be found in Section 6.5.12.

Starting at an arbitrary point on the line $y = x$ shown in Figure 6.4 and then repeatedly moving one to the right and one up leads to the top right point $(10, 10)$. Adding the vectors $\begin{pmatrix} 10 \\ 10 \end{pmatrix}$ and $\begin{pmatrix} 1 \\ 1 \end{pmatrix}$ brings us back to $\begin{pmatrix} 0 \\ 0 \end{pmatrix}$.

The line $y = x$ looks very similar in both the finite and the Euclidean case over the reals $\mathbb{R}$ to which one is used to. Other finite lines do not necessarily look familiar. Figure 6.5 shows the line $y = 6x + 2$ in Z_{11}. The slope $m = 6$ indicates to always move one to the right and six up, for example, starting in $(0, 2)$ and counting mod 11. After having arrived at the top where you can't go up further, one jumps to the bottom and continues to count. For example being located at the point $(3, 9)$ there's only one step to go up, and the second step is actually a jump to the bottom. The remaining four steps start at $(4, 0)$ and end at $(4, 4)$.

There's an alternative interpretation for the slope $m = 6$: Because in $\mathbb{Z}_{11}$ the equality $2 \cdot 6 = 1$ is true, one knows that 6 is the multiplicative inverse of 2; that is, the reciprocal value of 2. If one uses a fraction instead of the notation 2^{-1}, one

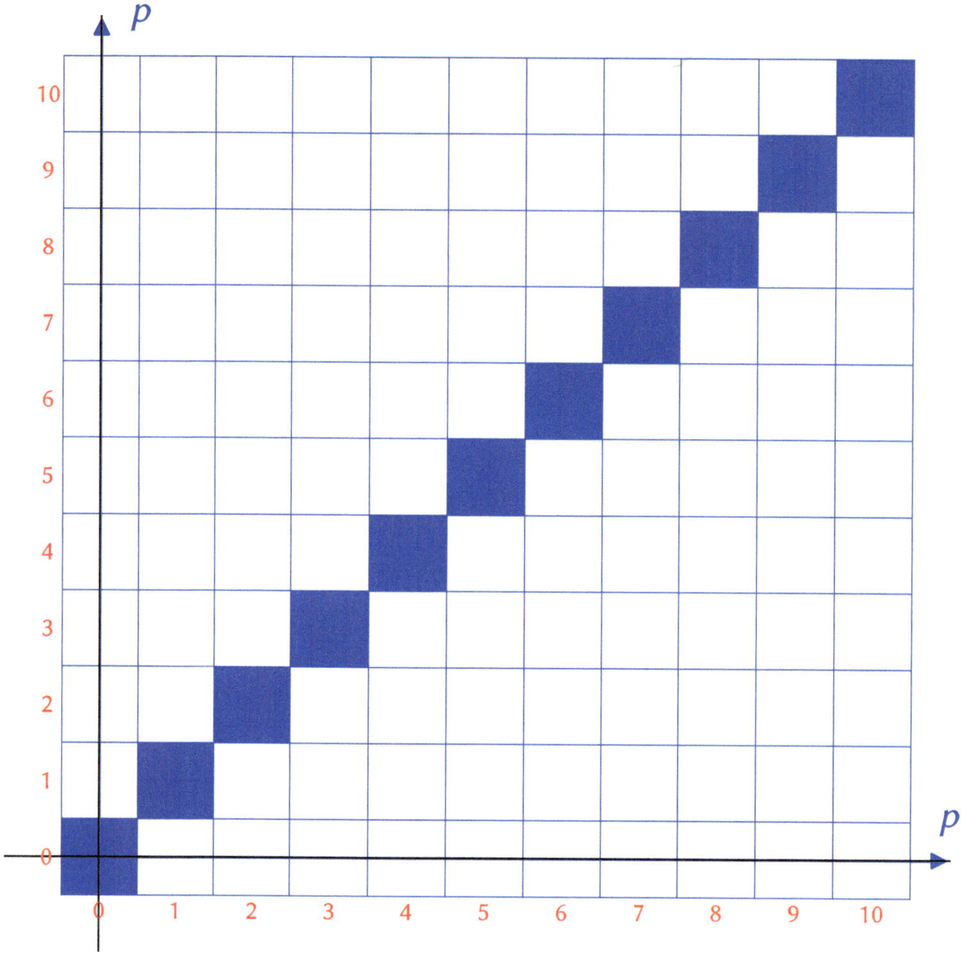

Figure 6.4 The line $y = x$ in the plane over the field Z_{11} with $p = q = 11$.

can think of $m = \frac{1}{2}$ instead of $m = 6$. Interpreting $\frac{1}{2}$ as "2 to the right and 1 up," the situation is consistent with the position of the blue squares of Figure 6.5.

Also, the connection between the three parts of the line $y = 6x + 2$ under consideration—the three higher squares, six in the middle and two at the bottom like in Figure 6.5—can be made clear easily: If one chooses the point $(10, 7)$ and wants to move two to the right and one up, this is not possible. Instead, one has to replace "1 right" by "all the way to the left to $p = 0$." Moving from $(10, 7)$ two to the right and one up then leads to the point $(1, 8)$, from which one can continue until $(5, 10)$. From there it is possible to go two to the right but not further up. Therefore, we replace "one up" by "all the way down to zero."

No matter what directional vector a line under consideration has, $\begin{pmatrix} 1 \\ 1 \end{pmatrix}$ or $\begin{pmatrix} 2 \\ 1 \end{pmatrix}$ or in general $\begin{pmatrix} x \\ y \end{pmatrix}$, the p-fold multiple of this vector is the zero vector because of the characteristic: $p \cdot \begin{pmatrix} x \\ y \end{pmatrix} = \begin{pmatrix} p \cdot x \\ p \cdot y \end{pmatrix} = \begin{pmatrix} 0 \\ 0 \end{pmatrix}$.

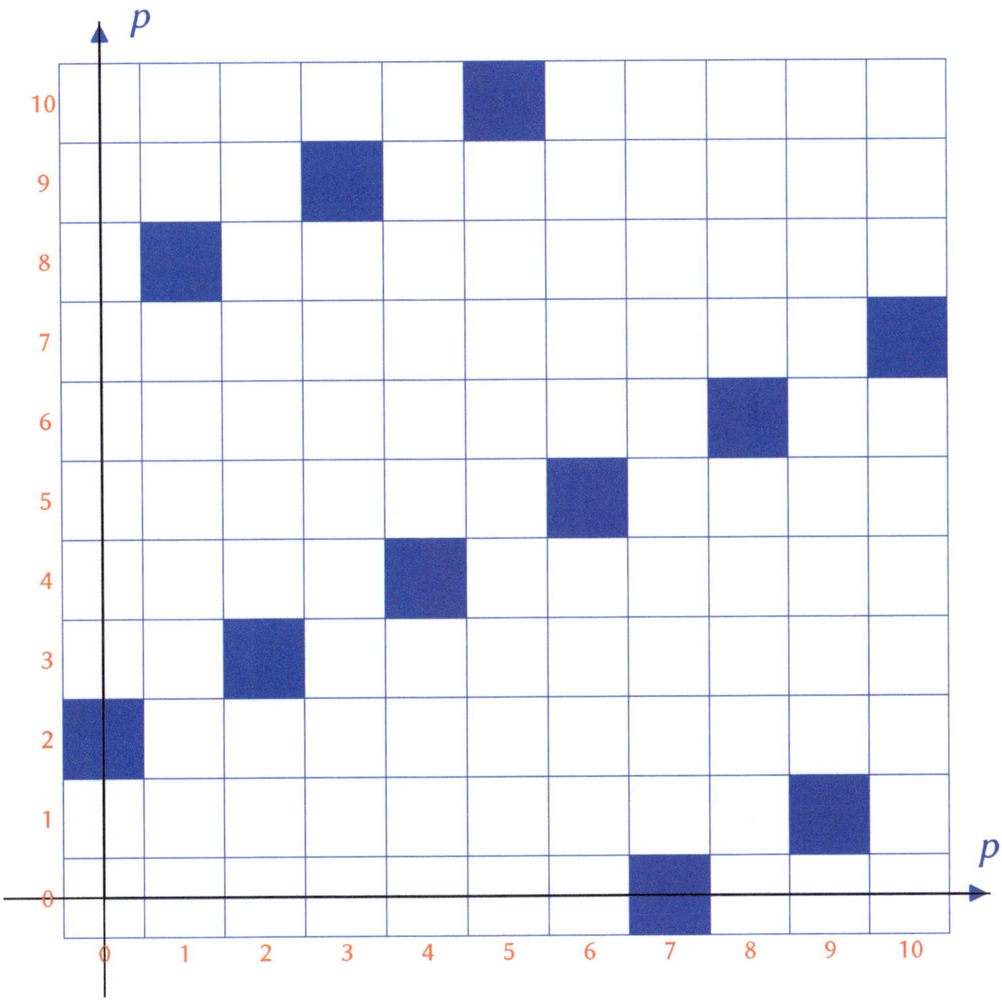

Figure 6.5 The line $y = 6x + 2$ in the finite plane over the field $\mathbb{Z}_{11}$.

Since we are in $\mathbb{Z}_p$ and p is prime, the k–fold multiple of every nonzero vector for $k < p$ is different from the zero vector. Therefore, we can conclude that every line of a finite plane has exactly p points.

6.5.4 Lines in the RSA Plane

One can try to carry over the concept of a line in a finite plane to the RSA plane. To do this, we begin with the analogon of the line through the origin that we plotted earlier in Figure 6.4. The set

$$\{(x, y) \in \mathbb{Z}_p \times \mathbb{Z}_q : y = x, x \in \mathbb{Z}_p\}$$

is a subset of the RSA plane with p elements; see Figure 6.6. We call the structure that is formed by this set an *RSA line* or just line. Formally we should write the RSA line more precisely like this:

$$\{(x, y) \in \mathbb{Z}_p \times \mathbb{Z}_q : y \equiv x \pmod{q}, x \in \mathbb{Z}_p\}$$

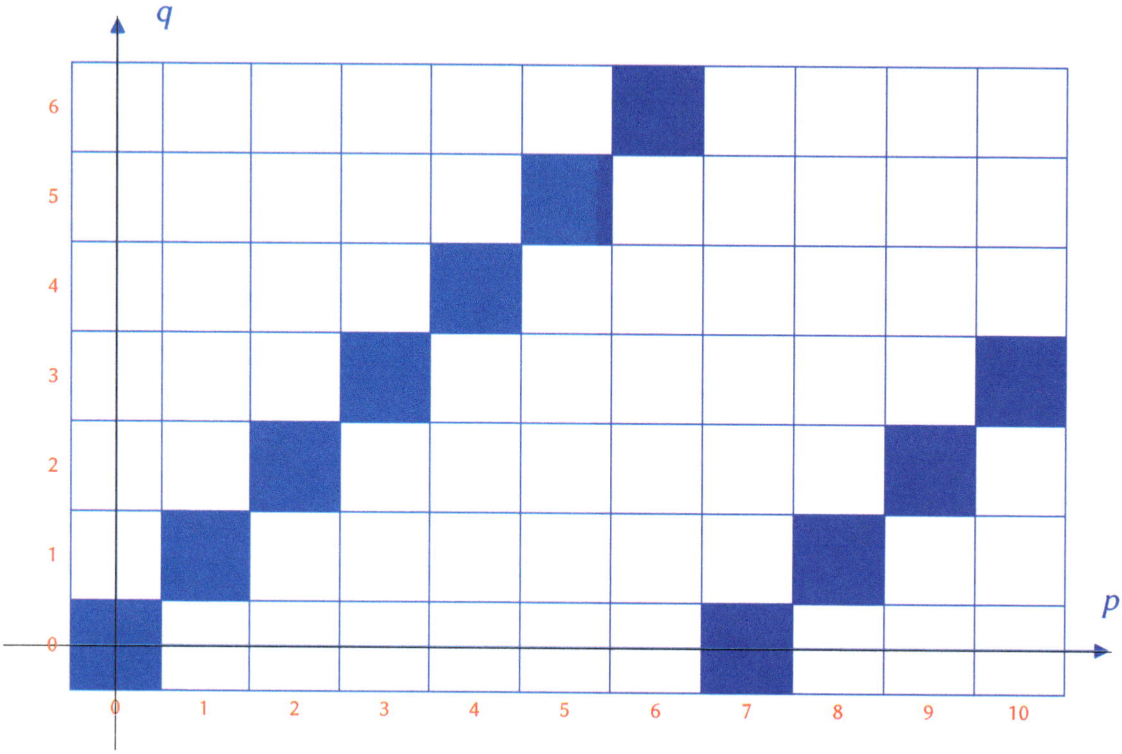

Figure 6.6 RSA line $y = x$ in the RSA plane for $p = 11$ and $q = 7$.

Our first interesting observation is that the linear equation $y = x$ has two different solutions in the $(11, 7)$-plane for some of the possible y values, which is not possible in a finite plane over a field. When choosing examples with $p > q$ and in the case $p = 11$ and $q = 7$ additionally $p < 2q$, then the equation $y = x$ has two solutions for $p - q = 4$ of the possible 7 $y-$values. In the case of $p > 2q$ there would be several solutions for every y, so just think of the pattern in Figure 6.6 extended to the right.

The vertical lines are a special case: They have exactly q points whereas all oblique lines consist of p points.

In Figure 6.7 one can see the analogon to Figure 6.5, the RSA line $y = 6x + 2$. This line has slope -1 since the number 6 is congruent to -1 modulo 7, resulting in the pattern "one to the right and one down."

There's another interesting difference between the situation in $\mathbb{Z}_p \times \mathbb{Z}_p$ (finite plane) and the situation in $\mathbb{Z}_p \times \mathbb{Z}_q$ (RSA plane; as before $p \neq q$) that we want to look at: The length of the series of points that we pass through if we repeatedly move one to the right and one up. In the case of the square one comes back to the starting point after p steps like in Figure 6.6. In the case of the rectangle one passes all $n = p \cdot q$ points before getting back to the starting point. Figure 6.3 shows this behavior, beginning in $(0, 0)$, then passing through all points until arriving at number 76 or $(10, 6)$. Compared to the finite plane, in the RSA plane case one passes $n = p \cdot q$ points and not only p before closing the loop.

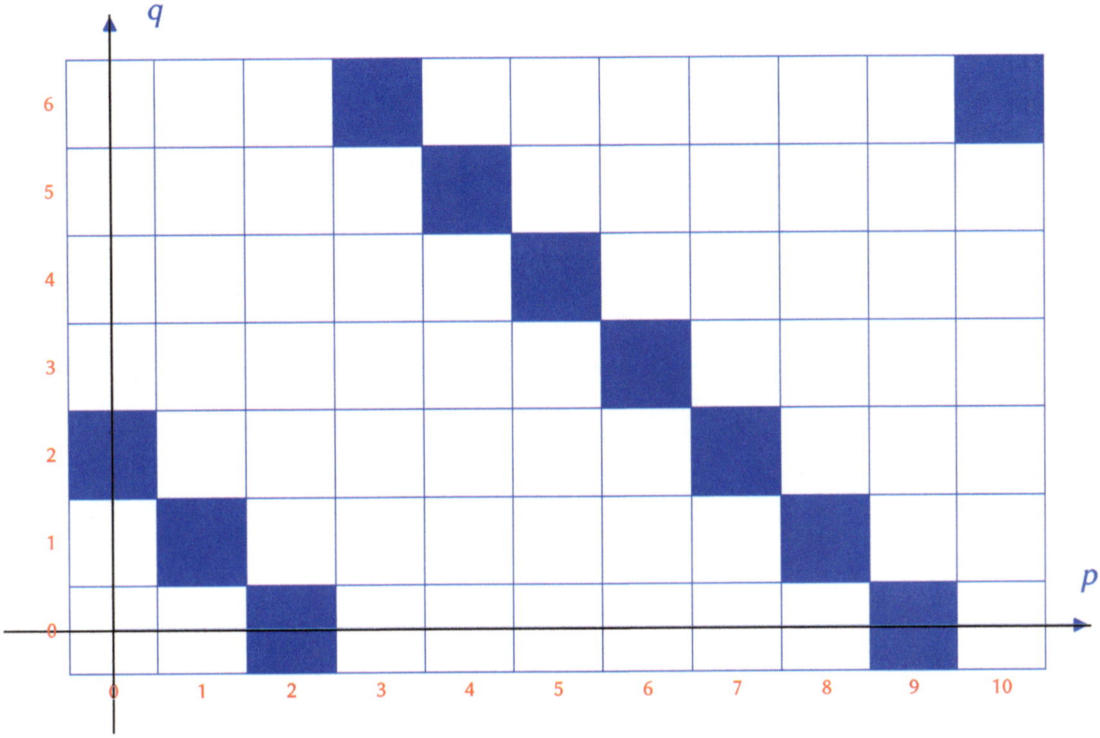

Figure 6.7 The RSA line $y = 6x + 2$ in the RSA plane with $p = 11$ and $q = 7$.

6.5.5 Alternative Choice of Representatives

Instead of $\{0, \ldots, n-1\}$ one can also choose the symmetric (or central with respect to 0) representatives $\left\{-\frac{(n-1)}{2}, \ldots, \frac{n-1}{2}\right\}$ for $\mathbb{Z}_n$. Figure 6.8 shows the usual representatives in blue brackets and the alternative representatives in black brackets, the latter being positioned symmetrically to the left and right of 0.

Choosing the representatives of $\mathbb{Z}_p$ and $\mathbb{Z}_q$ in this alternative way, one has the two sets$\left\{-\frac{(p-1)}{2}, \ldots, \frac{p-1}{2}\right\}$ and $\left\{-\frac{(q-1)}{2}, \ldots, \frac{q-1}{2}\right\}$.

In the two-dimensional version this gives us Figure 6.9. Here all the points are symmetrical to the p- and q-axis. In addition to that the last point when following the line starting in the origin and moving in the direction to the upper left is not the point $(10, 6)$, respectively, 76 like in Figure 6.3, but $(-1, -1)$ resp. -1.

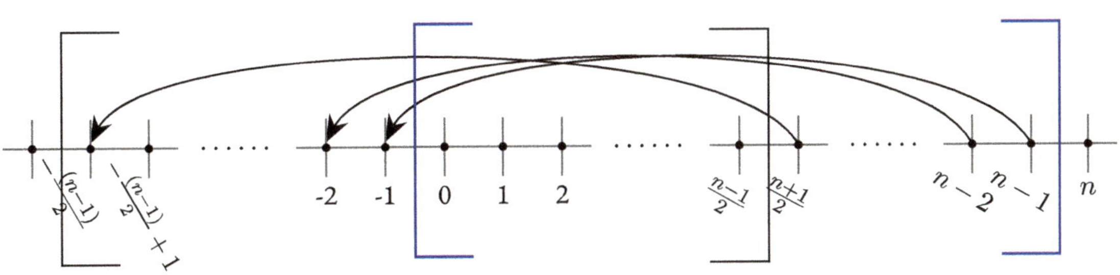

Figure 6.8 Translation of representatives mod n.

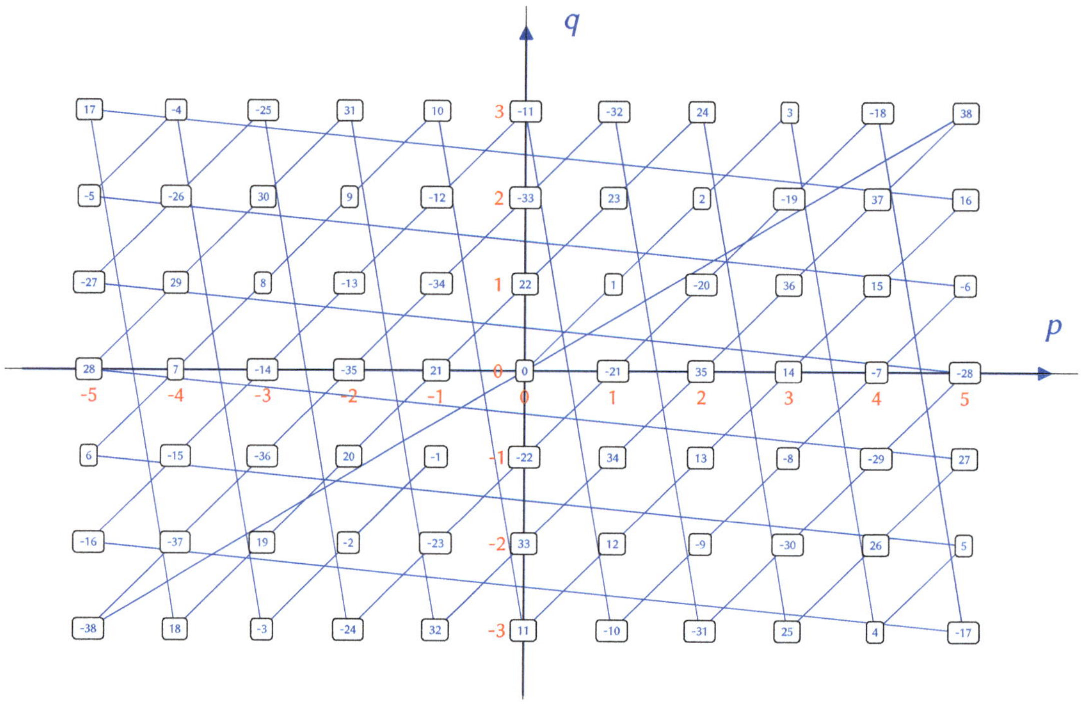

Figure 6.9 Rectangular two-dimensional pattern of the numbers from -38 to 38 for $p = 11$, $q = 7$, $n = 77$ with the origin in the center instead of in the bottom left corner.

Definition: We use the term *pq-plane* or *RSA plane* for both rectangular representations of $\mathbb{Z}_p \times \mathbb{Z}_q$ regardless of whether the origin is in the center or bottom left. When using specific numbers instead of the variables p and q we also write (p, q)-plane; for example, the $(11, 7)$-plane for Figure 6.6.

6.5.6 Points on the Axes and Inner Points

If one looks at Figure 6.9 it is immediately apparent that the points on the p-axis correspond to the numbers $\{0, \pm 7, \pm 14, \pm 21, \pm 28, \pm 35\}$, which are all multiples of $q = 7$. On the q-axis on the other hand there are the 2D images of multiples of $p = 11$ (i.e., $\{0, \pm 11, \pm 22, \pm 33\}$). In the other representation (Figure 6.3) these are the sets $\{0, 7, 14, \ldots, 70\}$ and $\{0, 11, 22, \ldots, 66\}$.

The CRT ensures that a number that has a common divisor with n has one coordinate $= 0$ if mapped to the corresponding 2-tuple in $\mathbb{Z}_p \times \mathbb{Z}_q$. Therefore points on the 2D axes correspond exactly to the numbers from 0 to $n-1$ (resp. from $-\frac{(n-2)}{2}$ to $\frac{n-1}{2}$) that have common divisors with n. These points are called *axis points* or *axis elements*. There are exactly $p + q - 1$ of them. The points not contained in the axes are called *inner points* or *inner elements*. Figure 6.10 shows the inner points (orange) and the axis points (violet), again for the case $p = 11$ and $q = 7$.

6.5.7 The Action of the Map $z \mapsto z^k$

Taking an element of $\mathbb{Z}_n$ to the power of k induces a well-defined map on the pq-plane.

$$f_k : \left\{ \begin{array}{ccc} \mathbb{Z}_n & \to & \mathbb{Z}_n \\ z & \mapsto & z^k \end{array} \right\}$$

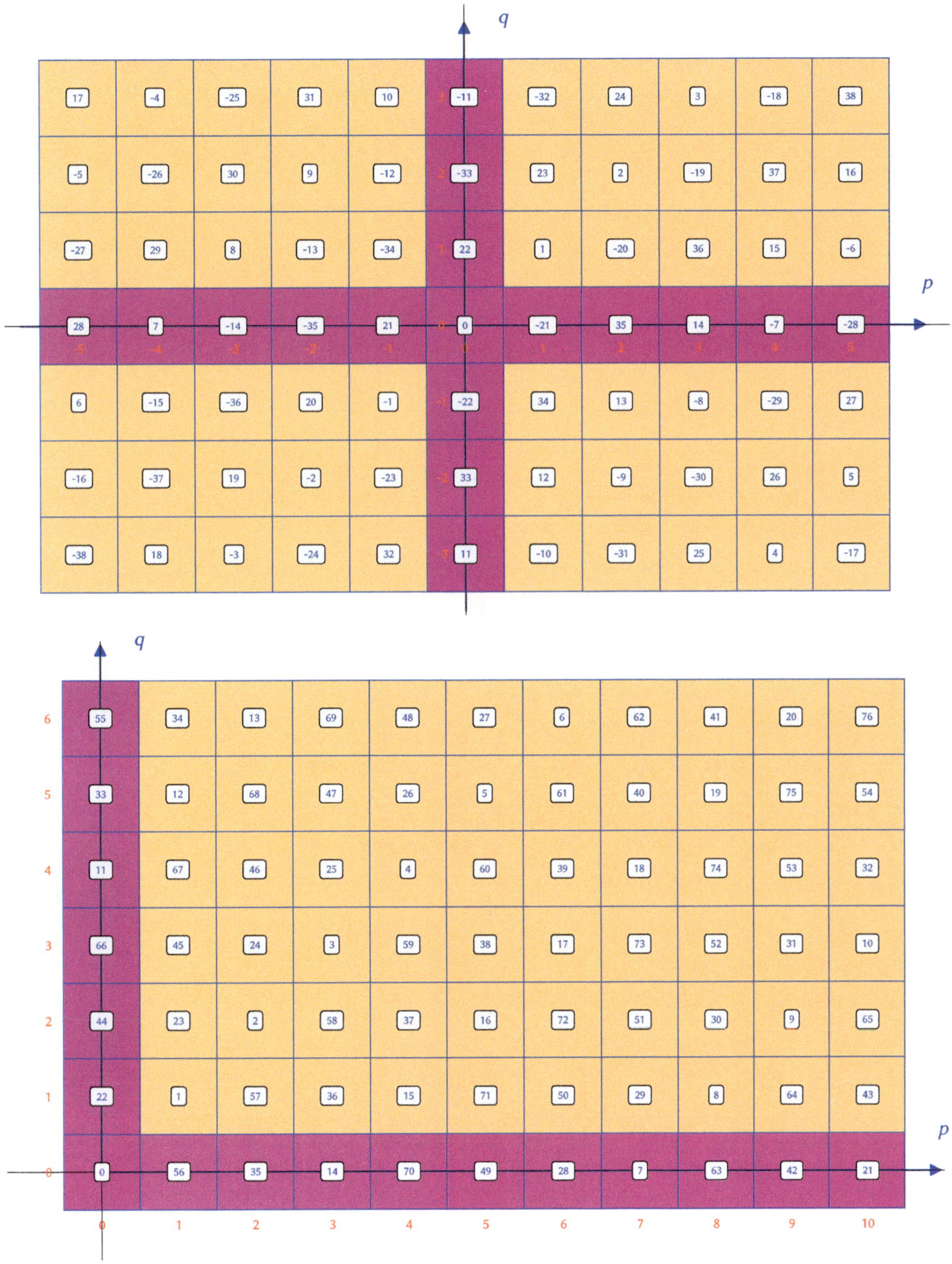

Figure 6.10 Axis points and inner points of the $(11, 7)$-plane; the upper picture has the origin in the middle while the lower picture has the origin in the lower left corner.

The map f_k is not $1:1$ (or bijective) in general. The $(11, 7)$-plane of Figure 6.11, for example, shows the image of f_2 consisting of all squares in $\mathbb{Z}_{77}$. One sees that only 24 of 77 elements of $\mathbb{Z}_{77}$ are squares, and only 15 of the 60 inner points are squares.

Next, one observes that axis elements cannot be moved away from the axis by f_k and inner elements remain inner elements. Take for instance a point $(0, y)$ on the q-axis, then the corresponding axis element in $\mathbb{Z}_n$ is of the form pr for some $r \in \mathbb{Z}$. It follows that $(pr)^k = p^k r^k \equiv 0 \pmod{p}$ is also an axis element.

If we choose a k that is relatively prime to $\phi(n)$, in our example (e.g., 7), then f_k is a $1:1$ map. In this case f_k is not only a map from one set to another, but

1. Restricted to inner points: an automorphism of the multiplicative group $\mathbb{Z}_n^*$;
2. Restricted to the p-axis without the origin: a $1:1$ map;
3. Restricted to the q-axis without the origin: also a $1:1$ map.

We still have to give an answer to why f_k is bijective for $\text{kgV}(k, \phi(n)) = 1$. Assume such an f_k to be not $1:1$. Then there would exist two different elements $z, w \in \mathbb{Z}_n$ such that $z^k = w^k$. For an inner element w there exists a multiplicative inverse and so it's possible to multiply this equation with w^{-k} leading to $z^k \cdot (w^{-1})^k = (z \cdot w^{-1})^k = 1$. So the element $(z \cdot w^{-1})$ has order k, and therefore k must divide the group order $\phi(n)$, a contradiction to our assumption.

We'll come back to the operation of f_k on the axes later in Section 6.5.9. There we also will outline why f_k for $\gcd(k, \phi(n))$ is $1:1$ also on the axis elements.

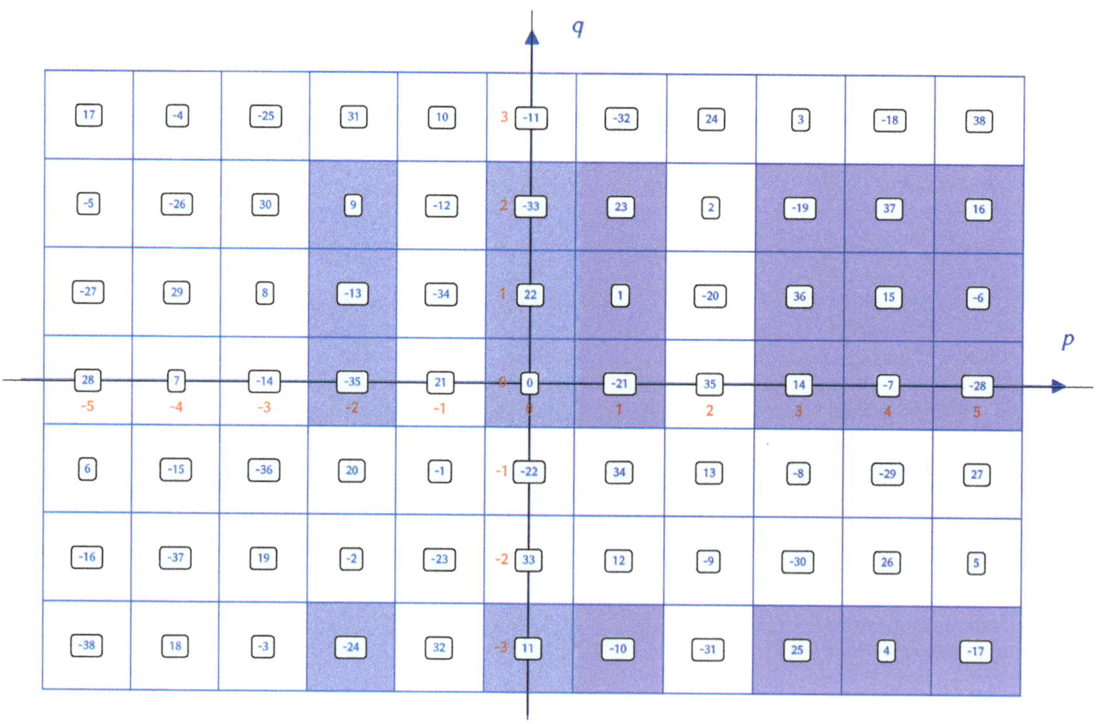

Figure 6.11 All squares (blue) of the $(11, 7)$-plane.

6.5.8 Orbits

We define the notion of an *orbit* or a *path* in such a way that it intuitively fits the geometric context. The usual group theoretic definition of an orbit would require the introduction of the operation of a group on a set. That would lead too far here.

For a $z \in \mathbb{Z}_n$ or $z \in \mathbb{Z}_p \times \mathbb{Z}_q$ the set $\langle z \rangle := \{z^k : k \in \mathbb{Z}\}$ is called the *orbit of* z or the path of z or also the (multiplicatively) generated set, z being called the *generator*. This path is finite and the number of its elements is called the *length of the orbit of* z. Later we also call $\langle z \rangle$ the *full orbit* of z, because then we will look at proper subsets of $\langle z \rangle$, the so called RSA orbits (see Section 6.5.8.3).

The orbits of different $z \in \mathbb{Z}_n$ are not necessarily disjoint, as can be seen in Figure 6.12.

Remark: In group theory, orbits are defined somewhat different from our definition here. Then they would have to be disjoint. That means their intersection would have to be empty.

The orbits of two points are disjoint if one point is an axis point and the other one is an inner point. The orbits of two inner points can never be disjoint since they always contain the point $(1, 1)$.

6.5.8.1 Examples of Orbits

For points on the axes, Figure 6.13 shows the orbits of the first multiples of 5 in the $(7, 5)$-plane. The lightest square is always the generator z, in the first picture we

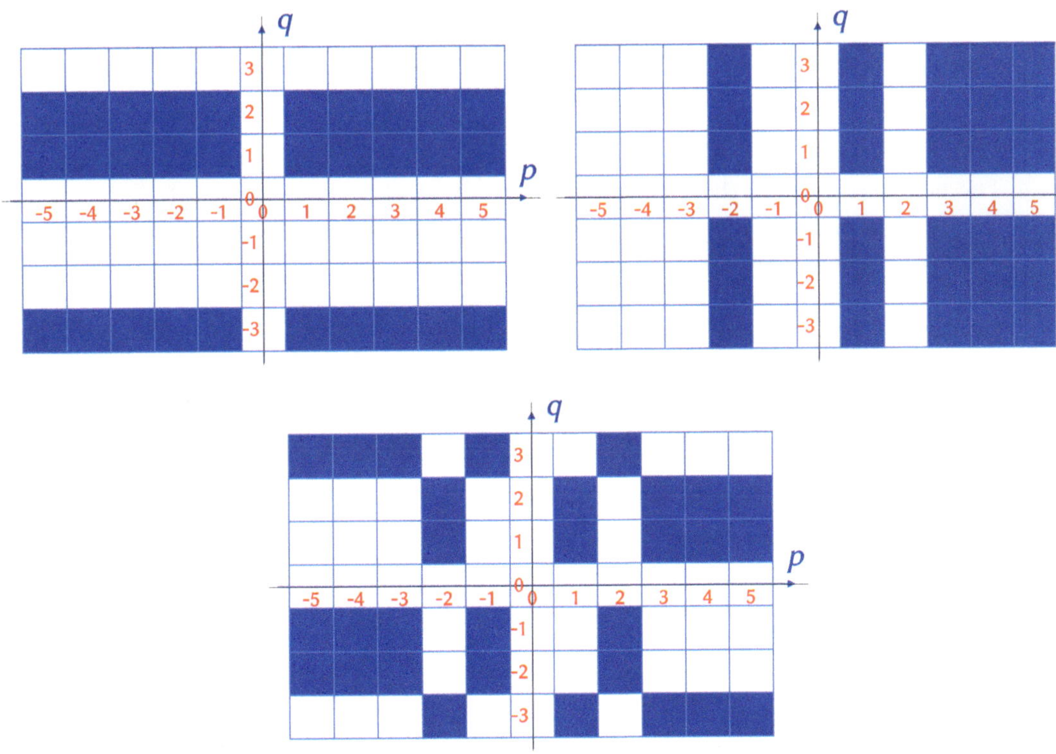

Figure 6.12 Orbit of 2 (top left), 3 (top right), and 17 (bottom) in the $(11, 7)$-plane.

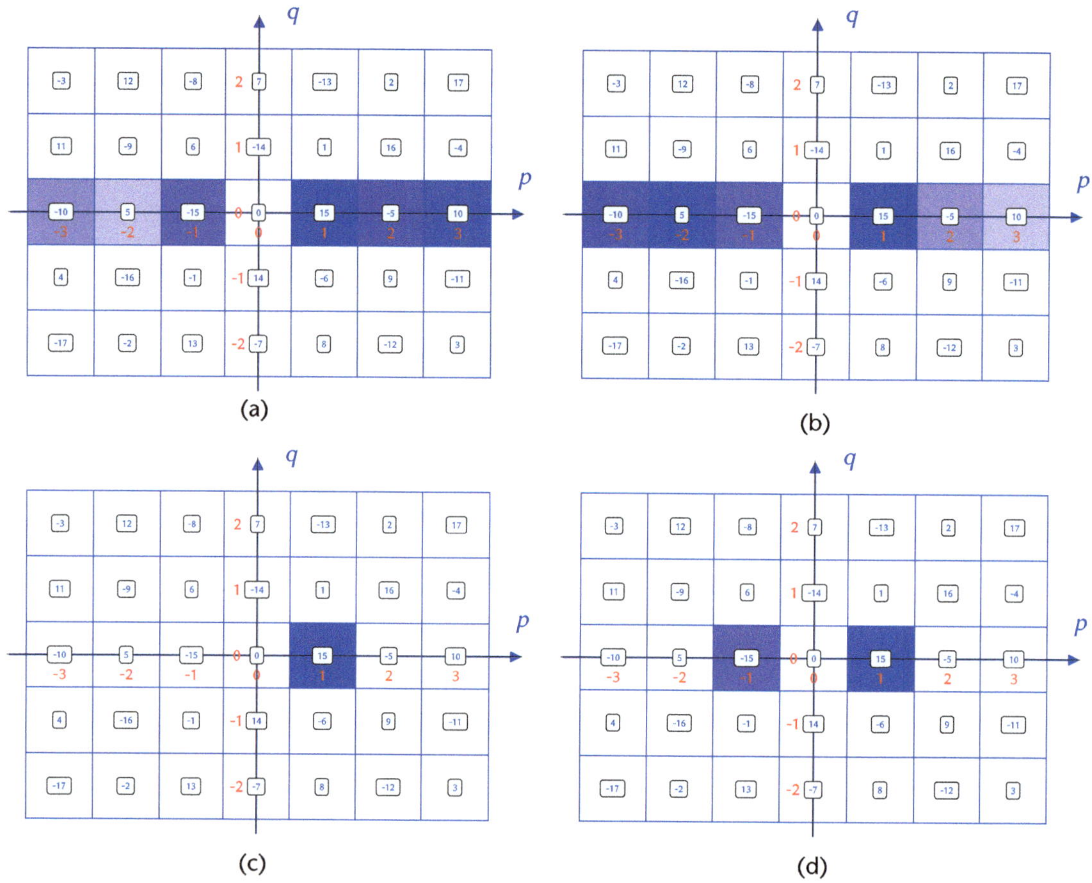

Figure 6.13 Orbits of multiples of 5 in the $(7, 5)$-plane: (a) Orbit of 5, (b) Orbit of 10, (c) Orbit of 15, and (d) Orbit of 20 or $20 - 35 = -15$.

have $z = 5$, in the second 10, in the third 15, and so forth, and in the fourth $z = 20$. The next darker field then is z^2 and so on, the darkest one is the last z^k that is not yet equal to z. This must be the square with p-coordinate $= 1$. In Section 6.5.9 we discuss why this is the case.

Let's have a closer look at Figure 6.13: The length of the path shown in Figure 6.13(a) and 6.13(b) is 6. The orbit on the top left has the generator $5 \in \mathbb{Z}_{35}$ or as 2D version $(-2, 0)$ because of $-2 \equiv 5 \bmod 7$ and $0 \equiv 5 \bmod 5$. On the top right the number of points in the orbit is also 6, but this time they are followed in a different order because the generator is 10, corresponding to the point $(3, 0)$ because of $10 \equiv 3 \bmod 7$ and $10 \equiv 0 \bmod 5$.

The lower two paths shown in Figure 6.13(c) and 6.13(d) have length 1 and 2.

Of course there is exactly one path of length 3 on the p-axis since 3 divides $6 = 7 - 1 = p - 1$ (see Figure 6.14). The order in which the points on the path are run through depends on the generator.

6.5.8.2 Lengths of Orbits

For an inner element z the smallest $k > 0$ such that $z^k = 1$ (and then $z^{k+1} = z$) is the *order of* z and the orbit of z is the cyclic subgroup of $\mathbb{Z}_n^*$ generated by z. The order of an inner element z is indeed the same as the length of the orbit of z.

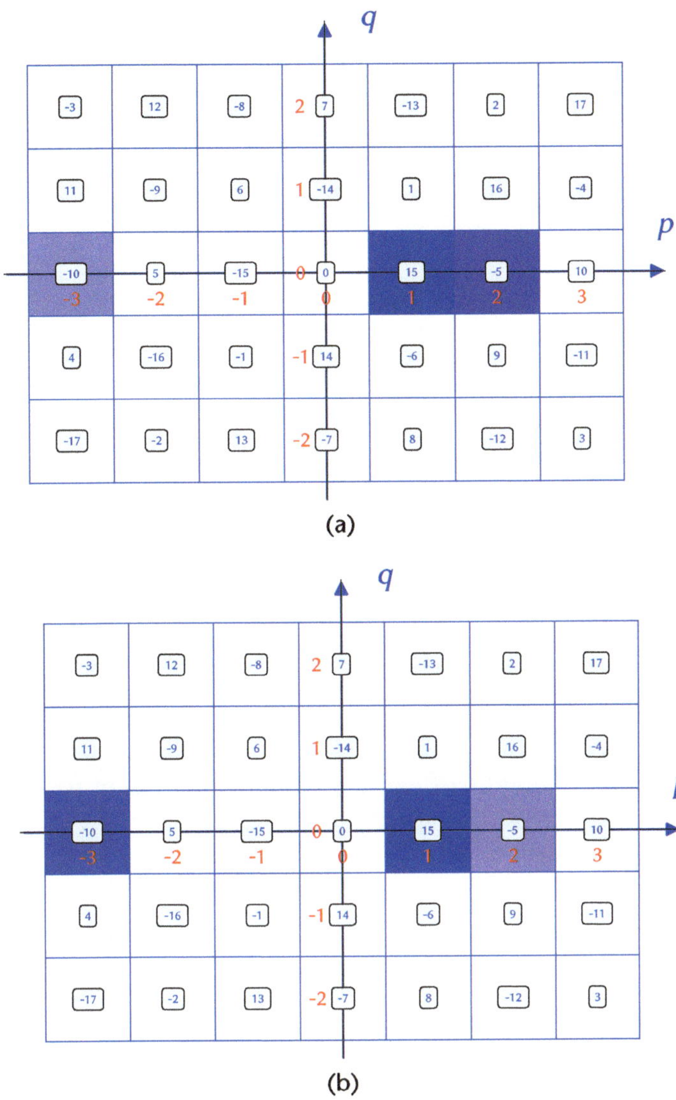

Figure 6.14 Orbit of 25 in the $(7, 5)$-plane; (a) orbit of 25, and (b) orbit of 30.

Since the order of any subgroup of a given group always divides the group order, we have:

The length of the orbits of inner elements are divisors of $\phi(n)$. The length of orbits of axis elements divides $p - 1$ or $q - 1$, depending on the element being on the p- or on the q-axis: Because the axes do not contain the element 1 or the point $(1, 1)$, respectively, they cannot have a path that forms a subgroup of $\mathbb{Z}_n^*$. But when thinking in projections—we will investigate projections in Section 6.5.9—one can identify the point $(x, 0)$ with $x \in \mathbb{Z}_p^*$ and $(0, y)$ with $y \in \mathbb{Z}_q^*$ $((x, y) \neq (0, 0))$. By doing this, one maps $1 : 1$ into a group, which implicates that the conditions on the divisors for subgroups are also satisfied by axis elements.

From theorems about the structure of groups [13, p. 42] it is further known that for any of the two axes there exist generators z_x or z_y for the two axes with maximal length $p - 1$ or $q - 1$, respectively. Among the inner points there exists an orbit of maximal length $\text{lcm}(p-1, q-1)$. Every orbit of an inner point corresponds to a cyclic subgroup of $\mathbb{Z}_n^*$. But $\mathbb{Z}_n^*$ itself is not cyclic. For if we assume it was cyclic, then there would exist an element of order $\phi(n)$ and there would also exist only one element of order 2, and that would be $x^{\phi(n)/2}$. The component $\mathbb{Z}_p^*$ has an element of order 2 and the component $\mathbb{Z}_q^*$ also has an element of order 2 because $q-1$ as well as $p-1$ are even numbers. The (multiplicative) group $\mathbb{Z}_n^*$ for $n = pq$ always has exactly three elements of order 2, and they are $\{(-1,1),(1,-1),(-1,-1)\}$. Together with $(1,1)$ they form a subgroup of $\mathbb{Z}_p \times \mathbb{Z}_q$ known as the *Klein four group*.

We want to illustrate the connection between the divisors of $\phi(n)$ and the lengths of possible orbits by looking at two examples: First, the case of only a few divisors, but also including large primes, and second, the case of many small prime divisors, but no large ones:

1. $p = 83,\ q = 59,\ n = 4897,\ \phi(n) = 4756,\ p-1 = 2\cdot 41,\ q-1 = 2\cdot 29$
2. $p = 97,\ q = 73,\ n = 7081,\ \phi(n) = 6912,\ p-1 = 2^5\cdot 3,\ q-1 = 2^3\cdot 3^2$

In the first case all possible orbits have the following lengths:

$$\{1,2,29,41,\underbrace{58}_{2\cdot 29},\underbrace{82}_{2\cdot 41},\underbrace{1189}_{29\cdot 41},\underbrace{2378}_{2\cdot 29\cdot 41}\}$$

The divisors $4, 116 = 4\cdot 29, 164 = 4\cdot 41$ and $4756 = 4\cdot 29\cdot 41 = \phi(4897)$ do not occur as lengths of existing orbits. The reason for this is that 4 divides all those divisors but 4 does divide neither $p-1$ nor $q-1$. Figure 6.15 shows the orbit of the element 2 with length 2378 (Figure 6.15(a)), the orbit of the element 60 with length 82 (Figure 6.15(b) with only one color and Figure 6.15(c) sorted by increasing darkness of the color) as well as the orbit of 117, also with length 82 (Figure 6.15(d)).

The lengths of all possible orbits in the second example with $p-1 = 2^5\cdot 3$ and $q-1 = 2^3\cdot 3^2$ are divisors of $\operatorname{lcm}(p-1,q-1) = \operatorname{lcm}(96,72) = 2^5\cdot 3^2$,

$$\{1,2,3,4,6,8,9,12,16,18,24,32,36,48,72,96,144,288\}$$

The divisors

$$\begin{array}{rclrclrcl}
27 &=& 3^3, & 216 &=& 2^3\cdot 3^3, & 864 &=& 2^5\cdot 3^3,\\
54 &=& 2\cdot 3^3, & 256 &=& 2^8, & 1152 &=& 2^7\cdot 3^2,\\
64 &=& 2^6, & 384 &=& 2^7\cdot 3, & 1728 &=& 2^6\cdot 3^3,\\
108 &=& 2^2\cdot 3^3, & 432 &=& 2^4\cdot 3^3, & 2304 &=& 2^8\cdot 3^2,\\
128 &=& 2^7, & 576 &=& 2^6\cdot 3^2, & 3456 &=& 2^7\cdot 3^3,\\
192 &=& 2^6\cdot 3, & 768 &=& 2^8\cdot 3, & 6912 &=& 2^8\cdot 3^3
\end{array}$$

do not occur as orbit lengths. These are exactly those divisors of $(p-1)(q-1)$ with a prime power divisor of the form 2^i with $i \geq 6$ or 3^j with $j \geq 3$. Those divide $(p-1)(q-1)$ but neither $p-1$ nor $q-1$. Figure 6.16 shows the orbits of 2, 5, and 811 with length 144, 288, and 3, respectively.

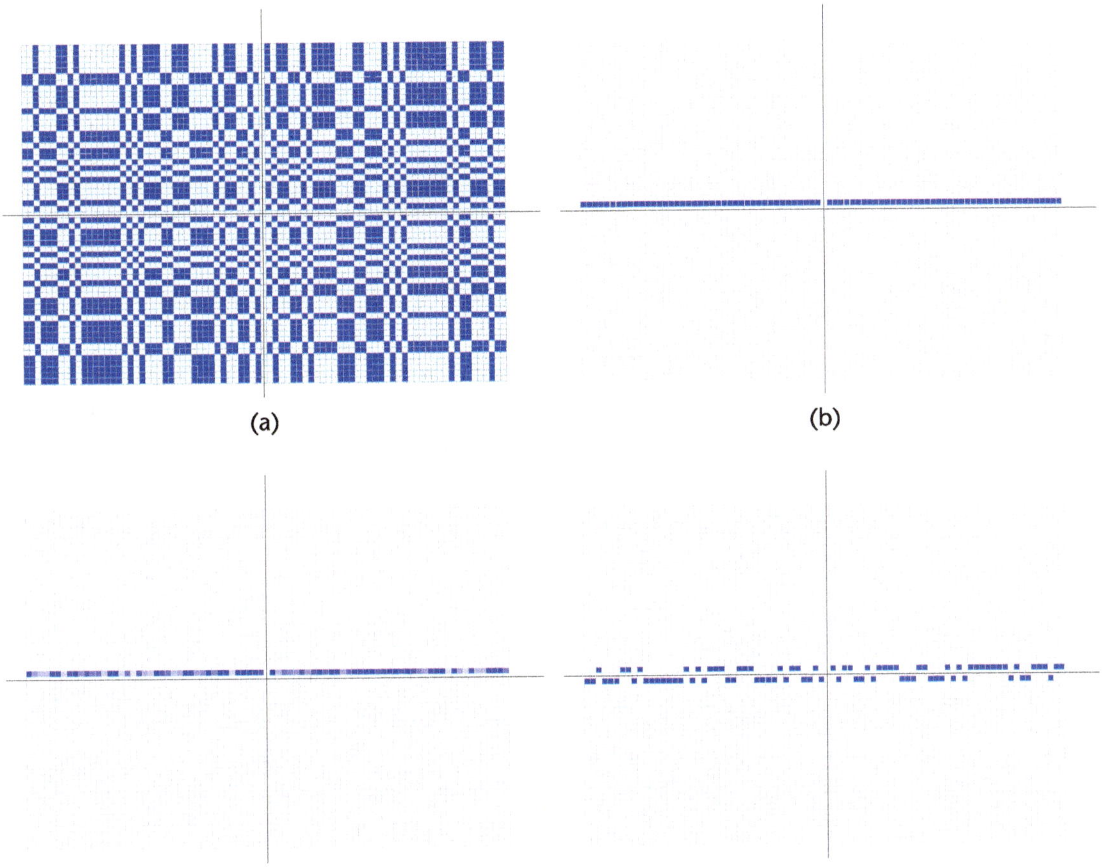

(a) (b) (c) (d)

Figure 6.15 Orbits in the $(83, 59)$-plane: (a) Orbit of 2, (b) orbit of 60, (c) orbit of 60 multicolor, and (d) orbit of 117.

Finally, we summarize: The orbits of the second case where there are only small prime divisors of $\phi(n)$ cannot be as long as in the first case where there are large prime divisors of $\phi(n)$.

This fact is, for example, used in the Pollard $p-1$ algorithm for factoring large numbers with only small prime divisors. We'll come back to this in Section 6.5.11.

6.5.8.3 RSA Orbits

We already mentioned in Section 6.5.7 that $f_k : z \mapsto z^k$ for $\gcd(k, \phi(n)) = 1$ is an automorphism of $\mathbb{Z}_n^*$.

First of all, this means that f_k is a 1:1 mapping. What cannot happen with an automorphism is similar to squaring, where two different points are mapped onto the same image point, as illustrated in Figure 6.11 (in fields of characteristic 2 squaring is bijective, for $p \neq 2$ it is not, but the case $p = 2$ is irrelevant for RSA).

If $\gcd(k, \phi(n)) = 1$ then also $\gcd(k, p-1) = 1$ and $\gcd(k, q-1) = 1$ because of $\phi(n) = (p-1)(q-1)$ for RSA. Therefore, the map f_k is also a 1:1 map on the axes. To make this clearer, have a look at Section 6.5.8.2 and associate the axes with groups of order $p-1$ or $q-1$.

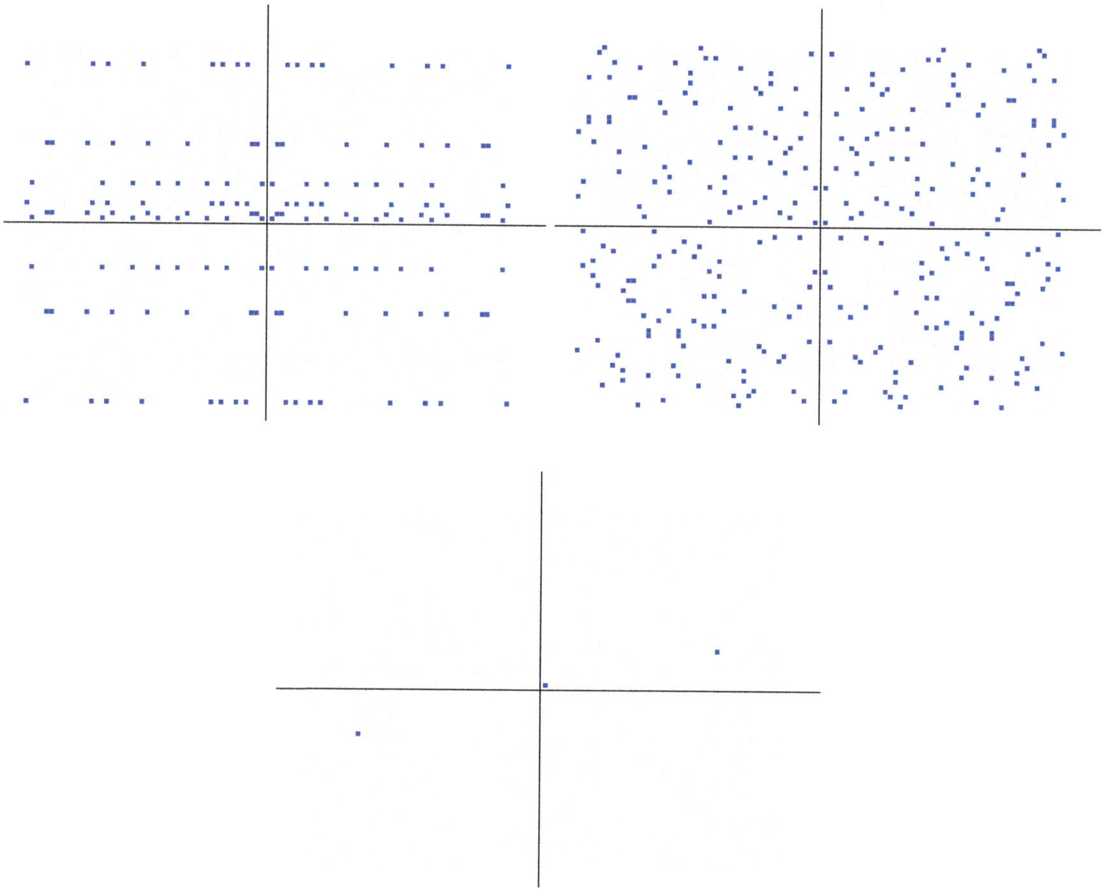

Figure 6.16 Orbit of 2 (upper left), 5 (upper right), and 811 (lower) in the $(97, 73)$-plane.

Until now, we looked at an orbit of an arbitrary but fixed z, where we thought of the set of all z^k for all possible k. Now we change our focus and choose some k relative prime to $\phi(n)$, think of k fixed and look at the set of all z^k for all possible z. Since f_k is one-to-one, f_k is a permutation of the points of the RSA plane and therefore the set of all such automorphisms is a subgroup of the symmetric group S_n on n elements. Every f_k leaves the 0 or the origin $(0, 0)$ invariant, also every axis invariant as a *set* but not *pointwise*. In general, a permutation $\sigma \in S_n$ leaves a subset $M \subseteq \{0, 1, \ldots, n-1\}$ invariant as a set if $\sigma(z) \in M$ for every $z \in M$. A permutation $\sigma \in S_n$ leaves a subset $M \subseteq \{0, 1, \ldots, n-1\}$ pointwise invariant if $\sigma(z) = z$ for every $z \in M$.

Furthermore, the point $(1, 1)$ is always a fixpoint—also called fixed point or invariant point—of any automorphism.

For encryption with the RSA algorithm the points of the RSA plane form the set of all possible plaintexts; the encrypted texts then consist of the images under f_k for some k relatively prime to $\phi(n)$. The points $(0, 0)$ and $(1, 1)$ apparently are not the best choice as a plaintext.

We now want to elaborate the composition of two automorphisms f_k and f_l as well as the inverse map of f_k, which is used for deciphering in the RSA cryptosystem. We restrict ourselves to the inner points or $\mathbb{Z}_n^*$.

The composition $f_l \circ f_k$ for two numbers $k, l \in \{1, \ldots, \phi(n)\}$ relatively prime to $\phi(n)$ yields:

$$(f_l \circ f_k)(z) = f_l(f_k(z)) = (z^k)^l \stackrel{(*)}{=} z^{k \cdot l} \stackrel{(*)}{=} z^{k \cdot l \bmod \phi(n)} = f_{k \cdot l \bmod \phi(n)}(z)$$

In the case of k and l not having common divisors with $\phi(n)$ the product kl also is relatively prime to $\phi(n)$. If kl is larger than $\phi(n)$, the remainder $r := kl \bmod \phi(n)$ with $0 < r < \phi(n)$ is also relatively prime to $\phi(n)$. Therefore, the composition $f_l \circ f_k$ is also a one-to-one map on the inner points.

We mention that the equalities marked with $(*)$ in the computation above hold only in the finite group $\mathbb{Z}_n^*$. Because if z^k or $z^{k \cdot l}$ would be natural numbers larger than n we would have to write $\cdots \equiv \ldots \bmod n$ instead.

For the inverse of f_k, the deciphering map, one uses the extended Euclidean algorithm: For every $k \in \{1, \ldots, \phi(n)\}$ there can be found a unique $k^{\text{inv}} \in \{1, \ldots, \phi(n)\}$ such that $k \cdot k^{\text{inv}} \equiv 1 \bmod \phi(n)$. Then $f_{k^{\text{inv}}} \circ f_k$ is the identity map and $(f_k)^{-1} = f_{k^{\text{inv}}}$ is the inverse map of f_k.

The set of all automorphisms of $\mathbb{Z}_n^*$ is usually denoted with $\text{Aut}(\mathbb{Z}_n^*)$. Note that not every automorphism of $\text{Aut}(\mathbb{Z}_n^*)$ is of the form $z \mapsto z^k$.

The Chinese remainder theorem gives us

$$\text{Aut}(\mathbb{Z}_n^*) \cong \text{Aut}(\mathbb{Z}_p^* \times \mathbb{Z}_q^*),$$

but because of $p - 1$ and $q - 1$ having common divisors, there is *no* isomorphism from the automorphism group of the product $\mathbb{Z}_p^* \times \mathbb{Z}_q^*$ to the product of the automorphism groups of $\mathbb{Z}_p^*$ and $\mathbb{Z}_q^*$:

$$\text{Aut}(\mathbb{Z}_p^* \times \mathbb{Z}_q^*) \ncong \text{Aut}(\mathbb{Z}_p^*) \times \text{Aut}(\mathbb{Z}_q^*)$$

The structure of the full automorphism group of finite abelian groups is described in the literature in, for example, `https://arxiv.org/pdf/math/0605185.pdf`.

For RSA we only need automorphisms that are power maps. Those form a subgroup of $\text{Aut}(\mathbb{Z}_n^*)$ that will be written as A or sometimes, to avoid confusion, as A_n or also $\text{Aut}_{\text{RSA}}(\mathbb{Z}_n^*)$. The corresponding automorphisms we call power automorphisms or RSA automorphisms. How many and which prime factors the number $\phi(n)$ has and also with which exponents they occur in $p - 1$ and $q - 1$ determines the number of possible power automorphisms of $\mathbb{Z}_n^*$. The following holds: $|A| = \phi(\text{lcm}(p-1, q-1))$

Now finally let's define the notion of an RSA orbit.

Definition: Let $n = pq$ be the product of two different primes and $A \subseteq \text{Aut}(\mathbb{Z}_n^*)$ the subgroup of power automorphisms of $\mathbb{Z}_n^*$. Further, let z be an arbitrary element of $\mathbb{Z}_n^*$. Then we call the set $z^A := \{z^a : a \in A\}$ the RSA orbit or the RSA path of z.

Every RSA orbit of an element z is contained in the full orbit of z, but RSA orbits are *not* groups themselves, because z^A cannot contain the unit 1 except, of course, if $z = 1$. If the orbit of z has length d for some divisor d of $\phi(n)$, then the RSA orbit of z has exactly $\phi(d)$ elements, whereas the normal orbit consists of d elements since as a set it is identical with the cyclic subgroup generated by z. But this is true only for exponents k that are relatively prime to the group order exponentiation to the power of k maps generators z of cyclic groups to generators of this group again.

Before we look at examples, we have to make sure that the notion of an RSA orbit can also be well-defined for axis elements. This could be shown with the CRT if one uses the theory of rings. Since we want to work without this theory, we will get back to this later when introducing projections in Section 6.5.9.

Now a first example: $n = 77, p = 11, q = 7, \phi(n) = 60, z = 2$

The top partial image in Figure 6.17 shows the full orbit of 2 in the $(11, 7)$-plane, the bottom one is its RSA orbit. The 2 generates the following subgroup of $\mathbb{Z}_{77}^*$ with 30 elements:

$$2, 4, 8, 16, 32, 64, 51, 25, 50, 23, 46, 15, 30, 60, 43, 9, 18, 36, 72, 67,$$
$$57, 37, 74, 71, 65, 53, 29, 58, 39, 1$$

The exponents that are relatively prime to 30 are 1, 7, 11, 13, 17, 19, 23, and 29, the corresponding RSA orbit then contains the following $\phi(30) = 8$ elements:

$$(*)\begin{cases} 2, \\ 51 \equiv 2^7 \bmod 77 = 128 \bmod 77, \\ 46 \equiv 2^{11} \bmod 77 = 2048 \bmod 77, \\ 30 \equiv 2^{13} \bmod 77 = 8192 \bmod 77, \\ 18 \equiv 2^{17} \bmod 77 = 131072 \bmod 77, \\ 72 \equiv 2^{19} \bmod 77 = 524288 \bmod 77, \\ 74 \equiv 2^{23} \bmod 77 = 8388608 \bmod 77, \\ 39 \equiv 2^{29} \bmod 77 = 536870912 \bmod 77 \end{cases}$$

For $\phi(11 \cdot 7) = 10 \cdot 6 = 60$ we have $2^2 \cdot 3 \cdot 5$ as prime factorization and $2 \cdot 3 \cdot 5 = 30$ as $\operatorname{lcm}(10, 6)$. Then we have

$$A_{77} = \mathrm{Aut}_{\mathrm{RSA}}(\mathbb{Z}_{77}^*) \cong \mathrm{Aut}_{\mathrm{RSA}}(\mathbb{Z}_{11}^* \times \mathbb{Z}_7^*) \cong \mathrm{Aut}_{\mathrm{RSA}}(C_{10} \times C_6)$$

with two cyclic groups of order 10 and 6. We use the standard notation C_N for a cyclic group of order N, this group being multiplicative if not told otherwise. Of course, we have an isomorphism between $(C_N, \cdot)$ and $(\mathbb{Z}_N, +)$.

Since $C_{10} \cong C_2 \times C_5$ and $C_6 \cong C_2 \times C_3$ with cyclic groups of order 2, 3, and 5 we also have:

$$\mathrm{Aut}_{\mathrm{RSA}}(C_{10} \times C_6) \cong \mathrm{Aut}_{\mathrm{RSA}}(C_2 \times C_2 \times C_3 \times C_5)$$

Since 2, 3, and 5 do not have common divisors, it follows that

$$\mathrm{Aut}_{\mathrm{RSA}}(C_2 \times C_2 \times C_3 \times C_5) \cong \mathrm{Aut}_{\mathrm{RSA}}(C_2 \times C_2) \times \mathrm{Aut}_{\mathrm{RSA}}(C_3) \times \mathrm{Aut}_{\mathrm{RSA}}(C_5).$$

The Klein four group $C_2 \times C_2$ does not allow power automorphisms because automophisms map generators of cyclic subgroups to generators again. However, each

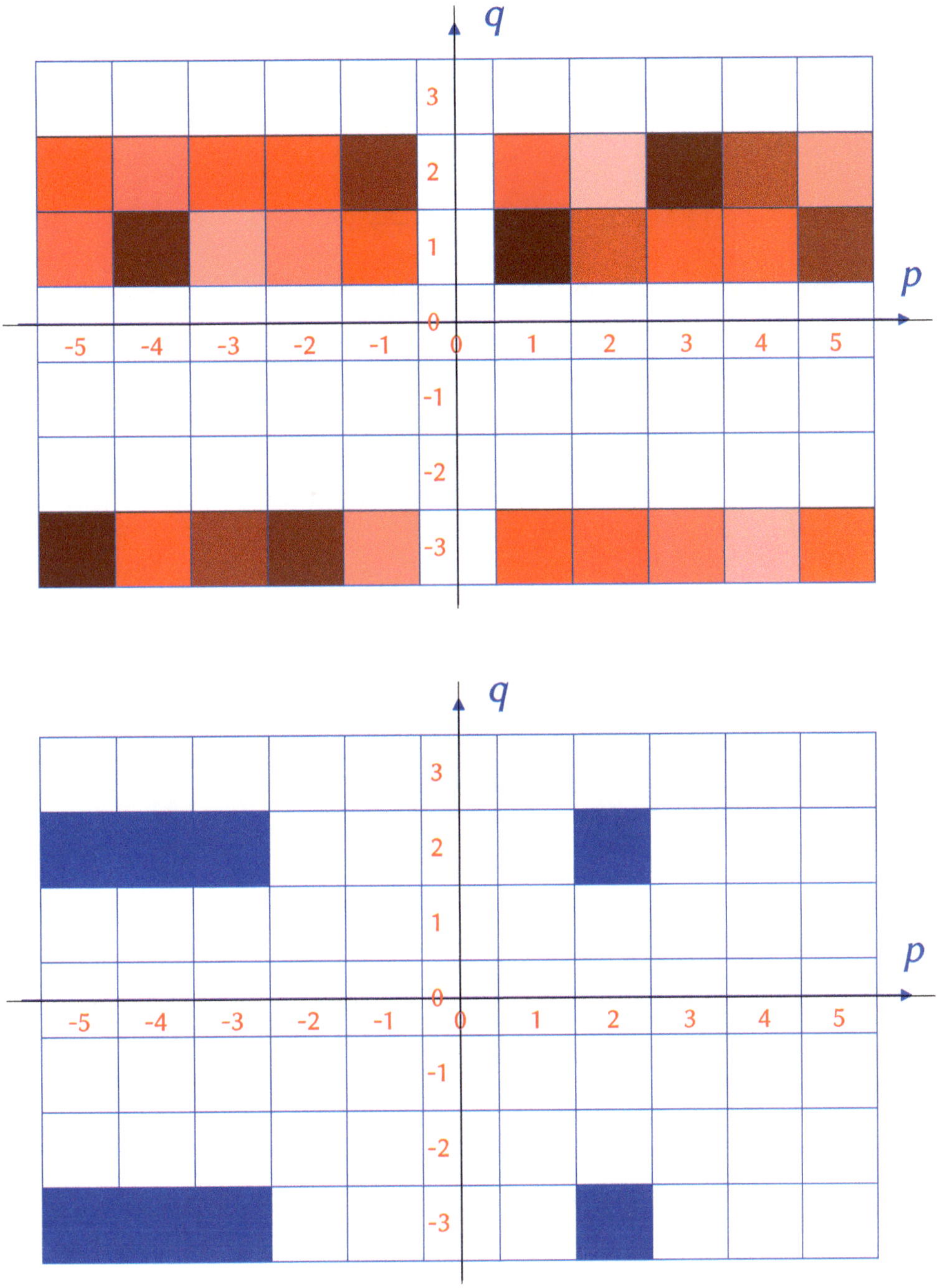

Figure 6.17 Full orbit of 2 (upper) and RSA orbit of 2 (lower), both in the $(11, 7)$-plane.

C_2-component has only a single generator and therefore does not allow a mapping to *another* generator of the same cyclic subgroup. This leaves:

$$A_{77} \cong \mathrm{Aut}_{\mathrm{RSA}}(C_2 \times C_2 \times C_3 \times C_5) \cong \mathrm{Aut}_{\mathrm{RSA}}(C_3) \times \mathrm{Aut}_{\mathrm{RSA}}(C_5).$$

The groups C_3 and C_5 have prime order, so there are no nontrivial (trivial subgroups of any group G are always the subgroup consisting of the unit element alone and the whole group G) subgroups and exactly $2 = 3 - 1$ and $4 = 5 - 1$ generators. From this we get $\mathrm{Aut}_{\mathrm{RSA}}(C_3) \cong C_2$ as well as $\mathrm{Aut}_{\mathrm{RSA}}(C_5) \cong C_4$, and then also $A_{77} \cong C_2 \times C_4$, and therefore finally $|A_{77}| = 8$.

The automorphism $\sigma := f_7 : z \mapsto z^7$, operating repeatedly on $2 \in \mathbb{Z}_{77}$, generates the following series:

$$2 \xrightarrow{\sigma} 2^7 \xrightarrow{\sigma} 2^{49} = 2^{19} \xrightarrow{\sigma} 2^{19\cdot 7} = 2^{13} \xrightarrow{\sigma} 2^{13\cdot 7} = 2$$

Using the automorphism $\tau := f_{11}$ we have $2 \rightarrow 2^{11} \rightarrow (2^{11})^{11} = 2^{121} = 2$.

For $\rho := f_{17}$ there are four steps again:

$$2 \xrightarrow{\rho} 2^{17} \xrightarrow{\rho} 2^{289} = 2^{19} \xrightarrow{\rho} 2^{19\cdot 17} = 2^{23} \xrightarrow{\rho} 2^{13\cdot 17} = 2$$

The last of the maps in the previous list $(*)$ we call $\lambda := f_{29}$. The map λ then is the multiplicative inverse $2^{-1} = 2^{29}$ of the element 2 in $\mathbb{Z}_{77}$. From this it follows that $\lambda^2 = f_1 = \mathrm{id}$.

We showed that $\{\sigma, \sigma^2 = \rho^2, \sigma^3, \tau, \rho, \rho^3, \lambda, \mathrm{id}\}$ form exactly the power automorphism group $\mathrm{Aut}_{RSA}(\mathbb{Z}_{77}^*)$, which is isomorphic to the additive group $\mathbb{Z}_2 \times \mathbb{Z}_4$.

This means the enciphering of the "plaintext" 2 via RSA with the modulus 77 can only be done in seven possible ways.

Second example: $n = 3097$, $p = 163$, $q = 19$, $p - 1 = 162 = 2 \cdot 3^4$, $q - 1 = 18 = 2 \cdot 3^2$, $\phi(n) = 2916 = 2^2 \cdot 3^6$, $\mathrm{lcm}(162, 18) = 162$.

The elements of $\mathbb{Z}_{3097}$ can have a path of the following length:

$$\{1, 2, 3, 6, 9, 18, 27, 54, 81, 162\}$$

Here are the first elements:

Element	Length of Path in $\mathbb{Z}_{3097}$
0	1
1	1
2	$162 = 2 \cdot 3^4$
3	162
4	$81 = 3^4$
5	$54 = 2 \cdot 3^2$
6	27
7	162
8	54
9	81
10	162
11	162
...	...
85	9
...	...

The lengths of the corresponding RSA orbits are as follows:

Element	Length of RSA Orbit in $\mathbb{Z}_{3097}$
0	1
1	1
2	54
3	54
4	54
5	18
6	18
7	54
8	18
9	54
10	54
11	54
...	...
85	6
...	...

This second example 2 was chosen similar to the preceding one (example 1) where also $\phi(n)$ does have only small prime divisors, here 2 and 3. We investigate the RSA automorphism group A_{3097} by decomposing like in the previous example:

$$\begin{aligned}
A_{3097} &\cong \text{Aut}_{RSA}(\mathbb{Z}^*_{163} \times \mathbb{Z}^*_{19}) \\
&\cong \text{Aut}_{RSA}(C_{2\cdot 3^4} \times C_{2\cdot 3^2}) \\
&\cong \text{Aut}_{RSA}(C_2 \times C_2 \times C_{3^2} \times C_{3^4}) \\
&\cong \text{Aut}_{RSA}(C_{3^4}) \\
&\cong C_{\phi(3^4)} \\
&\cong C_{54} \\
&\cong C_2 \times C_{3^3} \\
&\cong (\mathbb{Z}_2, +) \times (\mathbb{Z}_{3^3}, +)
\end{aligned}$$

So for $n = 3097 = 163 \cdot 19$ an RSA path has maximal length 54 and therefore every plaintext $m \in \mathbb{Z}_n$ cannot have more than 54 ciphertexts.

In Figure 6.18 we show as an example all paths of length 9 and in Figure 6.19 the corresponding RSA paths of the generating elements of those paths. The corresponding Table 6.4 consists of one cyclic group per row, denoted by $c_a, \ldots c_l$. The elements z^i are taken from the set of representatives $\{0, 1, \ldots, 3096\}$ with the generator z in the third column. For the pictures we chose the alternative set of representatives from $-(n-1)/2$ to $(n-1)/2$, see Section 6.5.5. SageMath Example 6.2 produces the LaTeX code for Table 6.4.

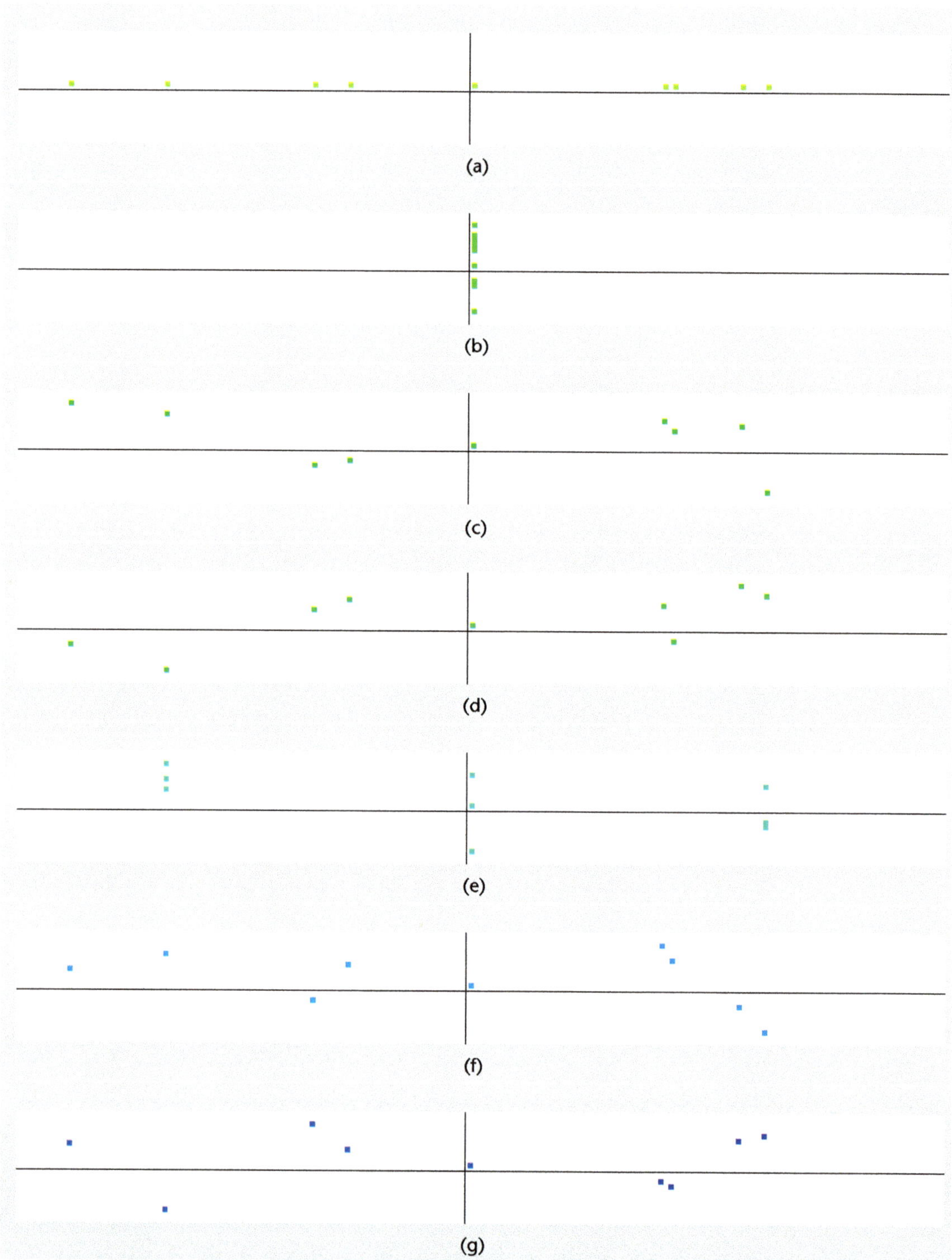

Figure 6.18 Paths of length 9 in the (163, 19)-plane: orbits of (a) 1996, (b) 327, (c) 2322, (d) 1600, (e) 593, (f) 574, (g) 2911, (h) 384, (i) 1505, (j) 2987, (k) 2648, and (l) 1926.

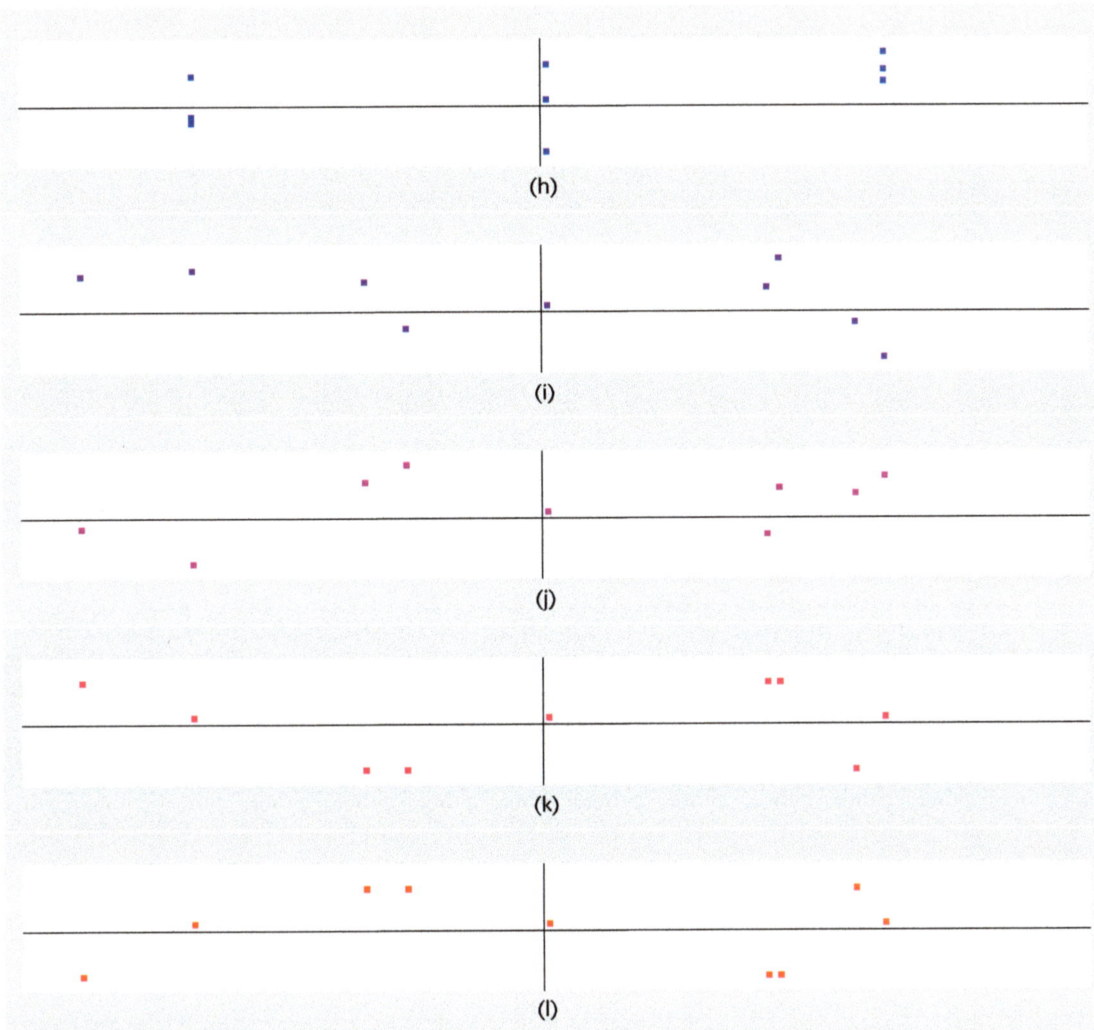

Figure 6.18 (*Continued*)

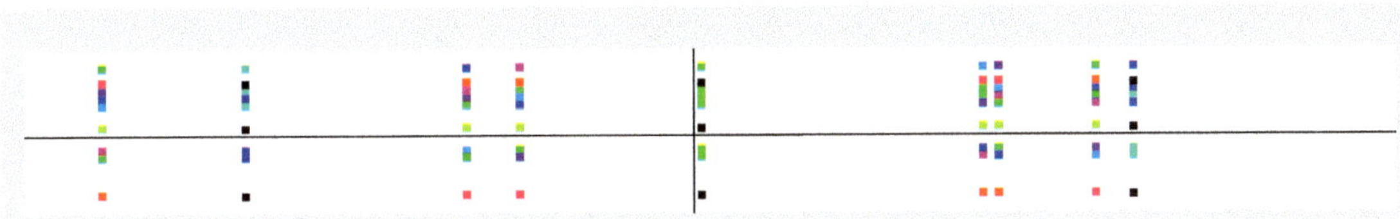

(a) All orbits of length 9 and the corresponding RSA orbits of length 6, common color for points in the same RSA orbit, subgroup of order 3 in black.

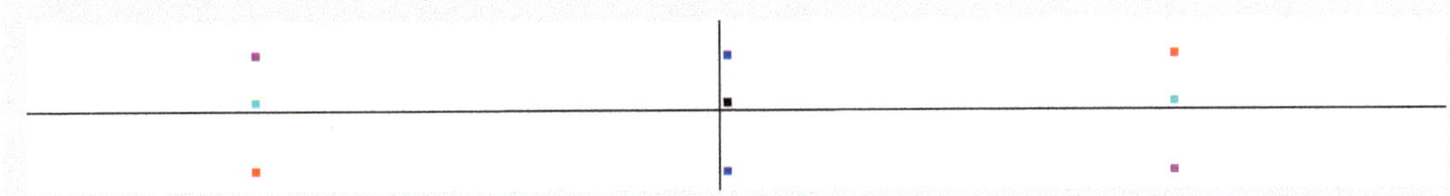

(b) All orbits of length 3 and the corresponding RSA orbits of length 2, common color for points in the same RSA orbit, unit element in black.

Figure 6.19 Orbits of length 9 and 3 in the $(163, 19)$-plane.

Table 6.4 Orbits of Inner Points of Length 9 in $\mathbb{Z}_{3097}$

	z^9	z	z^2	z^3	z^4	z^5	z^6	z^7	z^8
c_a	1	1996	1274	267	248	2585	58	1179	2661
c_b	1	327	1631	653	2935	2772	2120	2609	1468
c_c	1	2322	2904	919	85	2259	2177	690	1031
c_d	1	1600	1878	710	2498	1670	2386	2096	2646
c_e	1	593	1688	653	104	2829	2120	2875	1525
c_f	1	574	1194	919	1016	948	2177	1507	955
c_g	1	2911	529	710	1111	853	2386	2172	1715
c_h	1	384	1897	653	2992	3038	2120	2666	1734
c_i	1	1505	1118	919	1833	2335	2177	2856	2741
c_j	1	2987	2809	710	2422	3019	2386	785	366
c_k	1	2648	296	267	900	1607	58	1831	1683
c_l	1	1926	2367	58	216	1018	267	140	201

SageMath Example 6.2: Generation of Table 6.4 with SageMath

```
print("\n# CHAP06 -- Sage-Script-SAMPLE 010: =========")

n=3097; # n: RSA modulus
R=Zmod(n); U=R.unit_group(); # R: Ring of integers mod n, U: elements ▸
  ▸of R that do have
                                    # an multiplicative inverse mod n
oi=9 #order of interest

# define function that returns the multiplicative orbit of z in Zmod(n)▸
  ▸ as a list
def orbit(z,n):
  if gcd(z,n)!=1:
    return(print('multiplicative order does not exist!'))
  o=R(z).multiplicative_order()
  orbit=[]
  x=1
  for i in range(o):
    orbit.append(x)
    x=x*z%n
  return(orbit)

# define function returning a list of lists
# with all inner elements with order i, partitioned by cyclic group
def parti(i,n):
  l=[] # list of all orbits of order i as list, so list of lists
  s=Set() # set of all elements of order i
  for x in U:
    if x.multiplicative_order()==i:
      if R(x) not in s:
        l=l+[orbit(int(R(x)),n)]
        s=s+Set(Set(l[-1]))
  return(l)

# define function that returns the LaTeX source code for all the non-▸
  ▸header-rows of the table
def rows():
  orbs=parti(oi,n)
  s=''
  col0=[r'$c_a$',r'$c_b$',r'$c_c$',r'$c_d$',r'$c_e$',r'$c_f$', r'$c_g$▸
    ▸',r'$c_h$',r'$c_i$',r'$c_j$',r'$c_k$',r'$c_l$']
```

SageMath Example 6.2 (continued)

```
    for i in range(len(orbs)):
        zeile=LatexExpr(col0[i])+' & '+''.join(str(orbs[i][j])+' & ' for j in range(oi-1))+str(orbs[i][oi-1])+LatexExpr(r'\\')+'\n '
        s=s+zeile
    return(s)

# define string containing LaTeX source code for defining the structure and layout of the table
head=LatexExpr(r'\begin{tabular}{@{} r')+f'*{oi}'+LatexExpr(r'{>{$}r<{$}} @{}}\toprule')+'\n'
# define string containing LaTeX source code for the header line of the table
row0=LatexExpr(r'&z^9&\bm{z}&\bm{z^2}&z^3&\bm{z^4}&\bm{z^5}&z^6&\bm{z^7}&\bm{z^8}\\\midrule')+'\n'

# concatenate all strings for LaTeX table and append LaTeX code for ending the table
table=head+row0+rows()+LatexExpr(r'\bottomrule')+'\n'+LatexExpr(r'\end{tabular}')

# ------ main: write string >>table<< to file 'orbits-order9.txt' ------
filename = 'orbits-order' + str(int(oi)) + '.txt'
with open(filename, 'w') as file:
  try:
    file.write(table)
    print("File '" + filename + "' created.")
  except:
    print("File '" + filename + "' could not be opened.")
# after the "with" block is done, the file is guaranteed to be closed.
```

The smallest positive integer representing an element of order 9 in $\mathbb{Z}^*_{3097}$ is 85. It is contained in the third cyclic group c_c, but is not its generator, but its fourth power: $(2322^4 - 85) : 3097 = 9386585843$.

Table 6.4 was generated with SageMath. The loop that produces the rows follows the order in which SageMath orders the unit group (`U=R.unit_group()` and `for x in U`, see SageMath Example 6.2). We won't go into details here on how SageMath orders these units (it follows from the internal representation of abelian groups in SageMath). But this is the reason why on first sight the order of the rows of the table as well as the choice of the generating element z seem to be random. Instead of z, the generator of a cyclic group of order 9 can also be z^2, z^4, z^5, z^7, or z^8. These six elements then form an RSA path. For the RSA automorphism $f_k : z \mapsto z^k$ every $k \in \{1, 2, 4, 5, 7, 8\}$ is possible. But because of 2 dividing $(p-1)(q-1) = 2916$, there remain only $k = 1, 5$, and 7 at first thought. But as the following is valid:

$$
\begin{aligned}
5^2 &= 25 \equiv 7 \bmod 9 && \Longrightarrow f_5 \circ f_5 = f_7 \text{ reduced to } \langle z \rangle \text{ for some } z \text{ of order } 9\\
5^3 &\equiv 35 \equiv 8 \bmod 9 && \Longrightarrow f_5 \circ f_5 \circ f_5 = f_8 \qquad \dots\\
5^4 &\equiv 40 \equiv 4 \bmod 9 && \dots\\
5^5 &\equiv 20 \equiv 2 \bmod 9 && \dots\\
5^6 &\equiv 10 \equiv 1 \bmod 9, && \dots
\end{aligned}
$$

this means that—restricted to the cyclic groups of order 9 under consideration—also even exponents do occur.

The 12 different paths of length 9 can be further divided into four times three, the following triples having a subgroup of order 3 as intersection:

$$c_a \cap c_k \cap c_l = \{1, 267, 58\}$$
$$c_b \cap c_e \cap c_h = \{1, 653, 2120\}$$
$$c_c \cap c_f \cap c_i = \{1, 919, 2177\}$$
$$c_d \cap c_g \cap c_j = \{1, 710, 2386\}$$

If we strip away those three elements from such a cyclic group of order 9, the remaining set of elements forms an RSA orbit consisting of all the generators of this group of order 9.

This is illustrated in Figure 6.19. The upper part shows the union of all paths of length 9, and the lower part shows the four subgroups of order 3 (in the upper image (a) in black), these are exactly eight elements if the 1 (in the bottom figure (b) in black) is excluded. Points in the upper figure with the same color—not black—form exactly one RSA orbit of length 6.

6.5.9 Projections

Just as in the real plane $\{(x, y) : x, y \in \mathbb{R}\}$, one can also investigate projections onto the axes in finite planes. From the point of view of linear algebra, a two-dimensional vector space—a plane—is mapped onto a one-dimensional subspace with a line through the origin, so this map is a linear map of rank 1. The image of this map is the line that is projected onto, and the kernel of the map is identical with the line through the origin that is parallel to the direction of the projection, which must not be orthogonal to the image.

If the reader is not familiar with the terms *subspace*, *rank*, or *kernel*, please consult any introduction to linear algebra, such as [14].

Because in our model only orthogonal projections are considered, we use the short term *projection* instead of *orthogonal projection.*

The RSA plane is not a vector space in two dimensions; we already discussed this in Section 6.5.2. As a mapping of sets the projection map can be well-defined on the RSA plane.

The maps $\pi_x : (x, y) \mapsto (x, 0)$ and $\pi_y : (x, y) \mapsto (0, y)$ map points onto their shadows on the horizontal and vertical axes, so these are the two orthogonal projections that everyone knows from elementary geometry.

Following an inner point while stepping through its path, the shadows also follow those steps and move with the path. Vice versa, it is possible to reconstruct from the combined paths of the horizontal and vertical shadows the path of the point having those shadows.

We illustrate this with the example $p = 19, q = 7, n = 133, z = 12$.

First we show the path of 12 in Figure 6.20.

Figure 6.21 also shows the paths of the shadows.

For the path of 12 in the (19, 7)-plane, the length of the path of this inner point is for both axes equal to the length of the projected path. But this doesn't have to be

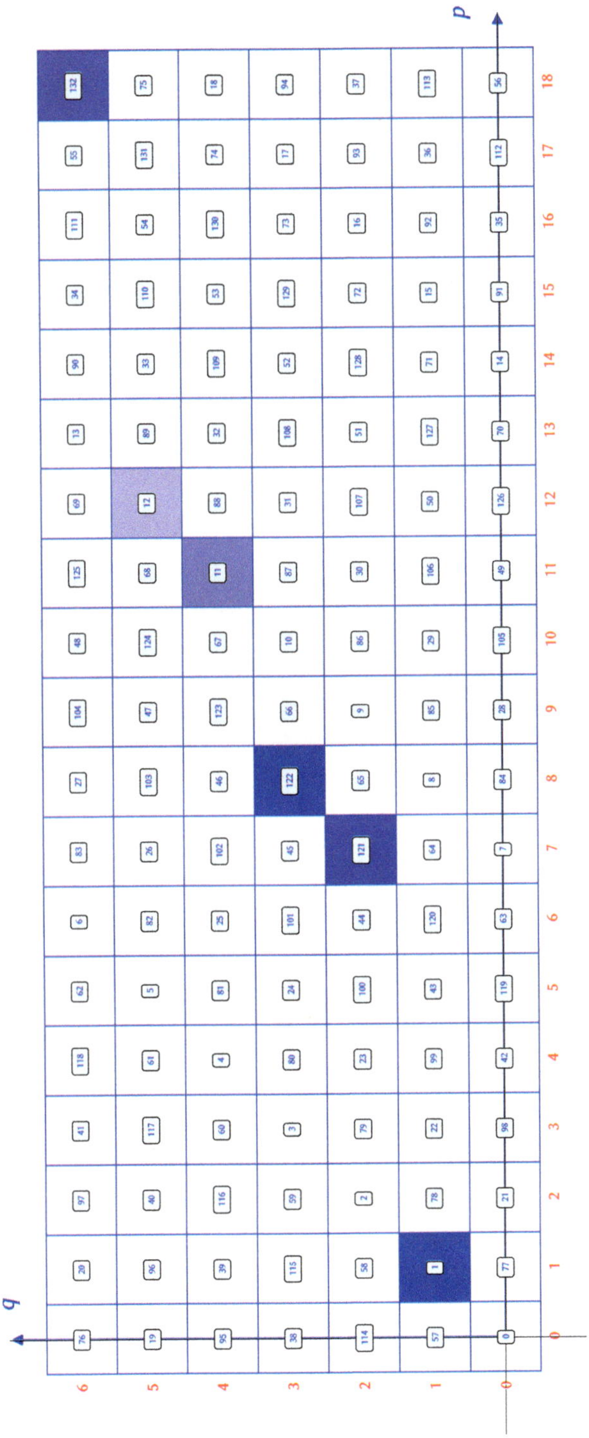

Figure 6.20 Orbit of 12 in the (19, 7)-plane.

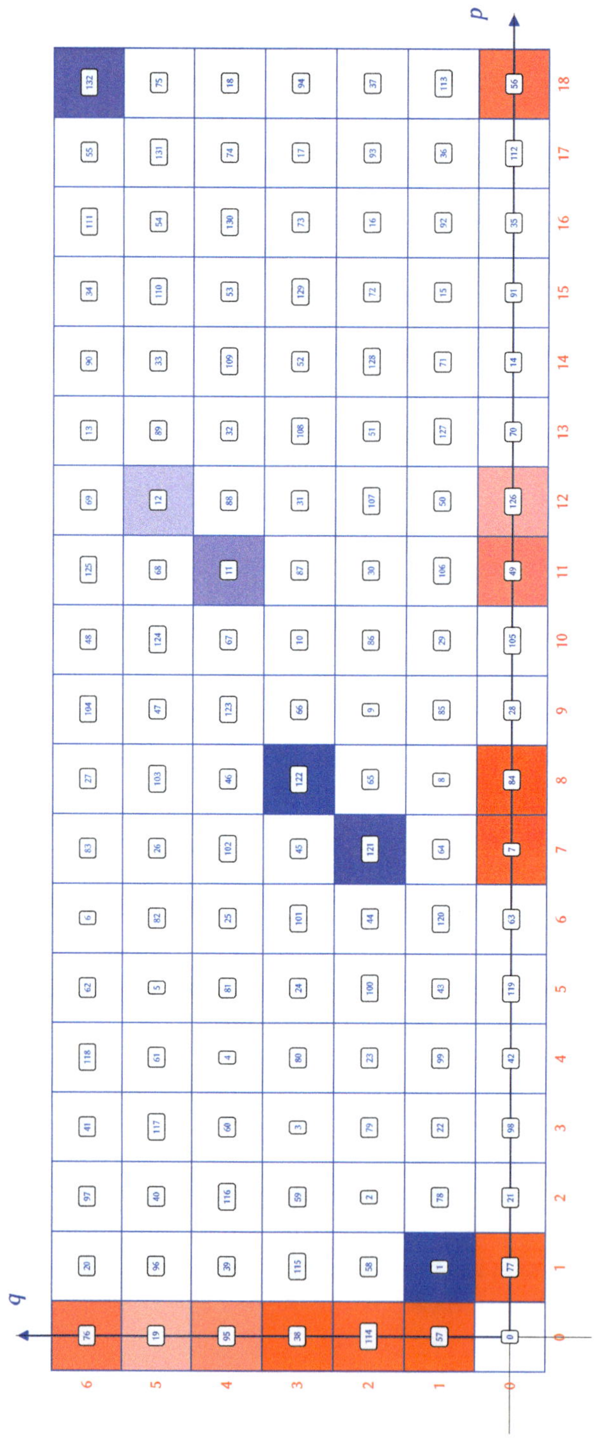

Figure 6.21 Orbit of 12 and its projections in the (19, 7)-plane.

this way; see the example of the path of 2 in Figure 6.22. It consists of 18 elements and the path projected onto the vertical axis has only three elements.

The projection of the path of an inner point onto one of the axes not only is a subset of the axis in question, but it inherits a part of the algebraic structure being projected.

The algebraic structure of any path is that of a multiplicative cyclic group with an order dividing the least common multiple of $p - 1$ and $q - 1$.

Formally, a projected path is not a multiplicative group because it does not contain the neutral element $(1, 1)$. But if one thinks *either* of just stripping away the one coordinate which equals 0, *or* of replacing the 0 with 1, the image of the projection map is identical or isomorphic to a subgroup of $\mathbb{Z}_p^*$ or $\mathbb{Z}_q^*$. Then π_x or π_y are inducing a map from $\mathbb{Z}_p \times \mathbb{Z}_q$ onto $\mathbb{Z}_p$ or $\mathbb{Z}_q$, respectively. For the sake of simplicity we denote this induced map also by π_x and π_y.

If one composes the map π defined in Section 6.5.1 with π_x or π_y one then gets $\mathbb{Z}_p$ or $\mathbb{Z}_q$ as image of $\mathbb{Z}_n$. In mathematics, such relationships are often described by commutative diagrams like in Figure 6.23.

Before moving to the next chapter, we want to point out that there is an interesting special case of RSA orbits: the fixed points. These are RSA orbits of length 1. In Section 5.17.7 one finds some information about fixed points.

Sometimes it is convenient not to look at projections onto the axes but onto the horizontal/vertical lines through $(0, 1)/(1, 0)$, because then we have a well-defined group homomorphism with $(1, 1)$ on both lines.

The definition of a homomorphism is given in Section 10.1, the chapter about the origin of the term "homomorphic." An isomorphism or automorphism are special cases of homomorphisms.

In the 2D model you can think of $\mathbb{Z}_q^*$ as all inner points $(1, y)$; that is, the vertical line without the axis point $(1, 0)$. The second projection then has all inner points $(x, 1)$ as image; that is, the horizontal line without the axis point $(0, 1)$. We then call those two lines—without their axis point—punched lines (see Figure 6.24).

The orbit of an element of one of those lines is necessarily contained in the corresponding line. This can be seen in a picture from before; see Figure 6.15(b) and 6.15(c). These pictures show the path of 60 in the $(83, 59)$-plane. The number 60 is congruent to 1 modulo 59, so it has coordinates $(60, 1)$.

6.5.10 Reflections

If one chooses the representation with the origin in the center, one can consider reflections and rotations of 180° as one knows them from elementary geometry. The reflections about an axis—at the horizontal as well as at the vertical axis—and their successive execution (i.e., the point reflection or 180° rotation) form a group that is isomorphic to the Klein four group. Also, the multiplicative group $\mathbb{Z}_n^*$ contains a subgroup isomorphic to the Klein four group, which we already addressed in Section 6.5.8.2. Despite the isomorphism, these groups are of course fundamentally different in our context: The reflections permute the elements of $\mathbb{Z}_p \times \mathbb{Z}_q$, but are not themselves elements of the RSA plane.

In mathematics, these reflections are special cases of involutions (i.e., mappings which, executed twice in succession, yield the identity). In geometry, an involution

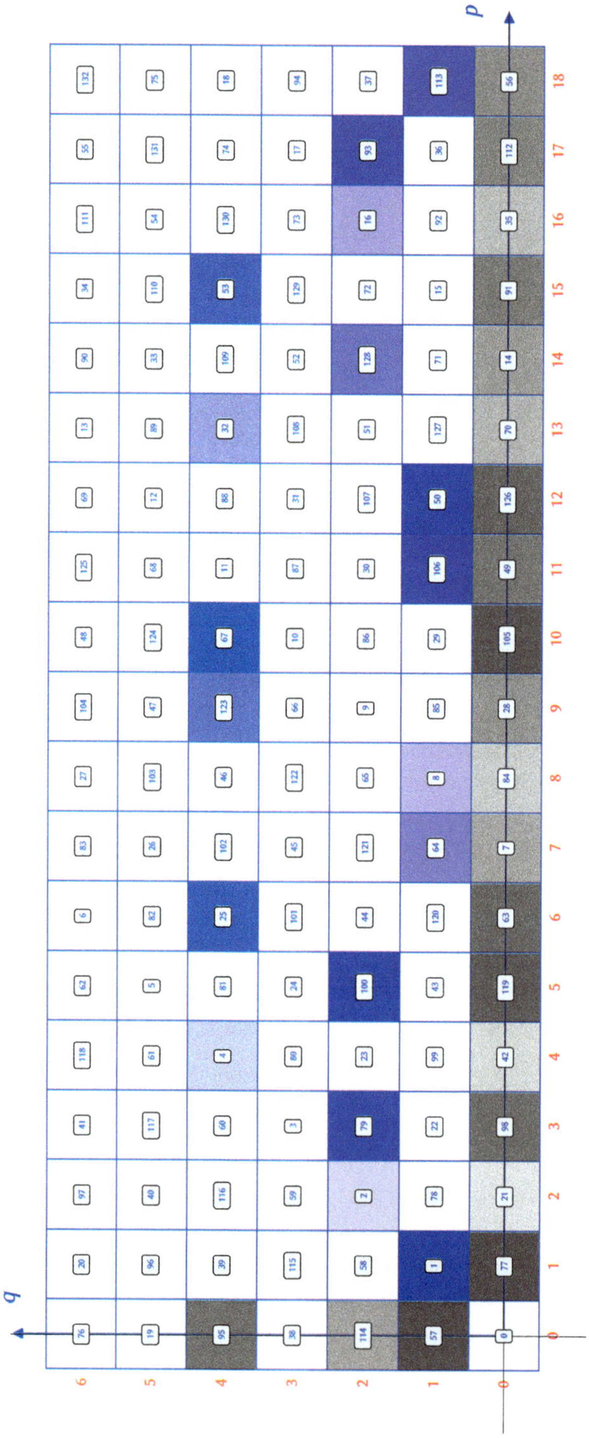

Figure 6.22 Orbit of 2 and its projections in the $(19, 7)$-plane.

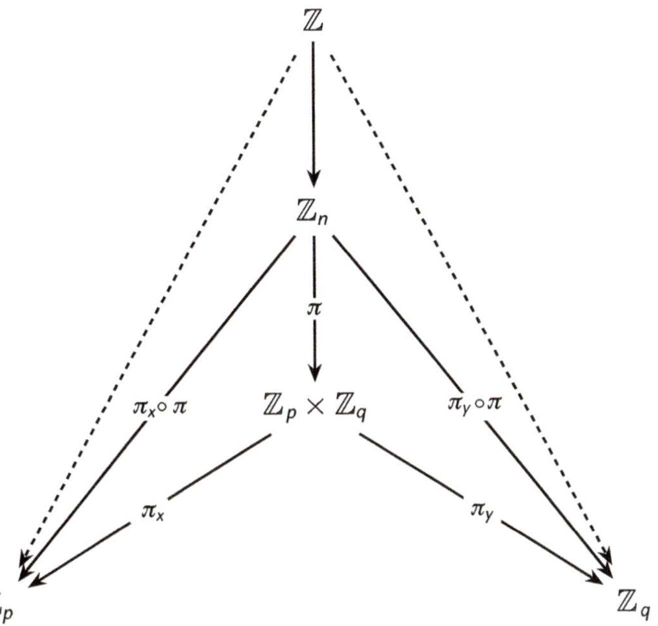

Figure 6.23 Projections π_x and π_y as part of a commutative diagram.

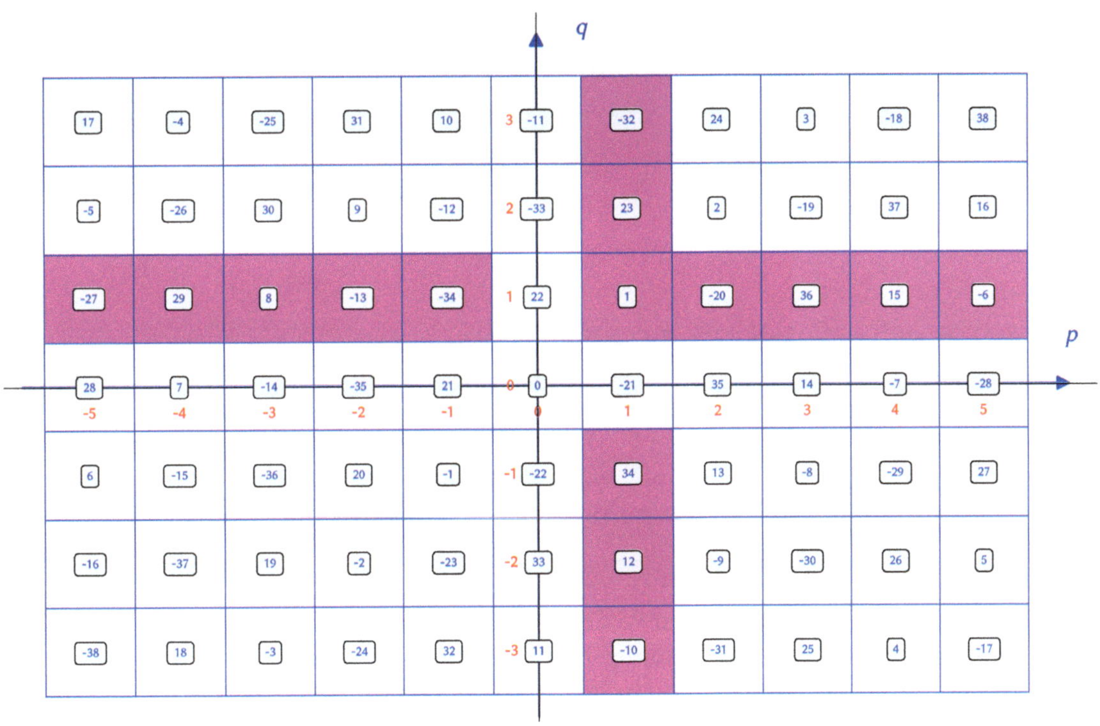

Figure 6.24 $\mathbb{Z}_{11}^*$ and $\mathbb{Z}_7^*$ as punched lines through $(1, 1)$ in the $(11, 7)$-plane.

is usually additionally required to respect the geometric structure; that is, to map straight lines back to straight lines and not to take a point out of a straight line and replace it by one not on the line. If one did not specify this condition, in principle every permutation on the given n points or numbers with order 2 would be an involution. Such permutations are then either transpositions or a product of disjoint transpositions.

A transposition on n elements swaps exactly two of these elements and fixes all others. Disjoint transpositions swap different elements; that is, element 1 with element 2 (noted by $(1,2)$) as well as element 3 with element 4, written $(3,4)$. The two transpositions $(1,2)$ and $(2,3)$ are not disjoint. If you compose these two, you get cycle $(1,3,2)$, which has not the order 2 but 3. More about permutations and transpositions can be found in Hungerford ([15]) on pages 46 ff.

We choose the following notations:

- σ: Reflection about the horizontal axis

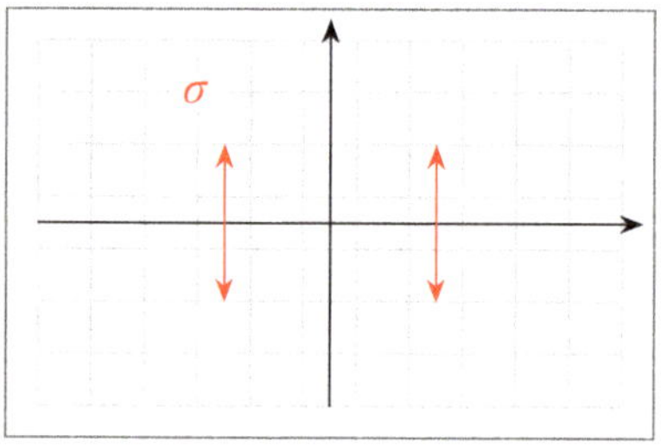

- τ: Reflection about the vertical axis

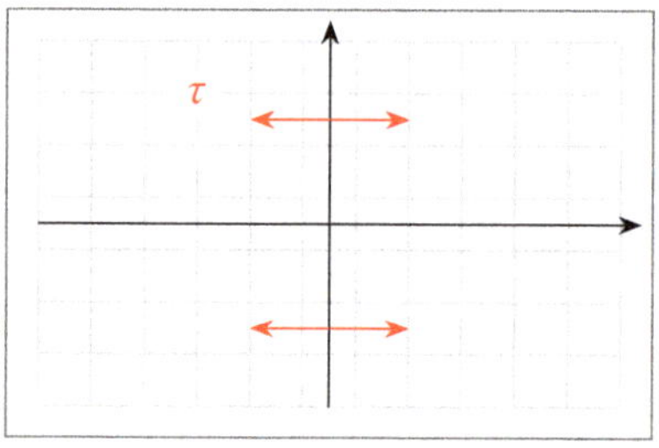

- ρ: Rotation of 180° $\sigma \circ \tau = \tau \circ \sigma = \rho$

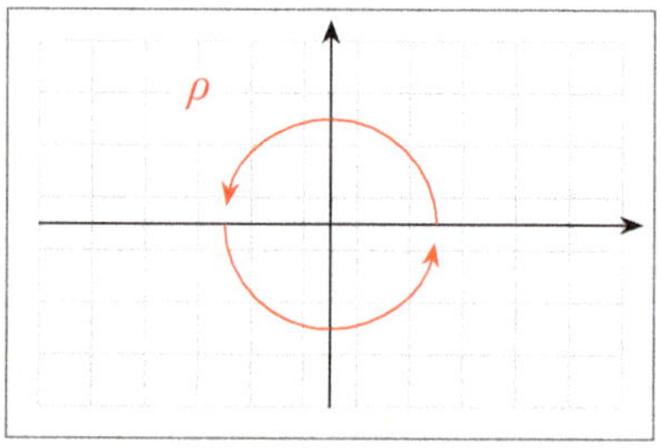

- κ: For the sake of simplicity, we sometimes use the letter κ for any involution from the set $K^* := \{\sigma, \tau, \rho\}$. When we speak of a reflection κ about an axis, either σ or τ is meant.

Although the question of whether there are other involutions operating on the RSA plane apart from σ, τ, and ρ is not uninteresting, we will not pursue it here.

It should be noted that the three reflections are not automorphisms of $\mathbb{Z}_n^*$: The neutral element of the multiplication, in the two-dimensional setting the point $(1,1)$, leaves its place under each of the three mappings σ, τ, and ρ, which cannot be the case for an automorphism. However, they are bijective and preserve the partition of $\mathbb{Z}_n$ into interior points and axis points; moreover, the origin is a fixed point. A fixed point is called an invariant.

As with all geometric mappings, one asks about the invariants. One distinguishes:

1. Single points, which remain invariant under κ, called *fixed points*; for example, the origin is fixed under κ.
2. Sets with several points, which remain *fixed pointwise*; that is, larger sets of fixed points. The axes of reflections are such sets.
3. Sets with several points, which remain *fixed setwise*, but not necessarily pointwise; any straight line perpendicular to the axis of a reflection κ forms such a set. Another classical example of larger sets that are fixed setwise is the circle line $\{z = e^{ix} : x \in \mathbb{R}\}$ of the complex number plane $\mathbb{C}$, when the inversion $z \mapsto z^{-1}$ is considered as an involution on $\mathbb{C}^*$.
4. Invariants that are not point sets but can be derived from geometric objects that are point sets, for example, area or orientation: For triangles, for example, there are two possible orientations, clockwise or counterclockwise; it is reversed by reflections, but is invariant under rotations by 180°. In two dimensions, the orientation can be defined with the help of the sign of the determinant by choosing one triangle vertex as origin and writing the vectors leading from this vertex to each of the other two vertices into a 2×2 matrix.

   However, we do not consider such derived invariants here, mainly because the RSA plane is not a plane in the strict mathematical sense (see 6.5.2) and therefore one has neither metrics nor the apparatus of linear algebra at hand and would therefore first have to examine how one could transfer the usual concepts such as determinant, surface content, angles, lengths, and so forth to the RSA plane at all.

We will now classify RSA planes whether they have orbits and/or RSA orbits that are invariant under one of these reflections. We will in Section 6.5.10.1 first classify the full orbits and then in Section 6.5.10.2 the RSA orbits.

6.5.10.1 Invariant Full Orbits

The existence of invariant full orbits is assured for each of the three involutions. Figure 6.12 shows an example of each one.

However, the question immediately arises whether such orbits exist for all p and q; after all one cannot draw general conclusions from $p = 11$ and $q = 7$. We now pursue this question, first for inner points and then for axis points.

Inner Points

We assume the orbit of an inner point $\langle(x, y)\rangle$ is symmetric. Then it contains the image of the point $(1, 1)$. Depending on the existing involution, this image is equal to the point $(-1, 1), (-1, -1)$ or $(1, -1)$, as can be seen in Figure 6.25.

If one of the points $(-1, 1), (-1, -1)$ or $(1, -1)$ lies on the path $\langle(x, y)\rangle$, then $\langle x\rangle \subseteq \mathbb{Z}_p$ or $\langle y\rangle \subseteq \mathbb{Z}_q$ or both must contain the element -1. Both -1 in the first component (i.e., taken as an element of the multiplicative group $\mathbb{Z}_p^*$) and -1 in the second component (i.e., taken as an element of the multiplicative group $\mathbb{Z}_q^*$), have order 2 as the only element in this respective group.

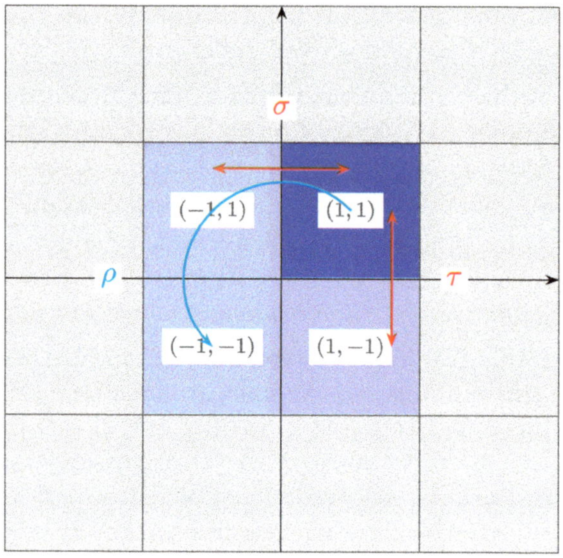

Figure 6.25 Possible images of the point $(1, 1)$ under the involutions σ, τ, or ρ.

Next, realize that a symmetric orbit can only contain exactly one of the points $\{(-1, 1), (-1, -1), (1, -1)\}$, because orbits are cyclic subgroups of $\mathbb{Z}_p^* \times \mathbb{Z}_q^*$ and these can have at most one element of order two.

Let us first consider the first component x of any orbit $\langle (x, y) \rangle$. It generates a cyclic subgroup of $\mathbb{Z}_p$ whose order $o(x)$ is a divisor of $p - 1$. If $o(x)$ is odd, the -1 is not in the path of x. Then none of the points $(-1, 1)$ and $(-1, -1)$ are contained in the orbit $\langle (x, y) \rangle$ and thus for odd $o(x)$ the reflections σ and ρ are already ruled out. Of course, the same reasoning applies to the second component y or $o(y)$. Since $o((x, y)) = \mathrm{lcm}(o(x), o(y))$, an orbit of odd length can have neither axis nor point symmetry.

Theorem 6.3 A necessary condition for the symmetry of an orbit is that the order of the element which generates it is even.

We now decompose the numbers $p - 1$ and $q - 1$ into an even and an odd part:

$$p - 1 = 2^a \cdot c \quad \text{and} \quad q - 1 = 2^b \cdot d,$$

where a and b are *positive* integers and c and d are *odd*. Since p and q are odd, $p - 1$ and $q - 1$ are even and therefore a and b positive.

Moreover, let u be a generating element of $\mathbb{Z}_p^*$ and v a generating element of $\mathbb{Z}_q^*$. Thus, for the pair (x, y) there are exponents r and s (these exponents are unique if one requires that they are divisors of $p - 1$ or $q - 1$ respectively) such that:

$$x = u^r, \qquad r \mid p - 1$$

$$y = v^s, \qquad s \mid q - 1$$

$$o(x) = \frac{p-1}{r} = \frac{2^a \cdot c}{r} := 2^\alpha \cdot \gamma, \qquad \gamma \quad \text{odd}, \ \alpha \geq 0$$

$$o(y) = \frac{q-1}{s} = \frac{2^b \cdot d}{s} := 2^\beta \cdot \delta, \qquad \delta \quad \text{odd}, \ \beta \geq 0$$
$$o((x, y)) = \text{lcm}(o(x), o(y)) = \text{lcm}(2^\alpha \gamma, 2^\beta \delta)$$

Four cases arise:

Case 1: $\alpha = \beta = 0$
Case 2: $\alpha = \beta > 0$
Case 3: $\alpha > \beta \geq 0$
Case 4: $\beta > \alpha \geq 0$

Case 1: If α and β are both equal to zero, the element orders of x and y are odd, so the element order of (x, y) is also odd, and thus the associated path is neither symmetric to any axis nor point-symmetric.

Case 2: For $\alpha = \beta \geq 1$, the orbit $\langle (x, y) \rangle$ is point-symmetric. This can be easily seen by representing (x, y) using the generating elements:

$$(x, y) = (u^r, v^s)$$

Considering the case $\alpha = \beta > 0$, we set $\alpha = \beta = m > 0$ and write $\text{lcm}(o(x), o(y)) = 2^m \cdot w$ with odd $w = \text{lcm}(\gamma, \delta)$. We also write $\gamma = t \cdot \gamma'$ and $\delta = t \cdot \delta'$, where t is the greatest common divisor of γ and δ, obtain $w = t\gamma'\delta'$, and then take the power as follows:

$$(x, y)^{2^{m-1} \cdot w}$$

This element is different from $(1, 1)$ and is of order 2 in $\mathbb{Z}_p^* \times \mathbb{Z}_q^*$. The same power applied to the individual components also yields an element of order 2, but in $\mathbb{Z}_p^*$ and $\mathbb{Z}_q^*$, respectively: $\mathbb{Z}_p^*$ or $\mathbb{Z}_q^*$:

$$x^{2^{m-1} \cdot w} = x^{2^{m-1} \cdot t\gamma'\delta'} = \left(x^{\delta'}\right)^{2^{m-1} \cdot \gamma}$$

$$y^{2^{m-1} \cdot w} = y^{2^{m-1} \cdot t\gamma'\delta'} = \left(y^{\gamma'}\right)^{2^{m-1} \cdot \delta}$$

Since $o(x)$ and δ' are coprime, x and $x^{\delta'}$ have the same order in $\mathbb{Z}_p^*$. If $(x^{\delta'})^{2^{m-1} \cdot \gamma}$ was already equal to 1, the element $x^{\delta'}$ and hence x would have an order equal to or a divisor of $2^{m-1} \cdot \gamma = o(x)$, a contradiction.

Arguing analogously for y, we find: $(-1, -1)$ is contained in the orbit of (x, y) if the maximal powers of two dividing $o(x)$ and $o(y)$ are identical.

Now we are nearly done: A trajectory containing $(-1, -1)$ is point-symmetric, because for each element (X, Y) from the orbit of (x, y), the point $(-1, -1) \cdot (X, Y) = (-X, -Y) = \rho((X, Y))$ is also a point on this orbit, since the orbit forms a cyclic group and therefore is multiplicatively closed. It follows $\rho(\langle (x, y) \rangle) \subset \langle (x, y) \rangle$.

It remains to explain why also $\langle (x, y) \rangle \subset \rho(\langle (x, y) \rangle)$ holds. So consider any element of $\langle (x, y) \rangle$, this has the form $(x, y)^k$ for a $k \in \mathbb{Z}$. If $(x, y)^k$ were not contained in $\rho(\langle(x, y)\rangle)$, then $(x, y)^k \neq (-1, -1) \cdot (x, y)^l$ for all $l \in \mathbb{Z}$. However, since an orbit is a group and equations are uniquely solvable in groups, there exists some $l \in \mathbb{Z}$ that solves the equation $(x, y)^k = (-1, -1) \cdot (x, y)^l$. Thus, we have shown:

Theorem 6.4 Let $n \in \mathbb{N}$ be a product of different odd primes p and q and let (x, y) be a point of the RSA plane associated with an element $z \in \mathbb{Z}_n^*$ via the Chinese remainder theorem. Further, let 2^a and 2^b be the respective maximal powers of two occurring as divisors of $p - 1$ and $q - 1$, respectively; that is, $p - 1 = 2^a \cdot c$ and $q - 1 = 2^b \cdot d$ with c and d odd. Then the following is true:

1. The orbit of an element $(x, y) \in \mathbb{Z}_p^* \times \mathbb{Z}_q^*$ is point-symmetric if and only if the order $o(x)$ in $\mathbb{Z}_p^*$ and the order $o(y)$ in $\mathbb{Z}_q^*$ have the same maximal power of two as divisors.
2. Every RSA plane contains at least one point-symmetric orbit, namely the one generated by the point $(-1, -1)$ and containing only the points $(-1, -1)$ and $(1, 1)$.
3. If an element $x \in \mathbb{Z}_p^*$ has order $o(x) = 2^\alpha \cdot k$ with $\alpha \geq 1$, $2^\alpha \cdot k \mid p - 1$ and odd l, then the orbit of (x, y) is point-symmetric. Therefore, for any power of two 2^α with $\alpha \geq 1$ on the one hand and $\alpha \leq \min(a, b)$ on the other hand, one can construct point-symmetric orbits of length $2^\alpha \cdot \mathrm{lcm}(k, l)$ by choosing x and y such that 2^α divides their order but $2^{\alpha+1}$ does not. (Such elements exist because, in general, in cyclic groups C of order N, for every divisor t of N there exists an element of that order t. For if ε is a generator of C, $\varepsilon^{\frac{N}{t}}$ has exactly the order t.)

Let us briefly consider the example of the $(11, 7)$-plane, which we took up before. Because of $11 - 1 = 10 = 2 \cdot 5$ and $7 - 1 = 6 = 2 \cdot 3$, α from point 3 of Theorem 6.4 can only be equal to 1. Then for point-symmetric orbits there are exactly the following combinations of possible element orders of the components x or y:

$o(x)$	$o(y)$	$o((x, y))$
2	2	2
2	6	6
10	2	10
10	6	$30 = \mathrm{lcm}(10, 6)$

Figure 6.26 shows the corresponding picture for the element of order 30, namely the 17.

The 17 corresponds to the point $(6, 3) = (-5, 3)$ in the $(11, 7)$-plane. The order of 6 in $\mathbb{Z}_{11}$ is in fact 10 and the order of 3 in $\mathbb{Z}_7$ is 6:

```
sage: for i in range(11):
....:     print(i,":",6^i%
....:
0 : 1
1 : 6
2 : 3
3 : 7
4 : 9
5 : 10
6 : 5
7 : 8
8 : 4
```

Figure 6.26 Orbit of 17 in the $(11, 7)$-plane.

```
9 : 2
10 : 1

sage: for i in range(7):
....:     print(i,":",3^i%
....:
0 : 1
1 : 3
2 : 2
3 : 6
4 : 4
5 : 5
6 : 1
```

For $(-1, 3)$, which corresponds to element 10 in $\mathbb{Z}_{77}$, Figure 6.27 shows on the top the associated path of length 6. The point $(-5, -1)$, which corresponds to element 6 in $\mathbb{Z}_{77}$, has a path of length 10, as can be seen in Figure 6.27 on the bottom. The point $(-1, -1)$ has the trivial, always existing path of length 2, which consists of $(-1, -1)$ and $(1, 1)$ (not shown in the figure, because it is trivial).

Case 3: $o(x) = 2^\alpha \cdot \gamma, o(y) = 2^\beta \cdot \delta, \quad \alpha > \beta \geq 0$
In this case, the orbit of (x, y) contains the element $(-1, 1)$ and is symmetric with respect to the vertical axis. Similar to case 2, for each pair $(2^\alpha, 2^\beta)$ with $\alpha > \beta \geq 0$ and $\alpha \leq a, \beta \leq b$, one can choose elements x, y of order $o(x) = 2^\alpha \cdot \gamma$ and $o(y) = 2^\beta \cdot \delta$ (γ and δ odd), respectively, which thus produce a symmetric orbit with respect to one of the two axes. This then has length $\operatorname{lcm}(o(x), o(y)) = 2^\alpha \cdot \operatorname{lcm}(\gamma, \delta)$. Elaborated proofs of this can be found in [16].

Case 4: $o(x) = 2^\alpha \cdot \gamma, o(y) = 2^\beta \cdot \delta, \quad 0 \leq \alpha < \beta$
Here you have symmetry about the horizontal axis, you argue analogous to case 3.

Axis Points

If we consider the horizontal axis, it is pointwise fixed by τ; that is, τ reduced to this axis, is the identical mapping. Moreover, ρ and σ have the same action when restricted to the horizontal axis, so it is sufficient to speak of symmetry, since there is no difference between point and axis symmetry in this case. The path of an element $(x, 0)$ on the horizontal axis is symmetric if and only if the order of x is even. So, again, for every even divisor t of $p - 1$, one can choose an element of order t and

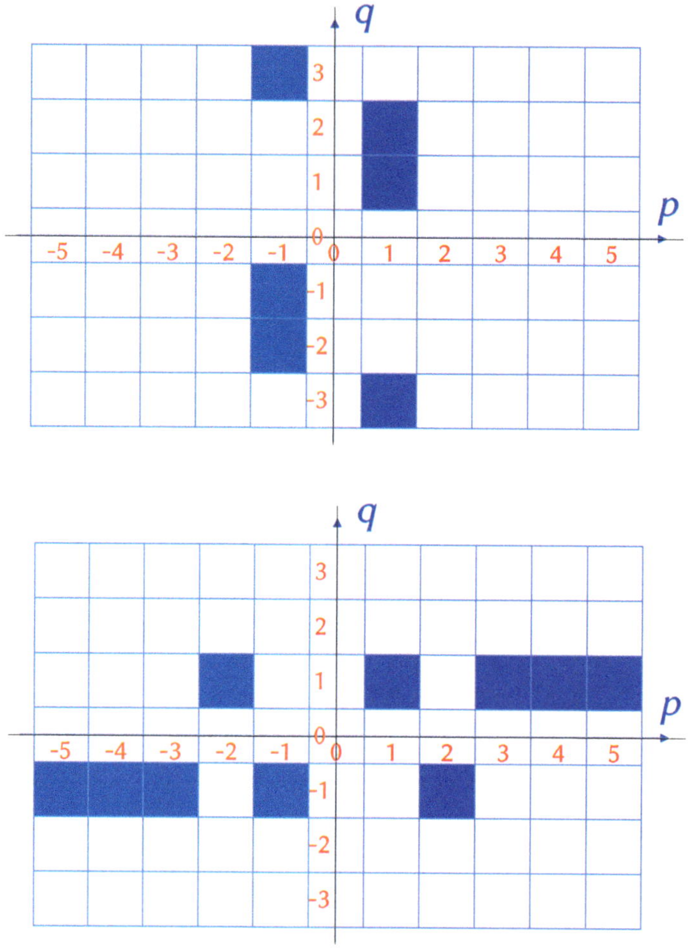

Figure 6.27 Point symmetric orbits of 10 (upper) and 6 (lower) in the $(11, 7)$-plane.

then construct a symmetric orbit of length t with it. The same is true for the vertical axis with elements of the form $(0, y)$ and even divisors of $q - 1$. This can be read in more detail in [16].

6.5.10.2 Invariant RSA Orbits

For symmetric RSA orbits, one finds that their existence is related to the divisibility of $(p-1) \cdot (q-1)$ by 4. For a proof of the Theorem 6.5 see [16].

Theorem 6.5 The RSA orbit of an inner point $(x, y) \in \mathbb{Z}_p^* \times \mathbb{Z}_q^*$ is symmetric with respect to a given reflection if and only if the full orbit of (x, y) is symmetric with respect to that reflection and the order of (x, y) is a multiple of 4.

The RSA orbit of an axis point $(x, 0) \in \mathbb{Z}_p^* \times \mathbb{Z}_q^*$ (different from the origin $(0, 0)$) is symmetric if and only if the full orbit of $(x, 0)$ is symmetric and the order of x in $\mathbb{Z}_p^*$ is a multiple of 4.

For $(0, y)$ the same is true, but the order of y of course has to be the order in $\mathbb{Z}_q^*$.

If we come back to our example with $p = 11$ and $q = 7$ we have to check that there are no symmetric RSA orbits at all, since the length of an orbit in this RSA plane is a divisor of $\mathrm{lcm}(11-1, 7-1) = 30$ and 4 is not a divisor of 30.

This leads to the partition of the RSA plane into disjoint RSA orbits shown in Table 6.5.

These 24 RSA orbits are shown in Figure 6.28 and, as can be seen, they are all nonsymmetric. The dark rectangle corresponds to the generator chosen for the respective RSA orbit. It always comes first in the list of 24 orbits.

An easy example where 4 divides $p - 1$ as well as $q - 1$ is $p = 13$ and $q = 5$. In this case we can observe point symmetry as well as symmetry with respect to the axes; see Figure 6.29. There you can see some—not all—symmetric orbits of the $(13, 5)$-plane.

Table 6.5 Partition of the (11,7)-Plane into Disjoint Nonsymmetric RSA Orbits

1 orbit:	$\{1\}$	13 orbit:	$\{17, 19, 24, -37, -25, -16, -9, -4\}$
2 orbit:	$\{2, 18, 30, -38, -31, -26, -5, -3\}$	14 orbit:	$\{20, 27, -29, -8\}$
3 orbit:	$\{3, 5, 26, 31, 38, -30, -18, -2\}$	15 orbit:	$\{21\}$
4 orbit:	$\{4, 9, 16, 25, 37, -24, -19, -17\}$	16 orbit:	$\{22\}$
5 orbit:	$\{6, 13, -36, -15\}$	17 orbit:	$\{23, -10\}$
6 orbit:	$\{7, 28, 35, -14\}$	18 orbit:	$\{32, -12\}$
7 orbit:	$\{8, 29, -27, -20\}$	19 orbit:	$\{33, -11\}$
8 orbit:	$\{10, -23\}$	20 orbit:	$\{34\}$
9 orbit:	$\{11, -33\}$	21 orbit:	$\{-34\}$
10 orbit:	$\{12, -32\}$	22 orbit:	$\{-22\}$
11 orbit:	$\{14, -35, -28, -7\}$	23 orbit:	$\{-21\}$
12 orbit:	$\{15, 36, -13, -6\}$	24 orbit:	$\{-1\}$

These correspond to the following points, respectively:

1 orbit:	$\{(1, 1)\}$
2 orbit:	$\{(2, 2), (-4, -3), (-3, 2), (-5, -3), (2, -3), (-4, 2), (-5, 2), (-3, -3)\}$
3 orbit:	$\{(3, 3), (5, -2), (4, -2), (-2, 3), (5, 3), (3, -2), (4, 3), (-2, -2)\}$
4 orbit:	$\{(4, -3), (-2, 2), (5, 2), (3, -3), (4, 2), (-2, -3), (3, 2), (5, -3)\}$
5 orbit:	$\{(-5, -1), (2, -1), (-3, -1), (-4, -1)\}$
6 orbit:	$\{(-4, 0), (-5, 0), (2, 0), (-3, 0)\}$
7 orbit:	$\{(-3, 1), (-4, 1), (-5, 1), (2, 1)\}$
8 orbit:	$\{(-1, 3), (-1, -2)\}$
9 orbit:	$\{(0, -3), (0, 2)\}$
10 orbit:	$\{(1, -2), (1, 3)\}$
11 orbit:	$\{(3, 0), (-2, 0), (5, 0), (4, 0)\}$
12 orbit:	$\{(4, 1), (3, 1), (-2, 1), (5, 1)\}$
13 orbit:	$\{(-5, 3), (-3, -2), (2, 3), (-4, -2), (-3, 3), (-5, -2), (2, -2), (-4, 3)\}$
14 orbit:	$\{(-2, -1), (5, -1), (4, -1), (3, -1)\}$
15 orbit:	$\{(-1, 0)\}$
16 orbit:	$\{(0, 1)\}$
17 orbit:	$\{(1, 2), (1, -3)\}$
18 orbit:	$\{(-1, -3), (-1, 2)\}$
19 orbit:	$\{(0, -2), (0, 3)\}$
20 orbit:	$\{(1, -1)\}$
21 orbit:	$\{(-1, 1)\}$
22 orbit:	$\{(0, -1)\}$
23 orbit:	$\{(1, 0)\}$
24 orbit:	$\{(-1, -1)\}$

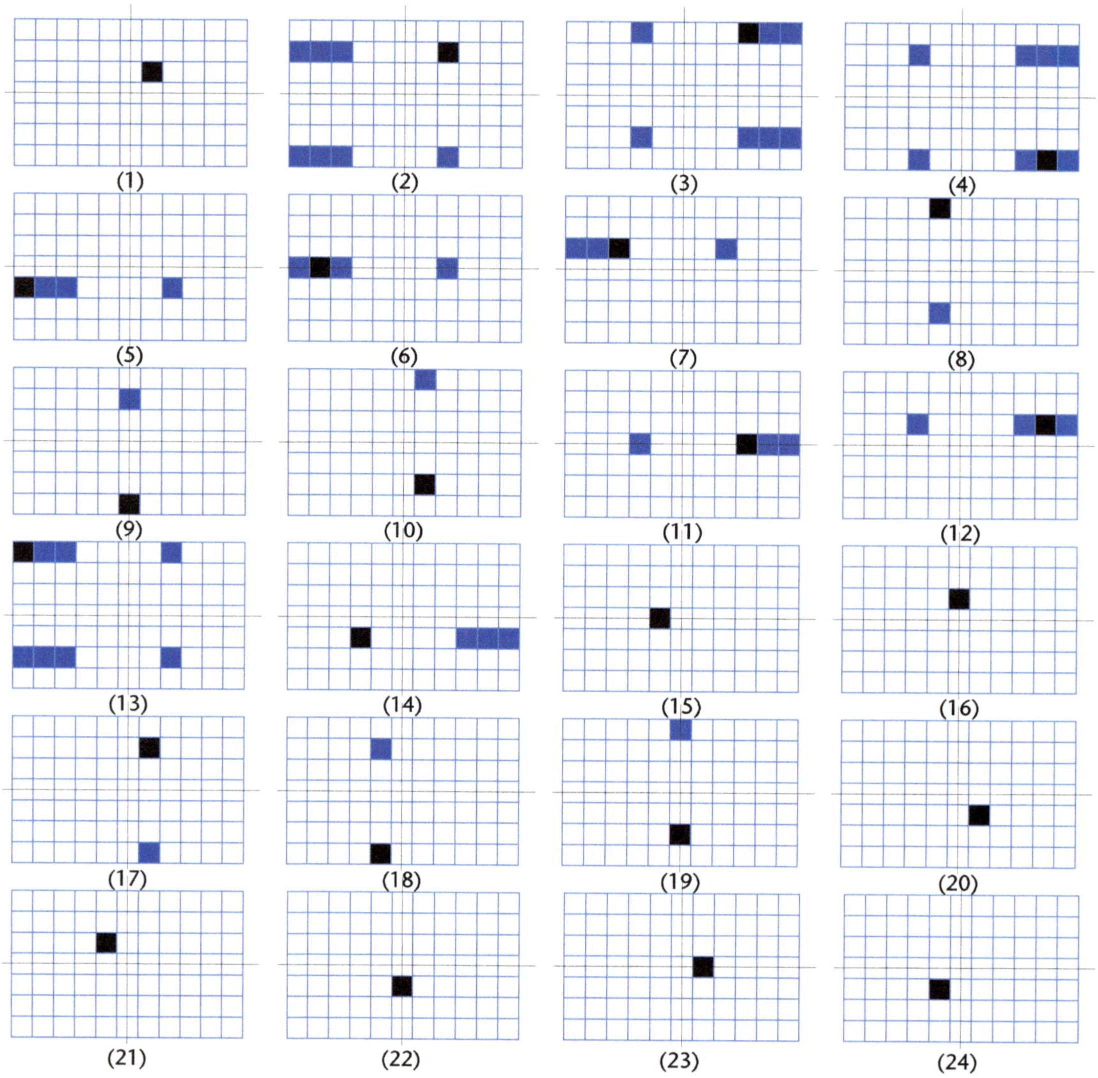

Figure 6.28 Partition of the $(11, 7)$-plane (except the origin) into RSA orbits. None of them have any kind of symmetry; the dark square is always the first element of the corresponding RSA orbit listed in Table 6.5.

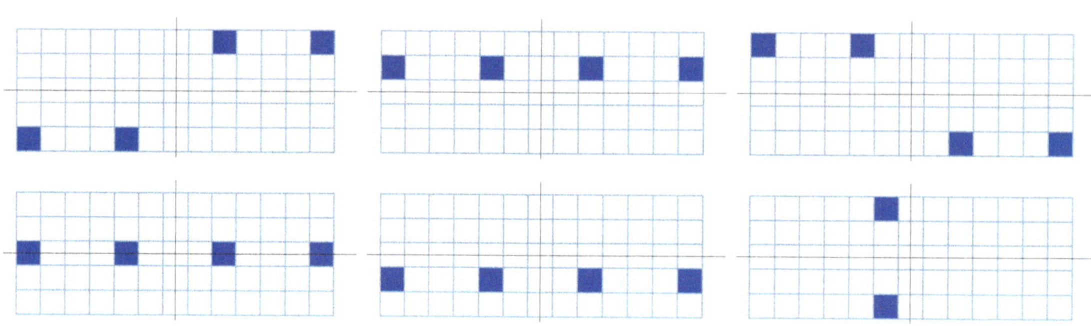

Figure 6.29 Some symmetric RSA orbits in the $(13, 5)$-plane.

After these geometric examinations of the inner workings of RSA with a pure mathematical focus, which are described nowhere else in the literature as far as we know, we now return to something widely known among cryptographers: The algorithm of Pollard.

6.5.11 The Pollard $p-1$ Algorithm for RSA in the 2D Model

If for some RSA modulus n, only small prime divisors of $\phi(n)$ exist, one can use the Pollard $p-1$ algorithm for factoring $n = pq$ in a "short" time. In [17] one can study what exactly *small* and *short* means in this context. We only want to illustrate the idea behind the algorithm in our geometric model.

Essentially, what is being done is this: One chooses an arbitrary point $z \in \mathbb{Z}_n^*$ and starts to compute not its path, but a part of its path, that being:

$$z, \quad z^2, \quad z^{2\cdot 3}, \quad z^{2\cdot 3\cdot 4}, \quad z^{2\cdot 3\cdot 4\cdot 5}, \quad \ldots \quad , \quad z^{k!}$$

Or more precisely:

$$z, \quad z^2 \bmod n, \quad z^{2\cdot 3} \bmod n, \quad z^{2\cdot 3\cdot 4} \bmod n, \quad z^{2\cdot 3\cdot 4\cdot 5} \bmod n, \quad \ldots \quad , \quad z^{k!} \bmod n$$

Now consider the p- and q-coordinate of such a z-power $z^{k!}$. If $z^{k!} \equiv x \bmod p$ and $z^{k!} \equiv y \bmod q$ we find that if we subtract 1, the number $z^{k!} - 1$ corresponds to the point that is located one step to the left and one step lower than the point representing $z^{k!}$; that is $(x-1, y-1)$. This should be clear if one recalls how the lines of Figures 6.3 or 6.9 were generated: Addition of 1 in the linear representation corresponds to "1 to the right and 1 up" in the two-dimensional model, except if a point lies on the upper or right border, then one has to jump to the bottom and then 1 right or to the left and then 1 up.

Every point lying one position lower than a point

- With p-coordinate $= 1$ is an axis point on the q-axis. These are all points corresponding to a multiple of p in the linear model. Therefore, p divides the number $z^{k!} - 1 \bmod n$ in this case. Then the Euclidean algorithm produces $\gcd(z^{k!} - 1 \bmod n, n) = p$.
- With q-coordinate $= 1$ is an axis point on the p-axis. These are all points corresponding to a multiple of q in the linear model. Therefore, q divides the $z^{k!} - 1 \bmod n$ in this case. Then the Euclidean algorithm produces $\gcd(z^{k!} - 1 \bmod n, n) = q$.

If $p-1$ as well as $q-1$ have only small prime factors, we can expect that for the most inner points z the series $z, z^2, z^{3!}, z^{4!}, \ldots$ will reach an element whose p- or q-coordinate is going to stabilize on 1. The latter follows quite easily from the isomorphism between $\mathbb{Z}_n$ and $\mathbb{Z}_p \times \mathbb{Z}_q$:

If $\pi(z) = (x, y) = (z \bmod p, z \bmod q)$, then also $\pi(z^{k!}) = (x^{k!} \bmod p, y^{k!} \bmod q)$. As soon as either $p-1$ or $q-1$ divides $k!$ for the first time, the p-coordinate or q-coordinate of $z^{k!}$ equals 1 because of $z^{k!} \equiv 1 \bmod p$ or $z^{k!} \equiv 1 \bmod q$. This follows from the fact that $\mathbb{Z}_p^*$ is a cyclic group of order $p-1$ (the same for q) in which the $p-1$th power of any element must be equal to the unit element. In other words, the shadow of an element path will meet 1 quite fast in most cases.

The groups $\mathbb{Z}_p^*$ and $\mathbb{Z}_q^*$ are cyclic, but the number of elements of maximal order $p-1$ or $q-1$ in these groups is small because $\phi(p-1)$ as well as $\phi(q-1)$ are small since $p-1$ and $q-1$ have only small prime divisors. So if one chooses z arbitrarily, the order of the element $\pi(z)$ will have only small divisors—say d—of $p-1$ or $q-1$. Then $\pi(z)$ will take 1 as p- or q-coordinate as soon as $d|k!$ is true.

To summarize, the $p-1$-algorithm of Pollard for the case $n = pq$ goes through these steps:

1. Choose an inner point z.
2. Compute successively the powers $z^{k!} \bmod n$ for $k = 1, 2, 3, \ldots$.
3. Test if $\gcd(z^{k!} - 1 \bmod n, n) \neq 1$ with the Euclidean algorithm.
4. As soon as the gcd is unequal to 1 for the first time, one has found the first point with p-coordinate or q-coordinate equal to 1 and therefore a nontrivial divisor $z^{k!} - 1 \bmod n$ of n.

We illustrate this with the example $p = 97$ and $q = 73$, which we already used in Figure 6.16.

We have $n = 97 \cdot 73 = 7081$, $p - 1 = 96 = 2^5 \cdot 3$, and $q - 1 = 72 = 2^3 \cdot 3^2$ and therefore only small prime divisors of n.

We choose $z = 3$ and first look at those points of the RSA plane that correspond to the set $\{z, z^2, z^{2\cdot 3}\} = \{3, 9, 729\}$. Since $\pi(3) = (3, 3)$, $\pi(9) = (9, 9)$ and $\pi(729) = (50, 72)$ we are not yet finished with $k = 3!$; see Figure 6.30 *without* the point $(96, 1)$.

If we add the next step, we get $\{z, z^2, z^{2\cdot 3}, z^{2\cdot 3\cdot 4}\} = \{3, 9, 729, 6498\}$. Since $\pi(6498) = (96, 1)$ we find a point lying exactly 1 above the horizontal axis, and so (cf. step 4 from above) we are done; see Figure 6.30 *with* point $(96, 1)$. Indeed, $\gcd(6497, 7081) = 73$.

Taking an even closer look at the orbit of 3, one has the following ordered list with $48 = 2^4 \cdot 3$ elements in $\mathbb{Z}_{7081}$:

3	9	27	81	243	729	2187	6561	5521	2401	122	366
1098	3294	2801	1322	3966	4817	289	867	2601	722	2166	6498
5332	1834	5502	2344	7032	6934	6640	5758	3112	2255	6765	6133
4237	5630	2728	1103	3309	2846	1457	4371	6032	3934	4721	1

The corresponding points in the 2D model then are:

(3, 3)	(9, 9)	(27, 27)	(81, 8)	(49, 24)	(50, 72)	(53, 70)	(62, 64)	(89, 46)	(73, 65)	(25, 49)	(75, 1)
(31, 3)	(93, 9)	(85, 27)	(61, 8)	(86, 24)	(64, 72)	(95, 70)	(91, 64)	(79, 46)	(43, 65)	(32, 49)	(96, 1)
(94, 3)	(88, 9)	(70, 27)	(16, 8)	(48, 24)	(47, 72)	(44, 70)	(35, 64)	(8, 46)	(24, 65)	(72, 49)	(22, 1)
(66, 3)	(4, 9)	(12, 27)	(36, 8)	(11, 24)	(33, 72)	(2, 70)	(6, 64)	(18, 46)	(54, 65)	(65, 49)	(1, 1)

Being projected onto the q-axis, the series $\pi(3^i)$ takes the y-coordinate (or q-coordinate) 1 every 12th time. By allowing only factorials as powers in the Pollard

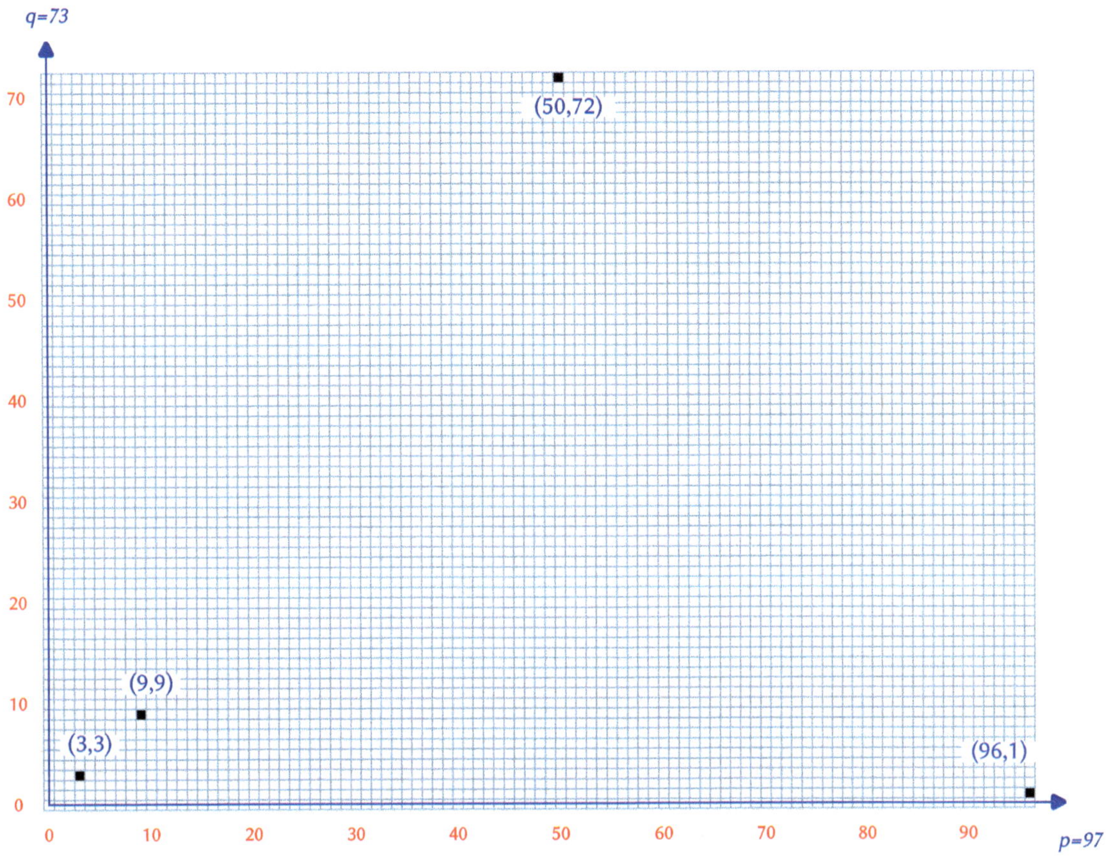

Figure 6.30 (97, 73)-plane with $3, 3^{2!}, 3^{3!}$, and $3^{4!}$.

$p-1$ algorithm one misses the first candidate $\pi(3^{12}) = (75, 1)$ and the first hit is $\pi(3^{23}) = (96, 1)$.

6.5.12 Final Remarks about the RSA Plane

There are other aspects of RSA that can be illustrated with the RSA plane. The fixed points have already been mentioned as special cases of orbits. However, we also see potential to illustrate other attacks, for example, Fermat's factorization method (see, e.g., [18]), which is applicable for the case where the difference of p and q is small.

If one interprets the RSA plane embedded in a two-dimensional lattice (on the topic of lattices see Chapter 11), one can also shed light on some interesting aspects of lattice cryptography, in particular ideas of Coppersmith [19]. For this purpose, however, it is necessary to consider the neighborhoods of the integer points in the Euclidean plane. The notion of "neighborhood" plays a major role in the mathematical subfield of topology. For example, if we represent the point $(1, 1)$ as a square, the square can also be taken to be the neighborhood $U = \{(x, y) \in \mathbb{Q} : \frac{1}{2} < x, y < \frac{3}{2}\}$ of $(1, 1)$ in the plane over the rational numbers. If one takes U as a subset of $\mathbb{Q}^2$, and not as a subset of $\mathbb{R}^2$, one can build on this to create a different, non-Euclidean geometry, the ultrametric plane. It allows certain structure-preserving mappings onto finite planes. See, for example, [20] or [21] for more background on this topic. Ultrametric planes are an example of compact planes. Since these mappings are defined not only on the integer lattice, but also on the neighborhood of the lattice points, we chose the representation with squares: A square symbolizes the lattice point together with its neighborhood, which contains infinitely many points. In comparison to points, this square model is extensible by topological aspects and therefore a more flexible model.

6.6 Outlook

Several other modern asymmetric methods exist. Some are presented in Chapters Elliptic Curves (Chapter 8), Homomorphic Ciphers (Chapter 10), Lattices (Chapter 11), Solving Discrete Logarithms and Factoring (Chapter 12), and Long-Term Cryptographic Perspectives (Chapter 13). Elliptic curves for instance provide useful groups for public-key encryption procedures, which offer shorter key lengths.

References

[1] Balcazar, J. L., J. Daaz, and J. Gabarr, *Structural Complexity I*, Berlin: Springer Verlag, 1998.

[2] Hesselink, W. H., *The Borderline between P and NP*, February 2001, https://www.cs.rug.nl/~wim/pub/whh237.pdf.

[3] Merkle, R., and M. Hellman, "Hiding Information and Signatures in Trapdoor Knapsacks," *IEEE Transactions on Information Theory*, Vol. 24, No. 5, 1978.

[4] Kellerer, H., U. Pferschy, and D. Pisinger, *Knapsack Problems*, Berlin: Springer, 2004.

[5] Shamir, A., "A Polynomial Time Algorithm for Breaking the Basic Merkle-Hellman Cryptosystem," in *Symposium on Foundations of Computer Science*, 1982, pp. 145–152.

[6] Adleman, L., "On Breaking the Iterated Merkle-Hellman Public-Key Cryptosystem," in *Advances in Cryptology, Proceedings of Crypto 82*, Plenum Press, 1983, pp. 303–308.

[7] Lagarias, J.C., "Knapsack Public Key Cryptosystems and Diophantine Approximation," *Advances in Cryptology Proceedings of Crypto 83*, Plenum Press, 1983.

[8] Brickell, E. F., "Breaking Iterated Knapsacks," in *Advances in Cryptology: Proceedings of CRYPTO'84*, Vol. 196, Berlin: Springer, 1985, pp. 342–358.

[9] Rivest, R. L., A. Shamir, and L. Adleman, "A Method for Obtaining Digital Signatures and Public-Key Cryptosystems," *Communications of the ACM*, Vol. 21, No. 2, April 1978, pp. 120–126.

[10] Stinson, D. R., *Cryptography–Theory and Practice*, Third Edition, Boca Raton, FL: Chapman & Hall/CRC, 2006.

[11] Schneier, B., *Applied Cryptography, Protocols, Algorithms, and Source Code in C, 20th Anniversary Edition*, Second Edition, New York: John Wiley, & Sons, 2015.

[12] *NMBRTHRY Archives*, https://listserv.nodak.edu/cgi-bin/wa.exe?A2=NMBRTHRY;fd743373.1912.

[13] Lang, S., *Algebra*, Third Edition, Reading MA: Addison-Wesley, 1993.

[14] Lang, S., *Introduction to Linear Algebra,*Second Edition, New York: Springer, 1986.

[15] Hungerford, T. W., *Algebra*, Springer Verlag, 1974.

[16] Quade, L., "RSA Studies in a Two-Dimensional Group-Theoretic Model," BA thesis, Universität Siegen, 2023.

[17] Pollard, J. M., "Theorems on Factorization and Primality Testing," in *Mathematical Proceedings of the Cambridge Philosophical Society*, Vol. 76, November 3, 1974, pp. 521–528.

[18] Erra, R., and C. Grenier, "The Fermat Factorization Method Revisited," *IACR Cryptology*, ePrint Archive, 2009, p. 318, https://eprint.iacr.org/2009/318.pdf.

[19] Coppersmith, D., "Small Solutions to Polynomial Equations, and Low Exponent RSA Vulnerabilities," *Journal of Cryptology*, Vol. 10, 1997, pp. 233–260, https://link.springer.com/article/10.1007/s001459900030.

[20] Wagner, D., "Ovale und ebene algebraische Kurven mit unendlicher Kollineationsgruppe," 2004. https://opus4.kobv.de/opus4-fau/frontdoor/index/index/year/2005/docId/168.

[21] Salzmann, H., et al., *Compact Projective Planes*, De Gruyter, 1995, https://doi.org/10.1515/9783110876833.

CHAPTER 7

Hash Functions, Digital Signatures, and Public-Key Infrastructures

This chapter mainly gives a high-level overview about hash functions and their application in digital signatures and certificates.

7.1 Hash Functions

A *hash function* maps a message of any length to a bit sequence of constant size (like 128 bit). The function value is called hash value or message digest. Outside of cryptography, hash functions are used to quickly retrieve stored data.

We consider here only cryptographic hash functions[1] fulfilling the requirements laid out in Section 7.1.1.

They have a multitude of uses such as in message authentication codes (MACs), pseudorandom number generators (PRNGs), digital signatures, blockchain, and cryptocurrencies. To store password verification values or derive keys from passwords, special hash functions or key derivation functions (KDFs) are used, which were created to slow brute-force searches and other guessing attacks. A current KDF recommendation is Argon2.

7.1.1 Requirements for Hash Functions

Cryptographically secure hash functions fulfill the following three requirements (the order is in a way that the requirements increase):

1. Cryptographic hash functions are implemented in CT at several places:
- The menus `CT1 Indiv. Procedures ▷ Hash` and `CT1 Analysis ▷ Hash` enable
  - To apply one of six hash functions to the content of the current window;
  - To calculate the hash value of a file;
  - To test how changes to a text change the corresponding hash value (in the "Hash Demonstration");
  - To calculate a key from a password according to the PKCS#5 standard;
  - To calculate HMACs from a text and a secret key;
  - To perform a simulation, how digital signatures could be attacked by a targeted search for hash value collisions.

- Using `CT2 Templates ▷ Hash Functions` you can find BLAKE, Grostl, Keccak, MD5, and attacks on them. Plus password-based key-derivation functions (PBKDF) like PKCS#5 and PBKDF2. Modern KDFs are PBKDF2, scrypt, or Argon2.
- `JCT Default Perspective ▷ Visuals ▷ Hash Sensitivity` demonstrates how sensitive hash values react to the smallest changes in the input.
- See the functions' lists within Sections A.2 and A.3.

1. *Resistance against first preimage attacks*:
   It should be practically impossible for a given hash value to find a message that has precisely the same hash value.
   Given (fix): hash value H'.
   Searched: message m, so that: $H(m) = H'$.
2. *Resistance against second preimage attacks*:
   It should be practically impossible for a given message to find another message has precisely the same hash value.
   Given (fix): message m_1 [and so the hash value $H_1 = H(m_1)$].
   Searched: message m_2, so that $H(m_2) = H_1$.
3. *Collision resistance*:
   It should be practically impossible to find any two messages with the same hash value—this is called a collision. Remark: Here, it doesn't matter what the hash value is.
   Searched: 2 messages m_1 and m_2, so that $H(m_1) = H(m_2)$.

There are several common concepts to build hash functions; for example, they can be based on block cipher components (like with SHA-1 and SHA-2) or sponge functions (like with Keccak), or they can use a so-called Merkle-Damgård construction to process an arbitrary-length message into a fixed-length output.

Figure 7.1 shows how a small change in the input ("Hello World!" becomes "Hello World.") drastically changes the output. This is the so-called avalanche effect.

7.1.2 Generic Collision Attacks

Hash functions always have collisions as they map messages of arbitrary size into the smaller set of hash values of fixed length.

Collisions of n bit hash functions can be expected for $O(2^{n/2})$ or more messages due to the birthday paradox. There is an algorithm that can find collisions with this effort without significant storage requirements and even in parallel [1, p. 383].

For this reason, hash functions need to have the double size of the corresponding symmetric encryption (i.e., SHA-512 for the security level of AES-256).

7.1.3 Attacks Against Hash Functions Drive the Standardization Process

So far, no formal proof has been found that perfectly secure cryptographic hash functions exist.

As early as 1996, Dobbertin found collisions in the compression function of MD5. This was not a full break of MD5, but had cast doubt on its security.

Until 2004, for several years no new attacks against hash algorithms came up, and so the candidates that had not yet shown any weaknesses in their structure in practice were trusted (e.g., SHA-1 or RIPEMD-160). RIPEMD-160 was developed in Europe and published in 1996. SHA-1 is a 160 bit hash function specified in FIPS 180 (by NIST), ANSI X9.30 Part 2, and [2]. SHA means secure hash algorithm, and is widely used (e.g., with DSA, RSA, or ECDSA).

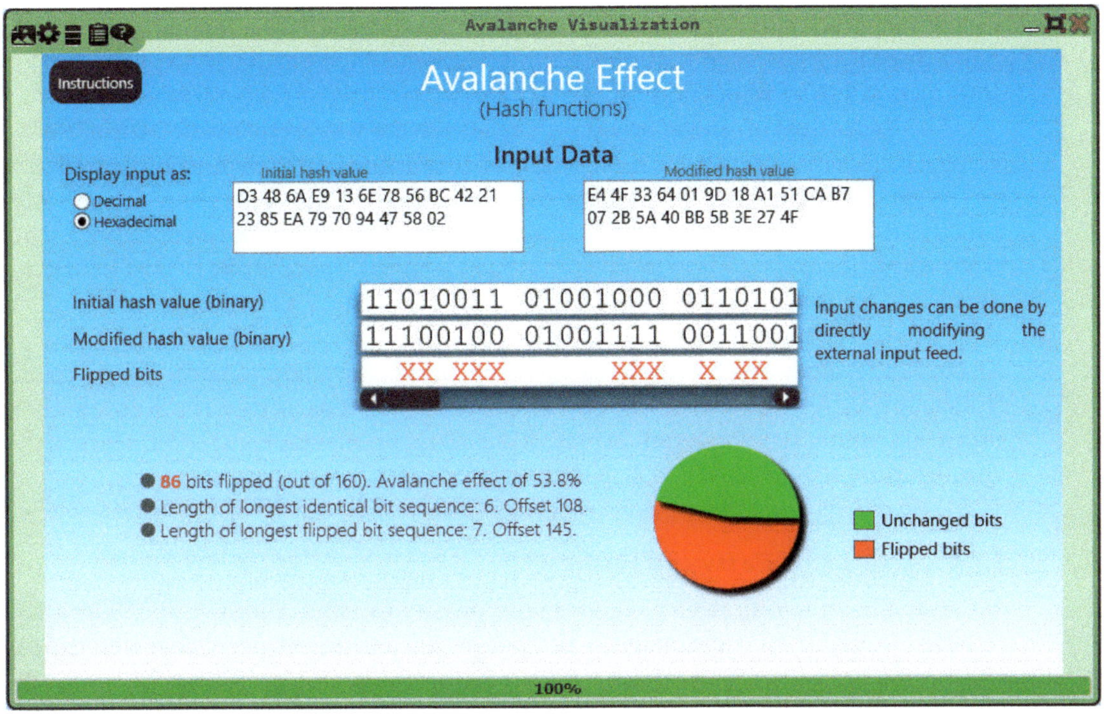

Figure 7.1 Avalanche effect with hash functions in CT2.[2]

The output length of the SHA algorithms was enhanced for the SHA-2 family because of the possibility of birthday attacks: these make n-bit AES and a 2n-bit hash roughly equivalent: 128-bit AES – SHA-256, 192-bit AES – SHA-384, 256-bit AES – SHA-512.[3]

At Crypto 2004 (August 2004) this safety-feeling was disputed: Chinese researchers published collision attacks against MD4, SHA-0, and parts of SHA-1. This globally caused new motivation to engage in new hash attack methods.

The initially published result reduced the expected complexity for one SHA-1 collision search from 2^{80} (brute-force) to 2^{69} [3]. Further announcements claimed to reduce the required effort to 2^{63} [4] and 2^{52} [5]. This would bring collision attacks into the practical realm, as similar efforts have been mastered in the past (see Section 1.2.2).

Already before Crypto 2004, the U.S. National Institute of Standards and Technology (NIST) decided to discontinue SHA-1 and to standardize new methods with longer hash values.

So in 2008 NIST opened a competition to develop a new cryptographic hash algorithm beyond the SHA-2 family [6]. In October 2012, Keccak was announced as "SHA-3."

2. `CT2 Templates ▷ Cryptoanalysis ▷ Modern ▷ Avalanche (Hash Functions).`
3. - Using `CT1 Analysis ▷ Hash ▷ Attack on the Hash Value of the Digital Signature` you can comprehend the birthday attack on digital signatures.
   - `CT2 Templates ▷ Hash Functions ▷ MD5 Collision Finder.`

SHA-3 (Secure Hash Algorithm 3) was released by NIST in August 2015. SHA-3 is a subset of the cryptographic family Keccak. This current NIST standard is described in FIPS Publication 202.

The Keccak algorithm is based on a relatively new approach called sponge construction. Given a state vector (sponge) of b bit, and a message in blocks of size $r < b$. The sponge then absorbs the message blockwise into a subset of the state, which is then transformed as a whole using a random permutation function f (this is the inputting part). Then the result is squeezed out (outputting). This construction leads to great flexibility. For SHA-3, four instances are defined: SHA3-224, SHA3-256, SHA3-384, and SHA3-512.

In CT2, there are 3 templates showing the Keccak component—as a stream cipher, for SHA-3, and as a PRNG. The individual steps within the Keccak algorithm are animated in detail in the Keccak component. Figure 7.2 shows the visualization of Keccak in CT2.

7.1.4 Attacks on Password Hashes

An attack has the objective to recover the password from a hash value. Thus, the resistance against first preimage is the password hashes property that determines its security.

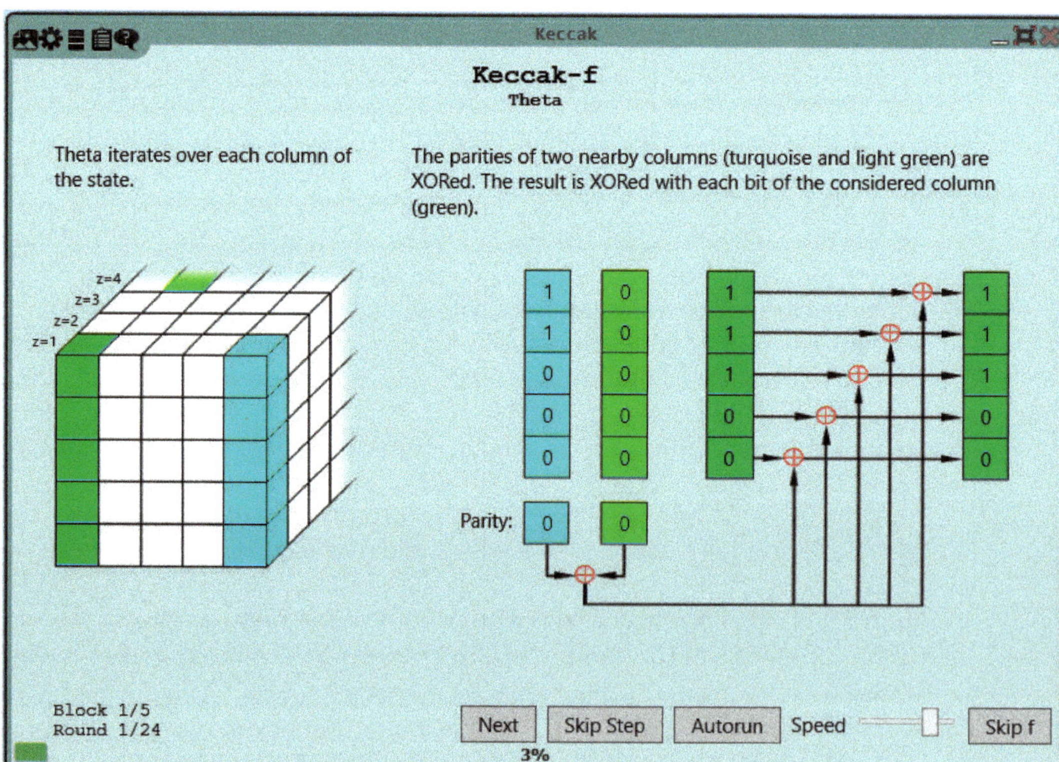

Figure 7.2 SHA-3: Keccak hash function: Theta part of the permutation function f in CT2.[4]

4. CT2 Templates ▷ Hash ▷ Keccak Hash (SHA-3).

In contrast to collision attacks, no practical 1st-pre-image attacks against common hash functions are known.

Successful password hash attacks guess trial passwords, hash them, and compare the result with one or more target hashes. A brute-force attack tries all possible passwords. Dictionary attacks try all passwords from a dictionary. Both attack methods can be modified (e.g., by appending digits or special characters).

The program Hashcat can test on an NVIDIA GeForce RTX 4090 GPU 164 billion passwords per second (164 GH/s, giga hashes per second). Table 7.1 shows the length of passwords that can be broken in one day with this setup.

A first countermeasure against those attacks hashes the password together with a salt. The salt is selected randomly for each user and stored together with the hash. This measure does not impact attacks on a single password hash. If the attack targets more then one password hash, then each of the hashes needs to be cracked individually, which is not required if no salt is used. At the same time, rainbow attacks are prevented. Like brute-force attacks, rainbow attacks hash all possible passwords. The result is then stored in compressed form. During the actual attack these precalculated values are used.

A second countermeasure artificially slows down password hashing by iterating a hash function like SHA-256 with the goal to use more computation time in order to slow down brute-force attacks. The number of iterations should be configurable and set such that, for example, one millisecond is required. As a result, attack speed is reduced to 1,000 tries per second. The attack can be sped up again arbitrarily by parallelization, however.

A downside of this measure is that it not only increases the effort for the attacker, but for the application verifying the passwords as well.

A third countermeasure prevents offline attacks on password hashes by using a secret key, also called pepper, in the hash calculation. This can be accomplished with an HMAC, for instance.

Similar to the salt, there is no significant slow down to be expected when using a pepper. This measure is fully effective only if attackers cannot gain access to the pepper. This can be accomplished by using a hardware security module (HSM).

7.2 Digital Signatures

The aim of digital signatures[5] is to guarantee the following two points:

- *User authenticity:* It can be checked whether a message really does come from a particular person.

5. Different variants of digital signatures can be found in CT:
- In CTO in the plugin "OpenSSL": `https://www.cryptool.org/en/cto/openssl`.
- Via the submenus of `CT1 Digital Signatures / PKI` or via `CT1 Indiv. Procedures` ▷ `RSA Cryptosystem` ▷ `Signature Demonstration (Signature Generation)` you can generate and check digital signatures.
- Using `CT2 Templates` ▷ `Cryptography` ▷ `Modern` ▷ `Asymmetric` ▷ `Blind Signature with RSA` and using `CT2 Templates` ▷ `Cryptography` ▷ `Modern` ▷ `Asymmetric` ▷ `Blind Signature with Paillier`.
- JCT (in the default and the algorithm perspective) contains newer signature variants from the field of post-quantum computing (Merkle, Winternitz, SPHINCS+), but also from the application area such as redactable signature schemes (RSS).

Table 7.1 A Brute-Force Attack with 164 GH/s Can Crack in One Day Passwords of Length l From a Set of a Characters

a	l
10	16
26	11
36	10
62	9
128	8
256	7

- *Message integrity:* It can be checked whether the message has been changed (en route).

An asymmetric technique is used again (see Chapter 6). Participants who wish to generate a digital signature for a document must possess a pair of keys. They use their secret key to generate signatures and the recipient uses the sender's public key to verify whether the signature is correct. As before, it must be impossible to use the public key to derive the secret key.

In detail, a *signature procedure* looks like this: Senders use their message (or document) and secret key to calculate the digital signature for the message. Compared to handwritten signatures, digital signatures have the advantage that they also depend on the document to be signed. Signatures from one and the same participant are different unless the signed documents are completely identical. Even inserting a blank in the text would lead to a different signature.

The document is sent to the recipient together with the signature. The recipient can then use the sender's public key in order to determine whether the received document fits with the received signature. So he/she checks whether the signature is correct. Thus, the recipient can detect any injury to the message integrity.

The procedure we just described has in practice, however, a decisive disadvantage. The signature of the message (like an encryption) would be approximately as long as the document itself. To prevent an unnecessary increase in data traffic, and also for reasons of performance, a cryptographic hash function is applied to the document before signing.

7.2.1 Signing the Hash Value of the Message

The signature procedure with hash functions is as follows: Rather than signing the actual document, the sender first calculates the hash value of the message and signs this. The recipient also calculates the hash value of the (received) message (he knows the used hash algorithm), then verifies whether the signature sent with the message is a correct signature of the hash value. If this is the case, the signature is verified to be correct. This means that the message is authentic, because we assumed that only the correct owner knows the private key.

Some digital signature schemes are based on asymmetric *encryption* procedures, the most prominent example being the RSA system, which can be used for signing by performing the same private key operation, but on the hash value of the document to be signed.

Other digital signature schemes where developed exclusively for this purpose, as the DSA (digital signature algorithm), and are not directly connected with a corresponding encryption scheme.

The two signature methods that are probably still most frequently used, RSA and DSA, are discussed in more detail in the following Sections 7.3 and 7.4. After that we explain shortly how digital signatures can be used to create the digital equivalent of ID cards. This is called public-key certification.

7.3 RSA Signatures

As mentioned at the end of Section 5.10.3, it is also possible to perform the RSA private and public-key operation in reverse order: If $H = hash(message)$, then raising H first to the power of $d \pmod N$ and then to the power of $e \pmod N$ yields H again. Therefore, RSA can be used as a signature scheme.

The RSA signature S for a message hash H is created by performing the private key operation:

$$S \equiv H^d \pmod N$$

In order to verify H is correct, the corresponding public-key operation is performed on the signature S and the result is compared with the message hash H:

$$S^e \equiv (H^d)^e \equiv (H^e)^d \equiv H \pmod N$$

If the result matches the message hash H, then the signature is accepted by the verifier; otherwise the message has been tampered with, or was never signed by the holder of d.

Figure 7.3 shows a step-by-step visualization creating RSA signatures with CT1.[6]

To prevent certain attacks on the RSA signature procedure (alone or in combination with encryption) it is necessary to format the hash value before doing the exponentiation, as described in the PKCS#1 (Public-Key Cryptography Standard #1 [7]). The fact that this standard had to be revised after several years of use can serve as an example of how difficult it is to define cryptographic details correctly.

7.4 DSA Signatures

In August of 1991, the U.S. NIST proposed a DSA, which was subsequently adopted as a U.S. Federal Information Processing Standard (FIPS 186 [2]).

The algorithm is a variant of the ElGamal scheme. Its security is based on the discrete logarithm problem (see Chapter 6). The DSA public and private key and its procedures for signature and verification are summarized in Crypto Procedure 7.1.

While DSA was specifically designed so that it can be exported from countries regulating export of encryption software and hardware (like the U.S. at the time

6. `CT1 Digital Signatures/PKI ▷ Signature Demonstration (Signature Generation).`

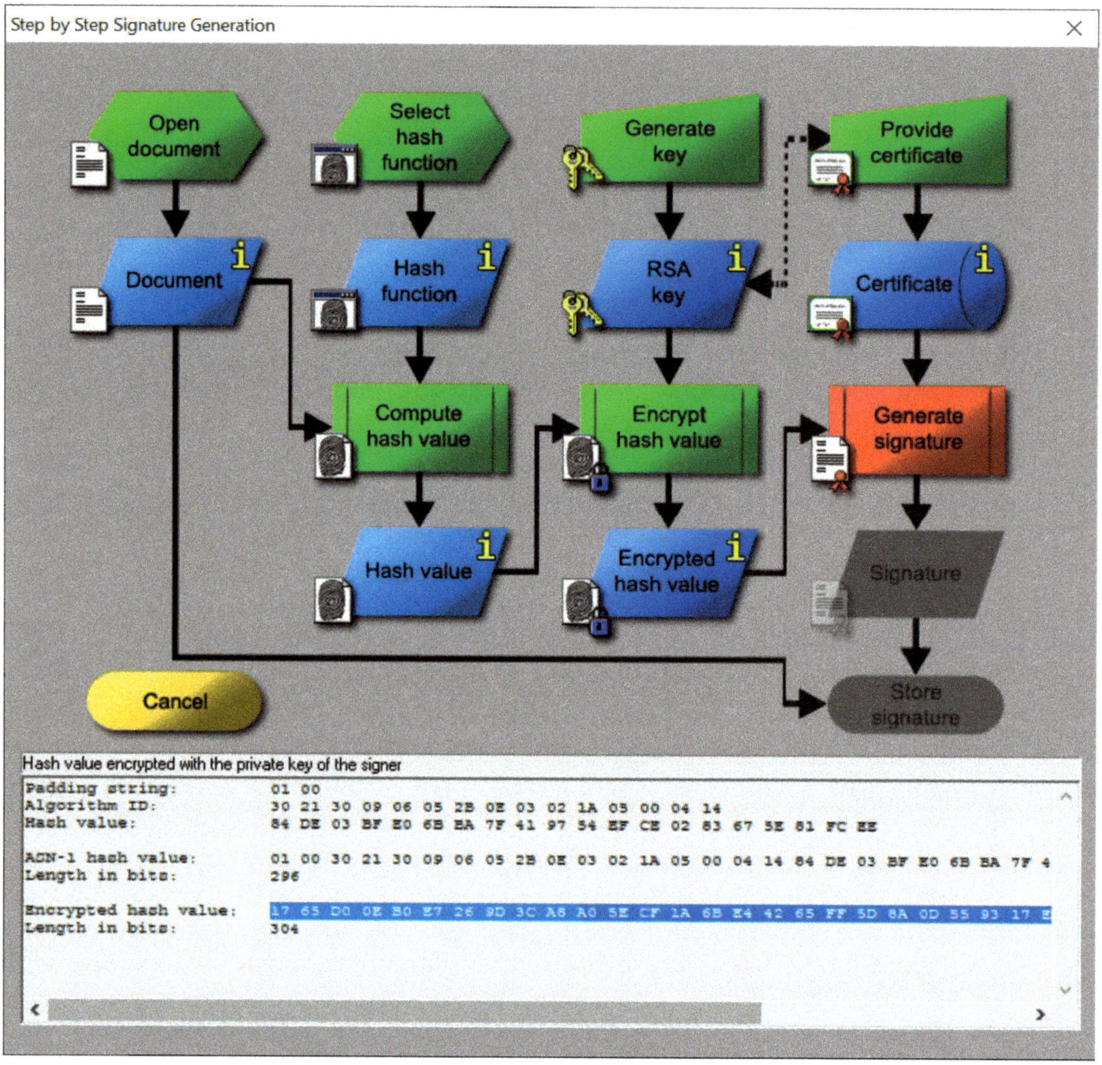

Figure 7.3 Workflow for generating an RSA signature in CT1.

when it was specified), it has been noted [8, p. 490], that the operations involved in DSA can be used to emulate RSA and ElGamal encryption.

Crypto Procedure 7.1: DSA Signature

Public key

p prime

q 160-bit prime factor of $p-1$

$g = h^{(p-1)/q} \bmod p$, where $h < p-1$ and $h^{(p-1)/q} > 1 \pmod p$

$y \equiv g^x \bmod p$

Remark: Parameters p, q, and g can be shared among a group of users.

Crypto Procedure 7.1 (continued)

Private key

$x < q$ (a 160-bit number)

Signing

m the message to be signed

k choose at random, less than q

$r = (g^k \mod p) \mod q$

$s = (k^{-1}(\text{SHA-1}(m) + xr)) \mod q$

Remark:

- (s, r) is the signature.
- The security of the signature depends not only on the mathematical properties, but also on using a good random source for k.
- SHA-1 is a 160-bit hash function.

Verifying

$w = s^{-1} \mod q$

$u_1 = (\text{SHA-1}(m)w) \mod q$

$u_2 = (rw) \mod q$

$v = (g^{u_1} y^{u_2}) \mod p) \mod q$

Remark: If $v = r$, then the signature is verified.

7.5 Public-Key Certification

Public-key infrastructure (PKI) is the term used to describe a system that issues, distributes, and verifies digital certificates. The certificates issued within a PKI can be used to enable computers and people in entire organizations to authenticate each other and to secure their communication.

The aim of public-key certification is to guarantee the connection between a public key and a user and to make it traceable for external parties. This trust anchor is confirmed by the so-called certificate. In cases in which it is impossible to ensure that a public key really belongs to a particular person, many protocols are no longer secure, even if the individual cryptographic components cannot be broken.

Another area using PKIs are internet-connected devices (IoT). The cybersecurity of IoT affects not only smart home and consumer electronics products, but also vehicles and industrial plants. PKI is particularly suitable for the provisioning of digital identities during production and operation.

7.5.1 Impersonation Attacks

Assume Charlie has two key pairs (PK_1, SK_1) and (PK_2, SK_2), where SK denotes the secret private key and PK the public key. Further, assume that he manages to palm off PK_1 on Alice as Bob's public key, and PK_2 on Bob as Alice's public key (by falsifying a public key directory).

Then he can attack as follows:

- Alice wants to send a message to Bob. She encrypts it using PK_1 because she thinks that this is Bob's public key. She then signs the message using her secret key and sends it.
- Charlie intercepts the message, removes the signature and decrypts the message using SK_1. If he wants to, he can then change the message in any way he likes. He then encrypts the message again, but this time using Bob's genuine public key, which he has taken from a public-key directory, signs the message using SK_2, and forwards it to Bob.
- Bob verifies the signature using PK_2 and will reach the conclusion that the signature is correct. He then decrypts the message using his secret key.

In this way, Charlie can listen in on communication between Alice and Bob and change the exchanged messages without them noticing. The attack will also work if Charlie only has one pair of keys.

The impersonation attack is a kind of "man-in-the-middle attack." Users are promised protection against this type of attack by public-key certification, which is intended to guarantee the authenticity of public keys. The most common certification method is the X.509 standard.

7.5.2 X.509 Certificate

X.509 [9] is an International Telecommunication Union (ITU) standard defining the format of public-key certificates.

X.509 certificates play a major role in TLS (for security between a browser and web server when accessing web pages via HTTPS) and in S/MIME (for signing and encrypting e-mails),[7] but also play a role in offline applications such as the electronic signature of documents. For a long time, most web server certificates were created using OpenSSL and the Microsoft CA. This share has dropped drastically since almost all domain-validated TLS certificates are created by Let's Encrypt. As of July 2023, the TLS certificates of 350 million websites come from Let's Encrypt [10]. You can get these in an automated process and free of charge.

Each participant who wants to have an X.509 certificate verifying that his public key belongs him/her as a real person consults what is known as a certification authority (CA). If the certificates are not only offered to a closed user group, the CA is called trust center or more generally trusted third party.

The participant proves his/her identity to this CA (for example by showing his/her ID) often via a registration authority (RA). This process is called registration. The CA then issues him/her an electronic document (certificate) that essentially contains the name of the certificate-holder and the name of the CA, the certificate-holder's public key, and the validity period of the certificate. The CA then signs the certificate using its private key.

Figure 7.4 shows the demonstration PKI in JCT. You can create key pairs and certificate signing requests for public keys. Afterwards, you can take a look at each

7. In Germany, for example, all university members can receive free email S/MIME certificates. These are issued by the DFN (`https://www.pki.dfn.de/ueberblick-dfn-pki/`) after an identification by the local university.

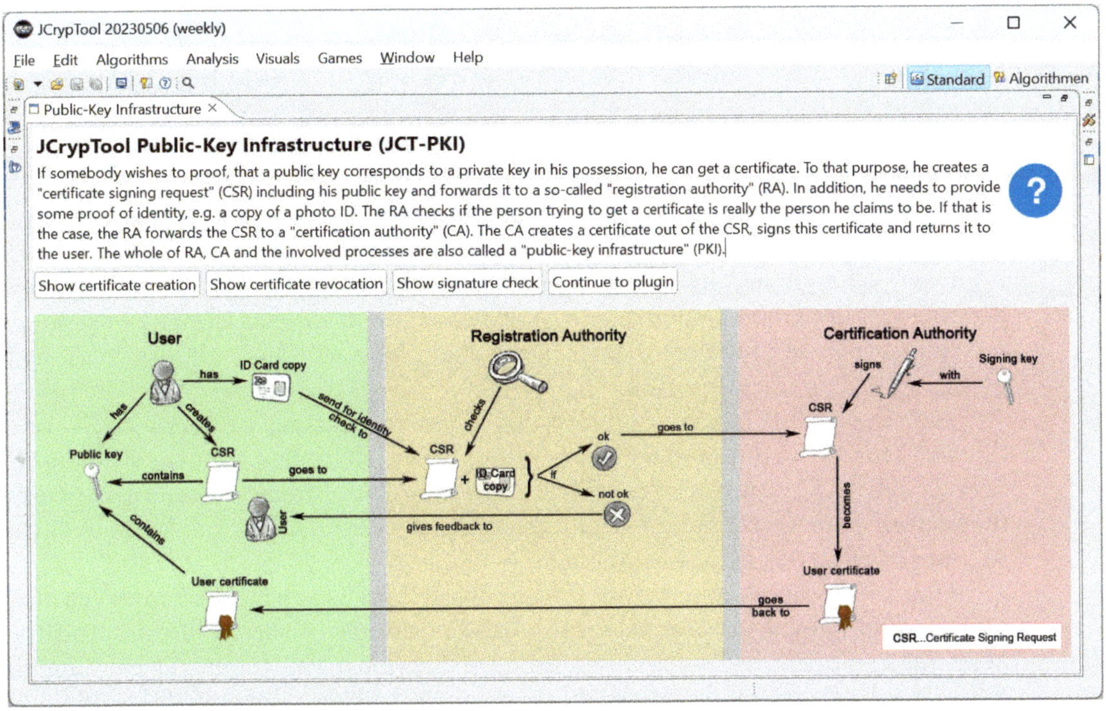

Figure 7.4 Demonstrations-PKI in JCT.[8]

step that your request needs to take through the different parts of the PKI. After the certificate was issued, you can create and check signatures for texts or files. Also, you can revoke your certificate. Every step is accompanied by additional explanations and online help.

A digital certificate is nothing more than binding an email address to a public key and is used for authentication. The term has nothing to do with the certificates (often risky products) from the financial world or with other official documents relating to a person.

Anyone can now use the CA's public key to verify whether a certificate is falsified. The CA therefore guarantees with the certificate that a public key belongs to a particular user.

This procedure is only secure as long as it can be guaranteed that the CA's public key is correct and that the registration process is handled seriously. For this reason, each CA has its public key certified by another CA that is superior in the hierarchy. In the uppermost hierarchy level (root CA) there is usually only one CA, which of course then has no higher CA to certify its key. It must therefore transfer its key securely in another way. In the case of many software products (such as the Microsoft, Mozilla, or Google web browsers), the certificates of these root CAs are permanently embedded in the program right from the start and cannot be changed by users at a later stage. However, (public) CA keys, particularly those of the root entity, can also be secured by means of making them available publicly (at websites or newspapers).

8. `JCT Default Perspective ▷ Visuals ▷ Public-Key Infrastructure.`

7.5.3 Signature Validation and Validity Models

Usually, a signature is invalid if the message has been modified (thus signature and message no longer match) or if a wrong signer key is used to check the signature. However, the verification algorithms also consider the validity timeframe of the certificates involved.

So, a signature can also become invalid, depending on how long the certificate of the signer or of the CAs are valid and when the verification takes place. The behavior is determined by so-called validity models. The validity model describes how the validity of the signature is to be evaluated depending on the date of signing, the validity period of the certificates in the chain between signer and root CA, and possibly the date of verification.

Although, while in some countries like Germany, various validity models were discussed for a long time and with almost ideological sharpness, only the shell model defined in RFC 5280 and used worldwide has prevailed. Attempts to enforce models that deviate from this standard (including by law) were unsuccessful: Users let them become meaningless by not using them.

Figure 7.5 shows a screenshot from JCrypTool[9]: The "Certificate Verification" plugin allows the user to set the validity period of the certificates and the times for signature creation and signature verification using a slider. Depending on the selected validity model, different (!) results can then be obtained. So this plugin allows one to playfully experience the effect of the parameters on the result

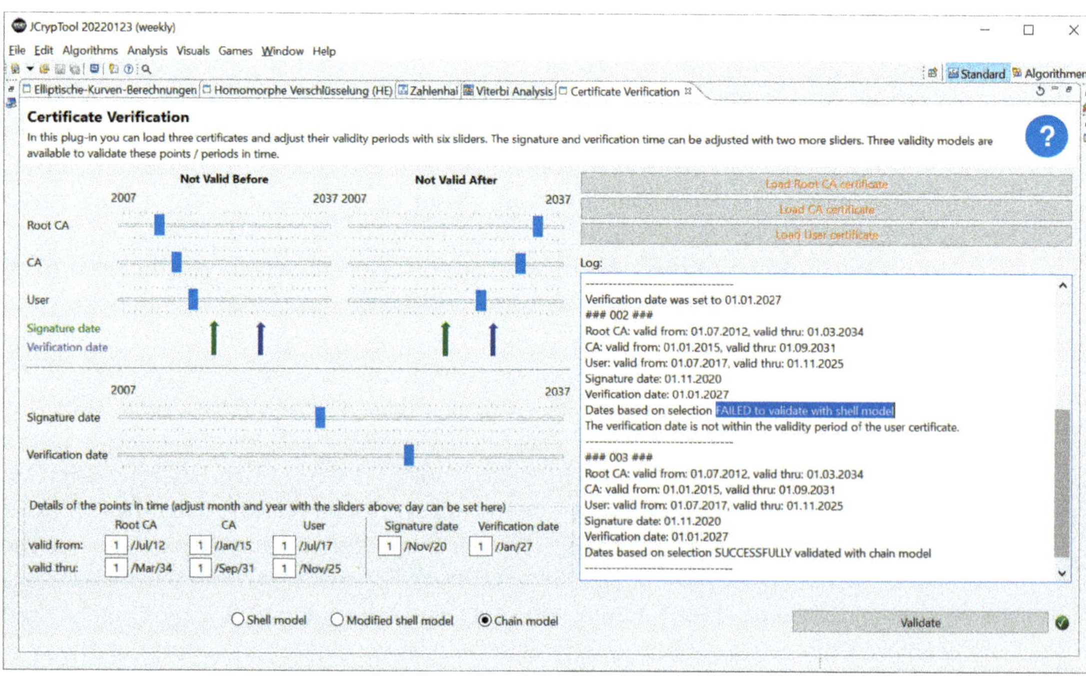

Figure 7.5 Signature validation and validity models in JCT.

9. JCT Default Perspective ▷ Visuals ▷ Certificate Verification.

Figure 7.6 Signature validation and validity models in CTO.

determined by the validity models. Figure 7.6 shows how the the same topic is implemented in CTO.[10]

You may notice that depending on the validity model, a signature can be evaluated as invalid if just the validation point of time is changed. It is astonishing that even if the message and the corresponding signature didn't change, you today can get the evaluation result that the signature is valid, and tomorrow that the signature is invalid. Especially with software signing, even signatures that become invalid in this way are often immediately associated with malware warnings by the Windows operating system. This is an example of counterproductive security warnings and false awareness. The developers should include the reason for the invalidation result in their warning to enable a user to take appropriate measures.

This is particularly relevant for contract documents, where it is often the case that—in contrast to e-mails and SSL—the signature is subject to legal regulations, and the verification of the signature (also called validation) can take place much later after the signature has been created.

References

[1] Menezes, A. J., P. C. van Oorschot, and S. A. Vanstone, *Handbook of Applied Cryptography*, 5th ed., Series on Discrete Mathematics and Its Application, CRC Press, 2001, https://cacr.uwaterloo.ca/hac/.

10. See "validation and validity models" `https://www.cryptool.org/en/cto/validitymodels`.

[2] *Digital Signature Standard (DSS)*, Federal Information Processing Standards (FIPS) 186-4, National Institute of Standards and Technology (NIST), Gaithersburg: U.S. Department of Commerce, July 19, 2013, https://nvlpubs.nist.gov/nistpubs/FIPS/NIST.FIPS.186-4.pdf; https://csrc.nist.gov/publications/fips; https://www.nist.gov/publications/digital-signature-standard-dss-2.

[3] Wang, X., Y. Yiqun, and H. Yu, "Finding Collisions in the Full SHA-1," in *Advances in Cryptology-Crypto, LNCS 3621*, 2005, pp. 17–36.

[4] Wang, X., A. Yao, and F. Yao, *New Collision Search for SHA-1*, Tech. rep., *Crypto 2005, Rump Session*, 2005, https://www.iacr.org/conferences/crypto2005/rumpSchedule.html.

[5] McDonald, C., P. Hawkes, and J. Pieprzyk, "Differential Path for SHA-1 with Complexity O(2^{52})," in *Cryptology ePrint Archive*, 2012, https://eprint.iacr.org/2009/259.

[6] Dang, Q. H., *Secure Hash Standard (SHS)*, Federal Information Processing Standards (FIPS) 180-4, National Institute of Standards and Technology (NIST), Gaithersburg: U.S. Department of Commerce, 2015, https://nvlpubs.nist.gov/nistpubs/FIPS/NIST.FIPS.180-4.pdf; https://csrc.nist.gov/publications/fips.

[7] RSA Labs, *PKCS #1 v2.1 Draft 3: RSA Cryptography Standard*, Tech. rep., RSA Laboratories, April 2002.

[8] Bruce Schneier. *Applied Cryptography, Protocols, Algorithms, and Source Code in C.*, 2nd ed., Wiley, 1996.

[9] ITU-T,"ITU-T Recommendation X.509 (1997 E): Information Technology—Open Systems Interconnection—The Directory: Authentication Framework," Tech. rep., *International Telecommunication Union ITU-T*, June 1997.

[10] Let's Encrypt, *Let's Encrypt Stats*, https://letsencrypt.org/stats/.

CHAPTER 8

Elliptic-Curve Cryptography

This chapter is about elliptic-curve cryptography[1] and describes elliptic curves in cryptography: They are an alternative to RSA and are particularly well suited for use on smart cards.

8.1 Elliptic-Curve Cryptography: A High-Performance Substitute for RSA?

In many business sectors, secure and efficient data transfer is essential. In particular, the RSA algorithm is used in many applications for encrypting and signing. Although the security of RSA itself is beyond doubt, the evolution in computing power has caused to raise the necessary key length several times. Today, 2048-bit RSA keys are standard but, for example, the BSI recommends the usage of 3000-bit keys from 2023 on (compare Section 5.12). The fact that most chips on smart cards cannot process keys extending around 2000 bit shows that there is a need for alternatives in the area of asymmetric cryptography. Elliptic-curve cryptography (ECC) is such an alternative. It is used widely on smart cards.

The efficiency of a cryptographic algorithm depends on the key length and the calculation effort that is necessary to provide a prescribed level of security. The major advantage of ECC compared to RSA is that it requires much shorter key lengths.

If we assume that there is no quantum computer and the computing power continues to increase by Moore's law (i.e., it doubles every 18 months), then the evolution of the key lengths for secure communication will be as in Figure 8.1, which was generated from Table 1 on page 32 in [1].[2] Moore's law formulates the empirical observation and the corresponding forecast that the number of components or transistors on an integrated circuit doubles every two years.

Creating a digital signature can be processed 10 times faster with ECC than with RSA. However, verification of a given signature is more efficient with RSA than with ECC. Refer to Figure 8.2 for a comparison. The reason is that RSA public keys can be chosen to be relatively small as long as the secret key is long enough.

Nevertheless, thin clients like smart cards usually have to store the (long) secret key and have to process a digital signature rather than verify one. Therefore, there is a clear advantage in using ECC in terms of efficiency.

1. We write elliptic-curve cryptography with a hyphen like public-key cryptography. In the literature this isn't used consistently.
2. Further information about key length comparison by Arjen Lenstra and Eric Verheul, plus more modern evaluations can be found in the interactive BlueKrypt website [1]. Also see Figure 13.1.

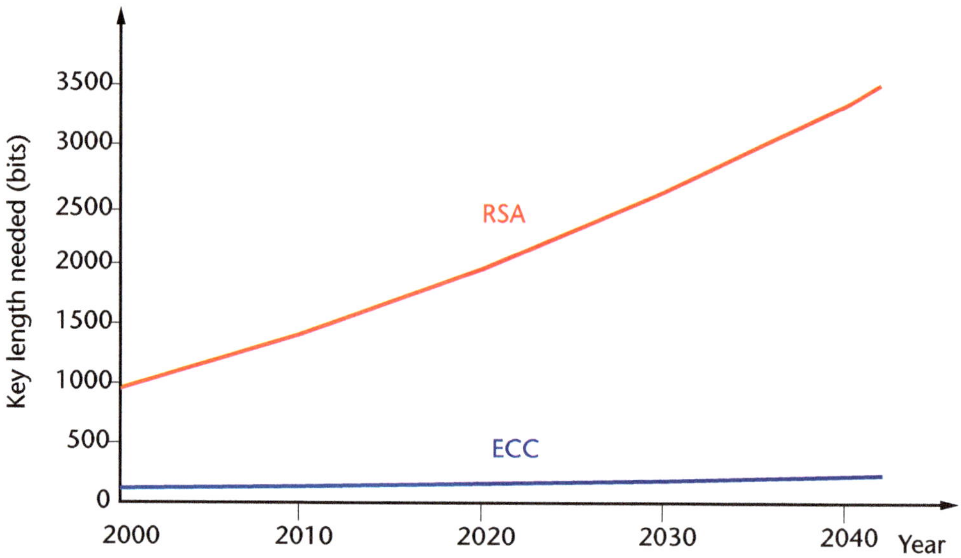

Figure 8.1 Prognosis of the key lengths regarded to be safe for RSA and for elliptic curves.

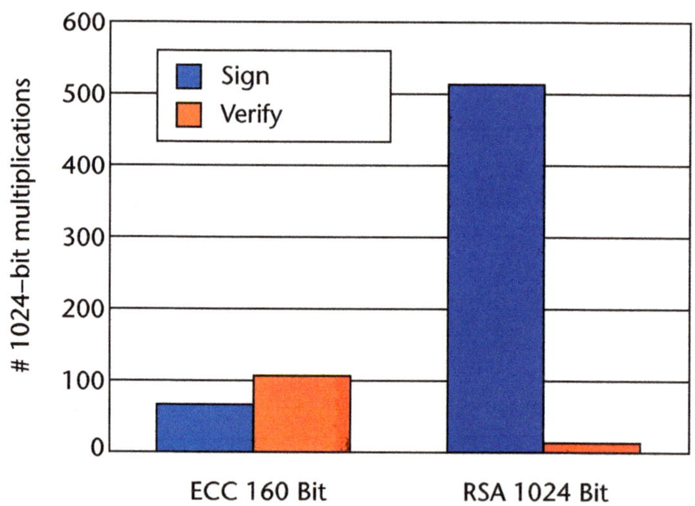

Figure 8.2 Comparison between signing and verification time for RSA and elliptic curves.

Nowadays, a major problem with ECC implementations is the lack of standardization. There is only one RSA algorithm, but there are many variants of ECC: One can work with different sets of numbers, different (elliptic) curves (described by parameters), and a variety of representations of the elements on the curve. Each choice has its advantages and disadvantages, and one can certainly construct the most efficient for each application. However, this causes problems in interoperability. But if all ECC tools should be able to communicate with each other, they will have to support all different algorithms, which might put the advantage of efficient computation and the need of less storage capacity to the contrary.

Therefore, international standardization organizations like IEEE (P1363), ASC (ANSI X9.62, X9.63), ISO/IEC as well as major players like RSA labs or Certicom have recently started standardization initiatives. While the IEEE only describes the different implementations, the ASC has explicitly stated 10 elliptic curves and recommends their usage. The advantage of the ASC approach is that one needs only a single byte to indicate which curve is meant. However, it is not yet clear whether the ASC curves will become a de facto standard.

Although there is no need to replace RSA in applications running today, one seriously should take the usage of ECC into consideration where key space is limited [3]. Current information about the security of the RSA algorithm can be found in Section 5.12 and in Chapter 12. Discussions about the security of ECC can be found in Chapter 12.

8.2 The History of Elliptic Curves

Mathematicians have been researching elliptic curves for over 100 years. Over the course of time, many lengthy and mathematically complex results have been found and published that are connected to elliptic curves. A mathematician would say that elliptic curves (or the mathematics behind them) are widely understood. This research was originally purely mathematical. That is to say, elliptic curves were investigated, for example, in the mathematical areas of number theory and algebraic geometry, which are generally highly abstract. Even in the recent past, elliptic curves played an important role in pure mathematics. In 1993 and in 1994, when the gaps in the first proof have been closed, Andrew Wiles published mathematical works that triggered enthusiasm far beyond the specialist audience. In these works, he proved a conjecture put forward in the 1960s. To put it short, this conjecture was concerned with the connection between elliptic curves and what are called module forms.

What is actually interesting for many people is that Wiles' work also proved the famous second or last theorem of Fermat (see Section 5.2). Mathematicians spent centuries trying to find a strict proof of this theorem. Understandably, Wiles' proof therefore got a good response. Fermat formulated his theorem as follows (written in the borders of a book from Diophantus) [4]:

> *Cubum autem in duos cubos, aut quadratoquadratum in duos quadratoquadratos, et generaliter nullam in infinitum ultra quadratum potestatem in duos ejusdem nominis fas est dividere: cujus rei demonstrationem mirabilem sane detexi. Hanc marginis exiguitas non caperet.*

With a free translation, using the denotation of modern mathematics, this means: No positive integers x, y, and z greater than zero exist such that $x^n + y^n = z^n$ for $n > 2$. I have found an amazing proof of this fact, but there is too little space within the confines of this book to include it.

This is truly amazing: A statement that is relatively simple to understand (we are referring to Fermat's second theorem here) could only be proved after such a long period of time, although Fermat himself claimed to have found a proof. What's more, the proof found by Wiles is extremely extensive (all of Wiles' publications connected with the proof made up a book in themselves). This should therefore make it obvious that elliptic curves are generally based on highly complex mathematics.

So much for the role of elliptic curves in pure mathematics. In 1985 Neal Koblitz and Victor Miller independently suggested using elliptic curves in cryptography. Elliptic curves have thus also found a concrete practical application. Another interesting area of application for elliptic curves is for factorizing integers: The RSA cryptographic system is based on the difficulty/complexity of finding prime factors of an extremely large number (see Section 5.12). In this area, procedures based on elliptic curves have been investigated and used since 1987 (see Section 8.8). There are also prime number tests based on elliptic curves.

So, elliptic curves are used differently in the various areas: Encryption procedures based on elliptic curves are based on the difficulty of the problem known as elliptic curve discrete logarithm. The factorization of natural composite numbers n uses the fact that numerous elliptic curves can be generated for n.

8.3 Elliptic Curves: Mathematical Basics

This section provides information about *groups* and *fields*. (See Section 5.7.) Didactically very well-prepared introductions into elliptic curves can be found in [5] and via the CTO plugin "Elliptic Curves" shown in Figure 8.5.

8.3.1 Groups

Because the term group is used differently in everyday language than in mathematics, we will, for reasons of completeness, begin by introducing the essential statement of the formal definition of a group:

- A group is a nonempty set G on which an operation "$\cdot$". The set G is closed under this operation, which means that for any two elements a, b taken from G, performing the operation on them gives an element in G (i.e., $ab = a \cdot b$ lies in G).
- For all elements a, b, and c in G: $(ab)c = a(bc)$ (associative law).
- There exists an element e in G that behaves neutrally with respect to the operation $\cdot$. That means that for all a in the set $G:\ ae = ea = a$.
- For each element a in G there exists a unique inverse element a^{-1} in G such that: $aa^{-1} = a^{-1}a = e$. The inverse is uniquely determined because if $x, y \in G$ are both inverse to a (i.e., $ax = xa = e$) and $ay = ya = e$, then $x = xe = x(ay) = (xa)y = ey = y$.

If, in addition, it applies $ab = ba$ (commutative law) for all a, b in G, then we call the group an *abelian* group.

Since we may define different operations on the same set, we distinguish them by giving them different names (e.g., $+$ addition or $\cdot$ multiplication).

The simplest example of an (abelian) group is the group of integers under the standard operation of addition. The set of integers denoted as $\mathbb{Z} = \{\cdots, -4, -3, -2, -1, 0, 1, 2, 3, 4, \cdots\}$ has an infinite number of elements. For example, the operation of $1+2$ lies in $\mathbb{Z}$, for $1+2=3$ and 3 lies in $\mathbb{Z}$. The neutral element in the group $\mathbb{Z}$ is 0. The inverse element of 3 is -3, for $3+(-3)=0$.

For our purpose, so-called *finite* groups play an important role. This means that there exists a set $\mathcal{M}$ with a fixed number of elements and an operation $+$ such that the previous conditions are fulfilled. One example is the group $\mathbb{Z}_n = \{0, 1, 2, 3, \cdots, n-1\}$ of the remainders of the division by $n \in \mathbb{N}$, and the operation is an addition mod n. So, for example, a and b in $\mathbb{Z}_n$ are subject to the operation $a+b \bmod n$.

8.3.1.1 Cyclic Groups

Cyclic groups are those groups G' that possess an element g from which the group operation can be used to generate all other elements in the group. This means that for each element a in G' there exists a positive integer i such that if g is subject to the operation i times (i.e., $g \cdot i$), $g+g+\cdots+g=a$ (additive group) or $g^i = g \cdot g \cdots g = a$ (multiplicative group). The element g is the *generator* of the cyclic group—each element in G' can be generated using g and the operation. In general, cyclic groups can be also endless like the additive group of the integer numbers. We consider here only finite cyclic groups.

8.3.1.2 Group Order

Now to the order of an element of the group: Let a be in G. The smallest positive integer r for which a, subject to the operation with itself r times is the neutral element of the group G' (i.e., $r \cdot a = a + a + \cdots + a = e$, respectively $a^r = e$), is called the *order* of a.

The order of the group is the number of elements in the set G. If G is cyclic and g is a generator, then the order of g equals the order of G. One easily shows that the order of an element of a group always divides the order of the group. In the special case of a group with prime order (i.e., the number of elements in the group is a prime), the group must be cyclic.

8.3.2 Fields

In mathematics, one is often interested in sets on which at least two (group) operations are defined, frequently called addition and multiplication. The most prominent are so-called fields.

A field is understood to be a set K with two operations (denoted as $+$ and $\cdot$) which fulfills the following conditions:

- The set K forms an abelian group together with the operation $+$ (addition), where 0 is the neutral element of the operation $+$.
- The set $K \setminus \{0\}$ also forms an abelian group together with the operation $\cdot$ (multiplication).
- For all elements a, b, and c in K, we have $c \cdot (a+b) = c \cdot a + c \cdot b$ and $(a+b) \cdot c = a \cdot c + b \cdot c$ (distributive law).

Fields may contain an infinite number of elements (e.g., the field of real numbers). They are called *infinite* fields. In contrast, we call a field *finite*, if it contains only a finite number of elements (e.g., $\mathbb{Z}_p = \{0, 1, 2, 3, \cdots, p-1\}$, where p is a prime. $\mathbb{Z}_p$ with addition mod p and multiplication mod p).

8.3.2.1 Characteristic of a Field

Let K be a field and 1 be the neutral element of K with respect to the multiplicative operation "$\cdot$". Then the characteristic of K is said to be the order of 1 with respect to the additive operation. This means that the characteristic of K is the smallest positive integer n such that

$$\underbrace{1 + 1 + \cdots + 1}_{n \text{ times}} = 0.$$

If there is no such n (i.e., if $1 + 1 + \cdots + 1 \neq 0$ no matter how many 1s we add) then we call K a field with characteristic 0.

Thus, fields with characteristic 0 are infinite since they contain the (pairwise distinct) elements $1,\ 1 + 1,\ 1 + 1 + 1, \ldots$. On the other hand, fields with finite characteristic may be either finite or infinite.

If the characteristic is finite, it has to be prime. This fact can easily be proved: Assume $n = pq$, $p, q < n$, is the characteristic of a field K. By definition of n, the elements $\bar{p} = \underbrace{1 + 1 + \cdots + 1}_{p \text{ times}}$, $\bar{q} = \underbrace{1 + 1 + \cdots + 1}_{q \text{ times}}$ of K are not equal to 0. Thus, there exist inverse elements $\bar{p}^{-1}, \bar{q}^{-1}$ with respect to multiplication. It follows that $(\bar{p}\bar{q})(\bar{p}^{-1}\bar{q}^{-1}) = 1$, which contradicts the fact that $\bar{p}\bar{q} = \bar{n} = \underbrace{1 + 1 + \cdots + 1}_{n \text{ times}} = 0$ and, hence, $\underbrace{(\bar{p}\bar{q})}_{=0}(\bar{p}^{-1}\bar{q}^{-1}) = 0$.

Comment

The field of real numbers has the characteristic 0; the field $\mathbb{Z}_p$ has the characteristic p. If p is not prime, $\mathbb{Z}_p$ is not a field at all.

The most simple field is $\mathbb{Z}_2 = \{0, 1\}$. It contains only two elements, the neutral elements with respect to addition and multiplication. In particular, we have $0+0 = 0$, $0+1 = 1+0 = 1$, $1+1 = 0$, $1 \cdot 1 = 1$, $0 \cdot 0 = 0 \cdot 1 = 1 \cdot 0 = 0$.

8.3.2.2 Finite Fields

As mentioned previously, each finite field has a characteristic $p \neq 0$, where p is a prime. On the other hand, given a prime p there is a field which has exactly p elements; that is $\mathbb{Z}_p$.

However, the number of elements of a field need not be prime in general. For example, it is not hard to construct a field with 4 elements: The set $K = \{0, 1, a, b\}$ fitted with the operations defined in the following table is a field:

+	0	1	a	b
0	0	1	a	b
1	1	0	b	a
a	a	b	0	1
b	b	a	1	0

and

$\cdot$	0	1	a	b
0	0	0	0	0
1	0	1	a	b
a	0	a	b	1
b	0	b	1	a

One can show that the order of any field is a prime power (i.e., the power of a prime number). We can construct a field with p^n elements for any given prime p and positive integer n and denote it by $GF(p^n)$ or by $\mathbb{F}_p^n$. Here GF stands for *Galois Field* to commemorate the French mathematician Galois.

Two fields that have the same number of elements cannot be distinguished in the following sense: If K, K' are fields with $k = p^n$ elements, then there is a one-to-one map $\varphi : K \to K'$, that respects the arithmetic of the field. Such a map is called an isomorphism. Isomorphic fields mathematically behave in the same way so that it makes no sense to distinguish between them. For example, $\mathbb{Z}_2$ and $K' = \{ZERO, ONE\}$ with zero-element $ZERO$ and one-element ONE are isomorphic. We note that mathematical objects are only defined by their mathematical properties.

The fields $GF(p)$ of prime order play a prominent role. They are called prime fields and are often denoted by $\mathbb{Z}_p$. For prime fields, both additive and multiplicative group are cyclic. Furthermore, each field $GF(p^n)$ contains a subfield that is isomorphic to the prime field $\mathbb{Z}_p$.

8.4 Elliptic Curves in Cryptography

In general, expressions of the form $P = \sum_{i_1,\ldots,i_n} a_{i_1\ldots i_n} x_1^{i_1} \ldots x_n^{i_n}$ with $i_1, \ldots, i_n \in \mathbb{N}$ with coefficients $a_{i_1\ldots i_n} \in K$ are called *polynomials* in n variables $x_1, \ldots, x_n$ with underlying field K, if $\deg P := \max\{i_1 + \cdots + i_n : a_{i_1\ldots i_n} \neq 0\}$ is finite [i.e., the sum has only finitely many nonzero terms (monomials)]. The sum of the exponents of the variables of each term of the sum is at most 3, and at least one term of the sum has a single variable with 3 as value of the corresponding exponent.

We now consider a curve given by the zeros of a *polynomial* F of degree 3 in 3 variables.

In cryptography, elliptic curves are a useful tool. Such curves are described as solutions of an equation of the form

$$F(x_1, x_2, x_3) = -x_1^3 + x_2^2 x_3 + a_1 x_1 x_2 x_3 - a_2 x_1^2 x_3 + a_3 x_2 x_3^2 - a_4 x_1 x_3^2 - a_6 x_3^3 = 0. \quad (8.1)$$

are especially useful. The variables x_1, x_2, x_3 and parameters $a_1, \ldots, a_4, a_6$ are elements of a given field K, which has certain properties that make it useful from the cryptographic point of view. The underlying field K might be the well known field of real numbers or some finite field (see Section 8.3.2). In order to obtain a curve that is useful for cryptography, the parameters have to be chosen in a way that the following conditions hold

$$\frac{\partial F}{\partial x_1} \neq 0, \quad \frac{\partial F}{\partial x_2} \neq 0, \quad \frac{\partial F}{\partial x_3} \neq 0.$$

We identify points on the curve that can be derived from each over by multiplying each component with some scalar. This makes sense since (x_1, x_2, x_3) solves (8.1) if and only if $\alpha(x_1, x_2, x_3)$ $(\alpha \neq 0)$ does. Formally, this means that we consider classes of equivalent points instead of single points, where points are called equivalent if one is the scalar multiple of the other one.

If we put $x_3 = 0$ in (8.1), then this equation collapses to $-x_1^3 = 0$, leading to $x_1 = 0$. Thus, the equivalence class which includes the element $(0, 1, 0)$ is the only one that contains a point with $x_3 = 0$. For all points on the curve that are not equivalent to $(0, 1, 0)$, we may apply the following transformation

$$K \times K \times (K \setminus \{0\}) \ni (x_1, x_2, x_3) \mapsto (x, y) := \left(\frac{x_1}{x_3}, \frac{x_2}{x_3}\right) \in K \times K\,,$$

which reduces the number of variables from three to two. We note that with the basic equation (8.1), $F(x_1, x_2, x_3) = 0$ was chosen in a way that this transformation leads to the famous so-called Weierstrass equation holds:

$$y^2 + a_1xy + a_3y = x^3 + a_2x^2 + a_4x + a_6 \qquad (8.2)$$

Since all but one point (i.e., equivalence class) of the elliptic curve can be described using (8.2), this equation is often called the elliptic equation, and its solutions written as

$$\mathrm{E} = \left\{(x, y) \in K \times K \mid y^2 + a_1xy + a_3y = x^3 + a_2x^2 + a_4x + a_6\right\} \cup \{\mathcal{O}\}.$$

Here, $\mathcal{O}$ represents the point $(0, 1, 0)$ that is loosely speaking mapped to infinity by the transformation (division by x_3) that reduces the three variables to two.

In contrast to Figure 8.3 only finite fields $K = GF(p^n)$ are used in elliptic-curve cryptography. The reason is that in modern communication engineering, data processing is always based on discrete data (simply because computers accept only discrete data).

For practical reasons, it turned out to be useful to take either $GF(p)$ with a large prime p or $GF(2^n)$ with a (large) positive integer n. Using $GF(p)$ has the advantage of providing a relatively simple arithmetic; on the other hand, $GF(2^n)$ allows a binary representation of each element that supports the way computers work. Other fields like $GF(7^n)$ do not have any of these advantages and are, thus, not considered, although there is no mathematical reason why they should not be.

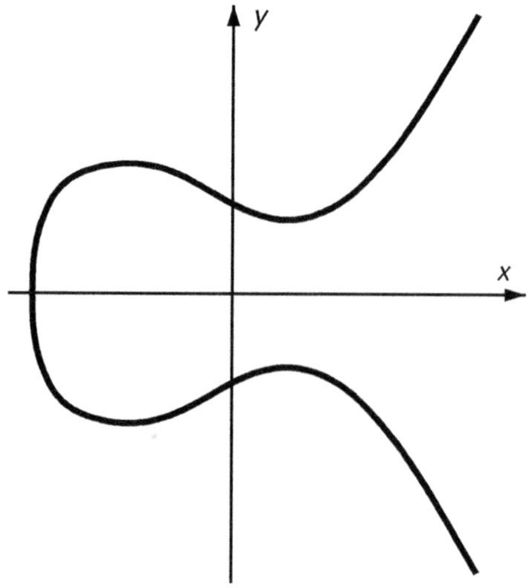

Figure 8.3 Example of an elliptic curve with the real numbers as an underlying field.

A coordinate transformation is combination of a rotation and a dilatation of the coordinate system without changing the elliptic curve itself. A coordinate transformation can result in a simpler version of the Weierstrass equation. Depending on whether $p > 3$, different transformations are used, and we obtain

- In case of $GF(p)$, $p > 3$, the elliptic curve equation of the form

$$y^2 = x^3 + ax + b \tag{8.3}$$

  with $4a^3 + 27b^2 \neq 0$.
- In case of $GF(2^n)$ the elliptic curve equation of the form

$$y^2 + xy = x^3 + ax^2 + b$$

  with $b \neq 0$.

These conditions on the parameters a, b ensure that the elliptic equation can be used in the context of cryptography. Formally we call such curves nonsingular.

The form (8.3) is called the standard form of the Weierstrass equation. If the characteristic of the field is 2 or 3, we obtain $4 = 0$, respectively, $27 = 0$, which means that the condition on parameters a, b collapse. Loosely speaking, this is the reason why the transformation to the standard form does not work in these cases.

Let $|E|$ denote the number of elements of an elliptic curve E given an underlying field $GF(k)$ (for practical reasons either $k = p$ with p prim or $k = 2^n$). Then Hasse's theorem [6] yields $||E| - k - 1| \leq 2 \cdot \sqrt{k}$. This inequality is equivalent to $k + 1 - 2\sqrt{k} < |E| < k + 1 + 2\sqrt{k}$. In particular, this means that the number of elements of an elliptic curve is approximately k (for large k).

8.5 Operating on the Elliptic Curve

In order to work with elliptic curves in practice, we define an operation (often written in an additive way +) on the set of points on the curve. If we have a curve over the field $GF(p)$, we define the commutative operation + by

1. $P + \mathcal{O} = \mathcal{O} + P = P$ for all $P \in E$;
2. For $P = (x, y)$ and $Q = (x, -y)$ we set $P + Q = \mathcal{O}$;
3. For $P_1 = (x_1, x_2)$, $P_2 = (x_2, y_2) \in E$ with $P_1, P_2 \neq \mathcal{O}$ and $(x_2, y_2) \neq (x_1, -y_1)$ we set $P_3 := P_1 + P_2$, $P_3 = (x_3, y_3)$ defined by

$$x_3 := -x_1 - x_2 + \lambda^2, \qquad y_3 := -y_1 + \lambda(x_1 - x_3)$$

   with the auxiliary quotient

$$\lambda := \begin{cases} \dfrac{y_1 - y_2}{x_1 - x_2} & \text{if } P_1 \neq P_2, \\[2ex] \dfrac{3x_1^2 + a}{2y_1} & \text{if } P_1 = P_2. \end{cases}$$

In particular, we obtain $-P = (x, -y)$ for $P = (x, y) \in E$.

If we deal with a curve over the field $GF(2^n)$, we define the operation $+$ in an analogous way by

1. $P + \mathcal{O} = \mathcal{O} + P = P$ for all $P \in E$;
2. For $P = (x, y)$ and $Q = (x, x + y)$ we set $P + Q = \mathcal{O}$;
3. For $P_1 = (x_1, x_2)$, $P_2 = (x_2, y_2) \in E$ with $P_1, P_2 \neq \mathcal{O}$ and $(x_2, y_2) \neq (x_1, x_1 + y_1)$ we set $P_3 := P_1 + P_2$, $P_3 = (x_3, y_3)$ defined by

$$x_3 := -x_1 + x_2 + \lambda + \lambda^2 + a\,, \qquad y_3 := y_1 + x_3 + \lambda(x_1 + x_3)$$

with auxiliary quotient

$$\lambda := \begin{cases} \dfrac{y_1 + y_2}{x_1 + x_2} & \text{if } P_1 \neq P_2, \\ x_1 + \dfrac{y_1}{x_1} & \text{if } P_1 = P_2. \end{cases}$$

In particular, we obtain $-P = (x, -y)$ for $P = (x, y) \in E$.

Note that $-(-P) = (x, x+(x+y)) = (x, 2x+y) = (x, y)$, since the underlying field has characteristic 2.

One can verify that $+$ defines a group operation on the set $E \cup \{\mathcal{O}\}$. In particular this means that the sum of two points is again a point on the elliptic curve. How his operation works is geometrically visualized in the following subsection.

How to Add Points on an Elliptic Curve

Figure 8.4 shows how points on an elliptic curve over the field of real numbers are summed up using affine coordinates: (a) doubles a point P, (b) sums up two different points P and Q. We note that the point infinity $\mathcal{O}$ cannot be shown in the affine plane.

8.5.1 Web Programs with Animations to Add Points on an Elliptic Curve

The best interactive animations about operations on elliptic curves can be found on the internet in the following:

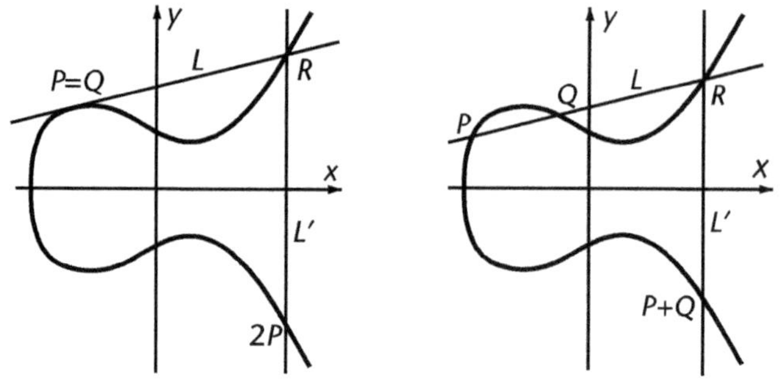

Figure 8.4 Operations on continuous elliptic curves.

- Certicom Online Tutorial [7];
- Tutorial with Java applets by Thomas Laubrock (German only) [8];
- ECC Tutorial by Johannes Bauer [5];
- Elliptic Curve Plotter by Stefan Kebekus [9];
- Elliptic curve point addition and multiplication tutorial by Andrea Corbellini [10];
- Animated Elliptic Curve by Michael Driscoll [11];
- Elliptic Curves plugin by CTO (see Figure 8.5).[3]

Figure 8.5 shows a screenshot from CTO: In the GUI, a curve type can be selected and different operations on the curve are available. In the screenshot, a chaining is performed: First, two points P and Q are added. Then the resulting point R is multiplied with a scalar, which leads to the point R_2. All resulting points are again on the curve.

8.6 Security of Elliptic-Curve Cryptography: The ECDLP

As mentioned in Section 8.4, we only consider elliptic curves over the finite fields $GF(2^n)$ or $GF(p)$ (for a large prime p). This means that all parameters that describe

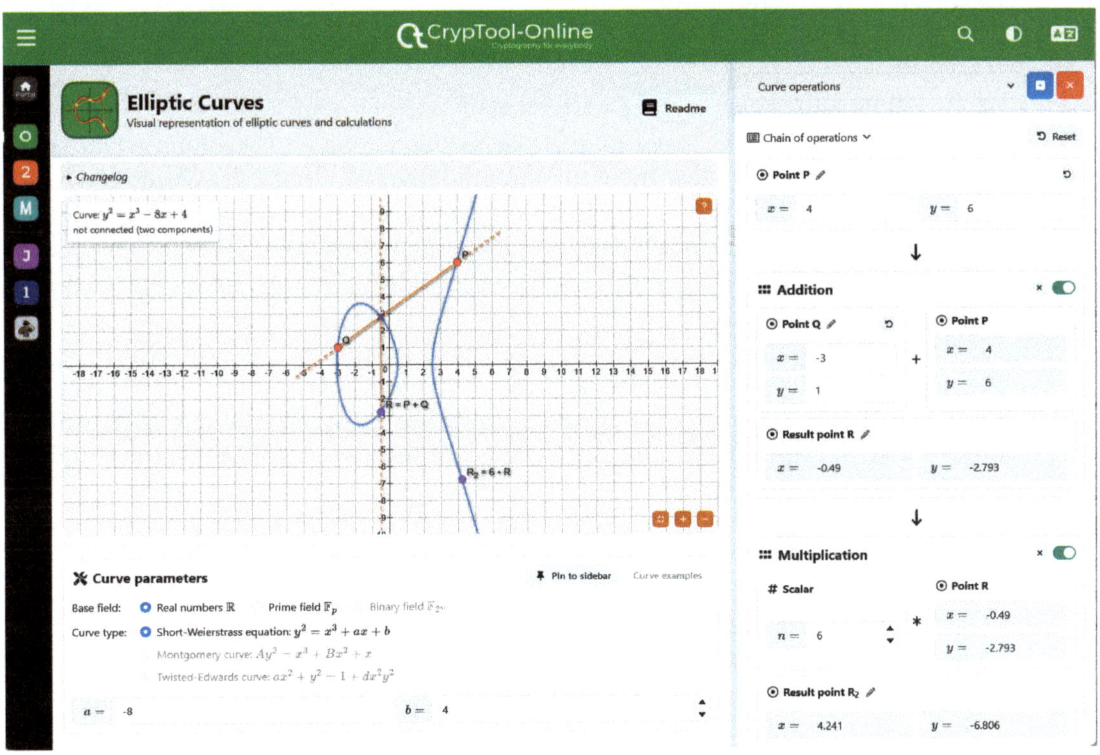

Figure 8.5 CrypTool-Online: Operations on elliptic curves.

3. CTO plugin "Elliptic Curves": `https://www.cryptool.org/en/cto/elliptic-curve`.

the curve are taken from this underlying field. If E is an elliptic curve over such a field and P is a point on the curve E, then we can derive for all positive integers m

$$mP := \underbrace{P + P + \cdots + P}_{m \text{ times}} .$$

Looking at this operation from the cryptographic point of view, it turns out to be very interesting because of the following reason: On the one hand, one needs only $\log m$ operations to calculate mP—one simply has to calculate P, $2P$, 2^2P, 2^3P, ..., write m in a binary form, and finally add all these multiples 2^kP of P with respect to the binary representation of m. On the other hand, it seems to be very hard to find m given P and $Q = mP$ on E. Of course, we may simply calculate $P, 2P, 3P, 4P, 5P, \ldots$ and compare each of them with Q. But this will take as much as m operations.

Yet there is no algorithm known that efficiently derives m given P and G. The best algorithms known so far need about $\sqrt{q}$ operations where q is the (largest) prime factor of $p - 1$, in case the underlying field is $GF(p)$; here m should be between 1 and q so that one needs at most $\log q$ operations to calculate mP. However, the quotient $\frac{\sqrt{q}}{\log q}$ tends to $+\infty$ very fast for large q.

If we choose sufficiently large parameters (for example, let p be prime and at least 160 bits long), a computer will easily be able to calculate mP (in less than a second). The *inverse problem*, however, to derive m from mP and P, can (still) not be solved in reasonable time.

This problem is known as the *elliptic curve discrete logarithm problem* (ECDLP).

In elliptic-curve cryptography we formally look at points on the elliptic curve as elements of a group with point addition $+$ as operation. Furthermore, we use only elliptic curves that have a sufficiently large number of points. However, in special cases curves may be weak and not useful due to other reasons. For such special cases the ECDLP can be much easier to solve than in the general case. This means that one has to look carefully at the parameters when choosing an elliptic curve for cryptographic applications.

Not useful for cryptography are *a-normal* (curves over $\mathbb{Z}_p$ for which the set E consists of exactly p elements) and *supersingular* curves (curves for which the ECDLP can be reduced to the "normal" discrete logarithms in another, smaller finite field). This means that there are cryptographically useful and nonuseful elliptic curves. Given the parameters a and b, it is possible to determine whether a curve is useful or not. In many publications one can find parameters that turned out to be useful for cryptography. The open (scientific) discussion guarantees that these results take into account latest research.

Given a secure curve, the time that is needed to solve the ECDLP is strongly correlated with parameter p in case $GF(p)$, respectively, n in case of $GF(2^n)$. The larger these parameters become, the more time an attacker needs to solve the ECDLP—at least with the best algorithms known so far. Experts recommend bit-lengths of 200 for p for secure curves. A comparison with the RSA modulus length shows why elliptic curves are so interesting for applications. We note that the computation effort for signing and encryption is closely related to the bit-length of the parameters. In addition, the initiation process (i.e., the generation of the private-public-key-pair) becomes more complicated the larger p is. Thus, one looks for the smallest parameters that still come along with the security required. It is remarkable that a length of 200 bits for p is sufficient to construct a good elliptic curve that is as secure as RSA with a 1024-bit RSA modulus (as far as we know today). In short, the reason for this advantage of ECC lies in the fact that the best algorithms known for solving the ECDLP need exponential time while the best algorithms for factorizing are subexponential (number field sieve quadratic sieve or factorizing with elliptic curves). Hence, the parameters for a cryptosystem that is based on the problem of *factorizing large integers* have to be larger than the parameters for a system based on ECDLP.

8.7 Encryption and Signing with Elliptic Curves

The elliptic curve discrete logarithm problem is the basis for elliptic-curve cryptography. Based on this problem, there are different signature schemes. In order to apply one of these, we need:

- An elliptic curve E with an underlying field $GF(p^n)$.
- A prime $q \neq p$ and a point G on the elliptic curve E with order q. This means that $qG = \mathcal{O}$ and $rG \neq \mathcal{O}$ for all $r \in \{1, 2, \ldots, q-1\}$. Thus, q is a factor of the group order (i.e., the number of elements) $\#E$ of E. Since q is prime, G generates a cyclic subgroup of E of order q.

The parameters mentioned are often called *domain* parameters. They describe the elliptic curve E and the cyclic subgroup of $\mathbf{E}$ on which the signature scheme is based.

8.7.1 Encryption

Using elliptic curves, one can construct a key exchange protocol based on the Diffie-Hellman protocol (see Section 6.4.2). The key exchanged can be used for a subsequent symmetric encryption. We note that in contrast to RSA there is no pair of private and public key that can be used for encryption and decryption!

In the notation of elliptic curves, the Diffie-Hellman protocol reads as follows: First, both partners (A and B) agree on a group G and an integer q. Then they choose $r_A, r_B \in \{1, 2, \ldots, q-1\}$ at random, derive the points $R_A = r_A G$, $R_B = r_B G$ on the elliptic curve, and exchange them (using an insecure channel). After that, A easily obtains $R = r_A R_B$; B gets the same point ($R = r_A r_B G$) by calculating $r_B R_A = r_B r_A G = r_A r_B G = R$. We note that R_A, R_B are easy to derive as long as r_A respectively r_B are known G. However, the inverse operation, to get R_A respectively R_B from r_A respectively r_B is hard.

Using the best algorithms known so far, it is impossible for any attacker to obtain R without knowing either r_A or r_B—otherwise he would have to solve the ECDLP.

In order to prohibit a man-in-the-middle attack, one may sign the values G, q, R_A, R_B as described in Section 7.5.1.

8.7.2 Signing

Using the DSA signature scheme, one can proceed as follows: The signing party chooses a (nontrivial) number $s \in \mathbb{Z}_q$, which will be the private key, and publishes q, G and $R = sG$. We note that s cannot be obtained from G and R sufficiently—a fact on which the security of the signature scheme is based.

Given the message m, which should be signed, one first constructs a digital fingerprint using a hash-algorithm h such that $h(m)$ has its values in $\{0, 1, 2, \ldots, q-1\}$. Thus, $h(m)$ can be considered as an element of $\mathbb{Z}_q$. Then the signing party chooses a random number $r \in \mathbb{Z}_q$ and derives $R = (r_1, r_2) = rG$. We note that the first component r_1 of R is an element of $GF(p^n)$. This component will then be projected onto $\mathbb{Z}_q$ (i.e., in the case of $n = 1$ it is interpreted as the remainder of an element of $\{0, 1, \ldots, p-1\}$ divided by q. This projection of r_1 onto $\mathbb{Z}_q$ is denoted by $\bar{r}_1$. Then one determines $x \in \mathbb{Z}_q$ such that

$$rx - s\bar{r}_1 - h(m) = 0.$$

The triple (m, r_1, x) is then published as the digital signature of message m.

8.7.3 Signature Verification

In order to verify a signature, one has to build $u_1 = h(m)/x$, $u_2 = \bar{r}_1/x$ in $\mathbb{Z}_q$ and derive

$$V = u_1 G + u_2 Q.$$

Since we have $Q = sG$, the point $V = (v_1, v_2)$ satisfies $v_1 = u_1 + u_2 s$. We note that this operation (addition) takes place in the field $GF(p^n)$. The projection of $GF(p^n)$ on $\mathbb{Z}_q$ mentioned previously should be chosen in such a way that $\bar{v}_1 = u_1 + u_2 s$ is an element of $\mathbb{Z}_q$. Then it follows that

$$\bar{v}_1 = u_1 + u_2 s = h(m)/x + \bar{r}_1 s/x = (h(m) + \bar{r}_1 s)/x = rx/x = r.$$

Since $R = rG$, we obtain $\bar{v}_1 = \bar{r}_1$ (i.e., R and V coincide modulo the projection onto $\mathbb{Z}_q$).

8.8 Factorization Using Elliptic Curves

There are factorization algorithms based on elliptic curves. The biggest factor found by factoring compound numbers with elliptic curves (GMP-ECM) has 83 decimal digits (state July 2023); see the ECMNET project [12]. More precisely, these procedures exploit the fact that elliptic curves can be defined over $\mathbb{Z}_n$ (n composite number). Elliptic curves over $\mathbb{Z}_n$ do not form a group, because not every point on such an elliptic curve has an inverse point. This is connected with the fact that, if n is a composite number, there exist elements in $\mathbb{Z}_n$ that do not have an inverse with respect to multiplication mod n. In order to add two points on an elliptic curve over $\mathbb{Z}_n$, we can calculate in the same way as on elliptic curves over $\mathbb{Z}_p$.

Addition of two points (on an elliptic curve over $\mathbb{Z}_n$), however, fails if and only if a factor of n has been found. The reason for this is that the procedure for adding points on elliptic curves gives elements in $\mathbb{Z}_n$ and calculates the inverse elements for these (with respect to multiplication mod n) in $\mathbb{Z}_n$. The extended Euclidean algorithm is used here. If the addition of two points (that lie on an elliptic curve over $\mathbb{Z}_n$) gives an element in $\mathbb{Z}_n$ that does not have an inverse element in $\mathbb{Z}_n$, then the extended Euclidean algorithm delivers a genuine factor of n.

Factorization using elliptic curves thus principally works as follows: Random curves over $\mathbb{Z}_n$ are selected, as well as random points (that lie on this curve), and added; you thus obtain points that also lie on the curve or find a factor of n. Factorization algorithms based on elliptic curves therefore work probabilistically. The opportunity of defining a large number of elliptic curves over $\mathbb{Z}_n$ allows you to increase the probability of finding two points that you can add to obtain a factor of n. These procedures are therefore highly suitable for parallelization.

8.9 Implementing Elliptic Curves for Educational Purposes

Besides the web programs listed in Section 8.5.1, there are not many free didactic programs offering especially ECC under a graphical user interface. Sections 8.9.1 and 8.9.2 explain shortly which corresponding functionality is available in CrypTool and in SageMath.

Remark: Another interesting educational program was ECvisual, which visualizes elliptic curves and finite fields, but has not been maintained since 2012. Currently, neither the static nor the dynamic version can be started on current Ubuntu versions [13]. This underlines once again how important it is that learning programs are maintained permanently and reliably so that students and teachers can build on them.

8.9.1 CrypTool

CT1 offers elliptic curves for the digital signature function[4] and for ECC-AES hybrid encryption.[5] Also implemented are the basic algorithms for group operations, for generating elliptic curves, and for importing and exporting parameters for elliptic curves over finite fields with p elements (p prime). The implementation in CT1 complies with draft number 8 of the IEEE P1363 work group Standard Specifications for Public-Key Cryptography [14]. The point addition on elliptic curves is visualized for two types of elliptic curves in CT1 and for 3 types in JCT.[6]

Figure 8.6 shows the visualization of the point addition in JCT (the implementation in JCT is much more mature than that in CT1). Figure 8.5 shows a visualization of elliptic curves in CrypTool-Online (CTO).

4. The dialog box in `CT1 Digital Signatures/PKI` ▷ `Sign Message` offers the EC methods ECSP-DSA and ECSP-NR (these Nyberg-Rueppel and DSA signatures are based on elliptic curves).
5. See `CT1 Encrypt/Decrypt` ▷ `Hybrid`.
6. - `CT1 Indiv. Procedures` ▷ `Number Theory -- Interactive` ▷ `Point Addition on Elliptic Curves`.
   - `JCT Default Perspective` ▷ `Visuals` ▷ `Elliptic Curve Calculations`.

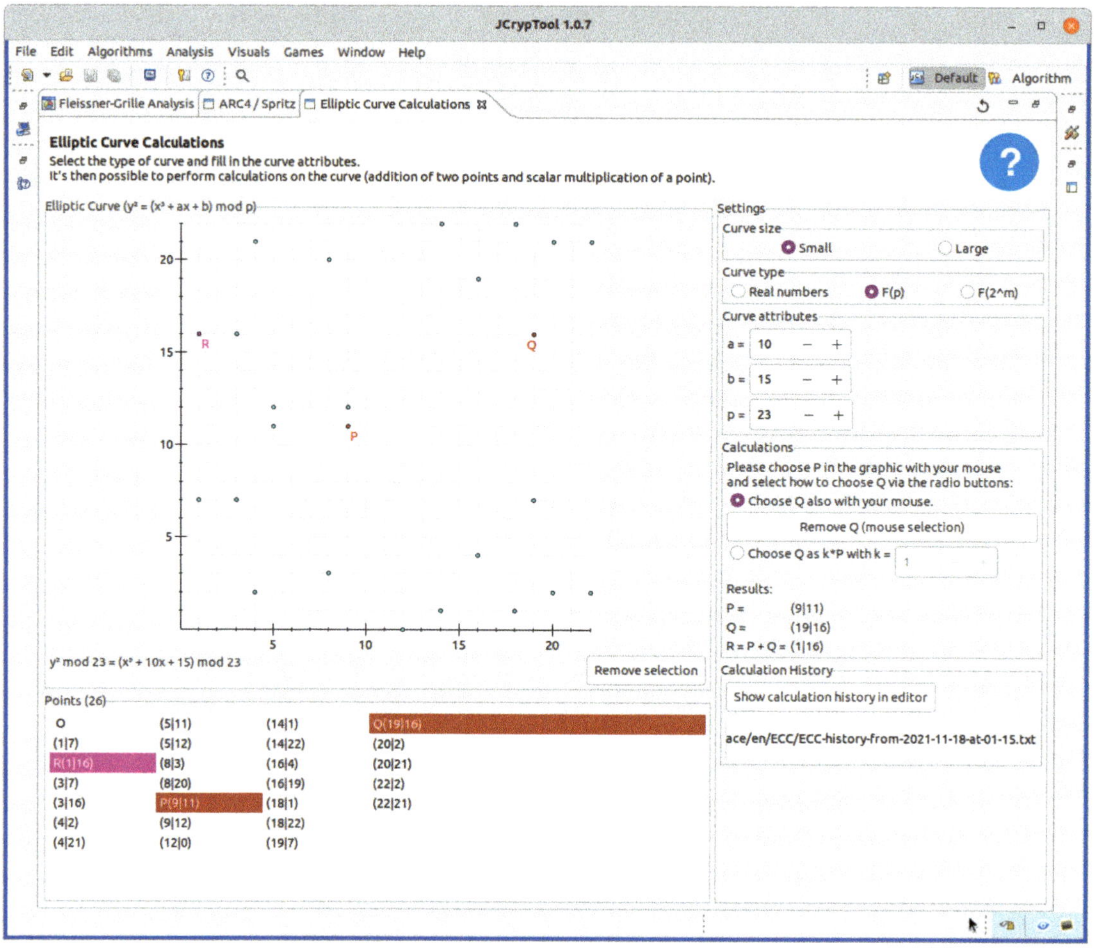

Figure 8.6 Addition on an $\mathbb{F}_p$ type of elliptic curve in JCT.

8.9.2 SageMath

There are many functions around elliptic curves in SageMath. See [15–20].

By Johannes Bauer there is a very nice tutorial (as of 2015) with SageMath and OpenSSL code [15].

8.10 Patent Aspects

If the field $GF(2^n)$ is used instead of the prime field $GF(p)$, one has to make substantial changes in the implementation. The advantage of $GF(2^n)$ lies in the fact that calculations in $GF(2^n)$ can be implemented very efficiently using the binary representation. In particular, divisions are much easier to process compared to $GF(p)$ (this is particularly important in the signature scheme mentioned previously where a division is needed for processing a signature as well as for the verification).

In order to achieve maximal gain in efficiency, one may choose a field that allows a special basis like a polynomial basis (useful for software implementations) or a normal basis (best for hardware implementations). For special n (like, for example, $n = 163179181$) one may even combine both advantages. However, they are still nonstandard.

Sometimes only the first component and one additional bit is used as representation of a point on the elliptic curve instead of the full two components. Since the first component together with the additional bit is sufficient to derive the full point, this representation minimizes the memory capacity needed. In particular, for a normal basis, this point compression can be implemented efficiently. In addition, the cryptographic protocols themselves become more effective. A disadvantage is, however, that point compression can be used for about half of all elliptic curves only and is protected under U.S. patent (US Patent 6141420, Certicon), causing additional costs. In the general case $GF(p^n)$ (and also in case $n = 1$) often so-called affine or projective coordinates are used. Depending on the application, these coordinates may result in a gain in efficiency as well.

A comprehensive description of all implementations and their advantages and disadvantages would go far beyond the scope of this book. We only want to state that there is a variety of possible implementations for elliptic-curve cryptography, much more than for RSA. Therefore, there are serious efforts to reduce this large number of implementation to a few standard implementations. Some standardization committees even try to reduce the complexity by focusing on some (prescribed) curves (ASC approach).

It is still not clear whether these standardization initiatives will be successful or not. However, without agreed standards, ECC is not likely to become a real alternative for RSA.

8.11 Elliptic Curves in Use

Today elliptic-curve cryptography is already broadly in use. Besides S/MIME or TLS, a prominent example is the information network Bonn-Berlin, used for the exchange of strictly confidential documents between different German federal governmental institutions in Berlin and Bonn. The main requirement was security, but not interoperability.

In Austria ECC has been massively launched for the bank card with digital-signature function.

Both examples show typical applications for elliptic-curve cryptography: For high security solutions and for implementations on smart cards in which the key length is crucial (because of lack of physical memory available).

References

[1] Lenstra, A. K., and E. R. Verheul, "Selecting Cryptographic Key Sizes (1999 + 2001)," *Journal of Cryptology*, Vol. 14, 2001, pp. 255–293, https://www.cs.ru.nl/E.Verheul/papers/Joc2001/joc2001.pdf.

[2] Merkle, J., Elliptic-Curve Cryptography Workshop. 2001.

[3] BSI,"Cryptographic Mechanisms: Recommendations and Key Lengths (Version 2022-01)," Tech. rep., Technical Guideline TR-02102-1, 2022, https://www.bsi.bund.de/SharedDocs/Downloads/EN/BSI/Publications/TechGuidelines/TG02102/BSI-TR-02102-1.pdf.

[4] Nagell, T., *Introduction to Number Theory*, Wiley, 1951.

[5] Bauer, J., "ECC-Tutorial (with Python and SagemathCode)," 2015, https://www.johannes-bauer.com/compsci/ecc/.

[6] Silverman, J.,*The Arithmetic of Elliptic Curves*, 2nd ed., Graduate Texts in Mathematics, Springer, Vol. 106, 2009.

[7] Certicom, "ECC Tutorial," https://www.certicom.com/content/certicom/en/ecc-tutorial.html.

[8] Laubrock, T.,"Tutorial mit Java Applets – Krypto-Verfahren basierend auf elliptischen Kurven," 1999, http://www.warendorf-freckenhorst.de/elliptische-kurven/frame.html (visited on07/26/2023).

[9] Kebekus, S., "Elliptic Curve Plotter," https://cplx.vm.uni-freiburg.de/storage/software/ellipticcurve/wasm/ellipticcurve.

[10] Corbellini, A., "Elliptic Curve Point Addition and Multiplication in R and F_p," https://andrea.corbellini.name/ecc/interactive/reals-add.html.

[11] Driscoll, M.,"The Animated Elliptic Curve," https://curves.xargs.org/ (visited on 07/26/2023).

[12] Zimmermann, P., "The ECMNET Project," 2023, https://members.loria.fr/PZimmermann/records/ecmnet.html.

[13] Shene, C.-K., "ECvisual," 2012, https://pages.mtu.edu/~shene/NSF-4/.

[14] Working Group P1363, The IEEE P1363 Home Page: Standard Specifications for Public-Key Cryptography, https://web.archive.org/web/20150405005140/; http://grouper.ieee.org/groups/1363/.

[15] The Sage Development Team,"Constructions: Elliptic Curves," https://doc.sagemath.org/html/en/constructions/elliptic_curves.html.

[16] Stein, W., "An Elliptic Curve Cryptography (ECC) Tutorial," 2006, http://www.williamstein.org/simuw06/notes/notes/node12.html.

[17] The Sage Development Team,"Thematic Tutorials: Elliptic Curves," https://doc.sagemath.org/html/en/thematic_tutorials/explicit_methods_in_number%20_theory/elliptic_curves.html.

[18] Stein, W., "Elliptic Curves in Sage—Presentation at Microsoft Research," 2011. url: https://www.youtube.com/watch?v=9j_Lj071vek.

[19] De Feo, L.,"Using Elliptic Curves and Isogenies in Sage," in *Jupyter* notebook, 2017, https://defeo.lu/jupyter/notebooks/example.html.

[20] Mastermath, "Elliptic Curves," worksheet with exercises, 2011, https://cocalc.com/share/public_paths/9ba3432cea2b7afaa75e400a2bf7815f8c6e64ee/3701-3801%2F3755-Elliptic%20curves.sagews.

[21] Giry, D., "BlueKrypt: Cryptographic Key Length Recommendation," Version 32.3, May 2020, https://www.keylength.com/.

CHAPTER 9

Foundations of Modern Symmetric Encryption

While number theoretic methods prevail for the construction and analysis of asymmetric encryption algorithms, modern symmetric encryption algorithms almost always rely on Boolean algebra, that is on the manipulation of bits. This involves a quite different kind of mathematics and might be unfamiliar to beginners. Therefore, in this chapter we attempt a smooth introduction into this mathematical subject. As previous knowledge, we assume elementary mathematical notions such as variable and function, and a small knowledge of elementary algebra and number theory.

Nevertheless, this chapter quickly becomes very mathematical—and is more mathematical than most of the other chapters in this book. We try to compensate for this by elaborating on the ideas and making them comprehensible by means of examples, so that readers can skip the mathematics and still understand the connections anyway. Extensive use is made of SageMath for this purpose. On the website (see Section 9.4) there is quite a bit more material for deepening.

Let us start with the description on how to interpret and process bits, and how to apply functions to them. Such functions are called Boolean functions, named after George Boole who formalized logic by introducing the elementary logical operations, and thereby made logic a part of mathematics (logical calculus). Most modern symmetric ciphers, as well as hash functions, can be expressed as systems of Boolean functions.

The focus of this chapter is on introducing the mathematical foundations of ciphers that operate on bits. We won't define single ciphers in detail but instead recommend the books by Menezes/Orschot/Vanstone [1], Oppliger [2, 3], Paar and Pelzl [4], Schmeh [5, 6], and Stamp [7].

A word on nomenclature: In the existing literature these ciphers usually are called *block ciphers* or *stream ciphers* without the prefix "bit." Sometimes this usage might cause a misunderstanding since, in particular for stream ciphers, ciphers could operate on other character sets (alphabets, letters) as their basic units. For clarity, in case of doubt it's better to make the "bits" explicit parts of the notations.

The explanations in this chapter are supported by almost 20 code examples in SageMath. These are available in separate SageMath scripts of the form `chap09_sample<nnn>.sage` that can be executed directly. Within them, functions and classes written by ourselves are used, which come from the small library `bitciphers.sage` which is included in the SageMath examples if necessary. See Section 9.4.

This being said, we could express the subject of this chapter—(bit)block ciphers as well as (bit)stream ciphers—in other words as symmetric encryption of information given by bits.

The mathematical foundations and methods belong to the domains of Boolean algebra and finite fields.

9.1 Boolean Functions

In Boolean functions, the independent variables x_i can only take the values 0 and 1 and the result is 1 bit. n different independent variables can be combined in at most 2^{2^n} different ways. If the result is a bit vector instead of one bit, one speaks of a vector Boolean function (like an S-box in cryptography).

9.1.1 Bits and Their Composition

On the lowest level, computers operate on bits, or small groups of bits, for example bytes, or words consisting of 32 or 64 bits depending on the processor architecture. This text assumes some familiarity with the bits 0 and 1 and with elementary logical operations such as AND, OR, NOT, and "exclusive or" (XOR). Nevertheless, we give a short description to make the terminology clear.

Bits have several distinct interpretations: logically as truth values "True" (T) and "False" (F), algebraically as objects 0 (corresponding to F) and 1 (corresponding to T). Mathematically they are the elements of the two-element set $\{0, 1\}$ that in this chapter is denoted by $\mathbb{F}_2$. Here is why:

Consider the residue class ring of $\mathbb{Z}$ modulo 2. This ring has two elements and is a field since 2 is a prime number. Addition in this field exactly corresponds to the logical composition XOR, and multiplication corresponds to the logical composition AND, as is seen in Table 9.1. Table 9.2 lists the transformation formulas between the elementary logical and algebraic operations.

Because this algebraic structure as a field plays a predominant role in cryptography, we use the common notation $\mathbb{F}_q$ for finite fields from algebra (often also noted as GF(q) for "Galois field" where q is the number of elements). In this context it also makes sense to use the algebraic symbols $+$ (for XOR) and $\cdot$ (for AND), and, as is common in mathematics, we often omit the multiplication dot. Cryptographers instead tend to use the symbols $\oplus$ and $\otimes$, that in mathematics are loaded

Table 9.1 The Most Important Compositions of Bits*

Logical					*Algebraic*			
Bits		*Composition*			*Bits*		*Composition*	
x	y	*OR*	*AND*	*XOR*	x	y	$+$	$\cdot$
F	F	F	F	F	0	0	0	0
F	T	T	F	T	0	1	1	0
T	F	T	F	T	1	0	1	0
T	T	T	T	F	1	1	0	1

* The logical XOR is identical with the algebraic +, the logical AND with the algebraic · (multiplication).

Table 9.2 Transformation of Algebraic Operations to Logical Ones and Vice Versa

	Algebraic to Logic	
$x + y$	$=$	$(x \vee y) \wedge (\neg x \vee \neg y)$
$x \cdot y$	$=$	$x \wedge y$
	Logic to Algebraic	
$x \vee y$	$=$	$x + y + x \cdot y$
$x \wedge y$	$=$	$x \cdot y$
$\neg x$	$=$	$1 + x$

with quite different meanings like direct sum and tensor product of vector spaces. Therefore, in this chapter we avoid them except in diagrams.

For clarification, we explicitly hint at some special aspects of algebraic calculations in the binary case (or in characteristic 2):

- Two equal summands in a sum cancel out, that is, together give 0. As a general rule: $x + x = 0$, or $2x = 0$.
- More generally, an even number of equal summands always gives 0 and an odd number of equal summands gives exactly this summand. As a general rule:

$$m\,x := \underbrace{x + \cdots + x}_{m} \quad = \quad \begin{cases} 0 & \text{for even } m \\ x & \text{for odd } m. \end{cases}$$

- For algebraic manipulations, a subtraction means exactly the same operation as an addition—plus and minus signs are arbitrarily interchangeable. As a general rule: $x + y = x - y$.
- All three binomial formulas, for $(x + y)^2$, $(x - y)^2$, $(x + y)(x - y)$, collapse to a single one:

$$(x + y)^2 = x^2 + y^2.$$

Since mixed term occurs twice, that results in a 0.

9.1.2 Description of Boolean Functions

Let us first define Boolean functions quite naively: A Boolean function is a rule (or an algorithm) that takes a certain number of bits and produces a new bit from them. Before rephrasing this naive definition more precisely in mathematical language (see Definition 9.1) we make it a little more vivid.

As a first simple example, consider AND as a Boolean function: It takes two bits and produces one new bit by the well-known rules shown in Table 9.1.

For a slightly more complex example take the function f_0 that produces the value

$$f_0(x_1, x_2, x_3) = x_1 \text{ AND } (x_2 \text{ OR } x_3) \tag{9.1}$$

from three bits x_1, x_2, x_3. The mechanism inside this black box can be described from several different points of view:

- *Mathematically* by a formula;
- *Informatically* by an algorithm;
- *Technically* by a circuit (or plugging diagram);
- *Pragmatically* by a truth table (that is the complete lookup table of its values).

Our sample function f_0 is mathematically defined by (9.1). The corresponding algorithm is also adequately specified by this formula since it has no branching points or conditional statements. As a circuit we visualize f_0 in Figure 9.1. The truth table is in Table 9.3.

The term truth table is motivated by the interpretation of the bits in logical calculus: 0 (= F) means false, 1 (= T) means true. The value $f(x_1, \ldots, x_n)$ of a Boolean function f indicates whether the complete expression is true or false whenever the single input bits $x_1, \ldots, x_n$ have the respective truth values.

The connection with electrical engineering—that is the connection between logical calculus and electric circuits—was essentially developed by Claude Elwood Shannon.

9.1.3 The Number of Boolean Functions

The truth table of f_0, Table 9.3, suggests an easy way of enumerating all Boolean functions: Three variables combine to $8 = 2^3$ different input triples, since each input bit may assume the values 0 or 1 independently of the other ones. Furthermore, a Boolean function f may assume the values 0 or 1 at each triple independently of the seven other triples. This makes 8 independent choices of 0 or 1, a total of 2^8. Therefore, the number of Boolean functions of three variables is $256 = 2^8$.

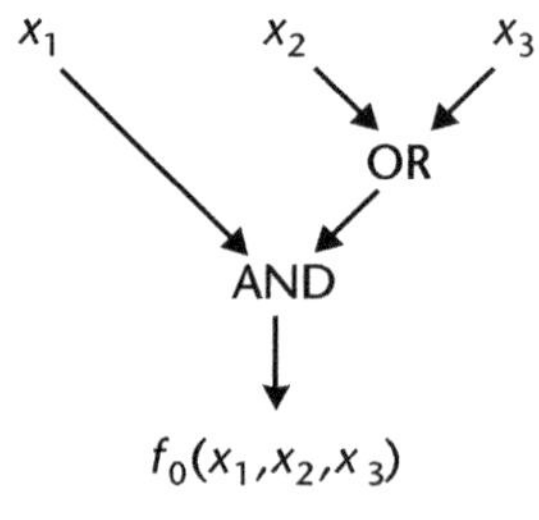

Figure 9.1 Example of a circuit.

Table 9.3 Example of a Truth Table

x_1	x_2	x_3	$f_0(x_1, x_2, x_3)$
0	0	0	0
0	0	1	0
0	1	0	0
0	1	1	0
1	0	0	0
1	0	1	1
1	1	0	1
1	1	1	1

In the general case we have $N = 2^n$ different allocations of the n input variables, and for each of these N input tuples the function may assume the values 0 or 1. This makes a total of 2^N different choices. Thus, the general formula is:

Theorem 9.1 The number of different Boolean functions of n variables is 2^{2^n}.

For four variables we have $2^{16} = 65536$ different functions. By the formula, the number grows superexponentially with n, and even the exponent grows exponentially.

All the 16 Boolean functions of two variables are listed in Table 9.4.

9.1.4 Bitblocks and Boolean Functions

Collections of bits are denoted by different names depending on the context: vectors, lists, (n-)tuples, ... For certain sizes we often use special denotations such as bytes or octets (for 8 bits), or words (for 32 or 64 bytes depending on the processor architecture). In this chapter we usually use the denomination bitblocks which is common in cryptography. Thus, a bitblock of length n is a list $(x_1, \ldots, x_n)$ of bits where the order matters. There are eight different bitblocks of length 3:

$$(0,0,0), (0,0,1), (0,1,0), (0,1,1), (1,0,0), (1,0,1), (1,1,0), (1,1,1).$$

If the danger of misunderstanding is negligible, we write them as bitstrings without parentheses or commas:

$$000, 001, 010, 011, 100, 101, 110, 111. \tag{9.2}$$

We often use the abbreviation x for $(x_1, \ldots, x_n)$. This short form highlights the fact that we consider bitblocks as objects of their own.

The 2^n different bitblocks of length n are the elements of the Cartesian product $\mathbb{F}_2^n = \mathbb{F}_2 \times \cdots \times \mathbb{F}_2$. This Cartesian product has a natural structure as a vector space over the field $\mathbb{F}_2$; bitblocks x and $y \in \mathbb{F}_2^n$ may be added or multiplied by scalars $a \in \mathbb{F}_2$:

$$(x_1, \ldots, x_n) + (y_1, \ldots, y_n) = (x_1 + y_1, \ldots, x_n + y_n),$$
$$a \cdot (x_1, \ldots, x_n) = (a \cdot x_1, \ldots, a \cdot x_n).$$

Now we can write down the mathematically exact definition:

Definition 9.1 A Boolean function of n variables is a map

$$f \colon \mathbb{F}_2^n \longrightarrow \mathbb{F}_2.$$

It takes a bitblock of length n as argument, and produces a single bit.

In this text we sometimes denote the set of all Boolean functions on $\mathbb{F}_2^n$ by $\mathcal{F}_n$. By Theorem 9.1 its cardinality is $\#\mathcal{F}_n = 2^{2^n}$.

Convention: If we describe a Boolean function by its truth table, we usually order the truth table lexicographically with respect to $x \in \mathbb{F}_2^n$, as in the previous example.

This order corresponds to the natural order of the integers $a = 0, \ldots, 2^n - 1$, if these are expanded in base 2

$$a = x_1 \cdot 2^{n-1} + \cdots + x_{n-1} \cdot 2 + x_n$$

and assigned to the corresponding bitblocks $(x_1, \ldots, x_n) \in \mathbb{F}_2^n$.

The *lexicographic order* orders strings (like a dictionary) by the value of their first characters—here 0 or 1 with $0 < 1$. If the first characters are equal, the order looks at the second character, and so on. The sequence 011, 100, 101 is in lexicographic order. A counterexample is the sequence 100, 101, 011. Here the third string begins with 0, which is smaller than the first character of the string preceding it. The sequence of the eight bitblocks of length 3 in (9.2) is in lexicographic order.

9.1.5 Logical Expressions and Conjunctive Normal Form

For describing Boolean functions in mathematical terms, that is by formulas, there are two approaches (beyond truth tables):

- In the logical approach, Boolean functions are expressed by disjunctions (the operation OR, also written as $\vee$), conjunctions (the operation AND, also written as $\wedge$), and negations (the operation NOT, also written as $\neg$). Compositions of these operations are called *logical expressions.*
- In the algebraic approach, Boolean functions are expressed by additions $+$ and multiplications $\cdot$ in the field $\mathbb{F}_2$. Compositions of these operations are called *(binary) polynomial expressions.*

In Theorem 9.2 and Theorem 9.3 we see that both approaches describe all Boolean functions, and that we even can require a certain structure as so-called normal forms. Of course there are algorithms to switch between the three representations (truth tables, logical expressions, and binary polynomial expressions); for all Boolean functions. But we cannot hope that these algorithms are efficient for large numbers n of variables—even writing down the truth table involves 2^n bits.

It seems that the algebraic approach allows a smoother handling of Boolean functions for cryptologic purposes due to its (yet to explore) more rigid structure. In contrast, the logical approach more easily leads to a realization in hardware by circuits since the elementary Boolean operations have direct analogs as circuit elements (gates).

Since in this subsection the logical approach plays a minor role we state the result on normal forms without further reasoning. The possibility of a logical representation (without normalization) will follow as a corollary in Theorem 9.5.

Theorem 9.2 Each Boolean function of n variables $x_1, \ldots, x_n$ has a representation of the form (conjunction)

$$f(x) = s_1(x) \wedge \ldots \wedge s_r(x)$$

with some index r where the $s_j(x)$ for $j = 1, \ldots, r$ each have the form (disjunctions)

$$s_j(x) = t_{j1}(x) \vee \ldots \vee t_{jn_j}(x)$$

with a certain number n_j of terms $t_{jk}(x)$ ($j = 1, \dots, r$ and $k = 1, \dots, n_j$), each of which in turn has the form x_i (an input bit) or $\neg x_i$ (a negated input bit) for some index i.

In particular, $n_j \leq n$ for $j = 1, \dots, r$. Each individual input bit x_i occurs in each of the $t_{jk}(x)$ either directly, negated, or not at all.

In other words, we can build each Boolean function by first forming a handful of expressions (the $s_j(x)$) as OR of some of the input bits or their negations, and then join these expressions by AND ("conjunction of disjunctions"). This normal form cleanly separates AND- and OR-compositions into two layers—there is no further intermixture. The example function f_0 from Section 9.1.2 was defined by the formula

$$f_0(x_1, x_2, x_3) = \underbrace{x_1}_{s_1(x)} \wedge \underbrace{(x_2 \vee x_3)}_{s_2(x)}$$

that already has the conjunctive form from Theorem 9.2 with

$$n_1 = 1, \ \ s_1(x) = t_{11}(x) = x_1, \quad n_2 = 2, \ \ t_{21}(x) = x_2, \ \ t_{22}(x) = x_3.$$

This is no longer true if we expand it:

$$f_0(x) = (x_1 \wedge x_2) \vee (x_1 \wedge x_3)$$

This example doesn't display negated input bits. However, in Table 9.4 we see some of them.

The form of a Boolean function according to Theorem 9.2 is called *conjunctive normal form (CNF)*. It is not unique.[1] Without further explanation we remark that there is a further simplification as a canonical CNF that guarantees a certain uniqueness. There is also an analogous disjunctive normal form (DNF) (a "disjunction of conjunctions").

9.1.6 Polynomial Expressions and Algebraic Normal Form

We consider (binary) polynomial expressions in the variables $x_1, \dots, x_n$, such as $x_1^2 x_2 + x_2 x_3 + x_3^2$. Since we work over the field $\mathbb{F}_2$ only the constants 0 and 1 occur as coefficients and these don't show up explicitly.

As $0^2 = 0$ and $1^2 = 1$, or more general $a^2 = a$ for all elements $a \in \mathbb{F}_2$, and even $a^e = a$ for all exponents $e \geq 1$, leads to another simplification of the expressions. As a consequence for binary polynomial expressions we need to consider the variables $x_1, \dots, x_n$ with exponents 0 and 1 only. Therefore, our sample expression may be written as $x_1 x_2 + x_2 x_3 + x_3$. Another example: $x_1^3 x_2 + x_1 x_2^2 = x_1 x_2 + x_1 x_2 = 0$.

In general, a *monomial expression* (or simply *monomial*) has the form

$$x^I := \prod_{i \in I} x_i \quad \text{with a subset } I \subseteq \{1, \dots, n\};$$

1. The SageMath class `sage.logic.boolformula.BooleanFormula` provides transformations of a logical expression to the CNF by the function `convert_cnf()`, and to the corresponding truth table by the function `truthtable()`.

in other words it is a product of some of the variables where the subset I specifies the choice of "some." Here is an illustrative example with $n = 3$:

$$I = \{2, 3\} \Longrightarrow x^I = x_2 x_3.$$

The total number of such monomial expressions is exactly 2^n, corresponding to the number of choices of products of n potential factors. Here the empty set corresponds to an empty product of 0 factors whose usual interpretation is 1. Note that empty sums are usually are interpreted as 0. Thus,

$$I = \emptyset \Longrightarrow x^I = 1$$

A monomial expression has an immediate interpretation as a Boolean function. At first sight we don't know whether all of these functions are distinct, but we'll see this in a few moments.

A *polynomial expression* is a sum of monomial expressions—remember that we are in the binary case where coefficients take the values 0 or 1 only. Thus, the most general (binary) polynomial expression has the form

$$\sum_{I \subseteq \{1,\ldots,n\}} a_I x^I,$$

where all coefficients a_I are 0 or 1. In other words we add a subset of the 2^n potential monomial expressions, and for this we have 2^{2^n} choices. All these expressions give different Boolean functions, but we have yet to prove this. First, we must prove that each Boolean function has a polynomial expression.

Theorem 9.3 (ANF) For each Boolean function $f\colon \mathbb{F}_2^n \longrightarrow \mathbb{F}_2$ there are coefficients $a_I \in \mathbb{F}_2$ (that is $= 0$ or 1), where I runs through all subsets of $\{1, \ldots, n\}$, such that f may be written as a polynomial expression in n variables of the form:

$$f(x_1, \ldots, x_n) = \sum_{I \subseteq \{1,\ldots,n\}} a_I x^I. \tag{9.3}$$

Proof

(Induction on n) Start with $n = 1$. The four Boolean functions of one variable x are the constants 0 and 1 and the functions given by x and $1 + x$ (= the negation of x). They all have the claimed form.

Now let $n \geq 1$. For $x = (x_1, \ldots, x_n) \in \mathbb{F}_2^n$ we abbreviate $(x_2, \ldots, x_n) \in \mathbb{F}_2^{n-1}$ as x'. Then we can also write $x = (x_1, x')$ instead of $x = (x_1, \ldots, x_n)$.

Now take a function $f \in \mathcal{F}_n$. For each fixed value b of the first variable x_1, the choices being $b = 0$ or $b = 1$, we consider the function $x' \mapsto f(b, x')$ of the $n - 1$ variables that x' consists of. By induction (for $b = 0$ as well as for $b = 1$) we know

$$f(b, x') = p_b(x') \qquad \text{for all } x' \in \mathbb{F}_2^{n-1}$$

where p_0, p_1 are polynomial expressions in x' of the desired form:

$$p_0(x') = \sum_{J \subseteq \{2,\ldots,n\}} b_J x^J, \qquad p_1(x') = \sum_{J \subseteq \{2,\ldots,n\}} c_J x^J.$$

Therefore,

$$f(x_1, x') = \begin{cases} p_0(x'), & \text{if } x_1 = 0, \\ p_1(x'), & \text{if } x_1 = 1, \end{cases} \qquad \text{for all } x = (x_1, x') \in \mathbb{F}_2^n$$

since x_1 assumes the values 0 or 1 only. We combine this conditional formula into

$$f(x_1, x') = (1 + x_1)p_0(x') + x_1 p_1(x') \qquad \text{for all } x \in \mathbb{F}_2^n, \tag{9.4}$$

To check, substitute $x_1 = 0$ or $x_1 = 1$ in (9.4). By expanding the right-hand side and eliminating repeated monomials we get a polynomial expression in x of the claimed form:

$$\begin{aligned} f(x_1, x') &= p_0(x') + x_1(p_0(x') + p_1(x')) \\ &= \underbrace{\sum_{J \subseteq \{2,\ldots,n\}} b_J x^J}_{\text{all monomials without } x_1} + \underbrace{\sum_{J \subseteq \{2,\ldots,n\}} (b_J + c_J) x_1 x^J}_{\text{all monomials with} x_1}. \end{aligned}$$

□

The wording of this theorem is mathematically compact. As an illustration look at the second column of Table 9.4, where the variables are x and y instead of x_1 and x_2, and the coefficients are a, b, c, d instead of $a_\emptyset$, $a_{\{1\}}$, $a_{\{2\}}$, $a_{\{1,2\}}$. Each row of the table describes a Boolean function of two variables. The corresponding polynomial expression is the sum of the terms 1, x, y, xy that have a coefficient 1 in the representation by (9.3), whereas terms with coefficients 0 don't show up explicitly.

Theorem 9.3 provides a representation of a Boolean function as a polynomial expression. This expression is called the *algebraic normal form (ANF)*.[2] The ANF is unique: Since the total number of polynomial expressions is 2^{2^n}, and since they represent all 2^{2^n} different Boolean functions, all these polynomial expressions must differ as functions, and furthermore this representation of a Boolean function as a polynomial expression must be unique. We have shown:

Theorem 9.4 The representation of a Boolean function in algebraic normal form is unique.

Definition 9.2 The *(algebraic) degree* of a Boolean function $f \in \mathcal{F}_n$ is the degree of its polynomial expression in algebraic normal form,

$$\deg f = \max\{\#I \mid a_I \neq 0\}.$$

It is always $\leq n$.

2. The transformation of ANF to truth table and vice versa is provided by the (internal) function `__convert()` of the class `BoolF()`, which comes from the file `bitciphers.sage`. SageMath's own module `sage.crypto.boolean_function` also provides initialization by a truth table or by a Boolean polynomial, and functions `algebraic_normal_form()` and `truth_table()` for the transformations.

Table 9.4 The 16 Operations on Two Bits (= Boolean Functions of 2 Variables), Using Table 9.2*

a	b	c	d	*ANF*	*Logical Operation*		*CNF*
0	0	0	0	0	False	constant	$x \wedge \neg x$
1	0	0	0	1	True	constant	$x \vee \neg x$
0	1	0	0	x	x	projection	x
1	1	0	0	$1+x$	$\neg x$	negation	$\neg x$
0	0	1	0	y	y	projection	y
1	0	1	0	$1+y$	$\neg y$	negation	$\neg y$
0	1	1	0	$x+y$	x XOR y	XOR	$(x \vee y) \wedge (\neg x \vee \neg y)$
1	1	1	0	$1+x+y$	$x \Longleftrightarrow y$	equivalence	$(\neg x \vee y) \wedge (x \vee \neg y)$
0	0	0	1	xy	$x \wedge y$	AND	$x \wedge y$
1	0	0	1	$1+xy$	$\neg(x \wedge y)$	NAND	$(\neg x) \vee (\neg y)$
0	1	0	1	$x+xy$	$x \wedge (\neg y)$		$x \wedge (\neg y)$
1	1	0	1	$1+x+xy$	$x \Longrightarrow y$	implication	$(\neg x) \vee y$
0	0	1	1	$y+xy$	$(\neg x) \wedge y$		$(\neg x) \wedge y$
1	0	1	1	$1+y+xy$	$x \Longleftarrow y$	implication	$x \vee (\neg y)$
0	1	1	1	$x+y+xy$	$x \vee y$	OR	$x \vee y$
1	1	1	1	$1+x+y+xy$	$\neg(x \vee y)$	NOR	$(\neg x) \wedge (\neg y)$

* The order of the first column is lexicographic if a, b, c, d are considered in reverse order.

The degree indicates how many different variables maximally occur in a monomial of the ANF.

Example Independently of the number of variables there are exactly two Boolean functions of degree 0: the two Boolean constants 0 and 1.

Functions of degree ≤ 1 are called affine functions. They are a sum of a constant and a Boolean linear form; see Section 9.1.9. If the degree is > 1 the function is called nonlinear, even though the denomination "nonaffine" would be more accurate.

Example The Boolean function given by $x \mapsto x_1x_2 + x_2x_3 + x_3$ has degree 2.

Remark Boolean functions have a high degree not by high powers of some variables but only by large products of different variables. Each single variable occurs with exponent at most 1 in each monomial of the ANF. Another way to express this fact is to say that all partial degrees (the degrees in the single variables x_i without regard for the other variables) are ≤ 1.

9.1.7 Boolean Functions of Two Variables

All the $2^4 = 16$ Boolean functions of two variables x and y are enumerated in Table 9.4, as polynomial expressions in algebraic normal form $a+bx+cy+dxy$, and as logical expressions. The parameters a_I from Theorem 9.3 translate as follows: $a = a_\emptyset$, $b = a_{\{1\}}$, $c = a_{\{2\}}$, $d = a_{\{1,2\}}$, the input variables as $x = x_1$, $y = x_2$.

We have already seen that each Boolean function admits a polynomial expression. To show that each Boolean function also admits a logical expression we only have to make sure that the algebraic operations $+$ and $\cdot$ have expressions by the logical operations $\vee$, $\wedge$, and $\neg$. To see this, look at the corresponding rows of Table 9.4. Thus, we have shown (as a weak form of the here unproven Theorem 9.2):

Theorem 9.5 Each Boolean function admits a logical expression, that is a representation by a composition of the logical operations $\vee$, $\wedge$, and $\neg$.

Hint In the algebraic interpretation the logical negation $\neg$ corresponds to the addition of 1.

Remark The analogous form[3] of the ANF for a Boolean function of three variables x, y, z is

$$(x, y, z) \mapsto a + bx + cy + dz + exy + fxz + gyz + hxyz.$$

Here we see eight coefficients $a, \ldots, h$. This fits the observations that:

- A Boolean function of three variables has up to $8 = 2^3$ monomials;
- The number of such functions is $2^{2^3} = 2^8 = 256$.

Example What is the ANF of the function f_0 from Section 9.1.2, written as $f_0(x, y, z) = x \wedge (y \vee z)$ and using the variables x, y, z? By Table 9.4 we have $(y \vee z) = y + z + yz$, whereas $\wedge$ simply is the multiplication in the field $\mathbb{F}_2$. Hence,

$$f_0(x, y, z) = x \cdot (y + z + yz) = xy + xz + xyz,$$

and by the way we see that, the degree of f_0 is 3.

Remark From Table 9.4 we might directly read off a naive algorithm for translating logical expressions into (binary) polynomial expressions, and vice versa.

9.1.8 Boolean Maps

Cryptographic algorithms usually produce several bits at once, not only single bits. An abstract model for this is a Boolean map, that is a map[4]

$$f : \mathbb{F}_2^n \longrightarrow \mathbb{F}_2^q$$

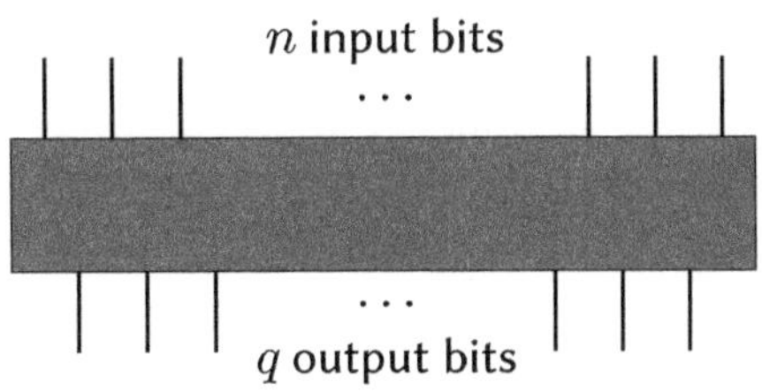

with natural numbers n and q.

The instances of f are bitblocks of length q. Decomposing them into their components,

$$f(x) = (f_1(x), \ldots, f_q(x)) \in \mathbb{F}_2^q,$$

3. In this formula, the letter f (in contrast with the common use in this text) denotes a coefficient, not a function. Mathematicians almost always use letters as symbols relative to the context, and only in exceptional cases with an absolute meaning. Such exceptions are the numbers e, i, and π. But even i often denotes (in contexts without complex numbers) some other object, for example an index in a sum. Or sometimes e is used as exponent, or coefficient.

4. The distinction between the concepts of function and map is somewhat arbitrary. Mathematicians often use them to indicate whether the values belong to a one-dimensional or multi-dimensional domain. Boolean maps (as systems of Boolean functions) are often denoted as vector valued Boolean functions or vectorial Boolean functions (VBF).

we see that we may interpret a Boolean map to $\mathbb{F}_2^q$ as a q-tuple (or system) of Boolean functions

$$f_1, \ldots, f_q : \mathbb{F}_2^n \longrightarrow \mathbb{F}_2.$$

Definition 9.3 The (algebraic) degree of a Boolean map $f : \mathbb{F}_2^n \longrightarrow \mathbb{F}_2^q$ is the maximum of the algebraic degrees of its components,

$$\deg f = \max\{\deg f_i \mid i = 1, \ldots, q\}.$$

Theorem 9.6 Each Boolean map f: $\mathbb{F}_2^n \longrightarrow \mathbb{F}_2^q$ has a unique representation as

$$f(x_1, \ldots, x_n) = \sum_{I \subseteq \{1,\ldots,n\}} x^I a_I$$

with $a_I \in \mathbb{F}_2^q$, and monomials x^I as in Theorem 9.3.

This representation of a Boolean map is also called algebraic normal form. It results from combining the algebraic normal forms of its component functions $f_1, \ldots, f_q$. Compared with Theorem 9.3 the x^I and a_I occur in reversed order. This follows the convention that usually scalars (here the $x^I \in \mathbb{F}_2$) precede vectors (here the $a_I \in \mathbb{F}_2^q$). The a_I are the q-tuples of the respective coefficients of the component functions.

Example

Define a Boolean map g: $\mathbb{F}_2^3 \longrightarrow \mathbb{F}_2^2$ by a pair of logical expressions in three variables x, y, z:

$$g(x, y, z) := \begin{pmatrix} x \wedge (y \vee z) \\ x \wedge z \end{pmatrix}$$

where the components are written below each other, in column form, for clarity. We recognize the function f_0 as the first component. The second component is the product $x \cdot z$. Hence, the ANF of g is

$$g(x, y, z) = \begin{pmatrix} xy + xz + xyz \\ xz \end{pmatrix} = xy \cdot \begin{pmatrix} 1 \\ 0 \end{pmatrix} + xz \cdot \begin{pmatrix} 1 \\ 1 \end{pmatrix} + xyz \cdot \begin{pmatrix} 1 \\ 0 \end{pmatrix}.$$

The algebraic degree is 3, and the value table is in Table 9.5. Here the values $g(x, y, z) \in \mathbb{F}_2^2$ of g are written as bitstrings of length 2.

9.1.9 Linear Forms and Linear Maps

A Boolean function $f : \mathbb{F}_2^n \longrightarrow \mathbb{F}_2$ is called a linear form if it has degree 1 and absolute term 0. This means that its algebraic normal form has linear terms only:

$$f(x) = \sum_{i=1}^{n} s_i x_i \qquad \text{for all } x = (x_1, \ldots, x_n) \in \mathbb{F}_2^n$$

Table 9.5 The Value Table of a Sample Boolean Map

x	y	z	$g(x,y,z)$
0	0	0	00
0	0	1	00
0	1	0	00
0	1	1	00
1	0	0	00
1	0	1	11
1	1	0	10
1	1	1	11

with $s_i \in \mathbb{F}_2$ for $i = 1, \dots, n$. Because the s_i are 0 or 1 a linear form is a partial sum

$$f(x) = \alpha_I(x) = \sum_{i \in I} x_i \qquad \text{for all } x = (x_1, \dots, x_n) \in \mathbb{F}_2^n$$

over a subset $I \subseteq \{1, \dots, n\}$ of all indices, namely

$$I = \{i \mid s_i = 1\}.$$

In particular there are exactly 2^n Boolean linear forms in n variables, and they correspond in a natural way to the power set $\mathfrak{P}(\{1, \dots, n\})$.

Other common notations are (for $I = \{i_1, \dots, i_r\}$):

$$f(x) = \alpha_I(x) = x[I] = x[i_1, \dots, i_r] = x_{i_1} + \cdots + x_{i_r}.$$

Theorem 9.7 relates the definition with the notion of linear forms from linear algebra:

Theorem 9.7 A Boolean function $f\colon \mathbb{F}_2^n \longrightarrow \mathbb{F}_2$ is a linear form if and only if the following two conditions hold:

(i) $f(x+y) = f(x) + f(y)$ for all $x, y \in \mathbb{F}_2^n$.

(ii) $f(ax) = af(x)$ for all $a \in \mathbb{F}_2$ and all $x \in \mathbb{F}_2^n$.

Proof

The representation by partial sums shows that each linear form meets the two conditions mentioned.

For the reverse direction let f be a Boolean function with (i) and (ii). Let $e_1 = (1, 0, \dots, 0)$, ..., $e_n = (0, \dots, 1)$ be the canonical unit vectors. Then each $x = (x_1, \dots, x_n) \in \mathbb{F}_2^n$ is a sum

$$x = x_1 e_1 + \cdots + x_n e_n.$$

Hence,

$$f(x) = f(x_1 e_1) + \cdots + f(x_n e_n) = x_1 f(e_1) + \cdots + x_n f(e_n)$$

is the partial sum of the x_i over the index set consisting of the i for which the constant value $f(e_i)$ is 1. Therefore, f is a linear form in the sense of the previous definition. □

A Boolean map $f\colon \mathbb{F}_2^n \longrightarrow \mathbb{F}_2^q$ is called linear if all of its component functions $f_1, \ldots, f_q$ are linear forms. As in the case $q = 1$ we can show Theorems 9.8 and 9.9.

Theorem 9.8 A Boolean map $f\colon \mathbb{F}_2^n \longrightarrow \mathbb{F}_2^q$ is linear if and only if the following two conditions hold:

(i) $f(x + y) = f(x) + f(y)$ for all $x, y \in \mathbb{F}_2^n$.
(ii) $f(ax) = af(x)$ for all $a \in \mathbb{F}_2$ and all $x \in \mathbb{F}_2^n$.

Theorem 9.9 A Boolean map $f\colon \mathbb{F}_2^n \longrightarrow \mathbb{F}_2^q$ is linear if and only if it has the form

$$f(x) = \sum_{i=1}^{n} x_i s_i$$

with $s_i \in \mathbb{F}_2^q$.

(Here again the x_i and s_i are written in reverse order.)

Affine (Boolean) maps are maps of algebraic degree ≤ 1. They result from adding linear maps and constants.

In the case $q = 1$, that is for functions, the only possible constants are 0 and 1. Adding the constant 1 effects a logical negation, that is a flipping of all bits. Therefore we can say the affine Boolean functions are the linear forms and their negations.

9.1.10 Systems of Boolean Linear Equations

Linear algebra over the field $\mathbb{F}_2$ is quite simple; many complications known from other mathematical areas boil down to trivialities. Such is the case for the solution of systems of linear equations, explicitly written as

$$\begin{array}{ccccccc} a_{11}x_1 & + & \cdots & + & a_{1n}x_n & = & b_1 \\ \vdots & & & & \vdots & & \vdots \\ a_{m1}x_1 & + & \cdots & + & a_{mn}x_n & = & b_m \end{array}$$

with given a_{ij} and $b_i \in \mathbb{F}_2$, and unknown x_j for which we search solutions. In matrix terms this system has an elegant expression as

$$Ax = b$$

where A is an $m \times n$ matrix, and x and b are column vectors, that is $n \times 1$ or $m \times 1$ matrices.

9.1.10.1 Systems of Linear Equations in SageMath

To clarify the relation with common linear algebra we consider an example of a system of linear equations over the rational numbers:

$$\begin{array}{ccccccc} x_1 & + & 2x_2 & + & 3x_3 & = & 0 \\ 3x_1 & + & 2x_2 & + & x_3 & = & -4 \\ x_1 & + & x_2 & + & x_3 & = & -1 \end{array}$$

and study how to handle this in SageMath. The complete solution is in SageMath Example 9.1. Here are the single steps:

1. Define the coefficient matrix $A = \begin{pmatrix} 1 & 2 & 3 \\ 3 & 2 & 1 \\ 1 & 1 & 1 \end{pmatrix}$.
2. Define the image vector $b = (0, -4, 1)$.
3. Let SageMath calculate a solution vector x. Since we wrote the left-hand side of the system as matrix product Ax we have to use the method `solve_right()`.
4. Our system of linear equations could admit several solutions. We find them all by solving the corresponding homogeneous system $Az = 0$ replacing the right-hand side b by 0. If z is a solution of the homogeneous system, then $A \cdot (x + z) = Ax + Az = b + 0 = b$, so $x + z$ is a solution of the original (inhomogeneous) system. In this way we get all solutions. This is because if $Ax = b$ and $Ay = b$, then $A \cdot (y - x) = 0$, hence the difference $y - x$ solves the homogeneous system. For the solution of the homogeneous system we use the SageMath method `right_kernel()`. In SageMath, the default kernel of a matrix A is the left kernel, (i.e., the space of vectors z, such that $zA = 0$). Therefore, the right kernel is needed here.
5. The output appears somewhat cryptic. It says that all solutions of the homogeneous system are multiples of the vector $z = (1, -2, 1)$. Since all coefficients were integers SageMath worked over $\mathbb{Z}$ (= `Integer Ring`).
6. We verify the solution $y = x - 4z$ by checking that $Ay = b$.

SageMath Example 9.1: Solution of a System of Linear Equations Over $\mathbb{Q}$

```
print("\n# CHAP09 -- Sage-Script-SAMPLE 010: =========")

A = Matrix([[1,2,3],[3,2,1],[1,1,1]])
b = vector([0,-4,-1])
x = A.solve_right(b)
print("x =  ", x)

LK = A.kernel()          # left kernel: just for information
print("LK = ", LK)

RK = A.right_kernel()    # right kernel: this is needed here
print("RK = ", RK)

## LEBM = A.kernel().matrix()
## print("lvvv", vector(LEBM))
## y = x - 4*vector(LEBM)
REBM = A.right_kernel().matrix()     # calculate REBM dynamically, ▸
  ▸instead
print("vec_REBM =", vector(REBM))    # of setting the known solution like
y = x - 4*vector(REBM)               # y = x - 4*vector([1,-2,1])

print("y =  ", y)
print("A*y =", A*y)
print("A*y==b :", A*y == b )
```

SageMath Example 9.1 (continued)

```
#----------------------------------
# CHAP09 -- Sage-Script-SAMPLE 010: =========
# x =    (-2, 1, 0)
# LK =  Free module of degree 3 and rank 1 over Integer Ring
#    Echelon basis matrix:
#    [ 1  1 -4]
# RK =  Free module of degree 3 and rank 1 over Integer Ring
#    Echelon basis matrix:
#    [ 1 -2  1]
# vec_REBM = (1, -2, 1)
# y =    (-6, 9, -4)
# A*y = (0, -4, -1)
# A*y==b : True
```

9.1.10.2 Systems of Linear Equations in the Boolean Case

In the general case (over an arbitrary field) the underlying algorithm for solving a system of linear equations is Gaussian elimination. This algorithm of course also hides in the SageMath method `solve\_right()`.

In the Boolean case (over the field $\mathbb{F}_2$), the solution of a system of linear equations by Gaussian elimination is extremely simple since all coefficients are 0 or 1, and multiplication and division are completely trivial. We don't need to deal with complicated coefficients (such as fractions over $\mathbb{Q}$), or inexact coefficients (such as floating point numbers over $\mathbb{R}$). So simple is the method that even for six unknowns, calculating by paper and pencil almost outperforms the feeding of the corresponding small SageMath program with the correct input values. The following example will illustrate this effect.

The idea of elimination is: reduce a system of m equations with n unknowns to a system with only $n-1$ unknowns, or "eliminate" one unknown.

Case 1 x_n only occurs with coefficients $a_{in} = 0$ for $i = 1, \ldots, m$. In other words, x_n doesn't occur at all. Then the system is already reduced.

Case 2 x_n has coefficient 1 in one of the equations. Then solve this equation for x_n, and substitute the resulting expression for x_n,

$$x_n = a_{i1}x_1 + \cdots a_{i,n-1}x_{n-1} + b_i,$$

in the other $m-1$ equations. Then, the remaining equations contain only the unknowns $x_1, \ldots, x_{n-1}$.

Continue recursively until there remains only one unknown or one equation. Now for the example that illustrates this simple procedure.

Example with $m = 5$ and $n = 6$

$$\begin{array}{cccccccc}
x_1 & & +x_3 & & & +x_6 & = & 1 \\
x_1 & +x_2 & & +x_4 & & +x_6 & = & 0 \\
 & x_2 & +x_3 & & +x_5 & +x_6 & = & 0 \\
x_1 & & & +x_4 & +x_5 & & = & 1 \\
 & x_2 & & +x_4 & +x_5 & & = & 1
\end{array}$$

From the first equation we get $x_6 = x_1 + x_3 + 1$ (using the rule that plus and minus are the same). Elimination results in a reduced system consisting of the equations 2 to 5 (note $x_1 + x_1 = 0$, etc.):

$$\begin{array}{cccccc} & x_2 & +x_3 & +x_4 & & = & 1 \\ x_1 & +x_2 & & & +x_5 & = & 1 \\ x_1 & & & +x_4 & +x_5 & = & 1 \\ & x_2 & & +x_4 & +x_5 & = & 1 \end{array}$$

Solving the second equation of the reduced system for x_5 and substituting $x_5 = x_1 + x_2 + 1$ in the other ones gives

$$\begin{array}{ccccc} & x_2 & +x_3 & +x_4 & = & 1 \\ & x_2 & & +x_4 & = & 0 \\ x_1 & & & +x_4 & = & 0 \end{array}$$

Now the last two equations yield $x_4 = x_2 = x_1$, and then the first one yields $x_3 = 1$. Thus, the complete solution is

$$x_1 = x_2 = x_4 = x_6 = a \quad \text{with } a \in \mathbb{F}_2 \text{ arbitrary}, \quad x_3 = 1, \quad x_5 = 1.$$

Since a may assume the values 0 and 1 our result consists of exactly two solutions: $(0, 0, 1, 0, 1, 0)$ and $(1, 1, 1, 1, 1, 1)$.

The Example in SageMath

SageMath Example 9.2 shows the solution in SageMath code. The SageMath method `solve_right()` gives the solution $(0, 0, 1, 0, 1, 0)$ only. To get all solutions we have to solve the homogeneous system. Its solutions are the multiples of the vector $v = (1, 1, 0, 1, 0, 1)$, that is, the two vectors $(0, 0, 0, 0, 0, 0) = 0 \cdot v$ and $(1, 1, 0, 1, 0, 1) = 1 \cdot v$. Thus, the second solution of the inhomogeneous system is $(0, 0, 1, 0, 1, 0) + (1, 1, 0, 1, 0, 1) = (1, 1, 1, 1, 1, 1)$.

SageMath Example 9.2: Solution of a System of Boolean Linear Equations

```
print("\n# CHAP09 -- Sage-Script-SAMPLE 020: =========")

M = MatrixSpace(GF(2), 5, 6) # GF(2) = field with two elements
A = M([[1,0,1,0,0,1],[1,1,0,1,0,1],[0,1,1,0,1,1],[1,0,0,1,1,0],\
    [0,1,0,1,1,0]])
print("A  :\n", A, sep="")  # sep="" prevents print adding a space ▸
  ▸between arguments
b = vector(GF(2),[1,0,0,1,1])
x1= A.solve_right(b)
print("x1 :", x1)
K = A.right_kernel()
print("K  :", K)
REBM = A.right_kernel().matrix()
x2 = x1 + vector(REBM)
print("x2 :", x2)
```

SageMath Example 9.2 (continued)

```
#-----------------------------------
# CHAP09 -- Sage-Script-SAMPLE 020: =========
# A  :
# [1 0 1 0 0 1]
# [1 1 0 1 0 1]
# [0 1 1 0 1 1]
# [1 0 0 1 1 0]
# [0 1 0 1 1 0]
# x1 : (0, 0, 1, 0, 1, 0)
# K  : Vector space of degree 6 and dimension 1 over Finite Field of ▸
  ▸size 2
#    Basis matrix:
#    [1 1 0 1 0 1]
# x2 : (1, 1, 1, 1, 1, 1)
```

9.1.10.3 Estimate of the Costs

What about the costs of solving a system of Boolean linear equations in general? Consider m equations with n unknowns. Then the matrix A of coefficients has size $m \cdot n$. The expanded matrix (A, b) has size $m \cdot (n+1)$.

We only aim at a coarse estimate and neglect possible optimizations of the procedure. For simplicity, we assume $m = n$. In the case $m > n$ we would ignore additional equations (of course, we must then check if the solutions we found also satisfy the additional equations). In the case that $m < n$ we would append null equations (of the kind $0 \cdot x_1 + \cdots + 0 \cdot x_n = 0$).

The elimination step, that is the reduction of the problem size from n to $n-1$, amounts to exactly one pass through all n rows of the *expanded* matrix:

- At first we search the first entry 1 in column n, consisting of the coefficients of x_n. This costs at most n single bit comparisons.
- Then we add the chosen row (containing the first entry 1 in column n) to all those rows below it that also contain a 1 in column n. This amounts (per row) to a single bit comparison and up to n bit additions—we ignore the nth entry since we know already that it becomes 0.

All in all this makes n bit comparisons and at most $n \cdot (n-1)$ bit additions, a total of at most n^2 bit operations. Let $N(n)$ be the number of bit operations for the complete solution of the system. Then we have the following inequality:

$$N(n) \leq n^2 + N(n-1) \quad \text{for all } n \geq 2.$$

Now $N(1) = 1$: We only have to check the one coefficient of the one unknown whether it is 0 or 1. From this we decide whether the equation has a unique solution (for coefficient 1), or whether it is never true (coefficient 0, right-hand side $b = 1$), or whether it is true for arbitrary values of the unknown (coefficient 0, right-hand side $b = 0$).

Then we conclude $N(2) \leq 2^2 + 1$, $N(3) \leq 3^2 + 2^2 + 1$, and so on. By induction, we immediately get

$$N(n) \leq \sum_{i=1}^{n} i^2.$$

The explicit value of this sum is well-known, and we have shown:

Theorem 9.10 The number $N(n)$ of needed bit comparisons and bit additions for solving a system of n Boolean linear equations with n unknowns is upper bounded by

$$N(n) \leq \frac{1}{6} \cdot n \cdot (n+1) \cdot (2n+1).$$

A somewhat more sloppy wording of this result expresses the cost as $\mathcal{O}(n^3)$. In any case it is polynomial of small degree in terms of the problem size n.

Remark The notation by "$\mathcal{O}$" obscures the difference with the cost over arbitrary fields that is generally bounded by $\mathcal{O}(n^3)$. The "felt" much better performance in the Boolean case is partly founded by the exact estimate in Theorem 9.10 that even in the worst case is about $\frac{1}{3} \cdot n^3$. Moreover, in the Boolean case we count simple bit operations only, and not arithmetic operations or floating point instructions that are significantly more expensive.

9.1.11 The Representation of Boolean Functions and Maps

We will discuss here what exactly is meant by the different representation forms of Boolean functions.

9.1.11.1 Various Interpretations of Bitblocks

We used the term bitblock for a variety of slightly different objects. A bitblock $b = (b_1, \ldots, b_n) \in \mathbb{F}_2^n$ describes:

- A vector $b \in \mathbb{F}_2^n$ written as a row or a column. This is the primary meaning of the term bitblock.
- An argument of a Boolean function or map of n variables, also used as row index of a value table (or truth table).
- A bitstring of length n.
- A subset $I \subseteq \{1, \ldots, n\}$ defined by b as indicator: $i \in I \Leftrightarrow b_i = 1$.
- A linear form α on $\mathbb{F}_2^n$ expressed as sum of the variables x_i with $b_i = 1$. The evaluation of α comes down to the scalar product of vectors: $\alpha(x) = b \cdot x$.
- A monomial in n variables $x_1, \ldots, x_n$ with all partial degrees ≤ 1. In this interpretation b_i specifies the exponent 0 or 1 of the variable x_i.
- An integer between 0 and $2^n - 1$ in binary representation (that is in the base-2 system). The sequence of binary digits (bits) is identical with the corresponding bitstring.[5] Conversely, the integer is the index (beginning with 0) of the bitstring when the bitstrings are lexicographically ordered in a list.

Of course there are further interpretations—after all, each piece of information has a binary coding. The bitblocks for $n = 3$ are listed in Table 9.6.

5. The SageMath method `binary()` transforms an integer to a bitstring, suppressing leading zeros. Example: `10.binary()` yields `'1010'`.

Table 9.6 Interpretations of Bitblocks of Length 3

Integer	*Bitstring*	*Subset*	*Linear Form*	*Monomial*
0	000	∅	0	1
1	001	{3}	x_3	x_3
2	010	{2}	x_2	x_2
3	011	{2, 3}	$x_2 + x_3$	x_2x_3
4	100	{1}	x_1	x_1
5	101	{1, 3}	$x_1 + x_3$	x_1x_3
6	110	{1, 2}	$x_1 + x_2$	x_1x_2
7	111	{1, 2, 3}	$x_1 + x_2 + x_3$	$x_1x_2x_3$

9.1.11.2 Representation of the Truth Table of a Boolean Function

Section 9.1.11.1 described (and Table 9.6 illustrated) how to interpret the bitblocks $x = (x_1, \ldots, x_n)$ of length n as integers $i(x) = 0, 1, \ldots, 2^n - 1$ in base-2 representation. The example in Table 9.7 suggests how to describe the truth table of a Boolean function $f\colon \mathbb{F}_2^n \longrightarrow \mathbb{F}_2$ in a parsimonious way by a bitblock $b = (b_0, \ldots, b_{2^n-1})$ of length 2^n: simply take the last column in the order given by the indices $i(x)$. The general procedure for arbitrary n runs as follows:

$$b_{i(x)} = f(x) \quad \text{where } i(x) = x_1 \cdot 2^{n-1} + \cdots + x_{n-1} \cdot 2 + x_n$$
$$\text{for} \quad x = (x_1, \ldots, x_n) \in \mathbb{F}_2^n.$$

This might look entangled, but it simply means: Interpret x as the base-2 representation of an integer $i(x)$, and set $f(x)$ as the bit at position $i(x)$ from the bitblock b. An additional column $i(x)$ in the truth table of the function f_0 (f_0 was defined in Formula 9.1) illustrates this procedure (see Table 9.7). The last column of this table, written in row form, is the bitblock b.

In this way, the bitblock $(0, 0, 0, 0, 0, 1, 1, 1)$ or, even more parsimoniously, the bitstring

$$00000111$$

of length $2^3 = 8$ completely specifies the truth table of f_0.

9.1.11.3 Representation of the Algebraic Normal Form

The algebraic normal form is a sum of monomials. Each monomial is a product of a subset of $\{x_1, \ldots, x_n\}$, and hence has a representation as an integer between

Table 9.7 An Extended Truth Table [for $f_0(x_1, x_2, x_3) = x_1 \wedge (x_2 \vee x_3)$] with $n = 3$ and $2^n = 8$

x_1	x_2	x_3	$i(x)$	$f_0(x_1, x_2, x_3)$
0	0	0	0	0
0	0	1	1	0
0	1	0	2	0
0	1	1	3	0
1	0	0	4	0
1	0	1	5	1
1	1	0	6	1
1	1	1	7	1

0 and $2^n - 1$. So the ANF has a characterization by 2^n bits: the coefficients of the 2^n different monomials (see Theorem 9.3). Therefore, we may view a bitblock $a = (a_0, \ldots, a_{2^n-1})$ as representation of the ANF of a Boolean function $f\colon \mathbb{F}_2^n \longrightarrow \mathbb{F}_2$ in the following way:

$$f(x) = \sum_{i=0}^{2^n-1} a_i x_1^{e_1(i)} \cdots x_n^{e_n(i)} \quad \text{where } i = e_1(i)\cdot 2^{n-1} + \cdots + e_n(i)$$

$$\text{with} \quad e_1(i), \ldots, e_n(i) = 0 \text{ or } 1.$$

This formula means: Interpret the n-tuple e of exponents of a monomial as the base-2 representation of an integer i. The ith element of the bitblock a indicates whether this monomial occurs in the ANF of f or not.

For the sample function f_0 (see Table 9.7) we already saw (or can easily check, because $f(1,1,1) = 1+1+1 = 1$ since we add mod 2) that the ANF is

$$f_0(x) = x_1x_3 + x_1x_2 + x_1x_2x_3.$$

It involves the monomials with exponent triples $101, 110, 111$ that correspond to the integers $5, 6, 7$. Therefore, we set the bits at the positions $5, 6, 7$ to 1, and the remaining bits to 0, and get the parsimonious representation of the ANF by a bitstring:

$$00000111.$$

Warning This is the same bitstring as for the truth table *by pure chance*—a special property of the function f_0! The function $f(x_1, x_2) = x_1$ has truth table 0011 (it takes the value 1 if and only if $x_1 = 1$, or if the argument has the form $x = (1, \text{any bit})$) and ANF 0010 (since it contains the single monomial x_1).

The SageMath class `BoolF()` has a method for calculating the ANF[6]. SageMath Example 9.3 demonstrates its application to f_0.

SageMath Example 9.3: A Boolean Function with Truth Table and ANF

```
print("\n# CHAP09 -- Sage-Script-SAMPLE 030: =========")

load("./bitciphers.sage")
bits = "00000111"
x = str2bbl(bits)
print("x :", x)
f = BoolF(x)
y = f.getTT()
print("y :", y)
z = f.getANF()
print("z :", z)
```

6. This transformation that converts a bitstring of length 2^n (the truth table) into another bitstring of length 2^n (the coefficient list of the ANF) is sometimes called Reed-Muller transformation or binary Moebius transformation.

SageMath Example 9.3 (continued)

```
#-----------------------------------
# CHAP09 -- Sage-Script-SAMPLE 030: =========
# x : [0, 0, 0, 0, 0, 1, 1, 1]
# y : [0, 0, 0, 0, 0, 1, 1, 1]
# z : [0, 0, 0, 0, 0, 1, 1, 1]
```

Remark Evaluating a Boolean function f at all arguments $x \in \mathbb{F}_2^n$ the naive way costs 2^n evaluations $f(x)$, each with at most 2^n summands, each of which needing at most $n - 1$ multiplications. Thus, the costs have an order of magnitude of about $n \cdot 2^n \cdot 2^n$. If we relate the costs to the input size $N = 2^n$ they are essentially quadratic: $N^2 \cdot \log_2(N)$. A common method, binary recursion, or "divide-and-conquer," divides a problem into two subproblems of half the input size, and leads to a significantly more efficient algorithm. Starting from (9.4), finally we achieve a reduction to almost linear costs $3N \cdot \log_2 N$. This algorithm, also denoted as fast binary Moebius transformation, an analogue of the fast Fourier transformation (FFT), is implemented in the class `BoolF()`.

9.2 Block Ciphers

In classical cryptography the weakness of simple monoalphabetic substitutions is remedied in two different ways: first by polygraphic substitutions that encrypt groups of letters at once, second by polyalphabetic substitutions that change the substitution alphabet depending on the position in the plaintext.

If we consider bits instead of letters, the two principles of hardening monoalphabetic substitutions can be transferred to two classes of useful encryption methods for binary encoded information:

- Block ciphers split bitstrings into blocks of a fixed length and encrypt one complete block per step.
- Stream ciphers encrypt bit by bit, each one by another substitution (so each single bit is either unchanged or flipped by a position-dependent rule).

No mathematically complete proof exists for the security of any block or stream cipher. Thus, the situation is even worse than for asymmetric ciphers where the proof of security often reduces to a well-studied, if not solved, mathematical problem. The best we can do is to consider a symmetric cipher as secure if none of the known attacks is significantly faster than a complete exhaustion of the key space (also known as "brute-force attack").

9.2.1 General Description

Block ciphers transform bitblocks of a fixed length n to bitblocks of the same length controlled by a key that itself is a bitblock of a certain length l. The extension of this cipher to input bitstrings of arbitrary lengths is the subject of Section 9.2.4. For the moment we neglect this aspect.

An adequate model of a block cipher is a Boolean map

$$F : \mathbb{F}_2^n \times \mathbb{F}_2^l \longrightarrow \mathbb{F}_2^n$$

often interpreted as a family $(F_k)_{k \in K}$ of Boolean maps

$$F_k : \mathbb{F}_2^n \longrightarrow \mathbb{F}_2^n \quad \text{for all } k \in K = \mathbb{F}_2^l$$

where $F_k(a) = F(a, k)$.

9.2.1.1 Choosing the Key Length

For the key length l we have an obvious criterion: l must be large enough to prevent an exhaustion of the key space, a "brute force attack." The key space is the set $\mathbb{F}_2^l$, so it contains 2^l different keys. We assume that the probabilities for all keys are the same, that is $1/2^l$. In other words, we assume that keys are chosen uniformly at random.

With these assumptions we have a lower bound of about 80 bits for a secure key length according to the state of the art [8]. Popular ciphers use keys of lengths 128 or more, so have a sufficient security margin. The outdated standard cipher DES used 56 bit keys. The technology of today breaks it relative quickly.

9.2.1.2 Choosing the Block Length

The block length n should be large enough to prevent analyses of patterns or frequencies. Even more it should prevent leaking any information about the plaintext into the ciphertext, for example the presence of repetitions.

If the attacker observes about $2^{n/2}$ ciphertexts corresponding to random plaintexts encrypted with the same key, the probability of a collision (by the birthday paradox) is about $\frac{1}{2}$. Therefore, this number $2^{n/2}$ should exceed the number of available memory cells. And the key should change frequently—long before this number of blocks is reached.

From this point of view the frequently used block length 64 is risky. Only frequent key change could justify it, and only if the plaintext contains few repetitions. A better cipher, as the current standard AES, uses blocks of 128 bits.

These considerations about key or block lengths are typical for the discussion of security in modern cryptography: We use large security margins and avoid any weaknesses, no matter how small they might be, even if there is no known practical attack that uses them. But since we have a broad choice of good and fast ciphers that provide large security margins there is no need to rely on a weaker cipher, even if this precaution seems paranoid.

9.2.2 Algebraic Cryptanalysis

Cryptanalysis deals with the analysis of ciphers in order to discover hidden secrets. Algebraic cryptanalysis is a method of cryptanalysis that mainly uses algebraic techniques such as equation-solving algorithms. Thus, it involves expressing the operations in the cipher as a system of equations and replacing some of the variables with known data to find the key [9].

An algebraic attack is similar to a brute-force attack in that it can theoretically break any symmetric cipher, but in practice it is completely useless against any reasonable cipher.

9.2.2.1 Attacks with Known Plaintext

An attack with known plaintext assumes that the attacker knows or guesses a small piece of plaintext, and then tries to deduce the key or some more plaintext that is unknown to her. For the present section we assume that the known plaintext is a complete bitblock.

Let a block cipher be given by a Boolean map

$$F: \mathbb{F}_2^n \times \mathbb{F}_2^l \longrightarrow \mathbb{F}_2^n.$$

By Theorem 9.6, F is an n-tuple $F = (F_1, \ldots F_n)$ of polynomial expressions in $n+l$ variables all of whose partial degrees are ≤ 1.

A known plaintext block $a \in \mathbb{F}_2^n$ with corresponding ciphertext block $c \in \mathbb{F}_2^n$ yields a system

$$F(a, x) = c$$

of n polynomial equations for the unknown key $x \in \mathbb{F}_2^l$.

Systems of equations of this type (over arbitrary fields) are subjects of algebraic geometry. The general theory is quite deep, in particular if we search for concrete solution procedures. However—couldn't the fact that our polynomials have all their partial degrees ≤ 1 simplify the problem?

Example 1 Let $n = l = 2$,

$$F(a_1, a_2, x_1, x_2) = (a_1 + a_2x_1, a_2 + a_1x_2 + x_1x_2),$$

$a = (0, 1)$, $c = (1, 1) \in \mathbb{F}_2^2$. The equations for the key $(x_1, x_2) \in \mathbb{F}_2^2$ are

$$\begin{pmatrix} 1 \\ 1 \end{pmatrix} = \begin{pmatrix} 0 + x_1 \\ 1 + 0 + x_1x_2 \end{pmatrix}.$$

The immediate solution is $x_1 = 1$, $x_2 = 0$.

Example 2 Linear maps: If F is a *linear* map, then the system of equations is accessible by the efficient solution algorithms of linear algebra; see Section 9.1.10. We have n linear equations for l unknowns. If $l < n$ the attacker needs some additional blocks of known plaintext, or she executes an exhaustion of the remaining $n - l$ key bits. For this method to work F needs to be linear only in x.

Example 3 Substitution: Often polynomial equations look complex at first sight but aren't so. Here is an example (over $\mathbb{F}_2$)

$$x_1x_2x_3 + x_1x_2 + x_1x_3 + x_2x_3 + x_2 + x_3 = 0.$$

By the substitutions $x_i = z_i + 1$ it is transformed to

$$z_1z_2z_3 + z_1 = 0$$

(for an easy proof look in the reverse direction). This has the solutions

$$z_1 = 0,\ z_2, z_3 \text{ arbitrary } or\ z_1 = z_2 = z_3 = 1.$$

Therefore the complete solution of the original equation is

$$x_1 = 1,\ x_2, x_3 \text{ arbitrary } or\ x_1 = x_2 = x_3 = 0.$$

There are two powerful general approaches for solving systems of (polynomial) equations over $\mathbb{F}_2$:

- *SAT solvers* [10]: SAT denotes the satisfiability problem of propositional logic. Consider a logical expression in Boolean variables $x_1, \ldots, x_n$ and ask if there exist values of the variables that make the expression True. In other words, consider a Boolean function f and ask if it assumes the value 1. A SAT solver is an algorithm that takes a logical expression in CNF and decides the satisfiability by finding a solution x, or showing there's no solution. The naive algorithm uses the truth table and exhausts the 2^n possible arguments. However, there are much faster algorithms, the most popular being the DPLL algorithm (after Davis, Putnam, Logemann, and Loveland) and BDD based algorithms (Binary Decision Diagram). The SageMath modules `sage.sat.solvers` and `sage.sat.boolean_polynomials` contain some of these algorithms.
- *Elimination using Groebner bases* See the dissertation [11], the textbooks [9, 12, 13], the script [14], or the paper [15].

Both methods work well for a few unknowns. With a growing number of unknowns their complexity becomes unmanageable (in fact, SAT was the first problem in history shown to be NP-complete). Of course, we always find a solution by searching through the complete value table. But this naive method is inefficient (exponential in the number of unknowns, and so hopeless for 80 or more unknowns). But also the costs of SAT solvers and Groebner-basis methods grow exponentially with the number of unknowns. Not even the fact that all partial degrees are ≤ 1 is of vital help.

9.2.2.2 The Complexity of the Algebraic Attack

The theoretical analysis of the cost for finding a solution leads to one of the central notions of complexity theory, NP-completeness.

Theorem 9.11 (Garey/Johnson): The problem of finding a solution for a system of polynomial equations over $\mathbb{F}_2$ is NP complete.

We won't explain the notion NP-complete but only mention that the (up to now unproven) **P ≠ NP** conjecture implies that an NP-complete problem admits no efficient algorithmic solution, or that there is no solution algorithm whose execution time grows at most polynomially with the number of input variables.

A common interpretation of this theorem is: For an appropriately chosen block cipher $F: \mathbb{F}_2^n \times \mathbb{F}_2^l \longrightarrow \mathbb{F}_2^n$ the attack with known plaintext (against the key $k \in \mathbb{F}_2^l$) is not efficient. However, from a strict mathematical point of view the theorem doesn't prove anything of practical relevance:

1. It relates to an algorithm for arbitrary polynomial equations (over $\mathbb{F}_2$). It doesn't contain any assertion for special classes of polynomials, or for a concrete system of equations.
2. Likewise, it gives a pure proof of (non)existence, and provides no hint as how to construct a concrete example of a difficult system of equations. Note that we know that some concrete systems admit easy solutions.
3. Even if we could find concrete examples of difficult systems the theorem would not make any assertion, whether only some rare instances (the worst cases) are difficult, or almost all (the generic cases)—and this is what the cryptologist wants to know. Maybe there is an algorithm that solves polynomial systems for almost all tuples of unknowns efficiently, and only fails for a few exceptional tuples.

Despite these critical comments the theorem raises hope that there are secure block ciphers, and the designers of block ciphers follow the:

Rule of thumb Systems of linear equations for bits admit very efficient solutions. Systems of nonlinear equations for bits in almost all cases admit no efficient solution.

9.2.3 The Structure of Block Ciphers

In an ideal world we would know how to reliably measure the security of a block cipher

$$F : \mathbb{F}_2^n \times \mathbb{F}_2^l \longrightarrow \mathbb{F}_2^n$$

for realistic values of the block length n and the key length l, say of an order of magnitude of 128 bit or more.

In fact, we know explicit measures of security, for example the linear potential, or the differential potential, that quantify the deviation from linearity, or the algebraic immunity, or others. Unfortunately all of these only give necessary, not sufficient, conditions for security; and moreover the efficient computability of these measures is limited to small block lengths n, about 8 or slightly larger.

Lacking a general efficient approach to security, the design of block ciphers usually relies on a structure that, although not obligatory, in practice seems to provide plausible security (according to verifiable criteria). Most of the generally approved standard ciphers, such as DES and AES, follow this approach.

This common design scheme starts by constructing Boolean maps of small dimensions and then extending them to the desired block length in several steps:

1. Define one or more Boolean maps of small dimension q (block length of the definition domain), say $q = 4, 6$, or 8, that are good for several security criteria. These maps are called *S-boxes*, and are the elementary building blocks of the cipher. ("S" stands for substitution.)
2. Mix the round input with some of the key bits and then apply m S-boxes in parallel (or apply the one S-box m times in parallel) to get a map with the desired input width $n = mq$.
3. Then permute the complete resulting bitblock over its total width.
4. These steps together are a *round* of the complete scheme. Assess the weaknesses of the round map, that mainly result from using S-boxes of small dimension. Then reduce these weaknesses in a reasonably controlled way by iterating the scheme over several rounds of the same structure but with a changing choice of key bits.
5. Don't stop as soon as the security measures give satisfying values but add some surplus rounds to get a wide security margin.

Figure 9.2 outlines the scheme for a single round.

The complete scheme is a special case of a somewhat more general proposal that goes back to Shannon who required two basic features of block ciphers:

Diffusion The bits of the plaintext block "smear" over all parts of the block. This is achieved by permutations.

Confusion (complex dependencies): The interrelation between plaintext block and key on the one hand, as well as ciphertext block on the other hand should be as complex as possible (in particular as nonlinear as possible). Basic building blocks for this are substitutions.

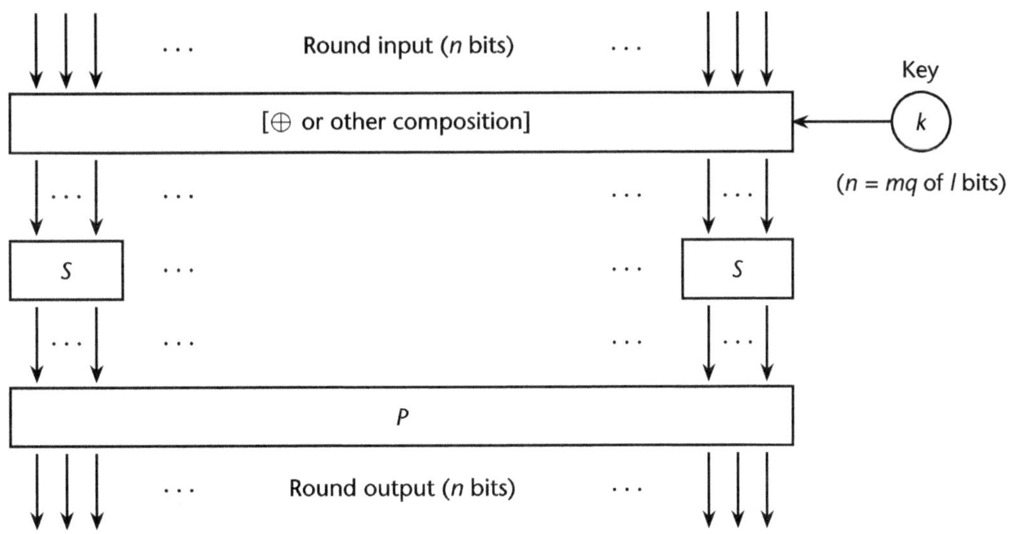

Figure 9.2 A single round of a block cipher (with S-boxes S, permutation P, and key k).

The overall effect of both requirements, taken together, should result in an unforeseeable change of ciphertext bits for a slight change of the key.

The attacker should have no means to recognize whether a guessed key is nearly correct.

For the construction of strong block ciphers Shannon proposed an alternating sequence of substitutions and transpositions (permutations), so-called SP networks:

$$\mathbb{F}_2^n \xrightarrow{S_1(\bullet,k)} \mathbb{F}_2^n \xrightarrow{P_1(\bullet,k)} \mathbb{F}_2^n \longrightarrow \dots$$
$$\dots \longrightarrow \mathbb{F}_2^n \xrightarrow{S_r(\bullet,k)} \mathbb{F}_2^n \xrightarrow{P_r(\bullet,k)} \mathbb{F}_2^n$$

depending on a key $k \in \mathbb{F}_2^l$. In this scheme

$$\begin{aligned} S_i &= i\text{th substitution} \\ P_i &= i\text{th permutation} \\ P_i \circ S_i &= i\text{th round} \end{aligned}$$

Altogether the encryption function consists of r rounds.

Note that the permutations are special linear maps $P: \mathbb{F}_2^n \longrightarrow \mathbb{F}_2^n$. Some recent block ciphers, the most prominent being AES, replace permutations by more general linear maps that provide an even better diffusion. However, the proper term LP network is not yet in use.

9.2.4 Modes of Operation

In this section the key plays no role: Therefore, we omit k in the notation. Consider a block cipher function $f: \mathbb{F}_2^n \longrightarrow \mathbb{F}_2^n$. If we want to apply it to longer or shorter bit sequences we must:

1. Split a bit sequence a into n-bit blocks $a_1, \dots, a_r$,
2. Fill ("pad") the last block a_r, if necessary, up to length n with
   - Zeros;
   - Random values;
   - Context information.

Then the most obvious encryption algorithm is encipher the blocks one by one. This is called ECB mode (for "Electronic Code Book"). In Figure 9.3 this is shown schematically.

ECB mode simply realizes a monoalphabetic substitution where the blocks in $\mathbb{F}_2^n$ are interpreted as letters. For a sufficiently large n this is secure from a ciphertext-only attack. However, the cipher leaks information on repeated blocks. For some plaintexts this is a real danger.

- For example, MS-Word files contain long sequences consisting of the bytes 00000000 and 00000001.

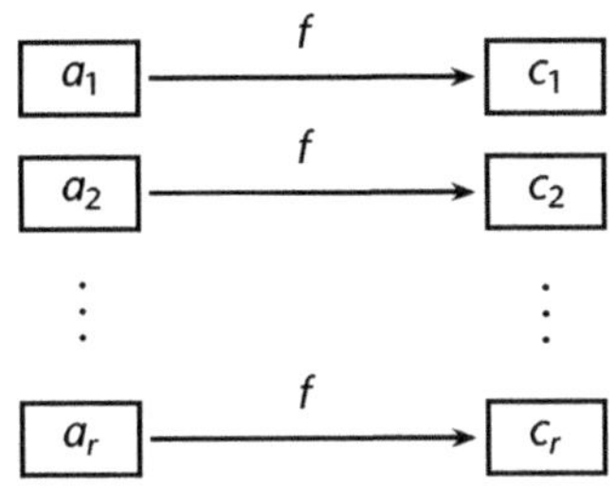

Figure 9.3 ECB mode.

- An even more alarming case is provided by image files with large single-color areas. They contain many identical blocks such that structures of the image may appear in the ciphertext file.[7]

In view of this weakness, generating some additional diffusion between the plaintext blocks seems a good idea. A simple but effective approach is CBC (cipher block chaining). Choose a random start value c_0 (also called IV for initialization vector). Then the procedure looks like it does in Figure 9.4.

The formula for encryption in CBC mode is

$$\begin{aligned} c_i &:= f(a_i + c_{i-1}) \quad \text{for } i = 1, \ldots, r \\ &= f(a_i + f(a_{i-1} + \cdots f(a_1 + c_0) \ldots)). \end{aligned}$$

Each ciphertext block depends on *all previous* plaintext blocks (diffusion), and identical plaintext blocks in general encrypt to different ciphertext blocks.

The formula for decryption is

$$a_i = f^{-1}(c_i) + c_{i-1} \quad \text{for } i = 1, \ldots, r.$$

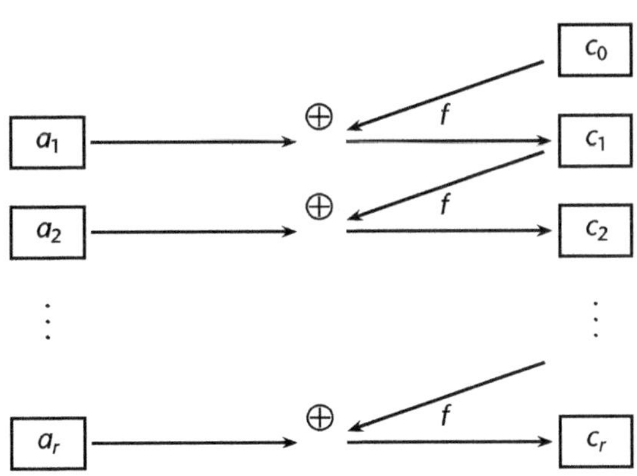

Figure 9.4 CBC mode.

7. - For a convincing example see the Wikipedia entry "Block cipher mode of operation."
- The template in `CT2 Startcenter ▷ Templates ▷ Cryptography ▷ Modern ▷ Symmetric ▷ Block Modes of Symmetric Ciphers` allows you to try this with your own pictures.

Question Does it make sense to keep the initialization vector c_0 secret and use it as an additional key component? (Then, for the example of DES, we had 56 proper key bit plus a 64-bit initialization vector, making a total of 120 key bit.)

Answer No!

Reason In the decryption process only a_1 depends on c_0. This means that keeping c_0 secret conceals known plaintext only for the first block. If the attacker knows the second or any later plaintext block, then she may attack the key as in ECB mode (by an attack with known plaintext).

There are several other modes of operation. Worth mentioning is that the modes OFB (output feedback) and CTR (counter) convert a block cipher into a stream cipher.

9.2.5 Statistical Analyses

For cryptanalyzing block ciphers we know some basic approaches:

1. Exhaustion = brute-force searching the complete key space;
2. Algebraic attack (see Section 9.2.2);
3. Statistical attacks against hidden linearity:
   (a) Linear cryptanalysis (LCA) (Matsui/Yamagishi 1992);
   (b) Differential cryptanalysis (DCA) (Murphy, Shamir, Biham 1990): Differential cryptanalysis was known at IBM and NSA as early as in 1974. Despite linear cryptanalysis is conceptually simpler, it was unknown to the designers of DES. Accordingly, the resistance of DES against linear cryptanalysis is suboptimal.
   (c) Generalizations and mixtures of (a) and (b).

All these statistical attacks hardly break a cipher in the sense of classical cryptanalysis. They usually assume lots of known plaintexts, much more than an attacker could gather in a realistic scenario. Therefore, a more adequate term is analysis instead of attack. The analyses make sense for finding measures for some partial aspects of security of block ciphers. They measure security for example by the number of known-plaintext blocks needed for the attack. If a cipher resists an attacker even with exaggerated assumptions on her capabilities, then we feel safe to trust it in real life.

Given an SP network, the analysis starts with the nonlinear components of the single rounds, in particular with the S-boxes. The next step is extending the potential attack over several rounds. This shows how the cost of the attack grows with the number of rounds. In this way we find criteria for the number of rounds for which the cipher is secure—at least from this special attack.

SageMath scripts applying linear cryptanalysis methods to Lucifer's S-box, to Mini-Lucifer, and to two other toy block ciphers can be found at `https://www.cryptool.org/en/documentation/ctbook/sagemath`.

In CT2, differential cryptanalysis can be reproduced on three toy ciphers in detail (each in an automatic and a tutorial mode).[8] The CT2 template "Differential Cryptanalysis Tutorial 3 (Automatic Mode)" shown in Figure 9.5 performs a fully-automatic attack on a 96-bit toy cipher:

- The DCA Oracle and ToyCipher components generate plaintext-ciphertext pairs needed by the DCA KeyRecovery component to recover the key using differential cryptanalysis.
- The DCA PathFinder component find characterstics and paths through the toy cipher needed by the DCA KeyRecovery component to perform the actual differential cryptanalysis.
- After five rounds and a total of 160,008 messages (see Figure 9.6), the used key could be recovered. This key is shown in the TextOutput component "Output of the key" of Figure 9.5. You can see that the correct key was found when comparing it to the key in the TextInput component "Input of the key," showing the key used to generate more than 160,000 plaintext-ciphertext pairs.

9.2.6 Security Criteria for Block Ciphers

To escape attacks, block ciphers, their round maps, or their S-boxes, should fulfill some requirements.

- *Balance:* All preimages have the same number of elements, or in other words, the values of the map are uniformly distributed. Irregularities of the distribution would provide hooks for statistical cryptanalysis.

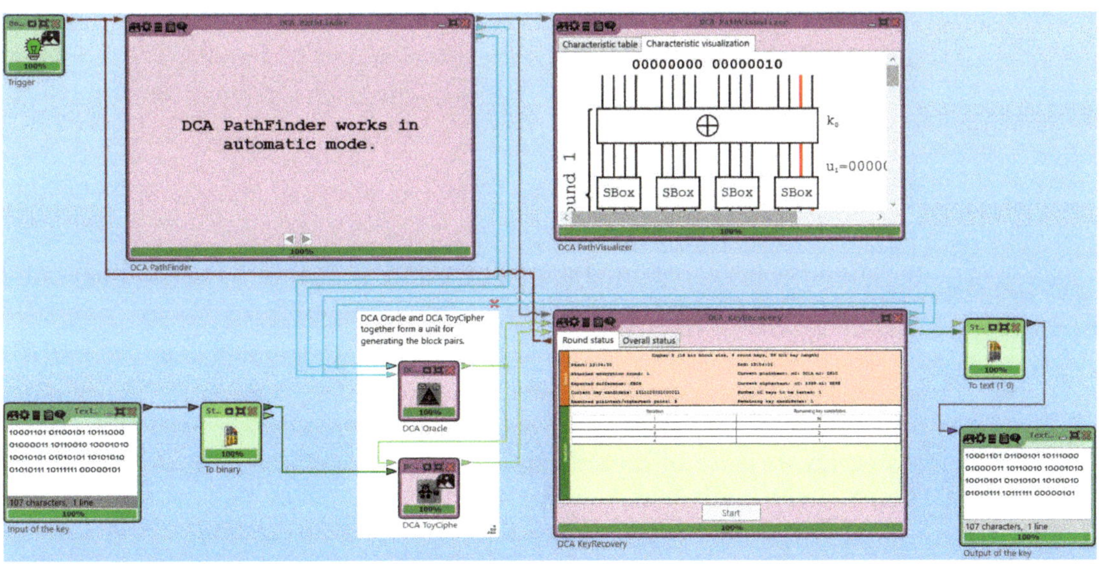

Figure 9.5 CT2 template "Differential Cryptanalysis Tutorial 3" performing DCA of a toy cipher.

8. See under `CT2 Templates` ▷ `Cryptanalysis` ▷ `Modern`.

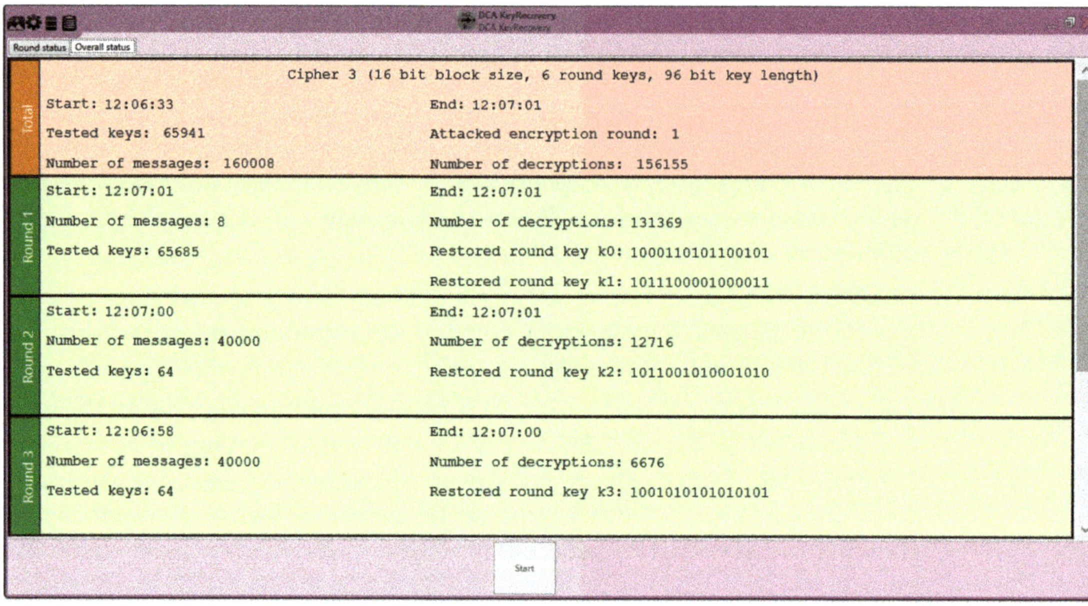

Figure 9.6 Presentation view of the DCA KeyRecovery component showing the final state after a successful differential cryptanalysis of a toy cipher (in CT2).

- *Diffusion/avalanche effect:* If a single plaintext bit changes, about 50% of the ciphertext bits change. This effect conceals similarity of plaintexts.
- *Algebraic complexity:* The determination of preimages or parts thereof should lead to equations whose solutions are as difficult as possible. This requirement is related to the algebraic degree of the map, but only in an indirect way. A suitable measure is algebraic immunity.
- *Nonlinearity:* We know several criteria that measure linearity, also hidden linearity, and are relatively easy to describe and to handle. For example, they quantify how susceptible Boolean maps are for linear or differential cryptanalysis [16].
    - The linear potential should be as low as possible, and the linear profile as balanced as possible.
    - The differential potential should be as low as possible, and the differential profile as balanced as possible.

Some of these criteria are compatible with each other; some criteria contradict other ones. Therefore, the design of a block cipher requires a balance between different criteria. Instead of optimizing a map for a single criterion the designer should aim at a uniformly high level for all criteria.

9.2.7 AES

The current standard for modern symmetric ciphers is AES (see Section 1.2.1).

Figures 9.7 and 9.8 show the design of AES[9] and the realization of the design principles derived in Section 9.2.3.

- The block length is $n = 128$, the key length, $l = 128$, 192, or 256, the number of rounds, $r = 10$, 12, or 14.
- At the beginning of each round and after the last round a partial key is added to the current bitblock. The complete algorithm involves $r + 1$ partial keys.
- The 128-bit "partial keys" $k^{(i)}$ are not partial keys in the proper sense but extracted from the master key k by a somewhat involved algorithm (key schedule). They are not independent.
- Each round starts by splitting the current 128-bit block into 16 parts each consisting of 8 bits. Each of these parts is fed into the same S-box $S\colon \mathbb{F}_2^8 \longrightarrow \mathbb{F}_2^8$. This S-box has a mathematically quite elegant description that however assumes some advanced knowledge of abstract algebra, hence it is omitted here. The linear potential of the S-box is $\frac{1}{64}$.
- The diffusion step consists of a permutation followed by a linear map. This step is slightly more complex than for a pure SP network as in Figure 9.2.

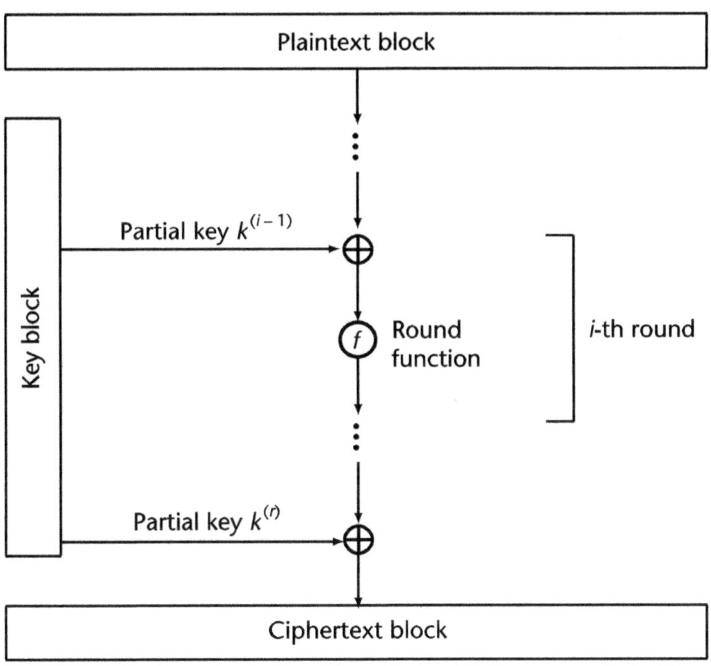

Figure 9.7 Structure of AES in the large.

9. - Using `CTO (cryptool-online.org)` in the browser, AES can be seen in 2 plugins: as animation and via "AES (step-by-step)."
- Using `CT1 Indiv. Procedures` ▷ `Visualization of Algorithms` ▷ `AES` you can find 3 visualizations for this cipher.
- Using the search string `AES` in `CT2 Startcenter` ▷ `Templates` you can find a plugin performing AES step-by-step.

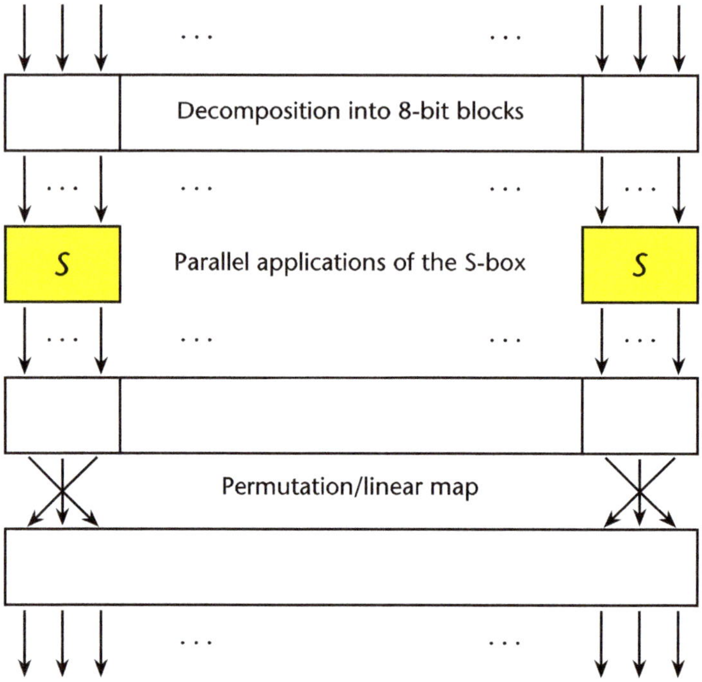

Figure 9.8 The round function f of AES.

9.2.8 Outlook on Block Ciphers

As we saw, linear cryptanalysis provides some evidence for the security of a cipher, in particular for choosing the number of rounds. But only some parts of the theory have a mathematically satisfying basis. Most existing publications only give ad-hoc analyses of concrete ciphers. For example, Matsui showed how for DES 2^{43} known plaintexts reveal 14 key bits with high certainty, reducing the exhaustion to the remaining $42 = 56 - 14$ key bits, a feasible task (at least if the analyst gets that many plaintexts). The treatment of linear cryptanalysis serves as an example of similar analyses.

Differential cryptanalysis as well as generalized and mixed variants follow similar lines of thought. For more information see [17] which also explicitly specifies the most important block ciphers DES and AES.

With the work of Gohr [18] an interesting research approach for cryptanalysis was brought to life in 2019: *differential-neural cryptanalysis* as the machine-learning assisted differential cryptanalysis of block ciphers. The idea is to distinguish ciphertext pairs that belong to a fixed plaintext difference from random ones. For this, neuronal networks were employed, leading to an 11-round key recovery attack on Speck32/64 which is at least competitive with classical differential cryptanalysis.

A lot of follow-up works have been published since. In [19], Gohr et al. examined different differential-neural distinguishers for a wide variety of ciphers and additionally provided general insights.

A very comprehensive report [20] with 200 pages, written by Leander et al. was published by the BSI in 2022. In the report, the applicability of machine-learning techniques was examined to the symmetric ciphers Simon and Speck, and the most promising ones were applied to the ciphers Present, Katan, ChaCha, and Skinny. For this purpose, round-reduced versions of these ciphers as well as versions with smaller block length were used.

9.3 Stream Ciphers

A stream cipher sequentially encrypts each single bit of a bitstring by an individual rule. The two possibilities are leave the bit unchanged or negate it. Leaving the bit unchanged is equivalent with (binary) adding 0; negating the bit is equivalent with adding 1. Thus, every stream cipher may be interpreted as an XOR encryption. We distinguish between

- Synchronous stream ciphers: where the key stream is generated independently of the plaintext;
- Asynchronous stream ciphers: where the key stream depends on the plaintext or other context parameters.

In this chapter we only treat synchronous stream ciphers.

9.3.1 XOR Encryption

The basic method of stream encryption is simply denoted by XOR. It interprets plaintexts[10] as sequences of bits. Also, the key is a bit sequence, called key stream. The encryption algorithm adds the current bit of the plaintext and the current bit of the key stream by XOR. Figure 9.9 illustrates the algorithm,[11] and Figure 9.10 shows an example.

In the 1920s, XOR ciphers were invented to encrypt teleprinter messages. These messages were written on five-hole punched tapes as in Figure 9.11. Another punched tape provided the key stream. Gilbert Vernam filed his U.S. patent in 1918. He used a key tape whose ends were glued together, resulting in a periodic key stream. Joseph Mauborgne immediately recognized that a nonperiodic key is obligatory for security.

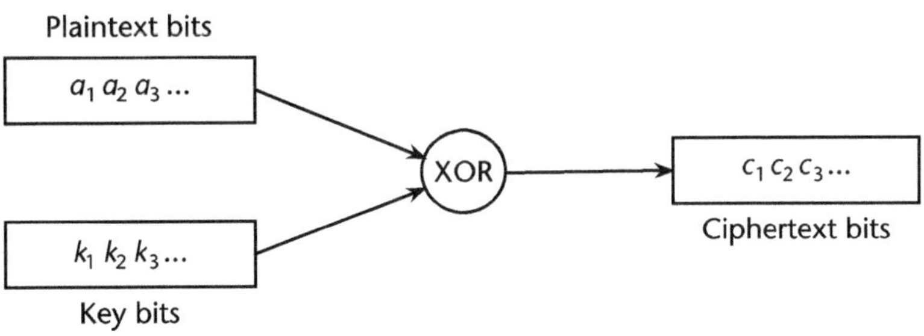

Figure 9.9 The principle of XOR encryption.

10. The default SageMath method `ascii_to_bin()` from the module `sage.crypto.util` converts ordinary texts to bitstrings. The inverse method is `bin_to_ascii()`. However, these bitstrings belong to the class `StringMonoidElement`, and are cumbersome to process further. Therefore, corresponding own functions are defined in the file `bitciphers.sage`.
11. In `CT2 Templates` ▷ `Cryptography` ▷ `Classical` ▷ `XOR Cipher` you can try this directly.

```
a: 01000100011101 ...
k: 10010110100101 ...
   -----------------
c: 11010010111000 ...
```

Figure 9.10 Example of XOR encryption.

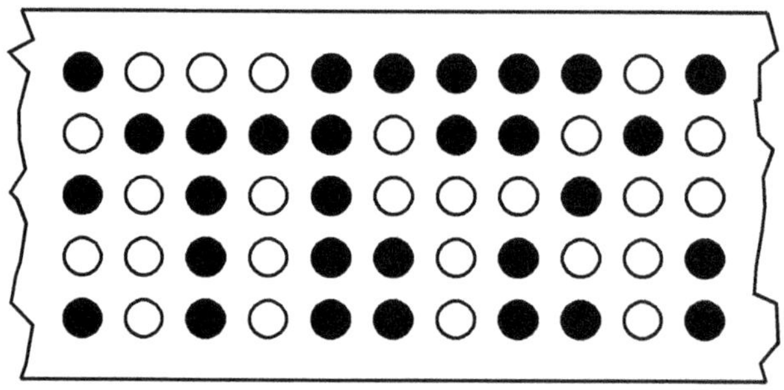

Figure 9.11 Punched tape—each column represents a five-bit character.

In its strongest form, the one-time pad (OTP), XOR encryption is an example for perfect security in the sense of Shannon. As algorithm A5 or E_0, XOR is used to secure mobile phones or the Bluetooth protocol for wireless data transmission. As RC4 it was part of the SSL protocol that encrypts client-server communication between browser and web server, and of the PKZIP compression software. There are many other current applications, not all of them fulfilling the expected security requirements.

The scope of XOR encryption ranges from simple ciphers that are trivially broken to unbreakable ciphers.

Advantages of XOR Ciphers

- Encryption and decryption are done by the same algorithm, since $c_i = a_i + k_i$ also $a_i = c_i + k_i$. Thus, decryption also consists of adding key stream and ciphertext (elementwise binary).
- The method is extremely simple to understand and to implement.
- It is very fast—provided that the key stream is available. For high transfer rates one may precompute the key stream at both ends of the line.
- If the key stream is properly chosen, the security is high.

Disadvantages of XOR Ciphers

- XOR ciphers are vulnerable for known-plaintext attacks, each correctly guessed plaintext bit reveals a key bit.
- If the attacker knows a piece of plaintext she also knows the corresponding piece of the key stream, and then is able to exchange this plaintext at will. For example, she might replace “I love you” with “I hate you,” or replace an amount of $1,000 with $9,999. In other words, the integrity of the message is poorly protected, so the sender has to implement additional procedures.

- XOR ciphers provide no diffusion in the sense of Shannon's criteria since each plaintext bit affects the one corresponding plaintext bit only. In the opposite, block ciphers were designed for using diffusion.
- Each reuse of a part of the key sequence (also in form of a periodic repetition) opens the door for an attack. The historical successes in breaking stream ciphers almost always used this effect, for example the attacks on encrypted teleprinters in World War II, or the project Venona during the Cold War.

A remark on the first disadvantage, the vulnerability for attacks with known plaintext: The common ISO character set for texts has a systematic weakness. The 8-bit codes of the lower-case letters a...z all start with 011, of the upper-case letters A...Z, with 010. A supposed sequence of six lower-case letters (no matter which) reveals $6 \cdot 3 = 18$ key bits. The occurrence of many zeros in the leading bits of the bytes is an important recognition feature for natural texts in European languages.

In other words we cannot prevent the attacker from getting or guessing a good portion of the plaintext. Thus, the security against an attack with known plaintext is a fundamental requirement for an XOR cipher, even more than for any other cryptographic procedure.

9.3.2 Generating the Key Stream

The main naive methods for generating the key stream are:

- Periodic bit sequence,
- Running-text,
- True random sequence.

A better method uses a

- Pseudorandom sequence

and leads to really useful procedures. The essential criterion is the quality of the pseudorandom generator.

9.3.2.1 Periodic Bit Sequences

Example: We generate a key stream of period 8 by repeating $k = 10010110$. Each letter is represented by bytes in the ISO character set. The plaintext here is in German.

```
       D    |   u    |        |   b    |   i    |   s    |
a: 01000100|01110101|00100000|01100010|01101001|01110011|
k: 10010110|10010110|10010110|10010110|10010110|10010110|
   -------- -------- -------- -------- -------- --------
c: 11010010|11100011|10110110|11110100|11111111|11100101|

       t    |        |   d    |   o    |   o    |   f
   01110100|00100000|01100100|01101111|01101111|01100110
   10010110|10010110|10010110|10010110|10010110|10010110
   -------- -------- -------- -------- -------- --------
   11100010|10110110|11110010|11111001|11111001|11110000
```

This encryption is easily done by hand, or by SageMath Example 9.4.

Reencoding the ciphertext bytes in the ISO-8859-1 character set the cryptogram looks like this:

Ò ã ¶ ô œ å â ¶ ò ù ù ð

This might bedazzle laypersons. An expert immediately notes that all characters are from the upper half of the possible 256 bytes. This observation suggests that the plaintext is in natural language, encrypted with a key whose leading bit is 1. If the attacker guesses that the conspicuous character ¶ = 10110110 corresponds to the space character 00100000, she derives the key as the difference 10010110. This breaks the cryptogram.

Known or probable plaintext easily breaks periodic XOR encryption.

SageMath Example 9.4: XOR Encryption in Python/SageMath

```
print("\n# CHAP09 -- Sage-Script-SAMPLE 130: =========")

load("./bitciphers.sage")  # for txt2bbl() + bbl2str()

plaintext = "Du bist doof"
bintext = txt2bbl(plaintext)  # array
binstr = bbl2str(bintext)
print(binstr)
testkey = [1,0,0,1,0,1,1,0]
keystr = bbl2str(testkey)
print("key:", keystr)
ciphertext = xor(bintext,testkey)
ciphstr = bbl2str(ciphertext)
print(ciphstr)

#-----------------------------------
# CHAP09 -- Sage-Script-SAMPLE 130: =========
# 010001000111010100100000011000100110100101110011
# 011101000010000001100100011011110110111101100110
# key: 10010110
# 110100101110001110110110111101001111111111100101
# 111000101011011011110010111110011111100111110000
```

9.3.2.2 MS Word and Periodic XOR

The following table (generated ad hoc by simple character counts) shows the frequencies of the most frequent bytes in MS Word files.

Byte (hexadecimal)	**Bits**	**Frequency**
00	00000000	7–70%
01	00000001	0.8–17%
20 (space)	00100000	0.8–12%
65 (e)	01100101	1–10%
FF	11111111	1–10%

Note that these frequencies relate to the binary files, heavily depend on the type of the document, and may change with every software version. The variation is large, we often find unexpected peaks, and all bytes 00–FF occur. But all this doesn't matter here since we observe *long chains of 00 bytes*.

For a Microsoft Word file that is XOR encrypted with a periodically repeated key, the ubiquity of zeros suggests an efficient attack. First, determine the length of the key. If the length of the period is unknown, you can determine it by the methods for periodic polyalphabetic substitutions from classical cryptanalysis named after Kasiski, Friedman, or Sinkov. Or simply try all possible lengths. Then split the stream of ciphertext bits into blocks corresponding to the length of the period and add the blocks pairwise. If one of the plaintext blocks essentially consists of zeros, then the sum is readable plaintext. Why? Consider the situation

	...	Block 1			...	Block 2			...
Plaintext:	...	a_1	...	a_s	...	0	...	0	...
Key:	...	k_1	...	k_s	...	k_1	...	k_s	...
Ciphertext:	...	c_1	...	c_s	...	c'_1	...	c'_s	...

where $c_i = a_i + k_i$ and $c'_i = 0 + k_i = k_i$ for $i = 1, \ldots, s$. Thus, the key reveals itself in block 2, however the attacker doesn't recognize this yet. But tentatively, pairwise adding all blocks she gets (among other things)

$$c_i + c'_i = a_i + k_i + k_i = a_i \quad \text{for } i = 1, \ldots, s,$$

that is, a plaintext block. If she realizes this (for example recognizing typical structures), then she sees the key $k_1, \ldots, k_s$.

Should it happen that the sum of two ciphertext blocks is zero then the ciphertext blocks are equal, and so are the corresponding plaintext blocks. The probability that both of them are zero is high. Thus, the key could immediately show through. To summarize: XOR encryption with a periodic key stream is quite easily broken for messages with a known structure.

This is true also for a large period, say 512 bytes = 4096 bits, in spite of the hyperastronomically huge key space of 2^{4096} different possible keys.

9.3.2.3 Running-Text Encryption

A classical approach to generating an aperiodic key is taking a data stream, file, or text, that has at least the length of the plaintext. In classical cryptography this method was called running-text encryption or book ciphers, and the keys were taken from books beginning at a certain position. The main method of breaking the cipher was finding or guessing the book. The same weakness affects the electronic analog that uses a file, a CD, or a DVD. As soon as the attacker knows the source of the key bits, the key space is much too small—exhausting a file of several gigabytes is easily done, the costs are linear in the size of the file.

But even when the attacker is unable to guess the source of the key bits ciphertext-only, cryptanalysis is possible: Plaintexts as well as keys contain structures that are not completely concealed by binary addition. We won't discuss this here[12] but summarize: XOR encryption with running-text keys is fairly easily broken.

9.3.2.4 True Random Sequence

The extreme choice for a key is a true random sequence of bits as key stream. Then the cipher is called (binary) one-time pad. In particular no part of the key stream must be repeated at any time. The notation pad comes from the idea of a tear-off calendar—each sheet is destroyed after use. This cipher is unbreakable, or perfectly secure. Shannon gave a formal proof of this in [17].

Without mathematical formalism the argument is as follows: The ciphertext divulges no information about the plaintext (except the length). It could result from *any* plaintext of the same length: simply take the (binary) difference of ciphertext and alleged plaintext as key. Consider the ciphertext $c = a + k$ with plaintext a and key k, all represented by bitstreams and added bit by bit as in Figure 9.9. For an arbitrary different plaintext b the formula $c = b + k'$ likewise shows a valid encryption using $k' = b + c$ as key.

This property of the OTP could be used in a scenario of forced decryption (rubber hose cryptanalysis) to produce an innocuous plaintext, as exemplified in Figure 9.12 or visualized in Figure 1.6.

If the one-time pad is perfect—why not use it in any case?

- The key management is unwieldy: Key agreement becomes a severe problem since the key is as long as the plaintext and awkward to memorize. Thus, the communication partners have to agree on the key stream (prior to transmitting the message) and store it. Agreeing on a key only just in time needs a secure communication channel—but if there was one why not use it to transmit the plaintext in clear?
- The key management is inappropriate for mass application or multiparty communication because of its complexity that grows with each additional participant.
- The problem of message integrity requires an extended solution for OTP like for any XOR cipher.

There is another, practical, problem when encrypting on a computer: how to get random sequences. True random bits arise from physical events like radioactive decay or thermal noise on an optical sensor. The apparently deterministic machine computer can also generate true random bits, for instance by special chips that produce usable noise. Moreover, many events are unpredictable, such as the exact mouse movements of the user, or arriving network packets that, although not completely random, contain random ingredients that may be extracted. On Unix systems these are provided by `/dev/random`.

However these random bits, no matter how true, are not that useful for encryption by OTP. The problem is on the side of the receiver that cannot reproduce the key. Thus, the key stream must be transmitted independently.

12. In `JCT Default Perspective` ▷ `Analysis` ▷ `Viterbi Analysis` an automatic recognition of the two plaintexts is offered, where only the running-key ciphertext is needed.

```
Plaintext bits and text:
01010100 01101000 01101001 01110011 00100000 01101101   This m
01100101 01110011 01110011 01100001 01100111 01100101   essage
00100000 01101001 01110011 00100000 01101000 01100001    is ha
01111010 01100001 01110010 01100100 01101111 01110101   zardou
01110011 00101110                                       s.

Key bits:
11001000 11010110 00110011 11000000 00111011 10001110
00001000 11101111 01001001 11100101 10111100 10111001
00010010 11000110 01110011 11010111 11000100 01100000
11100110 00010111 01101010 10111011 00010101 11011000
11110000 01000010

Ciphertext bits:
10011100 10111110 01011010 10110011 00011011 11100011
01101101 10011100 00111010 10000100 11011011 11011100
00110010 10101111 00000000 11110111 10101100 00000001
10011100 01110110 00011000 11011111 01111010 10101101
10000011 01101100

Pseudokey bits:
11001000 11010110 00110011 11000000 00111011 10001110
00001000 11101111 01001001 11100101 10111100 10111001
00010010 11000110 01110011 11010111 11000101 01101111
11110010 00011001 01111011 10101010 00010101 11011000
11110000 01000010

Pseudodecrypted bits and text:
01010100 01101000 01101001 01110011 00100000 01101101   This m
01100101 01110011 01110011 01100001 01100111 01100101   essage
00100000 01101001 01110011 00100000 01101001 01101110    is in
01101110 01101111 01100011 01110101 01101111 01110101   nocuou
01110011 00101110                                       s.
```

Figure 9.12 XOR encryption of a hazardous message, and an alleged alternative plaintext.

There are other, useful, cryptographic applications of true random bits: Generating keys for arbitrary encryption algorithms that are unpredictable for the attacker. Many cryptographic protocols rely on nonces that have no meaning except for being random; for example, the initialization vectors of the block cipher modes of operation, or the challenge for strong authentication (challenge-response protocol).

For XOR encryption—as approximation to the OTP—algorithmically generated bit sequences are much more practicable. But the attacker should have no means to distinguish them from true random sequences. This is the essence of the concept pseudorandomness, and generating pseudorandom sequences is of fundamental cryptologic relevance.

> XOR encryption with a pseudorandom key stream spoils the perfect security of the one-time pad. But if the pseudorandom sequence is cryptographically strong (see Section 9.3.8) the attacker has no chance to exploit this fact.

9.3.3 Pseudorandom Generators

Pseudorandom generators mimic true random processes by deterministic algorithms. Usually such an algorithm is called random generator, omitting the prefix "pseudo" if there is no danger of confusion. The main difference between a pseudorandom sequence and a true random sequence is its reproducibility.

Cipher designers hope to approximate the ideal properties of the one-time pad by using a pseudorandom bit sequence as key stream, treating the short start string as "effective key." Even for random generators of modest quality the resulting ciphertexts are immune against statistical analyses. The all-dominant problem is the security against known-plaintext attacks.

Thus, the critical question for a pseudorandom sequence and for a random generator is:

> Given a known chunk (maybe fragmented) of the sequence, is there a way to determine some more bits of the sequence, be it forwards or backwards?

For classical random generators that are popular in statistical applications and simulations the answer is yes (see Section 9.3.4). But we'll learn about random generators that (presumably) are cryptographically secure. The cipher designer faces the problem of finding a good trade-off between efficiency and security.

The two main serious methods of generating pseudorandom bit sequences, or key streams, are:

- (Feedback) shift registers (FSR) and combinations thereof, with theoretical foundations in Boolean algebra;
- Perfect random generators with theoretical foundations in number theory.

Figure 9.13 shows the schematic functionality of a random generator. It hides an inner state that changes with each step by a given algorithm. This algorithm is controlled by parameters, some of which are public, but some of which are secret

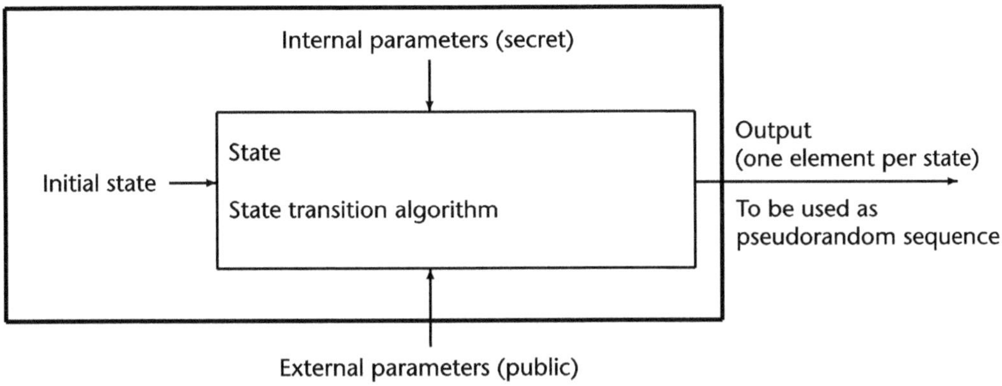

Figure 9.13 (Pseudo)random generator.

and serve as components of the key. The initial state (start value) is a true random value and likewise secret. With each step the random generator outputs a value, depending on its current inner state, until an exterior intervention stops it.

Thus, the random generator transforms a short, truly random, bit sequence, the initial state, into a long pseudorandom sequence. Cryptologists call this effect key expansion.

9.3.3.1 Feedback Shift Registers

Feedback shift registers are a classical and popular method of generating pseudorandom sequences. The method goes back to Solomon Golomb in 1955, but is often named after Tausworthe who picked up the idea in a 1965 paper. FSRs are especially convenient for hardware implementation.

An FSR of length l is specified by a Boolean function $f\colon \mathbb{F}_2^l \longrightarrow \mathbb{F}_2$, the feedback function. Figure 9.14 shows the mode of operation. The output consists of the rightmost bit u_0, all the other bits are shifted to the right by one position, and the leftmost cell is filled by the bit $u_l = f(u_{l-1}, \ldots, u_0)$. Thus, the recursive formula

$$u_n = f(u_{n-1}, \ldots, u_{n-l}) \quad \text{for } n \geq l \tag{9.5}$$

represents the complete sequence. For use with SageMath we define a general FSR with feedback function f by implementing a method `fsr()` for the class `BoolF`, see SageMath Example 9.5 (also contained in the file `bitciphers.sage`).

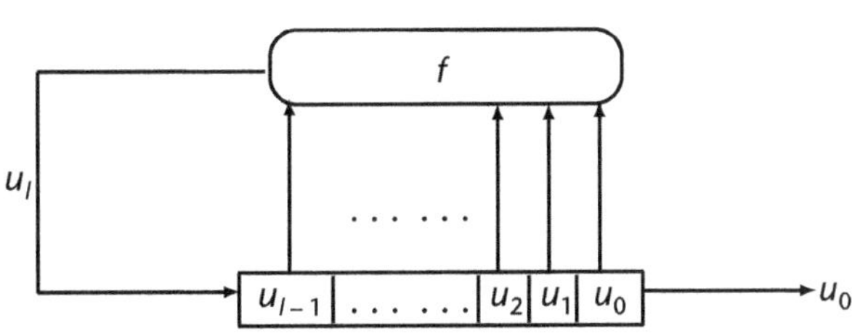

Figure 9.14 An FSR during the first iteration step. The Boolean function f calculates a new bit from the current state of the register. This new bit is slid in from the left.

In SageMath Example 9.6, the FSR is used to generate a bit sequence by a concrete sample feedback function $f : \mathbb{F}_2^4 \longrightarrow \mathbb{F}_2$.

The truth table of f is in short form given by the bitstring `bits = 1000010111001001`. This is converted to the bitblock `x`, and then used for instancing f. Table 9.8 shows the truth table in long form.

Table 9.9 illustrates the stepping of the register in this concrete example. The column "State" shows the actual state of the register. The initial state is `start = [0,1,1,1]`. In each step the rightmost bit of the state is output—indicated by an arrow in the column "Output." The actual value of f for the (now old) state is shifted into the register from the left—again indicated by an arrow in the column "Feedback."

The numbers (1) to (4) in the left column of the Table 9.8 indicate the order of the steps. Step 5 is identical with step 3, as is step 6 with step 4, and so on.

Note that the result doesn't look convincingly random take this as warning that the choice of the parameters needs significantly more care.

The bits u_0, ..., u_{l-1} form the start value. The key expansion transforms the short sequence $u = (u_0, \ldots, u_{l-1})$ (the effective key) of length l into a key stream $u_0, u_1, \ldots$ of arbitrary length. Additionally, in this context treating the internal parameters, that is the feedback function f or some of its coefficients, as components of the key makes sense. This makes the effective key length larger than l.

Table 9.8 Truth Table of the Boolean Sample Function f

	$a \in \mathbb{F}_2^4$	$f(a) \in \mathbb{F}_2$
	0000	1
	0001	0
	0010	0
	0011	0
	0100	0
(3)	0101	1
	0110	0
(1)	0111	1
	1000	1
	1001	1
(4)	1010	0
(2)	1011	0
	1100	1
	1101	0
	1110	0
	1111	1

Table 9.9 The Stepping of a Sample FSR

Feedback		*State*		*Output*
start		0111	$\longrightarrow$	1
$f(0111) = 1$	$\longrightarrow$	1011	$\longrightarrow$	1
$f(1011) = 0$	$\longrightarrow$	0101	$\longrightarrow$	1
$f(0101) = 1$	$\longrightarrow$	1010	$\longrightarrow$	0
$f(1010) = 0$	$\longrightarrow$	0101	from here on	periodic

In this respect the realization in hardware differs from a software implementation: Hardware allows using an adjustable feedback function only by complex additional circuits. Thus, in this case we usually assume an unchangeable feedback function, and (at least in the long run) we cannot prevent the attacker from figuring it out. In contrast, a software implementation allows a comfortable change of the feedback function at any time such that it may serve as part of the key.

SageMath Example 9.5: A Feedback Shift Register in Python/SageMath

```
def fsr(self,u,n):
  """Generate a feedback shift register sequence
  using the actual Boolean function (repres. by self).
  Parameters: start vector u, number n of output bits.
  Caution: The vector u is modified by this function."""
  outlist = []
  for i in range (0,n):
    b = self.valueAt(u)
    c = u.pop()
    u.insert(0,b)
    outlist.append(c)
  return outlist
```

SageMath Example 9.6: A (Very Poor) Pseudorandom Sequence in Python/SageMath

```
print("\n# CHAP09 -- Sage-Script-SAMPLE 140: =========")

load("./bitciphers.sage")  # for txt2bbl() and classes BoolF + LFSR

print("\n---------- Using f.fsr(): ----------")
bits = "1000010111001001" # truth table as bitstring
x = str2bbl(bits)         # truth table as list (= bitblock)
print("Input x (same as truth table of Bool function f):"); print(x)
f = BoolF(x)
# print("Truth table of Bool function f:")
# print(f.getTT())  # return truth table as bitlist
start = [0,1,1,1]   # initial state of the register
print("Input start:"); print(start)
bitlist = f.fsr(start, 32)  # output sequence
print("f.fsr() output:")
# print(bitlist)  #  array
print(bbl2str(bitlist))
print("New state of the register:", start)

print("\n---------- Using LFSR(): ----------")
# Sage Example 1.4 from https://www.staff.uni-mainz.de/pommeren/Cryptology/Bitstream/Bitstream.▸
 ▸pdf
coeff = [0,1,1,0,1,0,0,0,0,0,0,0,0,0,0,1]
print("Input coeff:"); print(coeff)
reg = LFSR(coeff)
## start = [0,1,1,1]  # AssertionError: LFSR_Error: Bitblock has wrong length (sl=4; __length=1▸
 ▸6).
start = [0,1,1,0,1,0,1,1,0,0,0,1,0,0,1,1]  # original
print("Input start:"); print(start)
reg.setState(start)
bitlist = reg.nextBits(20)
print("LFSR() output:")
print(bbl2str(bitlist))

print("\n-- Using f.fsr() with coeff from LFSR() as input x: --")
x = coeff
print("Input x:"); print(x)
f = BoolF(x)
```

SageMath Example 9.6 (continued)

```
start = [0,1,1,1]
## start = [0,1,1,0,1,0,1,1,0,0,0,1,0,0,1,1]  # AssertionError: boolF_Error: Block has false ▸
 ▸length (11=16; __dim=4).
print("Input start:"); print(start)
bitlist = f.fsr(start, 32)  # output sequence
print("f.fsr() output:")
print(bbl2str(bitlist))
print("New state of the register:", start)
```

9.3.3.2 The Period of a Finite-State Machine

In computer science a feedback shift register is a special case of a finite-state machine. Therefore, the sequence of its states is periodic. Here is why.

Let M be a finite set of $m = \#M$ elements. Imagine M as the collection of all possible states of a machine. Consider a map (transition)

$$g: M \longrightarrow M.$$

For each element (initial state) $x_0 \in M$ define a sequence $(x_i)_{i\geq 0}$ in M by the recursive formula $x_i = g(x_{i-1})$ for $i \geq 1$. After a previous period (preperiod) of length μ the sequence runs into a period of length ν (see Figure 9.15).

Since the set M is finite the states must eventually repeat. Thus, there are smallest integers $\mu \geq 0$ and $\nu \geq 1$ such that $x_{\mu+\nu} = x_\mu$. To see this, simply take μ as the first index such that the element x_μ reappears in the sequence at another position, and $\mu + \nu$ as the first index where this repetition occurs. Then also (by induction)

$$x_{i+\nu} = g(x_{i+\nu-1}) = g(x_{i-1}) = x_i \quad \text{for } i > \mu.$$

Here $0 \leq \mu \leq m-1$, $1 \leq \nu \leq m$, $\mu + \nu \leq m$. The values $x_0, \ldots, x_{\mu+\nu-1}$ are all different, and the values $x_0, \ldots, x_{\mu-1}$ never reappear in the sequence.

Definition: μ is called the (length of the) *preperiod*, ν, the (length of the) *period*.

> A pseudorandom generator in the sense of Figure 9.13 inevitably generates periodic sequences. The best we can hope for is a period so huge that the practical application never exhausts its size.

9.3.3.3 Linear Shift Registers

The simplest and best understood instances of FSRs are the linear feedback shift registers (LFSR). Their feedback functions f are linear. From Section 9.1.9 we know

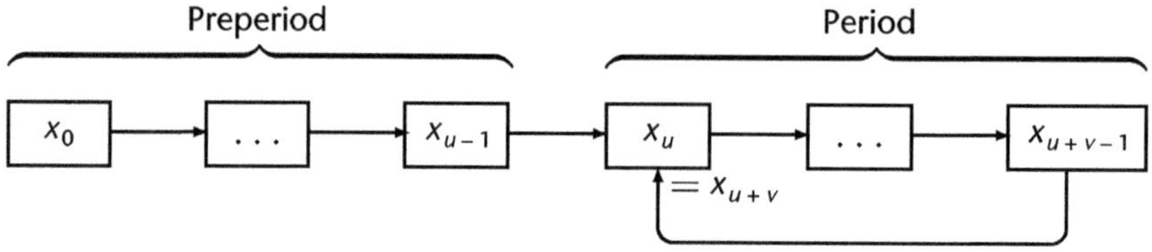

Figure 9.15 Period and preperiod.

that a linear function is simply a partial sum from an l-bit block:

$$f(u_{n-1}, \dots, u_{n-l}) = \sum_{j=1}^{l} s_j u_{n-j}, \tag{9.6}$$

where the coefficients s_j are 0 or 1. If I is the subset of indices j with $s_j = 1$, then the iteration formula (9.5) takes the form

$$u_n = \sum_{j \in I} u_{n-j}. \tag{9.7}$$

A simple graphical representation of an LFSR is shown in Figure 9.16. Here, the subset I defines the contacts (taps) that feed the respective cells into the feedback sum.

For a good choice of the parameters (that we won't discuss further) the sequence has a period of about 2^l, the number of possible different states of the register, and statistical tests are hardly able to distinguish it from a uniformly distributed true random sequence [21]. It is remarkable that such a simple approach generates pseudorandom sequences of fairly high quality. Of course the initial state $u = (0, \dots, 0)$ is inappropriate. For an initial state $\neq 0$ the maximum possible period is $2^l - 1$. Obtaining this period is easy. The necessary and sufficient condition is that the feedback polynomial $1 + s_1 x + s_2 x^2 + \cdots + s_l x^l$ is primitive as polynomial over the field $\mathbb{F}_2$. Caution: Don't confuse the feedback polynomial and the feedback function, the former being a (formal) polynomial in a single variable, the latter a Boolean linear form in l variables.

Using an LFSR for stream encryption the secret inner parameters, the coefficients $s_1, \dots, s_l$, as well as the initial state $u_0, \dots, u_{l-1}$, together constitute the key. In contrast, the length l of the register is assumed as known to the attacker.

SageMath Example 9.7 implements an LFSR as function `lfsr()`[13][14]; the output is a pseudorandom bitstream.

The class `LFSR` in the file `bitciphers.sage` also implements a linear feedback shift register. A sample call of this function for an LFSR of length 16, generating 1024 bits, is in SageMath Example 9.8. First the register `reg` is instantiated as an

Figure 9.16 Simple graphical representation of an LFSR.

13. Within, it uses the function `binScPr()` from the file `bitciphers.sage`, which defines the "scalar product" of two binary vectors.
14. For a more systematic approach, a class `LFSR` is defined in the file `bitciphers.sage`. Alternatively, we can reproduce the result of `lfsr()` by using `fsr()`. To this end we represent the linear feedback function by its truth table (or by its ANF), and then instantiate a Boolean function of the class `BoolF`. Instead of n bits (that define the taps) we need 2^n bits for the truth table. This is quite uneconomical. Thus, the separate definition of `lfsr()` makes sense. Another alternative is the default function `sage.crypto.lfsr.lfsr_sequence` of SageMath.

object of the class LFSR, the taps being defined by the bitblock coeff. Then we set the initial state of the register as the bitblock start, and generate 1024 bits of output. Figure 9.17 shows this output (printed without parantheses or delimiters).

We could apply a series of statistical tests to this bitstream, for example tests of uniform distribution, and would always see good results. Instead, we visualize the sequence in Figure 9.18 for optical inspection—of course an even more insufficient proof. However, the superficial impression shows a quite random sequence. The function visualize in SageMath Example 9.8 generated this picture.

Don't take offense at the sequence of nine (black) ones in the third to last row; the probability of nine ones in nine random bits is $(1/2)^9 = 1/512$. Therefore, in a random bitstream of length 1024 a run of this kind occurs with high probability.

Neither the usual statistical tests nor the visual impression are valid testimonials of the quality of a pseudorandom sequence.

As we'll see, the random properties of LFSR sequences are poor. Cryptanalysis detects deficiencies that evade standard statistical tests.

SageMath Example 9.7: Defining an LFSR in Python/SageMath

```
def lfsr(s,x,n):
  """Generate a linear feedback shift register sequence.
  Parameters: Coefficient vector s, start vector x, number n of output ▸
    ▸bits."""
```

```
11001000110101100011001111000000 00111011100011100000100011101111
01001001111001011011110010111001 00010010110001100111001111010111
11000100011000001110011000010111 01101010101110110001010111011000
11110000010000100010111100011110 10100111000001111000100001011000
01010101000101111110110011011101 11001001110111110001011000100010
11100100101111110011011001010011 00001100100001100110100011100100
11101000100101110110011011001010 11011100100110111001011100000011
00100010111101111000110000010001 01110100001110011111101000100101
00111010001111000100000000110110 10000101110101110001100000010001
11011011011110111001000110101001 10001111110110101010011111100001
11101110111101011001010110001010 00000100001001100110001110100110
00010100101110100000010101100100 10010110101011111110111111011101
11001010010100010010110111111110 10100101001111110110100100010001
10111100011001111001011111010110 01110111010100100010100101101111
01100111011000000111011111010000 11011101111111110000010001000100
10010111111110101011101110111111 01110010110000010001111001100111
```

Figure 9.17 A pseudorandom bit sequence from an LFSR.

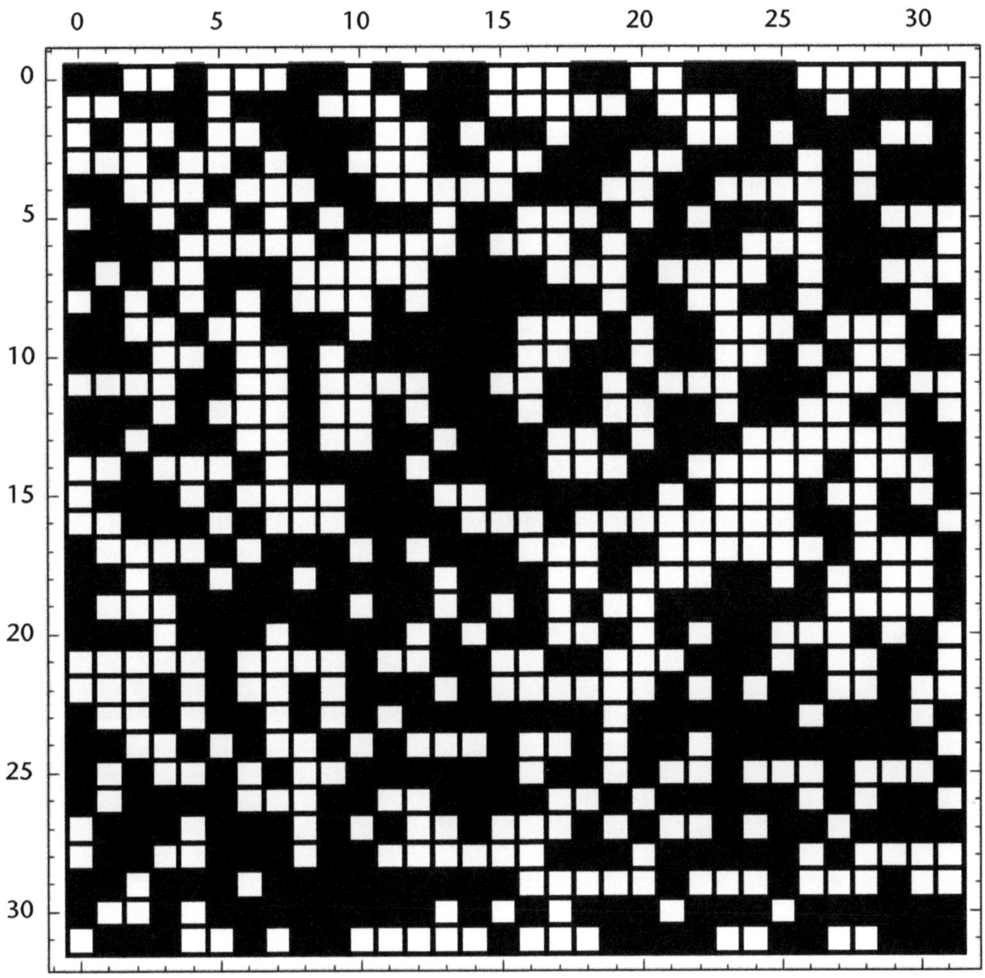

Figure 9.18 Visualization of the pseudorandom bit sequence from Figure 9.17, generated by SageMath Example 9.8 (1 = black, 0 = white).

SageMath Example 9.7 (continued)

```
l = len(s)
assert l == len(x), "lfsr_Error: Bad length of start vector."
u = x                                # in Python use u = x.copy()
outlist = []
for i in range (0,n):
  b = binScPr(s, u)
  c = u.pop()
  u.insert(0,b)
  outlist.append(c)
return outlist
```

SageMath Example 9.8: A Pseudorandom Bit Sequence in Python/SageMath

```
print("\n# CHAP09 -- Sage-Script-SAMPLE 150: =========")

load("./bitciphers.sage")  # for bbl2str() and class LFSR
```

SageMath Example 9.8 (continued)

```
print("\n---------- Using LFSR(): ----------")
coeff = [0,1,1,0,1,0,0,0,0,0,0,0,0,0,0,1] # define the taps of the LFSR
print("Input coeff:"); print(coeff)
reg = LFSR(coeff)                         # initialize the register
start = [0,1,1,0,1,0,1,1,0,0,0,1,0,0,1,1] # seed the register
print("Input start:"); print(start)
reg.setState(start)
bitlist = reg.nextBits(1024)              # ... and make it work
print("LFSR() output:")
print(bbl2str(bitlist))

print("\n---------- Visualize bit sequence: ----------")
def visualize(lst,w,h):
  """Arrange the list lst as a rectangle of width w and height h, row
  by row, as far as lst has entries, and discarding excess entries.
  Then plot a 'checkerboard' showing a white square for zero entries,
  and a black square otherwise. If lst has less than w*h entries,
  the remaining squares are left gray."""

  l = len(lst)
  p = polygon([(0,0), (w,0), (w,h), (0,h)], color = "grey")
  for j in range(h+1):
    p += line([(0,j),(w,j)], color="black")
  for i in range(w+1):
    p += line([(i,0),(i,h)], color="black")
  for j in range(h):
    for i in range(w):
      if (j*w + i < l):
        if (lst[j*w + i] == 0):
          p += polygon([(i,h-1-j),(i+1,h-1-j),(i+1,h-j),(i,h-j)], color = "white")
        else:
          p += polygon([(i,h-1-j),(i+1,h-1-j),(i+1,h-j),(i,h-j)], color = "black")
  p.axes(False)
  return p

p2 = visualize(bitlist,32,32)
# show(p2)                    # works in Jupyter notebook
# Display(p2, "LFSRsquare") # works in script called from CLI if Display() is defined
p2.save("LFSRseq.png")
print("Graphic saved to LFSRseq.png")
```

Figure 9.19 shows how LFSR is implemented in CT2. On the right of the screenshot you can see the settings for the LFSR component.

LFSRs have been implemented many times. A very nice implementation is available in the Python package "pylfsr" [22]. Section 9.3.3.3 shows an example with pylfsr. This code also contains the conversions between the input at CT2 and at pylfsr—because there is no generally accepted standard in which order the bits of the seed have to be entered and how to enter the coefficients of the feedback polynomial.

Python Example 9.1: LFSR with the pylfsr Package in Python

```
print("\n# CHAP09 -- Python-Script-SAMPLE 153: =========")
print("---------- Apply LFSR using Python package pyLFSR ----------")
# Output sequence by pyLFSR using state = [0,0,1,1,1,0,0,1] and fpoly =▸
 ▸ [8,6,4,3]
# - script creates the same result as CT2, which needs fpoly and state ▸
 ▸just in reversed order.
# - see https://lfsr.readthedocs.io/en/latest/dispViz.html

from pylfsr import LFSR
```

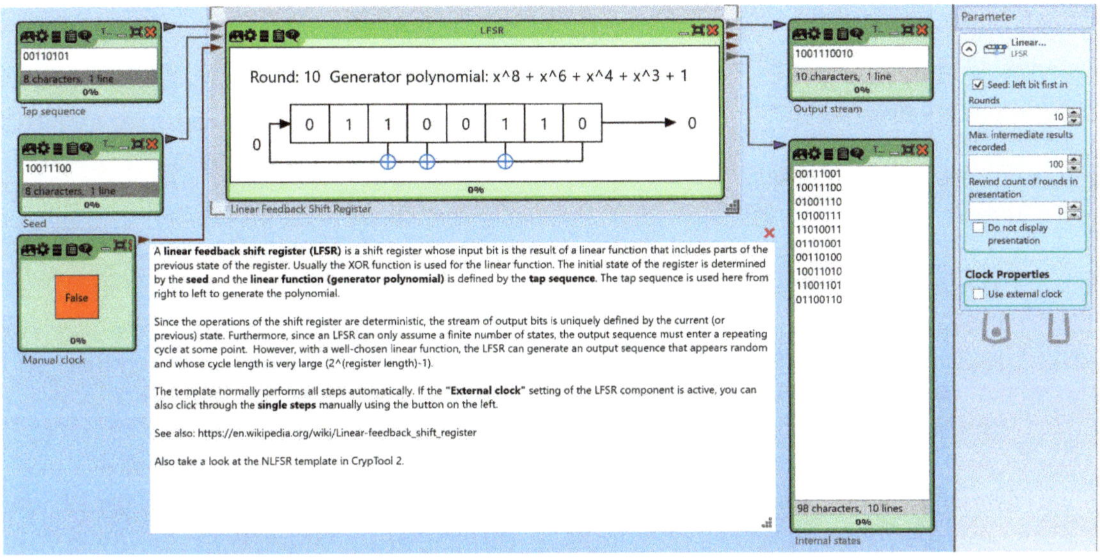

Figure 9.19 Sample LFSR $x^8 + x^6 + x^4 + x^3 + 1$ in CT2.

Python Example 9.1 (continued)

```
# Give bitsequence for seed as in CT2; reverse it and offer it as list for pyLFSR .......
seed = '10011100'  # seed here is a string. This bit sequence is given in CT2 as state.
width = len(seed)  # length of register
num = int(seed, 2)  # you also could enter it directly as 0b10011100 or 15 or 0x15 or 0xAAAA

output = [int(x) for x in '{:0{size}b}'.format(num, size=width)]  # width for potentially leeding bits
print('1a: given seed    =', output)
output.reverse()  # reverse, as pyLFSR needs state in reverse order as CT2
print('1b: reversed seed =', output)
state = output  # state = seed = initial value
statecheck = [0,0,1,1,1,0,0,1]; assert state == statecheck

# Give bitsequence for tap sequence as in CT2; create list of polynomial orders for pyLFSR .......
fpoly = '00110101'  # tap sequence
num = int(fpoly, 2)
output = [int(x) for x in '{:0{size}b}'.format(num, size=width)]
print('2a: fpoly = ', output)
fpoly = [i+1 for i, val in enumerate(output) if val]
fpoly.reverse()
print("2b: fpoly = ",fpoly)
fpolycheck = [8,6,4,3]; assert fpoly == fpolycheck

# Calculate the binary LFSR output sequence
L = LFSR(initstate=state,fpoly=fpoly,counter_start_zero=True)
print('-'*50)
print('count \tstate \t\t\t\toutbit \tseq')
print('-'*50)
for _ in range(15):
```

Python Example 9.1 (continued)

```
    print(L.count,L.state,'',L.outbit,L.seq,sep='\t')
    L.next()
print('-'*50)
print('Output:  ',L.seq)
```

9.3.4 Algebraic Attack on LFSRs

Even simple random generators such as LFSRs produce bit sequences that are virtually indistinguishable from true random sequences by statistical methods, and so provide no hooks for statistical methods of cryptanalysis. This is not true for attacks with known plaintext. The resulting equations for the key bits are accessible for algebraic cryptanalysis. If the key stream originates from a known source, trying to solve these equations promises success. In particular this holds for LFSRs.

Consider a key bitstream $u_0, u_1, \ldots$ generated by an LFSR by formulas (9.6) or (9.7). Assume a plaintext a is XOR encrypted using this key stream, resulting in the ciphertext c, where $c_i = a_i + u_i$ for $i = 0, 1, \ldots$ What are the prospects of an attacker who knows a chunk of the plaintext?

Assume she knows the first $l + 1$ bits of the plaintext. She immediately derives the corresponding bits $u_0, \ldots, u_l$ of the key stream, in particular the initial state of the LFSR. For the yet unknown coefficients s_i she knows a linear relation:

$$s_1 u_{l-1} + \cdots + s_l u_0 = u_l.$$

Each additional known plaintext bit yields one more relation, and having l relations, from $2l$ bits of known plaintext, the easy linear algebra over the field $\mathbb{F}_2$ (in nondegenerate cases) finds a unique solution. The $l \cdot l$ coefficient matrix of this linear equation system is essentially the matrix U of Section 9.3.4.1.

Theorem 9.12 An LFSR of length l is completely predictable from the first $2l$ bits for the cost of about $\frac{1}{3} \cdot l^3$ bit operations.

9.3.4.1 Prediction of LFSRs

Assume we know the first $2l$ bits $u_0, \ldots, u_{2l-1}$ from an LFSR of length l. For an elegant formulation of the linear algebra methods we introduce the state vectors

$$u_{(i)} = (u_i, \ldots, u_{i+l-1}) \quad \text{for } i = 0, 1, \ldots$$

The vector $u_{(i)}$ is the register content for step i (in reversed order compared with Figure 9.14). Thus, the analysis focuses on the states, not directly on the output. The recursion formula (9.6) in matrix form (for $n \geq l$) is

$$\begin{pmatrix} u_{n-l+1} \\ \vdots \\ u_{n-1} \\ u_n \end{pmatrix} = \begin{pmatrix} 0 & 1 & \ldots & 0 \\ \vdots & \vdots & \ddots & \vdots \\ 0 & 0 & \ldots & 1 \\ s_l & s_{l-1} & \ldots & s_1 \end{pmatrix} \begin{pmatrix} u_{n-l} \\ \vdots \\ u_{n-2} \\ u_{n-1} \end{pmatrix}$$

or more parsimoniously (the indices being substituted by $m = n - l + 1$)

$$u_{(m)} = S \cdot u_{(m-1)} \quad \text{for } m \geq 1$$

where S is the coefficient matrix. As a further step we collect l consecutive state vectors $u_{(i)}, \ldots, u_{(i+l-1)}$ in a state matrix

$$U_{(i)} = \begin{pmatrix} u_i & u_{i+1} & \ldots & u_{i+l-1} \\ u_{i+1} & u_{i+2} & \ldots & u_{i+l} \\ \vdots & \vdots & \ddots & \vdots \\ u_{i+l-1} & u_{i+l} & \ldots & u_{2l-2} \end{pmatrix}$$

and set $U = U_{(0)}$, $V = U_{(1)}$. This gives the formula

$$V = S \cdot U$$

that expresses the unknown coefficients $s_1, \ldots, s_l$ by the known plaintext bits $u_0, \ldots, u_{2l-1}$. Most notably it allows us to write down the solution immediately, provided that the matrix U is invertible:

$$S = V \cdot U^{-1}.$$

The matrix S explicitly displays the coefficients $s_1, \ldots, s_l$. We'll discuss the invertibility later on.

Example

Assume we are given a ciphertext:

```
10011100 10100100 01010110 10100110 01011101 10101110
01100101 10000000 00111011 10000010 11011001 11010111
00110010 11111110 01010011 10000010 10101100 00010010
11000110 01010101 00001011 11010011 01111011 10110000
10011111 00100100 00001111 01010011 11111101
```

We suspect that the cipher is XOR with a key stream from an LFSR of length $l = 16$. The context suggests that the text is in German and begins with the word "Treffpunkt" (meeting point). To solve the cryptogram we need 32 bits of plaintext, that is the first four letters only, presupposed that the theory applies. This gives 32 bits of the key stream:

```
01010100 01110010 01100101 01100110 = T r e f
10011100 10100100 01010110 10100110   cipher bits
-------- -------- -------- --------
11001000 11010110 00110011 11000000   key bits
```

SageMath Example 9.9 determines the coefficient matrix. Its last row tells us that all $s_i = 0$ except $s_{16} = s_5 = s_3 = s_2 = 1$.

Now we know the LFSR and the initial state, and can reconstruct the complete key stream—yes, it is the same as in Figure 9.17—and write down the plaintext (that by the way begins a bit differently from our guess).

SageMath Example 9.9: Determining a Coefficient Matrix

```
print("\n# CHAP09 -- Sage-Script-SAMPLE 160: =========")

l = 16
kbits = [1,1,0,0,1,0,0,0,1,1,0,1,0,1,1,0,0,0,1,1,0,0,1,1,1,1,0,0,0,0,0,▸
 ▸0]
ulist = []
for i in range(0,l):
  state = kbits[i:(l+i)]
  ulist.append(state)
U = matrix(GF(2),ulist)
print("det(U) =", det(U))
W = U.inverse()
vlist = []
for i in range(1,l+1):
  state = kbits[i:(l+i)]
  vlist.append(state)
V = matrix(GF(2),vlist)
S = V*W
print(S)

#-------------------------------------
# CHAP09 -- Sage-Script-SAMPLE 160: =========
# det(U) = 1
# [0 1 0 0 0 0 0 0 0 0 0 0 0 0 0 0]
# ...
# [1 0 0 0 0 0 0 0 0 0 0 1 0 1 1 0]
```

9.3.4.2 Proof of Theorem 9.12: How Many Bits Must be Known to Predict an LFSR

We showed that the coefficients are uniquely determined assuming the state matrix $U = U_{(0)}$ is invertible. As a consequence in this case the LFSR is completely known, and all output bits are predictable. We have yet to discuss the case where the matrix U is singular.

If one of the first l state vectors (= rows of the matrix U) is zero, then all following state vectors are zero too, and prediction is trivial.

Thus, we may assume that none of these vectors are zero, but that they are linearly dependent. Then there is a smallest index $k \geq 1$ such that $u_{(k)}$ is contained in the subspace spanned by $u_{(0)}, \ldots, u_{(k-1)}$, and we find coefficients $t_1, \ldots, t_k \in \mathbb{F}_2$ such that

$$u_{(k)} = t_1 u_{(k-1)} + \cdots + t_k u_{(0)}.$$

Then also $u_{(k+1)} = S \cdot u_{(k)} = t_1 S \cdot u_{(k-1)} + \cdots + t_k S \cdot u_{(0)} = t_1 u_{(k)} + \cdots + t_k u_{(1)}$, and by induction we get

$$u_{(n)} = t_1 u_{(n-1)} + \cdots + t_k u_{(n-k)} \quad \text{for all } n \geq k.$$

This formula predicts all the following bits.

The statement on the cost follows from Theorem 9.10.

Discussion

- For a singular state matrix this consideration yields a shorter LFSR (of length $k < l$) that generates exactly the same sequence. Then our method doesn't determine the coefficients of the original register but nevertheless correctly predicts the sequence.
- If the bits the attacker knows aren't just the first ones but $2l$ contiguous ones at a later position, then the theorem yields only the prediction of the following bits. In the main case of an invertible state matrix U the LFSR is completely known and may be run backwards to get the previous bits. For a singular state matrix we achieve the same effect using the shorter LFSR constructed previously.
- The situation where $2l$ bits of the key stream are known but at noncontiguous positions is slightly more involved. We get linear relations that contain additional (unknown) intermediate bits. If m is the number of these then we get $l + m$ linear equations for $l + m$ unknown bits.
- What if the length l of the LFSR is unknown? Exhaustively trying all values $l = 1, 2, 3, \dots$ is nasty but feasible. A better approach is provided by the Berlekamp-Massey[15] algorithm that is efficient also without knowledge of l. We won't treat it in this chapter.

Summary
Given a random generator as in Figure 9.13 cryptanalytic, targets are:

- Secret parameters
- Initial state
- Additional parts of the output (prediction problem)

given some parts of the output. As we saw for LFSRs the prediction problem has a solution even when the internal parameters remain unknown. Thus:

> Cryptanalysis of a random generator first of all means solving the prediction problem. A random generator is cryptographically secure if its prediction problem admits no efficient solution.
>
> Linear feedback shift registers are not cryptographically secure.

9.3.5 Approaches to Nonlinearity for Feedback Shift Registers

LFSRs are popular (in particular among electrical engineers and military) for several reasons:

- Very easy implementation;
- Extreme efficiency in hardware;

15. Berlekamp-Massey is contained in SageMath as `sage.crypto.lfsr.berlekamp_massey`. With CT2 `Templates` ▷ `Cryptanalysis` ▷ `Generic` ▷ `Berlekamp-Massey` you can try this algorithm directly.

- Good qualification as random generators for statistical applications and simulations;
- Unproblematic operation in parallel even in large quantities.

But unfortunately from a cryptological view they are completely insecure if used naively. To capitalize their positive properties while escaping their cryptological weakness there are several approaches.

Approach 1: Nonlinear Feedback

Nonlinear feedback follows the scheme from Figure 9.14 with a nonlinear Boolean function f. We won't pursue this approach here. We saw a very simple toy example in SageMath Example 9.6. There is a general proof that in realistic use cases NLFSRs (abbreviation for nonlinear feedback shift register) are cryptographically useless if used in the direct naive way [23].

NLFSRs have often been used within stream ciphers, but free implementations of the principle are rare. Figure 9.20 shows how a simple NLFSR is implemented in CT2.[16] On the right of the screenshot you can see the settings for the NLFSR component.

Approach 2: Nonlinear Output Filter

The nonlinear output filter (nonlinear feedforward) realizes the scheme from Figure 9.21. The shift register itself is linear, the Boolean function f, nonlinear. The nonlinear output filter is a special case of a nonlinear combiner.

Approach 3: Nonlinear Combiner

The nonlinear combiner uses a battery of n LFSRs—preferably of different lengths—operated in parallel. The output sequences of the LFSRs serve as input of a Boolean function $f\colon \mathbb{F}_2^n \longrightarrow \mathbb{F}_2$, see Figure 9.22.

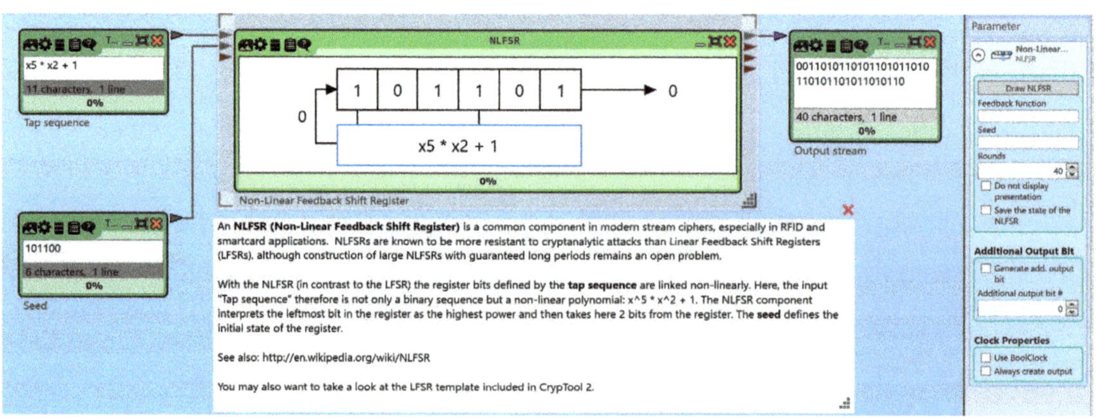

Figure 9.20 Simple NLFSR $x^5 * x^2 + 1$ in CT2.

16. With `CT2 Templates` ⊳ `Mathematics` ⊳ `LFSR or NLFSR` you can try this directly.

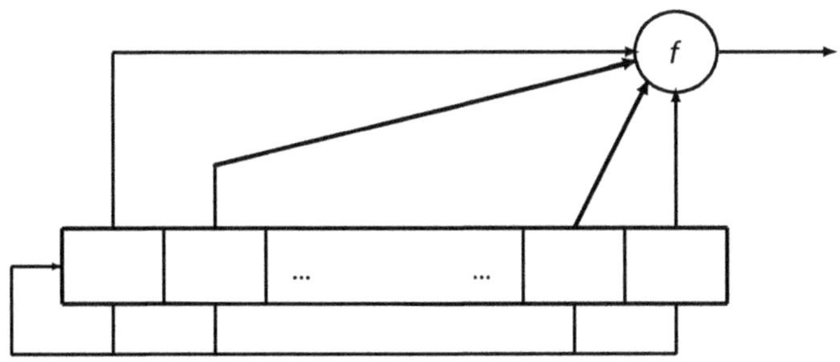

Figure 9.21 Nonlinear output filter for an LFSR.

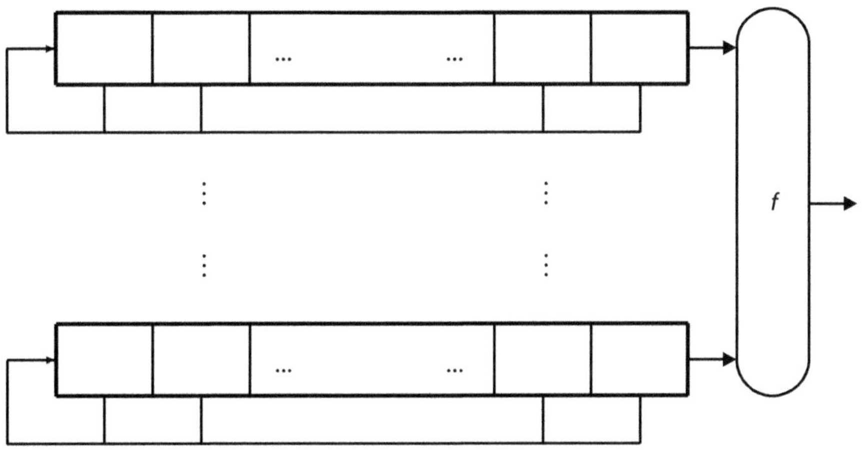

Figure 9.22 Nonlinear combiner.

Approach 4: Output Selection/Decimation/Clocking

There are different ways of controlling a battery of n parallel LFSRs by another LFSR:

- *Output selection* takes the current output bit of exactly one of the LFSRs from the battery, depending on the state of the auxiliary register, and outputs it as the next pseudorandom bit. More generally we could choose r from n.
- For *decimation* one usually takes $n = 1$ and outputs the current bit of the one battery register only if the auxiliary register is in a certain state, for example its own current output is 1. Of course this kind of decimation applies to arbitrary bit sequences in an analogous way.
- For *clocking* we look at the state of the auxiliary register and, depending on it, decide which of the battery registers to step in the current cycle (and by how many positions), leaving the other registers in their current states. This is reminiscent of the control logic of rotor machines in classical cryptography.

These methods turn out to be special cases of nonlinear combiners if properly rewritten. Thus approach 3 represents the most important method of making the best of LFSRs.

The encryption standard A5/1 for mobile communications uses three LFSRs of lengths 19, 22, and 23, each with maximum possible period, and slightly differently clocked. It linearly (by simple binary addition) combines the three output streams. The (even weaker) algorithm A5/2 controls the clocking by an auxiliary register. Both variants can be broken on a standard PC in real-time.

The Bluetooth encryption standard E_0 uses four LFSRs and combines them in a nonlinear way. This method is somewhat stronger than A5, but also too weak for real security [5].

Example: The Geffe Generator

The Geffe generator provides a simple example of output selection. Its description is in Figure 9.23. The output is x, if $z = 0$, and y, if $z = 1$. Expressed by a formula:

$$u = \begin{cases} x, & \text{if } z = 0, \\ y, & \text{if } z = 1 \end{cases}$$
$$= (1 - z)x + zy = x + zx + zy.$$

This formula shows how to interpret the Geffe generator as a nonlinear combiner with a Boolean function $f\colon \mathbb{F}_2^3 \longrightarrow \mathbb{F}_2$ of degree 2. For later use we implement f in SageMath Example 9.10.

SageMath Example 9.10: The Geffe Function

```
print("\n# CHAP09 -- Sage-Script-SAMPLE 170: =========")

load("./bitciphers.sage")
geff = BoolF(str2bbl("00011100"),method="ANF")
geff.printTT()

#------------------------------------
# CHAP09 -- Sage-Script-SAMPLE 170: =========
# Value at 000 is 0
# Value at 001 is 0
# Value at 010 is 0
# Value at 011 is 1
# Value at 100 is 1
```

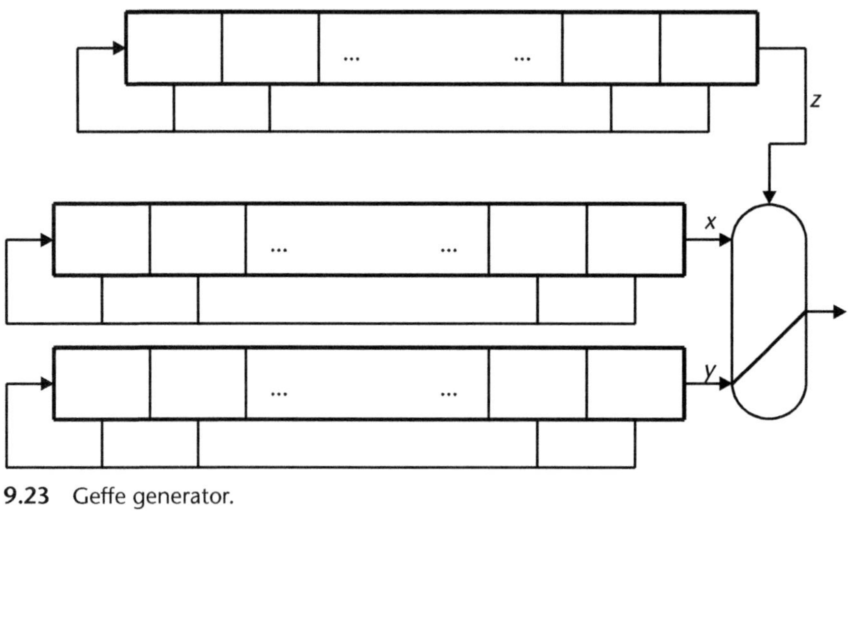

Figure 9.23 Geffe generator.

SageMath Example 9.10 (continued)

```
# Value at 101 is 0
# Value at 110 is 1
# Value at 111 is 1
```

9.3.6 Implementation of a Nonlinear Combiner with the Class LFSR

A nonlinear combiner uses several LFSRs, operated in parallel. This suggests an implementation of LFSRs as objects of a class `LFSR`.

Class `LFSR`:

Attributes:

- `length`: The length of the register;
- `taplist` (constant): The list of coefficients (or taps) that define the bits for feedback;
- `state` (variable): The state of the register.

Methods:

- `setLength`: Define the length (used only implicitly for initialization);
- `setTaps`: Define the list of taps (used only implicitly for initialization);
- `setState`: Set the state of the register;
- `getLength`: Output the length;
- `nextBits`: Generate a given number of output bits, and set the next state.

9.3.6.1 Example: Complete Geffe Generator

First we choose three LFSRs of lengths 15, 16, 17, whose periods are $2^{15} - 1 = 32767$, $2^{16} - 1 = 65535$, and $2^{17} - 1 = 131071$. These are pairwise coprime (see SageMath Example 9.11). Combining their outputs (in each step) as bit-blocks of length 3 yields a sequence with a period that has an impressive length of 281459944554495, about $300 \cdot 10^{12}$ (300 trillions for Americans are 300 billions for Europeans).

SageMath Example 9.12 defines the three LFSRs. The recursive formula for the third one, the control register `reg17`, is $u_n = u_{n-3} + u_{n-17}$, since exactly the taps 3 and 17 are active. We let each of the LFSRs generate a sequence of length 100 (see SageMath Example 9.13). The Geffe function combines them in SageMath Example 9.14. This is the complete sample.

SageMath Example 9.11: Calculating a Period

```
print("\n# CHAP09 -- Sage-Script-SAMPLE 180: =========")

n15 = 2**15 - 1; print("n15 =", n15, "=", n15.factor())
n16 = 2**16 - 1; print("n16 =", n16, "=", n16.factor())
n17 = 2**17 - 1; print("n17 =", n17, "=", n17.factor())
```

SageMath Example 9.11 (continued)

```
print("lcm    =", lcm([n15,n16,n17]))
period = n15 * n16 * n17; print("period =", period)

#------------------------------------
# CHAP09 -- Sage-Script-SAMPLE 180: =========
# n15 = 32767 = 7 * 31 * 151
# n16 = 65535 = 3 * 5 * 17 * 257
# n17 = 131071 = 131071
# lcm    = 281459944554495
# period = 281459944554495
```

SageMath Example 9.12: Three LFSRs

```
print("\n# CHAP09 -- Sage-Script-SAMPLE 190: =========")

load("./bitciphers.sage")

reg15 = LFSR([1,0,0,0,0,0,0,0,0,0,0,0,0,0,1])
reg15.setState([0,1,1,0,1,0,1,1,0,0,0,1,0,0,1])
print(reg15)
reg16 = LFSR([0,1,1,0,1,0,0,0,0,0,0,0,0,0,0,1])
reg16.setState([0,1,1,0,1,0,1,1,0,0,0,1,0,0,1,1])
print(reg16)
reg17 = LFSR([0,0,1,0,0,0,0,0,0,0,0,0,0,0,0,0,1])
reg17.setState([0,1,1,0,1,0,1,1,0,0,0,1,0,0,1,1,1])
print(reg17)

#------------------------------------
# CHAP09 -- Sage-Script-SAMPLE 190: =========
# Length: 15 | Taps: 100000000000001 | State: 011010110001001
# Length: 16 | Taps: 0110100000000001 | State: 0110101100010011
# Length: 17 | Taps: 00100000000000001 | State: 01101011000100111
```

SageMath Example 9.13: Three LFSR Sequences

```
print("\n# CHAP09 -- Sage-Script-SAMPLE 200: =========")

load("./bitciphers.sage")

reg15 = LFSR([1,0,0,0,0,0,0,0,0,0,0,0,0,0,1])
reg15.setState([0,1,1,0,1,0,1,1,0,0,0,1,0,0,1])

reg16 = LFSR([0,1,1,0,1,0,0,0,0,0,0,0,0,0,0,1])
reg16.setState([0,1,1,0,1,0,1,1,0,0,0,1,0,0,1,1])

reg17 = LFSR([0,0,1,0,0,0,0,0,0,0,0,0,0,0,0,0,1])
reg17.setState([0,1,1,0,1,0,1,1,0,0,0,1,0,0,1,1,1])

nofBits = 100
outlist15 = reg15.nextBits(nofBits)
print("outlist15:\n", outlist15)
outlist16 = reg16.nextBits(nofBits)
print("outlist16:\n", outlist16)
outlist17 = reg17.nextBits(nofBits)
print("outlist17:\n", outlist17)
```

SageMath Example 9.14: The Combined Sequence

```
print("\n# CHAP09 -- Sage-Script-SAMPLE 210: =========")

load("./bitciphers.sage")

reg15 = LFSR([1,0,0,0,0,0,0,0,0,0,0,0,0,0,1])
reg15.setState([0,1,1,0,1,0,1,1,0,0,0,1,0,0,1])

reg16 = LFSR([0,1,1,0,1,0,0,0,0,0,0,0,0,0,0,1])
reg16.setState([0,1,1,0,1,0,1,1,0,0,0,1,0,0,1,1])

reg17 = LFSR([0,0,1,0,0,0,0,0,0,0,0,0,0,0,0,0,1])
reg17.setState([0,1,1,0,1,0,1,1,0,0,0,1,0,0,1,1,1])

nofBits = 100
outlist15 = reg15.nextBits(nofBits)
outlist16 = reg16.nextBits(nofBits)
outlist17 = reg17.nextBits(nofBits)

geff = BoolF(str2bbl("00011100"),method="ANF")

outlist = []
for i in range(0,nofBits):
    x = [outlist15[i], outlist16[i], outlist17[i]]
    outlist.append(geff.valueAt(x))
print(outlist)

#------------------------------------
# CHAP09 -- Sage-Script-SAMPLE 210: =========
# [1, 1, 0, 1, 0, 0, 0, 1, 1, 1, 0, 0, 0, 1, 1, 0, 1, 1, 0, 1,
#  0, 0, 1, 0, 0, 1, 0, 0, 1, 1, 0, 0, 0, 0, 1, 1, 0, 0, 1, 1,
#  1, 0, 1, 0, 1, 1, 0, 0, 0, 0, 0, 0, 1, 0, 0, 0, 0, 1, 1, 0,
#  1, 1, 1, 1, 0, 1, 0, 0, 1, 0, 0, 0, 1, 1, 0, 1, 0, 1, 0, 1,
#  0, 0, 1, 1, 0, 1, 0, 0, 1, 1, 0, 1, 1, 0, 0, 0, 1, 0, 0, 1]
```

9.3.7 Design Criteria for Nonlinear Combiners

From the forgoing discussion we derive design criteria for nonlinear combiners:

- The battery registers should be as long as possible.
- The combining function f should have a low linear potential.

How long should the battery registers be? There are some algorithms for fast correlation attacks using the Walsh transformation, in particular against sparse linear feedback functions (that use only a few taps) [24]. These don't reduce the complexity class of the attack (exponential in the length of the shortest register) but reduce the cost by a significant factor. So they are able to attack registers whose feedback functions have up to 100 monomials with coefficients in their ANF. As a consequence

- The single LFSRs should have a length of at least 200 bits, and use about 100 taps each.

To assess the number n of LFSRs we bear in mind that the combining function should be correlation immune, in particular have a low linear potential. A well-chosen Boolean function of 16 variables should suffice. To us, there are no known recommendations in the literature.

Rainer Rueppel found an elegant way out to make the correlation attack break down: Use a time-dependent combining function, that is a family $(f_t)_{t\in\mathbb{N}}$. The bit u_t of the key stream is calculated by the function f_t. We won't analyze this approach here.

Observing that the correlation attack needs knowledge of the taps, the security could be somewhat better if the taps are secret. Then the attacker has to perform additional exhaustions that multiply the complexity by factors such as 2^{l_1} for the first LFSR alone. This scenario allows choosing LFSRs of somewhat smaller lengths. But bear in mind that for a hardware implementation the taps are parts of the algorithm, not of the key, so they are public parameters in the sense of Figure 9.13.

Efficiency

LFSRs and nonlinear combiners allow efficient realizations by special hardware that produces one bit per clock cycle. This rate can be enlarged by parallelization. From this point of view, estimating the cost of execution on a usual PC processor is somewhat inadequate. Splitting each of the $\geq$ 200 bit registers into 4 parts of about 64 bits shifting a single register requires at least 4 clock cycles, summing up to 64 clock cycles for 16 registers. Add some clock cycles for the combining function. Thus, one single bit would take about 100 clock cycles. A 2-GHz processor, even with optimized implementation, would produce at most $2 \cdot 10^9/100 = 20$ million bits per second.

As a summary we note: Using LFSRs and nonlinear combining functions, we can build useful and fast random generators, especially in hardware.

Unfortunately there is no satisfying theory for the cryptologic security of this type of random generators, even less a mathematical proof. Security is assessed by plausible criteria that—as for block ciphers—are related to the nonlinearity of Boolean functions.

9.3.8 Perfect (Pseudo)Random Generators

As we saw, the essential cryptologic criterion for random generators is unpredictability. In the 1980s cryptographers, guided by an analogy with asymmetric cryptography, found a way of modeling this property in terms of complexity theory: Prediction should boil down to a known hard algorithmic problem such as factoring or discrete logarithm. This idea established a new quality standard for random generators, much stronger than statistical tests, but eventually building on unproven mathematical hypotheses. Thus, the situation with respect to the security of random generators is comparable to asymmetric encryption.

As an interesting twist it soon turned out that in a certain sense unpredictability is a universal property. For an unpredictable sequence there is no efficient algorithm at all that distinguishes it from a true random sequence, a seemingly much stronger requirement [see Theorem 9.13 (Yao's theorem)]. This universality justifies the denomination perfect for the corresponding random generators. In particular there is no efficient statistical test that is able to distinguish the output of a perfect random generator from a true random sequence. Thus, on the theoretical side, we have a very appropriate model for random generators that are absolutely strong from a statistical viewpoint, and invulnerable from a cryptological viewpoint. In other words:

> Perfect random generators are cryptographically secure and statistically indistinguishable from true random sources.

> Presumably perfect random generators exist, but there is no complete mathematical proof of their existence.

The first concrete approaches to the construction of perfect random generators, the best known being the BBS generator (for Blum, Blum, Shub), yielded algorithms that were too slow for most practical uses (given the then current CPUs). But modified approaches soon provided random generators that are passably fast and nevertheless (presumably) cryptographically secure.

9.3.9 The BBS Generator

As with the RSA cipher we consider an integer modulus m that is a product of two large prime numbers. For the BBS generator we choose (for technical reasons not to be discussed here) Blum primes; these are primes $\equiv 3 \bmod 4$. A product of two Blum primes is called a Blum integer.

The BBS generator works in the following way: As a first step, we choose two large random Blum primes p and q, and form their product $m = pq$. As a second step, we choose a random integer seed s with $1 \leq s \leq m - 1$, and coprime with m. Remark 1: If we catch an s not coprime with m, we have factorized m by hazard. This might happen, but is extremely unlikely, and can easily be captured at initialization time. Remark 2: If $x_i < \sqrt{m}$, then $x_i^2 \bmod m = x_i^2$, the integer square, so x_{i+1}^2 has the same parity as x_i. In order to avoid a constant segment at the beginning of the output, often the boundary area $s < \sqrt{m}$, as well as $s > m - \sqrt{m}$, is excluded. However, if we really choose s as a true random value, the probability for s falling into these boundary areas is extremely low. But to be on the safe side we may require $\sqrt{m} \leq s \leq m - \sqrt{m}$.

Now we proceed with generating a pseudorandom sequence: Take $x_0 = s^2 \bmod m$ as initial state (so x_0 is a quadratic residue), and form the sequence of inner states of the random generator: $x_i = x_{i-1}^2 \bmod m$ for $i = 1, 2, 3, \ldots$ In each step, output that last significant bit of the binary representation; that is $u_i = x_i \bmod 2$ for $i = 0, 1, 2, \ldots$, or in other words, the parity of x_i.

Example

Of course an example with small numbers is practically irrelevant, but it illustrates the algorithm. Take $p = 7$, $q = 11$, $m = 77$, $s = 53$. Then $s^2 = 2809$, hence $x_0 = 37$, and $u_0 = 1$ since x_0 is odd. The naive SageMath Example 9.15 shows the beginning of the sequence of states:

i	0	1	2	3	...
x_i	37	60	58	53	...
u_i	1	0	0	1	...

SageMath Example 9.15: A (Much Too) Simple Example for BBS

```
print("\n# CHAP09 -- Sage-Script-SAMPLE 260: =========")

p = 7; q = 11; m = p*q; s = 53
print("m =", m, ", s =", s)
x0 = (s^2) % m; print("x0 =", x0)
x1 = (x0^2) % m; print("x1 =", x1)
x2 = (x1^2) % m; print("x2 =", x2)
x3 = (x2^2) % m; print("x3 =", x3)
```

Treating the Blum primes p and q as secret is essential for the security of the BBS generator. They serve for forming m only; afterward they may even be destroyed. In contrast with RSA there is no further use for them. Likewise, all the nonoutput bits of the inner states x_i must be secret.

The standard distribution of SageMath contains the BBS generator. It consists of the procedures:

- `random_blum_prime()` in the module `sage.crypto.util`. To generate a random Blum prime p with a given number k of bits (= digits of the binary representation) call it as `p = random_blum_prime(2**(k-1), 2**k)`. The correctness of this algorithm is only empirically founded: In fact there is always a prime between 2^{k-1} and 2^k, but this doesn't need to be a Blum prime. This is a special case of Bertrand's postulate, proved by Chebyshev in 1850: There is a prime between n and $2n$ (for all $n \geq 2$). Nevertheless, empiricism tells us that there are lots of Blum primes in this interval, namely about $2^k/(k \log(2))$. Thus, an attack by exhaustion will fail.
- `blum_blum_shub()` from `sage.crypto.stream`. To generate a sequence of r pseudorandom bits first generate two random Blum primes p and q and an initial value $x_0 = s^2 \bmod pq$, and then call the procedure as `blum_blum_shub(r,x_0,p,q)`.

SageMath Example 9.16 demonstrates the procedure. The intermediate results p, q, and x_0 are shown in Tables 9.10, 9.11, and 9.12, the result in Table 9.13. By convention s as well as the factors p and q must be kept secret. Moreover, there is no reason to reveal the product $m = pq$. However, considering the progress of factorization algorithms we should better use Blum integers of at least 2048 bit (see Section 9.3.10). And in any case s must be a true random value! We neglected this duty by choosing s as a pure power.

SageMath Example 9.16: Generating a Sequence of BBS Pseudorandom Bits

```
print("\n# CHAP09 -- Sage-Script-SAMPLE 270: =========")

from sage.crypto.util   import random_blum_prime
from sage.crypto.stream import blum_blum_shub

sl = 1000  # Number of bits to generate in the pseudo random bit ▸
  ▸sequence
print("Using random start values (1000 bit length):")
p = random_blum_prime(2^511, 2^512)
```

SageMath Example 9.16 (continued)

```
q = random_blum_prime(2^511, 2^512)
x0 = 11^248 % (p*q)              # s = 11^124 % (p*q)
print(blum_blum_shub(sl,x0,p,q))

print("\nUsing the fixed values from the given table as start values:")
no = "8 445 834 617 855 090 512 176 000 413 196 767 417 799 332\
626 936 992 170 472 089 385 128 414 279 550 732 184 808 226\
736 683 775 727 426 619 339 706 269 080 823 255 441 520 165\
438 397 334 657 231 839 251"
p = ZZ(no.replace("\n", "").replace("\r", "").replace(" ", ""))

no = "12 580 605 326 957 495 732 854 671 722 855 802 182 952 894\
  232 088 903 111 155 705 856 898 413 602 721 771 810 991 595\
  365 229 641 230 483 180 760 744 910 366 324 916 344 823 400\
  588 340 927 883 444 616 787"
q = ZZ(no.replace("\n", "").replace("\r", "").replace(" ", ""))

no = "1 842 408 460 334 540 507 430 929 434 383 083 145 786 026\
  412 146 359 363 362 017 837 922 966 741 162 861 257 645 571\
  680 482 798 249 771 263 305 761 292 545 408 040 659 753 561\
  970 871 645 393 254 757 072 936 076 922 069 587 163 804 708\
  256 246 366 137 431 776 175 309 050 064 068 198 002 904 756\
  218 898 942 856 431 647 438 473 529 312 261 281"
x0 = ZZ(no.replace("\n", "").replace("\r", "").replace(" ", ""))

s= str(blum_blum_shub(sl,x0,p,q))
bl=4 # blocklength
split = [s[i:i+bl] for i in range (0, len(s), bl)]
print("Number of blocks = len(split) =", len(split))

for i in range(len(split)):
  print("%s " % (split[i]), end = '' )
print()
```

Table 9.10 A Blum Prime p with 512 Bits (154 Decimal Places)

8 445 834 617 855 090 512 176 000 413 196 767 417 799 332
626 936 992 170 472 089 385 128 414 279 550 732 184 808 226
736 683 775 727 426 619 339 706 269 080 823 255 441 520 165
438 397 334 657 231 839 251

Table 9.11 A Blum Prime q with 512 Bits (155 Decimal Places)

12 580 605 326 957 495 732 854 671 722 855 802 182 952 894
232 088 903 111 155 705 856 898 413 602 721 771 810 991 595
365 229 641 230 483 180 760 744 910 366 324 916 344 823 400
588 340 927 883 444 616 787

Table 9.12 An Initial Value x_0

1 842 408 460 334 540 507 430 929 434 383 083 145 786 026
412 146 359 363 362 017 837 922 966 741 162 861 257 645 571
680 482 798 249 771 263 305 761 292 545 408 040 659 753 561
970 871 645 393 254 757 072 936 076 922 069 587 163 804 708
256 246 366 137 431 776 175 309 050 064 068 198 002 904 756
218 898 942 856 431 647 438 473 529 312 261 281

Table 9.13 1000 BBS Pseudorandom Bits

1010	0110	0011	0100	0000	0111	1111	0100	1111	0111	0010	1001
0000	0100	1111	0000	0010	1010	1011	1111	1000	0101	1110	0011
1110	1000	1001	1100	1000	1000	0110	0111	0011	0011	1010	0011
1100	1111	0011	1000	1011	0110	1011	1110	0110	1110	0111	1000
1101	0011	1101	0010	1000	1101	0000	1100	0100	1011	1110	0011
0110	0010	1011	0000	1010	1001	0110	0000	0011	1010	0011	1111
1010	0110	0101	1000	1011	0100	0100	1111	1010	1011	0001	1100
0000	0011	1101	1001	0001	0000	1111	1010	1001	0111	0111	0111
0000	1010	0101	0111	0111	0001	0110	1001	0011	1011	0000	0011
1000	0000	0111	0110	0110	1010	0110	0011	0111	1100	0010	0110
0011	1001	1010	1111	0001	0010	1111	0010	1100	1111	0110	0100
0001	1000	0101	0011	0000	0101	1111	1100	0101	0000	0100	0100
0100	0101	0010	1110	1010	1011	1011	0110	0101	1011	1111	1110
1100	1001	1011	0110	1001	0111	0111	1110	0101	0111	0011	0100
1101	1110	0011	1111	1101	0100	1111	1011	1010	0010	0111	1111
1010	1000	1100	1001	1010	1001	1010	0111	0100	0100	1010	0110
0011	0010	1110	0111	0101	0111	1101	0000	0110	0000	1110	1100
0101	1010	0111	1000	0101	1111	0010	1101	0110	0100	0010	1101
0000	1101	0111	1011	0010	1010	1000	0110	0100	0111	1100	0000
1101	0000	1011	1111	0101	1011	0011	1110	0010	1110	1101	0001
1110	1111	1000	0111	1010	0000	1100	0101	0110	0001		

9.3.10 Perfectness and the Factorization Conjecture

Informally, we define a pseudorandom generator (shortly: a random generator) as an efficient algorithm that takes a short bitstring $s \in \mathbb{F}_2^n$ and converts it into a long bitstring $s \in \mathbb{F}_2^r$.

The terminology of complexity theory allows us to give a mathematically exact (but not completely satisfying from a practical point of view) definition by considering parameter-dependent families of Boolean maps $G_n : \mathbb{F}_2^n \longrightarrow \mathbb{F}_2^{r(n)}$, and analyzing their behavior when the parameter n grows to infinity. Such an algorithm—represented by the family (G_n) of Boolean maps—can be efficient only if the expanding function $r : \mathbb{N} \longrightarrow \mathbb{N}$ grows at most polynomially with the parameter n, otherwise even writing down the output sequence efficiently way is impossible. Then we measure the cost somehow in a meaningful way; for example, count the number of needed bit operations that likewise must be at most polynomial with respect to the asymptotic behavior.

On the attacker's side we consider algorithms that predict further bits, or aim at detecting some other weaknesses of our random generator. We analyze the costs of these algorithms also as functions of n. In case the cost grows faster than any polynomial, say exponentially, we rate the attack as inefficient.

Pursuing this approach would require a lot of additional formalism including a model of probabilistic algorithms that are essential tools for the cryptanalyst. This would take us too far apart for the moment being. However, we bear in mind that there is a mathematically correct theory formalizing the intuitive idea of efficiency. Relying on this knowledge we don't hesitate to reasoning the naive way, and draft the following definition that in the given form is mathematically incorrect but might be made correct.

Definition 9.4 Consider a pseudorandom generator. A *next bit predictor* is an algorithm that takes a piece $u_0, \ldots, u_{r-1}$ from the beginning of the pseudorandom sequence and calculates the next bit u_r, without using the internal parameters of the pseudorandom generator.

The pseudorandom generator *passes the prediction test* if there is no efficient next bit predictor.

For example, LFSRs don't pass the prediction test: We constructed an efficient next bit predictor in Theorem 9.12.

Definition 9.5 Consider a pseudorandom generator. A *distinguisher* is an algorithm that decides whether a given sequence is purely random or is generated by the pseudorandom generator, without using the internal parameters of the pseudorandom generator.

The pseudorandom generator is *perfect* if there is no efficient distinguisher for it.

In particular no efficient statistical test is able to distinguish a perfect pseudorandom generator from a true random source. It is a bit of a surprise that the seemingly much weaker property of passing the prediction test already implies perfectness. In other words the prediction test is universal:

Theorem 9.13 (Yao's criterion) A pseudorandom generator is perfect if and only if it passes the prediction test.

Here stated without proof. Unfortunately, this approach only gives qualitative results, and so it is somewhat dissatisfying. However, as often in complexity theory, this is the best we can achieve.

9.3.10.1 The (Conjectured) Perfectness of the BBS Generator

The factorization hypothesis states that there is no efficient algorithm that decomposes large natural numbers into their prime factors. This hypothesis is the base of the security of RSA, as well of the perfectness of the BBS generator:

Theorem 9.14 (Blum/Blum/Shub/Vazirani/Vazirani) Assume the factorization hypothesis holds. Then the BBS generator is perfect.

We omit the proof (that is quite involved). Sloppily expressed, the theorem says:

> Whoever is able to predict a single bit of a BBS sequence given a partial sequence is also able to factor the modulus.

This statement assumes that the attacker knows the modulus m of the BBS generator. However, the modulus might also be secret, that is, considered as a part of the key. Assuming this, the cryptographic security of BBS should even be better—but no proof of this stronger statement seems to be known, not even an informal one.

9.3.11 Examples and Practical Considerations

We saw that the BBS generator is perfect under a plausible but unproven assumption, the factorization hypothesis. However, we don't know relevant concrete details, for example what parameters might be inappropriate. We know that certain initial states generate output sequences with short periods. Some examples of this effect are known, but we are far from a complete answer. However, the security proof (depending on the factorization hypothesis) doesn't require additional assumptions. Therefore, we may confidently use the BBS generator with a pragmatic attitude: randomly choosing the parameters (primes and initial state) the probability of hitting bad values is extremely low, much lower than finding a needle in a haystack, or even in the universe.

Nevertheless some questions are crucial for getting good pseudorandom sequences from the BBS generator efficiently:

- How large should we choose the parameter m?
- How many bits can we use for a fixed modulus and initial state without compromising the security?

The provable results, relative to the factorization hypothesis, are qualitative only, not quantitative. The recommendation to choose a modulus that escapes the known factorization methods also rests on heuristic considerations only, and doesn't seem absolutely mandatory for a modulus that itself is kept secret. The real quality of the pseudorandom bit sequence, be it for statistical or for cryptographic applications, can only be assessed by empirical criteria for the time being. We are confident that the danger of generating a bad pseudorandom sequence is extremely small, in any case negligible, for moduli that escape the presently known factorization algorithms, say at least of a length of 2048 bit, and for a true random choice of the modulus and the initial state. Remark: Émile Borel proposed an informal ranking of negligibility of extremely small probabilities: $\leq 10^{-6}$ from a human view; $\leq 10^{-15}$ from a terrestrial view; $\leq 10^{-45}$ from a cosmic view. By choosing a sufficiently large modulus m for RSA or BBS we easily undercut Borel's bounds by far.

For the length of the usable output sequence we only know the qualitative criterion "at most polynomially many" that is useless in a concrete application. But even if we only use "quadratically many" bits we wouldn't hesitate to take 4 millions bits from the generator with a ≥ 2000 bit modulus. If we need substantially more bits we would restart the generator with new parameters after every few millions of bits.

An additional question suggests itself: Are we allowed to output more than a single bit of the inner state in each iteration step to enhance the practical benefit of the generator? At least 2 bits?

Vazirani and Vazirani, and independently Alexi, Chor, Goldreich, and Schnorr gave a partial answer to this question, unfortunately also a qualitative one only: at least $\mathcal{O}(\log_2 \log_2 m)$ of the least significant bits are safe. Depending on the constants that hide in the $\mathcal{O}$, we need to choose a sufficiently large modulus, and trust empirical experience. A common recommendation is using $\lfloor \log_2 \log_2 m \rfloor$ bits per step. Then for a modulus m of 2048 bits, or roughly 600 decimal places, we can use 11 bits per step. Calculating $x^2 \bmod m$ for an n bit number m takes $(\frac{n}{64})^2$ multiplications of 64-bit integers and subsequently the same number of divisions of the type 128 bits by 64 bits.

For $n = 2048$ this makes a total of $2 \cdot (2^5)^2 = 2048$ multiplicative operations to generate 11 bits, or about 200 operations per bit. A well-established rule of thumb says that a modern CPU executes one multiplicative operation per clock cycle. Remark: Special processors that use pipelines and parallelism are significantly faster. Thus, on a 2-GHz CPU with 64-bit architecture we may expect roughly $2 \cdot 10^9/200 \approx 10$ million bits per second, provided the algorithm is implemented in an optimized way. This consideration shows that the BBS generator is almost competitive with a software implementation of a sufficiently secure nonlinear combiner of LFSRs, and is fast enough for many purposes if executed on a present day CPU.

The cryptographic literature offers several pseudorandom generators that follow similar principles as BBS:

The RSA Generator (Shamir) Choose a random modulus m of n bits as a product of two large primes p, q, and an exponent d that is coprime with $(p-1)(q-1)$, furthermore a random initial state $x = x_0$. The state transition is $x \mapsto x^d \bmod m$. Thus, we calculate $x_i = x_{i-1}^d \bmod m$, and output the least significant bit, or the $\lfloor \log_2 \log_2 m \rfloor$ least significant bits. If the RSA generator is not perfect, then there exists an efficient algorithm that breaks the RSA cipher. Since calculating dth powers is more expensive by a factor n than squaring, the cost is even higher for BBS: for a random d the algorithm needs $\mathcal{O}(n^3)$ cycles per bit.

The Index Generator (Blum/Micali) As modulus choose a random large prime p of n bits, and find a primitive root a for p. A primitive root for p is an integer whose powers run through all residue classes $\neq 0 \bmod p$, or in algebraic terms, a generating element of the multiplicative group mod p. Furthermore, choose a random initial state $x = x_0$, coprime with $p-1$. Then calculate $x_i = a^{x_{i-1}} \bmod p$, and output the most significant bit of x_i, or the $\lfloor \log_2 \log_2 p \rfloor$ most significant bits. The perfectness of the index generator relies on the hypothesis that calculating discrete logarithms mod$p$ is hard. The cost per bit also is $\mathcal{O}(n^3)$.

The Elliptic Index Generator (Kaliski) It works like the index generator, but replaces the group of invertible elements of the field $\mathbb{F}_p$ by an elliptic curve over $\mathbb{F}_p$ (such a curve is a finite group in a canonical way).

9.3.12 The Micali-Schnorr Generator

Micali and Schnorr proposed a pseudorandom generator that is a descendant of the RSA generator. Fix an odd number $d \geq 3$. The parameter set is the set of all products m of two primes p and q whose bit length differs by at most 1, and such that d is coprime with $(p-1)(q-1)$. For an n-bit number m let $h(n)$ be an integer $\approx \frac{2n}{d}$. Then the dth power of an $h(n)$-bit number is (approximately) a $2n$-bit number.

In the ith step calculate $z_i = x_{i-1}^d \bmod m$. Take the first $h(n)$ bits as the new state x_i, that is $x_i = \lfloor z_i/2^{n-h(n)} \rfloor$, and output the remaining bits, that is $y_i = z_i \bmod 2^{n-h(n)}$. Thus, the bits of the result z_i are partitioned into two disjoint parts: the new state x_i, and the output y_i. Figure 9.24 illustrates this scheme.

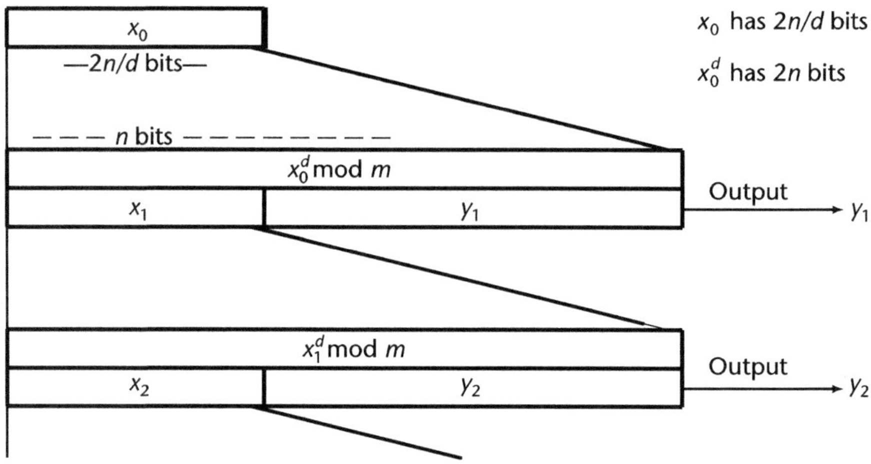

Figure 9.24 Micali-Schnorr generator.

But why may we hope that this random generator is perfect? This depends on the hypothesis: There is no efficient test that distinguishes the uniform distribution on $\{1, \ldots, m-1\}$ from the distribution of $x^d \bmod m$ for uniformly distributed $x \in \{1, \ldots, 2^{h(n)}\}$. If this hypothesis is true, then the Micali-Schnorr generator is perfect. This argument seems tautological, but heuristic considerations show a relation with the security of RSA and with factorization. Anyway we have to concede that this proof of security seems considerably more airy than that for BBS.

How fast do the pseudorandom bits tumble out of the machine? As elementary operations we again count the multiplication of two 64-bit numbers, and the division of a 128-bit number by a 64-bit number with 64-bit quotient. We multiply and divide by the classical algorithms. Remark: Multiplication by fast Fourier transformation (FFT) has an advantage only for much larger numbers. Thus, the product of s (64-bit) words and t words costs $s \cdot t$ elementary operations. The cost of division is the same as the cost of the product of divisor and quotient.

The specific recommendation by the inventors is: $d = 7$, $n = 512$. Today we would choose a larger n. The output of each step consists of 384 bits, withholding 128 bits as the new state. The binary power algorithm for a 128-bit number x with exponent 7 costs several elementary operations:

- x has 128 bits, hence 2 words.
- x^2 has 256 bits, hence 4 words, and costs $2 \cdot 2 = 4$ elementary operations.
- x^3 has 384 bits, hence 6 words, and costs $2 \cdot 4 = 8$ elementary operations.
- x^4 has 512 bits, hence 8 words, and costs $4 \cdot 4 = 16$ elementary operations.
- x^7 has 896 bits, hence 14 words, and costs $6 \cdot 8 = 48$ elementary operations.
- $x^7 \bmod m$ has ≤ 512 bits, and likewise costs $6 \cdot 8 = 48$ elementary operations.

This makes a total of 124 elementary operations; among them only one reduction mod m (for x^7). Our reward consists of 384 pseudorandom bits. Thus, we get about 3 bits per elementary operation, or, by the assumptions in Section 9.3.11, about 6

American billions (European milliards) bits per second. Compared with the BBS generator this amounts to a factor of about 1000.

Parallelization increases the speed virtually without limit: The Micali-Schnorr generator allows complete parallelization. Thus distributing the work among k CPUs brings a profit by the factor k since the CPUs can work independently of each other without need of communication.

9.3.13 Summary and Outlook on Stream Ciphers

Stream ciphers need cryptographically secure random generators. These probably exist, however their security (like the security of almost all ciphers) is mathematically not completely proven.

But implementing a useful stream cipher takes more than just a good random generator:

- Message integrity requires additional means such as a combination with a cryptographic hash function.
- The operational conditions must prevent the reuse of (parts of) the key stream reliably. This means that the key management requires utmost prudence. A possible approach is using a longtime general key that consists of certain inner parameters of the random generator, and use the remaining parameters including the initial state as one-time message key. So you should treat the start value as a part of the key (i.e., keep it secret and use it only once). This was a usual approach with the cipher machines of World War II.

In contrast with block ciphers where we have the accepted standard AES (and the outdated standard DES) for stream ciphers there is no established standard. Closest to standardization is the eSTREAM portfolio developed in a European project from 2004 until 2008. It recommends a bunch of several ciphers [5].

Unfortunately, several proprietary ciphers, mostly stream ciphers developed in back rooms, found their way into security critical applications. Despite the fact that they relied on security through obscurity, they could (easily) be analyzed by reverse engineering, and teared to shreds by cryptologists. Therefore, we finish this chapter with the advice that in an analogous form applies to all parts of cryptography:

> Never trust a random generator whose algorithm is kept secret, or for which no analysis results are publicly available. Statistical analyses are insufficient as security proofs, just as little as gargantuan periods, or a gigantic choice of initial states.

9.4 Table of SageMath Examples in This Chapter

Table 9.14 lists all SageMath scripts used in this chapter.

The files from the third column (including the small libraries `bitciphers.sage` and `FSR.sage`) can be downloaded from: `https://www.cryptool.org/en/documentation/ctbook/sagemath`.

Further samples not listed in this chapter can be found on the same website.

Table 9.14 SageMath Examples in This Chapter

#	*Title*	*File name*
9.1	Solution of a System of Linear Equations Over $\mathbb{Q}$	`chap09_sample010.sage`
9.2	Solution of a System of Boolean Linear Equations	`chap09_sample020.sage`
9.3	A Boolean Function with Truth Table and ANF	`chap09_sample030.sage`
9.4	XOR Encryption in Python/SageMath	`chap09_sample130.sage`
9.5	A Feedback Shift Register (FSR) in Python/SageMath	`bitciphers.sage`
9.6	A (Very Poor) Pseudorandom Sequence in Python/SageMath	`chap09_sample140.sage`
9.7	Defining an LFSR in Python/SageMath	`FSR.sage`
9.8	A Pseudorandom Bit Sequence in Python/SageMath	`chap09_sample150.sage`
9.9	Determining a Coefficient Matrix	`chap09_sample160.sage`
9.10	The Geffe Function	`chap09_sample170.sage`
9.11	Calculating a Period	`chap09_sample180.sage`
9.12	Three LFSRs	`chap09_sample190.sage`
9.13	Three LFSR Sequences	`chap09_sample200.sage`
9.14	The Combined Sequence	`chap09_sample210.sage`
9.15	A (Much Too) Simple Example for BBS	`chap09_sample260.sage`
9.16	Generating a Sequence of BBS Pseudorandom Bits	`chap09_sample270.sage`

References

[1] Menezes, A. J., P. C. van Oorschot, and S. A. Vanstone, *Handbook of Applied Cryptography*, 5th ed., Series on Discrete Mathematics and Its Application, CRC Press, 2001, https://cacr.uwaterloo.ca/hac/.

[2] Oppliger, R., *Contemporary Cryptography, Second Edition*, Norwood, MA: Artech House, 2011, https://www.esecurity.ch/Books/cryptography2e.html.

[3] Oppliger, R., *Cryptography 101: From Theory to Practice, First Edition*, Norwood, MA: Artech House, 2021, https://rolf.esecurity.ch/?page_id=465.

[4] Paar, C., and J. Pelzl, *Understanding Cryptography—A Textbook for Students and Practioners*, Springer, 2009, https://www.crypto-textbook.com/.

[5] Schmehl, K.,*Cryptography and Public Key Infrastructure on the Internet*, in German, the 6th edition was published in 2016. John Wiley, 2003.

[6] Schmeh, K., *Kryptographie—Verfahren, Protokolle, Infrastrukturen*, 6th ed., Written in German, dpunkt.verlag, 2016.

[7] Stamp, M., and R. M. Low, *Applied Cryptanalysis: Breaking Ciphers in the Real World.* Wiley-IEEE Press, 2007, https://www.cs.sjsu.edu/~stamp/crypto/.

[8] Lenstra, A. K., and E. R. Verheul, "Selecting Cryptographic Key Sizes," in *Lecture Notes in ComputerScience 558—PKC*, 2000, pp. 446–465.

[9] Bard, G. V., *Algebraic Cryptanalysis*, Springer, 2009.

[10] Garey, M. R., and D. S. Johnson, *Computers and Intractability*, Freeman, 1979.

[11] Brickenstein, M., "Boolean Gröbner Bases—Theory, Algorithms and Applications;" See also "BRiAl, the successor to PolyBoRi (Polynomials over Boolean Rings)," 2010, https://github.com/BRiAl/BRiAl.

[12] Cox, D., J. Little, and D. O'Shea, *Ideals, Varieties, and Algorithms*, 3rd ed., Springer, 2007.

[13] von zurGathen, J., and J. Gerhard, *Modern Computer Algebra*, Cambridge University Press, 1999.

[14] Segers, A. J. M., "Algebraic Attacks from a Gröbner Basis Perspective," TU Eindhoven, 2004, https://www.win.tue.nl/~henkvt/images/ReportSegersGB2-11-04.pdf.

[15] Lazard, D., "GröbnerBases, Gaussian Elimination and Resolution of Systems of Algebraic Equations," in *Lecture Notes in Computer Science 162*, EUROCAL '83, Springer, 1983, pp. 146–156.

[16] Pommerening, K., *Fourier Analysis of Boolean Maps—A Tutorial*, last revision on August 11, 2014; German equivalent: *Linearitätsmaße für BoolescheAbbildungen*, 2014, https://www.staff.uni-mainz.de/pommeren/Cryptology/Bitblock/Fourier/Fourier.pdf.

[17] Stinson, D. R., *Cryptography—Theory and Practice*, 3rd ed., Chapman & Hall/CRC, 2006.

[18] Gohr, A., "Improving Attacks on Round-Reduced Speck32/64 Using Deep Learning," in *Advancesin Cryptology–CRYPTO 2019: 39th Annual International Cryptology Conference, Santa Barbara, CA, USA, August 18–22, 2019, Proceedings, Part II*, Vol. 11693, Springer. 2019, pp. 150–179, https://eprint.iacr.org/2019/037.pdf.

[19] Gohr, A., G. Leander, and P. Neumann. "An Assessment of Differential-Neural Distinguishers," in *Cryptology ePrint Archive*, 2022, https://eprint.iacr.org/2022/1521.pdf.

[20] BSI, and Leander, et al., *AI-Supported Analysis Methods for Symmetric Cryptography*, Tech. rep. 2022, https://www.bsi.bund.de/SharedDocs/Downloads/EN/BSI/Publications/Studies/Cryptography/KIMSK_Final-Report.html.

[21] Golomb, S. W., *Shift Register Sequences*,revised edition, Aegean Park Press, 1982.

[22] Bajaj, N., *Nikeshbajaj/Linear_Feedback_Shift_Register: 1.0.7*, Version 1.0.7, January 2023, https://zenodo.org/record/7501241.

[23] Pommerening, K., "Cryptanalysis of Nonlinear Shift Registers," in *Cryptologia*, Vol. 40, No. 4, 2016, https://www.tandfonline.com/doi/abs/10.1080/01611194.2015.1055385.

[24] Meier, W., and O. Staffelbach, "Fast Correlation Attacks on Certain Stream Ciphers," in *Journal of Cryptology*, Vol. 1, 1989, pp. 159–176.

CHAPTER 10

Homomorphic Ciphers

Homomorphic ciphers are public-key cryptosystems with special properties. They allow performing certain arithmetic operations on encrypted ciphertexts, without knowing the corresponding plaintexts and without having to decrypt the ciphertexts first. These special properties have led to a huge amount of applications for homomorphic ciphers (e.g., in the domain of cloud computing). A very famous cryptosystem with homomorphic properties is the Paillier (1999) cryptosystem. But also some of the older and well established cryptosystems, such as ElGamal (1985) or RSA (1977), have homomorphic properties. In the meantime, we have reached the third fully homomorphic (FHE) generation of homomorphic methods. The typical representatives are FHEW (2014) and TFHE (2016).

This chapter attempts to introduce this topic in an easy-to-read manner and to elaborate on the ideas of this new crypto concept without going into more detail about the difficult mathematics of the newer methods.

10.1 Origin of the Term Homomorphic

We first clarify the meaning and the origin of the term homomorphic. This term in cryptography is derived from its counterpart in mathematics: In mathematics, a homomorphism is a structure-preserving map between two algebraic structures. In the common sense, this means that a homomorphism $f : X \to Y$ maps the structure of X to the structure of Y. Using an example, this can be easily illustrated: Let $(X, +)$ and $(Y, *)$ two algebraic groups with group operations $+$ and $*$, respectively. A homomorphism $f : X \to Y$ maps any given $x \in X$ to a value $y \in Y$, in a way that it holds (additive homomorphic system):

$$f(x_1 + x_2) = f(x_1) * f(x_2) \tag{10.1}$$

for any two x_1, x_2 in X. This means that for any two values x_1, x_2 it does not matter whether we first compute their sum (group operation of X) and then apply f (this is the left side of (10.1)) or, whether we first apply f to the values x_1, x_2, and then compute their product in Y, thus applying the group operation of Y. Please note that the operations $+$ and $*$ were chosen here only as an example; they always depend on the algebraic group they belong to. Naturally, the same relation holds for homomorphisms between groups with the same group operation.

Example: Let $X = \mathbb{Z}$ be the set of integer values. The set $\mathbb{Z}$ together with the addition operation forms an algebraic group $G_1 = (\mathbb{Z}, +)$. Similarly, the real values $\mathbb{R}$ without the value zero together with the multiplication operation form a group $G_2 = (\mathbb{R}\backslash\{0\}, *)$. The function $f : \mathbb{Z} \to \mathbb{R}\backslash\{0\}, z \to e^z$ is a homomorphism, since for all $z_1, z_2 \in \mathbb{Z}$ it holds: $f(z_1 + z_2) = e^{(z_1+z_2)} = f(z_1) * f(z_2)$. On the contrary, $f : \mathbb{Z} \to \mathbb{R}\backslash\{0\}, z \to z^2$ is an example of a function that is not a homomorphism.

10.2 Decryption Function Is a Homomorphism

In the remainder of this chapter we will consider public-key cryptosystems with a special property, namely that its decryption function is a homomorphism. A public-key cryptosystem with this property will be called homomorphic.

Let us for now assume, the described homomorphism f (see (10.1)) is the decryption function of a known cryptosystem. This means that we can perform certain algebraic operations in the ciphertext space, knowing which effects this will have on their plaintexts. Following the previous given example we have: Y corresponds to the set of cipher texts, X is the set of plaintexts. For two plaintexts x_1, x_2 with corresponding ciphertexts y_1, y_2 it holds (multiplicative homomorphic system):

$$f(y_1 * y_2) = f(y_1) + f(y_2) = x_1 + x_2 \tag{10.2}$$

The equation (10.2) can be interpreted as follows: If we multiply two ciphertexts y_1, y_2 with each other and subsequently decrypt their product, then we will obtain the sum of the originally encrypted values x_1 and x_2. Everybody can—without knowledge of the plaintexts, without having to decrypt, and even without knowing the private decryption key—compute a product of the two ciphertexts and know that upon decryption the owner of the private key will obtain the sum of the two originally encrypted plaintexts.

10.3 Classification of Homomorphic Methods

Homomorphic encryption can be performed by different encryption schemes. The schemes are classified as:

1. Partially homomorphic;
2. Somewhat homomorphic;
3. Leveled fully homomorphic;
4. Fully homomorphic encryption.

Fully homomorphic systems possess both additive and multiplicative homomorphic properties.

The first fully homomorphic method was presented by Craig Gentry in 2009 in his dissertation [1]. It was a lattice-based method (see Chapter 11).

The first cryptosystems that allowed fully-homomorphic encryption required far too much effort to be used in practice. According to their practicability, these FHE methods are divided into (so far four) generations, with only the first three representing a real breakthrough:

1. First-generation FHE (e.g., Gentry-Halevi from 2010). The idea of these systems is to first construct a somewhat homomorphic cryptosystem and then convert it to a fully homomorphic cryptosystem using bootstrapping.

2. Second-generation FHE (e.g., Brakerski-Gentry-Vaikuntanathan (BGV) from 2011). The security of most of these second-generation schemes is based on the hardness of the (ring) learning with errors (RLWE) problem.
3. Third-generation FHE (e.g., Gentry-Sahai-Waters (GSW) from 2013 [2]), FHEW from 2014 [3], and TFHE from 2016 [4]. They drastically reduced the bootstrapping time.
4. Fourth-generation FHE (e.g., Cheon-Kim-Kim-Song (CKKS) ab 2016 [5]).

Since 2017, the Homomorphic Encryption Standardization consortium (`https://homomorphicencryption.org/`) tries to standardize homomorphic encryption. Curated lists of open-source FHE libraries can be found in [6–8].

A very good overview including classification, not only about fully homomorphic encryption but also about the current state of techniques such as secure multiparty computation (MPC), is offered by the 100-page report "Encrypted Computing Compass," prepared by KIT and CISPA at the end of 2022 and published by the German cyber agency in 2023 [9].

Note: The cryptosystems based on lattice problems, which are used for post-quantum cryptography, are usually fully homomorphic.

10.4 Examples of Homomorphic Pre-FHE Ciphers

Here, the homomorphic properties of the three pre-FHE cryptosystems Paillier, RSA, and ElGamal are described as an introduction.

10.4.1 Paillier Cryptosystem

One of the oldest cryptosystems with homomorphic properties is the one by Paillier [10]. First we will see how the Paillier key generation process works. After that, we will show that the Paillier cryptosystem indeed has homomorphic properties.

10.4.1.1 Key Generation

First, we generate two random prime numbers p, q in a way that their product $n = pq$ forms a valid RSA modulus. Here n should have a bit length that corresponds to the current security standard for RSA modules (as of January 2022, 2048 bits). Using the prime values p and q, we can compute the value $\lambda = lcm(p-1, q-1)$. lcm here denotes the least common multiple. The RSA modulus n will now be the public key, while the private key is the value λ.

10.4.1.2 Encryption

Let m be the message which will be encrypted, where m is taken from the plaintext space $\mathbb{Z}_n$. For each encryption, we first choose a random element r from the plaintext space $\mathbb{Z}_n$. Subsequently, using the public key, we compute the ciphertext n as:

$$c = E(m, r) = (n+1)^m * r^n \bmod n^2$$

10.4.1.3 Decryption

Given the private key λ and a ciphertext $c \in \mathbb{Z}^*_{n^2}$, we first compute $S = c^\lambda \bmod n^2$ and subsequently $T = \phi(n)^{(-1)} \bmod n^2$, where ϕ denotes the Euler function.

Finally, we compute the plaintext $m = D(c) = (S-1)/n \cdot T \bmod n$.

10.4.1.4 Homomorphic Property

To prove the homomorphic property of the Paillier cryptosystem, we use E to denote the encryption and D to denote the decryption function. For simplicity, we set in the following $g := n+1$. From any two plaintexts m_1, m_2 and random values r_1, r_2 we obtain ciphertexts c_1, c_2 as

$$c_1 = g^{m_1} \cdot {r_1}^n \bmod n^2 \text{ and } c_2 = g^{m_2} \cdot {r_2}^n \bmod n^2,$$

respectively. Now it is easy to see that for the product $c_3 = c_1 \cdot c_2$ it holds

$$\begin{aligned} c_3 &= (g^{m_1} \cdot {r_1}^n \bmod n^2) \cdot (g^{m_2} \cdot {r_2}^n \bmod n^2) = g^{m_1+m_2} \cdot (r_1 \cdot r_2)^n \bmod n^2 \\ &= E(m_1 + m_2, r_1 \cdot r_2) \end{aligned}$$

Thus, the product of two given ciphertexts is in fact a valid ciphertext, namely the encryption of the sum of the originally encrypted messages. Now it is straightforward to see that the decryption function is a homomorphism. Given two plaintexts m_1, m_2 it holds

$$\begin{aligned} D(E(m_1, r_1) \cdot E(m_2, r_2)) &= D(E(m_1 + m_2, r_1 r_2)) = m_1 + m_2 \\ &= D(E(m_1, r_1)) + D(E(m_2, r_2)) \end{aligned}$$

10.4.2 Other Cryptosystems

Older public-key cryptosystems can also have homomorphic properties. Both the ElGamal cryptosystem and RSA constitute famous examples. We will show their homomorphic properties by means of some easy examples.

10.4.2.1 RSA

Let (e, n) be the public RSA key (e the public encryption exponent, n the RSA modulus). For any two messages m_1, m_2 we obtain the ciphertexts $c_1 = {m_1}^e \bmod n$ and $c_2 = {m_2}^e \bmod n$. For the product of these two ciphertexts, it holds: $c_1 \cdot c_2 = {m_1}^e \cdot {m_2}^e \bmod n = (m_1 \cdot m_2)^e \bmod n$. Thus, we obtain an encryption of the product of the two messages m_1 and m_2. As it is straightforward to see, this property holds for any two plaintexts m_1, m_2 and similar as for Paillier, the decryption function is a homomorphism. As we have seen here, RSA is an example for a homomorphism, where both groups have the same group operation.

10.4.2.2 ElGamal

Similar to RSA, we can also show the homomorphic properties of the ElGamal cryptosystem. Let (p, g, K) be the public key while the private key is k (thus, it holds

$g^k \bmod p = K$). For any two messages m_1, m_2 and random values r, s we obtain ciphertexts $(R, c_1) = (K^r \bmod p, m_1 \cdot g^r \bmod p)$ and $(S, c_2) = (K^s \bmod p, m_2 \cdot g^s \bmod p)$. As for RSA, we verify that their product $(R \cdot S, c_1 \cdot c_2)$ is an encryption of $m_1 \cdot m_2$. Again, it is straightforward to see that the decryption function is a homomorphism.

10.5 Applications

The homomorphic property can be used to add two encrypted values or to multiply any value under encryption with a known constant (note that the multiplication corresponds to the repeated application of the addition operation). This makes homomorphic ciphers important and easy to use base primitives in cryptographic applications.

1. One of these applications is *electronic voting*. Electronic voting allows many voters to submit their ballots in an encrypted form. This is important in situations where the voters cannot come together to the same location. This happens, for example, if the voters can only communicate over the internet via email. If the voting behavior of the single parties should remain secret, then the use of homomorphic ciphers is a good solution to this problem. The main principle of electronic voting using homomorphic ciphers is as follows.
   - All voters (on the left in Figure 10.1) encrypt the value 1 if they opt positive and the value 0, if opposed to the decision.
   - Using the homomorphic property, one can compute the sum of all encrypted ballots. Since this happens on encrypted values, the voting behavior of all participants remains secret.

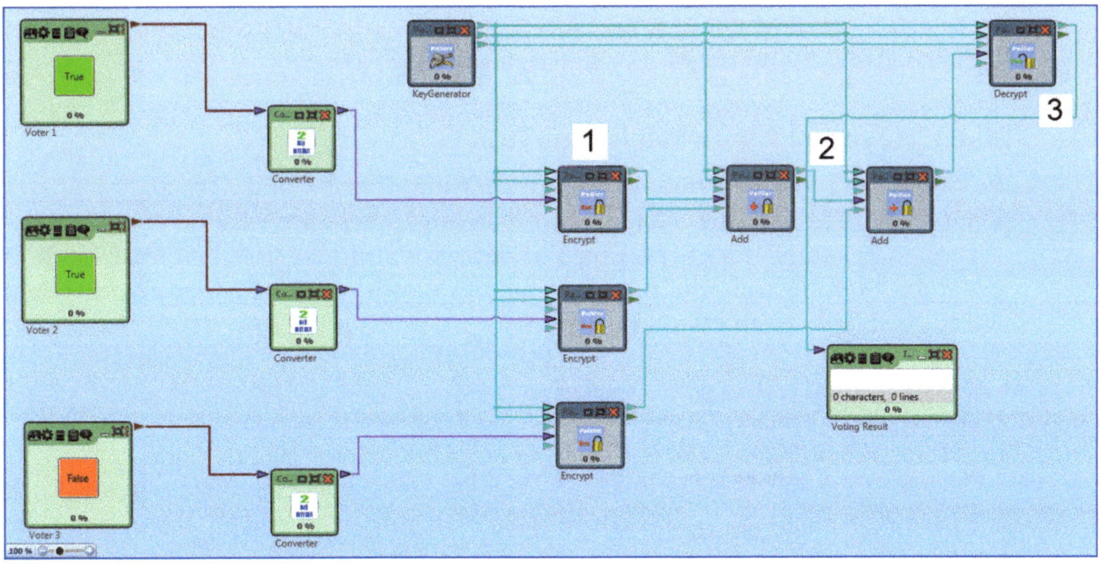

Figure 10.1 Voting example for Paillier.

- At the end, the result of the election is determined and published; this happens by decrypting the sum that was computed using the homomorphic property.

2. A second application of homomorphic ciphers is *secure multiparty computation*. Here, two or more parties can compute any commonly known function. Each of the parties provides one or more of the inputs for the function to be computed. The goal of the secure computations is to keep all private inputs secret, while only the result of the function is revealed. The use of homomorphic encryption helps to perform these computations on encrypted data. However, since the Paillier encryption only allows one to compute additions of encrypted values (e.g., no multiplications can be performed), a number of additional methods and techniques have to be applied. The Wikipedia page [11] offers a good start for reading more about this topic and more advanced techniques for secure multiparty computation.
3. Furthermore, it is expected that homomorphic encryption will provide great advantages in the area of cloud computing. Using fully-homomorphic encryption [12] it is possible to run applications on external servers only on encrypted data. For this, one needs to be able to perform both arithmetic operations, the addition, and the multiplication on encrypted data (in contrast to Paillier encryption, which only allows one to perform additions). Such a cryptosystem was first presented in 2009 [13] and since then continuously improved [12].

10.6 Homomorphic Methods in CrypTool

Homomorphic methods are contained in the three CrypTool variants CT2, JCT, and CTO. Further implementations in Python or SageMath can be found on the internet. A curated list of homomorphic encryption libraries can be found at `https://github.com/jonaschn/awesome-he`, for example the open-source libraries HEAAN, HElib, SEAL, TFHE, and PALISADE.

10.6.1 CrypTool 2 with Paillier and DGK

In CrypTool 2 you can find an implementation of the Paillier cryptosystem (see Figure 10.2). Among the ready-to-run templates, there are methods for key generation, an example for encryption and decryption of text with Paillier, as well as examples that illustrate the homomorphic properties of the Paillier cryptosystem (addition, blinding, and voting).[1]

1. Here are the paths to the corresponding templates in CT2:
   - `CT2 Cryptography ▷ Modern ▷ Asymmetric ▷ Blind Signature with Paillier;`
   - `CT2 Cryptography ▷ Modern ▷ Asymmetric ▷ Paillier Key Generator;`
   - `CT2 Cryptography ▷ Modern ▷ Asymmetric ▷ Paillier Cipher (Number Input);`
   - `CT2 Cryptography ▷ Modern ▷ Asymmetric ▷ Paillier Cipher (Text Input);`
   - `CT2 Cryptography ▷ Modern ▷ Asymmetric ▷ Paillier Cipher with Addition;`
   - `CT2 Cryptography ▷ Modern ▷ Asymmetric ▷ Paillier Cipher with Blinding;`
   - `CT2 Cryptography ▷ Modern ▷ Asymmetric ▷ Paillier Cipher (Voting).`

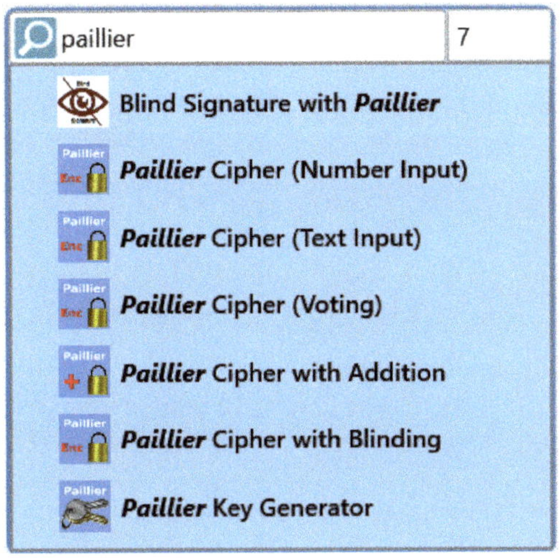

Figure 10.2 Templates for the Paillier cryptosystem in CrypTool 2 (CT2).

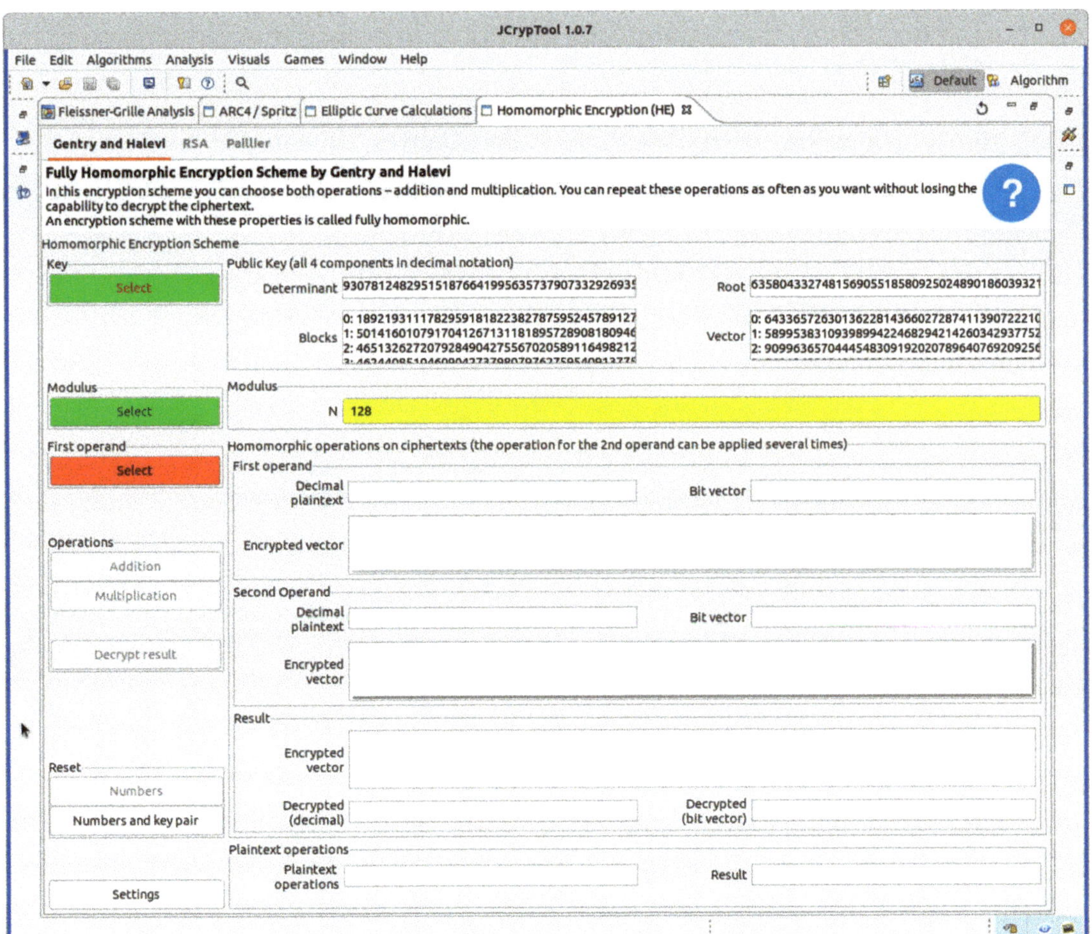

Figure 10.3 Visualization of homomorphic properties in JCrypTool (JCT).

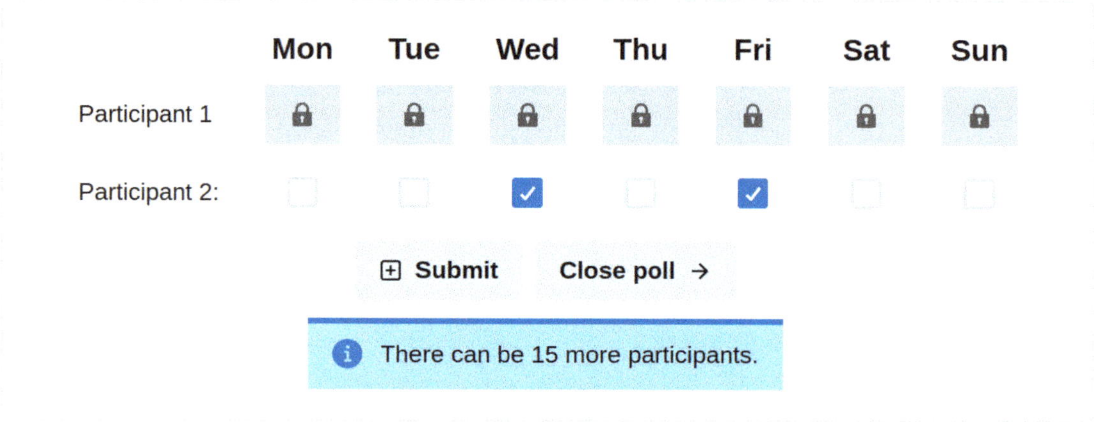

Figure 10.4 Poll demo in CTO using homomorphic encryption.

In 2008, Damgard et al. suggested a homomorphic encryption scheme (DGK) [14] that can be used as the basis of efficient and general secure MPC. They used it in a protocol for secure comparison of integers in order to improve the security of online auctions with this protocol.[2]

10.6.2 JCrypTool with RSA, Paillier, and Gentry/Halevi

In JCrypTool there is an implementation (see Figure 10.3), that visualizes the homomorphic properties of various cryptosystem:

- For RSA and Paillier it shows that, for each of them, either one operation of one type (multiplications for RSA and additions for Paillier) can be performed on encrypted values.
- For the fully-homomorphic cryptosystem by Gentry-Halevi it is possible to perform both multiplications and additions on encrypted values.[3]

10.6.3 Poll Demo in CTO Using Homomorphic Encryption

The poll demo in CTO visualizes, how a Doodle-like application can be built with a homomorphic encryption algorithm such that security by design is fulfilled. This implementation makes use of the fully homomorphic BFV scheme with the help of the library node-seal. When conducting a poll online this way, the inputs, calculations, and results are hidden from the aggregation server. Figure 10.4[4] shows what participant 2 can see after participant 1 entered his data.

References

[1] Gentry, C., *A Fully Homomorphic Encryption Scheme*, 2009, https://crypto.stanford.edu/craig/craig-thesis.pdf.

2. `CT2 ▷ Cryptography ▷ Modern ▷ Asymmetric ▷ DGK Cipher (Text Input).`
3. `JCT Default Perspective ▷ Visuals ▷ Homomorphic Encryption (HE).`
4. See "Poll Demo Using Homomorphic Encryption," `https://www.cryptool.org/en/cto/fhe-poll` and the corresponding thesis [15].

[2] Gentry, C., A. Sahai, and B. Waters, "Homomorphic Encryption from Learning with Errors: Conceptually-Simpler, Asymptotically-Faster, Attribute-Based," in *Cryptology ePrint Archive*, 2013, https://eprint.iacr.org/2013/340.

[3] Ducas, L., and D. Micciancio. "FHEW: Bootstrapping Homomorphic Encryption in Less Than a Second," in *Advances in Cryptology–EUROCRYPT 2015: 34th Annual International Conference on theTheory and Applications of Cryptographic Techniques, Sofia, Bulgaria, April 26-30, 2015, Proceedings, Part I 34*, Springer, 2015, pp. 617–640.

[4] Chillotti, I., et al., "TFHE: Fast Fully Homomorphic Encryption Over the Torus," *Journal of Cryptology*, Vol. 33, No. 1, 2020, pp. 34–91.

[5] Cheon, J. H., et al., "Homomorphic Encryption for Arithmetic of Approximate Numbers," in *International Conference on the Theory and Application of Cryptology and Information Security*, Springer, 2017, pp. 409–437.

[6] Wikipedia, *Homomorphic Encryption*, https://en.wikipedia.org/wiki/Homomorphic_encryption.

[7] Schneider, J., *Awesome Homomorphic Encryption*, https://github.com/jonaschn/awesome-he.

[8] FHE.org, *Libraries*, https://fhe.org/resources/libraries.

[9] Döttling, N., et al., *Encrypted Computing Compass, v 1.0*, Tech. rep., November 2022, https://www.cyberagentur.de/encrypted-computing-compass/.

[10] Paillier, P., "Public-Key Cryptosystems Based on Composite Degree Residuosity Classes," in *Advancesin Cryptology—EUROCRYPT '99*, 1999.

[11] Wikipedia contributors, *Secure Multiparty Computation*, https://en.wikipedia.org/wiki/Secure_multi-party_computation.

[12] Wikipedia contributors, *Homomorphic Encryption* and *Homomorphismus*, Wikipedia, The Free Encyclopedia, https://en.wikipedia.org/wiki/Homomorphic_encryption; https://de.wikipedia.org/wiki/Homomorphismus.

[13] Gentry. C., "Fully Homomorphic Encryption Using Ideal Lattices," in *41st ACM Symposium on Theory of Computing (STOC)*, 2009.

[14] Damgard, I., M. Geisler, and M. Kroigard. "Homomorphic Encryption and Secure Comparison," in: *International Journal of Applied Cryptography*, Vol. 1, No. 1, 2008, pp. 22–31.

[15] Bastian, M., *The State of Homomorphic Encryption*, 2023, https://www.cryptool.org/assets/ctp/documents/BA_Heep.pdf (visited on 08/09/2023).

CHAPTER 11

Lightweight Introduction to Lattices

In this chapter, our goal is to cover the basic theory behind lattices in a lightweight fashion. The theory covered is accompanied by many practical examples, SageMath code, and cryptographic challenges.

Sections 11.1 through 11.7 introduce the notation and methods needed to work with and understand lattices (this makes up about a third of this chapter). Sections 11.8 and 11.9 cover lattices in more detail and their application to attack RSA. Section 11.10 is intended as a deeper look, providing some algorithms for lattice basis reduction and their use to break cryptosystems. Section 11.12 contains screenshots of where lattice algorithms can be found in the CrypTool programs.

11.1 Preliminaries

You are not required to have an advanced background in any mathematical domain or programming language. Nevertheless, expanding your knowledge and learning new mathematical concepts and programming techniques will give you a great boost towards your goal as a future expert in cryptology. This chapter is self-contained—you are not required to read the previous chapters in the book. The examples and the practical scenarios are implemented in SageMath (a computer-algebra system that uses Python as scripting language; see `https://www.cryptool.org/en/documentation/ctbook/`). We believe that the code in this chapter is very intuitive and even if you don't have any experience with Python syntax you will understand the idea behind it.

11.2 Equations

Equations help us to mathematically describe a relationship between some objects. Equations are mathematical statements that can be correct or incorrect. These statements claim that the values of two mathematical expressions (on the left and on the right of the sign $=$) are equal.

The equations can be trivial (without an unknown variable involved)

$$
\begin{aligned}
0 &= 0 \\
1 &= 1 \\
1+1 &= 2 \\
1.\overline{9} &= 2
\end{aligned}
$$

or with one or more variables (also called indeterminates), leading to a nonempty solution set or no solution (unsolvable). The solution set is the set of assignments to these variables such that all statements are true. The solution set also depends on boundary conditions; that is, which values the variables can take, or in other words, over which basic set the variables are considered.

$$x + x = 10$$
$$x + y = 10$$
$$x + y = z$$
$$x_1 + x_2 + x_3 + \cdots + x_{10} = z$$

In some cases, the solution is straightforward and unique, but in some other cases we have a set of possible solutions. The domain is the set of input values for which the equation is defined. For example, the equation $x + 1 = -10$ has no solution over $\mathbb{N}$, but has one solution over $\mathbb{Z}$. From now until the end of this chapter, we will work only with the set of integers $\mathbb{Z}$ as the domain.

SageMath makes it easy to define variables. The following declaration defines the special symbol x as a variable:

```
sage: x = var('x')
```

Here is an example of a polynomial: If a coefficient is explicitly written down (only those different from ± 1, such as 5 in the following listing), the multiplication operator $*$ must be used between the coefficient and the variable term. The symbols $**$ mean "to the power of" in both SageMath and Python (i.e., exponentiation).

```
sage: pol = x + 5*x**2 + x**3
```

Now, we are ready to construct our equation. Let's say that we want to find the solution to the following equation $x + x^2 + x^3 = 100$. First, we need to define our left side of the equation. Using SageMath, the declaration is straightforward. We will refer to the left side of our equation as *leq*.

For our example we are using a term with no explicitly written coefficients (i.e., the coefficients are 1 or 0):

```
sage: leq = x + x**2 + x**3
```

We are ready to solve the equation and find the solutions.

```
sage: eq_sol = solve(leq==100, x)
```

The SageMath command `solve()` tries to find all x for which $x+x^2+x^3 = 100$. If you run this on your computer, you will see a list of possible solutions that don't look like integers. This is because we have defined x as a variable without any restrictions on its domain. So, let's define the symbol x as a variable in the domain of integers, ZZ in SageMath. Integers in SageMath can be further restricted to Boolean (true or false) or to integers modulo n (`IntegerModRing(n)` or `GF(n)` if n is prime).

```
sage: x = var('x', domain=ZZ)
```

Now, when we try to solve the equation, we get as set of solutions the empty set ([] in Python defines an empty list). This means that there is no such x that satisfies the defined equation.

Instead of using **solve()**, it is easier and more stable to work with the polynomial ring: **x = polygen(ZZ)**. This is equivalent to `polynomialRing(ZZ, 'x').gen()`.

```
sage: x = polygen(ZZ)
sage: p = x + x^2 + x^3
sage: (p - 10^2).roots(ZZ, multiplicities=False)
[]
sage: (p - 14).roots(ZZ, multiplicities=False)
[2]
```

Here is a reasoning that there is no such integer x that satisfies the defined equation `leq==100`. Let's see the values of the polynomial function $x \mapsto x+x^2+x^3$ for consecutive values of x:

```
sage: for i in range(-6,6):
....:         print(leq(x=i), "for", "x =", i)
-186  for x = -6
-105  for x = -5
 -52  for x = -4
 -21  for x = -3
  -6  for x = -2
  -1  for x = -1
   0  for x = 0
   3  for x = 1
  14  for x = 2
  39  for x = 3
  84  for x = 4
 155  for x = 5
 258  for x = 6
```

We can see two characteristics of our equation. First, the left-hand side of the equation becomes larger as the variable x increases. Second, the solution to our equation is a noninteger between 4 and 5.

Note: Normally, the term on the left-hand side also includes the number on the right-hand side if it's not 0. So the usual way to write this equation is as follows:

$$x^3 + x^2 + x - 100 = 0.$$

Challenge 11.1: Equations

We have found this strange list of (independent) equations. Can you find the hidden message? Think of the values found for the variable of each equation as the ASCII value of a character. The correct ASCII values form the correct word in the same order as the 8 equations that appear here. A copy-and-paste version of this system of equations is also available, at

Challenge 11.1 (continued)

https://www.cryptool.org/en/documentation/ctbook/sagemath

$$x_0^4 - 150x_0^3 + 4389\,x_0^2 - 43000\,x_0 + 131100 = 0$$

$$\begin{aligned}&x_1^{10} - 177\,x_1^9 + 9143\,x_1^8 - 228909\,x_1^7 + 3264597\,x_1^6 - 28298835\,x_1^5 + \\ &\quad +152170893\,x_1^4 - 502513551\,x_1^3 + 974729862\,x_1^2 - 995312448\,x_1 + 396179424 = 0\end{aligned}$$

$$\begin{aligned}&x_2^{10} - 196\,x_2^9 + 12537\,x_2^8 - 397764\,x_2^7 + 7189071\,x_2^6 - 77789724\,x_2^5 + \\ &+506733203\,x_2^4 - 1941451916\,x_2^3 + 4165661988\,x_2^2 - 4501832400\,x_2 + 1841875200 = 0\end{aligned}$$

$$x_3^5 - 153\,x_3^4 + 5317\,x_3^3 - 77199\,x_3^2 + 510274\,x_3 - 1269840 = 0$$

$$\begin{aligned}&x_4^8 - 194\,x_4^7 + 11791\,x_4^6 - 352754\,x_4^5 + 6011644\,x_4^4 - \\ &\qquad -61295576\,x_4^3 + 370272864\,x_4^2 - 1222050816\,x_4 + 1696757760 = 0\end{aligned}$$

$$x_5^6 - 169\,x_5^5 + 7702\,x_5^4 - 153082\,x_5^3 + 1477573\,x_5^2 - 6672349\,x_5 + 11042724 = 0$$

$$\begin{aligned}&x_6^8 - 202\,x_6^7 + 12936\,x_6^6 - 406082\,x_6^5 + 7170059\,x_6^4 - \\ &\qquad -74124708\,x_6^3 + 439747164\,x_6^2 - 1365683328\,x_6 + 1701311040 = 0\end{aligned}$$

$$\begin{aligned}&x_7^9 - 206\,x_7^8 + 13919\,x_7^7 - 467924\,x_7^6 + 8975099\,x_7^5 - 102829454\,x_7^4 + \\ &\qquad +699732361\,x_7^3 - 2673468816\,x_7^2 + 4956440220\,x_7 - 2888395200 = 0\end{aligned}$$

11.3 Systems of Linear Equations

We have already introduced the concepts of variables, equations, and the domain of an equation. We have shown how to declare variables in SageMath and how to automatically find solutions to single-variable equations using `solve()`. What if we have two different variables in our equation? Let's take the following equation as an example: $x + y = 10$. Again we will try to solve this equation using SageMath. This time we need the tuple (x, y) as the solution of `solve()`.

```
sage: x = var('x', domain=ZZ)
sage: y = var('y', domain=ZZ)
sage: solve(x+y==10, (x,y))
(t_0, -t_0 + 10)
```

We get as a solution $x = t_0$ and $y = -t_0+10$ and indeed $x+y = t_0+(-t_0+10) = t_0 - t_0 + 10 = 10$. The notation with t_0 is used by SageMath to show us that there are infinitely many integer solutions to the given equation. In correct mathematical notation, the solution is the following so-called parameterized set. The parameter t_0 is used by SageMath to indicate that there is only one degree of freedom, counting the first degree of freedom as number zero as usual. Outside SageMath we don't use this index and write just t:

$$\mathbb{L} = \{(x, y) \in \mathbb{Z}^2 \,|\, x = t,\ y = -t + 10,\ t \in \mathbb{Z}\} \text{ or } \mathbb{L} = \{(t, -t + 10) \in \mathbb{Z}^2 \,|\, t \in \mathbb{Z}\}$$

What if we have other constraints on x and y? For example, what if we know that they are equal? Equality defines another equation that is related to the first one, so we can form a system of equations (in a system of equations, the individual equations are not independent of each other):

$$\begin{cases} x + y = 10 \\ x = y \end{cases}$$

Let's solve this system of equations. We can easily organize all the equations from this two-equation system using a list array.

```
sage: x,y = var('x y', domain=ZZ)
sage: solve([x+y==10,x==y], (x,y))
[[x == 5, y == 5]]
```

As we can see, we have only the one solution $(x, y) = (5, 5)$.

A rich collection of mathematical problems can be solved by using systems of linear equations. For example, let's take the simple puzzle in Figure 11.1 and solve it using SageMath.

As usual, each row consists of three items and their total price. Usually, the goal in such puzzles is to find the price of each individual item. We have three different items. Let's define the price of each pencil as x, the price of each computer display as y, and the price of each bundle of servers as z. From the previous declarations we can write down the following system of linear equations:

$$\begin{cases} 2x + y = 15 \\ x + y + z = 20 \\ 3z = 30 \end{cases}$$

We can also solve this puzzle using only pen and paper. The last equation gives the value of $z = 10$. If we eliminate the variable z by replacing its value in the previous equations, we reduce the system to system of two unknown variables:

$$\begin{cases} 2x + y = 15 \\ x + y = 10 \\ \quad (z = 10 \text{ not unknown anymore}) \end{cases}$$

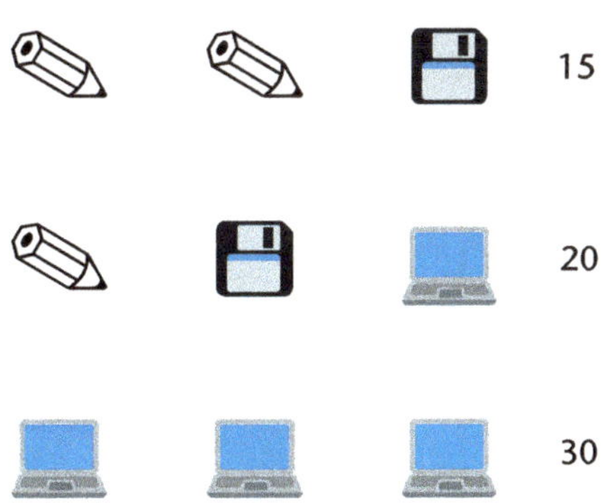

Figure 11.1 Visual puzzle.

We can now subtract the second equation from the first to get:

$$2x + y - (x + y) = 15 - 10 = 5$$
$$\Rightarrow \quad x = 5$$

Substituting this into the second equation $x + y = 10$, we get $y = 5$. So we have the following solution to the puzzle:

$$\begin{cases} x = 5 \\ y = 5 \\ z = 10 \end{cases}$$

Now let's try to solve the same puzzle using SageMath.

```
sage: x,y,z = var('x y z', domain=ZZ)
sage: solve([x + x + y == 15, x+y+z == 20, z+z+z == 30], (▸
  ▸x,y,z))
[[x == 5, y == 5, z == 10]]
```

Challenge 11.2: System of Linear Equations as a Picture

Can you find the hidden message in the picture puzzle in Figure 11.2? Each symbol represents a distinct decimal digit. There is a balance so that each left side is equal to the corresponding right side. Automate the process using SageMath. Hint: ASCII (American Standard Code for Information Interchange) is involved.

In the next sections, we will introduce the definition of matrices, which will help us describe a given system of linear equations in a much more compact way.

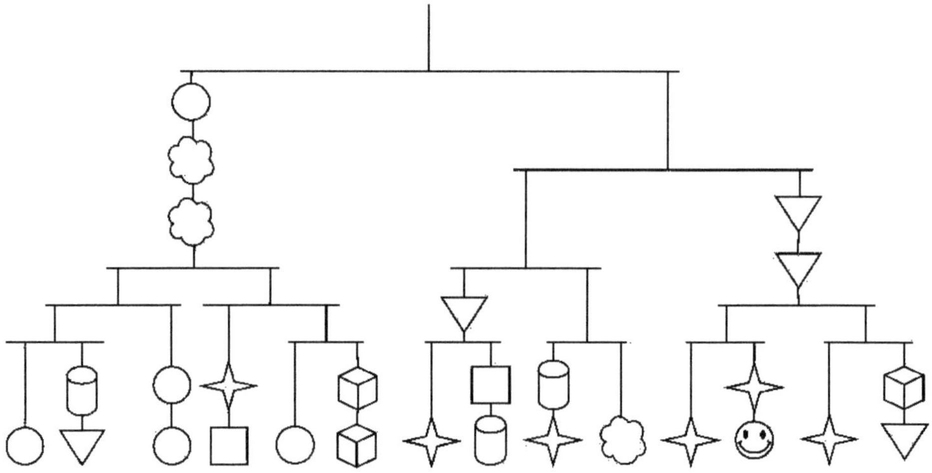

Figure 11.2 Puzzle challenge (picture created by the author).

11.4 Matrices

Is there a more convenient way to write large systems of linear equations? We will introduce this way by using augmented matrices. You can think of a matrix as a rectangular or square array of numbers. The numbers are arranged in rows and columns. For example, let's analyze the following matrix:

$$M = \begin{bmatrix} 1 & 2 & 3 \\ 3 & 4 & 5 \end{bmatrix}$$

We have two rows and three columns, and a total of six elements. We define an element as a_{ij} if we want to emphasize that the element is located in the ith row and jth column. For example, $a_{11} = 1$, $a_{13} = 3$, $a_{22} = 4$.

Note that here we are using indexing starting from 1, as usual in mathematics. However, later in the SageMath examples, the index of row and column starts from 0 (as usual in computer languages such as C or Python).

In the following system of linear equations, the independent variables are labeled a, b, c, d, e, and f.

$$\begin{cases} 6a + 7b + 11c + 18d + 4e + 7f = 5 \\ 8a + 14b + 2c + 13d + 2e + f = 19 \\ a + b + 3c + 4d + 4e + 7f = 15 \\ 3a + 4b + c + d + 14e + 17f = 1 \\ 5a + 5b + 2c + 2d + 2e + 6f = 2 \\ 11a + 17b + c + d + e + f = 9 \end{cases}$$

We can easily write this system of linear equations as a matrix. Let's write all the coefficients before the variable a in the first column of our new matrix, all the coefficients before the variable b in the second column, and so on. This is the coefficient matrix.

The right side of each equation forms another column—the last one. For clarity, we will separate it from the other columns with a vertical line. We call such a matrix an augmented matrix.

$$\left(\begin{array}{cccccc|c} 6 & 7 & 11 & 18 & 4 & 7 & 5 \\ 8 & 14 & 2 & 13 & 2 & 1 & 19 \\ 1 & 1 & 3 & 4 & 4 & 7 & 15 \\ 3 & 4 & 1 & 1 & 14 & 17 & 1 \\ 5 & 5 & 2 & 2 & 2 & 6 & 2 \\ 11 & 17 & 1 & 1 & 1 & 1 & 9 \end{array}\right)$$

Let's analyze the behavior of a system of linear equations. We can make the following observations:

- Swapping the positions of two equations doesn't affect the solution of the system of linear equations.

- Multiplying the equation by a nonzero number doesn't affect the solution of the system of linear equations.
- Adding a randomly chosen equation to another randomly chosen equation doesn't affect the solution of the system of the linear equations.

We can easily transform these properties into properties in the augmented matrix. Furthermore, these properties allow us to build a complete automatic system for finding solutions to a given augmented matrix. In linear algebra, Gaussian elimination (also known as row reduction) is an algorithm for solving systems of linear equations. It is usually understood as a sequence of operations performed on the corresponding augmented matrix. When all leading coefficients (the left-most nonzero entry in each row) are one, and each column containing a leading coefficient has zeros elsewhere, the matrix is said to be in reduced row echelon form.

Let's apply Gaussian elimination to the following system of linear equations:

$$\begin{cases} 4x + 8y + 3z = 10 \\ 5x + 6y + 2z = 15 \\ 9x + 5y + z = 20 \end{cases}$$

We can easily transform this system of linear equations into an augmented matrix.

$$\left(\begin{array}{ccc|c} 4 & 8 & 3 & 10 \\ 5 & 6 & 2 & 15 \\ 9 & 5 & 1 & 20 \end{array}\right)$$

Then we begin to transform the matrix into row echelon form. First, we divide the first row by 4.

$$\left(\begin{array}{ccc|c} 4 & 8 & 3 & 10 \\ 5 & 6 & 2 & 15 \\ 9 & 5 & 1 & 20 \end{array}\right) \to \left(\begin{array}{ccc|c} 1 & 2 & 0.75 & 2.5 \\ 5 & 6 & 2 & 15 \\ 9 & 5 & 1 & 20 \end{array}\right)$$

The reason for dividing the first row by 4 is simple—we need the first element of the first row to be equal to 1, which allows us to multiply the first row by 5 and 9 and to subtract it from the second and third rows, respectively. Let's recall that we are trying to transform the augmented matrix into the reduced row echelon form. Now, let's apply the previous observations.

$$\left(\begin{array}{ccc|c} 1 & 2 & 0.75 & 2.5 \\ 5 & 6 & 2 & 15 \\ 9 & 5 & 1 & 20 \end{array}\right) \to \left(\begin{array}{ccc|c} 1 & 2 & 0.75 & 2.5 \\ 0 & -4 & -1.75 & 2.5 \\ 9 & 5 & 1 & 20 \end{array}\right) \to \left(\begin{array}{ccc|c} 1 & 2 & 0.75 & 2.5 \\ 0 & -4 & -1.75 & 2.5 \\ 0 & -13 & -5.75 & -2.5 \end{array}\right)$$

We now divide the second row by −4. This will transform the second element of the second row to 1 and allow us to continue with our strategy of reducing the augmented matrix to the required row echelon form.

$$\left(\begin{array}{ccc|c} 1 & 2 & 0.75 & 2.5 \\ 0 & -4 & -1.75 & 2.5 \\ 0 & -13 & -5.75 & -2.5 \end{array}\right) \rightarrow \left(\begin{array}{ccc|c} 1 & 2 & 0.75 & 2.5 \\ 0 & 1 & 0.4375 & -0.625 \\ 0 & -13 & -5.75 & -2.5 \end{array}\right)$$

And again, following the previous strategy we applied to the first row, we multiply the second row by 2 and subtract it from the first row. Immediately after this operation, we multiply the second row by 13 and add it to the last row.

$$\left(\begin{array}{ccc|c} 1 & 2 & 0.75 & 2.5 \\ 0 & 1 & 0.4375 & -0.625 \\ 0 & -13 & -5.75 & -2.5 \end{array}\right) \rightarrow \left(\begin{array}{ccc|c} 1 & 0 & -0.125 & 3.75 \\ 0 & 1 & 0.4375 & -0.625 \\ 0 & -13 & -5.75 & -2.5 \end{array}\right)$$

$$\left(\begin{array}{ccc|c} 1 & 0 & -0.125 & 3.75 \\ 0 & 1 & 0.4375 & -0.625 \\ 0 & -13 & -5.75 & -2.5 \end{array}\right) \rightarrow \left(\begin{array}{ccc|c} 1 & 0 & -0.125 & 3.75 \\ 0 & 1 & 0.4375 & -0.625 \\ 0 & 0 & -0.0625 & -10.625 \end{array}\right)$$

We are almost done. Now we normalize the last row by dividing it by -0.0625.

$$\left(\begin{array}{ccc|c} 1 & 0 & -0.125 & 3.75 \\ 0 & 1 & 0.4375 & -0.625 \\ 0 & 0 & -0.0625 & -10.625 \end{array}\right) \rightarrow \left(\begin{array}{ccc|c} 1 & 0 & -0.125 & 3.75 \\ 0 & 1 & 0.4375 & -0.625 \\ 0 & 0 & 1 & 170 \end{array}\right)$$

We follow the same steps as in the previous operations. First, we multiply the last row by 0.125 and add it to the first row. Then, we multiply the last row by 0.4375 and subtract it from the second row.

$$\left(\begin{array}{ccc|c} 1 & 0 & -0.125 & 3.75 \\ 0 & 1 & 0.4375 & -0.625 \\ 0 & 0 & 1 & 170 \end{array}\right) \rightarrow \left(\begin{array}{ccc|c} 1 & 0 & 0 & 25 \\ 0 & 1 & 0.4375 & -0.625 \\ 0 & 0 & 1 & 170 \end{array}\right)$$

$$\left(\begin{array}{ccc|c} 1 & 0 & 0 & 25 \\ 0 & 1 & 0.4375 & -0.625 \\ 0 & 0 & 1 & 170 \end{array}\right) \rightarrow \left(\begin{array}{ccc|c} 1 & 0 & 0 & 25 \\ 0 & 1 & 0 & -75 \\ 0 & 0 & 1 & 170 \end{array}\right)$$

We have reduced the augmented matrix to the reduced row echelon form. Let's transform the problem back into a system of linear equations.

$$\begin{cases} 1 \cdot x + 0 \cdot y + 0 \cdot z = x = 25 \\ 0 \cdot x + 1 \cdot y + 0 \cdot z = y = -75 \\ 0 \cdot x + 0 \cdot y + 1 \cdot z = z = 170 \end{cases}$$

We now have a tool (algorithm) for solving a system of linear equations. How to do this with SageMath is described in Section 11.6.

Challenge 11.3: System of Linear Equations

We found a solution to the system of linear equations by just following an algorithm and using only three main operations. As an exercise, and using the newly discovered method of solving systems of linear equations, can you solve this challenge?

$$\begin{cases} 115\,b + 111\,h + 108\,f = 2209 \\ 118\,b + 101\,h + 115\,f = 2214 \\ 111\,b + 114h + 116\,f = 2286 \end{cases}$$

$$\begin{cases} 97\,q + 100\,m + 100\,a = 1582 \\ 111\,q + 110\,m + 101\,a = 1748 \\ 116\,q + 111\,m + 101\,a = 1786 \end{cases}$$

$$\begin{cases} 97\,r + \ 99\,n + 104\,t = 910 \\ 108\,r + 101\,n + 116\,t = 1005 \\ 116\,r + 101\,n + 114\,t = 1019 \end{cases}$$

Definition 11.1 Some, but not all, quadratic matrices have inverses; that is, for $A = (a_{i,j})$ there exits a matrix A^{-1} such that

$$A \cdot A^{-1} = \begin{pmatrix} 1 & 0 & \dots & 0 \\ 0 & 1 & \dots & 0 \\ \vdots & \vdots & \ddots & \vdots \\ 0 & 0 & \dots & 1 \end{pmatrix}$$

Then A^{-1} is called the inverse of A. If A has an inverse matrix, A is said to be invertible.

If A has such an inverse, the inverse is unique and the result of the product $A \cdot A^{-1} = A^{-1} \cdot A$ doesn't depend on the order of the factors. Note that matrix multiplication in general is not commutative.

Of course, we can easily calculate the inverse of a matrix in SageMath, if it exists:

```
sage: A=matrix([[0,2,0,0],[3,0,0,0],[0,0,5,0],[0,0,0,7]]);▸
 ▸ A
[0 2 0 0]
[3 0 0 0]
[0 0 5 0]
[0 0 0 7]
sage: A.inverse()
[  0 1/3   0   0]
[1/2   0   0   0]
[  0   0 1/5   0]
[  0   0   0 1/7]
```

```
sage: ~A
[  0 1/3   0   0]
[1/2   0   0   0]
[  0   0 1/5   0]
[  0   0   0 1/7]
sage: B=matrix([[1,0],[0,0]])
sage: B.inverse()
#... lines of error info, ending with:
ZeroDivisionError: matrix must be nonsingular
```

Definition 11.2 A diagonal matrix is a matrix $A = (a_{ij})$ where $a_{ij} = 0$ for all i, j with $i \neq j$. That is, all entries outside the main diagonal are zero.

Note that matrix multiplication, restricted to diagonal matrices, is commutative. Note also that transposition is trivial on diagonal matrices: $A^T = A$ for any diagonal matrix A.

To continue on our way defining lattices and their properties, we need to introduce some basic definitions and notations about vectors that, depending on the context, can sometimes be considered as special matrices with either only one column or only one row.

11.5 Vectors

A scalar is a one-dimensional measure of a quantity, such as temperature or mass. A vector has more than one number associated with it. A simple example is velocity. It has a magnitude, called speed, and a direction, like north or southwest or 10 degrees west of north. A vector can have more than two numbers associated with it [1]. We often draw a vector as an arrow, as shown in Figure 11.3 where a vector v is drawn starting at the origin $(0, 0)$ and ending at point $(1, 1)$. But how to write the vector?

Definition 11.3 A directed line from point $P(x_1, x_2)$ to point $Q(y_1, y_2)$ is a vector with the following components: $\overrightarrow{PQ} = \overrightarrow{OS} = (s_1, s_2) = (y_1 - x_1, y_2 - x_2)$.

The starting point of the vector $\overrightarrow{OP} = (x_1, x_2)$ is at the origin $O = (0, 0)$ and the end point is $P = (x_1, x_2)$ [2].

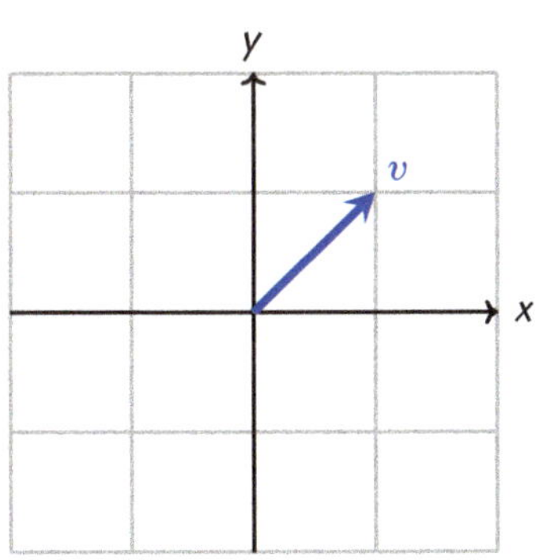

Figure 11.3 Example of a vector.

Let's express the vectors $\overrightarrow{PQ}$ and $\overrightarrow{RQ}$ with the three points $P(0,1)$, $Q(2,2)$, and $R(1.5,1.5)$ as shown in Figure 11.4.

We can easily do this by following the definition:

$$\overrightarrow{PQ} = (2-0, 2-1) = (2,1)$$
$$\overrightarrow{RQ} = (2-1.5, 2-1.5) = (0.5, 0.5)$$

Furthermore, if we define the origin as $O(0,0)$ and some random point $Z(x,y)$ we can easily define the vector $\overrightarrow{OZ} = (x-0, y-0) = (x,y)$. Using this observation we can easily calculate the desired vectors using SageMath.

```
sage: vOP = vector([0,1])
sage: vOQ = vector([2,2])
sage: vOR = vector([1.5,1.5])
sage: vPQ = vOQ - vOP
sage: vRQ = vOQ - vOR
sage: print(vPQ, vRQ)
(2,1) (0.5, 0.5)
```

Intuitively, we can easily check the results. $\overrightarrow{PQ} = (2,1)$ means that if we start at point P and move two times to the right and one time up, we will reach point Q. Following the same interpretation, $\overrightarrow{RQ} = (0.5, 0.5)$ means that if we start from point R and move 0.5 to the right and 0.5 up, we will reach point Q.

Challenge 11.4: Hidden ASCII

As an exercise, find a famous English quote hidden in Figure 11.5. Hint: 0xASCII, vectors.

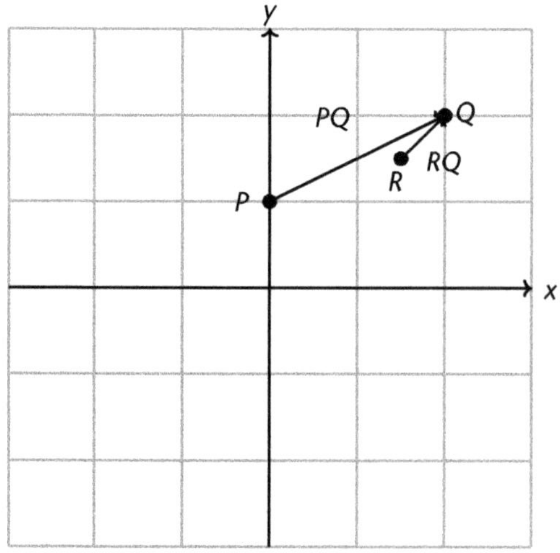

Figure 11.4 Finding vectors (defining them via start and end points).

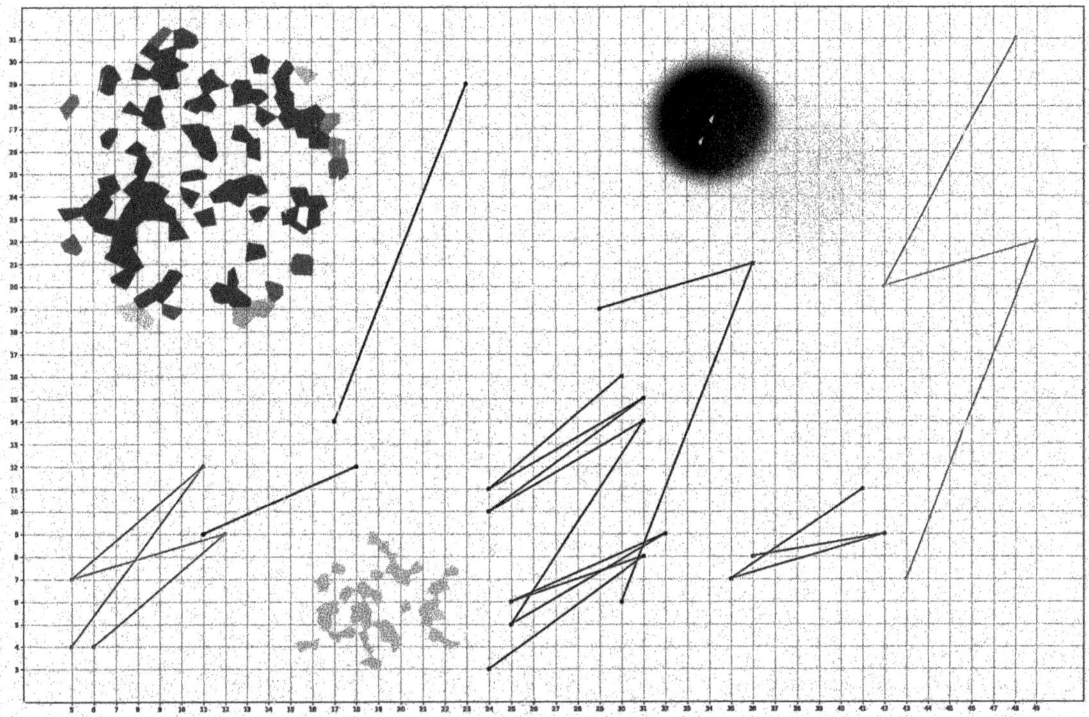

Figure 11.5 Puzzle challenge with vectors.

Definition 11.4 (Addition of vectors, multiplication of a scalar with a vector) For any two vectors $x = \begin{bmatrix} x_1 \\ x_2 \end{bmatrix}$, $y = \begin{bmatrix} y_1 \\ y_2 \end{bmatrix}$ in R^2 and a scalar k, the sum $x + y$ and the product kx are defined as follows: $x + y = \begin{bmatrix} x_1 + y_1 \\ x_2 + y_2 \end{bmatrix}$ and $kx = \begin{bmatrix} kx_1 \\ kx_2 \end{bmatrix}$.

Definition 11.5 The zero vector is a vector where all its components are equal to 0 (its origin is the origin of the coordinate system).

Definition 11.6 An ordered n-tuple of (real) numbers $(x_1, x_2, \ldots, x_n)$ is called an n-dimensional vector and can be written as

$$x = (x_1, x_2, \ldots, x_n) = \begin{bmatrix} x_1 \\ x_2 \\ x_3 \\ \ldots \\ x_n \end{bmatrix}$$

We call $x_1, x_2, \ldots, x_n$ the components of x.

We sum n-dimensional vectors and multiply n-dimensional vectors by some scalar in the same way as we did with the two-dimensional ones.

Definition 11.7 For vectors $v_1, v_2, \ldots, v_k$ in R^n and scalars $c_1, c_2, \ldots, c_k$,

$$x = c_1v_1 + c_2v_2 + \cdots + c_kv_k$$

is called a linear combination of vectors $v_1, v_2, \ldots, v_k$.

Definition 11.8 Given a vector $x = (x_1, x_2, \ldots, x_n)$ in R^n,

$$\|x\| = \sqrt{x_1^2 + x_2^2 + \cdots + x_n^2}$$

is called the norm (or length) of x.

You can easily calculate the norm of the vector by using SageMath:

```
sage: v = vector([3,6,2])
sage: v.norm()   # square root of (9 + 36  + 4)
7
```

Definition 11.9 For vectors $x = (x_1, x_2, \ldots, x_n)$, $y = (y_1, y_2, \ldots, y_n)$ in R^n,

$$x_1y_1 + x_2y_2 + \cdots + x_ny_n$$

is called the dot product, or scalar product, or inner product of x and y and is denoted by $x \cdot y$.

We will not use the Definitions 11.10 to 11.12 in the following sections, but they are useful consequences of the previous definitions.

Definition 11.10 For nontrivial vectors $x = (x_1, x_2, \ldots, x_n)$, $y = (y_1, y_2, \ldots, y_n)$ in R^n there exists θ with $0 \leq \theta \leq \pi$ or $0^\circ \leq \theta \leq 180^\circ$ and $\frac{x \cdot y}{\|x\| \cdot \|y\|} = \cos\theta$. Then θ is called the angle between x and y.

Definition 11.11 If $x \cdot y = 0$, then x is orthogonal to y. If x is a scalar multiple of y, then x is parallel to y.

We can easily calculate the inner product of two vectors using SageMath.

```
sage: x = vector([5,4,1,3])
sage: y = vector([6,1,2,3])
sage: x*y
45
sage: x.inner_product(y)
45
```

We can either use the multiply operator or be more strict and use the second syntax. In fact, $x \cdot y = 5 \cdot 6 + 4 \cdot 1 + 1 \cdot 2 + 3 \cdot 3 = 45$.

Now it's time to define the building blocks of a vector space.

Definition 11.12 For any arbitrary, nonzero vector $v \in R^n$, $u = \frac{1}{\|v\|} \cdot v$ is a unit vector. In R^n, unit vectors of the form:

$$e_1 = (1, 0, 0, \ldots, 0), e_2 = (0, 1, 0, \ldots, 0), \ldots, e_n = (0, 0, 0, \ldots, 1)$$

are called standard unit vectors or coordinate vectors.

Definition 11.12 allows us to express any arbitrary vector $x = (x_1, x_2, \ldots, x_n)$ of R^n as linear combination of the standard unit vectors:

$$x = x_1e_1 + x_2e_2 + \cdots + x_ne_n$$

11.6 Equations Revisited

We introduced the concept of matrices, more specifically the coefficient matrix and the augmented matrix (see Section 11.4). We studied the Gaussian elimination and how to use it to solve a system of linear equations.

So, let's solve the following system of linear equations using SageMath:

$$\begin{cases} 96x_1 + 11x_2 + 101x_3 = 634 \\ 97x_1 + 15x_2 + 99x_3 = 637 \\ 88x_1 + 22x_2 + 100x_3 = 654 \end{cases}$$

First, we define the coefficient matrix A.

```
sage: A = matrix([[96, 11, 101],[97, 15, 99],[88, 22, 100▸
 ▸]])
sage: A
[ 96  11 101]
[ 97  15  99]
[ 88  22 100]
```

Then, we define the right-hand sides of the equations as a vector b and construct the augmented matrix.

```
sage: b = vector([634, 637, 654])
sage: b
(634, 637, 654)
sage: B = A.augment(b)
sage: B
[ 96  11 101 634]
[ 97  15  99 637]
[ 88  22 100 654]
```

Now all we have to do is compute the reduced row echelon form using the command `rref`.

```
sage: B.rref()
[1 0 0 1]
[0 1 0 3]
[0 0 1 5]
```

As a final solution, we have $x_1 = 1, x_2 = 3, x_3 = 5$.

We have already defined operations for dealing with vectors (see Definition 11.4 and Definition 11.9). Let's use the same operations when dealing with matrices.

Definition 11.13 (Addition) Given two matrices $A = [a_{ij}]_{m \times n}$ and $B = [b_{ij}]_{m \times n}$, the sum of $A + B$ is defined by

$$A + B = [a_{ij} + b_{ij}]_{m \times n}$$

Definition 11.14 Given a matrix $A = [a_{ij}]_{m \times n}$ and a real number k, the scalar multiple kA is defined by

$$kA = [ka_{ij}]_{m \times n}$$

Let's try some examples with SageMath:

```
sage: A = matrix([[9,7,0],[0,5,6],[1,3,3]])
sage: B = matrix([[8,5,2],[8,2,2],[0,0,1]])
sage: A
[9 7 0]
[0 5 6]
[1 3 3]
sage: B
[8 5 2]
[8 2 2]
[0 0 1]
sage: A+B
[17 12  2]
[ 8  7  8]
[ 1  3  4]
sage: A-B
[ 1  2 -2]
[-8  3  4]
[ 1  3  2]
sage: 2*A
[18 14  0]
[ 0 10 12]
[ 2  6  6]
sage: A.row(0)   # get first row of a matrix
(9, 7, 0)
sage: A.column(0)   # get first col of a matrix
(9, 0, 1)
```

Definition 11.15 Given two matrices $A = [a_{ij}]_{m \times p}$ and $B = [b_{ij}]_{p \times n}$, we define the product AB of A and B, so that $AB = [c_{ij}]_{m \times n} = C$, where

$$c_{ij} = a_{i1}b_{1j} + a_{i2}b_{2j} + a_{i3}b_{3j} + \cdots + a_{ip}b_{pj}$$

Note that the number of columns of the first factor must be equal to the number of rows in the second factor.

```
sage: A*B
[128  59  32]
[ 40  10  16]
[ 32  11  11]
```

From the definition of a product of two matrices, we can make the following observations: The inner product (see Definition 11.9) of the ith row vector of A and the jth column vector of B is the (i, j) entry of $C = AB$. To demonstrate this observation we first must introduce another simple definition:

Definition 11.16 The transpose of a matrix is a new matrix whose rows are the columns of the original; that is, if $A = [a_{ij}]_{m \times n}$, then A^T, the transpose of A, is $A^T = [a_{ji}]_{n \times m}$.

```
sage: B.transpose()  # alternative command doing the same:▶
 ▶ B.T
[8 8 0]
[5 2 0]
[2 2 1]
```

Using this handy SageMath method, we can easily calculate the product of two matrices:

```
sage: for a in A:
....:     for b in B.transpose():
....:         print(a*b)
....:     print()
....:
128
59
32

40
10
16

32
11
11
```

Note the preference of SageMath operators:
`A*B.transpose() = A*(B.transpose())` and not `(A*B).transpose()`

In order to introduce the concept of a lattice, we need some more definitions. Let's look at the next system of equations and try to express each of the variables x, y, and z as an expression of the coefficients $a, b, c, d, e, f, g, h, i, r_1, r_2, r_3$.

$$\begin{aligned} ax + by + cz &= r_1 \\ dx + ey + fz &= r_2 \\ gx + hy + iz &= r_3 \end{aligned} \tag{11.1}$$

Let's multiply the first equation in (11.1) by ei, the second equation by hc, and the last equation by bf.

$$\begin{aligned} aeix + beiy + ceiz &= eir_1 \\ dhcx + ehcy + fhcz &= hcr_2 \\ gbfx + hbfy + ibfz &= bfr_3 \end{aligned} \tag{11.2}$$

Again, using (11.1), we multiply the first line by fh, the second line by bi, and the last line by ce.

$$\begin{aligned} afhx + bfhy + cfhz &= fhr_1 \\ dbix + ebiy + fbiz &= bir_2 \\ gcex + hcey + icez &= cer_3 \end{aligned} \tag{11.3}$$

Now we derive a new equation by subtracting from the sum of all (11.2) and the sum of all (11.3):

$$\begin{aligned} &(aei + dhc + gbf - afh - dbi - gce) \cdot x + \\ &\quad + (bei + ehc + hbf - bfh - ebi - hce) \cdot y + \\ &\quad + (cei + fhc + ibf - cfh - fbi - ice) \cdot z = \\ &\quad = (ei - fh) \cdot r_1 + (hc - bi) \cdot r_2 + (bf - ce) \cdot r_3 \end{aligned}$$

Simplify the equation by removing the equal expressions:

$$\begin{aligned} &(aei + dhc + gbf - afh - dbi - gce) \cdot x \\ &\quad = (ei - fh) \cdot r_1 + (hc - bi) \cdot r_2 + (bf - ce) \cdot r_3 \end{aligned}$$

We are now ready to express x:

$$x = \frac{(ei - fh) \cdot r_1 + (hc - bi) \cdot r_2 + (bf - ce) \cdot r_3}{aei + dhc + gbf - afh - dbi - gce} \tag{11.4}$$

Following the same procedure, and carefully choosing the coefficients with which to multiply the equations, we can also express y and z. But what if we have a system of 100 equations with 100 variables? Moreover, how does one use a more elegant way to recover the variables? It's time to introduce the definitions of minors and determinants.

Definition 11.17 A minor M_{ij} of a square matrix A of size n is the $(n-1) \times (n-1)$ matrix formed by the rows and columns of A excluding the ith row and the jth column.

Example

Let's have a matrix $A = \begin{pmatrix} a & b & c \\ d & e & f \\ g & h & i \end{pmatrix}$. By the definition of minors we have $M_{11} = \left(\begin{smallmatrix} e & f \\ h & i \end{smallmatrix}\right)$, $M_{12} = \left(\begin{smallmatrix} d & f \\ g & i \end{smallmatrix}\right)$ or $M_{22} = \left(\begin{smallmatrix} a & c \\ g & i \end{smallmatrix}\right)$.

Let's take the general case of a matrix A with size n.

$$A = \begin{pmatrix} a_{1,1} & a_{1,2} & a_{1,3} & \dots & a_{1,j} & \dots & a_{1,n} \\ a_{2,1} & a_{2,2} & a_{2,3} & \dots & a_{2,j} & \dots & a_{2,n} \\ a_{3,1} & a_{3,2} & a_{3,3} & \dots & a_{3,j} & \dots & a_{3,n} \\ \dots & \dots & \dots & \dots & \dots & \dots & \dots \\ a_{i,1} & a_{i,2} & a_{i,3} & \dots & a_{i,j} & \dots & a_{i,n} \\ \dots & \dots & \dots & \dots & \dots & \dots & \dots \\ a_{n,1} & a_{n,2} & a_{n,3} & \dots & a_{n,j} & \dots & a_{n,n} \end{pmatrix}$$

Then the minor M_{ij} of A is equal to

$$M_{ij} = \begin{pmatrix} a_{1,1} & a_{1,2} & a_{1,3} & \dots & a_{1,j-1} & a_{1,j+1} & \dots & a_{1,n} \\ a_{2,1} & a_{2,2} & a_{2,3} & \dots & a_{2,j-1} & a_{2,j+1} & \dots & a_{2,n} \\ a_{3,1} & a_{3,2} & a_{3,3} & \dots & a_{3,j-1} & a_{3,j+1} & \dots & a_{3,n} \\ \dots & \dots & \dots & \dots & \dots & \dots & \dots & \\ a_{i-1,1} & a_{i-1,2} & a_{i-1,3} & \dots & a_{i-1,j-1} & a_{i-1,j+1} & \dots & a_{i-1,n} \\ a_{i+1,1} & a_{i+1,2} & a_{i+1,3} & \dots & a_{i+1,j-1} & a_{i+1,j+1} & \dots & a_{i+1,n} \\ \dots & \dots & \dots & \dots & \dots & \dots & \dots & \\ a_{n,1} & a_{n,2} & a_{n,3} & \dots & a_{n,j-1} & a_{n,j+1} & \dots & a_{n,n} \end{pmatrix}.$$

Now that we have the definitions of minors, we can finally define the determinant of a matrix.

Definition 11.18 Let A be a square matrix with real elements and some fixed integers $r \in \{1, \dots, n\}$ and $c \in \{1, \dots, n\}$. Then its determinant, $det(A) = \left|A\right|$ is a real number that can be calculated either by column c or by row r:

$$\left|A\right| = \sum_{i=1}^{n} (-1)^{i+c} a_{i,c} \left|M_{ic}\right| \qquad \text{(expansion along column } c\text{)}$$

$$\left|A\right| = \sum_{i=1}^{n} (-1)^{r+i} a_{r,i} \left|M_{ri}\right| \qquad \text{(expansion along row } r\text{)}$$

Let's calculate the determinants of a 2×2 matrix B and a 3×3 matrix C by using minors on row 1 and column 2, respectively.

Example

Expansion along row 1:

$$det(B) = \begin{vmatrix} a & b \\ c & d \end{vmatrix} = (-1)^{1+1} \cdot a \cdot \left|d\right| + (-1)^{1+2} \cdot b \cdot \left|c\right| = ad - bc$$

Note that in the above calculation, $|d|$ and $|c|$ do *not* denote the absolute values of the numbers d and c, but the determinants of the 1×1 matrices consisting of d or c, the minors $d = M_{11}$ and $c = M_{12}$ of the matrix B. Note also that $\det(x) = x$ for every 1×1 matrix with entry x, so $|d| = d$ and $|c| = c$ in the last step.

Example

Expansion along column 2:

$$det(C) = \begin{vmatrix} a & b & c \\ d & e & f \\ g & h & i \end{vmatrix}$$

$$= (-1)^{1+2} \cdot b \cdot \begin{vmatrix} d & f \\ g & i \end{vmatrix} + (-1)^{2+2} \cdot e \cdot \begin{vmatrix} a & c \\ g & i \end{vmatrix} + (-1)^{3+2} \cdot h \cdot \begin{vmatrix} a & c \\ d & f \end{vmatrix}$$

$$= -b(di - fg) + e(ai - cg) - h(af - cd) =$$

$$= aei + dhc + gbf - afh - dbi - gce$$

The determinant of this example is exactly equal to the denominator of the right side of (11.4). What about the numerator? We can easily verify that the numerator is equal to the determinant of the matrix $B_1 = \begin{vmatrix} r_1 & b & c \\ r_2 & e & f \\ r_3 & h & i \end{vmatrix}$. If we define the matrices $B_2 = \begin{vmatrix} a & r_1 & c \\ d & r_2 & f \\ g & r_3 & i \end{vmatrix}$ and $B_3 = \begin{vmatrix} a & b & r_1 \\ d & e & r_2 \\ g & h & r_3 \end{vmatrix}$, we can easily calculate the solutions x, y and z of the given system of linear equations:

$$x = \frac{det(B_1)}{det(C)}, \quad y = \frac{det(B_2)}{det(C)}, \quad z = \frac{det(B_3)}{det(C)} \tag{11.5}$$

So we can use determinants to solve systems of linear equations. SageMath provides an easy way to compute a determinant, as well as submatrices of our choice.

```
sage: M = matrix([[1,2,3], [4,5,6], [7,8,9]])
sage: M
[1 2 3]
[4 5 6]
[7 8 9]
sage: M.determinant()
0
sage: det(M)
0
sage: M.det()
0
```

```
# consider row 1 and 3, and col 1 and 3 (row 1 has index 0▶
  ▶]
sage: M.matrix_from_rows_and_columns([0,2],[0,2])
[1 3]
[7 9]
```

$$M_{22} = \begin{pmatrix} 1 & 3 \\ 7 & 9 \end{pmatrix}$$

The constructor `matrix_from_rows_and_columns` takes two lists as arguments. The first list defines which rows of a matrix A should be taken to construct the new matrix, while the second list defines the columns. For example, to construct the minor (see Definition 11.17) M_{ij} of the matrix A with size n, we call:

```
A.matrix_from_rows_and_columns([0,\dots ,i-1,i+1,\dots
,n-1],[0,\dots ,j-1,j+1,\dots ,n-1])
```

Now, with all this information we can automate the process of solving a large system of equations.

Let's look at the following system of equations:

$$\begin{cases} x + 9y + 3z = 61 \\ 2x + 4y + 8z = 94 \\ 5x + 7y + 6z = 128 \end{cases}$$

We can convert the same system of equations into a matrix equation:

$$\begin{pmatrix} 1 & 9 & 3 \\ 2 & 4 & 8 \\ 5 & 7 & 6 \end{pmatrix} \cdot \begin{pmatrix} x \\ y \\ z \end{pmatrix} = \begin{pmatrix} 61 \\ 94 \\ 128 \end{pmatrix}$$

We solve this equation using SageMath:

```
sage: M = matrix([[1,9,3], [2,4,8], [5,7,6]])
sage: M
```

$$\begin{pmatrix} 1 & 9 & 3 \\ 2 & 4 & 8 \\ 5 & 7 & 6 \end{pmatrix}$$

```
sage: r = matrix([[61],[94],[128]])
sage: r
```

$$\begin{pmatrix} 61 \\ 94 \\ 128 \end{pmatrix}$$

```
sage: M.solve_right(r)
```

$$\begin{pmatrix} 13 \\ 3 \\ 7 \end{pmatrix}$$

This gives the final solutions $x = 13$, $y = 3$, and $z = 7$.

Challenge 11.5: Leetspeak

Alice and Bob have come up with an interesting (but insecure) encryption scheme. Alice creates n equations with n variables and sends them to Bob over an insecure channel using two packets. The first packet consists of all the coefficients used in the equations in the form of a matrix without any changes. The second packet, however, consists of all the right-hand sides of the equations in a scrambled order. Their shared secret key consists of the original indeces of the scrambled right-hand sides of the equations.

With the secret key, Bob can unscramble the right-hand sides of the equations and recover the unknown variables. Then, he multiplies all the recovered variables and the final number is the decrypted message. They used the `leet` language to create or read the final number. For example, the word `sage` in leet language is `5463`. Eve captured the following two packets P_1 and P_2:

$$P_1 = \begin{pmatrix} 33 & 79 & 29 & 41 & 47 \\ 79 & 27 & 39 & 79 & 44 \\ 90 & 83 & 58 & 1 & 90 \\ 38 & 32 & 13 & 15 & 96 \\ 72 & 82 & 88 & 83 & 23 \end{pmatrix}; \qquad P_2 = [73300,\ 167887,\ 243754,\ 254984,\ 458756]$$

Can you reconstruct the original message?

11.7 Vector Spaces

We need to introduce another important building block of linear algebra, *vector spaces*. First, let's define what a vector space is. This definition can be found in any undergraduate book on linear algebra:

Definition 11.19 A vector space over the real numbers $\mathbb{R}$ consists of a set V and two operations $\oplus$ and $\odot$, subject to the following conditions/properties/axioms:

1. *Closure under addition*: $\forall \vec{x}\ \forall \vec{y} \in V : \vec{x} \oplus \vec{y} \in V$
2. *Commutativity of addition*: $\forall \vec{x}\ \forall \vec{y} \in V : \vec{x} \oplus \vec{y} = \vec{y} \oplus \vec{x}$
3. *Associativity of addition*: $\forall \vec{x}\ \forall \vec{y}\ \forall \vec{z} \in V : (\vec{x} \oplus \vec{y}) \oplus \vec{z} = \vec{x} \oplus (\vec{y} \oplus \vec{z})$
4. *Neutral element under addition*: $\exists \vec{o} \in V : (\forall \vec{x} \in V : \vec{x} \oplus \vec{o} = \vec{x})$
   We call $\vec{o}$ the *zero vector*.
5. *Additive inverse*: $\forall \vec{x} \in V\ \exists \vec{y} \in V : \vec{x} \oplus \vec{y} = \vec{o}$
   We call $\vec{y}$ the *additive inverse or inverse element of addition* of $\vec{x}$, and vice versa.

6. *Closure in scalar multiplication*: $\forall r \in \mathbb{R}, \forall \vec{x} \in V : r \odot \vec{x} \in V$
7. *Distributivity 1*: $\forall r, s \in \mathbb{R}, \forall \vec{x} \in V : (r+s) \odot \vec{x} = r \odot \vec{x} \oplus s \odot \vec{x}$
8. *Distributivity 2*: $\forall r \in \mathbb{R}, \forall \vec{x}, \vec{y} \in V : r \odot (\vec{x} \oplus \vec{y}) = r \odot \vec{x} \oplus r \odot \vec{y}$
9. *Associativity*: $\forall r, s \in \mathbb{R}, \forall \vec{x} \in V : (rs) \odot \vec{x} = r \odot \ (s \odot \vec{x})$
10. *Neutral action of scalar multiplication with* 1: $\forall \vec{x} \in V : 1 \odot \vec{x} = \vec{x}$

Usually, we do not use the symbol $\odot$, but just the multiplication point or no dot at all. But sometimes, as in this definition, we want to point out that the scalar multiplication is a mapping from $\mathbb{R} \times V$ to V, while the "regular" multiplication is something else (i.e., a mapping from $\mathbb{R} \times \mathbb{R}$ to $\mathbb{R}$). Also note that there is some danger of confusion if we use the same multiplication symbol (or no symbol at all) for both the scalar multiplication and the scalar product (vector product, inner product). This can cause some problems when looking at more complicated formulas that contain both types of products.

We can also define vector spaces over larger sets, such as the set of complex numbers. It is also possible to define structures like the one above over smaller sets, for example the set of integers. In this case—when the set of scalars is a so-called ring and not a so-called field—these algebraic structures are not called vector spaces, but *modules*. Our vectors, as well as our choice of the two operators $\oplus$ and $\odot$ play an important role in determining whether our space is a vector space or not.

Example Let's define the set M of all 2×2 matrices with entries of real numbers. Furthermore, we choose the operator $\oplus$ as a regular additive operator on matrices; that is,

$$\begin{pmatrix} a_1 & a_2 \\ a_3 & a_4 \end{pmatrix} \oplus \begin{pmatrix} b_1 & b_2 \\ b_3 & b_4 \end{pmatrix} = \begin{pmatrix} a_1+b_1 & a_2+b_2 \\ a_3+b_3 & a_4+b_4 \end{pmatrix}.$$

We choose the operation $\odot$ to be the already known scalar multiplication of matrices; that is,

$$r \odot \begin{pmatrix} a_1 & a_2 \\ a_3 & a_4 \end{pmatrix} = \begin{pmatrix} ra_1 & ra_2 \\ ra_3 & ra_4 \end{pmatrix}.$$

We can easily check that all the conditions hold and that this is indeed a vector space, in which the zero vector is $\left(\begin{smallmatrix} 0 & 0 \\ 0 & 0 \end{smallmatrix}\right)$.

Example The set P of polynomials with real coefficients is a vector space with the operator $\oplus$ defined as the regular additive operator on polynomials and the operator $\odot$ defined via $r \odot (\sum a_i x^i) := \sum r a_i x^i$. For example, if $a_i, b_i, r \in \mathbb{R}$, then:

$$\begin{aligned} (a_0 + a_1 x + \cdots + a_n x^n) \quad &\oplus \quad (b_0 + b_1 x + \cdots + b_n x^n) = \\ &= \quad (a_0 + b_0) + (a_1 + b_1)x + \cdots + (a_n + b_n)x^n \\ r \odot (a_0 + a_1 x + \cdots + a_n x^n) \quad &= \quad (ra_0) + (ra_1)x + \cdots + (ra_n)x^n \end{aligned}$$

Definition 11.20 For any vector space V with operations $\oplus$ and $\odot$, a subspace U is a subset of V which is itself a vector space over the same field under the inherited operations $\oplus$ and $\odot$. This means that U is closed under addition $\oplus$ and scalar multiplication $\odot$.

Example A trivial subspace of $\mathbb{R}^2$ is the 1-element set of the zero vector $\{(0, 0)\}$.

Example Let's define the vector space of cubic polynomials

$$C_{\mathrm{u}} = \{a + bx + cx^2 + dx^3 \mid a, b, c, d \in \mathbb{R}\}$$

and the vector space of linear polynomials $L_{\mathrm{i}} = \{e + fx \mid e, f \in \mathbb{R}\}$. Then, L_{i} is a subspace of C_{u}.

Definition 11.21 The span (or linear closure) of a non-empty subset S of a vector space V is the set of all linear combinations of vectors from S.

In short, we can write down the span of a subset S of a vector space V as follows:

$$span(S) = \{c_1\vec{v_1} \oplus \cdots \oplus c_n\vec{v_n} \mid c_i \in \mathbb{R},\ \vec{v_i} \in S\}$$

Note that S itself need not be a subspace, but span(S) is always a subspace of V. If span$(S) = U$, we say that S generates (or spans) U.

Example For any nonzero vector $\vec{x} \in \mathbb{R}^3$, the span of $\vec{x}$ is a line through the origin $(0, 0, 0)$, or, more precisely, a one-dimensional subspace containing the zero vector. Note that a point (e.g., the origin) is not the same as a vector (e.g., the zero vector).

Example Let's define $S = \left\{\binom{2}{2}, \binom{2}{-2}\right\}$. We will show that $span(S) = \mathbb{R}^2$.

If this is the case, then every vector $\binom{x}{y} \in \mathbb{R}^2$ can be represented as a linear combination of vectors in S. So $r_1, r_2 \in \mathbb{R}$ exist, such that

$$r_1 \cdot \begin{pmatrix} 2 \\ 2 \end{pmatrix} + r_2 \cdot \begin{pmatrix} 2 \\ -2 \end{pmatrix} = \begin{pmatrix} x \\ y \end{pmatrix}$$

So we can express every possible vector $\binom{x}{y} \in \mathbb{R}^2$ by choosing r_1 and r_2 such that $r_1 = \frac{x+y}{4}$ and $r_2 = \frac{x-y}{4}$. For example, if $x = 9$ and $y = 1$, we have $r_1 = \frac{5}{2}$ and $r_2 = 2$.

$$\frac{5}{2} \cdot \begin{pmatrix} 2 \\ 2 \end{pmatrix} + 2 \cdot \begin{pmatrix} 2 \\ -2 \end{pmatrix} = \begin{pmatrix} 5 \\ 5 \end{pmatrix} + \begin{pmatrix} 4 \\ -4 \end{pmatrix} = \begin{pmatrix} 9 \\ 1 \end{pmatrix}$$

The previous example is one way of spanning $\mathbb{R}^2$. Can the set $\mathbb{R}^2$ be spanned by three or more vectors? Sure, we can just duplicate one of the elements in the previous example, namely we can take the set $\left\{\binom{2}{2}, \binom{2}{-2}, \binom{1}{-1}\right\}$. But can $\mathbb{R}^2$ be spanned by only one vector? The answer is of course no, because one vector can only span a 1-dimendional space.

Definition 11.22 A subset of a vector space is said to be linearly independent if none of its elements is a linear combination of the others. Otherwise it is said to be linearly dependent.

Definition 11.23 A basis of a vector space is a set B of vectors that is linearly independent and spans the space. If $|B| = n$ for some $n \in \mathbb{N}$, we define the dimension of span(B) as $dim(\text{span}(B)) := n$. If $|B| = \infty$, the notion of dimension is also well defined, but the theory is a bit more complicated because of different types of infinity in mathematics. We will not go into detail here.

Example We have already shown that the set $\left\{\binom{2}{2}, \binom{2}{-2}\right\}$ is a basis of $\mathbb{R}^2$. Another one is $\left\{\binom{1}{0}, \binom{0}{1}\right\}$.

Example We can easily construct the basis ξ_n of $\mathbb{R}^n$ for any n:

$$\xi_n = \left\{ \begin{pmatrix} 1 \\ 0 \\ 0 \\ \vdots \\ 0 \end{pmatrix}, \begin{pmatrix} 0 \\ 1 \\ 0 \\ \vdots \\ 0 \end{pmatrix}, \ldots, \begin{pmatrix} 0 \\ 0 \\ 0 \\ \vdots \\ 1 \end{pmatrix} \right\} =: \{e_1, \ldots, e_n\}$$

We say that this is the *standard* or *canonical* basis of $\mathbb{R}^n$.

Every invertible matrix $B = (b_{ij})$ induces a mapping from this standard basis to another basis $\{b_1, \ldots, b_n\}$ with $b_i = \begin{pmatrix} b_{1i} \\ \vdots \\ b_{ni} \end{pmatrix}$; that is, each column of the matrix B can be seen as an image of one of the vectors of the canonical basis. Using matrix multiplication, we have $B \cdot e_i = b_i$.

A special class of mappings between bases are permutations. If we just change the order of the canonical basis vectors e_i, this is called a permutation and the corresponding matrix has only 0 or 1 entries, exactly one 1 per row and column. For example, mapping e_1 to e_3, e_3 to e_2, and e_2 to e_1 is done by

$$\begin{pmatrix} 0 & 1 & 0 \\ 0 & 0 & 1 \\ 1 & 0 & 0 \end{pmatrix}.$$

Such a matrix is called a permutation matrix. In algebra, permutations are sometimes written in parentheses, such as $(1, 3, 2)$, which means to map the first element to the third, the third element to the second, and the second element to the first. This is the notation used by SageMath: `[1,3,2]`. But be careful! The tuple $(1, 3, 2)$ can also mean map the first element to the first element, the second element to the third element, and the third element to the second element.

```
sage: Permutation([3, 1, 2])
[3, 1, 2]
sage: per=Permutation((1,3,2))
sage: per  # alternative: per.show() displays the ▶
  ▶permutation as a drawing
[3, 1, 2]
sage: per.reverse()  # not available: per.transpose() and ▶
  ▶per.T
```

```
[2, 1, 3]
sage: per.inverse()
[2, 3, 1]
sage: per.complement()
[1, 3, 2]
sage: matrix(per)
[0 1 0]
[0 0 1]
[1 0 0]
sage: grelt=PermutationGroupElement([1,3,2])
sage: grelt.matrix()
[1 0 0]
[0 0 1]
[0 1 0]
```

An interesting thing about these permutation matrices is that their inverse is identical to their transposed matrix:

```
sage: matrix(per).inverse()==matrix(per).T
True
```

Challenge 11.6: Vector Challenge

Inspired by the concept of a basis of a vector space, Alice and Bob invent yet another cryptosystem. This time they use an encoding encode, which first encodes each letter into a number. This number is equal to the index of the corresponding letter in the English alphabet. For example, the word **Bob** is encoded in an array `[ 2, 15, 2 ]`.

Alice and Bob have carefully chosen a set of private keys K that they share as common prior knowledge. Depending on the length of the message to be sent, a different key is used. Let's define the key $k_m \in K$ as the key used to encryt messages of length m.

Each key k_m is an $m \times m$ matrix generated by the following rules:

- Each element of the matrix is either 0 or a prime number between 100 and 999.
- The row vectors of the matrix are linearly independent.
- There are exactly m numbers other than 0.

Now, the encryption E of a message M of length m is a straightforward procedure:

1. Alice encodes M, so she has the encoded message $\mathsf{encode}(M)$.
2. Alice encrypts $\mathsf{encode}(M)$ with the corresponding key k_m; that is,

$$E(\mathsf{encode}(M), k_m) = k_m \cdot \mathsf{encode}(M).$$

3. The message $E(\mathsf{encode}(M), k_m)$ is sent to Bob over the insecure channel.

Challenge 11.6 (continued)

Then, the decryption D of a message $E(\mathsf{encode}(M), k_m)$ is performed by Bob following these simple steps:

1. The length of the encrypted message uniquely defines the key k_m that Bob should use.
2. Bob constructs the decryption key d_m by replacing every element greater than zero in the secret key k_m with its reciprocal value.
3. Bob then performs the decryption; that is, $D(E(\mathsf{encode}(M), k_m)) = E(\mathsf{encode}(M), k_m) \cdot d_m$.
4. Bob decodes the decrypted message to recover the original message.

Can you verify the correctness of this encryption scheme? Why does this decryption work or why does it not? Can you recover the following encrypted ciphertext (one word):
(6852, 3475, 17540, 3076, 12217, 6383, 745, 1347, 661, 6088, 15354, 2384, 2097, 11415, 3143)

Note the third step from Bob in the Challenge 11.6: The matrix d_m is not the inverse of k_m. The matrix k_m can be written as a product $k_m = P \cdot D$ with a diagonal matrix D and a permutation matrix P. Then $k_m^{-1} = (P \cdot D)^{-1} = D^{-1} \cdot P^{-1} = D^{-1} \cdot P^T$. If we now construct the matrix d_m by replacing each nonzero element in k_m by its reciprocal, we get $d_m = P \cdot D^{-1}$. If we look at the transposed matrix d_m^T, we have $d_m^T = (P \cdot D^{-1})^T = (D^{-1})^T \cdot P^T = D^{-1} \cdot P^{-1} = (P \cdot D)^{-1} = k_m^{-1}$. Instead of using the matrix d_m^T on a column vector from the left, we can let the matrix d_m operate on a row vector from the right, since in general for matrices A and column vectors v and b there is an equivalence $Av = b \Leftrightarrow (Av)^T = b^T \Leftrightarrow v^T A^T = b^T$. This corresponds to the notation from above, where in the encryption process, $E(\mathsf{encode}(M), k_m) = k_m \cdot \mathsf{encode}(M)$ means $k_m \cdot \mathsf{encode}(M)$ as well as $\mathsf{encode}(M)$ is treated as a column vector while in the decryption process, d_m is written on the right of $E(\mathsf{encode}(M), k_m)$, so in this case $E(\mathsf{encode}(M), k_m)$ as well as $E(\mathsf{encode}(M), k_m) \cdot d_m$ is treated as a row vector.

11.8 Lattices

Now we have all the building blocks to introduce the concept of lattices.

Definition 11.24 Let $v_1, \dots, v_n \in \mathbb{Z}^m$, $m \geq n$ be linearly independent vectors. An integer lattice L spanned by $\{v_1, \dots, v_n\}$ is the set of all integer linear combinations of $v_1, \dots, v_n$, such that:

$$L = \left\{ v \in \mathbb{Z}^m \mid v = \sum_{i=1}^{n} a_i v_i,\ a_i \in \mathbb{Z} \right\} \tag{11.6}$$

Note: Linear combination means that all a_i are integers. Integer lattice means that all v_{ij} (components of vectors v_i) are integers, and thus all points in the infinite graph have integer coordinates.

SageMath uses the notion of an integral lattice, which is a slightly more complicated concept. Roughly speaking, the components of the vectors $v_1, \ldots, v_n$ are not restricted to $\mathbb{Z}$, but can be arbitrary real numbers, while the allowed linear combinations still have to be integer linear combinations; that is, the coefficients a_i in $\sum_{i=1}^{n} a_i v_i$ have to be integers. Every integer lattice is an integral lattice.

The set of vectors $B = \{v_1, \ldots, v_n\}$ is called a basis of the lattice L. We also say that $L = L(B)$ is spanned by the vectors of the basis B.

We define the dimension of L as $\dim(L) := n$.

In the case where $n = m$, we can canonically construct a quadratic matrix from the vectors of a lattice basis by writing them down row by row (or column by column). If we denote this matrix by M, we can compute the product $M \cdot M^T$, which is sometimes called the *Gram matrix of the lattice*. If this Gram matrix has only integer entries, the lattice is integral. Note that to go into detail here we would have to introduce some more mathematics, in particular the theory of quadratic forms, symmetric bilinear forms, and so on, which are a kind of generalization of the vector product introduced earlier.

The example in Figure 11.6(a) shows a two-dimensional lattice with

$$B_a = \{v_1, v_2\} = \{(1,2), (-1,1)\},$$

while the example in Figure 11.6(b) shows a two-dimensional lattice with

$$B_b = \{v_1, v_2\} = \{(-1,-2), (-1,-1)\}.$$

Let's see how we can construct a lattice using SageMath:

```
sage: M = matrix(ZZ, [[1,2], [-1,1]])
sage: M
```

$$\begin{pmatrix} 1 & 2 \\ -1 & 1 \end{pmatrix}$$

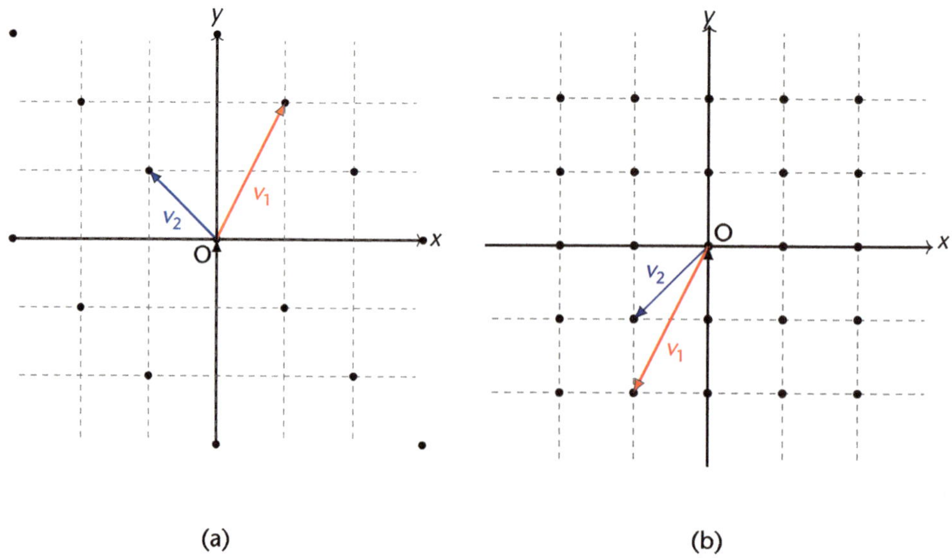

Figure 11.6 Example of 2D lattices with different bases: (a) b_a and (b) B_b.

Now we can easily check if a given point (z_1, z_2) belongs to the lattice. If $(z_1, z_2) \in L$, then it belongs to the span of M.

```
#  The span of a matrix in SageMath is by default the span▶
 ▶ of the row
#  vectors of the matrix.
sage: vector([1,1]) in span(M)
False

sage: vector([1,2]) in span(M)  # true because [1,2]=1*x+0▶
 ▶*y
True

sage: vector([-1,2]) in span(M)
False

sage: vector([-101,5]) in span(M)
True
```

What if we want to see the exact linear combination that produces the point $z = (z_1, z_2)$? That is, we want to know how this point z is uniquely constructed by the given basis vectors x=[1,2] and y=[-1,1].

```
sage: M.solve_left(vector([-101,5]))
(-32, 69)
```

And indeed:

```
sage: -32*M[0] + 69*M[1]
(-101, 5)
```

We can define the same lattice using different bases. For example, let's introduce a lattice over the matrix $M2$ with the following basis:

```
sage: M2 = matrix(ZZ, [[1,2], [0,3]])
sage: M2
```

$$\begin{pmatrix} 1 & 2 \\ 0 & 3 \end{pmatrix}$$

Using SageMath's span function, we can easily compare the identity of two objects defined by different bases.

```
sage: span(M) == span(M2)
True
```

Screenshots from CT2 showing different lattices can be found in Section 11.12.2.

11.8.1 Merkle-Hellman Knapsack Cryptosystem

Now let's create another definition before we take a look at the Merkle-Hellman knapsack cryptosystem. See [3], pages 472 and on.

Definition 11.25 Any set of distinct nonzero natural numbers is called a knapsack. Furthermore, if this set can be arranged in an increasing list in such a way that each number is greater than the sum of all previous numbers, we call this list a superincreasing knapsack.

Challenge 11.7: Superincreasing Knapsacks

Inspired by the definition of the superincreasing knapsack, Alice and Bob constructed another insecure cryptosystem of their own. Can you find the hidden word in this intercepted message, shown in the following sequence of numbers? Hint: Is the knapsack superincreasing? Why or why not? Each number contains a secret bit. (A copy-and-paste version of these numbers is also available at `https://www.cryptool.org/en/documentation/ctbook/sagemath`.)

0, 0, 1, 2, 3, 6, 12, 25, 49, 98, 197, 394, 787, 1574, 3148, 6296, 12593, 25185, 50371, 100742, 201484, 402967, 805935, 1611870, 3223740, 6447479, 12894959, 25789918, 51579835, 103159670, 206319340, 412638681, 825277361, 1650554722, 3301109445, 6602218890, 13204437779, 26408875558, 52817751117, 105635502233, 211271004467, 422542008933, 845084017867, 1690168035734, 3380336071467, 6760672142934, 13521344285869, 27042688571737, 54085377143475

The Merkle-Hellman knapsack cryptosystem is another asymmetric cryptosystem that is theoretically interesting because it basically allows sensitive information to be sent over an insecure channel.

It consists of two knapsack keys:

- Public key, used only for encryption. It's called a hard knapsack.
- Private key, used only for decryption. It consists of a superincreasing knapsack, a multiplier, and a modulus. The multiplier and modulus can be used to transform the superincreasing knapsack into the hard knapsack.

The key generation algorithm performs the following steps:

1. First, we create a superincreasing knapsack $W = [w_1, w_2, \cdots, w_n]$.
2. We choose an integer q that is greater than the sum of all elements in W; that is,

$$q > \sum_{i=1}^{n} w_i$$

   We define q as the modulus of our cryptosystem.

3. We choose an integer r; that is, $r \in [1, q)$ and $(r, q) = 1$, where (r, q) is the notation for the greatest common divisor (gcd) of r and q and $[1, q) = \{1, 2, \ldots, q-1\}$. We define r as the multiplier of our cryptosystem.
4. The private key of the cryptosystem consists of the tuple (W, r, q).
5. We generate the sequence $H = [h_1, h_2, \cdots, h_n]$, s.t. $h_j = w_j * r \bmod q$, for $1 \leq j \leq n$. We define H as the public key of the cryptosystem.

If we want to encrypt a message m, we first take its bit representation $B_m = \overline{m_1 m_2 \ldots m_n}$, where m_i denotes the ith bit; that is, $m_i \in \{0, 1\}$. To ensure the correctness of the algorithm, our superincreasing knapsack K should have at least n elements. Let's define it as $W = [w_1, w_2, \ldots, w_n, \ldots]$. After the key generation procedure, we generate its corresponding public key H; that is, $H = [h_1, h_2, \ldots, h_n, \ldots]$, with some appropriate q and r. Then, the encryption c of m is the sum

$$c = \sum_{i=1}^{n} m_i h_i.$$

If we want to decrypt the message c, we first compute $c' = c \cdot r^{-1} \bmod q$, where r^{-1} is the modular inverse of $r \bmod q$. Then, we start a decomposition procedure of c' by selecting the largest elements in W that are less than or equal to the remaining value being decomposed. Finally, we recover $m = \overline{m_1 m_2 \ldots m_n}$ by replacing m_j with 1 if the element w_j was selected in the previous step. Otherwise, m_j is 0.

We intentionally describe the pure algorithm only as a cryptographic scheme (because of its weakness against lattice attacks), and do not discuss practical implementation issues such as ensuring that the length of B_m is less than or equal to the length of H, or how and when to apply padding.

Example Let's assume that Alice wants to encrypt and send the message crypto to Bob using the Merkle-Hellman knapsack cryptosystem. Throughout this example, each letter is treated independently. Thus, n is always 8, because each letter has an 8-bit binary representation.

First, Bob must generate his private and public keys. Bob initiates the process of generating the private key by first generating a superincreasing knapsack W:

$$W = [11, 28, 97, 274, 865, 2567, 7776, 23253]$$

Then, Bob generates the corresponding modulus q and multiplier r:

$$q = 48433 > \sum_{i=1}^{n} w_i = 34871$$

$$r = 2333 < q$$

$$(2333, 48433) = 1$$

So, Bob composes the private key $P_r = (W, r, q)$:

$$P_r = ([11, 28, 97, 274, 865, 2567, 7776, 23253], 2333, 48433)$$

The final step for Bob is to generate the hard knapsack H and the public key $P_u = (H)$ and deliver it to Alice:

$$H = [25663, 16891, 32569, 9613, 32292, 31552, 27466, 4289]$$
$$P_u = ([25663, 16891, 32569, 9613, 32292, 31552, 27466, 4289])$$

Before encrypting the message M = crypto, Alice divides the message into individual letters and replaces each letter with its own bit representation; that is:

$$c = 01100011$$
$$r = 01110010$$
$$y = 01111001$$
$$p = 01110000$$
$$t = 01110100$$
$$o = 01101111$$

Now, Alice computes the corresponding encrypted number for the bit representation of each letter using the public key H. Thus, the algorithm must be applied six times. Finally, the list of encrypted numbers C of the word crypto is:

$$C = [81215, 86539, 95654, 59073, 90625, 145059]$$

When Bob receives C, he first calculates C' using r and q from P_r.

$$C' = [31154, 8175, 24517, 399, 2966, 34586]$$

Then, using W from P_r, he represents each element in C' as a sum of elements in W, following the above algorithm. For example, let's decompose 31154. The sign ✓ will denote the elements in W that are part of the decomposition of 31154, and the sign ✗ will denote those that aren't. The sign * will denote the unknowns.

$$\begin{array}{l}[11, 28, 97, 274, 865, 2567, 7776, 23253] \\ [\;* \quad * \quad * \quad * \quad * \quad * \quad * \quad *\;], 31154\end{array}$$

The largest number in W less than 31154 is 23253. We mark it as an element used in the decomposition of 31154, and we continue with the decomposition of the remaining value $7901 = 31154 - 23253$:

$$\begin{array}{l}[11, 28, 97, 274, 865, 2567, 7776, 23253] \\ [\;* \quad * \quad * \quad * \quad * \quad * \quad * \quad \checkmark\;], 7901\end{array}$$

The largest element less than 7901 is 7776. We continue with this algorithm until we reach 0.

$$
\begin{array}{cccccccccl}
[\,11, & 28, & 97, & 274, & 865, & 2567, & 7776, & 23253\,] & & \\
[\ \ast & \ast & \ast & \ast & \ast & \ast & \checkmark & \checkmark &], & 125 \\
[\ \ast & \ast & \ast & \ast & \ast & ✗ & \checkmark & \checkmark &], & 125 \\
[\ \ast & \ast & \ast & \ast & ✗ & ✗ & \checkmark & \checkmark &], & 125 \\
[\ \ast & \ast & \ast & ✗ & ✗ & ✗ & \checkmark & \checkmark &], & 125 \\
[\ \ast & \ast & \checkmark & ✗ & ✗ & ✗ & \checkmark & \checkmark &], & 28 \\
[\ \ast & \checkmark & \checkmark & ✗ & ✗ & ✗ & \checkmark & \checkmark &], & 0
\end{array}
$$

Thus, at the end, 31154 is decomposed into 01100011, which is the bit representation of the letter **c**. By applying the same decryption algorithm to all the elements of C, Bob finally recovers the encrypted message **crypto**.

Challenge 11.8: Encryption with Knapsacks

There are some risks involved in encrypting long messages by repeatedly using hard knapsacks of small length. In the next puzzle, you must recover the encrypted message that Alice sent to Bob. The private and public keys are different from those generated in the previous example. However, you know that the length of H is the same as before: $n = 8$. Can you recover the message even without knowing the public key?

333644, 560458, 138874, 389938, 472518, 394128, 138874, 472518, 560458,
138874, 465914, 384730, 550286, 138874, 462498, 472518, 638226, 560458,
138874, 634810, 389938, 138874, 628828, 472518, 465914, 384730, 550286,
628828, 472518, 465914, 551060, 478500, 560458, 138874, 394128, 550286,
389938, 550286, 394128, 138874, 465914, 634810, 138874, 394128, 550286,
472518, 462498, 551060, 465914, 633018, 295184, 138874, 465914, 384730,
550286, 633018, 138874, 472518, 394128, 550286, 138874, 468480, 634810,
465914, 138874, 478500, 550286, 394128, 465914, 472518, 551060, 468480,
295184, 138874, 472518, 468480, 383956, 138874, 472518, 560458, 138874,
389938, 472518, 394128, 138874, 472518, 560458, 138874, 465914, 384730,
550286, 633018, 138874, 472518, 394128, 550286, 138874, 478500, 550286,
394128, 465914, 472518, 551060, 468480, 295184, 138874, 465914, 384730,
550286, 633018, 138874, 383956, 634810, 138874, 468480, 634810, 465914,
138874, 394128, 550286, 389938, 550286, 394128, 138874, 465914, 634810,
138874, 394128, 550286, 472518, 462498, 551060, 465914, 633018, 301166

A screenshot visualizing the Merkle-Hellman knapsack cryptosystem is shown in Figure 11.19.

A screenshot of CT2 (Figure 11.15) shows a ready-to-run lattice-based attack against the Merkle-Hellman knapsack cryptosystem.

11.8.2 Lattice-Based Cryptanalysis

Encrypting a message using a hard knapsack that is at least as long as the message is much more secure, but still vulnerable. We will demonstrate this vulnerability using a specially designed lattice.

Given a public key with a hard knapsack H and an encrypted message c, we can represent each element of H as a vector in a $|H|$-dimensional lattice. We need $|H|$ dimensions in order to guarantee that they form a basis of the lattice L. In order to guarantee that they are linearly independent, we simply augment the transpose H to the identity matrix with dimension $|H| - 1$.

As an example, let's take H with length 8:

$$H = [h_1, h_2, \cdots, h_8]$$

Then the constructed lattice has the form:

$$L = \begin{pmatrix} 1 & 0 & 0 & 0 & 0 & 0 & 0 & 0 & h_1 \\ 0 & 1 & 0 & 0 & 0 & 0 & 0 & 0 & h_2 \\ 0 & 0 & 1 & 0 & 0 & 0 & 0 & 0 & h_3 \\ 0 & 0 & 0 & 1 & 0 & 0 & 0 & 0 & h_4 \\ 0 & 0 & 0 & 0 & 1 & 0 & 0 & 0 & h_5 \\ 0 & 0 & 0 & 0 & 0 & 1 & 0 & 0 & h_6 \\ 0 & 0 & 0 & 0 & 0 & 0 & 1 & 0 & h_7 \\ 0 & 0 & 0 & 0 & 0 & 0 & 0 & 1 & h_8 \end{pmatrix}$$

All rows are linearly independent. Furthermore, we add another row to the lattice by inserting the encrypted number c as the last element.

$$L = \begin{pmatrix} 1 & 0 & 0 & 0 & 0 & 0 & 0 & 0 & h_1 \\ 0 & 1 & 0 & 0 & 0 & 0 & 0 & 0 & h_2 \\ 0 & 0 & 1 & 0 & 0 & 0 & 0 & 0 & h_3 \\ 0 & 0 & 0 & 1 & 0 & 0 & 0 & 0 & h_4 \\ 0 & 0 & 0 & 0 & 1 & 0 & 0 & 0 & h_5 \\ 0 & 0 & 0 & 0 & 0 & 1 & 0 & 0 & h_6 \\ 0 & 0 & 0 & 0 & 0 & 0 & 1 & 0 & h_7 \\ 0 & 0 & 0 & 0 & 0 & 0 & 0 & 1 & h_8 \\ 0 & 0 & 0 & 0 & 0 & 0 & 0 & 0 & c \end{pmatrix}$$

Again, all the rows are linearly independent. However, we know that c is an exact sum of some h's. Our strategy is to find another basis of this lattice that contains at least one vector with a last element equal to 0. Moreover, since it can be represented as a linear combination of the vectors of the current basis, we know that this particular vector will only have elements equal to 0 or -1. A value of 0 in column i tells us that h_i doesn't participate in the decomposition of c, while -1 indicates that h_i is used in the construction of c.

But how to find such a basis? The following algorithm will help us:

Theorem 11.1 (Lenstra, Lenstra, Lovász [4, 5]) Let $L \in \mathbb{Z}^n$ be a lattice spanned by $B = \{b_1, \ldots, b_n\}$. The L^3 algorithm returns a reduced lattice basis $\{v_1, \ldots, v_n\}$ with

$$\|v_i\| \leq 2^{\frac{n(n-1)}{4(n-i+1)}} \det(L)^{\frac{1}{n-i+1}} \text{ for } i = 1, \ldots, n \tag{11.7}$$

in time polynomial in n and in the bit size of the entries of the basis matrix B.

In other words, the L^3 algorithm will produce another basis of the lattice consisting of vectors with restrained norms given by the inequality in Theorem 11.1. The L^3 algorithm is already built into SageMath.

Example Let's say Eve intercepts a message between Alice and Bob that is encrypted using the Merkle-Hellman knapsack cryptosystem. Since everyone has access to the public key of the cryptosystem, Eve also has it. The intercepted message C is:

$$C = [318668, 317632, 226697, 388930, 357448, 297811,$$
$$344670, 219717, 388930, 307414, 220516, 281175]$$

The corresponding public key hard knapsack H is the vector:

$$H = [106507, 31482, 107518, 60659, 80717, 81516, 117973, 87697]$$

To recover the message, Eve must decrypt each element c in C. For example, let's start with $c = 318668$.

```
sage: H = [106507, 31482, 107518, 60659,
           80717, 81516, 117973, 87697]
sage: c = 318668
```

Then we start to construct the lattice by first building the identity matrix:

```
sage: I = identity_matrix(8)
sage: I
```

$$\begin{pmatrix} 1 & 0 & 0 & 0 & 0 & 0 & 0 & 0 \\ 0 & 1 & 0 & 0 & 0 & 0 & 0 & 0 \\ 0 & 0 & 1 & 0 & 0 & 0 & 0 & 0 \\ 0 & 0 & 0 & 1 & 0 & 0 & 0 & 0 \\ 0 & 0 & 0 & 0 & 1 & 0 & 0 & 0 \\ 0 & 0 & 0 & 0 & 0 & 1 & 0 & 0 \\ 0 & 0 & 0 & 0 & 0 & 0 & 1 & 0 \\ 0 & 0 & 0 & 0 & 0 & 0 & 0 & 1 \end{pmatrix}$$

We add another row full of zeros:

```
sage: I = I.insert_row(8, [0 for x in range(8)])
sage: I
```

$$\begin{pmatrix} 1 & 0 & 0 & 0 & 0 & 0 & 0 & 0 \\ 0 & 1 & 0 & 0 & 0 & 0 & 0 & 0 \\ 0 & 0 & 1 & 0 & 0 & 0 & 0 & 0 \\ 0 & 0 & 0 & 1 & 0 & 0 & 0 & 0 \\ 0 & 0 & 0 & 0 & 1 & 0 & 0 & 0 \\ 0 & 0 & 0 & 0 & 0 & 1 & 0 & 0 \\ 0 & 0 & 0 & 0 & 0 & 0 & 1 & 0 \\ 0 & 0 & 0 & 0 & 0 & 0 & 0 & 1 \\ 0 & 0 & 0 & 0 & 0 & 0 & 0 & 0 \end{pmatrix}$$

Finally, we add the last column with H transposed and c. However, we will flip the sign of c; so the first vector of the reduced basis should have a last element equal to 0 and all other elements equal to 1 (instead of -1).

```
sage: L_helper = [[x] for x in H]   # vector of vectors
sage: L_helper.append([-c])
sage: L = I.augment(matrix(L_helper))
sage: L
```

$$\begin{pmatrix} 1 & 0 & 0 & 0 & 0 & 0 & 0 & 0 & 106507 \\ 0 & 1 & 0 & 0 & 0 & 0 & 0 & 0 & 31482 \\ 0 & 0 & 1 & 0 & 0 & 0 & 0 & 0 & 107518 \\ 0 & 0 & 0 & 1 & 0 & 0 & 0 & 0 & 60659 \\ 0 & 0 & 0 & 0 & 1 & 0 & 0 & 0 & 80717 \\ 0 & 0 & 0 & 0 & 0 & 1 & 0 & 0 & 81516 \\ 0 & 0 & 0 & 0 & 0 & 0 & 1 & 0 & 117973 \\ 0 & 0 & 0 & 0 & 0 & 0 & 0 & 1 & 87697 \\ 0 & 0 & 0 & 0 & 0 & 0 & 0 & 0 & -318668 \end{pmatrix}$$

To reduce the basis, we will now apply the L^3 algorithm by simply calling the SageMath `LLL()` function.

```
sage: L.LLL()
```

$$\begin{pmatrix} 0 & 1 & 0 & 0 & 0 & 1 & 1 & 1 & 0 \\ -1 & 1 & 0 & 1 & -1 & -2 & -2 & 2 & 1 \\ 3 & 1 & 2 & -1 & 1 & 1 & -1 & 1 & 1 \\ 1 & -1 & -2 & -1 & -3 & -1 & 1 & 1 & 1 \\ 2 & -2 & -1 & 1 & 0 & 2 & -3 & 1 & 1 \\ 0 & 0 & 3 & -4 & -2 & 1 & 0 & 0 & 0 \\ -1 & 3 & -1 & 3 & 0 & 0 & -1 & -3 & 2 \\ 0 & -1 & 1 & 4 & 0 & 0 & 0 & 0 & 4 \\ -2 & -1 & -2 & -3 & 1 & -1 & 2 & 1 & 3 \end{pmatrix}$$

The first candidate (the shortest vector in the reduced basis) is the one we were looking for:

```
sage: L.LLL()[0][:-1].dot_product(vector(H))
318668
```

So the binary representation of the encrypted character is 01000111. Using SageMath again, we can check the corresponding letter:

```
sage: bs = ''.join([str(x) for x in L.LLL()[0][:-1]])
sage: bs
'01000111'

sage: chr(int(bs,2))
'G'
```

Challenge 11.9: LLL

In the example in Section 11.8.2 above, G is the first letter of the recovered text. Using SageMath, lattices, and the LLL algorithm, can you recover the rest of the text?

The CT2 screenshots in Figures 11.12 and 11.13 show a mouse-driven visualization of reducing a two-dim basis with Gauss and a ready-to-run LLL implementation for reducing the basis of higher-dimensional bases.

11.9 Lattices and RSA

RSA is one of the first asymmetric cryptosystems. The inner workings of RSA have been thoroughly explained and demonstrated in Chapter 5 of this book. This section assumes that you are already familiar with how the RSA cryptosystem works. However, we will briefly review the basics of key generation for the RSA algorithm using SageMath. Then we show how RSA can be attacked using lattices.

11.9.1 Textbook RSA

The raw RSA method (without padding) is called textbook RSA and is not suitable for practical use as either an encryption or signature method [6]. It consists of the following steps (see Section 5.10.2 for more details).

- Two large distinct primes p and q are generated.
- Their product $n = pq$ is called the modulus.
- Then, we pick a number e, such as e is relatively prime to $\phi(n)$, Euler's totient function. We define e as the public-key exponent.
- We compute d as the modular multiplicative inverse of e modulo $\phi(n)$. We define d as the private-key exponent.
- The pair (n, e) is the public key.
- The pair (n, d) is the private key.

To avoid some known attacks on RSA, we need to choose our parameters wisely. Some of the requirements and recommendations can be found in [7].

Now let's encrypt the word asymmetric using SageMath and the RSA cryptosystem. First, we need to think about the encoding strategy (i.e., the translation of strings into numbers). Throughout this section, we will use the following encoding procedure:

- Let's denote the string to be encoded as $S = s_1 s_2 \cdots s_n$.
- We replace each symbol s_i in the string with its decimal ASCII code representation. For example, the symbol "g" is replaced by "103."
- Then each decimal ASCII code is replaced by its binary representation. For reversibility purposes, as long as the length of the binary representation is less than 8, we append at the beginning as many 0s as necessary. For example, the binary representation of 103 is 1100111. However, the length of the binary representation is seven, so we add another zero at the beginning to get 01100111.
- We form an $8n$-bit string by concatenating the n strings of 8 bit each, starting with the 8-bit representation of the first letter of S from the left.
- Finally, we convert S to an integer in decimal representation.

For example, let's encode the word *asymmetric*. First, we replace each symbol of S with its corresponding decimal ASCII value:

```
sage: S = "asymmetric"
sage: S_ascii = [ord(x) for x in S]
sage: S_ascii
[97, 115, 121, 109, 109, 101, 116, 114, 105, 99]
```

Then we replace each element in `S_ascii` with its binary equivalent. To get rid of the leading `0b` of the binary strings, we use `[2:]`.

```
sage: S_bin = [bin(x)[2:].zfill(8) for x in S_ascii]
sage: S_bin
['01100001', '01110011', '01111001', '01101101', '01101101▸
 ▸',
 '01100101', '01110100', '01110010', '01101001', '01100011▸
 ▸']
```

Finally, we concatenate all the elements in `S_bin` and convert this concatenation to a decimal number:

```
sage: SS = Integer(''.join(S_bin),2)
sage: SS
460199674176765747685731
```

To check the reversibility of the encoding procedure, let's decode the result back:

```
sage: SS_bin = bin(SS)[2:]
sage: SS_bin
'1100001011100110111100101101101011011010110010101110100011▸
 ▸1100100110100101100011'
sage: len(SS_bin)
79
```

```
sage: while len(SS_bin) %
....:     SS_bin = '0' + SS_bin
sage: SS_ascii = [chr(int(SS_bin[x*8:8*(x+1)],2))
....:      for x in range(len(SS_bin)/8)]
sage: ''.join(SS_ascii)
'asymmetric'
```

When we are done with the encoding procedure, we initialize the RSA parameter generation step and generate p, q, and n:

```
sage: b = 512
sage: p = random_prime(2**b-1, lbound=2**(b-1)+2**(b-2))
```

In the previous example, we generated a random prime number in the interval $[2^{b-1} + 2^{b-2}, 2^b - 1]$. Let's say we have two primes in this interval; that is,

$$p = 2^{b-1} + 2^{b-2} + \rho_1$$
$$q = 2^{b-1} + 2^{b-2} + \rho_2$$

for some ρ_1 and ρ_2. When we multiply the primes, we have:

$$\begin{aligned} p \cdot q &= (2^{b-1} + 2^{b-2} + \rho_1)(2^{b-1} + 2^{b-2} + \rho_2) = \\ &= 2^{2b-2} + 2^{2b-3} + 2^{b-1}\rho_2 + 2^{2b-3} + 2^{2b-4} + \\ &+ 2^{b-2}\rho_2 + \rho_1 2^{b-1} + \rho_1 2^{b-2} + \rho_1\rho_2 = \\ &= 2^{2b-2} + 2 \cdot 2^{2b-3} + \Omega = \\ &= 2^{2b-2} + 2^{2b-2} + \Omega = \\ &= 2 \cdot 2^{2b-2} + \Omega = 2^{2b-1} + \Omega > 2^{2b-1} \end{aligned}$$

This guarantees that the bit length of their product is $2b$. The method `nbits()` returns the bit length of a number.

```
sage: p.nbits()
512
sage: q = random_prime(2**b-1, lbound=2**(b-1)+2**(b-2)); ▸
  ▸q.nbits()
512
sage: N = p*q; N.nbits()
1024
```

It's time to choose the public exponent e. A common choice of value for e is $2^{16} + 1$.

```
sage: e = 2**16 + 1; e
65537
```

SageMath has a built-in function `euler_phi()`. However, if we directly type `euler_phi(N)`, SageMath will try to factor $N = pq$. It's better to manually calculate the number of positive integers that are relatively prime to N and not greater than N (this is easy as we know the factors of N and can use the formula—see Section 5.8.2):

```
sage: phi_N = (p-1)*(q-1)
```

Now that we have $\phi(n)$, we can calculate d using the built-in function `inverse_mod()`:

```
sage: d = inverse_mod(e,phi_N)
```

Let's make sure that $ed \equiv 1 \mod \phi(n)$:

```
sage: assert d < phi_N
sage: e*d %
1
```

We are ready to encrypt the encoding `SS` of the message "asymmetric." The encryption can be calculated directly using `SS**e%N`. However, we will use the built-in function `power_mod()`, which is much faster than the direct calculation. (On a 2020 Mac Mini, the direct calculation took 62 ms of CPU time, and `power_mod` took less than 1 ns. "Much faster" is indeed true, though not in our human experience of time.)

```
sage: encrypted = power_mod(SS,e,N)
```

To decrypt the message:

```
sage: decrypted = power_mod(encrypted,d,N)
sage: decrypted
460199674176765747685731
```

Challenge 11.10: RSA

Alice and Bob again decided to use their own encoding scheme and RSA implementation to secure their communication. The encoding scheme used is quite simple: Alice translates each letter from the plaintext to its decimal ASCII representation. In this way, Alice sends as many encrypted messages as the length of the original unencrypted message.

This scheme has one major drawback: When large plaintexts are encrypted, the collection of intercepted encrypted messages is vulnerable to frequency analysis attacks. To work around this, Alice and Bob renew their RSA keys when they reach a threshold number of messages sent.

However, they made another mistake. The RSA public key is:

$N = 6842176325842682031847125968264734689729927045799136522752318721517927993776878211790146955615938091152726743120686152933384202585716854144646470442805080811450030171938063091890893578048911727269235209816411041382264267029865784731222580175578439986459497511654781585601147479342995641817727780618783610106 1$

$e = 127$

Eve intercepted the following 11 messages (sent in this order):

$c1 = 20842273115788965044434568911351496365025941709963878891654635864614247250595415337877670412884297380645809556224513056164981861313077151294657843754553657687729573741274326907928991221246371764225295669345864765449163254397423969283552347 12078183$

$c2 = 20893454506753506195646780042379588087816651548061669824679147298111722465210531962697364936758882446841738989887223022907938140873068062126260148654403891017812919490588535501594235621529922408698085740859622639421733168633622772463221300 97359903249313$

Challenge 11.10 (continued)

$c3 =$ 15351988337964905099701960376347076532439043589617467841763029200910162258670 08339499789389756629361807066970996087640185785892959509059061040916309825290667 8 94027034857864536708707934721612923379017354322495378040804021369349050072497908 4 761978285646999197843456

$c4 =$ 27904208858505919506689042950498726643957813119400643293322771534900822430923 58584878184845560361208122657510080122657048421202659485825208921784032883790670 8 27601630611484289723657470143474224631114266432824789017052059285116164747098348 9 359620795002699

$c5 =$ 14562438053393942865563565506076319964337086564418653275680839364685346358348 26387270812896842341268168773581646273040911274525651721561895389722762725689853 3 45485804529793195837639495561047186775624449872519165568427413465770079493980103 1 70176004536034918 4

$c6 =$ 37370666535363066168961624931547694539547602092327751979114535130065929115448 53295308247797277717029030440472567012693658669860452964879358165926306097054693 8 25994483895291117047826544861482249517767722025270434054525178543495547662794471 7 241828329

$c7 =$ 57018830572739461491984788995673407709179639540390781558806685226845173001582 25274094629999215259149699283194431626990778523591567618587926423246578367287634 2 03463688598234376481269695823515506081211968626320267283411565778900665855308128 3 5468253729909927010 71

$c8 =$ 45667206025566800148122417818312587397117202643844088800399634507595677539812 53180438980063337356363520302653029580826718653786950185499999758581316561045994 5 09932304144989007625800895390336098999862209881749752726145591849769024710472559 4 1225650820350576211757 73

$c9 =$ 16862273135393186478865050591393354467469966242203319758781467127096457948108 88946761963350628222465151134851313061316471360362284419753231478405415985364477 2 39725795743107788714671289354822510203766481055710075778057712258940862586529599 5 70943303841410139029504

$c10 =$ 34184916515805352683250586319278293122416282275978861283289174865948602610 67 37910383076021602844922593096922368623753010447251025423582399396141954459468286 7 68770464472831982875580635659918156592765109749350304246573018358473129678989

$c11 =$ 51124105796698521835346083828711155727857040260880863977784321358860541909 82 58110942799596774222498729431095652955025535198037264822351140404848680838205182 1 39572299518969847143074424801281971337942843849336646216609681813575205566735348 8 388471305370330810546875

Can you decrypt each of these messages and reconstruct the original message?

Screenshots from CT1 of a ready-to-run implementation of an attack against textbook RSA can be found in Section 11.12.1.

11.9.2 Lattices versus RSA

In [8] and [9] a whole new family of attacks on RSA is published, attacks that use lattices and lattice reduction algorithms.

As we showed earlier, it is easy (using SageMath) to find the roots of a polynomial in a single variable over the integers. However, finding the roots of a modular polynomial is hard; that is,

$$f(x) \equiv 0 \mod N$$

Let N be a large composite integer of unknown factorization, and let us have a univariate integer polynomial (a polynomial in a single variable x) f of degree r; that is,

$$f(x) = x^r + a_{r-1}x^{r-1} + a_{r-2}x^{r-2} + \cdots + a_1 x + a_0$$

Suppose further that there exists an integer solution x_0 for the modular equation $f(x) \equiv 0 \mod N$, such that $x_0 < N^{\frac{1}{r}}$. D. Coppersmith showed how we can recover this value in polynomial time by using Theorem 11.2 of Howgrave-Graham, which we cite in the form of Alexander May's survey (see [10, p. 6]).

Theorem 11.2 Let $n, m \in \mathbb{N}$ and $g(x)$ be a univariate polynomial with n monomials. If we have some constraint X for $|x|$ and the following equations hold:

$$g(x_0) \equiv 0 \mod N^m \text{ and } |x_0| \le X$$

$$||g(xX)|| < \frac{N^m}{\sqrt{n}} \tag{11.8}$$

Then $g(x_0) = 0$ over the integers.

Remarks:

- A monomial is a single summand of a polynomial.
- Normally one would take $n - 1$ as the degree of the polynomial g. But the only condition is that g has at most n nonzero summands, so the degree can be higher than $n - 1$.
- The term $||g(xX)||$ is known as the polynomial norm, defined for any $f \in \mathbb{R}[x]$ with $f(x) = \sum_{i=0}^{k} c_i x^i$ to be $||f|| = \sqrt{\sum_{i=0}^{k} c_i^2}$. Thus, if we change the notation of $g(xX)$ to, say, $f(x) = g(xX)$, then the graph of the function f is created by horizontally compressing the graph of g by a factor of X (we had $X > 1$). The norm for $g(xX)$ is then $||f|| = ||g(xX)|| = \sqrt{\sum (a_i X^i)^2}$ and $a_i X^i = c_i$.

The reasons for using lattices:

- If we have some polynomials that have the same root x_0 over N^m, we can represent each of them as a row of a lattice. Then any linear combination of rows from the lattice will yield another polynomial with a root x_0.
- Then, by using the LLL algorithm on the specially designed lattice, we can find another reduced lattice basis in polynomial time, such that the norm of the shortest vector from the reduced basis successfully satisfies the inequality 11.2 from Theorem 11.2.
- Let's define the shortest vector in the reduced basis as $v = (v_0, v_1, \cdots, v_n)$. We construct the polynomial $g(x)$, such that,

$$g(x) = v_0 + \frac{v_1}{X}x + \frac{v_2}{X^2}x^2 + \cdots + \frac{v_n}{X^n}x^n = \sum_{i=0}^{N} v_i \left(\frac{x}{X}\right)^i \tag{11.9}$$

Since $g(x)$ has $n + 1$ momomials and is on the lattice, we have:

$$g(x_0) \equiv 0 \mod N^m$$

$$|x_0| \le X$$

$$\deg(g) = n$$

$$||g(xX)|| < \frac{N^m}{\sqrt{n+1}}$$

Following the results of Theorem 11.2, we can conclude that $g(x_0) = 0$ over the integers.

We can easily create polynomials that have the same root x_0 over N^m (we follow [9, p. 69, p. 80]). Consider the family of polynomials $g_{i,j}(x)$, such that

$$g_{i,j}(x) = x^j N^{m-i} f^i(x) \text{ for } i \in \{0, \ldots, m-1\}, j \in \{0, \ldots, \deg f - 1\} \qquad (11.10)$$

By construction, they all have the same root x_0 over N^m; that is, $g_{i,j}(x_0) \equiv 0 \mod N^m$. The larger the value of m, the more polynomials we can construct. The more polynomials we construct, the larger the lattice, and the longer it will take to reduce the lattice.

Now imagine that Eve intercepted a series of plaintext messages between Alice and Bob. The messages were:

The password for AES usage is: `4{8dXY!`
The password for AES usage is: `31kTbwj`
The password for AES usage is: `2rr#ETh`
...
The password for AES usage is: `&H,45zU`

Then, Alice and Bob start exchanging AES-encrypted files using the communicated password. When a new password is received, they immediately start using it. However, they realize that this is completely insecure and increase their security by using RSA.

They use the same encoding procedure that was demonstrated at the beginning of this section. As we showed, the word `asymmetric` is encoded into the decimal number `460199674176765747685731`.

Let's say Alice wants to send an RSA-encrypted string message `S` to Bob. She first encodes it to the decimal integer D. To denote the message, we use the uppercase D rather than the more common m, since m is the exponent of N in our context. Also, an uppercase M is not a good choice because it will be used later in the code examples for a matrix. Then she encrypts the number D using Bob's public key (N, e), that is $c = (D^e) \mod N$, and sends the encrypted message c over the insecure channel. Bob recovers the original message D using his private exponent; that is, $c^d = D \mod N$. Eve intercepts c.

Bob's public key has parameters $(N, 3)$, where the bit length of N is 512. The predictive nature of the message (popularly called "stereotyped messages") can lead to a devastating attack. Eve knows that the structure of the string message S is

$$S = \texttt{"The password for AES usage is: } C_1C_2C_3C_4C_5C_6C_7\texttt{"}$$

for some characters C_i. Before encrypting, Alice must translate each character to its ASCII binary string representation. Let's denote the binary translation function as T_1; that is, $T_1(\texttt{"xy"}) = T_1(\texttt{"x"})||T_1(\texttt{"y"})$, where the symbol || denotes the concatenation of strings.

With this in mind, we can write:

$$S' = T_1(S) = T_1(\texttt{"The password for AES usage is: "})||T_1(\texttt{"C}_1\texttt{C}_2\cdots\texttt{C}_7\texttt{"})$$

After this translation, Alice reads the final binary string as a decimal number. Let's call this function $T_2(S')$.

Each ASCII decimal representation of C_i is in the interval $[0, 255]$. Let's call the symbol with ASCII decimal representation 0 C_{00}, and the symbol with ASCII decimal representation 255 C_{ff}. So we choose indices in hexadecimal notation. For simplicity, let's denote

$$B = \texttt{"The password for AES usage is: "}.$$

With the encoding procedure in mind, we can conclude that:

$$T_2(T_1(B||C_{00}C_{00}\cdots C_{00})) < T_2(T_1(B||C_1C_2\cdots C_7)) < T_2(T_1(B||C_{ff}C_{ff}\cdots C_{ff}))$$

Let's introduce two new variables: a and X, such that:

$$a = T_2(T_1(B||C_{00}C_{00}\cdots C_{00}))$$
$$X = T_2(T_1(C_{ff}C_{ff}\cdots C_{ff}))$$

Since Eve knows c and a, she can reconstruct D if she is able to find a positive integer $x < X$ that satisfies the equation

$$(a + x)^3 \equiv c \mod N.$$

So we search for x such that

$$(a + x)^3 - c \equiv 0 \mod N$$

In fact, x denotes the difference between

$$T_2(T_1(C_1C_2\cdots C_7)) \quad \text{and} \quad T_2(T_1(C_{00}C_{00}\cdots C_{00})).$$

Let's pause for a moment and implement the current polynomial using *SageMath*. First, we introduce the `encode()` function—it is equivalent to $T_2(T_1(D))$. Here is an example of how to call this function and what it outputs:
`encode("A"): 65, encode("AB"): 16706, encode("ABC"): 4276803`

```
sage: def encode(D):
....:     return Integer(''.join([bin(ord(x))[2:].
....:     zfill(8) for x in D]),2)
```

We introduce the expected starting characters of the encrypted message.

```
sage: B = "The password for AES usage is: "
```

Now, we insert the values of $C_{00}C_{00}\cdots C_{00}$ and $C_{ff}C_{ff}\cdots C_{ff}$.

```
sage: padding = ''.join(['\x00' for x in range(7)])
sage: X_str = ''.join(['\xff' for x in range(7)])
```

We continue by calculating the values of a and X:

```
sage: a_str = B + padding
sage: a_const = encode(a_str)
sage: X_const = encode(X_str)
```

We also have to define $e = 3, c = 533\ldots455$, and $\mathtt{N}_{\mathtt{const}} = 871\ldots499$. You can take the rather long values from the Challenge 11.11 challenge or from the helper scripts on our website see `https://www.cryptool.org/en/documentation/ctbook/sagemath`.

```
sage: e=3
sage: c=533...
sage: N_const=871...
```

We introduce the polynomial f with the three variables X, N, and a, which will be replaced later by `X_const`, `N_const`, and `a_const`.

```
sage: R.<X,N,a> = ZZ[]
```

Now we are ready to construct the polynomial $f(X)$:

```
sage: f = (X+a)**3 - c
sage: f
X^3 + 3*X^2*a + 3*X*a^2 + a^3 - c
```

We don't know x_0. However, we do know a good upper bound for x_0; that is, $x_0 < X$. Since $e = 3$, the degree of our polynomial is 3. For this particular case, let's set m to the smallest possible value; that is, $m = 1$:

```
sage: f.degree()
3
sage: m = 1
```

Our lattice will be of dimension 4—we have exactly 3 polynomials $g_{i,j}$, as well as the final polynomial f.

```
sage: dim = 4
sage: M = matrix(dim,dim)
```

We construct the polynomials as in according to Theorem 11.2. Following the strategy of the lattice construction, we have 3 polynomials $g_{0,j} (j = 0, 1, 2)$ and f in the last row, so we have to define the following lattice:

$$\begin{pmatrix} N & 0 & 0 & 0 \\ 0 & NX & 0 & 0 \\ 0 & 0 & NX^2 & 0 \\ a^3 - c & 3a^2X & 3aX^2 & X^3 \end{pmatrix}$$

To do this, we first need to define the helper function `get_ext_monoms()`. It will help us to extract all monomials from a given polynomial, but with the coefficients included.

```
sage: def get_ext_monoms(ff):
....:     ff_m = ff.monomials()
....:     ff_coefs = [ff.monomial_coefficient(x) for x in ▶
  ▶ff_m]
....:     ff_monoms_ext = [ff_m[x]*ff_coefs[x]
....:     for x in range(len(ff_m))]
....:     return ff_monoms_ext
```

For example:

```
sage: get_ext_monoms(f)
[X^3, 3*X^2*a, 3*X*a^2, a^3, -c]
```

However, there is a problem here, because later we sort by powers of X, but here a^3 and $-c$ are treated as separate monomials. That's why we substitute `N_const` for N and `a_const` for a just before calling `get_ext_monoms()`:

```
sage: for i in range(m):
....:     for j in range(e):
....:       g = X**j * N**(m-i) * (f**i)
....:       g = g.subs({N:N_const, a:a_const})
....:       g_monoms_ext = get_ext_monoms(g)
....:       for monom in g_monoms_ext:
....:         row_pos = e*i+j
....:         column_pos = monom.degree()
....:         M[row_pos,column_pos] = monom.subs({X:X_const▶
  ▶})
```

Note that we don't need the first line of code in the listing above since $m = 1$ at the moment. However, in the example that follows later, we will need it. The same goes for the first line of the next listing.

Finally, we append the final row of the lattice – the values of the corresponding monoms of f after substituting `N_const` for N, `a_const` for a, and `X_const` for X.

```
sage: fg=f**m
sage: fg = fg.subs({N:N_const, a:a_const})
sage: fg_monoms_ext = get_ext_monoms(fg)
sage: for fg_monom in fg_monoms_ext:
....:     pos = fg_monom.degree()
....:     M[dim-1, pos] = fg_monom.subs({X:X_const})
```

Our lattice is ready. We can start the lattice reduction algorithm:

```
sage: B = M.LLL()
```

The shortest vector `B[0]` in our reduced basis contains the coefficients we need to construct the polynomial g over the rational ring. We introduced $g(x) = v_0 + \frac{v_1}{X}x + \frac{v_2}{X^2}x^2 + \cdots + \frac{v_n}{X^n}x^n = \sum_{i=0}^{N} v_i \left(\frac{x}{X}\right)^i$, in (11.9) after Theorem 11.2. We can easily construct it using SageMath:

```
sage: R.<x> = QQ[]
sage: g = sum([B[0][i]*(x**i)/(X_const**i) for i in range(▸
  ▸dim)])
```

According to Theorem 11.2, the last polynomial should have a solution over the integers. And indeed it does:

```
sage: sol = g.roots(ring=ZZ)[0][0]
sage: type(sol)
<type 'sage.rings.integer.Integer'>
```

Now let's define another helper function `decode()`, which translates back an encoded message $Z \in \mathbb{N}$; that is, $T_1^{-1}(T_2^{-1}(Z))$:

```
sage: def decode(n):
....:   nn = str(bin(n)[2:])
....:   while len(nn) %
....:     nn = '0' + nn
....:     return ''.join([chr(int(nn[x*8:8*(x+1)],2))
....:     for x in range(len(nn)/8)])
```

Our last step is simply to decode the solution:

```
sage: decode(sol)
```

Challenge 11.11: RSA Attack for Small Exponents

Eve has inspected Bob's public key and intercepted the encrypted message c:
$N = 87105263120665488502276714807618005091167547295414522403403$
$85826044593797820258419597692701154128696972650359076718923667$6
$742077646358458219594112627604 99$
$e = 3$
$c = 53324798259879463395628746557109686362316082380119849133012$
$62447142261322575224549371305566272165061124930469733249577503 4$
$76282414453312278092919951644 55$
As an exercise, can you recover the original message using the above lattice attack?

After this first case, which we could solve with $m = 1$ (see (11.10) after Theorem 11.2), we will now look at another example where we need a larger m.

Challenge 11.12: Harder RSA Attack for Small Exponents

Soon after Eve's successful lattice-reduction attack, Alice and Bob were aware of the insecure scheme they were using to encrypt their correspondence. However, they thought that this was possible because they were using passwords that were too short. They increased the length of their passwords from 7 to 13 characters.

Bob's public key was left intact. The newly intercepted message c is:

Challenge 11.12 (continued)

$c =$ 74875898628081924295230958863232737993265641657662214847638
48360348800688681255935199851002238854425476872148791894788351 4
84389959862897996873514056282948

Can you recover the newly exchanged password? Here are a few things to consider:

- First, let's try with $m = 1$. Why does the attack fail?
- Now let's try with $m = 2$. The dimension of the new lattice will be $em + 1$.
- The polynomial f is still the same. However, we need more helper polynomials $g_{i,j}$ to construct our larger lattice. Equation (11.10) gives 6 polynomials g_{ij} and the coefficients of the expanded polynomial $(f(x))^2$ in the last row. Finally, you should construct the following lattice:

$$\begin{pmatrix} N^2 & 0 & 0 & 0 & 0 & 0 & 0 \\ 0 & N^2X & 0 & 0 & 0 & 0 & 0 \\ 0 & 0 & N^2X^2 & 0 & 0 & 0 & 0 \\ N(a^3-c) & 3Na^2X & 3NaX^2 & NX^3 & 0 & 0 & 0 \\ 0 & N(a^3-c)X & 3Na^2X^2 & 3NaX^3 & NX^4 & 0 & 0 \\ 0 & 0 & N(a^3-c)X^2 & 3Na^2X^3 & 3NaX^4 & NX^5 & 0 \\ f_0 & f_1 & f_2 & f_3 & 15a^2X^4 & 6aX^5 & X^6 \end{pmatrix}$$

$$f_0 = (a^3 - c)^2$$

$$f_1 = 6a^2(a^3 - c)X$$

$$f_2 = a(15a^3 - 6c)X^2$$

$$f_3 = (20a^3 - 2c)X^3$$

Challenge 11.13: Even Harder RSA Attack for Small Exponents

For the next stereotype challenge, you have the following parameters:

$N =$ 11225354852531229312705821542018938144842129865964887302659
52745410910072681138663483074618935128265451387560973724847297 0
85037894275160093985827338655154551777903941595546130947578089 8
54083283079940232287825301027638695687835609359030774683694898 7
21093343101189799502070711082802196203627374677603082274488 37

$e = 7$

$c =$ 10670654096244930306696108877164811975817784621106090830133
61442402896883715423232034163629274021482627819113678709672437 6
91954131729343929285737922272207153114174438757138142540189592 4
31327528606195874021248932484514678389202737947583108275543928 4
15736794504415568836663027223190290104631408291835053910921 71

This time the degree of the polynomial f is 7 and the password consists of 14 ASCII characters. You will need to try different values of m to construct a large enough, but still compact lattice.

In general, if Alice uses an encryption exponent of e, then for this type of attack to work, Eve must know $\frac{e-1}{e}$ of Bob's message; see [9, p. 96].

Screenshots of CT1 (Figures 11.9 to 11.11) and CT2 (Figure 11.16) of ready-to-run lattice-based implementations of attacks against RSA can be found in Sections 11.12.1 and 11.12.2.

11.10 Lattice Basis Reduction

This chapter has thus far given a deeper, mathematical challenging outlook providing some algorithms for lattice basis reduction and their usage to break cryptosystems. At the end, we will briefly discuss the lattice-based procedures of the NIST standardization for PQC.

A given lattice has infinitely many different bases. The main goal of lattice basis reduction is to find (by using some lattice basis as an input) a basis that consists of short vectors, or, equivalently, a basis consisting of vectors that are pairwise nearly orthogonal. Thus, the reduced basis may lead to a solution of an underlying problem, like breaking a knapsack cryptosystem, as we have already shown in Section 11.8.2.

Let's first introduce the notion of Gram-Schmidt orthogonalization named after the mathematicians Jørgen Pedersen Gram and Erhard Schmidt.

Definition 11.26 With an ordered lattice basis $b_1, \cdots, b_m \in \mathbb{R}^n$ we associate the Gram-Schmidt orthogonalization $\hat{b}_1, \ldots, \hat{b}_m \in \mathbb{R}^n$ which can be computed from $b_1, \cdots, b_m$ together with the Gram-Schmidt coefficients $\mu_{i,j} = \frac{b_i \cdot \hat{b}_j}{\hat{b}_j \cdot \hat{b}_j}$ by the recursion

$$\begin{aligned} \hat{b}_1 &= b_1 \\ \hat{b}_i &= b_i - \textstyle\sum_{j=1}^{i-1} \mu_{i,j} \hat{b}_j \quad i = 2, \ldots, m \end{aligned}$$

Let $\text{span}(b_1, \ldots, b_{i-1})^\perp$ be the set of all vectors orthogonal to $\text{span}(b_1, \ldots, b_{i-1})$; that is,

$$\text{span}(b_1, \ldots, b_{i-1})^\perp = \{v \in \mathbb{R}^n \;\Big|\; v \cdot \sum_{j=1}^{i-1} x_j b_j = 0 \;\forall x_j \in \mathbb{R}\}.$$

The orthogonal projections of vectors b_j to $\text{span}(b_1, \ldots, b_{i-1})^\perp$ are named π_i
$\pi_i : \mathbb{R}^n \rightarrow \text{span}(b_1, \ldots, b_{i-1})^\perp, \quad \pi_i(b_j) := \sum_{t=i}^{j} \mu_{j,t} \hat{b}_t, \quad i = 1, \cdots, m.$

We have $\mu_{i,i} = 1$ and $\mu_{i,j} = 0$ for $i < j$. If the basis $b_1, \cdots, b_m$ is integral, then the vectors $\hat{b}_1, \cdots, \hat{b}_m$ and the Gram-Schmidt coefficients are rational. We can write the previous equations in matrix notation as:

$$(b_1, \cdots, b_m) = (\hat{b}_1, \cdots, \hat{b}_m)(\mu_{i,j})^T_{1 \le i,j \le m}$$

Definition 11.27 The ith successive minimum λ_i of a lattice L is defined as the minimum radius $r \in \mathbb{R}$ of an n-dimensional sphere B with center O that contains i linearly independent lattice vectors:

$$\lambda_i(L) = \min\{r \in \mathbb{R} \mid \dim(\text{span}\, L \cap B_{r,0}) \ge i\}.$$

Obviously λ_1 is the norm of the shortest nonzero lattice vector.

One of the strongest notions of lattice reduction is based on the work of Hermite [11], Korkine and Zolotarev [12–14], hence the notion HKZ-reduced:

Definition 11.28 A lattice basis $b_1, \ldots, b_m$ is called reduced in the sense of Hermite, Korkine, and Zolotarev or short HKZ-reduced if the following holds:

1. $|\mu_{i,j}| \leq \frac{1}{2}$ for $1 \leq j < i \leq m$,
2. $||\hat{b}_i|| = \lambda_1(L(\pi_i(b_i), \ldots, \pi_i(b_m))$ for $1 \leq i \leq m$.

The first vector of any HKZ-reduced lattice basis is a shortest lattice vector.

Let's take a look at two-dimensional lattices where we can easily find a shortest lattice vector with respect to the Euclidean norm by using the Gauss reduction algorithm. The process is similar to the process of calculating the greatest common divisor of two integers by applying the Euclidean algorithm.

Crypto Procedure 11.1: Gauß

input lattice basis $\{a, b\}$

repeat

$\{a, b\} = \{b - \lceil \frac{a \cdot b}{a \cdot a} \rfloor \cdot a, a\}$

until $||a|| \leq ||b|| \leq ||a - b||$

output Gauss reduced lattice basis $\{a, b\}$

For any real number x, $\lceil x \rfloor$ denotes the closest integer; that is, $\lceil x \rfloor = \lfloor x + 0.5 \rfloor$.

Example Let's run the Gauß reduction algorithm on a basis $B = \{a, b\} = \{(1, 7), (-1, 1)\}$ of a given lattice L in $\mathbb{Z}^2$.

Input: Lattice basis $\{a, b\} = \{(1, 7), (-1, 1)\}$

$$\begin{aligned}\{a, b\} &= \{b - \left\lceil \frac{a \cdot b}{a \cdot a} \right\rfloor \cdot a, a\} \\ &= \{(-1, 1) - \left\lceil \frac{(1, 7) \cdot (-1, 1)}{(1, 7) \cdot (1, 7)} \right\rfloor \cdot (1, 7), (1, 7)\} \\ &= \{(-1, 1) - \left\lceil \frac{6}{50} \right\rfloor \cdot (1, 7), (1, 7)\} \\ &= \{(-1, 1), (1, 7)\}\end{aligned}$$

Since $||b|| = \sqrt{50} > \sqrt{40} = ||a - b||$ we need to run another iteration:

$$\begin{aligned}\{a, b\} &= \{b - \left\lceil \frac{a \cdot b}{a \cdot a} \right\rfloor \cdot a, a\} \\ &= \{(1, 7) - \left\lceil \frac{(-1, 1) \cdot (1, 7)}{(-1, 1) \cdot (-1, 1)} \right\rfloor \cdot (-1, 1), (-1, 1)\}\end{aligned}$$

$$= \{(1,7) - \left\lceil \frac{6}{2} \right\rfloor \cdot (-1,1), (-1,1)\}$$
$$= \{(1,7) - 3 \cdot (-1,1), (-1,1)\}$$
$$= \{(4,4), (-1,1)\}$$

Now $||a|| = \sqrt{32} > \sqrt{2} = ||b||$ and we need another iteration:

$$\{a, b\} = \{b - \left\lceil \frac{a \cdot b}{a \cdot a} \right\rfloor \cdot a, a\}$$
$$= \{(-1,1) - \left\lceil \frac{(4,4) \cdot (-1,1)}{(4,4) \cdot (4,4)} \right\rfloor \cdot (4,4), (4,4)\}$$
$$= \{(-1,1) - \left\lceil \frac{0}{32} \right\rfloor \cdot (-1,1), (4,4)\}$$
$$= \{(-1,1), (4,4)\}$$

Since $||a|| = \sqrt{2} < \sqrt{32} = ||b|| < \sqrt{34} = ||a - b||$, the algorithm ends.

Output: $\{a, b\} = \{(-1,1), (4,4)\}$

The vector $(-1,1)$ is a shortest nonzero vector in L and $(4,4)$ is a shortest vector of L that is linear independent of $(-1,1)$.

In higher dimensions the calculation of an HKZ-reduced basis (see Definition 11.28) is very inefficient (no polynomial time algorithm is known). We define a hierarchy of notions that approximate HKZ-reduced bases in reasonable time.

Definition 11.29 An ordered lattice basis $b_1, \cdots, b_m \in \mathbb{R}^n$ is called *size-reduced* if $|\mu_{i,j}| \leq \frac{1}{2}$ for $1 \leq j < i \leq m$. An individual basis vector b_i is *size-reduced* if $|\mu_{i,j}| \leq \frac{1}{2}$ for $1 \leq j < i$.

A size-reduced lattice basis consists of vectors that are almost orthogonal to each other.

In Section 11.8.2 we already used the LLL algorithm in SageMath in order to reduce a basis and solve a knapsack example. We now give a formal definition of LLL-reduction as well as the details of the algorithm as this is used as a subroutine in further algorithms described later in this chapter.

Definition 11.30 Let δ be a constant, $0 < \delta \leq 1$. We call a basis $b_1, \ldots, b_m \in \mathbb{R}^n$ *LLL-reduced with* δ if it is size-reduced, such that

$$\delta ||\hat{b}_{k-1}||^2 \leq ||\hat{b}_k + \mu_{k,k-1}\hat{b}_{k-1}||^2 \text{ for } k = 2, \cdots, m$$

The two algorithms for calculating respectively the ordered lattice basis and LLL-reduced δ basis, are summarized in the Crypto Procedures 11.2 and 11.3.

Crypto Procedure 11.2: Algorithm for Size Reduction of Basis Vector b_k

input $b_1, \dots, b_m \in \mathbb{R}^n$ (lattice basis)
    $\mu_{i,j}$ for $1 \leq j < i \leq m$ (its Gram-Schmidt coefficients)
for $j = k-1, \dots, 1$ **do**
  if $|\mu_{k,j}| > \frac{1}{2}$ **then**
    $b_k \leftarrow b_k - \lceil \mu_{k,j} \rfloor b_j$
    for $i = 1, \dots, k-1$ **do**
      $\mu_{k,i} \leftarrow \mu_{k,i} - \lceil \mu_{k,j} \rfloor \cdot \mu_{j,i}$
output $b_1, \dots, b_m \in \mathbb{R}^n$ (lattice basis where b_k is size-reduced)
    $\mu_{i,j}$ for $1 \leq j < i \leq m$ (its Gram-Schmidt coefficients)

Crypto Procedure 11.3: Algorithm for LLL Reduction

input $b_1, \dots, b_m \in \mathbb{R}^n$ (lattice basis), δ with $0 \leq \delta \leq 1$
Step 1
$k \leftarrow 2$ (k is the stage. When entering stage k, the basis $b_1, \dots, b_{k-1}$ is already L^3-reduced with δ, the Gram-Schmidt coefficients $\mu_{i,j}$ are calculated for $1 \leq j < i < k$ as well as the normsquares $c_i = ||\hat{b}_i||_2^2$ for $i = 1, \dots, k-1$)
Step 2
while $k \leq m$ **do**
  for $j = 1, \dots, k-1$ **do**
    $\mu_{k,j} \leftarrow \frac{b_k \cdot b_j - \sum_{i=1}^{j-1} \mu_{j,i} \mu_{k,i} c_i}{c_j}$
  $c_k \leftarrow b_k \cdot b_k - \sum_{j=1}^{k-1} \mu_{k,j} c_j$
  Step 3 (size-reduce b_k)
  for $j = k-1, \dots, 1$ **do**
    $\mu \leftarrow \lceil \mu_{k,j} \rfloor$
    for $i = 1, \dots, j-1$ **do**
      $\mu_{k,i} \leftarrow \mu_{k,i} - \mu \mu_{j,i}$
    $\mu_{k,j} \leftarrow \mu_{k,j} - \mu$
    $b_k \leftarrow b_k - \mu b_j$
  Step 4
  if $\delta c_{k-1} > c_k + \mu_{k,k-1}^2 c_{k-1}$ **then**
    exchange b_k and b_{k-1}
    $k \leftarrow \max(k-1, 2)$
    $k \leftarrow k+1$

Crypto Procedure 11.3 (continued)

output basis $b_1, \ldots, b_m$ which is LLL-reduced with δ
Gram-Schmidt coefficients $\mu_{i,j}$ for $1 \leq j < i \leq m$
normsquares $c_i = ||\hat{b}_i||_2^2$ for $i = 1, \ldots, m$

In practice, replacement of step 4 with the deep insertion rule proposed by Schnorr and Euchner [15] proved to be more efficient (by still being polynomial in n, m and input length) for any fixed value t:

Crypto Procedure 11.4: t Deep Insertions

```
Step 4, alternative
c ← ||b_k||_2^2, T ←min(t, k − 1), i ← 1
while i < T do
    if δc_i > c then
        (b_1, ..., b_k) ← (b_1, ..., b_{i−1}, b_k, b_i, ..., b_{k−1})
        k ← max(i − 1, 2)
        goto Step 2
        c ← c − μ_{k,i}^2 c_i
        i ← i + 1
if δc_{k−1} > c_k + μ_{k,k−1}^2 c_{k−1} then
    exchange b_k and b_{k−1}
    k ← max(k − 1, 2)
    k ← k + 1
```

Furthermore, Schnorr and Euchner [15] invented the notion of blockwise Korine-Zolotarev-reduced bases with block size β. For a lattice basis $b_1, \ldots, b_m$ and $\beta = 2$ the notion is equivalent to a LLL-reduced basis and it is equivalent to the notion of HKZ-reduced bases for $\beta = m$.

Definition 11.31 Let $\beta \geq 2$ be an integer and $\delta \in (0, 1]$ be real. A basis $b_1, \ldots, b_m$ of a lattice $L \subseteq \mathbb{R}^n$ is called (β, δ)−block reduced, if the following holds for $i = 1, \ldots, m$:

1. $|\mu_{i,j}| \leq \frac{1}{2}$ for all $j < i$,
2. $\delta||\hat{b}_i||^2 \leq \lambda_1^2(L(\pi_i(b_i), \ldots, \pi_i(b_{\min(i+\beta-1,m)})))$.

Remark: In the literature, the notion of BKZ-reduced bases is also used for block-reduced bases.

Although there is no proven polynomial bound for the number of operations of any algorithm to calculate a (β, δ)–block-reduced basis for $\beta > 2$ (except for $\beta = 3$ and $\delta \in [\frac{1}{2}, \frac{1}{2}\sqrt{3})$; see [16]), the following algorithm proved to be efficient in practice for small bock sizes ($\beta \leq 30$). Its core component is the enumeration algorithm verb!enum(j,k)! which finds an integer, nonzero minimum $(u_j, \ldots, u_k)$ of the following term:

$$c_j(\tilde{u}_j, \ldots, \tilde{u}_k) := ||\pi_j(\sum_{i=j}^{k} \tilde{u}_i b_i)||_2^2, \ (\tilde{u}_j, \ldots, \tilde{u}_k) \in \mathbb{Z}^{k-j+1}$$

Before going into the details of `enum(j,k)` let's have a look at the block-reduction algorithm in the Crypto Procedure 11.5. It cyclically iterates over all positions j, ensures that the basis is size-reduced, and that it is enforced for all j:

$$\delta||\hat{b}_j||^2 \leq \lambda_1^2(L(\pi_j(b_j), \ldots, \pi_j(b_{min(j+\beta-1,m)})))$$

Crypto Procedure 11.5: Algorithm for (β, δ) Block Reduction

input basis $b_1, \ldots, b_m \in \mathbb{R}^n$ of L, $\beta \in \mathbb{N}$, $2 \leq \beta \leq m$, $\delta \in \mathbb{R}$, $0 \leq \delta \leq 1$

Step 1

LLL-reduce $b_1, \ldots, b_\beta$, $j \leftarrow m$, $z \leftarrow 0$

Step 2

while $z < m - 1$ **do**

  $j \leftarrow j + 1$

  if $j = m$ **then**

    $j \leftarrow 1$

  $k \leftarrow \min(j + \beta - 1, m)$

  enum(j, k) outputs integer coefficients $(u_j, \ldots, u_k)$ of a lattice vector $b_j^{\text{new}} = \sum_{i=j}^{k} u_i b_i$ and $\bar{c}_j := ||\pi_j(b_j^{\text{new}})||^2 = \lambda_1^2(L(\pi_j(b_j), \ldots, \pi_j(b_k)))$

  $h \leftarrow \min(k + 1, m)$

  if $\bar{c}_j < \delta c_j$ **then**

    extend $b_1, \ldots, b_{j-1}, b_j^{\text{new}}$ to a basis $b_1, \ldots, b_{j-1}, b_j^{\text{new}}, \ldots, b_k^{\text{new}}, b_{k+1}, \ldots, b_m$ of L,

    LLL-reduce $b_1, \ldots, b_h^{\text{new}}$

    $z \leftarrow 0$

    LLL-reduce $b_1, \ldots, b_h$

    $z \leftarrow z + 1$

output (β, δ) block reduced basis $b_1, \ldots, b_m$

The extension of $b_1, \ldots, b_{j-1}, b_j^{\text{new}}$ to a basis $b_1, \ldots, b_{j-1}, b_j^{\text{new}}, \ldots, b_k^{\text{new}}$, $b_{k+1}, \ldots, b_m$ of L is done with the algorithm in Crypto Procedure 11.6:

Crypto Procedure 11.6: Algorithm BASIS

input basis $b_1, \ldots, b_m, (u_j, \ldots, u_k)$

Step 1

$b_j^{\text{new}} = \sum_{i=j}^{k} u_i b_i$

Crypto Procedure 11.6 (continued)

Step 2

$g \leftarrow \max\{t : j \le t \le k, u_t \ne 0\}$

Step 3

while $|u_g| > 1$ **do**

  $i \leftarrow \max\{t : j \le t < g : u_t \ne 0\}$

  $q \leftarrow \lceil u_g/u_i \rfloor$

  $u_i \leftarrow u_g - q \cdot u_i$

  $u_g \leftarrow u_i^{old}$

  $b_g \leftarrow q \cdot b_g + b_i$

  $b_i \leftarrow b_g^{old}$

Step 4

for $i = g, \ldots, j+1$ **do**

  $b_i \leftarrow b_{i-1}$

Step 5

$b_j \leftarrow b_j^{\text{new}}$

output $b_1, \ldots, b_m$

By introducing the naming conventions $\tilde{c}_t := ||\pi_t(\sum_{i=t}^{k} \tilde{u}_i b_i||^2$ and $c_t := ||\hat{b}_t||^2 = ||\pi_t(b_t)||^2$, we get $\tilde{c}_t = \tilde{c}_{t+1} + (\tilde{u}_t + \sum_{i=t+1}^{k} \tilde{u}_i \mu_{i,t})^2 c_t$. For fixed $(\tilde{u}_{t+1}, \ldots, \tilde{u}_k)$ we can easily enumerate all integers $\tilde{u}_t$, such that corresponding values of $\tilde{c}_t$ are nondecreasing, starting with $\tilde{u}_t = \left\lceil -\sum_{i=t+1}^{k} \tilde{u}_i \mu_{i,t} \right\rfloor$. The (basic) variant of algorithm `enum` in Crypto Procedure 11.7 traverses the resulting search tree in depth-first search order. Other variants (e.g., traversing the tree in breadth-first search order or incomplete—pruned—traversals) are given in [16].

Crypto Procedure 11.7: Algorithm `enum(j,k)`

input j, k, c_i for $i = j, \ldots, k$ and $\mu_{i,t}$ for $j \le t < i \le k$

Step 1

$s \leftarrow t \leftarrow j, \bar{c}_j \leftarrow c_j, \tilde{u}_j \leftarrow u_j \leftarrow 1,$

$v_j \leftarrow y_j \leftarrow \Delta_j \leftarrow 0, \delta_j \leftarrow 1,$

for $i = j+1, \ldots, k+1$ **do**

  $\tilde{c}_i \leftarrow u_i \leftarrow \tilde{u}_i \leftarrow v_i \leftarrow y_i \leftarrow \Delta_i \leftarrow 0, \delta_i \leftarrow 1$

Step 2

while $t \le k$ **do**

  $\tilde{c}_t \leftarrow \tilde{c}_{t+1} + (y_t + \tilde{u}_t)^2 c_t$

  if $\tilde{c}_t < \bar{c}_j$ **then**

    if $t > j$ **then**

      $t \leftarrow t - 1, y_t \leftarrow \sum_{i=t+1}^{s} \tilde{u}_i \mu_{i,t},$

Crypto Procedure 11.7 (continued)

$\tilde{u}_t \leftarrow v_t \leftarrow \lceil -y_t \rfloor, \Delta_t \leftarrow 0$
if $\tilde{u}_t > -y_t$ **then**
$\delta_t \leftarrow -1$
$\delta_t \leftarrow 1$
$\overline{c}_j \leftarrow \tilde{c}_j, u_i \leftarrow \tilde{u}_i$ for $i = j, \ldots, k$
$t \leftarrow t + 1$
$s \leftarrow \max(s, t)$
if $t < s$ **then**
$\Delta_t \leftarrow -\Delta_t$
if $\Delta_t \delta_t \geq 0$ **then**
$\Delta_t \leftarrow -\Delta_t + \delta_t$
$\tilde{u}_t \leftarrow v_t + \Delta_t$
output $(u_j, \ldots, u_k), \overline{c}_j$

11.10.1 Breaking Knapsack Cryptosystems Using Lattice Basis Reduction Algorithms

For given natural numbers $n, a_1, \ldots, a_n$ and s, a knapsack problem consists of either finding a vector x with $x = (x_1, \ldots, x_n) \in \{0, 1\}^n$ such that $\sum_{i=1}^{n} x_i a_i = s$ or to prove that no such solution exists. $x = (x_1, \ldots, x_n)$ is called a solution of the knapsack problem $(n, a_1, \ldots, a_n, s)$. As the corresponding decision problem is NP-complete, several cryptosystems based on the knapsack problem were proposed. In Section 11.8.1 we already described and attacked the Merkle-Hellman knapsack cryptosystem. In the following subsections, we sketch the attacks on cryptosystems proposed by Chor and Rivest [17] and by Orton [18].

11.10.1.1 Breaking the Chor-Rivest Cryptosystem

Chor and Rivest construct n special weights a_i and code a binary message $x = x_1 \ldots x_n$ with q 1s and $n-q$ 0s by $s := \sum_{i=1}^{n} x_i a_i$. Let's have a look at the following lattice basis B:

$$B := \begin{pmatrix} b_1 \\ \\ \vdots \\ \\ b_{n+1} \end{pmatrix} := \begin{pmatrix} 1 & q & q & \ldots & q & n^2 s & n^2 q \\ 0 & n & 0 & \ldots & 0 & n^2 a_1 & n^2 \\ 0 & 0 & n & & 0 & n^2 a_2 & n^2 \\ \vdots & & \ddots & \ddots & \vdots & \vdots & \vdots \\ 0 & \ldots & & & n & n^2 a_n & n^2 \end{pmatrix}$$

Any lattice vector $v = (v_0, \ldots, v_{n+2}) = \sum_{i=1}^{n+1} u_i b_i$ with $v_0 = \pm 1$ and $v_{n+1} = v_{n+2} = 0$ decodes the message x in case exactly q of the coefficients v_j have value $v_0 \cdot (q - n)$ and $n - q$ coefficients have value $v_0 \cdot q$. In this case we get

$$x_i = \begin{cases} 1, & \text{if } v_i = v_0 \cdot (q - n), \\ 0, & \text{if } v_i = v_0 \cdot q \end{cases}$$

In his diploma thesis, H. Hörner uses variants of the block-reduction algorithm to find such vectors for parameters $(n, q) = (103, 12)$ and $(151, 16)$. Main results are published in [19]. We're not going into detail here.

11.10.1.2 Breaking the Orton Cryptosystem

In [18] Orton proposed the following public-key cryptosystem based on so-called dense, compact, and modular knapsack problems:

Public Parameters:

Natural numbers r, n, s. (Messages consist of n blocks with s bit each, r is the number of rounds to create the keys.)

Secret Key:

Integers $a_i^{(0)}$ with $a_1^{(0)} = 1, a_i^{(0)} > (2^s - 1) \sum_{j=1}^{i-1} a_j^{(0)}$ for $i = 1, \ldots, n$ and natural numbers $q_2, p^{(k)}, w^{(k)}$ for $k = 1, \ldots, r$, where $q_1 := p^{(r)}/q_2$ is an integer.

The part $\{a_i^{(0)}\}$ of the secret key represents an easy knapsack. It is transformed in a hard knapsack by the following transformations:

$$a_i^{(k)} := a_i^{(k-1)} w(k) \mod p^{(k)} \text{ for } i = 1, \ldots, n+k-1, \qquad a_{n+k}^{(k)} := -p^{(k)},$$

$$f_i^{(k)} := 2^{-\text{prec}(k)} \left\lfloor a_i^{(k)} 2^{\text{prec}(k)} / p^{(k)} \right\rfloor \text{ for } i = 1, \ldots, n+k-1, \qquad k = 1, \ldots, r,$$

$$a_{i,j} := a_i^{(r)} \mod q_j \text{ for } i = 1, \ldots, n+r-1, \qquad j = 1, 2.$$

The cryptosystem uses the secret trapdoor $q_2, p^{(k)}, w^{(k)}$ $(k = 1, \ldots, r)$. $\text{prec}(k)$ is the number of precision bits for calculating the quotients $f_i^{(k)}$ in round k. Orton proposed to use $\text{prec}(k) = s + log_2 n + k + 2$ in order to ensure unique decryption and prevent known attacks by Brickell [20] and Shamir [21].

Public Key:

- Natural numbers $q_1, \text{prec}(k)$ for $k = 1, \ldots, r-1$;
- Nonnegative integers $a_{i,j}$ for $i = 1, \ldots, n+r-1, \quad j = 1, 2$;
- Rational numbers $f_i^{(k)} \in 2^{-\text{prec}(k)}[0, 2^{\text{prec}(k)})$ for $k = 1, \ldots, r-1, \quad i = 1, \ldots, n+k-1$.

Crypto Procedure 11.8: Encryption by Orton

input public key, message $(x_1, \ldots, x_n) \in [0, 2^s)^n$

1: $x_{n+k} \leftarrow \left\lfloor \sum_{i=1}^{n+k-1} x_i f_i^{(k)} \right\rfloor$ for $k = 1, \ldots, r-1$

2: $y_1 \leftarrow \sum_{i=1}^{n+r-1} x_i a_{i,1} \mod q_1, \qquad y_2 \leftarrow \sum_{i=1}^{n+r-1} x_i a_{i,2}$

output encrypted message (y_1, y_2)

Crypto Procedure 11.9: Deyption by Orton

input public and secret key, encrypted message (y_1, y_2)

1: recombine $y^{(r)} \equiv y_j \mod q_j$ (j=1,2) using Chinese remainder theorem: $y^{(r)} \leftarrow q_2((y_1 - y_2)q_2^{-1} \mod q_1) + y_2$

2: $y^{(k-1)} \leftarrow y^{(k)}(w^{(k)})^{-1} \mod p^{(k)}$ for $k = r, \dots, 1$

3: solve $\sum_{i=1}^{n} x_i a_i^{(0)} = y^{(0)}$ with $x_i \in [0, 2^s)$ this can easily be done since $a_i^{(0)} > (2^s - 1)\sum_{j=1}^{i-1} a_j^{(0)}$)

output decrypted message $(x_1, \dots, x_n)$

In the following, by using lattice algorithms we show how to reconstruct a message encrypted by the Orton cryptosystem. We first construct a lattice basis $b_1, \dots, b_{m+2} \in \mathbb{Z}^{m+r+2}$ s.t. the original message can easily be recovered from any lattice vector with l_∞-norm 1. The l_∞-norm $||v||_\infty$ of a vector $v = (v_1, \dots, v_n)$ is defined as the maximal absolute value of its coefficients v_i.

$$||v||_\infty = \max(|v_1|, \dots, |v_n|), \quad v \in \mathbb{R}^n$$

We then show how such a lattice vector can be found efficiently.

The decryption problem is stated as follows:

Given the public parameters (r, n, s), the public key $(q_1, \text{prec}(k), a_{i,j}, f_i^k)$, and the encrypted message (y_1, y_2), find the plaintext message $(x_1, \dots, x_n)$; that is, find integers $x_1, \dots, x_n \in [0, 2^s)$, $x_{n+k} \in [0, 2^{s+k+\log_2 n+1})$ satisfying the following equations:

$$\sum_{i=1}^{n+r-1} x_i a_{i,1} = y_1 \mod q_1 \tag{11.11}$$

$$\sum_{i=1}^{n+r-1} x_i a_{i,2} = y_2 \tag{11.12}$$

$$x_{n+k} = \left\lfloor \sum_{i=1}^{n+k-1} x_i f_i^{(k)} \right\rfloor \quad \text{or } k = 1, \dots, r-1 \tag{11.13}$$

Let's transform these equations into a set of $r + 1$ integer linear equations with m 0-1-unknowns, where $m := ns + (r-1)(r/2 + s + \lceil \log_2 n \rceil - 1) + \sum_{k=1}^{r-1} \text{prec}(k)$. Since $f_i^{(k)} 2^{\text{prec}(k)} \in [0, 2^{\text{prec}(k)})$ is integral we can write (11.13) as

$$x_{n+k} 2^{\text{prec}(k)} = \sum_{i=1}^{n+k-1} x_i f_i^{(k)} 2^{\text{prec}(k)} - x_{n+r+k-1} \text{ for } k = 1, \dots, r-1, \tag{11.14}$$

where the additional variables $x_{n+r+k-1}$ are integers in $[0, 2^{\text{prec}(k)})$.

With

$$a_{i,k+2} := \begin{cases} f_i^{(k)} 2^{\text{prec}(k)} & \text{for } i = 1, \ldots, n+k-1 \\ -2^{\text{prec}(k)} & \text{for } i = n+k \\ 0 & \text{for } i = n+k+1, \ldots, n+r+k-2 \\ -1 & \text{for } i = n+r+k-1 \\ 0 & \text{for } i = n+r+k, \ldots, n+2r-2 \end{cases}$$

equations in (11.14) simplify to:

$$\sum_{i=1}^{n+2r-2} x_i a_{i,k+2} = 0 \text{ for } k = 1, \ldots, r-1 \tag{11.15}$$

The unique solution of (11.11), (11.12), (11.15) directly transforms into the unique solution of (11.11) - (11.13). To get $0-1$-variables we use the binary representation of the integer variables.

We set

$$d_i := \begin{cases} s & \text{for } 1 \le i \le n \\ s + i + \lceil \log_2 n \rceil - n - 1 & \text{for } n+1 \le i \le n+r-1 \\ \text{prec}(i - (n+r-1)) & \text{for } n+r \le i \le n+2r-2 \end{cases}$$

and $D_i := \sum_{j=1}^{i-1} d_j$.

Let $t_{D_i+1}, \ldots, t_{D_i+d_i} \in \{0,1\}$ be the binary representation of x_i; that is,

$$x_i = \sum_{l=0}^{d_i-1} t_{D_i+l+1} 2^l,$$

and set

$$A_{D_i+l+1,j} := a_{i,j} 2^l \text{ for } i = 1, \ldots, n+2r-2,\ j = 1, \ldots, r+1,\ l = 0, \ldots, d_i - 1,$$
where $a_{i,1} := a_{i,2} := 0$ for $i > n+r-1$.

With $y_3 := \cdots := y_{r+1} := 0$ equations (11.11), (11.12), (11.15) simplify to

$$\begin{aligned} \sum_{i=1}^{m} t_i A_{i,1} &= y_1 + z q_1 \\ \sum_{i=1}^{m} t_i A_{i,j} &= y_j \text{ for } j = 2, \ldots, r+1, \\ & t_i \in \{0,1\},\ z \in \mathbb{Z} \end{aligned} \tag{11.16}$$

The row vectors $b_1, \ldots, b_{m+2} \in \mathbb{Z}^{m+r+2}$ of the following matrix form the basis of lattice L:

$$\begin{pmatrix} 0 & 2 & 0 & \cdots & 0 & NA_{1,1} & NA_{1,2} & \cdots & NA_{1,r+1} \\ 0 & 0 & 2 & \ddots & 0 & NA_{2,1} & NA_{2,2} & \cdots & NA_{2,r+1} \\ \vdots & \vdots & \ddots & \ddots & \vdots & \vdots & \vdots & \vdots & \\ 0 & 0 & \ddots & 0 & 2 & NA_{2,1} & NA_{2,2} & \cdots & NA_{2,r+1} \\ 0 & 0 & \ddots & 0 & 0 & Nq_1 & 0 & \cdots & 0 \\ 1 & 1 & \cdots & 1 & 1 & Ny_1 & Ny_2 & \cdots & Ny_{r+1} \end{pmatrix} \tag{11.17}$$

For every integer $N \geq 2$ we can obtain the unique solution $t_1, \ldots, t_m$ of (11.16) from each vector $v = (v_0, \ldots, v_{m+r+1}) = \sum_{i=1}^{m+2} c_i b_i$ with l_∞-norm 1:

The vector v has the form $\{\pm1\}^{m+1} \times \{0\}^{r+1}$, where $c_{m+2} \in \{\pm1\}$, $c_{m+1} \in \mathbb{Z}$ and $c_1, \ldots, c_m \in \{0, -c_{m+2}\}$. The zeros in the last $r+1$ coefficients imply

$$\sum_{i=1}^{m} c_i A_{i,1} + c_{m+2} y_1 = 0 \mod q_1$$

$$\sum_{i=1}^{m} c_i A_{i,j} + c_{m+2} y_j = 0 \quad \text{for} j = 1, \ldots, r+1.$$

With $t_i := |c_i| = (|v_i - v_0|)/2$ for $i = 1, \ldots, m$ we obtain the unique solution of (11.16) and we directly get the original message from v:

$$x_i := \sum_{j=0}^{s-1} |v_{s(i-1)+j+1} - v_0| 2^{j-1} \text{ for } i = 1, \ldots, n.$$

To find a vector with l_∞-norm 1 we modify algorithm `enum` in order to search for short vectors in l_∞ norm instead of the Euclidean norm $||.||_2$. To do that we make use of *Hölder's inequality* [22, p. 347]:

$$|x \cdot y| \leq ||x||_\infty ||y||_1 \text{ for all } x, y \in \mathbb{R}^n.$$

The expression $||y||_1$ is defined to be the l_1-norm of y, given by $||y||_1 := \sum_{i=1}^{n} |y_i|,\ y \in \mathbb{R}^n$.

For $t = m, \ldots, 1$ we define the following functions w_t, $\overline{c}_t$ with integer arguments $\tilde{u}_t, \ldots, \tilde{u}_m$ (using the notions of Definition 11.26):

$$w_t := w_t(\tilde{u}_t, \ldots, \tilde{u}_m) := \pi_t\left(\sum_{i=t}^{m} \tilde{u}_i b_i\right) = w_{t+1} + \left(\sum_{i=t}^{m} \tilde{u}_i \mu_{i,t}\right)\hat{b}_t$$

$$\tilde{c}_t := \tilde{c}_t(\tilde{u}_t, \ldots, \tilde{u}_m) := ||w_t||_2^2 = \tilde{c}_{t+1} + \left(\sum_{i=t}^{m} \tilde{u}_i \mu_{i,t}\right)^2 ||\hat{b}_t||_2^2$$

Let's have a look into the algorithm `enum` described previously. It enumerates in depth-first search order all nonzero integer vectors $(\tilde{u}_t, \ldots, \tilde{u}_m)$ for $t = m, \ldots, 1$ satisfying $\tilde{c}_t(\tilde{u}_t, \ldots, \tilde{u}_m) < \overline{c}_1$, where $\overline{c}_1$ is the current minimum for the function $\tilde{c}_1(\tilde{u}_1, \ldots, \tilde{u}_m)$. In order to find a shortest lattice vector with respect to the l_∞-norm we modify this and recursively enumerate all nonzero integer vectors $(\tilde{u}_t, \ldots, \tilde{u}_m)$ satisfying $\tilde{c}_t(\tilde{u}_t, \ldots, \tilde{u}_m) < n \cdot \overline{B}^2$, where $\overline{B}$ is the current minimal l_∞-norm of all lattice vectors w_1 enumerated so far. The resulting enumeration area is illustrated in Figure 11.7. We enumerate all vectors w_t inside the sphere B with radius $\sqrt{n} \cdot \overline{B}$ centered at the origin. We can then stop the enumeration using the following observations:

Since, for fixed $\tilde{u}_t, \ldots, \tilde{u}_m$ we can only reach lattice vectors in the hyperplane H orthogonal to w_t, we can prune the enumeration as soon as this hyperplane doesn't intersect with the set M of all points with l_∞-norm less or equal $\overline{B}$. Using Hölder's inequality we get $\tilde{c}_t > \overline{B}||w_t||_1$ whenever the intersection is empty. The inequality can be tested in linear time and restricts the enumeration to the shaded area U; that is, the union of all balls with radius $\frac{1}{2}\sqrt{n}\overline{B}$ centered in $\{\pm\overline{B}/2\}^n$.

The number of vectors w_t to be enumerated and therefore the running time of the enumeration can roughly be approximated by the volume of the area that needs to be traversed. As a consequence the running time of the pruned enumeration algorithm $\texttt{enum}_\infty$ in the Crypto Procedure 11.11 is faster by the factor volume(U)/ volume(B). For dimension 2 this factor is exactly $\frac{\pi+2}{2\pi}$ and in dimension n it is approximately $(\frac{\pi+2}{2\pi})^{n-1}$. This means that $\texttt{enum}_\infty$ is faster by a factor exponential in the dimension of the lattice. For more details see [16].

We are now able to formulate the attack algorithm:

Crypto Procedure 11.10: Algorithm ATTACK-Orton

input public key, encrypted message y_1, y_2

1: Build the basis $b_1, \ldots, b_{m+2}$ with $N := n^2$ according to matrix (11.17) from before

2: LLL-reduce $b_1, \ldots, b_{m+2}$ with $\delta = 0.99$

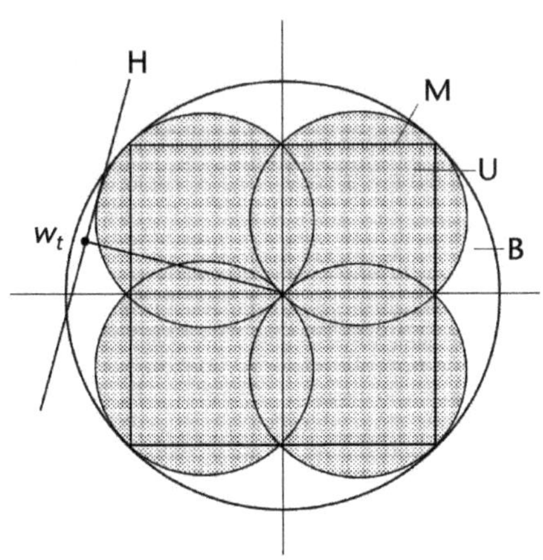

Figure 11.7 Pruning based on Hölder's inequality.

Crypto Procedure 11.10 (continued)

3: Call **enum**$_\infty$; we get a vector v with $||v||_\infty = 1$

4: $x_i \leftarrow \sum_{l=0}^{s-1} |v_{s(i-1)+l+1} - v_0| 2^{l-1}$ for $i = 1, \ldots, n$

output original message $x_1, \ldots, x_n$

Crypto Procedure 11.11: Algorithm $\mathtt{enum}_\infty$

input $\hat{b}_i$, $c_i \leftarrow ||\hat{b}_i||_2^2$, $\mu_{i,t}$ for $1 \leq t \leq i \leq m$

Step 1

$s \leftarrow t \leftarrow 1$

$\tilde{u}_1 \leftarrow u_1 \leftarrow 1$

$\overline{b} \leftarrow b_1$

$\overline{c} \leftarrow n||b_1||_\infty^2$

$\overline{B} \leftarrow ||b_1||_\infty$

$v_j \leftarrow y_j \leftarrow \Delta_j \leftarrow 0$

$\delta_j \leftarrow 1$

for $i = 1, \ldots, m+1$ **do**

  $\tilde{c}_i \leftarrow u_i \leftarrow \tilde{u}_i \leftarrow v_i \leftarrow y_i \leftarrow \Delta_i \leftarrow 0$

  $\eta_i \leftarrow \delta_i \leftarrow 1$

  $w_i \leftarrow (0, \ldots, 0)$

Step 2

while $t \leq m$ **do**

  $\tilde{c}_t \leftarrow \tilde{c}_{t+1} + (y_t + \tilde{u}_t)^2 c_t$

  if $\tilde{c}_t < \overline{c}$ **then**

    $w_t \leftarrow w_{t+1} + (y_t + \tilde{u}_t)\hat{b}_t$

    if $t > 1$ **then**

      if $\tilde{c}_t \geq \overline{B}||w_t||_1$ **then**

        if $\eta_t = 1$ **then**

          increase_t()

          $\eta_t \leftarrow 1, \Delta_t \leftarrow -\Delta_t$

          if $\Delta_t \delta_t \geq 0$ **then**

            $\Delta_t \leftarrow \Delta_t + \delta_t$

          $\tilde{u}_t \leftarrow v_t + \Delta_t$

        $t \leftarrow t-1, \eta_t \leftarrow \Delta_t \leftarrow 0, y_t \leftarrow \sum_{i=t+1}^{s} \tilde{u}_i \mu_{i,t}, \tilde{u}_t \leftarrow v_t \leftarrow$
        $\lceil -y_t \rfloor$

      if $\tilde{u}_t > -y_t$ **then**

        $\delta_t \leftarrow -1$

        $\delta_t \leftarrow 1$

Crypto Procedure 11.11 (continued)

```
            if $||w_1||_\infty < \overline{B}$ then
                $\overline{b} \leftarrow w_1, \overline{c} \leftarrow n||\overline{b}||_\infty^2$,
                $u_i \leftarrow \tilde{u}_i$ for $i = 1, \ldots, m$
        increase_t()
    output $(u_j, \ldots, u_k), \overline{b}$
```

Crypto Procedure 11.12: Subroutine **increase_(t)**

```
$t \leftarrow t + 1$
$s \leftarrow \max(t, s)$
if $\eta_t = 0$ then
    $\Delta_t \leftarrow -\Delta_t$
    if $\Delta_t \delta_t \geq 0$ then
        $\Delta_t \leftarrow \Delta_t + \delta_t$
    $\Delta_t \leftarrow \Delta_t + \delta_t$
$\tilde{u}_t \leftarrow v_t + \Delta_t$
```

With the following modifications of $\mathtt{enum}_\infty$ we can further improve the running time of the attack:

Since $||v||_2^2 = m + 1$ and $||v||_\infty = 1$, we initialize $\overline{c} := m + 1.0001, \overline{B} := 1.0001$ and stop the algorithm as soon as we have found v. We also cut the enumeration for $\tilde{u}_t$ as soon as there is an index $j \in [0, m]$ with $b_{i,j} = 0$ for $i = 1, \ldots, t-1$ and $b_{t,j} \neq 0$, $|w_{t,j}| \neq 1$. We don't miss the solution since $w_{1,j} = w_{t,j} \neq \pm 1$ for all choices of $\tilde{u}_1, \ldots, \tilde{u}_{t-1}$. As the original basis vectors $b_1, \ldots, b_{m+1}$ only depend on the public key, we can precompute the LLL-reduced basis $b'_1, \ldots, b'_{m+1}$ of $b_1, \ldots, b_{m+1}$ once for every public key we want to attack. For all messages which are encrypted with the same public key we use the precomputed vectors $b'_1, \ldots, b'_{m+1}$ together with b_{m+2} instead of the original basis. More details on the attack including practical results may be found in [23] and [16].

11.10.2 Factoring

Many public-key cryptosystems are based on the assumption that factoring large natural numbers is hard. In 1993, C. P. Schnorr [24] proposed to use lattice basis reduction to factorize natural numbers:

Crypto Procedure 11.13: Factoring

input N (a natural number with at least two prime factors),

$\alpha, c \in \mathbb{Q}$ with $\alpha, c > 1$

Step 1

calculate the list $p_1, \ldots, p_t$ of the first t primes, $p_t = (\ln N)^\alpha$

Crypto Procedure 11.13 (continued)

Step 2

Use lattice basis reduction in order to find $m \geq t+2$ pairs $(u_i, v_i) \in \mathbb{N}^2$ with

$u_i = \prod_{j=1}^{t} p_j^{a_{i,j}}$ *with* $a_{i,j} \in \mathbb{N}$

and

$|u_i - v_i N|$ can be factorized over prime factors $p_1, \ldots, p_t$

Step 3

Factorize $u_i - v_i N$ over primes $p_1, \ldots, p_t$ and $p_0 = -1$.

Let $u_i - v_i N = \prod_{j=0}^{t} p_j^{b_{i,j}}$, $b_i = (b_{i,0}, \ldots, b_{i,t})$ and $a_i = (a_{i,0}, \ldots, a_{i,t})$ with $a_{i,0} = 0$

Step 4

Find a `0-1`-solution $(c_1, \ldots, c_m) \neq (0, \ldots, 0)$ of equation

$$\sum_{i=1}^{m} c_i(a_i + b_i) = 0 \pmod 2$$

Step 5

$$x \leftarrow \prod_{j=0}^{t} p_j^{\sum_{i=1}^{m} c_i(a_{i,j}+b_{i,j})/2} \pmod N$$

$$y \leftarrow \prod_{j=0}^{t} p_j^{\sum_{i=1}^{m} c_i b_{i,j}} \pmod N = \prod_{j=0}^{t} p_j^{\sum_{i=1}^{m} c_i a_{i,j}} \pmod N$$

(this construction implies $x^2 = y^2 \pmod N$)

Step 6

If $x \neq \pm y \pmod N$, then output $\gcd(x+y, N)$ and stop, else goto step 4 and find another solution $(c_1, \ldots, c_m)$

In [25], enumeration of short lattice vectors in l_1-norm (similar to ENUM_∞) is used to find the solutions more efficiently. However, those algorithms are still far away from being efficient for large numbers.

11.10.3 Usage of Lattice Algorithms in Post-Quantum Cryptography and New Developments (Eurocrypt 2019)

As it is hard to find the shortest vector in a high-dimensional lattice (in cases when no special structures exist, like those found in the Chor-Rivest and Orton cryptosystem), several cryptosystems based on the shortest vector problem are proposed.

The basic idea for constructing lattice-based public-key encryption schemes is to use a well-formed high-dimensional lattice basis B as secret key and a scrambled version P of B as public key. For encryption, the sender of a message m maps the message to a point $\overline{m}$ in the lattice, by using the public basis P, and then adds a random error to $\overline{m}$; such that the resulting point c is still closer to $\overline{m}$ than to any other point in the lattice. Then, c is sent to the receiver who can use the well-formed basis B in order to find $\overline{m}$ efficiently and obtain the original message.

The security of the scheme is based on the assumption that an attacker who is not in the possession of the well-formed basis B needs to spend an infeasible amount of computational time in order to decipher the message, even with an aid of quantum computers. However, the security of lattice-based schemes against quantum-computer attacks is not yet well-understood. For example at Eurocrypt 2019, several aspects of post-quantum cryptography based on lattices were discussed:

- A. Pellet-Mary, G. Hanrot, and D. Stehlé [26] describe an algorithm to solve the approximate shortest vector problem for lattices corresponding to ideals of integers of an arbitrary number field K. The running time is still exponential in the input size, but improved compared to previous results.
- C. Ba etu, F. B. Durak, L. Huguenin-Dumittan, A. Talayhan, and S. Vaudenay [27] describe misuse attacks on several post-quantum cryptosystems submitted to the National Institute of Standards and Technology (NIST), including several lattice-based schemes.
- M.R. Albrecht, L. Ducas, G. Herold, E. Kirshanova, E.W. Postlethwaite, and M. Stevens [28] propose a sieve method in order to find a shortest lattice vector, or a lattice vector nearest to a given (nonlattice) vector as an alternative to the enumeration algorithms described in this chapter. It would be interesting to check the performance of the enumeration algorithms on modern computers, rather than the implementations on machines as of the late nineties; see [16].

11.11 PQC Standardization

In 2016, NIST launched a competition to identify and eventually standardize suitable alternative methods for the current generation of crypto methods (like RSA or ECDSA). This next generation of cryptographic algorithms is called post-quantum cryptography (PQC). Overall, 82 proposals were submitted. In July 2022, as a result of the third round, NIST announced which methods it wants to standardize [29]:

- For public-key encryption and key-exchange: CRYSTALS-Kyber;
- For digital signatures: CRYSTALS-Dilithium, Falcon, SPHINCS+.

There, NIST recommends using CRYSTALS-Kyber as the encryption algorithm and CRYSTALS-Dilithium for signatures for most use cases. Both methods belong to the group of lattice-based algorithms. Falcon has shorter signatures than CRYSTALS-Dilithium; SPHINCS+ is hash-based.

A very good overview about the post-quantum cryptography standardization organized by NIST can be found in [30, 31].

11.12 Screenshots and Related Plugins in the CrypTool Programs

Sections 11.12.1 to 11.12.3 contain screenshots from CrypTool 1 (CT1), CrypTool 2 (CT2), and JavaCrypTool (JCT). These show both plugins dealing with lattices in a didactical manner and plugins with attacks (like the attacks implemented in CT1 in Section 5.12.2). All of these CrypTool programs continue to be maintained; the vast majority of the further software development takes place in CT2 and CTO.

All functions in all CrypTool programs are listed at `https://www.cryptool.org/en/documentation/functionvolume`. Specifying a category or a filter string or unboxing one of the four programs allows one to search for a special function.

Figure 11.8 shows the result of the search, when the selection was restricted to the two programs CT1 and CT2, and the filter string lattice was set.

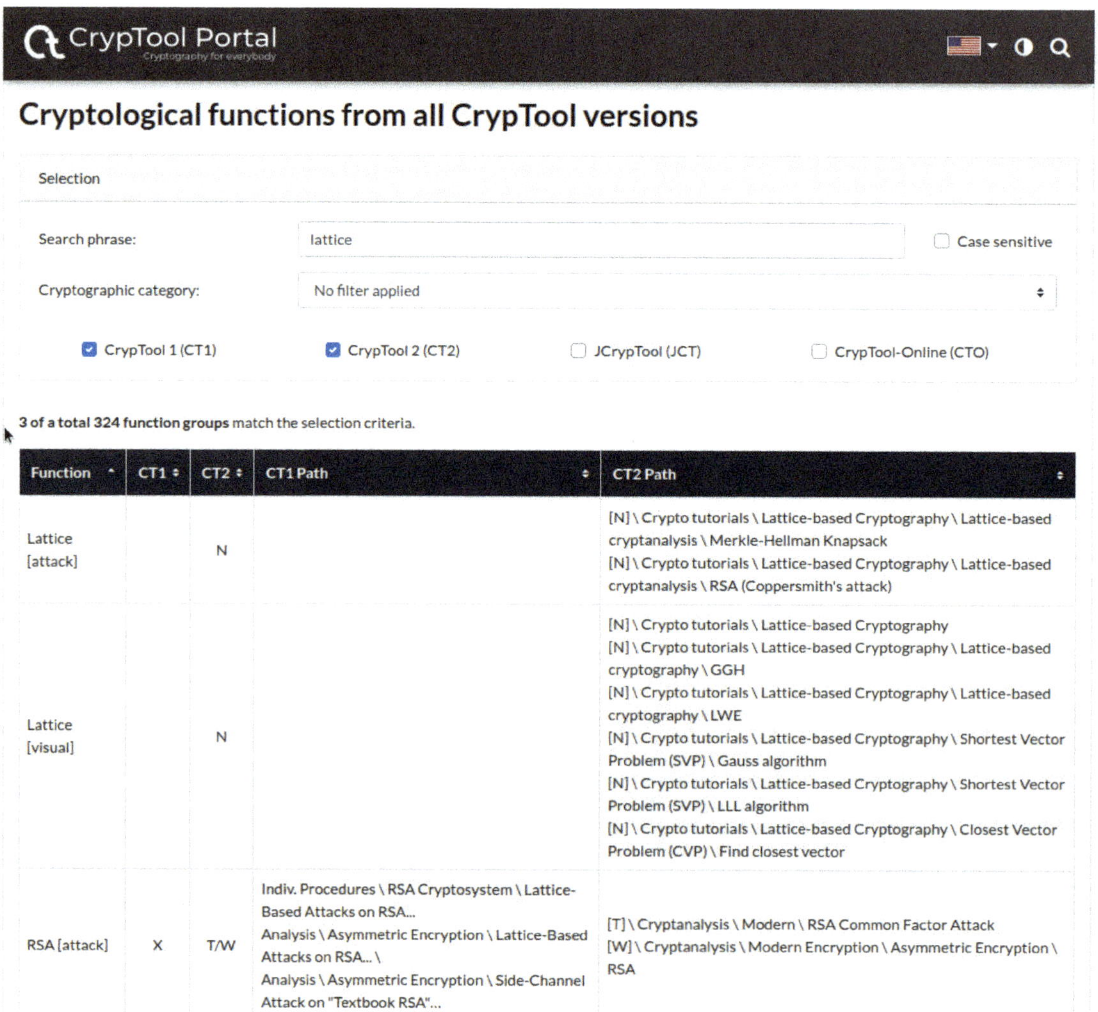

Figure 11.8 Restricted selection from the overview of all CrypTool functions.

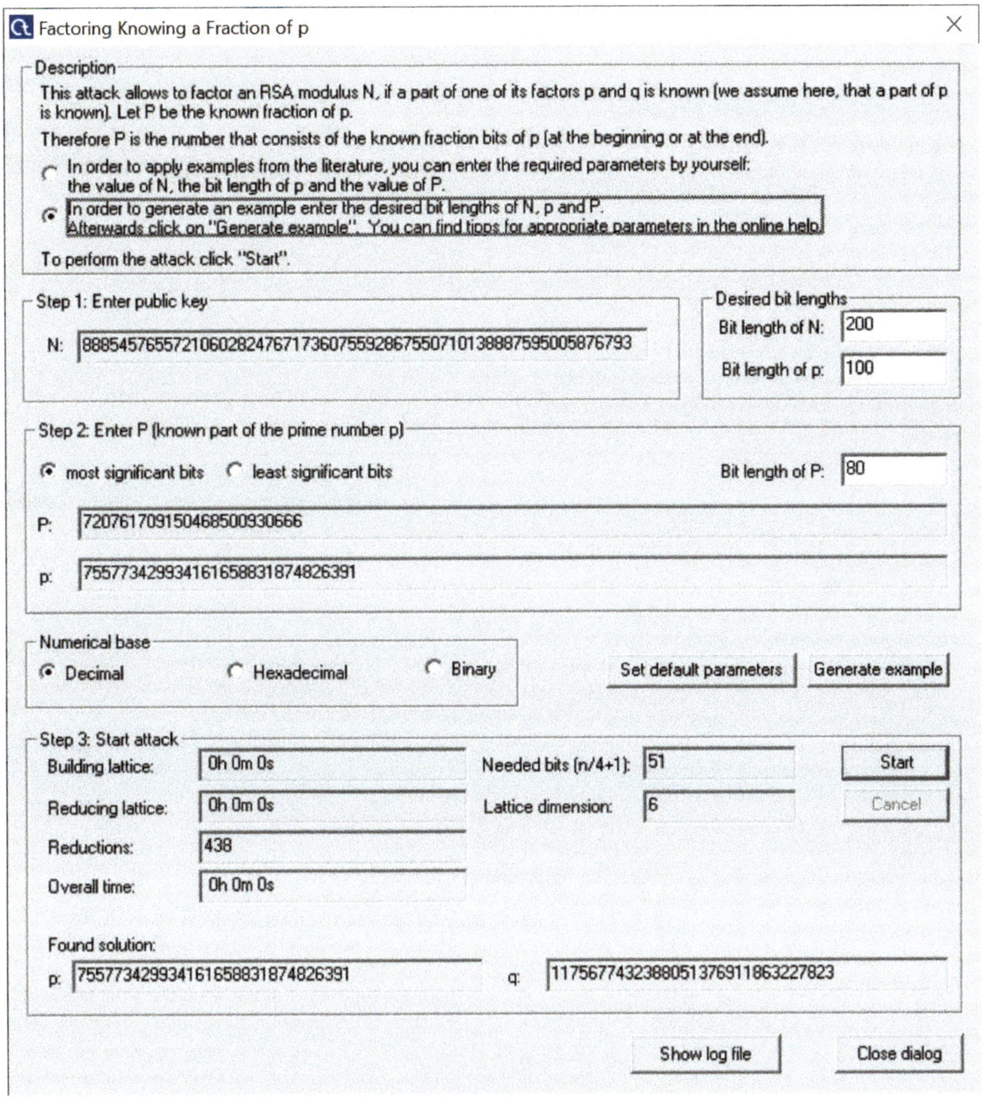

Figure 11.9 CT1 dialog: Factoring N with a hint (you know a fraction of p).

11.12.1 Dialogs in CrypTool 1 (CT1)

CT1 contains 4 dialogs dealing with attacks on RSA: The first one is a typical oracle attack (made possible by missing padding in plain textbook RSA implementations). The next three use lattices to attack RSA under certain circumstances[1]:

- Factoring with a hint;
- Attack on stereotyped messages;
- Attack on small secret exponents.

1. These three attacks can be found either below the menu:
`CT1 Indiv. Procedures ▷ RSA Cryptosystem ▷ Lattice-Based Attacks`
or below the menu
`CT1 Analysis ▷ Asymmetric Encryption ▷ Lattice-Based Attacks on RSA.`

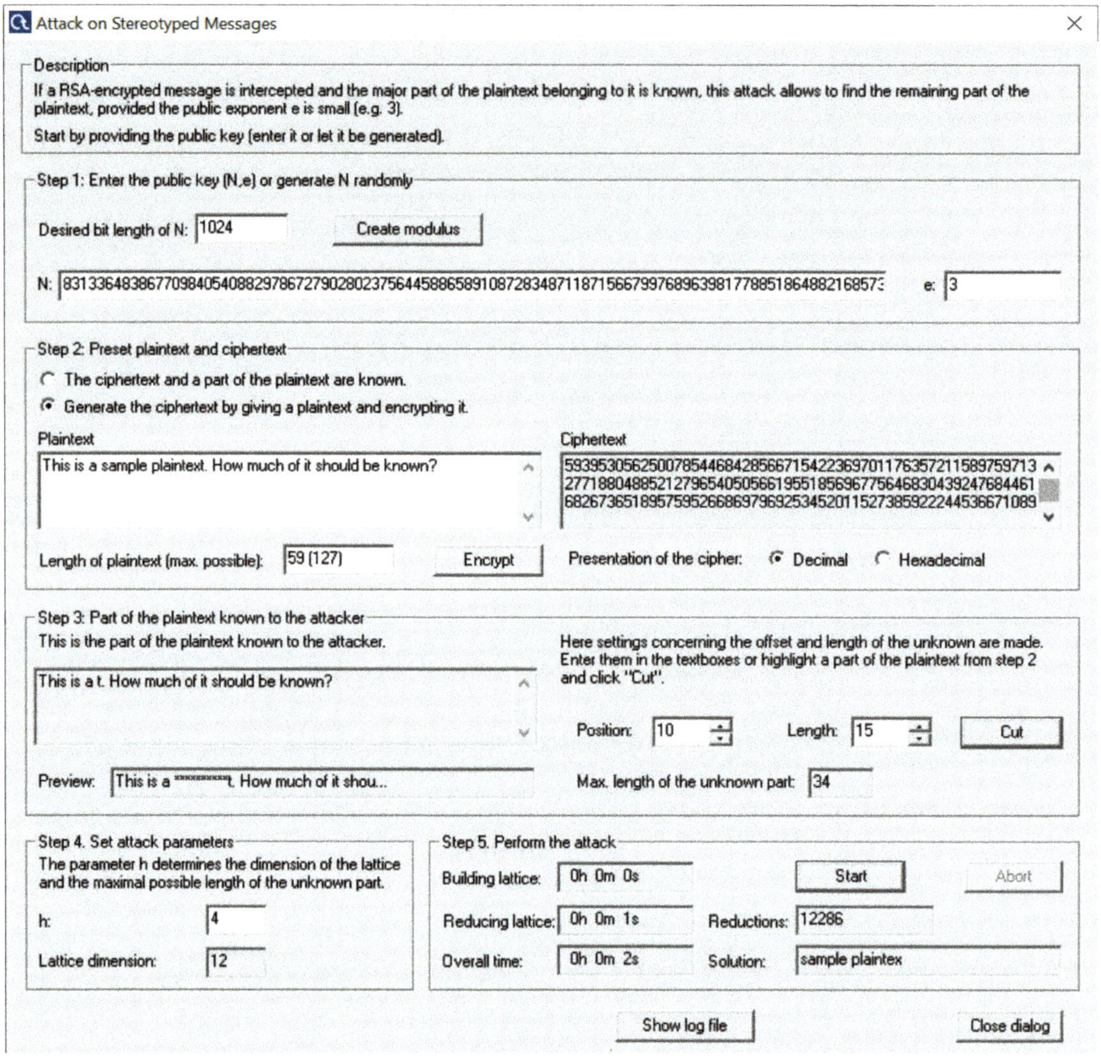

Figure 11.10 CT1 dialog: Attack on stereotyped messages (you know a part of the plaintext message).

11.12.2 Lattice Tutorial in CrypTool 2 (CT2)

The plugin Lattice-Based Cryptography[2] offers the following introductory programs:

- Algorithms to reduce lattice basis for shortest vector problem (SVP):
  - Gauß (nice visualization in 2-dim);
  - LLL.
- Closest vector problem (CVP):

2. `CT2 Crypto Tutorials ▷ Lattice-based cryptography`
Most of the plugins in CT2 appear in the workspace manager as components to be started as templates from the **Startcenter**. On the opposite, the **Crypto tutorials** used here are started from the CT2 main menu.

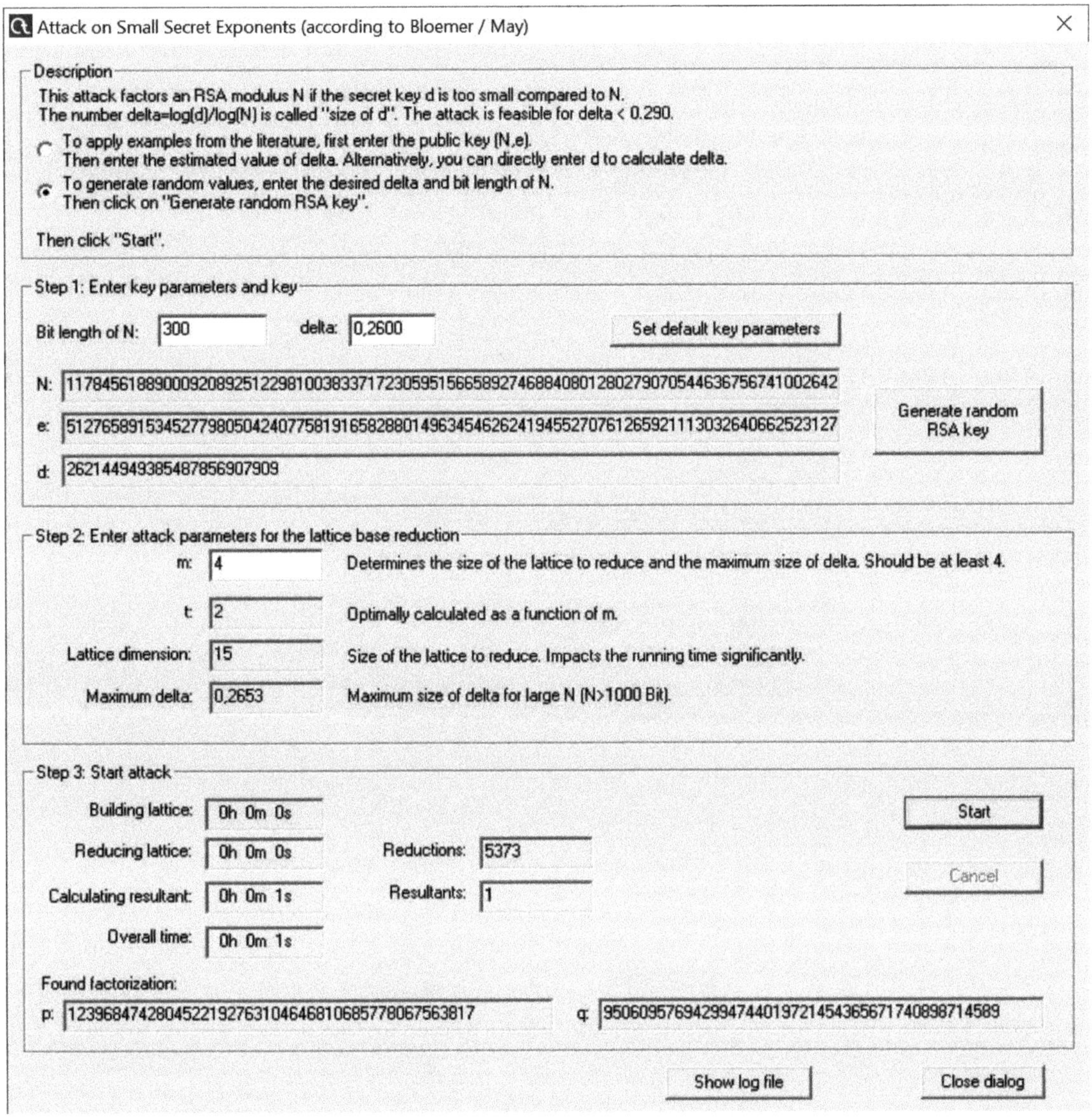

Figure 11.11 CT1 dialog: Factoring N when the private exponent/key is too small (Bloemer/May, 2001).

- Find closest vector (nice visualization in two-dim).

- Lattice-based attacks against:
  - Merkle-Hellman knapsack;
  - RSA (Coppersmith attack).
- Lattice-based cryptography:
  - GGH (Goldreich-Goldwasser-Halevi);
  - LWE (learning with errors).

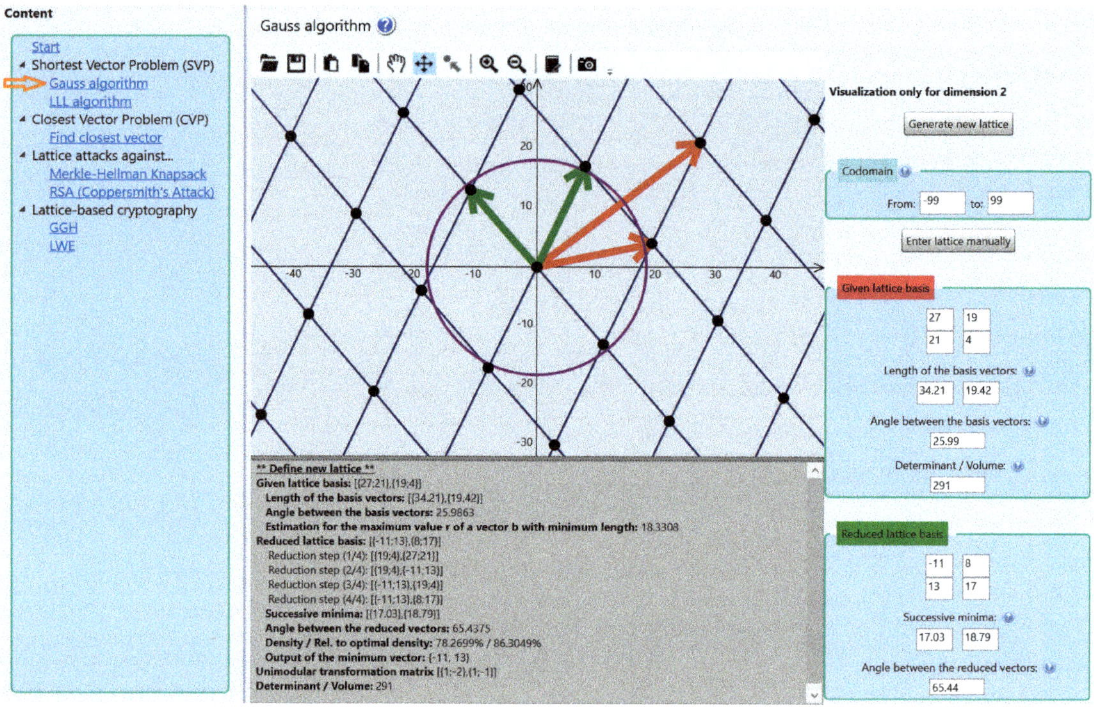

Figure 11.12 CT2 tutorial *Lattice-based cryptography*: SVP via Gauss.

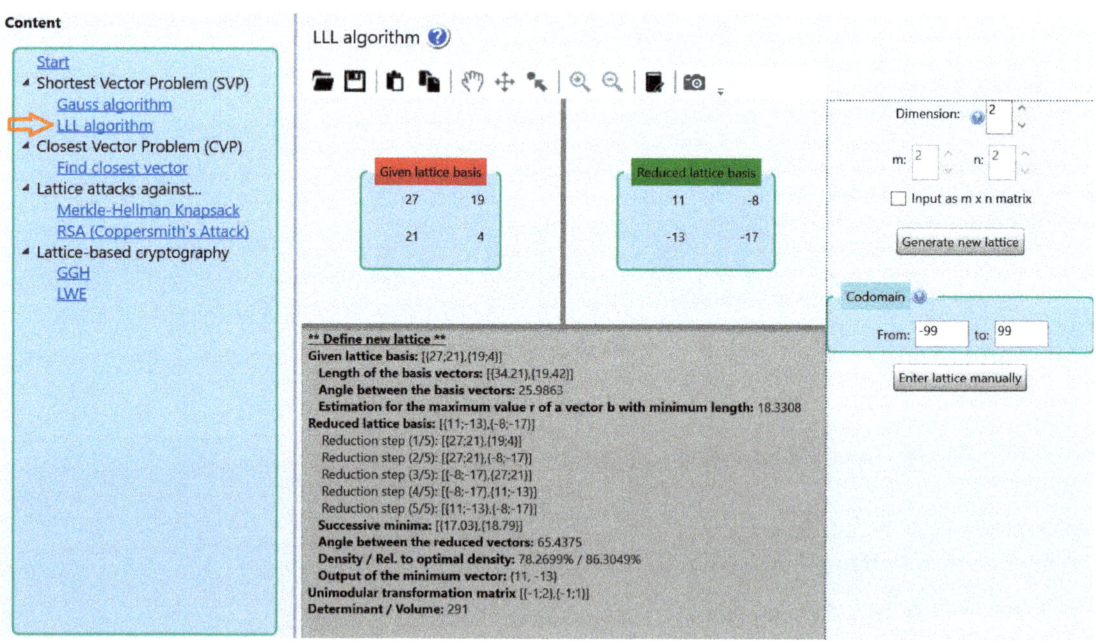

Figure 11.13 CT2 tutorial *Lattice-based cryptography*: SVP via LLL algorithm.

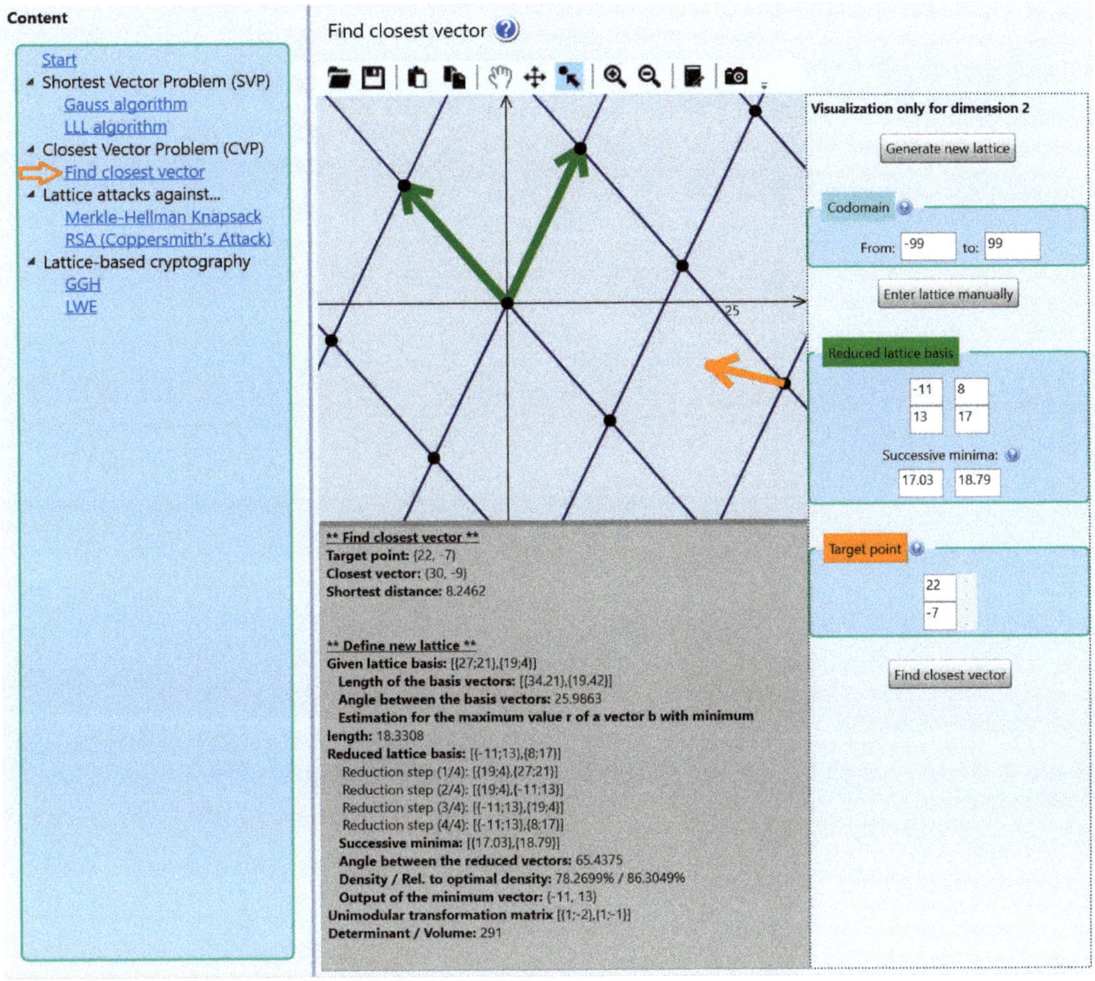

Figure 11.14 CT2 tutorial *Lattice-based cryptography*: CVP, Find closest vector.

11.12.3 Plugin in JCrypTool (JCT)

JCT contains a visualization of the Merkle-Hellman knapsack cryptosystem.[3] This plugin is just a didactical visualization showing all the necessary steps for private keys with maximum 20 elements. The Merkle-Hellman knapsack cryptosystem is vulnerable to Shamir's lattice reduction attack [32].

3. `JCT Default Perspective ▷ Visuals ▷ Merkle-Hellman Knapsack Cryptosystem.`

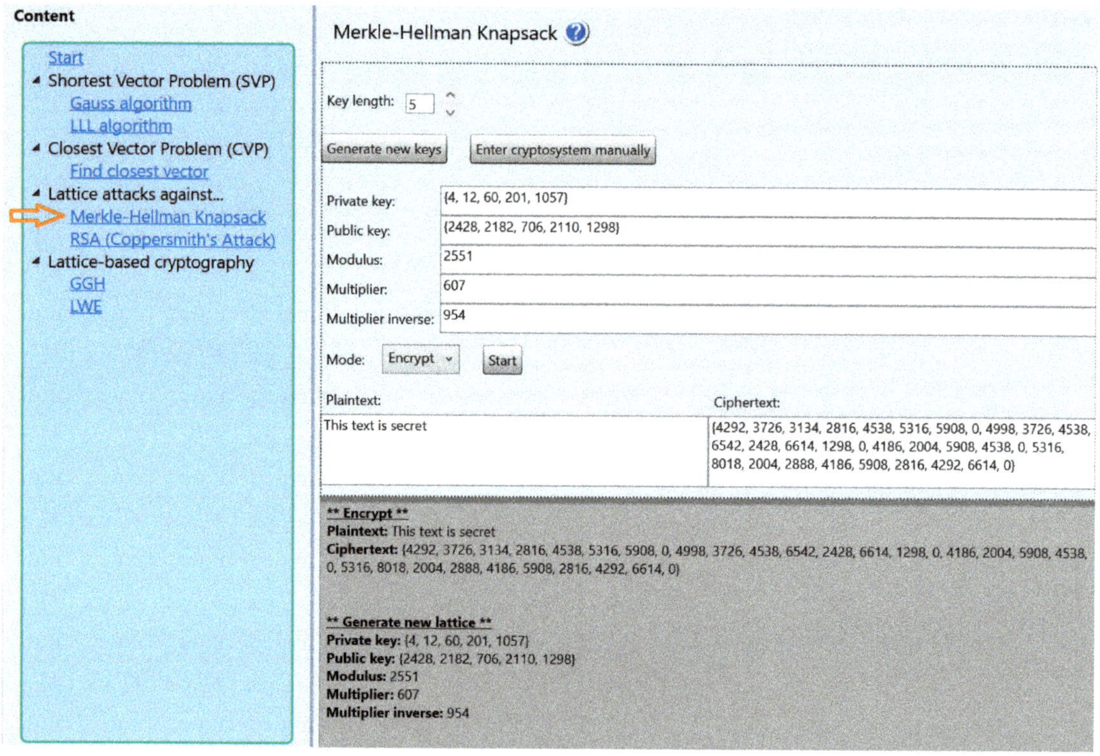

Figure 11.15 CT2 tutorial *Lattice-based cryptography*, attack against the Merkle-Hellman knapsack cryptosystem.

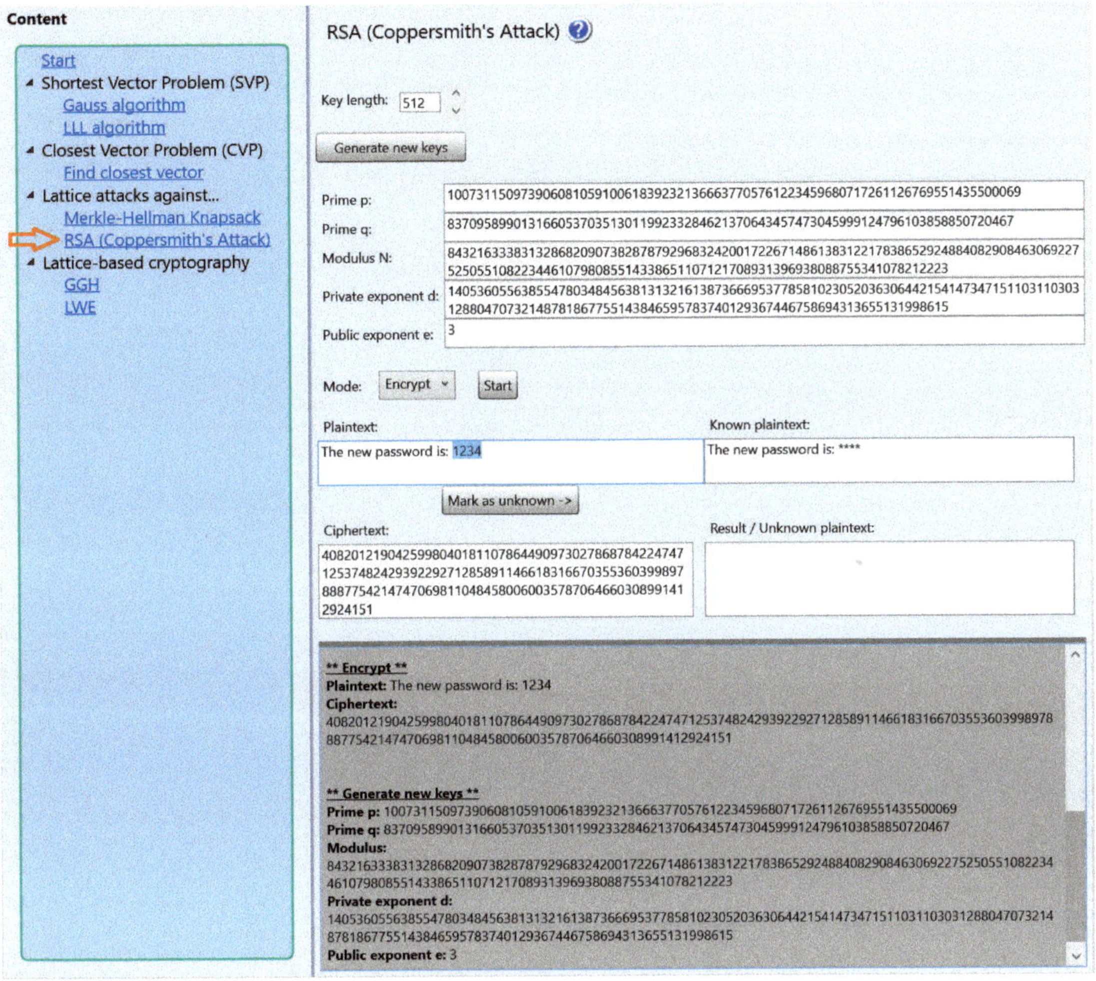

Figure 11.16 CT2 tutorial *Lattice-based cryptography*, attack against RSA (Coppersmith).

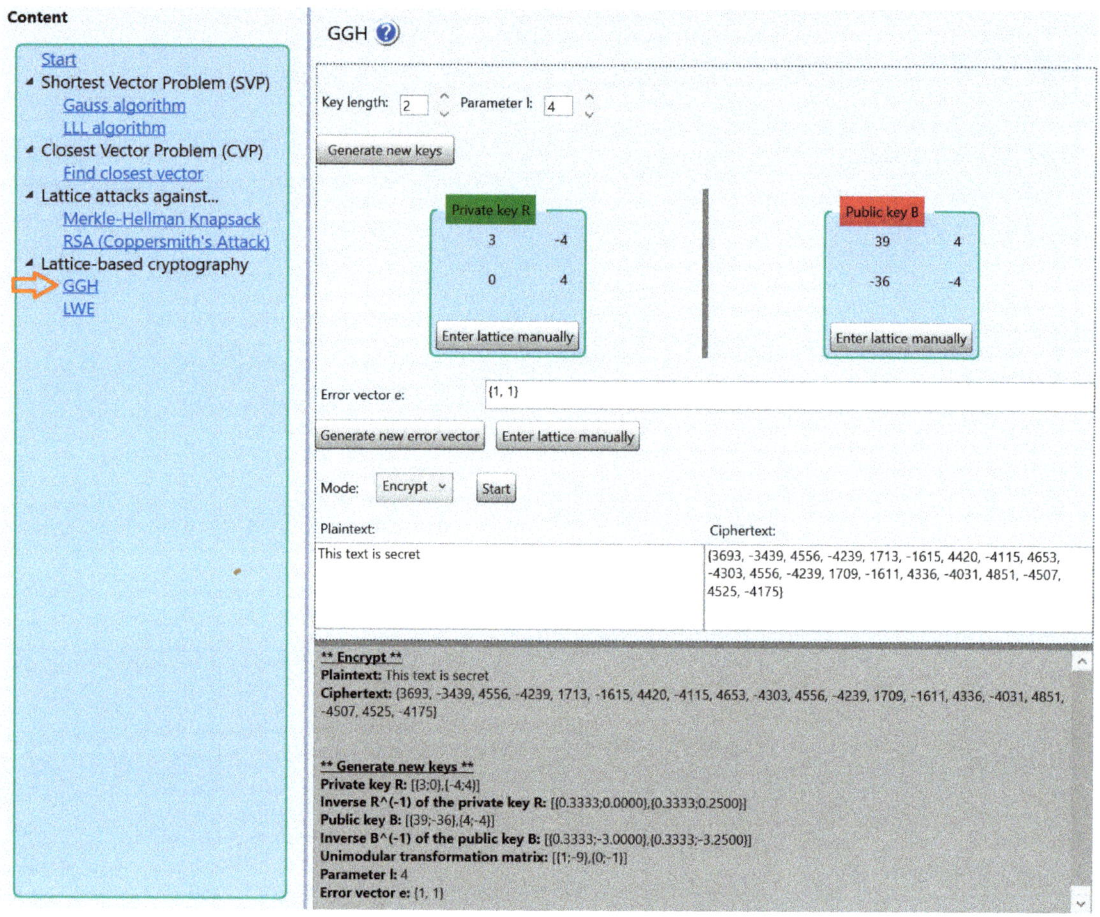

Figure 11.17 CT2 tutorial *Lattice-based cryptography*, the GGH cryptosystem.

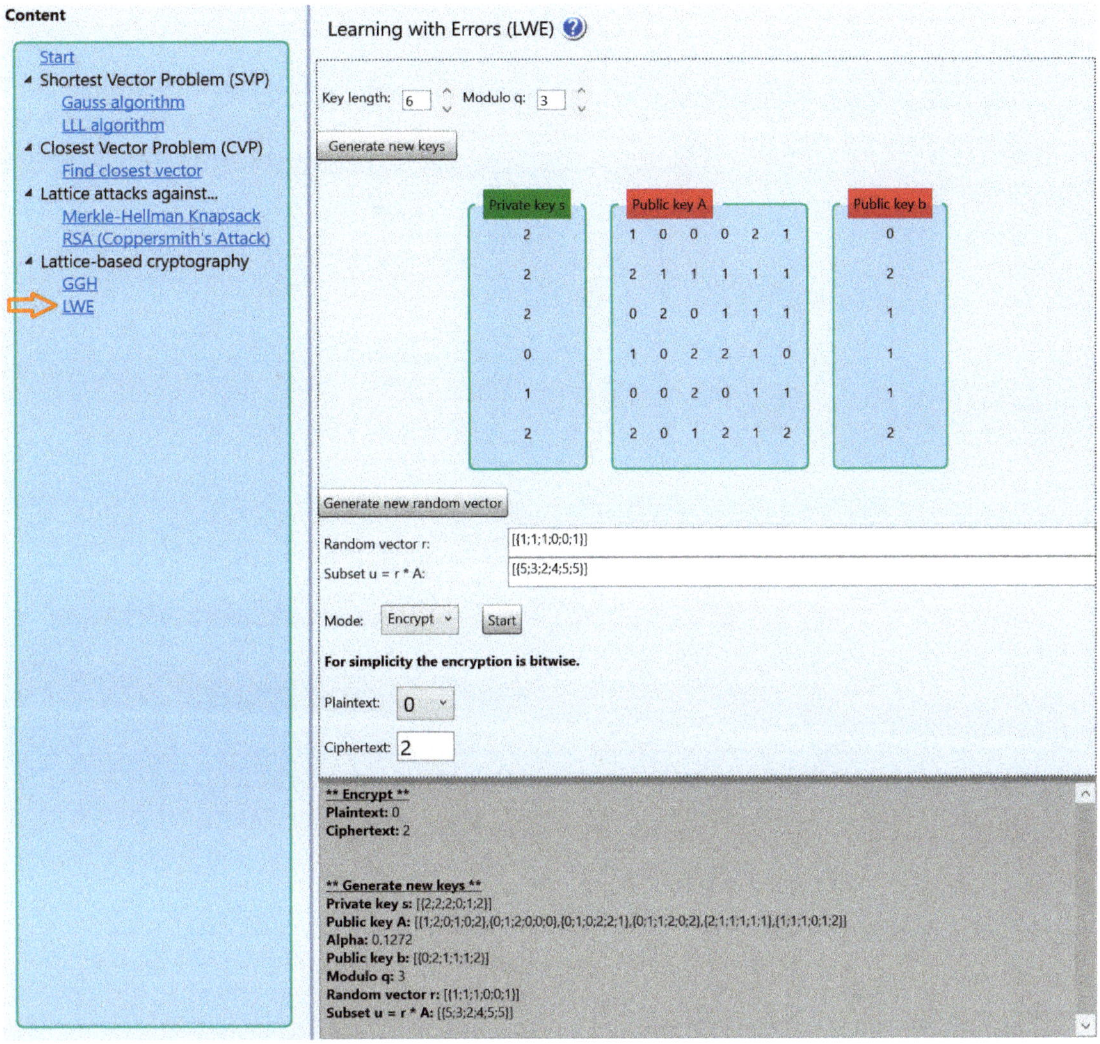

Figure 11.18 CT2 tutorial *Lattice-based cryptography*, the LWE cryptosystem.

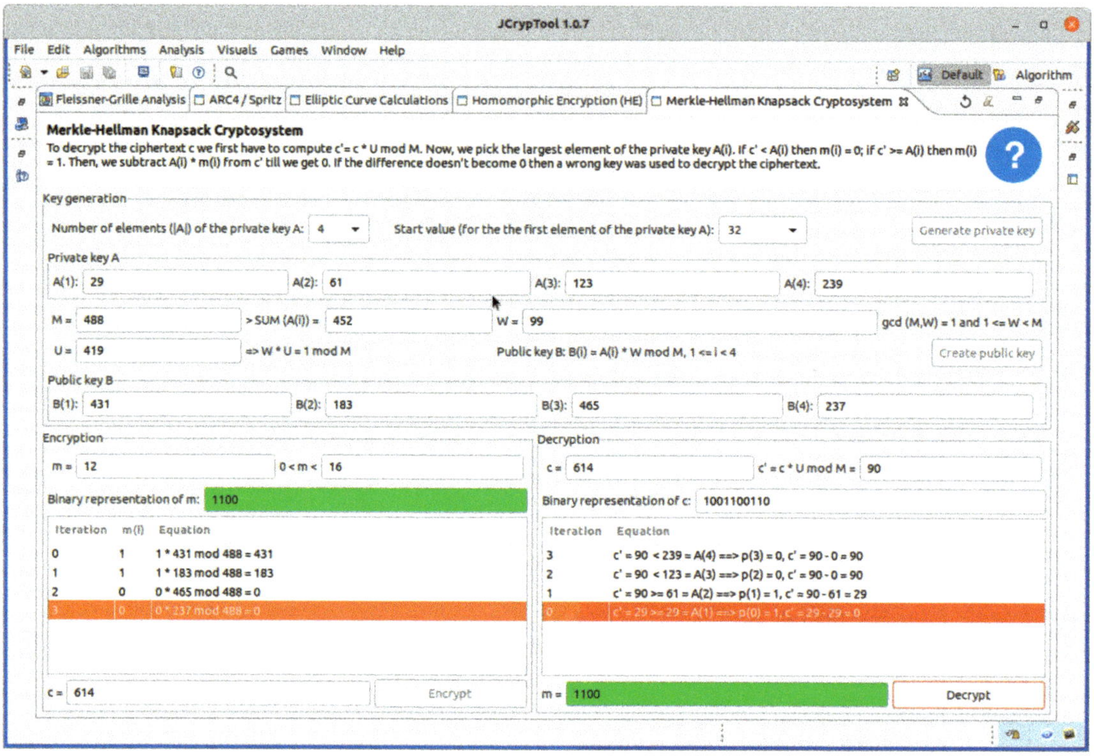

Figure 11.19 JCT plugin: Merkle-Hellman knapsack cryptosystem, step-by-step calculations.

References

[1] Ask the Van, *The Physics Van*, https://van.physics.illinois.edu/ask/listing/14225.

[2] Lee, S.-G., *Linear Algebra with Sage*, Kyobo Books, 2018, https://www.researchgate.net/publication/327362474_Linear_Algebra_seonhyeongdaesuhag_e-_book_-_2018_version.

[3] Kellerer, H., U. Pferschy, and D. Pisinger, *Knapsack Problems*, Springer, 2004.

[4] Lenstra, A. K., H. W. Lenstra, and L. Lovász. "Factoring Polynomials with Rational Coefficients," in *Mathematische Annalen*, Vol. 261, No. 4, 1982, pp. 515–534.

[5] May, A., *Using LLL-Reduction for Solving RSA and Factorization Problems. The LLLAlgorithm*, Springer, 2009, pp. 315–348.

[6] Boneh, D., "Twenty Years of Attacks on the RSA Cryptosystem," in *Notices of the American Mathematical Society (AMS)*, Vol. 46, No. 2, 1999, pp. 203–213, https://crypto.stanford.edu/%7Edabo/papers/RSA-survey.pdf.

[7] *Digital Signature Standard (DSS)*, Federal Information Processing Standards (FIPS) 186-4, National Institute of Standards and Technology (NIST), Gaithersburg: U.S. Department of Commerce, July 19,2013, https://nvlpubs.nist.gov/nistpubs/FIPS/NIST.FIPS.186-4.pdfurl2; https://csrc.nist.gov/publications/fipsurl3; https://www.nist.gov/publications/digital-signature-standard-dss-2.

[8] Coppersmith, D., "Small Solutions to Polynomial Equations, and Low Exponent RSA Vulnerabilities," in *Journal of Cryptology*, Vol. 10, 1997, pp. 233–260, https://link.springer.com/article/10.1007/s001459900030.

[9] Howgrave-Graham, N. A., "Computational Mathematics Inspired by RSA," PhD thesis, University of Bath, 1998, https://cr.yp.to/bib/1998/howgrave-graham.pdf.

[10] May, A., "Using LLL-Reduction for Solving RSA and Factorization Problems," February 2009, pp. 315–348, doi: 10.1007/978-3-642-02295-1_10.

[11] Hermite, C., "26. Extraits de lettres de M. Ch. Hermite à M. Jacobi sur différentsobjets de de la théoriedes nombres, deuxièmeletter," in *Journal für die reine und angewandte Mathematik*, Vol. 40, 1850, pp. 279–290.

[12] Korkine, A., and G. Zolotarev, "Sur les formesquadratiques positives quaternaires," in *Mathematische Annalen*, Vol. 5, 1872, pp. 581–583.

[13] Korkine, A., and G. Zolotarev, "Sur les formesquadratiques," in *Mathematische Annalen*, Vol. 6, 1873, pp. 366–389.

[14] Korkine, A., and G. Zolotarev, "Sur les formes quadratiques positives," in *Mathematische Annalen*, Vol. 11, 1877, pp. 242–292.

[15] Schnorr, C.-P., and M. Euchner, "Lattice Basis Reduction: Improved Practical Algorithms and Solving Subset Sum Problems," in *Mathematical Programming*, Vol. 66, No. 1–3, 1994, pp. 181–199.

[16] Ritter, H., "Aufzählung von kurzen Gittervektoren in allgemeiner Norm," PhD thesis, Johan Wolfgang Goethe-Universität Frankfurt, 1997.

[17] Chor, B., and R. L. Rivest, "A Knapsack Type Public-Key Cryptosystem Based on Arithmeticin Finite Fields," *IEEE Transactions on Information Theory*, Vol. 34, No. 5, 1988, pp. 901–909.

[18] Orton, G., A., "A Multiple-Iterated Trapdoor for Dense Compact Knapsacks," in *EUROCRYPT*, Vol. 950, 1994, pp. 112–130.

[19] Schnorr, C.-P., and H. H. Hörner, "Attacking the Chor-Rivest Cryptosystem by Improved Lattice Reduction," in *EUROCRYPT*, Vol. 921, 1995, pp. 1–12.

[20] Brickell, E. F., "Breaking Iterated Knapsacks," in *Proc. CRYPTO 84*, 1984, pp. 342–358.

[21] Shamir, A., "On the Cryptocomplexity of Knapsack Systems," in *Proc. 11th ACM Symp. on Theory ofComputing*, 1979, pp. 118–129.

[22] Heuser, H., *Lehrbuch der Analysis, Teil 1*, 11th ed., Stuttgart: Teubner, 1994.

[23] Ritter, H., "Breaking Knapsack Cryptosystems by l_∞-Norm Enumeration," in *Proceedings of the 1st International Conference on the Theory and Applications of Cryptology*, Prague, Czech Republic: CTU Publishing House, 1996, pp. 480–492.

[24] Schnorr, C.-P., "Factoring Integers and Computing Discrete Logarithms via Diophantine Approximations," in *Advances of Computational Complexity, DIMACS Series in Discrete Mathematicsand Theoretical Science*, Vol. 13, 1993, pp. 171–182.

[25] Ritter, H., and C. Rössner, *Factoring via Strong Lattice Reduction Algorithms*, Tech. rep., 1997., https://www.researchgate.net/publication/2266562_Factoring_via_Strong_Lattice_Reduction_Algorithms.

[26] Pellet-Mary, A., G. Hanrot, and D. Stehlé, *Approx-SVP in Ideal Lattices with Preprocessing*, Cryptology ePrint Archive, Report 2019/2015, 2019, https://eprint.iacr.org/2019/215.

[27] Ba etu, C., et al., *Misuse Attacks on Post-Quantum Cryptosystems*, Cryptology ePrint Archive, Report 2019/525, 2019, https://ia.cr/2019/525.

[28] Albrecht, M. R., et al., *The General Sieve Kernel and New Records in Lattice Reduction*, Cryptology ePrint Archive, Report 2019/089, 2019, https://ia.cr/2019/089.

[29] *Report on the Third Round of the NIST Post-Quantum Cryptography Standardization Process*, National Institute of Standards and Technology (NIST), July 2022 (updated 2022-09-26), https://nvlpubs.nist.gov/nistpubs/ir/2022/NIST.IR.8413.pdf.

[30] Wikipedia, *NIST Post-Quantum Cryptography Standardization*, https://en.wikipedia.org/wiki/NIST_Post-Quantum_Cryptography_Standardization.

[31] NIST Computer Security Resource Center CSRC, *Post-Quantum Cryptography PQC*, https://csrc.nist.gov/projects/post-quantum-cryptography/post-quantum-cryptography-standardization.

[32] Stamp, M., *Lattice Reduction Attack on the Knapsack*, http://www.cs.sjsu.edu/faculty/stamp/papers/topics/topic16/Knapsack.pdf.

CHAPTER 12

Solving Discrete Logarithms and Factoring

Recent algorithmic developments for solving discrete logarithms in finite fields of small characteristics led to some uncertainty among cryptographic users fueled by the media about the impact for the security of currently deployed cryptographic schemes (see for instance the discussion in [1] using the catchword "cryptocalypse").

This chapter provides a broader picture about the currently best algorithms for computing discrete logarithms in various groups and about the status of the factorization problem.

The subject requires a very mathematical presentation—more mathematical than most of the other chapters in this book. We try to compensate this by working out the ideas as well.

Our goal is to clarify what currently can be done algorithmically and what cannot be done without further major breakthroughs. In particular, we currently do not see a way how to extend the current algorithmic progress for finite fields of small characteristic to either the case of large characteristic finite fields or to the integer factorization problem. This means that there is no danger for the methods currently used (RSA, DHKE, DS, ECC) if the parameters are correctly selected and there are no breakthroughs either in algorithms or in quantum computers.

The recommendations also consider the possibility of embedding trapdoors in cryptographic schemes by governmental organizations.

12.1 Generic Algorithms for the Discrete Logarithm Problem in Any Group

The hardiness of the discrete logarithm problem depends on the group over which it is defined. In this chapter we review cryptanalytical algorithms that work for any group. From a cryptographic point of view it is desirable to identify groups for which one is unable to find better algorithms. One candidate for these groups are elliptic curve groups.

In this chapter, we describe general cryptanalytical algorithms that apply for any finite abelian group. That means, any group used in cryptography (e.g., multiplicative groups of finite fields or of elliptic curves) are susceptible to this kind of algorithm. We will see that we can always compute a discrete logarithm in a group of order n in $\mathcal{O}(\sqrt{n})$ steps by Pollard's rho method. This in turn means that for achieving a security level of 2^k one has to choose a group of order at least 2^{2k}. For example, for achieving a security level of 80 bit, one has to choose a group of order at least 160 bit. This explains why in practice we usually take elliptic curve groups with at least 160 bit order.

Moreover, let G be a group of order n and let $n = p_1^{e_1} \cdot \ldots \cdot p_\ell^{e_\ell}$ be the prime factorization of n. Then we will see that discrete logarithms in G can be computed in time $\mathcal{O}(e_1\sqrt{p_1} + \ldots + e_\ell\sqrt{p_\ell})$. Notice that this bound is equal to Pollard's bound $O(\sqrt{n})$ if and only if n is a prime. Otherwise, the complexity of computing the discrete logarithm is mainly determined by the size of the largest prime divisor of its group order. This explains why, for example, Schnorr/DSA signatures are implemented in groups which contain by construction a prime factor of size at least 160 bit. This also explains why usually elliptic curve groups have prime order or order containing only a very small *smooth cofactor*.

12.1.1 Pollard Rho Method

Let G be a finite abelian group. Let g be a generator of some large subgroup $G' = \{g, g^2, \ldots, g^n\} \subseteq G$ (e.g., g could generate G itself). Let $y = g^x$. Then the discrete logarithm problem is to find on input g and y the output $x \mod n$. We write $x = \mathrm{dlog}_g(y)$.

Pollard's rho method tries to generate elements $g^{a_i}y^{b_i} \in G'$ with $a_i, b_i \in \mathbb{N}$ in a pseudorandom but deterministic fashion. Let us assume for simplicity that we generate random elements from the n elements in G'. Then by the birthday paradox, we expect to find after only $\mathcal{O}(\sqrt{n})$ steps two elements which are identical. In our case, this means that

$$g^{a_i}y^{b_i} = g^{a_j}y^{b_j}.$$

This can be rewritten as $g^{\frac{a_i - a_j}{b_j - b_i}} = y$. This in turn implies that we can recover our discrete logarithm as $x \equiv \frac{a_i - a_j}{b_j - b_i} \mod n$.

Hence, with Pollard's rho method one can compute discrete logarithms in any finite abelian group of order n in $\mathcal{O}(\sqrt{n})$ steps. By using so-called cycle-finding techniques, one can also show that Pollard's rho method can be implemented within constant space.

Moreover, it is also possible to improve the efficiency of square root algorithms when multiple discrete logarithms in the same group are desired: When computing L distinct logarithms, one can reduce the global cost from $\mathcal{O}(L\sqrt{n})$ to $\mathcal{O}(\sqrt{Ln})$ [2].

12.1.2 Silver-Pohlig-Hellman Algorithm

As before let $y = g^x$ for a generator g of order n. We have to compute the discrete logarithm $x \bmod n$. Moreover, let $n = p_1^{e_1} \cdot \ldots \cdot p_\ell^{e_\ell}$ be the prime factorization of n. Then by the Chinese remainder theorem $x \bmod n$ is uniquely defined by the system of congruences

$$\begin{aligned} x &\equiv x_1 \bmod p_1^{e_1} \\ &\vdots \\ x &\equiv x_\ell \bmod p_\ell^{e_\ell}. \end{aligned} \tag{12.1}$$

The algorithm of Silver-Pohlig-Hellman computes all discrete logarithms $x_i \bmod p_i$ in the subgroups of order p_i in $\mathcal{O}(\sqrt{p_i})$ steps by using Pollard's rho method. Then it is quite easy to find a logarithm modulo the prime power $x_i \bmod p_i^{e_i}$ by a Hensel lifting process that performs e_i calls to the discrete logarithm procedure modulo p_i. In a Hensel lifting process, we start by a solution $x_i \bmod p_i$, and then consecutively compute $x_i \bmod p_i^2$, $x_i \bmod p_i^3$, and so on until $x_i \bmod p_i^{e_i}$ (see [3] for Hensel's formula).

Finally, one computes the desired discrete logarithm $x \bmod n$ from the system of (12.1) by Chinese remaindering. In total, the running time is mainly determined by computing $x_i \bmod p_i$ for the largest prime factor p_i. That is, the running time is roughly $\mathcal{O}(\max_i\{\sqrt{p_i}\})$.

12.1.3 How to Measure Running Times

Throughout this chapter, we want to measure the running time of analysis algorithms for discrete logarithms as a function of the bit-size of n. Note that any integer n can be written with (roughly) $\log n$ bit, where log is to base 2. Thus, the bit-size of n is $\log n$.

For expressing our running times we use the notation $L_n[b, c] = \exp^{c \cdot (\ln n)^b (\ln \ln n)^{1-b}}$ for constants $b \in [0, 1]$ and $c > 0$. Notice that $L_n[1, c] = e^{c \cdot \ln n} = n^c$ is a function that is for constant c a polynomial in n. Therefore, we say that $L_n[1, c]$ is *polynomial* in n. Also notice that $L_n[1, c] = n^c = (2^c)^{\log_2 n}$ is a function that is exponential in $\log n$. Therefore, we say that $L_n[1, c]$ is *exponential* in the bit-size $\log n$ of n. So our Pollard's rho algorithm achieves exponential running time $L[1, \frac{1}{2}]$.

On the other end, $L_n[0, c] = e^{c \cdot \ln \ln n} = (\ln n)^c$ is polynomial in the bit-size of n. Notice that the first parameter b is more important for the running time than the second parameter c, since b interpolates between polynomial and exponential running time. We shortly denote $L_n[b]$ if we do not want to specify the constant c.

Some of the most important algorithms that we discuss in the subsequent sections achieve a running time of $L_n[\frac{1}{2} + o(1)]$ or $L_n[\frac{1}{3} + o(1)]$ (where the $o(1)$-part vanishes for $n \to \infty$), which is a function that grows faster than any polynomial but slower than exponential. For cryptographic schemes, such attacks are completely acceptable, since the desired security level can be easily achieved by a moderate adjustment of the key sizes.

However, the recent algorithm of Joux et al. for computing discrete logarithms in finite fields of small characteristic achieves a running time of $L_n[o(1)]$, where $o(1)$ converges to 0 for $n \to \infty$. This means that these algorithms are quasi polynomial time, and the underlying fields are no longer acceptable for cryptographic applications. A finite field $\mathbb{F}_{p^n}$ has small characteristic if p is small, that is, the base field $\mathbb{F}_p$ is small and its extension degree n is usually large. In the recent algorithms we need a small p, since the algorithms enumerate over all p elements in the base field $\mathbb{F}_p$.

12.1.4 Insecurity in the Presence of Quantum Computers

In 1995, Shor published an algorithm for computing discrete logarithms and factorizations on a quantum computer. He showed that computing discrete logarithms in *any* group of order n can be done in polynomial time which is almost $\mathcal{O}(\log n^2)$.

The same running time holds for computing the factorization of an integer n. This running time is not only polynomial, but the attacks are even more efficient than the cryptographic schemes themselves! This in turn means that the problem cannot be fixed by just adjusting key sizes.

Thus, if we face the development of large-scale quantum computers in the next decades, then all classical dlog- and factoring-based cryptography has to be replaced. However, one should stress that the construction of large quantum computers with many qubits appears to be way more difficult than its classical counterpart, since most small quantum systems do not scale well and face decoherence problems.

Recommendation: It seems hard to predict the developments in constructing quantum computers. But experts in quantum physics currently do not see any major obstacle that would hinder the development of large quantum computers in the long term. It seems crucial to keep track of current progress in this area, and to have some alternative quantum-resistant cryptosystems ready to enroll within the next 15 years.

References and further reading: We recommend reading the books of Menezes, van Oorschot, and Vanstone [4], Joux [5], and Galbraith [6] for a survey of cryptanalytic techniques. An introductory course in cryptanalysis is provided by May's lecture notes on cryptanalysis [7, 8] (German). An introduction to quantum algorithms can be found in the books of Homeister [9] (German) and Mermin [10].

The algorithms of this section were originally presented in the superb works of Pollard [11, 12] and Shor [13]. Generic algorithms for multiple dlogs have recently been studied in [2].

12.2 Best Algorithms for Prime Fields $\mathbb{F}_p$

Prime fields $\mathbb{F}_p$ are (besides elliptic curves) the standard group for the discrete logarithm problem. There has been no significant algorithmic progress for these groups in the last 20 years. They are still a good choice for cryptography.

In Section 12.1, we learned that in any finite abelian group of order n, we can determine discrete logarithms in $\mathcal{O}(\sqrt{n})$ steps. Notice that both the Pollard rho method and the Silver-Pohlig-Hellman algorithm from Section 12.1 used no other property of representations of group elements than their uniqueness. In these methods, one simply computes group elements by group operations and checks for equality of elements. Algorithms of this type are called generic in the literature.

It is known that generic algorithms cannot compute discrete logarithms in time better than the Silver-Pohlig-Hellman algorithm [14]. Thus, the algorithms of Section 12.1 can be considered optimal if no further information about the group elements is used.

However, when we specify our group G as the multiplicative group of the finite field $\mathbb{F}_p$, where p is a prime, we can actually exploit the representation of group elements. Natural representatives of $\mathbb{F}_p$ are the integer $0, \ldots, p-1$. Thus, we can, for example, use the prime factorization of these integers. This is done in the so-called *Index Calculus type* discrete logarithm algorithms. This type of algorithm currently forms the class with the best running times for discrete logarithm over prime fields, prime extensions (Section 12.3) and for the factorization problem (Section 12.4).

We will now illustrate an index calculus algorithm with a very easy example.

12.2.1 An Introduction to Index Calculus Algorithms

An index calculus algorithm consists of three basic steps.

Factor base: Definition of a factor base $F = \{f_1, \ldots, f_k\}$. We want to express group elements as powers of elements of the factor base.

Relation finding: Find elements $z_i := g^{x_i} \in G$ for some integer x_i that can be written in the factor base; that is

$$g^{x_i} = \prod_{j=1}^{k} f_j^{e_{ij}}.$$

When we write this equality to the base g, we obtain a relation

$$x_i \equiv \sum_{j=1}^{k} e_{ij}\mathrm{dlog}_g(f_j) \bmod n,$$

where n is the order of g. A relation is a linear equation in the k unknowns

$$\mathrm{dlog}_g(f_1), \ldots, \mathrm{dlog}_g(f_k).$$

Once we have k linear independent relations of this type, we can compute these unknowns by linear algebra. This means we actually first compute all discrete logarithms of the factor base elements before we compute our desired individual logarithm of y.

Dlog computation: Express $yg^r = g^{x+r} = \prod_{j=1}^{k} f_j^{e_j}$ in the factor base for some integer r. This gives us another relation

$$x + r \equiv \sum_{j=1}^{k} e_j\mathrm{dlog}_g(f_j) \bmod n,$$

which can be easily solved in the only unknown $x = \mathrm{dlog}_g y$.

Let us provide an easy example for an index calculus algorithm that computes $x = \mathrm{dlog}_2(5)$ in $\mathbb{F}_{11}^*$. Since 2 generates the multiplicative group $\mathbb{F}_{11}^*$, the order of 2 is 10.

Factor base: Define $F = \{-1, 2\}$.

Relation finding: $2^1 = (-1)^0 2^1$ gives us a first trivial relation

$$1 \equiv 0 \cdot \mathrm{dlog}_2(-1) + 1 \cdot \mathrm{dlog}_2(2) \bmod 10.$$

If we compute $2^6 = 64 \equiv -2 \bmod 11$ we obtain a second relation

$$6 \equiv 1 \cdot \text{dlog}_2(-1) + 1 \cdot \text{dlog}_2(2) \bmod 10.$$

Therefore, we can solve the system of linear equations

$$\begin{pmatrix} 0 & 1 \\ 1 & 1 \end{pmatrix} \cdot \begin{pmatrix} \text{dlog}_2(-1) \\ \text{dlog}_2(2) \end{pmatrix} \equiv \begin{pmatrix} 1 \\ 6 \end{pmatrix} \bmod 10.$$

We obtain as the unique solution $\text{dlog}_2(-1) \equiv 5$ and $\text{dlog}_2(2) \equiv 1$.

Dlog computation: Since $5 \cdot 2^1 = 10 \equiv -1 \bmod 11$ we obtain that

$$x + 1 \equiv 1 \cdot \text{dlog}(-1) + 0 \cdot \text{dlog}(2) \bmod 10.$$

This leads to the solution $x \equiv 4 \bmod 10$.

Runtime: Choosing a large factor base makes it easier to find relations, since it increases the likelihood that a certain number splits in the factor base. On the other hand, for a large factor base we have to find more relations in order to compute the dlogs of all factor base elements. An optimization of this tradeoff leads to a running time of $L_p[\frac{1}{2}]$ for the relation finding step and also $L_p[\frac{1}{2}]$ for performing the individual discrete logarithm computation in step 3.

Let us briefly discuss the advantages and disadvantages of the simple index calculus algorithm from a cryptanalyst's point of view.

Advantages:

- For $g^{x_i} = \prod_{j=1}^{k} f_j^{e_{ij}}$ it is trivial to compute the discrete logarithm on the left-hand size.

Disadvantages:

- We need to factor relatively large numbers g^{x_i} over the integers. One can show that this intrinsically leads to a running time of $L_p[\frac{1}{2}]$, and there is no hope to get below the constant $\frac{1}{2}$.
- We need to compute all discrete logarithms of the factor base elements. This is inherent to all index calculus algorithms.

We will eliminate the first disadvantage by allowing factorizations over number fields. The second disadvantage is eliminated by choosing a factor base with very efficient discrete logarithm computations of its elements.

12.2.2 The Number Field Sieve for Calculating the Dlog

A number field $\mathbb{Q}[\alpha]$ is a k-dimensional vector space over $\mathbb{Q}$ and can be obtained by adjoining a root α of some irreducible degree-k polynomial f to $\mathbb{Q}$. This means we can write every element of $\mathbb{Q}[\alpha]$ as $a_0 + a_1\alpha + \ldots a_{k-1}\alpha^{k-1}$ with $a_i \in \mathbb{Q}$. If we restrict the a_i to integers we are in the ring $\mathbb{Z}[\alpha]$.

Remark: When calculating the dlog there is only the term *number field sieve* and no distinction between general versus special. This is in the opposite to the number field sieve for factorization in Section 12.4.1.

The number field sieve is also an index calculus algorithm. Compared to the previous approach it has the advantage to involve smaller numbers. This is done by choosing a specific representation of the prime field $\mathbb{F}_p$, which is implicitly defined as a finite field where two polynomials of small degree with small coefficients possess a common root. There are several methods that allow one to construct such polynomials with a common root modulo p. In particular, for primes of a special form (i.e., with a sparse representation), it is possible to construct polynomials which are much better than in the general case. One typical construction that works well is to choose a number m and write p in basis m as $\sum_{i=0}^{t} a_i m^i$. We then find that $f_1(X) = X - m$ and $f_2(X) = \sum_{i=0}^{t} a_i m^i$ have m as a common root modulo p.

Equipped with two polynomials f_1 and f_2 of this form, with m as their common root modulo p, we obtain the commutative diagram in Figure 12.1:

Let r_1, r_2 be roots of f_1, f_2, respectively. Then we are working with the number fields $\mathbb{Q}[r_1] \simeq \mathbb{Q}[X]/(f_1(X))$ and $\mathbb{Q}[r_2] \simeq \mathbb{Q}[X]/(f_2(X))$.

Factor base: Consists of small-norm prime elements in both number fields.

Relation finding: The basic principle of the number field sieve consists of sending elements of the form $a + bX$ to both sides of the diagram and to write a relation when both sides factor into the factor base. Technically, this is quite challenging, because we need to introduce several tools to account for the fact that the left and right sides are not necessarily unique factorization domains. As a consequence, we need to factor elements into ideals and take care of the obstructions that arise from the class groups and unit groups. This procedure gives us the discrete logarithms of the factor base elements.

Discrete log computation: Express the desired logarithm as a linear combination of the factor base elements.

Runtime: The number field sieve is the most efficient currently known algorithm for the large characteristic discrete logarithm problem. In the general case, which means that p is not of a special form (e.g., close to a prime power), its complexity is $L_p[\frac{1}{3}, (\frac{64}{9})^{1/3}]$.

References and further reading: For an introduction to index calculus and the involved mathematical tools see May's lecture notes on number theory [3]

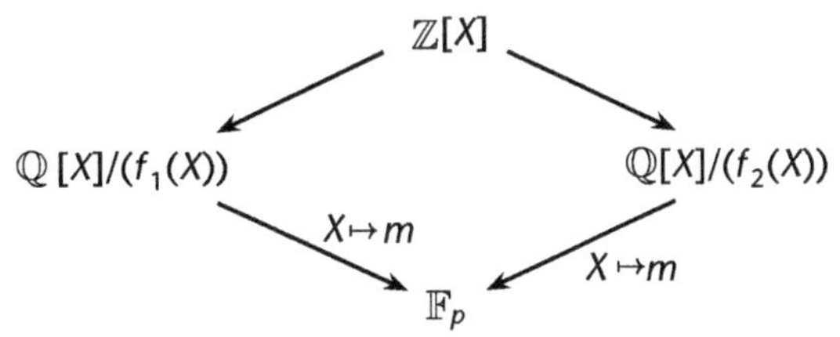

Figure 12.1 NFS in $\mathbb{F}_p$ with two polynomials and common roots, shown as a commutative diagram.

(in German) and the number theory book by Müller-Stach, Piontkowski [15]. For gaining a deep understanding of the number field sieve, one has to study the book of Lenstra and Lenstra [16] that contains all original works that led to the development of the number field sieve algorithm in the late 1980s and early 1990s.

As a good start for understanding the number field sieve, we recommend to first study its predecessors that are described in the original works of Adleman [17], Coppersmith [18], and Pomerance [19, 20].

12.3 Best Known Algorithms for Extension Fields $\mathbb{F}_{p^n}$ and Recent Advances

The groups over extension fields are attacked by the new algorithms of Joux et al. Before the invention of these attacks, the security of extension field groups appeared to be similar to the prime order groups from the last chapter. The new attacks render these groups completely insecure. However, the new attacks do not affect the security of prime order groups.

First, we will discuss the former best algorithm from 2006 (due to Joux and Lercier) that achieves a running time of $L_n[\frac{1}{3}]$. We will then describe the recent developments that led to the dramatic improvement in the running time down to $L_n[o(1)]$, which is quasi-polynomial time.

12.3.1 The Joux-Lercier Function Field Sieve

Any finite field $\mathbb{F}_{p^n}$ can be represented by a polynomial ring $\mathbb{F}_p[x]/f(x)$, where $f(x)$ is an irreducible polynomial over $\mathbb{F}_p$ with degree n. Thus, any element in $\mathbb{F}_{p^n}$ can be represented by a univariate polynomial with coefficients in $\mathbb{F}_p$ of degree less than n. An addition of two elements is the usual addition of polynomials, where the coefficients are reduced modulo p. Multiplication of two elements is the usual multiplication of polynomials, where the result is reduced modulo $f(x)$ in order to again achieve a polynomial of degree less than n.

It is important to notice that the description length of an element is $n\mathcal{O}(\log p)$. Thus, a polynomial time algorithm achieves a running time which is polynomial in n and $\log p$. We will also consider fields of small characteristic p, where p is constant. Then polynomial running time means polynomial in n.

It is known that for any p there are always polynomials $f(x)$ of degree n that are irreducible over $\mathbb{F}_p$. Usually, there are many of these polynomials, which in turn means that we obtain different representations of a finite field when choosing different polynomials $f(x)$. However, it is also known that all of these representations are isomorphic, and the isomorphisms are efficiently computable.

This fact is used in the algorithm of Joux and Lercier, who exploit different representations $\mathbb{F}_p[x]/f(x)$ and $\mathbb{F}_p[y]/g(y)$ of the same field. This is illustrated in the commutative diagram in Figure 12.2.

Factor base: We choose all degree-1 polynomials $x-a$ and $y-b$ from $\mathbb{F}_p[x] \cup \mathbb{F}_p[y]$. Thus, the factor base has size $2p$.

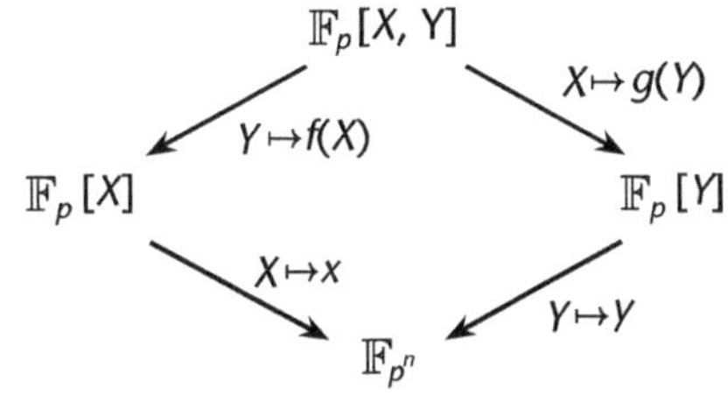

Figure 12.2 Isomorphic representations in $\mathbb{F}_{p^n}$ as a commutative diagram.

Relation finding: On both sides, that is for polynomials h from $\mathbb{F}_p[x]/f(x)$ and from $\mathbb{F}_p[y]/g(y)$, we try to factor into the linear factors from the factor base. This can be done by an easy gcd computation $\gcd(h, x^p - x)$ in time $\mathcal{O}(p)$ for each polynomial. It can be shown that the number of polynomials that have to be tested is bounded by $L_{p^n}[\frac{1}{3}]$.

Discrete log computation: This step is done by writing a polynomial as a linear combination of polynomials of smaller degree and by repeating recursively, until degree-1 is found. This recursion is called a (degree) decent and requires running time $L_{p^n}[\frac{1}{3}]$, just like the relation finding step.

12.3.2 Recent Improvements for the Function Field Sieve

The first recent improvement upon the Joux-Lercier FFS was presented at Eurocrypt 2013 by Joux, who showed that it is possible to drastically lower the complexity of finding relations by replacing the classical sieving approach with a new technique based on a linear change of variables called *pinpointing*.

At the Crypto Conference 2013, Göloglu, Granger, McGuire, and Zumbrägel presented another approach, related to pinpointing that works very efficiently within a characteristic-2 subfield. Their paper was considered so important by the cryptographic community that they received the best paper award.

The new results hold for finite fields $\mathbb{F}_{q^n}$ of characteristic two; that is, $q = 2^\ell$. Notice that we use the standard convention that denotes primes by p and prime powers by $q = p^\ell$. For these fields $\mathbb{F}_{q^n}$ the relation finding step in the Joux-Lercier algorithm simplifies, since one can construct polynomials that split with a higher probability than generic polynomials of the same degree.

Let us give a high-level description of the ideas of their improvement.

Factor base: All degree-1 polynomials as in the Joux-Lercier algorithm.

Relation finding: Göloglu, Granger, McGuire, and Zumbrägel show that one can construct a special type of polynomials over $\mathbb{F}_q[x]$ (the so-called Bluher polynomials) that by construction split over $\mathbb{F}_q[x]$. So similar to our simple version of index calculus for integers in Section 12.2.1, we obtain one side of the equation for free. The cost for splitting the polynomials in $\mathbb{F}_q[y]$ is roughly $\mathcal{O}(q)$ and the cost for finding the discrete logarithms of the factor base elements is roughly $\mathcal{O}(n \cdot q^2)$. We will explain why this gives us the discrete logarithms of the factor base in polynomial time for properly chosen parameters.

Discrete log computation: The individual discrete logarithm computation is similar to the Joux-Lercier algorithm.

Runtime: We are computing in a field $\mathbb{F}_{q^n}$, where $q = 2^\ell$. Hence, a polynomial time algorithm would require running time polynomial in the parameters n and $\log q$. However, the relation finding above takes time $\mathcal{O}(n \cdot q^2)$, which is polynomial in n but exponential in $\log q$. So actually the algorithm performs very poorly with respect to the size of the base field $\mathbb{F}_q = \mathbb{F}_{2^\ell}$.

The trick to work around this is to decrease the size of the base q to q' while slightly increasing the extension degree n to n'. Our goal is that the new base field size q' roughly equals the new extension degree n'; that is $q' \approx n'$. In this case, we again obtain a running time which is polynomial in n' and q', but now q' is also polynomially bounded by n'. So, in total, for step 2 our running time is polynomially bounded by n'.

Let us give a simple example of how this can be done for concrete parameters. Assume that we wish to compute a discrete logarithm in $\mathbb{F}_{(2^{100})^{100}}$. Then we would lower the base field to $q' = 2^{10}$ and at the same time increase the extension degree to $n' = 1000$; that is, compute in $\mathbb{F}_{(2^{10})^{1000}}$. Notice that this can always be done by using the efficiently computable isomorphisms between finite fields of the same cardinality.

Warning: One might be tempted to bypass the above with the selection of exponents that do not split appropriately; that is, by choosing $\mathbb{F}_{2^p}$ with prime p. However, we can always embed our finite field in some larger field—as well as the respective discrete logarithms. Hence, finite fields with small characteristic have to be considered insecure, independently of the special form of the extension degree n.

While the relation finding in step 2 of Göloglu, Granger, McGuire, and Zumbrägel can be done in polynomial time, the individual log computation is still time-consuming. If one does it naively, step 3 is even more time-consuming than in Joux-Lercier because of the increased extension degree n'. If one balances out the running times of step 2 and 3, one ends up with an improved overall running time of $L_{q^n}[\frac{1}{3}, (\frac{4}{9})^{\frac{1}{3}}]$.

12.3.3 Quasi-Polynomial Dlog Computation of Joux et al.

In the previous section, it was shown that the discrete logarithms of all elements of a factor base can be computed in polynomial time. However, it remained a hard problem to use that fact for computing individual logarithms.

This problem has been recently solved by Joux [21] and Barbulesu, Gaudry, Joux, and Thomé [22]. In the paper of Joux, it was shown that the individual logarithm step can be performed in $L[\frac{1}{4}]$. Shortly after, this was improved by Barbulescu, Gaudry, Joux, and Thomé to $L[o(1)]$, which is a function that grows slower than $L[\epsilon]$ for any $\epsilon > 0$. So they achieve quasi-polynomial time.

Let us briefly describe the modifications of these two papers to the Function Field Sieve (FFS) algorithm.

Factor base: Consists of degree-1 polynomials as before.

Relation finding: One starts with the trivial initial polynomial

$$h(x) = x^q - x = \prod_{a \in \mathbb{F}_q} (x - a)$$

that obviously factors in the factor base. Now, one applies linear and rational transformations (called homographies) to $h(x)$, which preserve its property to split over the factor base. One can show that there are sufficiently many independent homographies in order to construct sufficiently many relations. So out of one trivial polynomial $h(x)$, we obtain for free all $\mathcal{O}(q)$ relations. This enables us to compute the discrete logarithms of the factor base elements in time $\mathcal{O}(q)$.

Discrete log computation: Barbulescu et al. present an efficient degree decent algorithm that on input of a polynomial $p(x)$ of degree n outputs a linear relation between the discrete log of $p(x)$ and $\mathcal{O}(nq^2)$ polynomials of degree $\frac{n}{2}$ in time polynomial in q and D. This implies that we get a tree of polynomials, where the degree drops in every level by a factor of two, which in turns implies a tree depth of $\log n$. This results in a running time of $\mathcal{O}(q^{\mathcal{O}(\log n)})$.

Runtime: As in Section 12.3.2 let us assume that the size q of the base field is of the same size as the extension degree n; that is, $q = \mathcal{O}(n)$. Then step 2 runs in time $\mathcal{O}(q) = \mathcal{O}(n)$, which is polynomial in n. Step 3 runs in time $\mathcal{O}(q^{\mathcal{O}(\log n)}) = \mathcal{O}(n^{\mathcal{O}(\log n)}) = L_{q^n}[o(1)]$. Notice that $n^{\log n} = 2^{\log^2 n}$ grows faster than any polynomial function in n but slower than any subexponential function 2^{n^c} for some $c > 0$.

12.3.4 Conclusions for Finite Fields of Small Characteristic

To give some examples what the theoretical quasi-polynomial run time of the previous results implies in practice, we illustrate in Table 12.1 what can currently be achieved in computing discrete logarithms.

Recommendation: The use of small characteristic fields for discrete log-based is completely insecure, no matter which key sizes are used. Fortunately, we are not aware of such a usage in actual applications in wide-spread/standardized cryptographic schemes.

Table 12.1 Small Characteristic Records

Date	*Field*	*Bitsize*	*Cost (CPU hours)*	*Algorithm*
2012/06/17	$3^{6 \cdot 97}$	923	895 000	[23]
2012/12/24	p^{47}	1175	32 000	[24]
2013/01/06	p^{57}	1425	32 000	[24]
2013/02/11	2^{1778}	1778	220	[21]
2013/02/19	2^{1778}	1991	2200	[25]
2013/03/22	2^{4080}	4080	14 100	[21]
2013/04/11	2^{6120}	6120	750	[21]
2013/05/21	2^{6168}	6168	550	[21]

12.3.5 Do These Results Transfer to Other Index Calculus Type Algorithms?

From a crypto user's point of view, one could worry that the current breakthrough results that drop the complexity for discrete log computations in small characteristic fields from $L[\frac{1}{3}]$ to $L[o(1)]$ apply to discrete logarithms in other groups as well. For instance, one might be concerned by the actual security level of discrete log based cryptography in finite fields $\mathbb{F}_p$ of *large* characteristic.

Conjecture: We believe that the new techniques do not carry over to large-characteristic finite fields or elliptic curves that currently comprise the standard for cryptographic constructions.

Let us briefly collect some reasons why the current techniques do not carry over to these groups, and which problems have to be solved before we see any significant progress in the running time for these groups.

- *Runtime:* Notice that all index calculus algorithms described in this section are polynomial in the base field size q and thus exponential in the bit-length $\mathcal{O}(\log q)$. So the hardness of the discrete logarithm problem seems to stem from the hardness in the base field, whereas the extension degree n does not contribute to make the problem significantly harder.

  In particular, we note that each equation—constructed from the polynomial $x^q - x$ as done in the new small characteristic algorithms—contains at least q terms. Thus, whenever q becomes bigger than $L[1/3]$, even writing a single equation of this type would cost more than the full complexity of the number field sieve from Section 12.2.2.

  Notice that there is a similar situation for discrete logarithms in elliptic curve groups. When we use an elliptic curve over $\mathbb{F}_q$ in general the best known algorithm is the generic Pollard rho algorithm from Section 12.1 with running time $\mathcal{O}(\sqrt{q})$. However, Gaudry's algorithm (Section 12.5.2) requires for elliptic curves over $\mathbb{F}_{q^n}$ only running time $q^{2-\frac{2}{n}}$, which is way better than the generic bound $\mathcal{O}(q^{\frac{n}{2}})$. Like the algorithms in this chapter, Gaudry's algorithm is of the index calculus type. And similar to the algorithms in this chapter, the complexity of the discrete logarithm problem seems to be concentrated in the parameter q rather than the parameter n.

- *Polynomials vs numbers:* Notice that the current results make heavy use of polynomial arithmetic and of subfields of $\mathbb{F}_{q^n}$. However, neither is polynomial arithmetic available for $\mathbb{F}_p$ nor do there exist subfields for prime order groups. We would like to argue that many problems are efficiently solvable for polynomials, whereas they appear to be notoriously hard for integers. For instance, it is known that polynomials over finite fields and over the rationals can be efficiently factored by the algorithms of Berlekamp and Lenstra-Lenstra-Lovasz, whereas there is no equivalent algorithm for the integers. There is also an efficient algorithm for finding the shortest vectors in polynomial rings due to von zur Gathen, where its integer lattice counterpart is known to be NP-hard.

  What makes integers intrinsically harder than polynomials is the effect of carry bits. When we multiply two polynomials, we know by the convolution product exactly which coefficients contribute to which coefficients in the product, which is not true for integer multiplication due to the carry bits.

- *Complexity of steps 2 and 3:* Any algorithmic breakthrough for index calculus type discrete logarithms would have to efficiently solve the discrete logarithms of a well-defined factor base and express the desired logarithm in terms of this factor base. But currently, we do not have an efficient method for either step in the case of large prime fields $\mathbb{F}_p$.

References and further reading: Coppersmith's algorithm [26] from the mid-1980s was for a long time the reference method for computing discrete logarithms in small characteristic fields. The Joux-Lercier function field sieve was introduced 2006 in [23].

The recent advances started at Eurocrypt 2013 with Joux's pinpointing technique [24]. At Crypto 2013, Göloglu, Granger, McGuire, and Jens Zumbrägel [25] already improved the constant c in the $L[\frac{1}{3}, c]$ running time. The improvement to running time $L[\frac{1}{4}]$ was then presented in the work of Joux [21]. Eventually, Barbulescu, Gaudry, Joux, and Thomé [22] proposed an algorithm for the descent that led to running time $L[o(1)]$.

12.4 Best Known Algorithms for Factoring Integers

The best algorithm for factoring shows close similarity to the best algorithm for computing discrete logarithms in prime order groups. It seems that the new attacks do not help to improve any of the two algorithms.

The best algorithm for computing the prime factorization of integers, the so-called number field sieve, is very similar to the best algorithm for computing discrete logarithm in $\mathbb{F}_p$ from Section 12.2.2, and much less similar to the algorithm for $\mathbb{F}_{q^n}$ from Section 12.3.

In a nutshell, all known, sophisticated algorithms that factor RSA moduli $n = pq$ for primes p, q of the same size rely on the same basic simple idea. Our goal is to construct $x, y \in \mathbb{Z}/n\mathbb{Z}$ such that

$$x^2 \equiv y^2 \bmod n \text{ and } x \not\equiv \pm y \bmod n.$$

This immediately yields the factorization of n, since n divides the product $x^2 - y^2 = (x + y)(x - y)$ by the first property, but n does neither divide $x + y$ nor $x - y$ by the second property. Thus, one prime factor of n has to divide $x + y$, whereas the other one has to divide $x - y$. This in turn means that $\gcd(x \pm y, n) = \{p, q\}$.

The factorization algorithms only differ in the way in which these x, y are computed. The intention is to compute x, y with $x^2 \equiv y^2 \bmod n$ in an independent way. If this independence is given, it is easy to show that $x \not\equiv \pm y \bmod n$ holds with probability $\frac{1}{2}$, since every square in $\mathbb{Z}/n\mathbb{Z}$ has 4 square roots by the chinese remainder theorem—two different roots modulo p and two different roots modulo q.

12.4.1 The Number Field Sieve for Factorization

Remark: The term number field sieve here always means the general number field sieve (GNFS). In the context of factorization there is a difference between a special and a general number field sieve—this is in the opposite to Section 12.2.2.

Let $n \in \mathbb{N}$ be the integer that we want to factor. In the number field sieve algorithm we start by constructing two polynomials f, g that share a common root

m modulo N. Usually this is done by simply defining $g(X) = X - m \bmod n$ and constructing some low degree polynomial $f(X)$ with $f(m) \equiv 0 \bmod n$ (e.g., by expanding n in base m as in Section 12.2.2).

Since f and g are different, they define different rings $\mathbb{Z}[X]/f(X)$ and $\mathbb{Z}[X]/g(X)$. But since f and g share the same root m modulo n, both rings are isomorphic to $\mathbb{Z}/n\mathbb{Z}$; and this isomorphism can be explicitly computed by the mapping $X \mapsto m$. This is illustrated in the commutative diagram in Figure 12.3.

Factor base: Consists of small-norm prime elements in both number fields.

Relation finding: We look for arguments $\tilde{x}$ such that simultaneously $\pi_f := f(\tilde{x})$ splits in $\mathbb{Q}[X]/(f(X))$ and $\pi_g := g(\tilde{x})$ splits in $\mathbb{Q}[X]/(g(X))$ into the factor base. Such elements are called relations.

Linear Algebra: By linear algebra, we search for a product of the elements π_f which is a square and whose corresponding product of the π_g is also a square. If we send these elements via our homomorphism $X \mapsto m$ to $\mathbb{Z}/n\mathbb{Z}$, we obtain elements $x^2, y^2 \in \mathbb{Z}/n\mathbb{Z}$ such that $x^2 \equiv y^2 \bmod n$. If we first compute the square roots of π_f and π_g in their respective number fields before applying the homomorphism, we obtain $x, y \in \mathbb{Z}/n\mathbb{Z}$ with $x^2 \equiv y^2 \bmod N$, as desired. The independence of x, y here stems from the different representations in both number fields.

Runtime: The above algorithm is up to some details (e.g., the square root computation in the number field) identical to the algorithm of Section 12.2.2 and shares the same running time $L[\frac{1}{3}, (\frac{64}{9})^{1/3}]$.

12.4.2 Relation to the Index Calculus Algorithm for Dlogs in $\mathbb{F}_p$

Firstly, we know that computing discrete logarithms in composite order groups $\mathbb{Z}/n\mathbb{Z}$ is at least as hard as factoring $n = pq$. This in turn means that any algorithm that computes discrete logarithms in $\mathbb{Z}/n\mathbb{Z}$ computes the factorization of n:

$$\text{Dlogs in } \mathbb{Z}/n\mathbb{Z} \Rightarrow \text{Factoring } n.$$

Let us briefly give the idea of this relation. We compute the order $k = \text{ord}(a)$ for an arbitrary $a \in \mathbb{Z}/n\mathbb{Z}$ by our dlog algorithm; that is, we compute the smallest positive integer k such that $a^k \equiv 1 \bmod n$. If k is even, then $a^{\frac{k}{2}} \not\equiv 1$ is a square root of 1. We have $a^{\frac{k}{2}} \not\equiv -1$ with probability at least $\frac{1}{2}$, since 1 has 4 square roots modulo n. Set

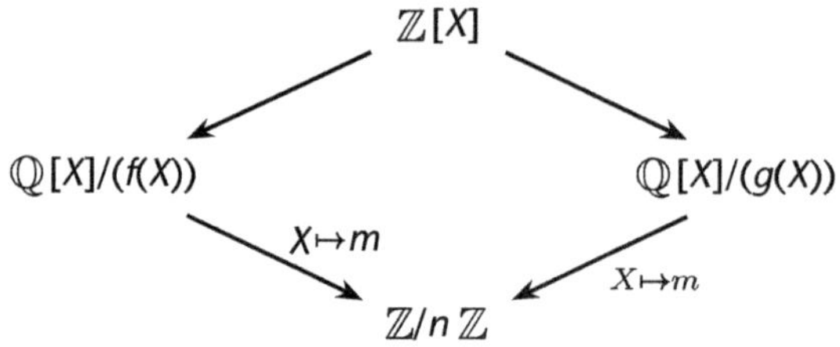

Figure 12.3 Isomorphic to $\mathbb{Z}/n\mathbb{Z}$ as a commutative diagram.

$x \equiv a^{\frac{k}{2}} \bmod n$ and $y = 1$. Then we obtain $x^2 \equiv 1 \equiv y^2 \bmod n$ and $x \not\equiv \pm y \bmod n$. By the discussion at the beginning of the chapter, this allows us to factor n.

Secondly, we also know that both problems factoring and computing discrete logarithms in $\mathbb{F}_p$ are together at least as hard as computing discrete logarithms in $\mathbb{Z}/n\mathbb{Z}$. In short

$$\text{Factoring + Dlogs in } \mathbb{F}_p \Rightarrow \text{Dlogs in } \mathbb{Z}/n\mathbb{Z}.$$

This fact can be easily seen by noticing that factoring and dlogs in $\mathbb{F}_p$ together immediately give an efficient version of the Silver-Pohlig-Hellman algorithm from Section 12.1. We first factor the group order n in prime powers $p_i^{e_i}$, and then compute the discrete logarithms in $\mathbb{F}_{p_i}$ for each i. Just as in the Silver-Pohlig-Hellman algorithm we lift the solution modulo $p_i^{e_i}$ and combine these lifted solutions via Chinese remaindering.

We would like to stress that these two known relations do not tell much about whether there is a reduction

$$\text{Factoring} \Rightarrow \text{Dlog in } \mathbb{F}_p \qquad \text{or} \qquad \text{Dlog in } \mathbb{F}_p \Rightarrow \text{Factoring}.$$

Both directions are a long-standing open problem in cryptography. Notice however that the best algorithms for factoring and dlog in $\mathbb{F}_p$ from Sections 12.2.2 and 12.4.1 are remarkably similar. Historically, algorithmic progress for one problem always immediately implied progress for the other problem as well. Although we have no formal proof, it seems to be fair to say that both problems seem to be closely linked from an algorithmic perspective.

12.4.3 Integer Factorization in Practice

Given the current state of the art of academic integer factorization research, even moderately sized (but properly chosen) RSA moduli offer a reasonable amount of protection against open community cryptanalytic efforts. The largest RSA challenge number factored in 2009 by a public effort had just 768 bit [27] and required the equivalent of about 2,000 years of computing on a single 2-GHz core (the current records are listed in Table 5.12). Attacking a 1024-bit RSA modulus is about a thousand times harder. Such an effort must be expected to be out of reach for academic efforts for several more years. Doubling the size to 2048-bit moduli increases the computational effort by another factor of 10^9. Without substantial new mathematical or algorithmic insights, 2048-bit RSA must be considered to be out of reach for at least two more decades (from 2013).

12.4.4 Relation of Key Size versus Security for Dlog in $\mathbb{F}_p$ and Factoring

The running time of the best algorithm for a problem defines the security level of a cryptosystem. For example, for 80-bit security, we want that the best algorithm requires at least 2^{80} steps.

As we already noted, the best running time for discrete logs in $\mathbb{F}_p$ and for factoring is $L[\frac{1}{3}, (\frac{64}{9})^{1/3}]$. The most accurate way to use this formula is to actually measure the running time for a large real world factorization/dlog computation, and then extrapolate to large values. Assume that we know that it took time T to factor a number n_1, then we extrapolate the running time for some $n_2 > n_1$ by the formula

$$T \cdot \frac{L_{n_1}\left[\frac{1}{3}, \left(\frac{64}{9}\right)^{1/3}\right]}{L_{n_2}\left[\frac{1}{3}, \left(\frac{64}{9}\right)^{1/3}\right]}.$$

So, we use the L-formula to estimate the relative factor that we have to spend in addition. Notice that this (slightly) overestimates the security, since the L-formula is asymptotic and thus becomes more accurate in the numerator than in the denominator—the denominator should include a larger error term. So, in practice, one obtains (only slightly) less security than predicted by this formula.

We computed the formula for several choices of the bit-size of an RSA number n, respectively a dlog prime p, in Table 12.2. Recall from Section 12.4.1 that the running time of the number field sieve algorithm for factoring is indeed a function of n and not of the prime factors of n.

We start with RSA-768 that has been successfully factored in 2009 [27]. In order to count the number of instructions for factoring RSA-768, one has to define what an instruction unit is. It is good practice in cryptography to define as a unit measure the time to evaluate DES in order to obtain comparability of security levels between secret and public key primitives. Then by definition of this unit measure, DES offers 56-bit security against brute-force key attacks.

In terms of this unit measure, the factorization of RSA-768 required $T = 2^{67}$ instructions. From this starting point, we extrapolated the security level for larger bit-sizes in Table 12.2.

We successively increase the bit-size by 128 up to 2048 bit. We see that in the beginning, this leads to roughly an increase of security of 5 bit per 128-bit step, whereas in the end we only have an increase of roughly 3 bit per 128-bit step.

By Moore's law the speed of computers doubles every 1.5 years. Hence after $5 \cdot 1.5 = 7.5$ years we have an increase of 2^5, which means that currently we should roughly increase our bit-size by 128 bit every 7.5 years; and when we come closer to 2000 bit our increase of 128-bit steps should be in intervals of no later than 4.5 years. For more conservative choices that also anticipate some algorithmic progress rather than just an increase in computers' speed see the recommendations in Section 12.7.

Table 12.2 Bitsize of n, p Versus Security Level

Bitsize	*Security*
768	67.0
896	72.4
1024	77.3
1152	81.8
1280	86.1
1408	90.1
1536	93.9
1664	97.5
1792	100.9
1920	104.2
2048	107.4

References and further reading: An introduction to several factorization algorithms including the quadratic sieve (the predecessor of the number field sieve) can be found in May's lecture notes on number theory [3]. We recommend Blömer's lecture notes on algorithmic number theory [28] as an introduction to the number field sieve. The development of the number field sieve is described in the textbook of Lenstra and Lenstra [16] that includes all original papers. The relation of discrete logarithms and factoring has been discussed by Bach [29]. Details of the current factorization record for RSA-768 can be found in [27].

12.5 Best Known Algorithms for Elliptic Curves E

Elliptic curves are the second standard group for the discrete logarithm problem. The new attacks do not affect these groups; their security remains unchanged.

We would like to discuss elliptic curves $E[p^n]$ over finite extension fields $\mathbb{F}_{p^n}$ and elliptic curves $E[p]$ over prime fields $\mathbb{F}_p$. The latter are usually used for cryptographic purposes. The reason to discuss the former too is to illustrate (similar to the previous sections) the vulnerabilities of extension fields $\mathbb{F}_{p^n}$ as opposed to prime field $\mathbb{F}_p$. However, we would like to point out that we assume in the following (in contrast to the previous section) that n is fixed. This is because as opposed to the algorithm of Joux et al., the algorithms for $E[p^n]$ have complexities that depend exponentially on n.

We present two different approaches for elliptic curves over extension fields: cover (or Weil descent) attacks introduced by Gaudry, Hess, and Smart (GHS), and decomposition attacks proposed by Semaev and Gaudry. In some cases, it is possible to combine the two approaches into an even more efficient algorithm as shown by Joux and Vitse [30].

12.5.1 The GHS Approach for Elliptic Curves $E[p^n]$

This approach introduced by Gaudry, Hess, and Smart aims at transporting the discrete logarithm problem from an elliptic curve E defined over an extension field $\mathbb{F}_{p^n}$ to a higher genus curve defined over a smaller field, for example $\mathbb{F}_p$. This can be done by finding a curve H over $\mathbb{F}_p$ together with a surjective morphism from H to E. In this context, we say that the curve H is a cover of E. Once such a curve H is obtained, it is possible using the so called conorm technique to pull back a discrete logarithm problem on E to a discrete logarithm problem on the Jacobian of H. If the genus g of the target curve is not too large, this can lead to an efficient discrete logarithm algorithm. This uses the fact that there exists an index calculus algorithm on high genus curve of genus g over $\mathbb{F}_p$ with complexity $\max(g!\,p, p^2)$. This was introduced by Enge, Gaudry, and Thomé [31].

Ideally, one would like the genus g to be equal to n. However, this is not possible in general. Classifying the possible covers for elliptic curve seems to be a difficult task.

12.5.2 The Gaudry-Semaev Algorithm for Elliptic Curves $E[p^n]$

Let $Q = \alpha P$ be a discrete logarithm on an elliptic curve $E[p^n]$. So the goal is to find the integer $\alpha \in \mathbb{N}$ such that k times the point $P \in E[p^n]$ added to itself is equal to the point $Q \in E[p^n]$.

Gaudry's discrete logarithm algorithm is of index calculus type. We briefly outline the basic steps.

Factor base: Consists of all points (x, y) on the elliptic curve $E[p^n]$ such that $x \in \mathbb{F}_p$. That is x lies in the ground field $\mathbb{F}_p$ rather than in the extension.

Relation finding: Given a random point $R = aP$, with $a \in \mathbb{N}$, we try to write R as a sum of exactly n points from the factor base, where n is the extension degree. This is achieved by using the nth Semaev polynomial f_{n+1}. This polynomial is a symmetric polynomial of degree 2^{n-2} in $n+1$ unknowns $x_1, \ldots, x_{n+1}$ which encodes the fact that there exists points with respective abscissae $x_1, \ldots, x_{n+1}$ that sum to zero. Of course, the coefficients of f depend on the curve E. Replacing x_{n+1} by the abscissa of R, we can find a decomposition of R as a sum of points from the factor base by searching for a solution $(x_1, \cdots, x_n)$ in the base field $\mathbb{F}_p$. In order to do this, one first rewrites f as a multivariate system of n equations by decomposing the constants that appear in the polynomial over some basis of $\mathbb{F}_{p^n}$ over $\mathbb{F}_p$. This system of n equations in n unknowns can be solved using a Groebner basis computation.

Individual discrete log computation: To compute the discrete logarithm of Q, it suffices to find one additional relation that expresses a random multiple of Q, namely $R = aQ$ in terms of the points in the factor base. This is done in the exact same way as the generation of relations in the previous step.

Runtime: The factor base can be computed in time $\mathcal{O}(p)$. Every R can be written as a sum of n factor base elements; that is, yields a relation, with probability exponentially small in n (but independent of p). If it yields a solution, the running time of a Groebner basis computation is also exponential in n (but polynomial in $\log p$). In total, we need roughly p relations which can be computed in time linearly in p and exponentially in n. Since we assumed n to be fixed, we do not care about the bad behavior in n. The linear algebra step on a $(p \times p)$-matrix can then be performed in $\mathcal{O}(p^2)$, since the matrix is sparse—every row contains exactly n nonzero entries. With additional tricks one achieves a running time of $\mathcal{O}(p^{2-\frac{2}{n}})$ for Gaudry's algorithm.

This should be compared to the generic bound of $\mathcal{O}(p^{\frac{n}{2}})$ that we achieve when using Pollard's rho algorithm from Section 12.1. Similar to Section 12.3, almost the whole complexity of the problem seems to be concentrated in the size of the base field p, and not in the extension degree n. Notice that as in Section 12.3, Gaudry's algorithm is exponential in $\log p$.

12.5.3 Best Known Algorithms for Elliptic Curves $E[p]$ Over Prime Fields

Generic discrete log solving: In general, the best algorithm that we know for arbitrary elliptic curves $E[p]$ is Pollard's rho method with a running time of $\mathcal{O}(\sqrt{p})$. For the moment, it seems that nobody knows how to exploit the structure of an elliptic curve group or its elements in order to improve over the generic bound.

We would also like to point out that random elliptic curves; that is, where the elliptic curve parameters a, b in the defining Weierstrass equation $y^2 \equiv x^3 + ax +$

$b \bmod p$ are chosen in a uniformly random manner, are among the hard instances. To further harden elliptic curves, one chooses for standardization only those curves that have (almost) prime order. This means that the cofactor of the largest prime in the group order is usually 1, which abandons the use of Silver-Pohlig-Hellman's algorithm.

Embedding $E[p]$ into $\mathbb{F}_{p^k}$: It is known that in general elliptic curves $E[p]$ can be embedded into a finite field $\mathbb{F}_{p^k}$, where k is the so-called embedding degree. In $\mathbb{F}_{p^k}$ we could use the number field sieve for discrete logarithm computations. Hence such an embedding would be attractive if $L_{p^k}[\frac{1}{3}]$ is smaller than $\sqrt{p}$, which is the case only if the embedding degree k happens to be very small. However, for almost all elliptic curves the embedding degree is known to be huge, namely comparable to p itself.

Some constructions in cryptography (e.g., those that make use of bilinear pairings), exploit the advantages of a small embedding degree. Thus, in these schemes elliptic curves are explicitly chosen with a small embedding degree (e.g., $k = 6$), balances out the hardness of the discrete logarithm problem on $E[p]$ and in $\mathbb{F}_p^k$.

The xedni calculus algorithm: In 2000, Silverman published his *xedni calculus algorithm* (read xedni backwards) that uses the group structure of $E[p]$ for discrete logarithm computations, and thus is the only known non-generic algorithm that works directly on $E[p]$. However, it was soon after his publication discovered that the so-called lifting process in Silverman's algorithm has a negligible probability of succeeding in computing a discrete logarithm.

12.5.4 Relation of Key Size versus Security for Elliptic Curves $E[p]$

Similar to the discussion in Section 12.4.4 about key sizes for dlog in $\mathbb{F}_p$ and for factoring, we want to evaluate how key sizes have to be adapted for elliptic curves $E[p]$ in order to guard against an increase in computer speed. For elliptic curves, such an analysis is comparably simple. The best algorithm that we know for the dlog in $E[p]$ is Pollard's rho method with running time

$$L_p[1, \frac{1}{2}] = \sqrt{p} = 2^{\frac{\log p}{2}}.$$

This means that for achieving a security level of k bit, we have to choose a prime p with $2k$ bit. In other words, increasing the bit-size of our group by 2 bit leads to increase of 1 bit in security. By Moore's law we loose 1 bit of security every 1.5 years just from an increase of a computer's speed. In order to guard against this loss over 10 years, it thus suffices to increase the group-size by just $7 \cdot 2 = 14$ bit. Notice that as opposed to the case of dlog in $\mathbb{F}_p$ and factoring in Section 12.4.4 this increase is linear and independent of the starting point. That means to guard against technological speedups over 20 years, an increase of 28 bit is sufficient.

Of course, this analysis only holds if we do not have to face any major breakthrough in computer technology or algorithms. For a more conservative choice see the advice in Section 12.7.

12.5.5 How to Securely Choose Elliptic Curve Parameters

A comprehensive description on how to choose elliptic curve domain parameters over finite fields can be found in RFC 5639 "ECC Brainpool Standard Curves and Curve Generation" by Manfred Lochter and Johannes Merkle [32, 33]. This RFC defines a publicly verifiable way of choosing pseudorandom parameters for elliptic curve parameters, and thus it excludes the main source for embedding a trapdoor in the definition of a group. The authors discuss all known properties of a curve $E[p]$ that might potentially weaken its security:

- *A small embedding degree* for the embedding into a finite field: This would allow for the use of more efficient finite field algorithms. Especially, the requirement excludes supersingular curves of order $p + 1$.
- *Trace one curves* that have order $|E[p]| = p$: These curves are known to be weak by the discrete logarithm algorithms of Satoh-Araki [34], Semaev [35], and Smart [36].
- *Large class number*: This excludes that $E[p]$ can be efficiently lifted to a curve defined over some algebraic number field. This requirement is quite conservative, since even for small class numbers there is currently no efficient attack known.

Moreover, the authors insist on the following useful properties:

- *Prime order*: This simply rules out subgroup attacks.
- *Verifiable pseudorandom number generation*: The seeds for a pseudorandom number generator are chosen in a systematic way by Lochter and Merkle, who use in their construction the first seven substrings of length 160 bit of the fundamental constant $\pi = 3.141\ldots$.

In addition, Lochter and Merkle specify a variety of curves for p's of bit-lengths in the range 160 to 512. For TLS/SSL there is also a new set of proposed Brainpool curves available [37].

The work of Bos, Costello, Longa, and Naehrig [38] gives a valuable introduction for practitioners on how to choose elliptic curve parameters that are secure and also allow for efficient implementation in various coordinate settings (Weierstrass, Edwards, Montgomery). Additionally, Bos et al. focus on side-channel resistance against timing attacks by proposing constant-time scalar multiplications.

We highly recommend the SafeCurve project by Daniel Bernstein and Tanja Lange [39] that provides an excellent overview for several selection methods, their benefits and drawbacks. The goal of Bernstein and Lange is to provide security of elliptic curve cryptography, rather than just strength of elliptic curves against discrete logarithm attacks. Therefore, they take into account various types of side-channels that may leak secrets in an implementation.

References and further reading: For an introduction to the mathematics of elliptic curves and their cryptographic applications we refer to the textbooks of Washington [40], Galbraith [6], and Silverman [41].

This section described the results of the original works of Gaudry, Hess, Smart [42], Gaudry [43], Semaev [44], and the xedni algorithm of Silverman [41].

12.6 Possibility of Embedded Backdoors in Cryptographic Keys

All cryptography seems to offer the possibility of embedding backdoors. Dlog schemes offer some advantage over factoring-based schemes in the sense that carefully chosen system-wide parameters protect all users.

The possibility of embedding trapdoors in cryptographic schemes to bypass cryptography and thus to decrypt/sign/authenticate without the use of a secret key is a long recognized problem that has been intensively discussed in the cryptographic community (e.g., at the panel discussion at Eurocrypt 1990). However, the wide-spread use of NSA's backdoors as described by Edward Snowden has recently renewed the interest in this topic.

It appears that by construction some schemes are way more vulnerable than others. For example, for discrete-log based schemes the definition of the group parameters is a system-wide parameter that is used by any user in the scheme. Thus, a party that is able to manipulate the definition of a group in such a way that enables this party to compute discrete logarithms in this group efficiently, can decrypt all communication. On the other hand, a carefully specified secure group also offers security for all users.

Currently, there is some speculation whether the NSA influenced NIST, the U.S. standardization agency, to standardize certain elliptic curves. But the definition of a group is not the only way to embed backdoors. All cryptographic schemes rely inherently on a good source of (pseudo)random bits. It is well known that so-called semantic security of encryption schemes cannot be achieved without randomness, and every cryptographic secret key is assumed to be randomly chosen. Thus, a weak pseudorandom generator opens the door for bypassing cryptography. Such a weak pseudorandom generator was standardized by NIST as Special Publication 800-90, although there have been warnings by the cryptographic community.

For factoring-based schemes the situation is slightly different from discrete log-based schemes. As opposed to discrete log schemes, there are no system-wide parameters that define a group. Nevertheless, there are known ways to embed, for example, information about the factorization of the RSA modulus N in the RSA public exponent e. Moreover, recent attacks on RSA public key infrastructures [45, 46] show that it appears to be a difficult problem to generate RSA public keys with different primes in the public, mainly due to bad initialization of pseudorandom generators. This of course does only affect badly chosen keys of individuals as opposed to all users of cryptographic scheme.

Recommendation: Dlog-based schemes seem to be easier to control from a cryptodesigners perspective, since here all users have to take the same system-wide parameters.

We do not discuss the possibility of malware here—which may render obsolete any cryptographic protection method—or how to protect against it. But we would like to stress the following (somewhat trivial) warning that addresses a crucial point in practice.

Warning: Cryptography can only protect data if it is properly implemented and does not leak its (imminent) secret. So in addition to the mathematical hardness

Table 12.3 Security Level 100 Bit

System	*Key Size in Bits*
Dlog in $\mathbb{F}_p$	2000 until 2019, then 3000
Factoring	2000 until 2019, then 3000
Dlog in $E[p]$	224 until 2015, then 250

Source: BSI [49], ANSSI [50].

of the underlying problems, we also have to trust in the implementor of a cryptographic scheme. This trust does not only include that the cryptographic scheme is implemented in the way it was originally designed (without embedding of any backdoors), but also that the implementor does not reveal the generated secret keys to a third party.

It seems that in the NSA affair, some companies were forced to reveal secret keys. Thus, one has to keep in mind that one has to buy cryptographic schemes from a completely reliable company that has not been compromised.

References and further reading: For a nice discussion of how to embed undetectable backdoors in various cryptographic schemes, see the original works of Young and Yung [47, 48]. See [45] for a current attack on a significant portion of RSA keys in practice due to bad pseudorandom number generation.

12.7 Conclusion: Advice for Cryptographic Infrastructure

Despite recent discrete logarithm attacks, discrete logarithm-based schemes over prime order groups and elliptic curve groups remain secure. The same holds for factoring-based schemes. All discrete logarithm-based groups with small characteristics are completely insecure. Our suggestion is to choose elliptic curve groups.

12.7.1 Suggestions for Choice of Scheme

As we saw in the previous sections of this chapter, discrete log-based schemes in $\mathbb{F}_p$ and over $E[p]$ remain secure, as well as factoring-based schemes. In this subsection, we suggest key sizes for these schemes that provide a sufficient security level for the next two decades under the assumption that no major algorithmic breakthrough occurs.

Our preference is to use elliptic curve groups $E[p]$ since they offer the following advantages:

- Algorithms for discrete logarithms in $\mathbb{F}_p$ and factoring are closely linked. So any progress in one of these two might imply some progress for the other. But such progress is unlikely to affect the security of elliptic curve groups.
- The best algorithms for $E[p]$ are those of generic type from Section 12.1, which are inferior to the best algorithms for prime order discrete logarithm and factoring with $L[\frac{1}{3}]$ running time. This in turn means that the key growth that compensates technological progress of faster computers is much smaller for $E[p]$—roughly 2 bit every 1.5 years according to Moore's law.

- Getting algorithmic progress by using the group structure of $E[p]$ seems to be harder than for $\mathbb{F}_p$ since, as opposed to $\mathbb{F}_p$, we do not even have an initial starting group-structure index calculus algorithm that we could improve.
- If an elliptic curve $E[p]$ is properly chosen (i.e., the group is computationally hard and backdoor-free, then all users profit from the hardness of the discrete logarithm problem. Notice that this choice is crucial: If the group is not secure, then all users also suffer from its insecurity.

Warning: One should keep in mind that the suggestions above only hold in a world without large quantum computers. It seems crucial to keep track of current progress in this area, and to have some alternative quantum-resistant cryptosystems ready to enroll within the next 15 years.

References and further reading: For a good and conservative choice of key sizes we highly recommend following the suggestions of the Bundesamt für Sicherheit in der Informationstechnik (BSI) [49] and the Agence nationale de la sécurité des systèmes d'information [50]. Both sources also provide various valuable recommendations how to correctly implement and combine different cryptographic primitives.

12.7.2 Year 2023: Conclusion Remarks

Since the first advice in April 2014, quite a lot of things have changed (there have been new records in dlog finite fields and some marginal improvements of the L(1/3) algorithms in some contexts). However, this does not affect the overall conclusion that (only) small characteristic finite fields are no longer secure.

The recommendations of this chapter are still valid. See:

- BSI: "TR-02102-1: Cryptographic Mechanisms: Recommendations and Key Lengths" in 2022 [49];
- NIST: "SP 800-186 (Final) Recommendations for Discrete Logarithm-Based Cryptography: Elliptic Curve Domain Parameters" in February 2023 [51];
- NIST: "SP 800-56B Recommendation for Pair-Wise Key-Establishment Schemes Using Integer Factorization Cryptography" in 2019 [52].

In April 2022, Fabrice Boudot et al. published the very good article "The State of the Art in Integer Factoring and Breaking Public-Key Cryptography" in IEEE Security & Privacy. There they review the three number-theoretic problems of integer factorization, discrete logarithms in finite fields, and discrete logarithms over elliptic curves, and come to very similar results [53].

References

[1] Ptacek, T., et al., "The Factoring Dead—Preparing for the Cryptopocalypse," in *Black Hat Conference* (2013).

[2] Fouque, P.-A., A. Joux, and C. Mavromati, "Multi-User Collisions: Applicationsto Discrete Logarithm, Even-Mansour and Prince," in *Cryptology ePrint Archive*, 2014, https://eprint .iacr.org/2013/761.

[3] May, A., *Vorlesungsskript Zahlentheorie*, 2013, https://www.cits.ruhr-uni-bochum.de/imperia/md/content/may/13/ss13/zahlenss13/zahlentheorie.pdf.

[4] Menezes, A. J., P. C. van Oorschot, and S. A. Vanstone, *Handbook of Applied Cryptography*, 5th ed., Series on Discrete Mathematics and Its Application, CRC Press, 2001, https://cacr.uwaterloo.ca/hac/.

[5] Joux, A., *Algorithmic Cryptanalysis*, CRC Cryptography and Network Security Series, Chapman & Hall, 2009.

[6] Galbraith, S. D., *Mathematics of Public Key Cryptography*, Cambridge University Press, 2012.

[7] May, A., *Vorlesungsskript Kryptanalyse 1*, 2008, https://www.cits.ruhr-uni-bochum.de/imperia/md/content/may/pkk08/skript.pdf.

[8] May, A., *Vorlesungsskript Kryptanalyse 2*, 2012, https://www.cits.ruhr-uni-bochum.de/imperia/md/content/may/12/ws1213/kryptanal12/kryptanalyse_2013.pdf.

[9] Homeister, M., *Quantum Computer Science: An Introduction*, Vieweg + Teubner Verlag, 2007.

[10] Mermin, D. N., *Quantum Computing Verstehen*, Cambridge University Press, 2008.

[11] Pollard, J. M., "A Monte Carlo Method for Factorization," in *BIT Numerical Mathematics 15*, Vol. 3, 1975, pp. 331–334.

[12] Pollard, J. M., "Kangaroos, Monopoly and Discrete Logarithms," in *J. Cryptology*, Vol. 13, No. 4, 2000, pp. 437–447.

[13] Shor, P. W., "Algorithms for Quantum Computation: Discrete Logarithms and Factoring," in *FOCS*, 1994, pp. 124–134.

[14] Shoup, V., "Lower Bounds for Discrete Logarithms and Related Problems," in *EUROCRYPT*, 1997, pp. 256–266.

[15] Müller-Stach, and Piontkowski, *Elementare und Algebraische Zahlentheorie*, Vieweg Studium, 2011.

[16] Lenstra, A. K., and H. W. Lenstra Jr., *The Development of the Number Field Sieve*, Lecture Notes in Mathematics, Springer, Verlag, 1993.

[17] Adleman, L. M., "A Subexponential Algorithm for the Discrete Logarithm Problem with Applicationsto Cryptography (Abstract)," in *FOCS*, 1979, pp. 55–60.

[18] Coppersmith, D., A. M. Odlyzko, and R. Schroeppel, "Discrete Logarithms in GF(p)," in *Algorithmica*, Vol. 1, No. 1, 1986, pp. 1–15, http://dx.doi.org/10.1007/BF01840433.

[19] Pomerance, C., "The Quadratic Sieve Factoring Algorithm," in *Proceedings of Crypto '84, LNCS 196*, G.R. Blakley and D. Chaum (eds.), Springer, 1984, pp. 169–182.

[20] Pomerance, C., "A Tale of Two Sieves," in *Notices Amer. Math. Soc*, Vol. 43, 1996, pp. 1473–1485.

[21] Joux, A., "A New Index Calculus Algorithm with Complexity L(1/4+o(1)) in Very Small Characteristic," in *IACR Cryptology ePrint Archive 2013*, 2013, p. 95.

[22] Barbulescu, R., et al., "A Quasi-Polynomial Algorithm for Discrete Logarithm in Finite Fields of Small Characteristic," in *CoRR*, 2013, abs/1306.4244.

[23] Joux, A., and R. Lercier, "The Function Field Sieve in the Medium Prime Case," in *EUROCRYPT*, 2006, pp. 254–270.

[24] Joux, A., "Faster Index Calculus for the Medium Prime Case Application to 1175-bit and 1425-bit Finite Fields," in *EUROCRYPT*, 2013, pp. 177–193.

[25] Göloglu, F., et al., "On the Function Field Sieve and the Impact of Higher Splitting Probabilities—Application to Discrete Logarithms," in *CRYPTO (2)*, 2013, pp. 109–128.

[26] Coppersmith, D., "Evaluating Logarithms in GF(2n)," in *STOC*, 1984, pp. 201–207, https://dl.acm.org/doi/10.1145/800057.808682.

[27] Kleinjung, T., et al., "Factorization of a 768-Bit RSA Modulus," in *CRYPTO*, 2010, pp. 333–350, http://dx.doi.org/10.1007/978-3-642-14623-7_18.

[28] Blömer, J., *Vorlesungsskript Algorithmische Zahlentheorie*, 1999.

[29] Bach, E., *Discrete Logarithms and Factoring*, UCB/CSD-84-186, June 1984, https://www2.eecs.berkeley.edu/Pubs/TechRpts/1984/5973.html; https://www2.eecs.berkeley.edu/Pubs/TechRpts/1984/CSD-84-186.pdf.

[30] Joux, A., and V. Vitse, "Cover and Decomposition Index Calculus on Elliptic Curves Made Practical. Application to a Seemingly Secure Curve Over Fp6," in *IACR Cryptology ePrint Archive*, 2011, p. 20.

[31] Enge, A., P. Gaudry, and E. Thomé, "An $L(1/3)$ Discrete Logarithm Algorithm for Low Degree Curves," in *J. Cryptology*, Vol. 24, No. 1, 2011, pp. 24–41.

[32] Lochter, M., and J. Merkle, *Elliptic Curve Cryptography (ECC) Brainpool Standard Curves and Curve Generation*, RFC 5639, 2010, https://datatracker.ietf.org/doc/html/rfc5639.

[33] Lochter, M., and J. Merkle, *ECCBrainpool Standard Curves and Curve Generation v. 1.0*, 2005, https://www.teletrust.de/fileadmin/files/oid/oid_ECC-Brainpool-Standard-curves-V1.pdf.

[34] Satoh, T., and K. Araki,"Fermat Quotients and the Polynomial Time Discrete Log Algorithm for Anomalous Elliptic Curves," in *Commentarii Mathematici Universitatis Sancti Pauli 47*, 1998.

[35] Semaev, I., "Evaluation of Discrete Logarithms on Some Elliptic Curves," in *Mathematics of Computation* 67, 1998.

[36] Smart, N., "The Discrete Logarithm Problem on Elliptic Curves of Trace One," in *Journal of Cryptology 12*, 1999.

[37] Lochter, M., and J. Merkle, *Elliptic Curve Cryptography (ECC) Brainpool Curves for Transport Layer Security (TLS)*, RFC 7027, 2013, https://datatracker.ietf.org/doc/html/rfc7027.

[38] Bos, J. W., et al., *Selecting Elliptic Curves for Cryptography: An Efficiency and Security Analysis*, 2014, https://www.microsoft.com/en-us/research/wp-content/uploads/2016/02/selecting.pdf.

[39] Bernstein, D., and T. Lange, *SafeCurves: Choosing Safe Curves for Elliptic-Curve Cryptography*, 2014, https://safecurves.cr.yp.to.

[40] Washington, L. C., *Elliptic Curves: Number Theory and Cryptography*, Discrete Mathematics and its Applications, Chapman and Hall/CRC, 2008.

[41] Silverman, J. H., "The Xedni Calculus and The Elliptic Curve Discrete Logarithm Problem," in *Designs, Codes and Cryptography*, Vol. 20, 1999, pp. 5–40.

[42] Gaudry, P., F. Hess, and N. P. Smart, "Constructive and Destructive Facets of Weil Descenton Elliptic Curves," in *J. Cryptology*, Vol. 15, No. 1, 2002, pp. 19–46.

[43] Gaudry, P., "Index Calculus for Abelian Varieties of Small Dimension and the Elliptic Curve Discrete Logarithm Problem," in *J. Symb. Comput.*, Vol. 44, No. 12, 2009, pp. 1690–1702.

[44] Semaev, I., "Summation Polynomials and the Discrete Logarithm Problem on Elliptic Curves," in *IACR Cryptology ePrint Archive*, 2004, p. 31.

[45] Lenstra, A. K., et al., "Public Keys," in *CRYPTO*, 2012, pp. 626–642, http://dx.doi.org/10.1007/978-3-642-32009-5_37.

[46] Heninger, N., et al., "Mining Your Ps and Qs: Detection of Widespread Weak Keys in NetworkDevices," in *Proceedings of the 21st USENIX Security Symposium*, August 2012, https://factorable.net/paper.html.

[47] Young, A. L., and M. Yung, "The Dark Side of Black-Box Cryptography, or: Should We TrustCapstone?" in *CRYPTO*, 1996, pp. 89–103.

[48] Young, A. L., and M. Yung. "Kleptography: Using Cryptography Against Cryptography," in *EUROCRYPT*. 1997, pp. 62–74.

[49] BSI,*Technical Guideline TR-02102-1, Cryptographic Mechanisms: Recommendations and Key Lengths (Version 2022-01)*, Tech. rep., 2022, https://www.bsi.bund.de/SharedDocs/Downloads/EN/BSI/Publications/TechGuidelines/TG02102/BSI-TR-02102-1.pdf.

[50] Agencenationale de la sécurité des systèmes d'information, *Référentiel général de sécurité Version 2.02*, 2013, https://www.ssi.gouv.fr/administration/reglementation/confiance-numerique/le-referentiel-general-de-securite-rgs/.

[51] Chen, L., et al., *Recommendations for Discrete Logarithm-based Cryptography: Elliptic Curve DomainParameters*, Special Publication (NIST SP), National Institute of Standards and Technology, 2023, https://csrc.nist.gov/publications/detail/sp/800-186/final.

[52] Barker, E., et al., *Recommendation for Pair-Wise Key-Establishment Schemes Using Integer Factorization Cryptography*, Special Publication (NIST SP), National Institute of Standards and Technology, 2019, https://nvlpubs.nist.gov/nistpubs/SpecialPublications/NIST.SP.800-56Br2.pdf.

[53] Boudot, F., et al., "The State of the Art in Integer Factoring and Breaking Public-Key Cryptography," in *IEEE Security & Privacy*, Vol. 20, No. 2, 2022, pp. 80–86, https://ieeexplore.ieee.org/document/9740707.

CHAPTER 13

Future Use of Cryptography

Cryptography is a fundamental building block of all IT security solutions. But, for how long will the cryptographic tools we use today remain secure? Is that long enough to ensure the confidentiality of medical data? Even in the short term, the potential for havoc is great if certain keys are broken. Consider the digital signatures that protect the authenticity of automatic updates for the Windows operating system or for critical business applications.

At the same time, the cryptographic community is anticipating future advances and providing methods that can withstand quantum computers [post-quantum cryptography (PQC)] or that enable trustworthy computing in the cloud (MPC).

13.1 Widely Used Schemes

In 1978, Rivest, Shamir, and Adleman proposed the RSA public-key encryption and signature schemes [1]. RSA is still the most widely used public-key scheme. The security of RSA depends on the difficulty of factoring so-called RSA moduli which are products of two large prime numbers. In their 1978 paper, the inventors of RSA suggested using RSA moduli with 200 decimal places for long-term security. Later, the company RSA Security published a list of RSA moduli of increasing size, the RSA Challenge. RSA Security offered a total of $635,000 in prizes for factoring these numbers; see Section 5.12.4.

In 2005, 27 years after the invention of RSA, Bahr, Boehm, Franke, and Kleinjung of the University of Bonn succeeded in factoring a 200-decimal-digit RSA challenge number (see Section 5.12.4). A key of this size, originally thought to be secure for a very long time, was broken with a calculation that took them only five months. This illustrates the tremendous progress factoring technology has made in the 30 years since the invention of the RSA algorithm. This progress is based on breakthrough mathematical ideas—such as the number field sieve proposed by John Pollard as well as significant developments in computer hardware and software implementation technology. Recent cryptanalytic results against RSA and Dlog were discussed in Chapter 12 and Section 5.12.

In 2000, Lenstra and Verheul [2] developed an extrapolation formula to help us predict the security that can be achieved with RSA and other important cryptographic schemes in the long run. The formula suggests using 850-decimal-digit RSA moduli if you want to protect data until 2038 (this corresponds to a 3072-bit RSA key). RSA-2048 has an effective security of about 88 bit, making it secure until about 2023 if you follow the Lenstra/Verheul equations from 2000; if you follow the Lenstra equations from 2004, it has an effective security of about 95 bit, making it secure until about the year 2040. So the experts have adjusted their opinions over time.

These results and the recommendations of seven other authorities are dynamically processed on the Bluekrypt website [3]. See Figures 13.1 and 13.2.

However, even a well thought-out extrapolation formula is no guarantee of security. At any time, a brilliant mathematical idea can allow us to easily factor large numbers, and destroy the security of RSA. In 1996, Peter Shor showed that a quantum computer—a new type of computer that leverages the laws of quantum mechanics to speed up certain types of computation—could in principle be used to quickly factor large numbers [4]. If Shor's algorithm could be practically applied,

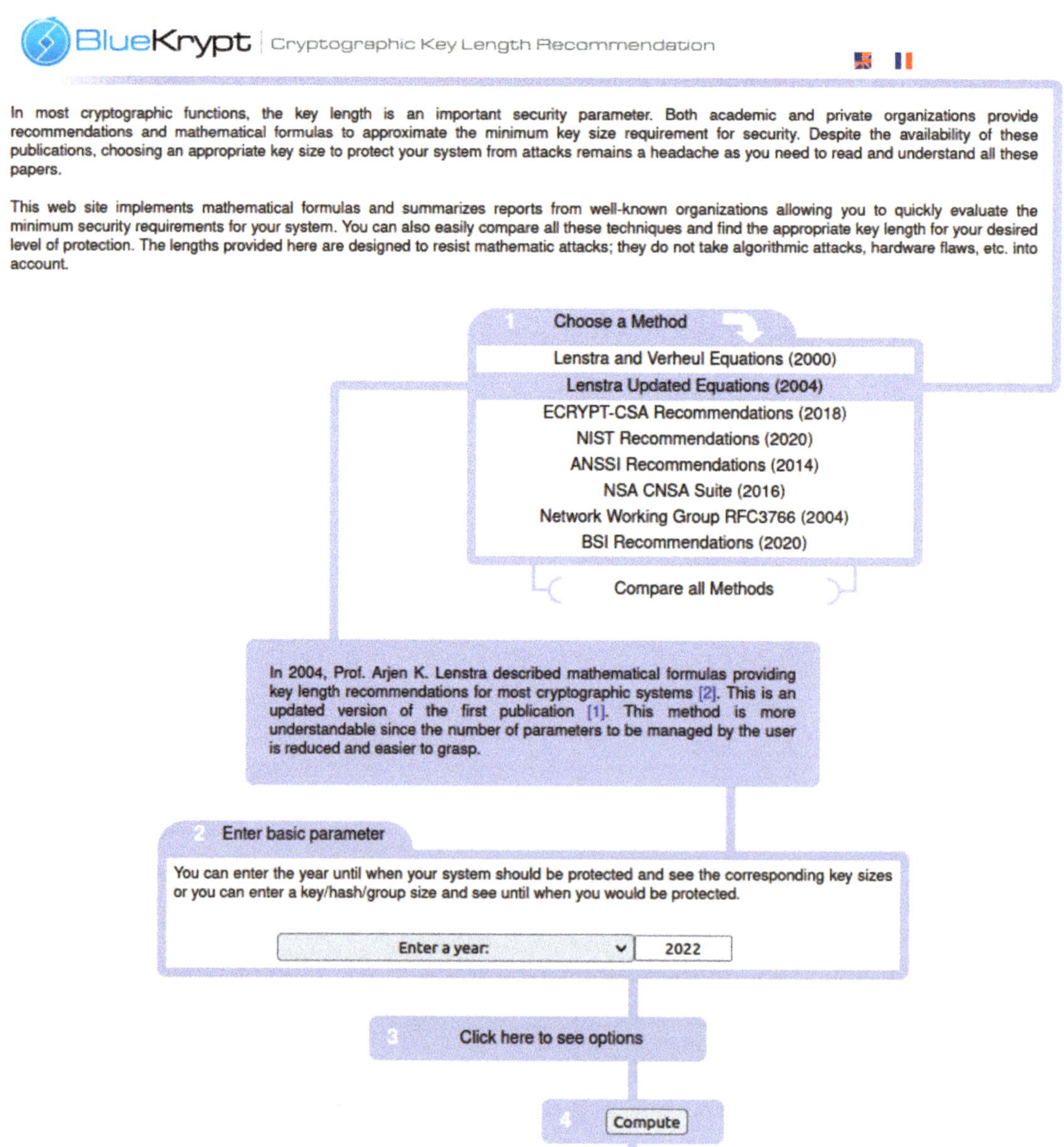

Figure 13.1 A graph to determine secure key length until a given year (from BlueKrypt).

All key sizes are provided in bits. These are the minimal sizes for security.

Year	Symmetric	Factoring (modulus) Optimistic	Factoring (modulus) Conservative	Discrete Logarithm Key	Discrete Logarithm Group	Elliptic Curve	Hash
2020	82	1387	1568	163	1387	163	163
2021	82	1416	1614	164	1416	164	164
2022	**83**	**1446**	**1660**	**166**	**1446**	**166**	**166**
2023	84	1476	1708	167	1476	167	167
2024	84	1507	1756	168	1507	168	168

Results in this table have to be considered with caution.

Figure 13.2 Secure key sizes: result in BlueKrypt for the year 2022.

one would have to double the bit length of an RSA key to achieve the same level of security. Despite intensive research in this area, it is still too early to say whether we will ever be able to build quantum computers of sufficient capacity to apply Shor's algorithm to numbers of relevant size. See Section 5.12.3.

Early announcements by D-Wave about the performance of their quantum computer were met with a lot of skepticism, even ridicule. As large companies have invested heavily in QC, the skepticism has turned into hype.

The development of attacks on another widely used scheme called Digital Signature Algorithm and the elliptic curve cryptography class of schemes is analogous to those on RSA. The security of these schemes depends on the difficulty of computing discrete logarithms. Even today, significant algorithmic progress is being made. Quantum computers would render these schemes insecure.

And what's the status of symmetric (so called secret-key) encryption schemes? In 1977, DES was introduced as Data Encryption Standard [5]. Twenty-one years later, the Electronic Frontier Foundation built Deep Crack, a specialized machine that took only 56 hours to break a DES key. The problem with DES was that it used keys that were too short. It seems that the inventors of DES did not anticipate the speed of hardware development. The Advanced Encryption Standard [6], the successor to DES, is currently considered secure, although there are interesting, though still inefficient, methods to attack AES using algebraic methods.

AES is the gold standard for all symmetric ciphers—and because of more powerful and cheaper chips, it is now even used in low-power, resource-constrained devices such as sensors.

13.2 Preparing for Tomorrow

Is the security of today's cryptography adequate for its growing importance? Experience shows: carefully designed and implemented cryptographic schemes have a lifetime of five to twenty years. Those who use RSA, ECC, or AES for short-term data protection can feel secure. It is also possible to achieve long-term authenticity, integrity, and nonreputability of data, for example, by using multiple signature schemes.

However, current schemes cannot guarantee long-term confidentiality. And what about twenty years from now? What should we do if, virtually overnight, unexpected mathematical progress renders an important cryptographic scheme insecure? Three things are needed to prepare us for this event:

- A pool of secure alternative cryptographic schemes;
- Infrastructures that allow us to easily and quickly replace one cryptographic scheme with another (agile APIs);
- Methods that ensure long-term confidentiality.

There is intensive research in post-quantum cryptography searching for cryptographic schemes that will remain secure even if powerful quantum computers are built. Good overviews of the current state of the art can be found in [7] and the ENISA report [8].

The security of public-key cryptography has traditionally been based on the difficulty of solving certain mathematical problems. Today, the following alternatives to the factorization and discrete logarithm problems are extensively discussed: the decoding problem, the shortest and closest vector problem in lattices, and the problem of solving large systems of multivariate quadratic equations. It is suggested that quantum computers offer little advantage in trying to solve these problems efficiently.

13.3 New Mathematical Problems

Let us take a closer look at these alternatives. The first encryption scheme based on the decoding problem was proposed by McEliece [9].[1] Background: Error-correcting codes are used to transmit or store electronic data in such a way that it remains undistorted even if a few bits are changed during transmission or on the storage media. This property is used, for example, in compact discs (CDs). The data on a CD can be reconstructed even if the disc is slightly scratched.

In a code-based encryption scheme, a message is encrypted by adding a fixed number of errors to the encrypted message (i.e., flipping a fixed number of bits). Decryption requires knowledge of an approriate decryption procedure that efficiently eliminates these errors. This procedure is called the secret key. Code-based encryption is generally very efficient. Research is currently underway to determine which codes lead to secure encryption methods with the smallest possible keys.

Encryption based on lattice problems is similar to encryption based on error-correcting codes. Lattices are regular structures of points in space. For instance, the points where the lines cross on a square piece of paper form a 2-dimensional lattice. For cryptographic use, the dimension of the lattice is chosen to be much larger. Encryption works as follows: The plaintext is used to construct a lattice point which is then slightly distorted so that it is no longer a lattice point, but close to one. Whoever knows a secret about the lattice is able to find this lattice point in the vicinity of the given point in space. The lattice point in turn yields the plaintext. Chapter 11 gave a lightweight introduction to lattices.

1. McEliece can be found in `JCT Algorithm Perspective` and in `JCT Default Perspective ▷ Visuals ▷ McEliece Cryptosystem`.

13.4 New Signatures

In 1979, Ralph Merkle proposed a remarkable framework for new signature schemes in his PhD thesis [10]. Unlike all other signature schemes, its security is not based on the difficulty of a number-theoretic, algebraic, or geometric problem. It requires only what other signature schemes require anyway: a cryptographically secure hash function and a secure pseudorandom number generator. Each new hash function leads to a new signature algorithm. As a result, the Merkle scheme has the potential to solve the problem of long-term availability of digital signature schemes.

Merkle uses so-called one-time signatures in his construction: Each new signature requires a new signing key and a new verification key. Merkle's idea was to use a hash tree to reduce the validity of many verification keys to the validity of a unique public hash value. When generating keys for the Merkle scheme, one must determine in advance the number of signatures that can be made with them. For a long time, this seemed to be a significant drawback. In [11], however, a variant the Merkle scheme was proposed that allows 2^{40} signatures to be computed with a single key pair.[2]

Another new signature scheme uses multivariate cryptography. This asymmetric scheme uses multivariate polynomials over a finite field.[3]

13.5 Quantum Cryptography: A Way Out of the Dead End?

From the point of view of the current state of cryptography, the problem of long-term confidentiality remains unsolved: There is no practical way to keep an encrypted message secret for a very long time.

Quantum cryptography can provide a way out here: These quantum technologies establish a secure channel to enable the exchange of keys (e.g., very long keys for one-time pads). Their security is guaranteed by the laws of quantum mechanics; see [13]. However, the known methods of quantum cryptography are currently rather inefficient and allow only symmetric methods. Governments, for example, can use them to exchange top-secret information. For many applications such as signatures, symmetric cryptography alone is not sufficient. Note that quantum cryptography should not be confused with post-quantum cryptography.

13.6 Post-Quantum Cryptography

Today's cryptography provides tools to ensure short- and medium-term security. Software developers can use these tools in their applications with a clear conscience

2. Under `JCT Default Perspective` ▷ `Visuals` you can find several components and variants of this: the one-time signature WOTS+, the normal Merkle signature (MSS), the extended Merkle signature scheme (XMSS), and the multitree Merkle signature scheme (XMSS_MT).

   In addition, the SPHINCS+ signature is extensively visualized. SPHINCS+ was one of the second track candidates in the NIST post-quantum computing contest in round three (2020).

   Many variants are offered in the `JCT Algorithm Perspective`, delivered by the BouncyCastle library.

3. In `JCT Default Perspective` ▷ `Visuals` ▷ `Multivariate Cryptography`, the rainbow signature variant by Jintai Ding and Dieter Schmidt [12] is used, which utilizes several layers of multivariate linear equation systems.

as long as they ensure that components can be quickly exchanged when they become insecure.

To ensure IT security in the future, we need to build a portfolio of secure cryptographic schemes. This portfolio must include schemes that are suitable for the world of ubiquitous computing with many less powerful computers. It also needs to include schemes that will remain secure in the event that powerful quantum computers are built. Several promising candidates have been discussed in this chapter. The question of how to ensure long-term confidentiality still remains an open research problem.

In 2016, NIST launched a competition to identify suitable alternatives to the current generation of cryptographic methods (such as RSA or ECDSA). This next generation of cryptographic algorithms is called "post-quantum cryptography."

In July 2022, as a result of the third round, NIST announced which methods it wants to standardize [14]:

- For public-key encryption and key exchange: CRYSTALS-Kyber (see Section 11.11);
- For digital signatures: CRYSTALS-Dilithium, Falcon, SPHINCS+.

13.7 Conclusion

Cryptography is important and a lot of work, but cryptography is also intellectually challenging and fun.

For the users (both private and business) cryptography is mostly an invisible part of IT security and of corporate risk management as outlined in Figure 13.3.[4]

We are seeing more and more end-to-end encryption and sophisticated protocols in products. Messengers are a good example: Signal[5] was the first widely used protocol for postcompromise security, and its successor Messaging Layer Security (MLS) will make even chat groups secure. MLS is an emerging standard that supports end-to-end encryption in messaging applications, and was published as RFC 9420 in July 2023 [15].

IT security is now less at risk from bad cryptographic algorithms than from:

- Attackers who just need to find one weak link in the chain. For example, one server with weak password hashing, one computer on a network without updates, one misconfigured router, one outdated component or library, and so on;
- Users who mainly want speed and good usability, but don't care about security (awareness, backups, and common sense are needed even on the computer);
- Monocultures and digital dependencies: This includes hardware with subsystems such as "management engine" or "secure technology" and operating

4. See "CrypTool for Awareness" `https://www.cryptool.org/en/education/awareness`.
5. In `JCT Default Perspective ▷ Visuals ▷ Signal Encryption`, the double ratchet scheme of the Signal protocol is visualized step by step.

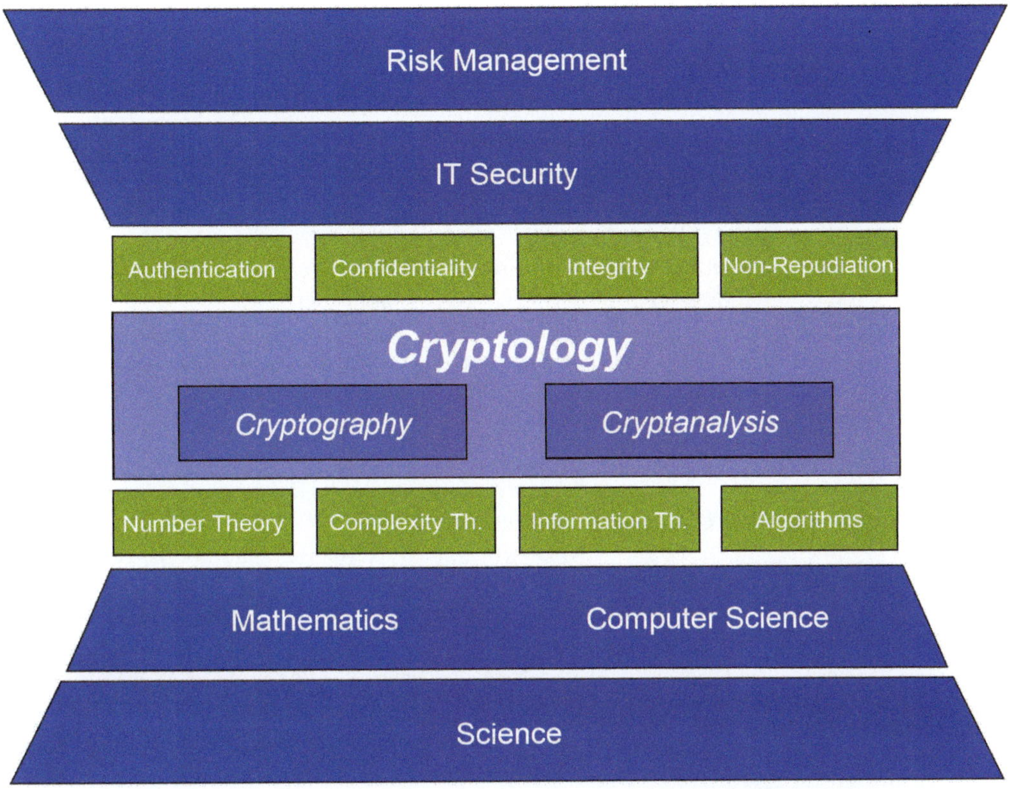

Figure 13.3 Embedding cryptology between corporate risk management and science.

systems or antivirus software that are always online and may send back sensitive data or have backdoors;

- Last but not least, law enforcement and surveillance: Despite all the legitimate reasons, government agencies have never been able to keep the data they collect to themselves. According to an unnamed intelligence source, all the zero-day exploits they collected end up in the hands of organized crime after an average of two years.

References

[1] Rivest, R. L., A. Shamir, and L. Adleman, "A Method for Obtaining Digital Signatures and Public-Key Cryptosystems," in *Communications of the ACM*, Vol. 21, No. 2, April 1978, pp. 120–126.

[2] Lenstra, A. K., and E. R. Verheul, *Selecting Cryptographic Key Sizes (1999 + 2001)*, in *Journal of Cryptology*, Vol. 14, 2001, pp. 255–293, https://www.cs.ru.nl/E.Verheul/papers/Joc2001/joc2001.pdf.

[3] Giry, D., *BlueKrypt: Cryptographic Key Length Recommendation*, Version 32.3, May 2020, https://www.keylength.com/.

[4] Shor, P. W., "Polynomial Time Algorithms for Prime Factorization and Discrete Logarithms on a Quantum Computer," in *SIAM Journal on Computing*, Vol. 26, No. 5, 1997, pp. 1484–1509.

[5] *Data Encryption Standard (DES)*, Federal Information Processing Standards (FIPS) 46. National Bureau of Standards, National Technical Information Service, Springfield, Virginia: U.S. Department of Commerce, 1977.

[6] Dworkin, M. J., et al., *Advanced Encyption Standard (AES)*, Federal Information Processing Standards(FIPS) 197, National Institute of Standards and Technology (NIST). Gaithersburg: U.S. Departmentof Commerce, November 26, 2001, https://nvlpubs.nist.gov/nistpubs/FIPS/NIST.FIPS.197.pdf.

[7] Bernstein, D., and T. Lange, *Post-Quantum Cryptography—Dealing with the Fallout of Physics Success*, in *Nature*, 2017, http://www.readcube.com/articles/10.1038/nature23461; https://eprint.iacr.org/2017/314.pdf.

[8] Beullens, W., et al., *Post-Quantum Cryptography: Current State and Quantum Mitigation*, Tech. rep., 2021, https://www.enisa.europa.eu/publications/post-quantum-cryptography-current-state-and-quantum-mitigation/@@download/fullReport.

[9] McEliece, R. J., "A Public Key Cryptosystem Based on Algebraic Coding Theory," in *DSN Progress Report 42–44*, 1978, pp. 114–116.

[10] Merkle, R. C., "Secrecy, Authentication, and Public Key Systems," PhD thesis, Department of Electrical Engineering, Stanford University, 1979.

[11] Buchmann, J., et al., "CMSS—An Improved Merkle Signature Scheme," in *7th International Conference on Cryptology in India—Indocrypt'06*, R. Barua and T. Lange (eds.), lecture notes incomputer science 4392, Springer-Verlag, 2006, pp. 349–363.

[12] Ding, J., and D. Schmidt, "Rainbow, a New Multivariable Polynomial Signature Scheme," in *Applied Cryptography and Network Security*, J. Ioannidis, A. Keromytis, and M. Yung (eds.), Springer, 2005, pp. 164–175.

[13] C. H. Bennett and G. Brassard. "An Update on Quantum Cryptography," in *Advances in Cryptology—CRYPTO '84*, G. R. Blakley, and D. Chaum (eds.), Vol. 196, lecture notes in computer science, Springer-Verlag, 1985, pp. 475–480.

[14] *Report on the Third Round of the NIST Post-Quantum Cryptography Standardization Process*, National Institute of Standards and Technology (NIST), July 2022 (updated Sept. 2022), https://nvlpubs.nist.gov/nistpubs/ir/2022/NIST.IR.8413.pdf.

[15] IETF, *The Messaging Layer Security (MLS) Protocol RFC 9420*, https://datatracker.ietf.org/doc/rfc9420/ (visited on 08/02/2023).

APPENDIX A
Software

Sections A.1 to A.4 briefly describe the four CT variants CT1, CT2, JCT, and CTO.[1] For each, the functions offered (via menus, templates, or plugins) are shown.

A.1 CrypTool 1 Menus

On the internet, a list with all functions offered by CrypTool 1 (CT1) can be created with: `https://www.cryptool.org/en/documentation/functionvolume?ctversion=ct1`.

The main menu of CT1 contains both generic service functions in the six main menu items

- File;
- Edit;
- View;
- Options;
- Window;
- Help.

and the actual crypto functions in the following four main menus:

- Encrypt / decrypt;
- Digital signature / PKI;
- Individual procedures;
- Analysis.

Within `Individual Procedures` you find visualizations of single algorithms and of protocols. Some procedures are implemented both for a fast performance (mostly under the main menu `Encrypt/Decrypt`) and for a step-by-step visualization.

Which of the menu items in CrypTool 1 are active (that means not grayed) depends on the type of currently active document window: The brute-force analysis for DES, for example, is only available if the active window is opened in the hexadecimal view. On the other hand, the menu item "Generate Random Numbers..." is always available (even if no document is opened).

Screenshots from CT1 can be found at `https://www.cryptool.org/en/ct1/screenshots/screenshots`.

1. From 2011, changes for CrypTool 1 were limited to bugfixes and pure maintenance. However, regularly new developments went into the two CT1 successors CrypTool 2 and JCrypTool (**JCT**). In the meantime, their functional range is bigger than the one of CT1. From 2023, JCT comes to pure maintenance. CT2 is still actively developed. The web version CrypTool-Online (CTO) was and further will be expanded considerably.
   - CT1: `https://www.cryptool.org/en/ct1/documentation/features`;
   - JCT: `https://www.cryptool.org/en/jct/documentation/resources`;
   - CT2: `https://www.cryptool.org/en/ct2/resources`;
   - CTO: `https://www.cryptool.org/en/cto/`.

A.2 CrypTool 2 Templates and the WorkspaceManager

When you start CT2 it first shows the Startcenter (see Figure A.1).[2]

Figure A.2 shows the beginning of a list with all functions offered by CrypTool 2. This list was created with `https://www.cryptool.org/en/documentation/functionvolume?ctversion=ct2`.

Besides the information on how you can get in the web a list of all functions in CT2, this appendix contains information about the templates (graphical programs included in CT2) and about the graphical editor ("WorkspaceManager") of CT2. When CT2 is started, the Startcenter opens first.

In the Startcenter, you have the choice to open CT2 templates in two different ways:

- Via the Wizard (second icon with magic wand, below "Main functions"), which guides you to the provided templates.
- Via the template tree (window in the center of the Startcenter), from which you can select ready-made cryptographic workflows.

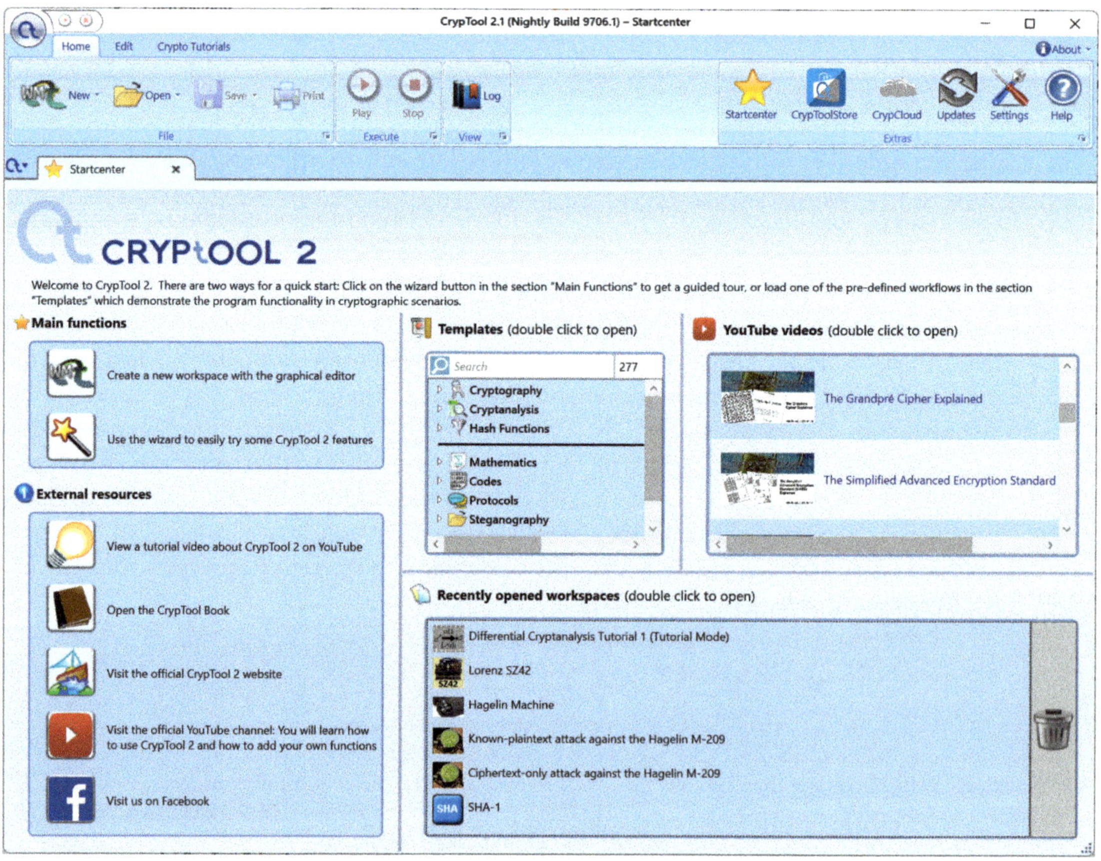

Figure A.1 Startcenter in CT2 (Nightly Build, October 2023).

2. The current CT2 release is CT 2.1 (release 2023.1 from June 2023). Each day a new "nightly build" is generated.

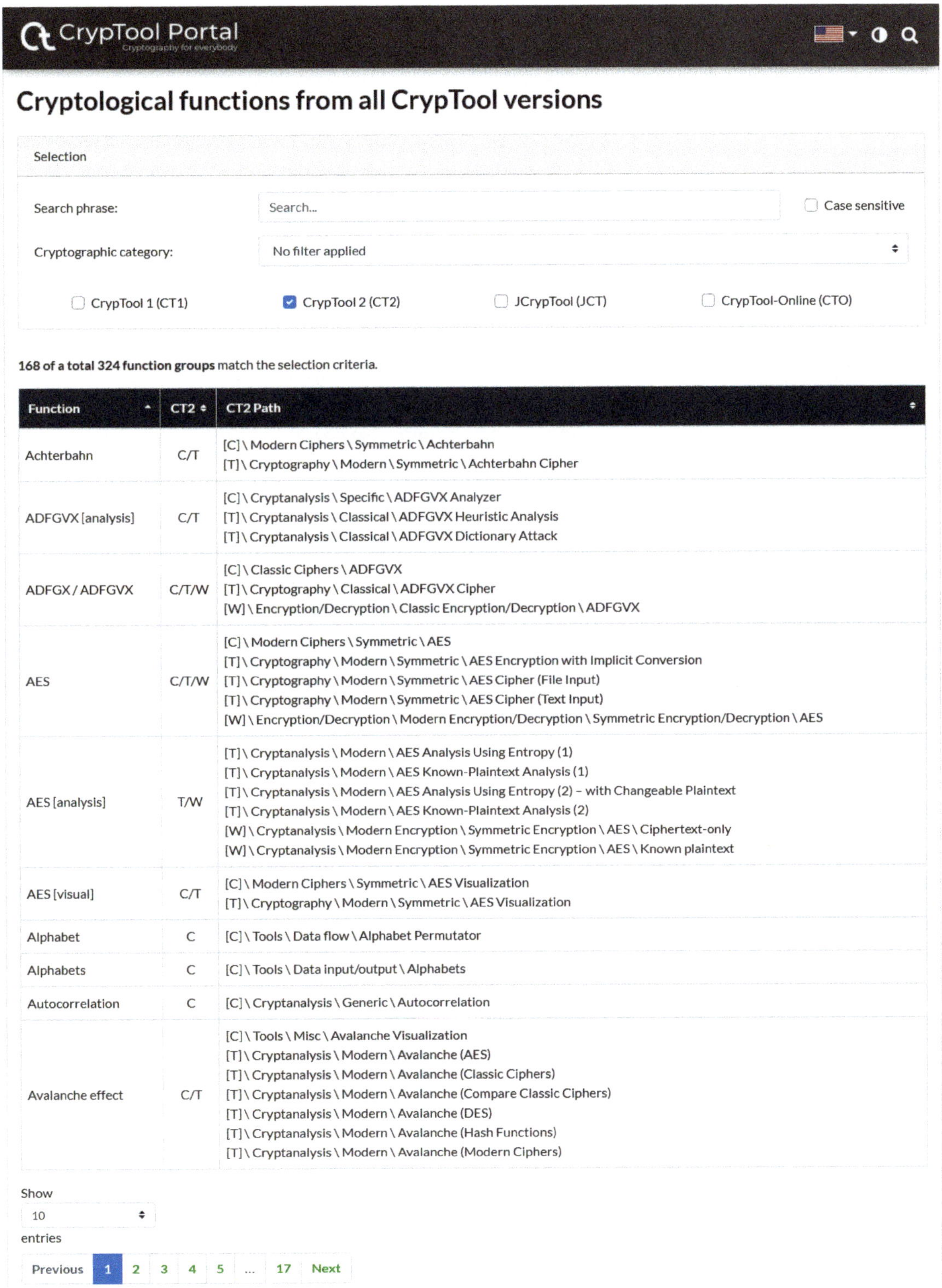
CrypTool Portal
Cryptography for everybody

Cryptological functions from all CrypTool versions

Selection

Search phrase: Search... ☐ Case sensitive

Cryptographic category: No filter applied

☐ CrypTool 1 (CT1) ☑ CrypTool 2 (CT2) ☐ JCrypTool (JCT) ☐ CrypTool-Online (CTO)

168 of a total 324 function groups match the selection criteria.

Function	CT2	CT2 Path
Achterbahn	C/T	[C] \ Modern Ciphers \ Symmetric \ Achterbahn [T] \ Cryptography \ Modern \ Symmetric \ Achterbahn Cipher
ADFGVX [analysis]	C/T	[C] \ Cryptanalysis \ Specific \ ADFGVX Analyzer [T] \ Cryptanalysis \ Classical \ ADFGVX Heuristic Analysis [T] \ Cryptanalysis \ Classical \ ADFGVX Dictionary Attack
ADFGX / ADFGVX	C/T/W	[C] \ Classic Ciphers \ ADFGVX [T] \ Cryptography \ Classical \ ADFGVX Cipher [W] \ Encryption/Decryption \ Classic Encryption/Decryption \ ADFGVX
AES	C/T/W	[C] \ Modern Ciphers \ Symmetric \ AES [T] \ Cryptography \ Modern \ Symmetric \ AES Encryption with Implicit Conversion [T] \ Cryptography \ Modern \ Symmetric \ AES Cipher (File Input) [T] \ Cryptography \ Modern \ Symmetric \ AES Cipher (Text Input) [W] \ Encryption/Decryption \ Modern Encryption/Decryption \ Symmetric Encryption/Decryption \ AES
AES [analysis]	T/W	[T] \ Cryptanalysis \ Modern \ AES Analysis Using Entropy (1) [T] \ Cryptanalysis \ Modern \ AES Known-Plaintext Analysis (1) [T] \ Cryptanalysis \ Modern \ AES Analysis Using Entropy (2) – with Changeable Plaintext [T] \ Cryptanalysis \ Modern \ AES Known-Plaintext Analysis (2) [W] \ Cryptanalysis \ Modern Encryption \ Symmetric Encryption \ AES \ Ciphertext-only [W] \ Cryptanalysis \ Modern Encryption \ Symmetric Encryption \ AES \ Known plaintext
AES [visual]	C/T	[C] \ Modern Ciphers \ Symmetric \ AES Visualization [T] \ Cryptography \ Modern \ Symmetric \ AES Visualization
Alphabet	C	[C] \ Tools \ Data flow \ Alphabet Permutator
Alphabets	C	[C] \ Tools \ Data input/output \ Alphabets
Autocorrelation	C	[C] \ Cryptanalysis \ Generic \ Autocorrelation
Avalanche effect	C/T	[C] \ Tools \ Misc \ Avalanche Visualization [T] \ Cryptanalysis \ Modern \ Avalanche (AES) [T] \ Cryptanalysis \ Modern \ Avalanche (Classic Ciphers) [T] \ Cryptanalysis \ Modern \ Avalanche (Compare Classic Ciphers) [T] \ Cryptanalysis \ Modern \ Avalanche (DES) [T] \ Cryptanalysis \ Modern \ Avalanche (Hash Functions) [T] \ Cryptanalysis \ Modern \ Avalanche (Modern Ciphers)

Show 10 entries

Previous 1 2 3 4 5 ... 17 Next

Figure A.2 Display in CTP (`https://www.cryptool.org/en/documentation/functionvolume`): the first functions offered by CT2.

The Wizard offers thematically nested choices for the desired cryptographic scenario, for example, "Encryption/Decryption ▹ Classical Encryption/Decryption ▹ Caesar," and then finally leads the user to the corresponding template. The selected scenario with the own inputs can be opened afterwards also as a graphical program in the WorkspaceManager (small WorkspaceManager symbol with plus sign on the top right of the respective last Wizard page) and can be stored in an own cwm file (own template).

Alternatively to the provided templates, you can create your own graphical programs. The WorkspaceManager is there for this purpose: It provides a workspace where you can assemble the components (e.g., an encryption function, a text input function) yourself using the visual programming language. The WorkspaceManager can be called in the Startcenter by means of the first icon under main functions. On the empty workspace you can drag and drop all components from the left navigation bar and then connect them as desired. The implemented crypto functionality is contained in these components (e.g. Enigma, AES).

In the template tree in the Startcenter there is at least one template for almost every component. The offered templates contain immediately executable cryptographic workflows. For example, if you change your input in the template for AES, you can see dynamically and immediately how outputs change accordingly (e.g., how padding adds a block or what the effect of chaining is).

Figure A.3 shows an extract from the template tree of the Startcenter of CT2.

Screenshots from CT2 can be found at `https://www.cryptool.org/en/ct2/screenshots`.

Resources and developer information about CT2 can be found at `https://www.cryptool.org/en/ct2/resources`.

A.3 JCrypTool Functions

When you start JCT[3] the first time it comes up with the welcome window (see Figure A.4).

Figure A.5 shows the beginning of a list with all functions in JCrypTool. This list was created with `https://www.cryptool.org/en/documentation/functionvolume?ctversion=jct`.

After pressing "Start JCT" you can directly use the different functions. The functions implemented in JCT are presented in two different perspectives:

- Default perspective;
- Algorithm perspective.

All functions of the default perspective can be found both in the menus and in the navigation bar called "Crypto Explorer" (at the right side). The default perspective contains all important methods like classic transposition or modern AES, and many visualizations (e.g., Diffie-Hellman key exchange or calculations on elliptic curves).

3. The current JCT release is JCT 1.0.9 (July 2023). Occasionally a new weekly build is generated. You can find further information about JCT at: `https://www.cryptool.org/en/jct/volunteer`.

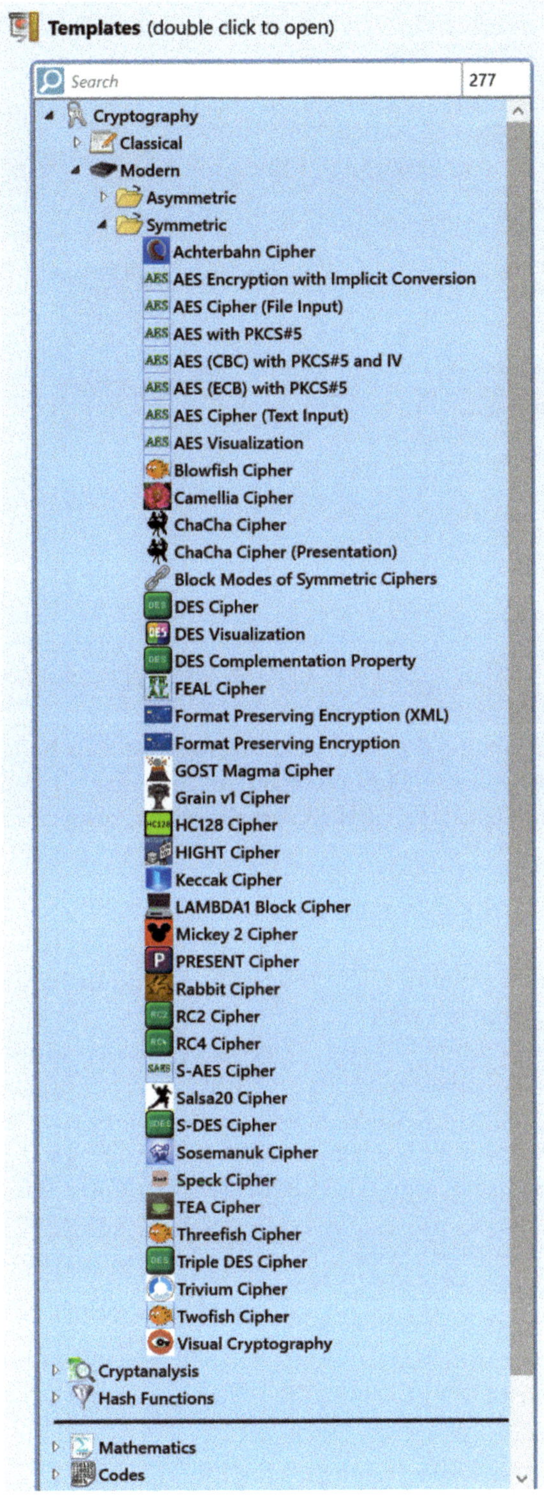

Figure A.3 Extract of the expanded template tree in CT2.

Figure A.4 Welcome screen in JCT (version 1.0.7, October 2021).

All functions of the algorithm perspective can be found in the navigation bar called "Algorithms." This perspective contains all detail settings of the various algorithms; it especially offers post-quantum computing algorithms.

A.4 CrypTool-Online Functions

On the starting page of CTO (Figure A.6) you can choose via text search or icon click which plugin to start.

CrypTool-Online (`https://www.cryptool-online.org`) is a website with applications (so-called plug-ins) for testing, learning, and discovering ancient and modern cryptography. Current web technologies such as React, Chakra UI, Bootstrap, and WebAssembly are used. The technological aim is a responsive design for all device sizes, and simultaneously a common full-screen like appearance for desktop monitors.

Figure A.7 shows the beginning of a list with all functions offered by CTO. This list was created with `https://www.cryptool.org/en/documentation/functionvolume?ctversion=cto`.

As the overall function list at the CTP (CrypTool portal) is only updated twice a year, the most current list of CTO plugins can be found on the CTO starting page (see Figure A.8).

Parts of CTO are:

- Simple ciphers like Caesar and ADFGVX;
- Homophonic substitution solver for both manual and automatic cryptanalysis;

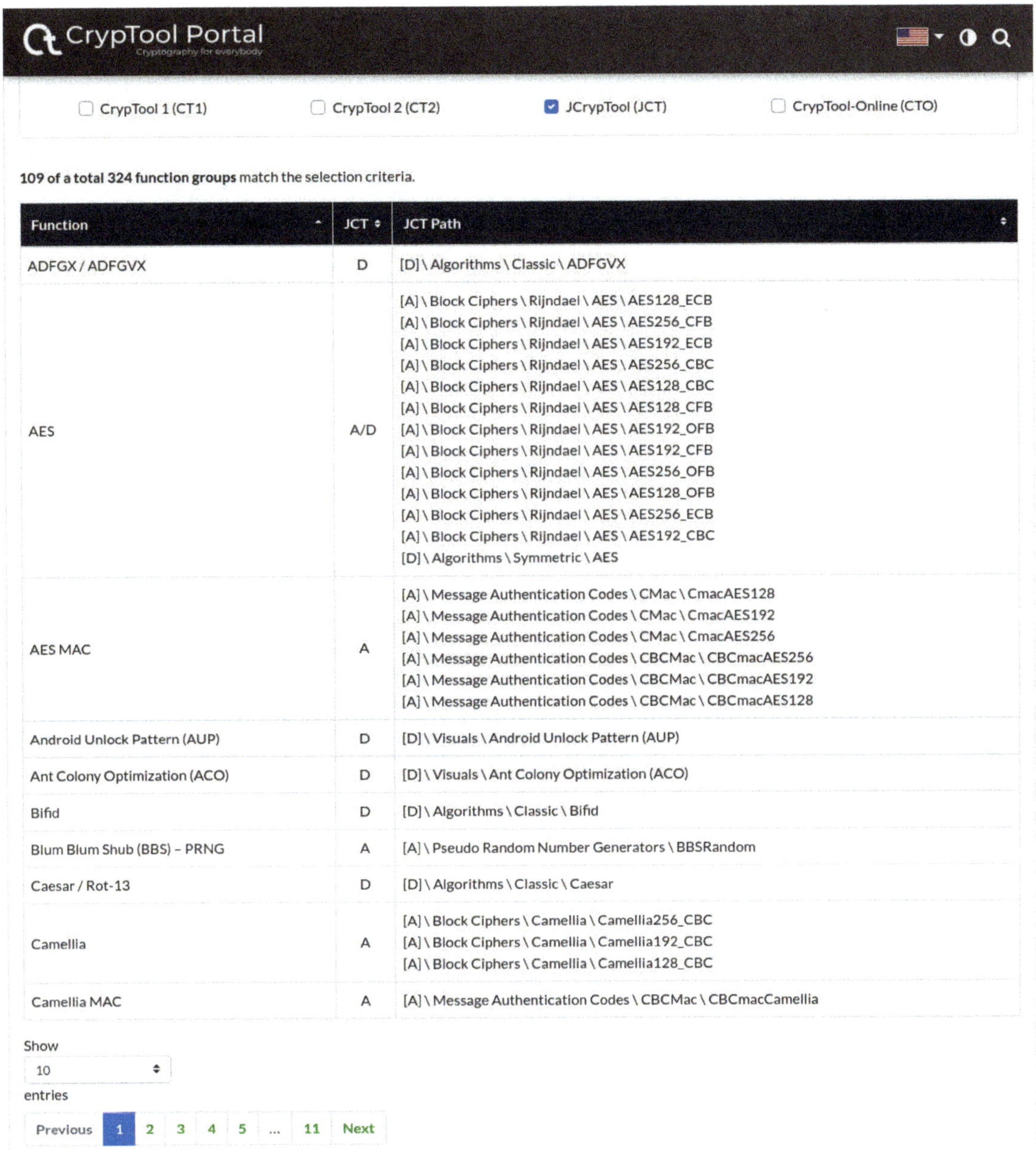

CrypTool Portal
Cryptography for everybody

CrypTool 1 (CT1) | CrypTool 2 (CT2) | JCrypTool (JCT) | CrypTool-Online (CTO)

109 of a total 324 function groups match the selection criteria.

Function	JCT	JCT Path
ADFGX / ADFGVX	D	[D] \ Algorithms \ Classic \ ADFGVX
AES	A/D	[A] \ Block Ciphers \ Rijndael \ AES \ AES128_ECB [A] \ Block Ciphers \ Rijndael \ AES \ AES256_CFB [A] \ Block Ciphers \ Rijndael \ AES \ AES192_ECB [A] \ Block Ciphers \ Rijndael \ AES \ AES256_CBC [A] \ Block Ciphers \ Rijndael \ AES \ AES128_CBC [A] \ Block Ciphers \ Rijndael \ AES \ AES128_CFB [A] \ Block Ciphers \ Rijndael \ AES \ AES192_OFB [A] \ Block Ciphers \ Rijndael \ AES \ AES192_CFB [A] \ Block Ciphers \ Rijndael \ AES \ AES256_OFB [A] \ Block Ciphers \ Rijndael \ AES \ AES128_OFB [A] \ Block Ciphers \ Rijndael \ AES \ AES256_ECB [A] \ Block Ciphers \ Rijndael \ AES \ AES192_CBC [D] \ Algorithms \ Symmetric \ AES
AES MAC	A	[A] \ Message Authentication Codes \ CMac \ CmacAES128 [A] \ Message Authentication Codes \ CMac \ CmacAES192 [A] \ Message Authentication Codes \ CMac \ CmacAES256 [A] \ Message Authentication Codes \ CBCMac \ CBCmacAES256 [A] \ Message Authentication Codes \ CBCMac \ CBCmacAES192 [A] \ Message Authentication Codes \ CBCMac \ CBCmacAES128
Android Unlock Pattern (AUP)	D	[D] \ Visuals \ Android Unlock Pattern (AUP)
Ant Colony Optimization (ACO)	D	[D] \ Visuals \ Ant Colony Optimization (ACO)
Bifid	D	[D] \ Algorithms \ Classic \ Bifid
Blum Blum Shub (BBS) – PRNG	A	[A] \ Pseudo Random Number Generators \ BBSRandom
Caesar / Rot-13	D	[D] \ Algorithms \ Classic \ Caesar
Camellia	A	[A] \ Block Ciphers \ Camellia \ Camellia256_CBC [A] \ Block Ciphers \ Camellia \ Camellia192_CBC [A] \ Block Ciphers \ Camellia \ Camellia128_CBC
Camellia MAC	A	[A] \ Message Authentication Codes \ CBCMac \ CBCmacCamellia

Show 10 entries

Previous 1 2 3 4 5 ... 11 Next

Figure A.5 Display of functions volume in CTP: the first functions offered by JCT.

- Sophisticated visualizations (like AES with PixiJS in `https://www.cryptool.org/en/cto/aes-animation`);
- Taxman game (`https://www.cryptool.org/en/cto/taxman`);
- Password meter (`https://www.cryptool.org/en/cto/password-meter`);
- Demonstration of the DP-3T cryptographic protocol (`https://www.cryptool.org/en/cto/corona-tracing` and `https://corona-tracing.cryptool.org/`);

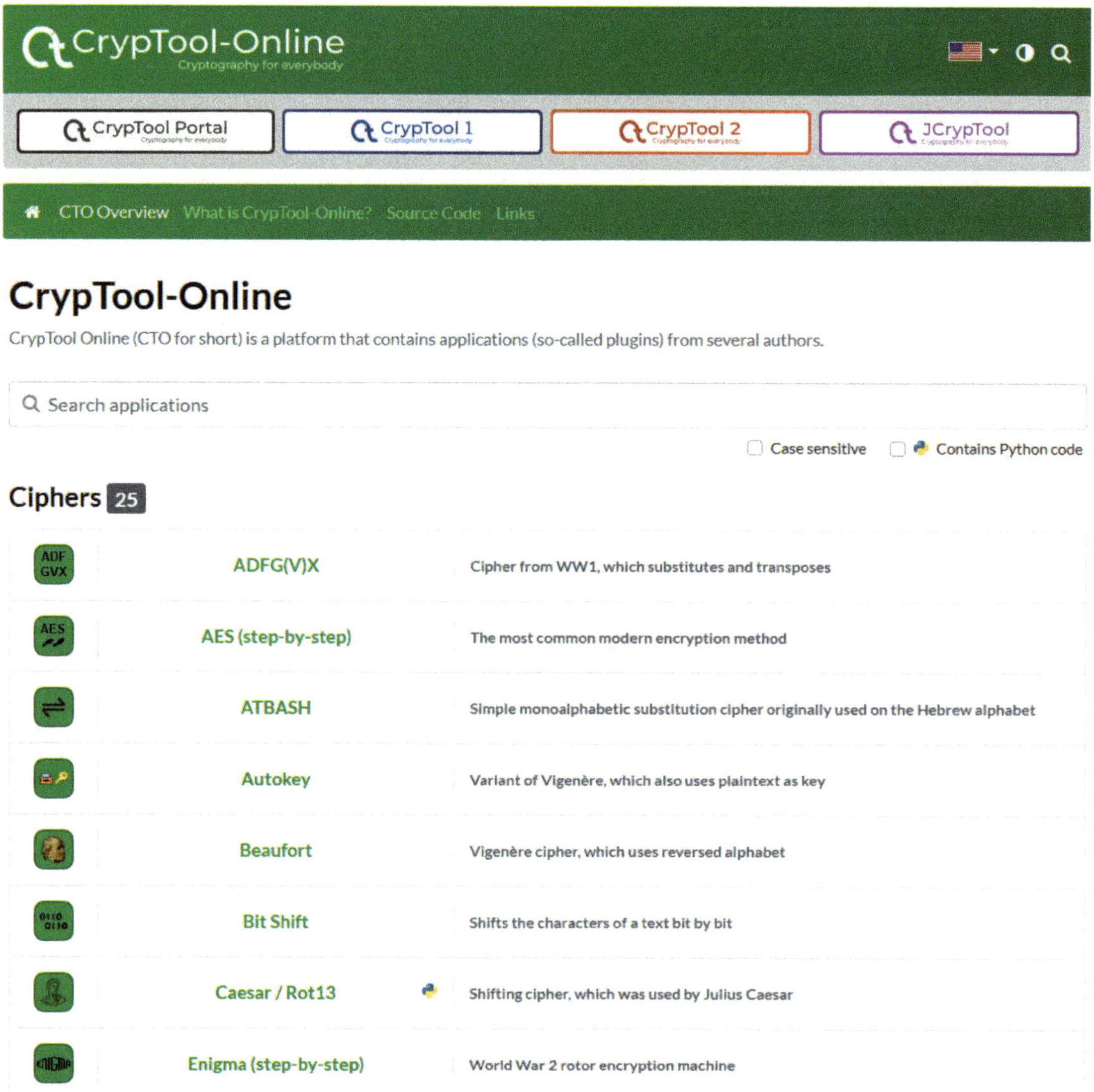

Figure A.6 Starting page of CTO (February 2022).

- Didactic version of RSA that is often used by teachers (`https://www.cryptool.org/en/cto/rsa-visual`);
- Browser-based implementation of CryptoBrief (former Sunset/FFapl). Sunset/FFapl is a simplified programming language specially designed for cryptography. It can be used to easily write down the code for protocols and public-key procedures similar to the notation used in textbooks, because the interpreter ensures the algebraic compatibility of the objects. (`https://www.cryptool.org/en/cto/cryptobrief`);
- Machine-learning based encryption type detection just by entering a short ciphertext (`https://www.cryptool.org/en/cto/ncid`);

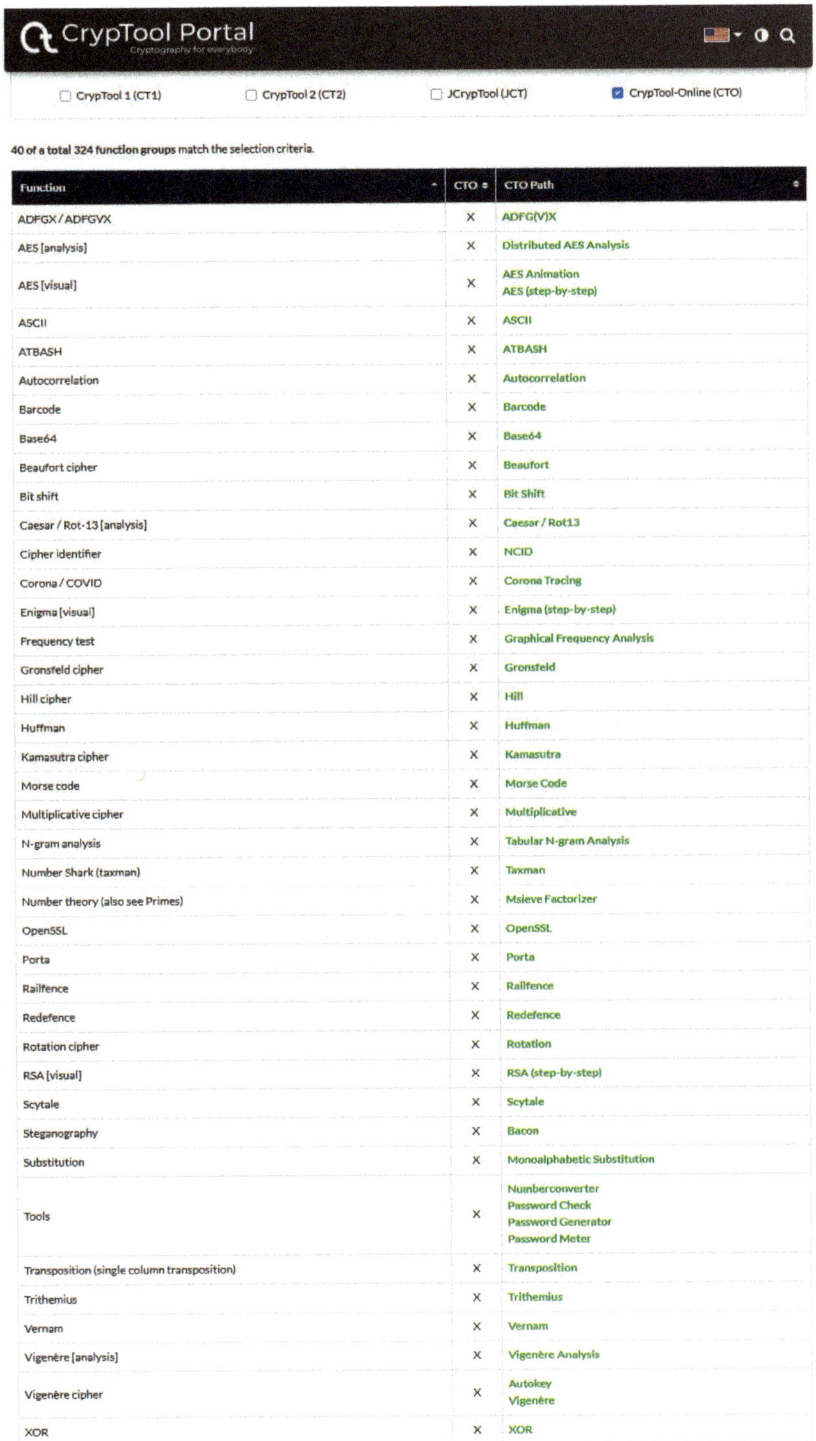

CrypTool Portal
Cryptography for everybody

☐ CrypTool 1 (CT1) ☐ CrypTool 2 (CT2) ☐ JCrypTool (JCT) ☑ CrypTool-Online (CTO)

40 of a total 324 function groups match the selection criteria.

Function	CTO	CTO Path
ADFGX / ADFGVX	X	ADFG(V)X
AES [analysis]	X	Distributed AES Analysis
AES [visual]	X	AES Animation AES (step-by-step)
ASCII	X	ASCII
ATBASH	X	ATBASH
Autocorrelation	X	Autocorrelation
Barcode	X	Barcode
Base64	X	Base64
Beaufort cipher	X	Beaufort
Bit shift	X	Bit Shift
Caesar / Rot-13 [analysis]	X	Caesar / Rot13
Cipher identifier	X	NCID
Corona / COVID	X	Corona Tracing
Enigma [visual]	X	Enigma (step-by-step)
Frequency test	X	Graphical Frequency Analysis
Gronsfeld cipher	X	Gronsfeld
Hill cipher	X	Hill
Huffman	X	Huffman
Kamasutra cipher	X	Kamasutra
Morse code	X	Morse Code
Multiplicative cipher	X	Multiplicative
N-gram analysis	X	Tabular N-gram Analysis
Number Shark (taxman)	X	Taxman
Number theory (also see Primes)	X	Msieve Factorizer
OpenSSL	X	OpenSSL
Porta	X	Porta
Railfence	X	Railfence
Redefence	X	Redefence
Rotation cipher	X	Rotation
RSA [visual]	X	RSA (step-by-step)
Scytale	X	Scytale
Steganography	X	Bacon
Substitution	X	Monoalphabetic Substitution
Tools	X	Numberconverter Password Check Password Generator Password Meter
Transposition (single column transposition)	X	Transposition
Trithemius	X	Trithemius
Vernam	X	Vernam
Vigenère [analysis]	X	Vigenère Analysis
Vigenère cipher	X	Autokey Vigenère
XOR	X	XOR

Figure A.7 Display in CTP: functions offered by CTO (Jan 2022).

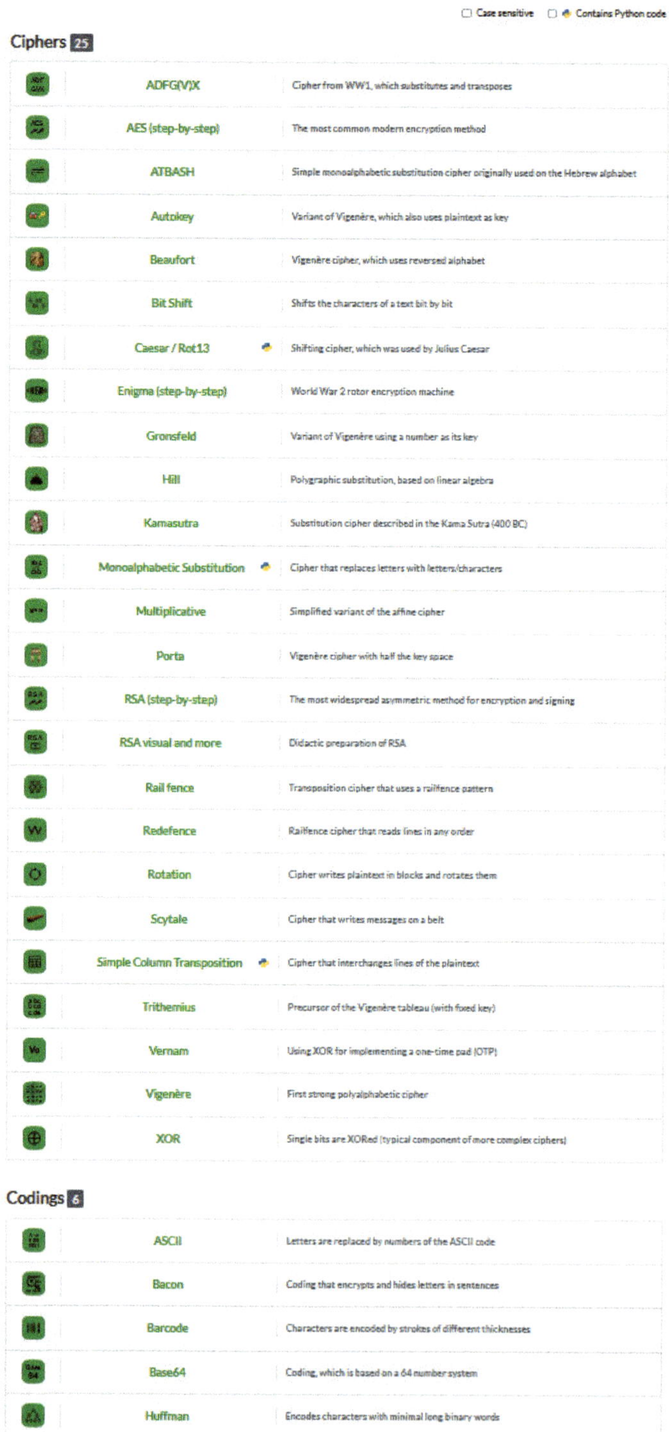

Figure A.8 Display of the first functions in CTO on the CTO starting page (February 2022).

- Various WebAssembly applications, all of which run purely locally in the browser:
  - Python development environment (Pyodide), used, for example, in `https://www.cryptool.org/en/cto/monoalpha`,
  - Port from Msieve to wasm (`https://www.cryptool.org/en/cto/msieve`);
  - Demonstration of a poll-like Doodle based on a second-generation FHE algorithm from the wasm library node-seal (`https://www.cryptool.org/en/cto/fhe-poll`);
  - First port of OpenSSL 3 to wasm, called "OpenSSL for Web" (`https://www.cryptool.org/en/cto/openssl` and `https://wiki.openssl.org/index.php/Binaries`).

APPENDIX B

Miscellaneous

B.1 Movies and Fictional Literature with Relation to Cryptography

Cryptographic applications—classical as well as modern ones—have been used in literature and movies. In some media they are only mentioned and are a pure admixture; in others they play a primary role and are explained in detail; and sometimes the purpose of the story, which forms the framework, is primarily to transport this knowledge and achieve better motivation. Here is the beginning of an overview.

B.1.1 For Grownups and Teenagers

***The Gold Bug*, Edgar Allan Poe, 1843.**

In this short story Poe tells as first-person narrator about his acquaintanceship with the curious Mr. Legrand. They detect the fabulous treasure of captain Kidd via a gold bug and a vellum found at the coast of New England. The cipher consists of 203 cryptic symbols, and it proves to be a general monoalphabetic substitution cipher (see Section 2.2.1). The story tells how they solve the riddle step by step using a combination of semantic and syntax analysis (frequency analysis of single letters in English texts).

In this novel the code breaker Legrand says the famous statement: "Yet it may be roundly asserted that human ingenuity cannot concoct a cipher which human ingenuity cannot resolve—given the according dedication." Poe not only was a well-known writer, but also a talented cryptographer. His story is also told in the book *Code Breaking* [1].

Mathias Sandorf, Jules Verne, 1885.

This is one of the most famous novels of the French author Jules Verne (1828–1905), who was called "Father of Science Fiction." In *Mathias Sandorf* he tells the story of the freedom fighter Earl Sandorf, who is betrayed to the police, but finally he can escape.

The whistleblowing worked, because his enemies captured and decrypted a secret message sent to him. For decryption, they needed a special grille, which they stole from him. This turning grille was a quadratic piece of jig with 6 × 6 squares, of which one-quarter (nine) were holes (see the turning grille in Section 2.1.1).

***Kim*, Rudyard Kipling, 1901.**

Rob Slade's review [2] of this novel says: "Kipling packed a great deal of information and concept into his stories, and in *Kim* we find The Great Game: espionage and spying. Within the first twenty pages we have authentication by something you have, denial of service, impersonation, stealth, masquerade, role-based authorization (with ad hoc authentication by something you know), eavesdropping, and trust based on data integrity. Later on we get contingency planning against theft and cryptography with key changes."

The book is out of copyright [3].

***The Adventure of the Dancing Men*, Arthur Conan Doyle, 1905.**

In this Sherlock Holmes short story (first published in 1903 in the *Strand Magazine*, and then in 1905 in the collection *The Return of Sherlock Holmes* the first time in book form), Sherlock Holmes has to solve a cipher that at first glance looks like a harmless kid's picture.

But it proves to be the monoalphabetic substitution cipher (see Section 2.2.1) of the criminal Abe Slaney. Sherlock Holmes solves the riddle using frequency analysis.

***Have His Carcase*, Dorothy L. Sayers, Harper/Victor Gollancz Ltd., 1932.**

In this novel the writer Harriet Vane finds a dead body at the beach. The police believe the death is suicide. Harriet Vane and the elegant amateur sleuth Lord Peter Wimsey together clear of the disgusting murder in this second of Sayers's famous Harriet Vane mystery series.

This requires them to solve a cryptogram. Surprisingly the novel not only describes the Playfair cipher in detail, but also the cryptanalysis of this cipher (see Playfair in Section 2.2.3).

***And Jimmy Went to the Rainbow (original title: Und Jimmy ging zum Regenbogen)*, Johannes Mario Simmel, Knaur Verlag, 1970.**

The novel plays between 1938 and 1967 in Vienna. The main character Manual Aranda uncovers step by step the past of his murdered father. Important for the plot is an encrypted manuscript, which is decrypted in Chapter 33. In the novel the cipher is called "25-fold Caesar cipher." It is actually a Vigenère cipher with a 25-character key. A movie of the novel appeared in 1971.

***Sphere*, Michael Crichton, Pan Books, 1987.**

A team of different scientists is sent to the ground of the ocean in order to investigate a highly developed 900m long spaceship. The human peculiarities and psychological problems of the researchers surface more and more because of life-threatening events and isolation. There are many mysteries: While the space ship lies on the ground for 300 years, it has English markings and a life of its own, and materializing of the researcher's imaginations appear. On a computer screen a cipher text appears, which is completely printed in the book. The genius mathematician Harry deciphers the simple helical substitution code.

***House of Cards*, Directed by Paul Seed, 1990.**

In this movie Ruth tries to solve the secret that made her daughter fall silent. Here two young people suffering from autism communicate via 5- and 6-digit primes (see Chapter 4). After more than 1 hour the movie contains the following encrypted two series of primes:

$$21383, 176081, 18199, 113933, 150377, 304523, 113933;$$
$$193877, 737683, 117881, 193877$$

Compare the story *The Dialogue of the Sisters*.

***Sneakers*, Directed by Phil Alden Robinson, Universal Pictures Film, 1992.**

In this movie the "sneakers," computer experts under their boss Martin Bishop, try to get back the deciphering box SETEC from the "bad guys." SETEC, invented by a genius mathematician before he was killed, allows decrypting all codes from any nation. In the movie the code is not described in any way.

Leonard Adleman (the "A" within RSA) worked as mathematical consultant for "sneakers." He describes the funny story about his contribution at his homepage `https://theworld.com/ reinhold/math/sneakers.adleman.html`. It is assumed that the cipher used everywhere is RSA. So on the chip a fast, unknown factorization method is implemented.

***Total Control*, David Baldacci, Mass Market Paperback, 1997.**

Jason Archer, executive with a technology company suddenly disappears. Sidney Archer tries to find out about her husband's surprising death. She gets a clue how the global financial system is abused and that the real control belongs to those with the most money. Here even good passwords don't help.

***Cube*, Directed by Vincenzo Natali, Mehra Meh Film, 1997.**

In this Canadian low-budget-movie, seven complete strangers of widely varying personality characteristics are involuntarily placed in a Kafkaesque maze of cubical rooms containing deadly traps.

To get out, the persons have to move through these rooms. To find out which rooms are dangerous, mathematics is crucial: Each cubic room has at its entrance a numerical marking consisting of three sets of three digits. First they deduce that all rooms marked at their entrance with at least one prime number are trapped. Later it comes out that a trapped room can also be marked by a number which is a power of a prime (so traps are p^n, e.g., $128 = 2^7$ or $101 = 101^1 = prime$, but not $517 = 11 * 47$).

***Mercury Rising*, Directed by Harold Becker, Universal Pictures Film, 1998.**

The NSA developed a new cipher, which is pretended to be uncrackable by humans and computers. To test its reliability some programmers hide a message encrypted with this cipher in a puzzle magazine.

Simon, a nine-year-old autistic boy, cracks the code. Instead of fixing the code, a government agent sends a killer. FBI agent Art Jeffries (Bruce Willis) protects the boy and sets a snare for the killers.

The code is not described in any way.

***Digital Fortress*, Dan Brown, E-Book, 1998.**

Dan Brown's first novel was published in 1998 as an e-book, but it was largely unsuccessful then.

The National Security Agency uses a huge computer, that enables them to decrypt all messages (needless to say only of criminals and terrorists) within minutes even if they use the most modern encryption methods.

An apostate employee invents an unbreakable code and his computer program Diabolus forces the super computer to do self-destructing operations. The plot, where also the beautiful computer expert Susan Fletcher has a role, is rather predictable.

The idea that the NSA or another secret service is able to decrypt any code is currently a popular topic. In *Digital Fortress* the super computer has 3 million processors—nevertheless from today's view this is by no means sufficient to hack modern ciphers.

***The Dialogue of the Sisters*, C. Elsner, c't, Heise, 1999.**

In this short story, the sisters confidentially communicate using a variant of RSA (see Section 5.10). They are residents of a madhouse being under permanent surveillance.

The PDF file is displayed in CT1 if you there enter, for example, the search term "sisters" in the online help.

***Cryptonomicon*, Neal Stephenson, Harper, 1999.**

This very thick novel deals with cryptography both in WW2 and today. The two heroes from the 1940s are the excellent mathematician and cryptanalyst Lawrence Waterhouse, and the overeager and morphine-addicted U.S. marine Bobby Shaftoe. They both are members of the special allied unit 2702, which tries to hack the enemy's communication codes and at the same time to hide its own existence.

This secretiveness also happens in the present plot, where the grandchildren of the war heroes—the dedicated programmer Randy Waterhouse and the beautiful Amy Shaftoe—team up.

Cryptonomicon is notably heavy for nontechnical readers in parts. Several pages are spent explaining in detail some of the concepts behind cryptography. Stephenson added a detailed description of the solitaire cipher (see Section 2.4), a paper-and-pencil encryption algorithm developed by Bruce Schneier which is called "Pontifex" in the book. Another, modern algorithm called "Arethusa" is not explained in detail.

***The Chinese Labyrinth*, C. Elsner, c't, Heise, 2001, Updated 2020.**

In this short story, which is included in the CrypTool package as a PDF file, Marco Polo has to solve problems from number theory within a competition to become a major consultant of the Great Khan. All solutions are included and explained.

The new version (with lots of SageMath code) can be found at `https://www.cryptool.org/assets/ctp/documents/cttc/chinlab-en.pdf`.

***Artemis Fowl*, Eoin Colfer, Viking, 2001.**

In this book for young people the 12-year-old Artemis, a genius thief, gets a copy of the top secret *Book of the Elfs*. After he decrypted it with his computer, he finds out things that men never should have known.

The used code is not described in detail or revealed.

***A Beautiful Mind*, Directed by Ron Howard, 2001.**

This is the film version of Sylvia Nasar's biography of the game theorist John Nash. After the brilliant but asocial mathematician accepts secret work in cryptography, his life takes a turn to the nightmarish. His irresistible urge to solve problems becomes a danger for himself and his family. Nash is—within his belief—a most important hacker working for the government. Details of his way of analyzing code are not described in any way.

***Enigma*, Directed by Michael Apted, 2001.**

This is the film version of Robert Harris' historical fiction *Enigma* (Hutchinson, London, 1995) about the World War II code-breaking work at Bletchley Park in early 1943, when the actual inventor of the analysis Alan Turing (after Polish pre-work) already was in the United States. So the fictional mathematician Tom Jericho is the lead character in this spy-thriller. Details of his way of analyzing the code are not described.

***The Museum of the Stolen Memories (original title: Das Museum der gestohlenen Erinnerungen)*, Ralf Isau, Thienemann-Verlag, 1997/2003.**

In this exciting novel the last part of the oracle can only be solved with the joined help of the computer community. The book got several awards and exists in eight different languages, but not in English yet.

***The Da Vinci Code*, Dan Brown, Doubleday, 2003.**

The director of the Louvre is found murdered in his museum in front of a picture of Leonardo da Vinci, and the symbol researcher Robert Langdon is involved in a conspiracy.

The plot mentions different classic codes (substitution like Caesar or Vigenère, as well as transposition and number codes). Also, there are hints about Schneier and the sunflower. The second part of the book contains a lot of theological considerations.

This book has become one of the most widely read books of all time.

***Final Solution*, Scott McBain, manuscript not published by Harper Collins, 2004 (German version was published in 2005).**

In a near future, politicians, chiefs of military, and secret services of many different countries take over all the power. With a giant computer network called "Mother" and complete surveillance, they want to cement their power and commercialization of life forever. Humans are only assessed according to their credit rating, and globally acting companies elude of any democratic control. Within the thriller, the obvious injustice, but also the realistic likelihood of this development, are considered again and again.

With the help of a cryptographer, a code to destroy was built into the super computer "Mother": In a race several people try to start the deactivation (Lars Pedersen, Oswald Plevy, the female American president, the British prime minister, and an unknown Finnish person named Pia, who wants to take revenge for the death of her brother). On the opposite side a killing group acts under the special guidance of the British foreign minister and the boss of the CIA.

***The Cryptographer*, Tobias Hill, Faber & Faber, 2003.**

London 2021: The company SoftMark developed and established an electronic currency that guarantees the highest security standards by an unbreakable code. The inventor and company founder, called the cryptographer because of his mathematical talent, has become the richest man in the world. But the code was hacked, and in a worldwide economic crisis his company goes bankrupt. Additionally, the tax investigator Anna Moore is set on him.

***Tyrannosaur Canyon*, Douglas Preston, Forge Books, 2005.**

A very exciting thriller that also struggles with the question of why the dinosaurs died off.

Archeologist Stem Weathers is shot in a canyon. Before his murderer appears he gives his notebook to Tom Broadbent, a local animal doctor, coming by accidentally.

The notebook contains on 60 pages only digits. Therefore Tom takes it to Wyman Ford an ex-CIA cryptanalyst, who now lives in a nearby abbey, after his wife was killed in action. Wyman first declines and says that self-invented codes are "idiot ciphers," devised by an idiot and easily crackable by each idiot. The notebook then proves to be not that easy. After intensive analysis he finds out that the digits are no code but the output of an Earth radar device showing the picture of a well-preserved *T. rex*.

After around 250 pages of endless chases, a surprising turn comes up: Masago, head of a so-called black-detachment unit of the CIA. He explains, new weapons invented once have always been used. Mankind will kill herself, but it's his task to postpone that as far as possible. As head of the LS480 department he will prevent by any means possible that terrorists get any new dangerous biological weapon.

When scanning the dead body of Weathers, the murderer only found some rock cuttings he took. These rocks are investigated by a young researcher named Melody Crookshank, although she doesn't know where the rock cuttings come from. She finds within them a very special kind of virus apparently coming from outer-space.

***Heidelberg Lies (original title: Heidelberger Lügen)*, Wolfgang Burger, Piper, 2006.**

This detective story playing in the Rhein-Neckar area in Germany has several independent strands and local stories, but mainly it is about police officer Gerlach from Heidelberg. On page 207, the cryptographic reference for one strand is shortly explained: The soldier Hörrle had copied circuit diagrams of a new digital NATO decryption device and the murdered man had tried to sell his perceptions to China.

***The Black Sun*, James Twinig, HarperCollins, 2006.**

A history-based thriller with some artificially constructed elements, dealing also with a treasure hunt to get the hidden uranium of the Nazis, and naturally the future of the world depends on today's bad guys being stopped in time.

Heros are Tom Kirk, a London-based ex-CIA agent and former professional art thief, and Dominique de Lecourt, who loves challenges including riddles and codes.

The only cryptographic parts are a "Sprungcode" (the criminals use this method to communicate via newsletter adverts), steganography (used to hide the Enigma key), and an Enigma message (containing the encrypted coordinates of the treasure).

At the beginning of the plot an Enigma device is stolen with high efforts which is necessary to let the story play in the constructed way. But in the reality today such a theft is completely needless, as there are great software emulators for the Enigma.

***Kryptum*, Agustin Sanchez Vidal, Dtv, 2006.**

The first novel of the Spanish professor of art history has some similarities with Dan Brown's *The Da Vinci Code* from 2003, but allegedly Vidal started his writing of the novel already in 1996. Vidal's novel is a mixture between historic adventure and mystery thriller. It was a huge success in Spain and Germany. There is currently no English version available.

In the year 1582, Raimundo Randa is waiting to be condemned to death—he was trying to solve a mystery all his life. This mystery is about a parchment with cryptic characters, where a unique power is behind. Around 400 years later the American scientist Sara Toledano is fascinated by this power until she vanishes in Antigua. Her colleague, the cryptographer David Calderon, and her daughter Rachel are searching for her, and simultaneously they try to solve the code. But also secret organizations like the NSA chase after the secret of the last key. They don't hesitate to kill for it.

***Perdition (original title: Flickan som lekte med elden)*, Stieg Larsson, 2006.**

The author was posthumously awarded in 2006 with the Scandinavian thriller award. The superhero Lisbeth Salander uses PGP and occupies herself with mathematical riddles like the Fermat theorem.

***The Judas Documents (original title: Die Judas-Papiere)*, Rainer M. Schröder, Arena, 2008.**

In the year 1899 Lord Pembroke has three men and one woman in his grip. So they have to follow his order to try to decipher the encrypted messages in the notebook of his dead brother Mortimer and to find the missing gospel according to Judas, which could shock the whole of Christendom. The four people therefore have to solve riddles at many places in the world. The story explains some classic ciphers like Polybius and Freemason.

***A King for Germany (original title: Ein König für Deutschland)*, Andreas Eschbach, Lübbe, 2009.**

The novel deals with manipulations of electronic voting machines.

Vincent Merrit, a young American programmer, is blackmailed to write such a program. Besides commercially oriented blackmailers, massively multiplayer online role-playing games (MMORPGs) and live action role playing (LARP), have a role. Because Merrit assumed that his program will be misused, he installed a trapdoor: If a party with the name VWM participates at the election, it automatically gets 95% of the votes. The fictional story line is based on many verifiable and well researched facts, which are referenced in footnotes. While the cryptographic protocols themselves could be made secure, their implementation and their organizational management stays susceptible against misuse.

Currently there is no English translation of the book.

***Tetraktys*, Ari Juels, Emerald Bay Books, 2009.**

The plot exposes the vulnerability of modern computer based identity, authenticity, and security interweaving modern cryptography with classical art and literature. Cryptographer and classicist Ambrose Jerusalem is a University of California Berkeley graduate with a beautiful girlfriend and a comfortable future, until the NSA recruits him to track a strange pattern of computer break-ins. Many small pieces provide disturbing evidence that someone has broken RSA encryption. Even more bizarre, a secret cult of latter-day followers of Pythagoras, the great Greek mathematician and philosopher who believed reality could be understood only through a mystical system of numbers, appears to be behind the attacks.

***Daemon*, Daniel Suarez, Penguin Books, 2009**

This is considered as one of the most exciting books during the last few years—it's a near-science-fiction thriller combining developments in the real world and possibilities coming from current research like from the Google X Lab (augmented reality head-mounted displays (HMD) like Google glass, self-driving cars, and 3-D printers) to a plausible story.

After the computer genius and game developer Matthew Sobol died, a daemon starts acting on the internet, which seemingly ruthlessly manipulates and trains more and more humans and companies.

By ruling the data everybody seems to be a helpless victim. All the communication of his mercenary soldiers is affected by high-tech and encryption—also the communication between the distributed instances of his incarnation. Core is an MMORPG game which reminds many of WoW. Here also encryption is used, for example, to advertise the best players: m0wFG3PRCoJVTs7JcgBwsOXb3U7yPxBB.

The plot is without redundancy, complex, manifold, very fascinating, and with its critique of the plutocrats it also contains concrete social elements. The end is open. And the ideas seem to be realizable in the very next future.

***Freedom (TM)*, Daniel Suarez, Penguin Books, 2010**

"The propulsive, shockingly plausible sequel to the bestseller *Daemon*." *Freedom (TM)* (*Daemon* #2) patches a number of holes the writer left in the first book. The prose is tighter, the descriptions more direct, the characters are fleshed out, especially Loki. Having laid the groundwork in *Daemon*, Suarez uses this foundation in order to explore a new concept of social organization based on empowering information technology and the reasoning why and how the battle runs between the old potentates and the daemon society, which also evolves further during the story. Cryptography is a natural part of modern technology and modern warfare as described in this book. The new society emerging in *Freedom (TM)* is based on the darknet, an alternative to the internet using fast wireless meshes in order to increase the durability and availability of the network. Despite the story being shocking in some parts, it appears to be realistic and not far away from the parallel usage of modern technology integrated into our modern lives as a virtual world overlaying our real world.

***Rafael 2.0*, Karl Olsberg, Thienemann Verlag, 2011**

Michael and Rafael Ogilvy are talented twins who get along very well. Before the terminally ill Rafael dies, his father developed a virtual computer effigy of him, an artificial intelligence (AI). This is a good kept secret until Michael one day finds out what his father is hiding before him. However, his first horror soon turns into joy. So he still has something that reminds him of his brother.

But this computer system is also interesting for the military. One day Michael's father is kidnapped and the company, and thus also the computer program Rafael 2.0, falls into the wrong hands. Michael is banished by his uncle in a boarding school, from which he can flee. Henceforth, Michael and his friends try their best to find his father, whom they assume was abducted by a competing company. From there the story gets really exciting. Michael learns that there is another artificial intelligence, Metraton, which is not so well-disposed to the people. Nothing is too engrossed; young teenagers are the target audience. Nevertheless, depth and substance are created when, for instance, the machinations in acquisitions are discussed.

From a crypto perspective, the section about factoring is thrilling: With a variant Michael can detect whether the computer is cheating.

***The Fifth Murderer (original title: Der fünfte Mörder)*, Wolfgang Burger, Piper, 2011.**

Location and time of the story: Germany / Heidelberg, 1990 to 2009. Episode 7 of the Alexander-Gerlach series. Inspector Alexander Gerlach almost became a victim of a bomb blast when the sport utility vehicle (SUV) of a Bulgarian panderer exploded. Gerlach starts investigating because he wants to prevent a gang warfare, but then his bosses call him off. When the journalist Machatschek supports Gerlach, he communicates with him only via Skype using an add-on encryption program which he believes is the most secure in the world.

***Master of the Universe: Master of all Staff (original title: Herr aller Dinge)*, Andreas Eschbach, Lübbe, 2011.**

This novel deserved a much broader audience: The idea in it of the "most terrific of all crimes," which is the origin of the whole story, is new and almost revolutionary, but also infinitely sad. Along the failing partnership of Hiroshi (inventor genius) and Charlotte, important topics like justice, human wealth, and power are dealt with.

From a crypto perspective, Hiroshi uses distributed calculations and developed an encryption and backup system which misleads the government which bugged him.

***Blackout – Tomorrow is too Late (original title: Blackout – Morgen ist es zu spät)*, Marc Elsberg, Blanvalet, 2012.**

During a cold day in winter, all power supply networks in Europe break down. Agencies, energy suppliers, and security companies are in the dark and unable to solve the problem. The Italian computer scientist Piero Manzano believes that this is caused by terrorists using hackers: All customers use smart meters, electricity meters controlled by software that was manipulated. Despite the integrated security and encryption components, they have been hacked, and are out of order by wrong control sequences. The terrifying consequences happening at various locations are described realistically and excitingly, as are the reactions of the human beings.

***The Eigths Revelation (original title: Die achte Offenbarung)*, Karl Olsberg, Aufbau Taschenbuch, 2013.**

Can a message from the past change our future? An ancient, encrypted manuscript fell into the hands of historian Paul Brenner. The more he decodes the text, the more puzzling the content is: Because the book tells with remarkable precision events years ahead of the time of its presumed creation. While highly dangerous genetic material disappears from a U.S. laboratory, someone tries to prevent, at any price, Paul from deciphering the last (the eighth) revelation. A gripping thriller about a shockingly realistic apocalypse with many human aspects.

As a reader, you can participate in the deciphering of the manuscript.

The experiments of Paul to make the right persons aware of his discovery and to correct it later, are described very excitingly—even chief editors have a dilemma with conspiracy.

The cipher on the last book page is offered as a challenge in the crypto competition MTC3: `https://mysterytwister.org/challenges/level-1/the-last-note`.

***ZERO – They Know What You Are Doing (original title: ZERO – Sie wissen, was du tust)*, Marc Elsberg, Blanvalet Verlag, 2014.**

London. In a pursuit a boy is shot. His death takes the journalist Cynthia Bonsant to the acclaimed internet platform Freemee. Freemee collects and analyzes data, and thus promises its millions of users (rightly) a better life and more success. There is only one who warns about Freemee and about the power that the online newcomer could give just a few: ZERO, the most searched online activist in the world. As Cynthia begins precisely to research, she's becoming the quarry. And in a world of cameras, headsets, and smartphones there is no escape.

Highly topical and menacing: the transparent person under control. The novel takes place in the near future (fiction) and contains many contemporary references such as PRISM, predictive analytics, and gamification. By the way, references to well-known science fiction media like *The Running Man*, *Monkey Wrench Gang*, *V as Vendetta* (V wears a Guy Fawkes mask, now the hallmark of Anonymous), *Network*, and *Body Snatchers* are processed.

Technologically / cryptologically the protagonists move on the highest level, which is not further explained: Alice Kinkaid communicates with a Raspberry Pi. Cynthia's daughter Vi uses mesh networks.

***Genocide of One*, Kazuaki Takano, 2014. (Orginal in Japanese: *Jenosaido*, 2011; as paperback in English again under the title *Extinction*, 2016)**

The cover text of the English version (Mulholland Books, 2014) says: "He is a new kind of human. He may mean the end for the rest of us... One bright morning in Washington D.C., the U.S. President learns of a terrifying new threat to national security. Soon afterward, American mercenary Jonathan Yeager is asked to lead a team into the Congo to eliminate a mysterious enemy—a job which will help him pay for treatment for his dying son. But when they reach Africa, the threat turns out to be a three-year-old child named Akili: the next step in human evolution. The soldiers are under orders to kill the boy before his full potential can be realized. Yet Akili's advanced knowledge might be the only hope Yeager has to save his son's life... With time running out to choose a side, Yeager must decide whether to follow his orders or to save a creature who may not be as harmless or innocent as he appears. Because Akili is already the smartest being on the planet, with the power to either save humanity—or destroy it."

This is a very exciting book. After having overcome the first 100–200 pages you'll be awarded with surprising insights. According to the recensions, it's very well researched, but not for superficial readers.

From a crypto perspective, RSA and OTP are direct drivers of the story and are explained correctly. Breaking RSA by factorization is so important that the CIA wouldn't accept that this knowledge isn't in their ownership.

***The Girl in the Spider's Web*, David Lagercrantz, Quercus, 2015.**

This is the fourth novel in the *Millennium* series, and the first not written by Stieg Larsson. While Mikael Blomkvist's print medium is struggling to survive, the reader gets more and more insight in the inner structures and the combinations of publishers, secret services, public agencies, organized crime, and industrial espionage. Here, no care is taken for single humans, and normal humans would have no chance against this mix of interests. However, the special skills of Lisbeth Salander make a difference, and so the NSA is informed that parts of it are led and misused by organized crime. The characters of the *Millenium* trilogy have been developed further in a credible way. Very exiting.

From a crypto perspective, Lisbeth and August deal with elliptic curves to crack RSA.

Remark 1:

A long list of (partly commented) samples of cryptology in fictional literature can be found on the following German web page: `https://www.staff.uni-mainz.de/pommeren/Kryptologie/Klassisch/0_Unterhaltung/`. For some older authors (e.g., Jules Verne, Karl May, Arthur Conan Doyle, and Edgar Allen Poe) there are even links to the original and relevant text pieces.

Remark 2:

You can find title pages of some of these books on the website of Tobias Schrödel, who collects classic books about cryptography: `https://cryptobooks.org/`.

Remark 3:

If you know of further books and movies, where cryptography has a major role then we would be very glad if you could send us the exact title and a short explanation about the movie/book's content. We will insinuate your possible enthusiasm for a title. Thanks a lot.

B.1.2 For Kids and Teenagers

The following list contains movies and children's books. The children books contain both stories, and collections of simple ciphers, prepared in a didactic and exciting manner (please send us similar English children books and children movies, because at the moment our list contains mostly German children books).

***Top Secret – The Book for Detectives and Spies (original title: Streng geheim – Das Buch für Detektive und Agenten)*, author unknown, Edition moses, year unknown.**

This is a thin book for small kids with Inspector Fox and Dr. Chicken.

***The Three Investigators: The Secret Key (Original German title: Die 3 ???: Der geheime Schlüssel nach Alfred Hitchcock (volume 119)*, Robert Arthur, Kosmos-Verlag (from 1960).**

The three detectives Justus, Peter, and Bob have to decrypt covered and encrypted messages within this story to find out what is behind the toys of the Copperfield company.

***Ciphers (original title: Geheimschriften)*, Karl-Heinz Paraquin, Ravensburger Taschenbuch Verlag, 1988 (1st edition 1977).**

On 125 pages filled with a small font this mini format book explains many methods that children can apply directly to encrypt or hide their messages. A little glossary and a short overview about the usage of encryption methods in history complete this little book.

Right at page 6 it summarizes for beginners in an old fashion style "The Important Things First" about paper-and-pencil encryption (compare Chapter 2):

- "It must be possible to encrypt your messages at any place and at any location with the easiest measures and a small effort in a short time."
- "Your cipher must be easy to remember and easy to read for your partners. But strangers should not be able to decrypt them. Remember: Fastness before finesse, security before carelessness."
- "Your message must always be as short and precise as a telegram. Shortness outranks grammar and spelling. Get rid of all needlessness like salutations or punctuation marks. Preferably use only small or only capital letters."

***The Manual for Detectives. Everything You Need to Know About Ciphers, Codes, Reading Tracks and the Biggest Detectives of the World (original title: Das Handbuch für Detektive. Alles über Geheimsprachen, Codes, Spurenlesen und die großen Detektive dieser Welt)*, Matthias Müller-Michaelis, Südwest, 2002.**

A small collection on 62 pages.

***Top Secret! – How to Encrypt Messages and to Hack Codes (original title: Streng geheim! – Wie man Botschaften verschlüsselt und Zahlencodes knackt)*, Rudolf Kippenhahn, rororo, 2002.**

In this novel, a grandpa, an expert for secret writings, teaches his four grandchildren and their friends how to encrypt messages that nobody should read. Because there is someone who hacks their secrets, the grandpa has to teach them more and more complicated methods.

Within the framework of this story, the most important classic encryption methods and their analyses are explained in a manner exciting and appropriate for children.

***Top Secret. The Big Book for Detectives (original title: Streng geheim. Das große Buch der Detektive)*, Corinna Harder and Jens Schumacher, Moses, 2003.**

A collection on 118 pages.

***Your Mission in the Weird Villa. Riddle Thriller (original title: Dein Auftrag in der unheimlichen Villa. Kennwort Rätselkrimi)*, Helga Talke and Milena Baisch, Loewe, 2003.**

From 4th form. `http://www.antolin.de`.

Young detectives solve simple ciphers and codes during their missions.

***The Three Investigators: Manual for Secret Messages (original title: Die 3 ???: Handbuch Geheimbotschaften)*, Bernd Flessner, Kosmos, 2004.**

On 127 pages you learn in an easy and exciting manner, structured by the method types, which secret languages (like the one of the Navajo Indians or dialects) and which secret writings (real encryption or hiding via technical or linguistic steganography) existed and how simple methods can be decrypted.

The author tells where in history the methods were used and in which novels authors used encryption methods [like in Edgar Allan Poe's *The Gold Bug*, like with Jules Verne's hero Mathias Sandorf, or like with Astrid Lindgren's master detective Blomquist who used the ROR language (similar inserting ciphers are the spoon or the B language)].

This is a didactically excellent introduction for younger teens.

***The Treasure of the White Hawks (original title: Der Schatz der weißen Falken)*, directed by Christian Zübert, 2005.**

This exciting adventure movie for kids ties in with the tradition of classics like Mark Twain's *The Adventures of Tom Sawyer and Huckleberry Finn* or Enid Blytons "The Famous Five." The plot happens in summer 1981. In an old half tumbledown villa three young kids find the treasure map of the "White Hawks," which they decrypt with the help of a computer. Traced by another gang they aim to go to an old castle.

***The Three Investigators: Secret Messages (German version: Die 3 ???: Geheimnisvolle Botschaften) (volume 160)*, Christoph Dittert, Kosmos, 2011.**

In the house of Professor Mathewson an old hand-made book was stolen. The three detectives Justus, Peter, and Bob are getting attacked by a ruthless opponent, who seems to be always a step ahead. A major part in this story is played by a palimpsest, an ancient manuscript page, written upon newly. Using X-rays they can make visible again the old text below. It's not only the story that is exciting, but also the way, how the instruction for a treasure hunt is encrypted. Despite using the simple railfence cipher it's not easy to solve it, as the message is distributed onto two slips and the printed symbols don't mean single letters.

Remark 1:

You can find title pages of many of these kid books on the website of Tobias Schrödel, who collects classic books about cryptography: `https://cryptobooks.org/`.

Remark 2:

If you know of further books that address cryptography in a didactic and, for children, adequate way, then we would be very glad if you could send us the exact book title and a short explanation about the book's content. Thanks a lot.

B.1.3 Code for the Light Fiction Books

Section B.1.1 lists *The Gold Bug* by E.A. Poe as the first book.

Using Python Example B.1 [4] you can decrypt the ciphertext of Captain Kidd (see the original text of *The Gold Bug* in `http://pinkmonkey.com/dl/library1/gold.pdf`, page 21).

The code already contains the ASCII characters of the ciphertext and the correlated alphabets for the plaintext and the ciphertext of this monoalphabetic cipher (MASC). Alternatively, you could use SageMath Example 2.1 which also uses a self-defined alphabet.

The easiest way to perform the decryption is using the SageMathCell server (`https://sagecell.sagemath.org/`) in a browser: There you can switch between the programming languages Sage and Python. The code can be executed by inserting it with "copy-and-paste" and then pressing "Evaluate."

Python Example B.1: Decryption of the Gold-Bug Ciphertext from the Novel of E.A. Poe (with Python)

```
print("\n# Appendix_B--SAMPLE 010: =========")

# Chap. B.1.3 Code for the light fiction books
# Decryption of the Gold-Bug ciphertext from the novel of E.A. Poe
# Usage on terminal: python appB1_sample01.py  (needs Python 3)

PA = 'ETHSONAIRDGLBVPFYMUC'
print('Plaintext alphabet PA: ', PA, ' Length of PA', len(PA))
CA = "8;4)+*56(!302'.1:9?-"
print('Ciphertext alphabet CA:', CA, ' Length of CA', len(CA))

codetableC2P = str.maketrans(CA,PA) # the strings CA and PA must have ▸
  ▸the same length

C = '''53++!305))6*;4826)4+.)4+);806*;48!8'60))85;1+(;:+*8!83(88)5*!;46
(;88*96*?;8)*+(;485);5*!2:*+(;4956*2(5*-4)8'8*;4069285);)6!8)4++;1(+9;4
8081;8:8+1;48!85;4)485!528806*81(+9;48;(88;4(+?34;48)4+;161;:188;+?;'''
P = C.translate(codetableC2P);
print('\nKidd decrypted:')
print(P)

# if str contains symbols not in the translation, they are left ▸
  ▸unchanged
intab  = "aeiou"
```

Python Example B.1 (continued)

```
outtab = "12345"
trantab = str.maketrans(intab, outtab)
stri = "this is string example..AE..wow!!!"
print("\nTest substituting only lower-case vocals:", stri.translate(▸
 ▸trantab))

#------------------------------------
# Appendix_B--SAMPLE 010: =========
# Plaintext alphabet PA:  ETHSONAIRDGLBVPFYMUC  Length of PA 20
# Ciphertext alphabet CA: 8;4)+*56(!302'.1:9?-  Length of CA 20
#
# Kidd decrypted:
# AGOODGLASSINTHEBISHOPSHOSTELINTHEDEVILSSEATFORTYONEDEGREESANDTHI
# ▸
 ▸RTEENMINUTESNORTHEASTANDBYNORTHMAINBRANCHSEVENTHLIMBEASTSIDESHOOTFROMTH▸
 ▸
# ELEFTEYEOFTHEDEATHSHEADABEELINEFROMTHETREETHROUGHTHESHOTFIFTYFEETOUT
#
# Test substituting only lower-case vocals: th3s 3s str3ng 2x1mpl2..AE▸
 ▸..w4w!!!
```

Remark 1:

When printing the ciphertext, Poe or his publisher "cheated," similarly to the author of the Python code who used only ASCII characters.

In the archive of an original publication (e.g., at `https://archive.org/details/goldbug00poegoog` at page 95) you can see that Poe used characters that were common in the letterpress printing (and most of them are also part of the ASCII set). It is very unlikely that an untaught pirate would use just such characters for his ciphertext.

Remark 2:

The sample code uses the Python string functions "maketrans" and "translate."

So both alphabets (for the plaintext and the ciphertext) are inserted as a simple string, and "maketrans" creates a mapping table. The actual encryption is done by "translate." For the decryption you just have to switch the arguments of "maketrans" for the two alphabets. The otherwise necessary transformations between characters and their ASCII numbers (using "str" and "ord") can be avoided. This is ideal for monoalphabetic ciphers—especially for lessons at the junior high school.

It's evident how less code is needed with Python 3 or SageMathfor such tasks. In the sample there were only 7 lines of code really necessary.

B.2 Recommended Spelling within the CrypTool Book

As a guide for the authors and because the internet and marketing ads often deviate from the official spelling, we list the recommendations from IEC (International Electrotechnical Commission), and so on.

The LaTeX package siunitx [5] allows you to conveniently enter numbers and units and display them consistently throughout the document. The package documentation is very good and easy to understand.

Bit: Upper/Lower Case and Abbreviation

The "bit" is not defined in the International System of Units (SI). However, the International Electrotechnical Commission issued standard IEC 60027, which specifies that the symbol for binary digit should be "bit," and this should be used in all multiples, such as "kbit," for kilobit. However, the lower-case letter "b" is widely used as well and was recommended by the IEEE 1541 Standard (2002). In contrast, the upper-case letter "B" is the standard and customary symbol for byte.

So as a unit in formulas, we write "bit" in lower case and without the plural "s."

Remark: The unit for quantum information is qubit.

Byte: Upper/Lower Case and Abbreviation

The unit symbol for the byte was designated as the upper-case letter "B" by the IEC and by the Institute of Electrical and Electronics Engineers (IEEE).

1000B = 1 kB = 1 kilobyte
1024B = 1 KiB = 1 kibibyte = 1 KB [sometimes wrongly as 1 kilobyte]

Hyphens

Public-key cryptography: Hyphen if the two words are used like one adjective.

Brute-force attack: Hyphen if the two words are used like one adjective.

`https://www.scribendi.com/academy/articles/hyphenation.en.html.`

`https://dictionary.cambridge.org/grammar/british-grammar/hyphens.`

References

[1] Kippenhahn, R., *VerschlüsselteBotschaften: Geheimschrift, Enigma und Chipkarte*, 1st ed, Rowohlt, 1997.

[2] Slade, R., *REVIEW: "Kim," Rudyard Kipling*, 2006, http://catless.ncl.ac.uk/Risks/24.49.html%5C#subj12.

[3] Kipling, R., *Kim*, https://kipling.thefreelibrary.com/Kim.

[4] Witten, H., I. Letzner, and R.-H. Schulz. "RSA & Co. in der Schule: ModerneKryptologie, alteMathematik, raffinierteProtokolle, Teil 1: Sprache und Statistik". In: *LOG IN* 3/4, 1998, pp. 57–65, https://informatik.schule.de/krypto/.

[5] Wright, J., *Siunitx—A comprehensive (SI) Units Package*, 2023, https://ctan.org/pkg/siunitx.

About the Author

Bernhard Esslinger

Initiator of the CrypTool project, editor, and main author of this book. Professor for IT security and cryptography at the University of Siegen. He is the former CISO of SAP AG, and former head IT security at Deutsche Bank.

Email: bernhard.esslinger@gmail.com.

Contributors

Doris Behrendt

Author of Section 6.5 ("The RSA Plane"). Mathematician, member of CT Team since 2018. Took over the project lead of the CT project in 2023 at Bundeswehr University, Munich. Email: doris.behrendt@unibw.de.

Matthias Büger

Contributor to Chapter 8 ("Elliptic-Curve Cryptography"). Research analyst at Deutsche Bank.

Miroslav Dimitrov

First author of Chapter 11 ("Lightweight Introduction to Lattices"). Bulgarian Academy of Sciences. Email: mirdim@math.bas.bg.

Bartol Filipovic

Original author of the CT1 elliptic curve implementation and of Chapter 8 ("Elliptic-Curve Cryptography").

Martin Franz

Original author of Chapter 10 ("Homomorphic Ciphers"). Works and carries out research in the area of applied cryptography.

Henrik Koy

Main developer and coordinator of CT1 development version 1.3 and 1.4. Book reviewer and TEX guru. Cryptographer and project leader IT at Deutsche Bank.

Vasily Mikhalev

Author of Section 2.5 ("Hagelin Machines as a Models for Precomputer Ciphers"), coauthor of Section 1.7 ("Best Known Attacks on Given Ciphers"), and coauthor of Chapter 3 ("Historical Cryptology"). Postdoctoral researcher at the University of Siegen.

Roger Oyono

First implementer of the CT1 factorization dialog and original author of Chapter 6 ("The Mathematical Ideas Behind Modern Asymmetric Cryptography").

Klaus Pommerening

Original author of Chapter 9 ("Foundations of Modern Symmetric Encryption"). Former professor of mathematics and computer science at Johannes-Gutenberg-Universität Mainz.

Harald Ritter

Contributor to Chapter 11 ("Lightweight Introduction to Lattices"). Member of IACR; PhD thesis on lattice basis reduction at the University of Frankfurt. Senior Consultant at NOVOSEC AG, Frankfurt/Main.

Email: harald.ritter@novosec.com.

Jörg Cornelius Schneider

Design and long-term support of CrypTool. Crypto enthusiast. IT architect and senior project leader IT at Deutsche Bank.

Christine Stötzel

Contributor to Chapter 2 ("Paper-and-Pencil and Precomputer Ciphers").

Johannes Buchmann

Coauthor of Chapter 13 ("Future Use of Cryptography"). Prof. Johannes Buchmann held the Chair for Theoretical Computer Science (Cryptography and Computer Algebra) at the department of Computer Science of the Technische Universität Darmstadt TUD). Retired.

Alexander May

Coauthor of Chapter 12 ("Solving Discrete Logarithms and Factoring") and of Chapter 13 ("Future Use of Cryptography"). Full professor at the department of mathematics (chair for cryptology and IT Security) of the Ruhr-Universität Bochum, and member of the Horst-Görtz Institute for IT Security. His research focuses on algorithms for cryptanalysis, especially on methods for attacking the RSA cryptosystem.

Erik Dahmen

Coauthor of Chapter 13 ("Future Use of Cryptography"). Researcher at the Chair for Theoretical Computer Science (Cryptography and Computer Algebra), Department of Computer Science, Technische Universität Darmstadt, Germany.

Ulrich Vollmer

Coauthor of Chapter 13 ("Future Use of Cryptography"). Researcher at the Chair for Theoretical Computer Science (Cryptography and Computer Algebra), Department of Computer Science, Technische Universität Darmstadt, Germany.

Antoine Joux

Coauthor of Chapter 12 ("Solving Discrete Logarithms and Factoring"). Antoine Joux is the holder of the Cryptology chair of the Foundation of the University Pierre et Marie Curie (Paris 6) and a senior security expert at CryptoExperts, Paris. He worked in various fields of cryptanalysis, and he is a key player in the recent advances in computing discrete logarithms in fields of small characteristic.

Arjen Lenstra

Coauthor of Chapter 12 ("Solving Discrete Logarithms and Factoring"). Arjen Lenstra is a full professor at École Polytechnique Fédérale de Lausanne (EPFL) and head of the laboratory for cryptological algorithms. He is one of the inventors of the best algorithm currently available for factoring integers (the number field sieve). He was involved in many practical factoring records.

Beáta Megyesi

Coauthor of Chapter 3 ("Historical Cryptology"). Professor of computational linguistics, Uppsala University, Sweden. PI of the DECRYPT project. Email: beata.megyesi@gmail.com.

Alicia Fornés

Coauthor of Chapter 3 ("Historical Cryptology"). Computer Vision Center, Universitat Autònoma de Barcelona, Spain.

Benedek Láng

Coauthor of Chapter 3 ("Historical Cryptology"). Historian of science, Eötvös Loránd University, Budapest, Hungary.

Michelle Waldispühl

Coauthor of Chapter 3 ("Historical Cryptology"). Associate professor of German linguistics and language acquisition at the University of Gothenburg, Sweden.

Nils Kopal

Coauthor of Chapter 3 ("Historical Cryptology") and Section 1.7 ("Best Known Attacks on Given Ciphers"). Leader of the development of the software CrypTool 2. Computer scientist and cryptanalyst working as a postdoctoral researcher at the University of Siegen. Email: nils@kopaldev.de.

Ralph Simpson

Coauthor of Section 1.6 ("Key Spaces: A Theoretical and Practical View"). See www.CipherHistory.com.

Minh Van Nguyen

SageMath developer and documentation quality reviewer.

Index

D

E

G

H

I

J

K

S

T

Recent Titles in the Artech House Computer Security Series

Rolf Oppliger, Series Editor

Bluetooth Security, Christian Gehrmann, Joakim Persson, and Ben Smeets

Computer Forensics and Privacy, Michael A. Caloyannides

Computer and Intrusion Forensics, George Mohay, et al.

Contemporary Cryptography, Second Edition, Rolf Oppliger

Cryptography 101: From Theory to Practice, Rolf Oppliger

Cryptography for Security and Privacy in Cloud Computing, Stefan Rass and Daniel Slamanig

Defense and Detection Strategies Against Internet Worms, Jose Nazario

Demystifying the IPsec Puzzle, Sheila Frankel

Developing Secure Distributed Systems with CORBA, Ulrich Lang and Rudolf Schreiner

Electric Payment Systems for E-Commerce, *Second Edition,* Donal O'Mahony, Michael Peirce, and Hitesh Tewari

Engineering Safe and Secure Software Systems, C. Warren Axelrod

Evaluating Agile Software Development: Methods for Your Organization, Alan S. Koch

Implementing Electronic Card Payment Systems, Cristian Radu

Implementing the ISO/IEC 27001 Information Security Management System Standard, Edward Humphreys

Implementing Security for ATM Networks, Thomas Tarman and Edward Witzke

Information Hiding, Stefan Katzenbeisser and Fabien Petitcolas, editors

Internet and Intranet Security, Second Edition, Rolf Oppliger

Introduction to Identity-Based Encryption, Luther Martin

Java Card for E-Payment Applications, Vesna Hassler, Martin Manninger, Mikail Gordeev, and Christoph Müller

Learning and Experiencing Cryptography with CrypTool and SageMath, Bernhard Esslinger

Lifecycle IoT Security for Engineers, Kaustubh Dhondge

Modern Vulnerability Management: Predictive Cybersecurity, Michael Roytman and Ed Bellis

Multicast and Group Security, Thomas Hardjono and Lakshminath R. Dondeti

Non-repudiation in Electronic Commerce, Jianying Zhou

Outsourcing Information Security, C. Warren Axelrod

The Penetration Tester's Guide to Web Applications, Serge Borso

Privacy Protection and Computer Forensics, Second Edition, Michael A. Caloyannides

Role-Based Access Control, Second Edition, David F. Ferraiolo, D. Richard Kuhn, and Ramaswamy Chandramouli

Secure Messaging with PGP and S/MIME, Rolf Oppliger

Securing Information and Communications Systems: Principles, Technologies and Applications, Javier Lopez, Steven Furnell, Sokratis Katsikas, and Ahmed Patel

Security Fundamentals for E-Commerce, Vesna Hassler

Security Technologies for the World Wide Web, Second Edition, Rolf Oppliger

Techniques and Applications of Digital Watermarking and Content Protection, Michael Arnold, Martin Schmucker, and Stephen D. Wolthusen

User's Guide to Cryptography and Standards, Alexander W. Dent and Chris J. Mitchell

For further information on these and other Artech House titles, including previously considered out-of-print books now available through our In-Print-Forever® (IPF®) program, contact:

Artech House
685 Canton Street
Norwood, MA 02062
Phone: 781-769-9750
Fax: 781-769-6334
e-mail: artech@artechhouse.com

Artech House
16 Sussex Street
London SW1V HRW UK
Phone: +44 (0)20 7596-8750
Fax: +44 (0)20 7630-0166
e-mail: artech-uk@artechhouse.com

Find us on the World Wide Web at: www.artechhouse.com